Humanized Mice for HIV Research

Larisa Y. Poluektova • J. Victor Garcia
Yoshio Koyanagi • Markus G. Manz
Andrew M. Tager
Editors

Humanized Mice for HIV Research

Springer

Editors
Larisa Y. Poluektova
Department of Pharmacology and Experimental Neuroscience
University of Nebraska Medical Center
Omaha
Nebraska
USA

J. Victor Garcia
Center for AIDS Research
The University of North Carolina at Chapel Hill
Chapel Hill
North Carolina
USA

Yoshio Koyanagi
Research Center for AIDS Laboratory of Viral Pathogenesis
Kyoto University Institute for Virus Research
Sakyo-ku
Japan

Markus G. Manz
Division of Hematology
University Hospital Zürich
Zürich
Switzerland

Andrew M. Tager
Center for Immunology and Inflammatory Diseases
Massachusetts General Hospital, Harvard Medical School
Charlestown
Massachusetts
USA

ISBN 978-1-4939-1654-2 ISBN 978-1-4939-1655-9 (eBook)
DOI 10.1007/978-1-4939-1655-9
Springer New York Heidelberg Dordrecht London

Library of Congress Control Number: 2014956932

Printed on acid-free paper

Springer is part of Springer Science+Business Media (www.springer.com)

Foreword

As a clinician and researcher involved in human immunodeficiency virus (HIV) disease since the beginning of the epidemic, I have huge respect for the contributions of work with humanized mice. As one of the only animal models that facilitates working with HIV as opposed to other lenitviruses, work with humanized mice has encompassed the entire spectrum of HIV pathogenesis research from transmission to immune dysregulation and the impact of preventive and therapeutic interventions. Finding suitable animal models has been a major impediment to HIV pathogenesis work since the beginning, as naturally occurring rodent cells are completely refractory to HIV infection. Even animals closely related to humans, such as chimpanzees, that can be infected with HIV do not develop the same immunodeficiency and disease. In addition, HIV has very limited tropism and so any kind of in vivo modeling must use a very similar organism, the most common one being simian immunodeficiency virus (SIV). While extremely useful, there are important genetic and biological differences between HIV and SIV, just as there are between humans and chimpanzees.

Over the last few decades, enormous strides have been made to improve the "humanization" of mouse models, particularly in the area of HIV research. Humanized mice have evolved into an invaluable alternative to SIV-based nonhuman primate models, as they are simpler, less costly, and also highly susceptible to HIV infection. Mouse models have been employed in basic pathogenesis research, preclinical and clinical testing of compounds with potential antiretroviral activity, and more recently, HIV biomedical prevention. For example, a humanized mouse model demonstrated that human breast milk has antiretroviral properties and may protect infants against oral transmission, thus helping to inform the debate about breast feeding for infected mothers without access to safe alternatives. Humanized mouse models are also being used to provide efficacy data about protection against rectal and vaginal infections with an array of regimens that might be used for pre-exposure prophylaxis. The models have helped to define the limits of protection for various dosing schedules, and are increasingly being used to investigate key pharmacologic parameters.

Reports of at least two individuals being cured of HIV infection, and several more with apparent functional cures (defined as long-term health in the absence of antiretroviral therapy) have renewed interest and excitement in this area. An important challenge is the difficulty of quantifying virus at extremely low levels in patients, but this will need to be overcome in future to be able to establish whether or not an infected individual has truly been cleared of any virus. Humanized mice have already been used in this context to demonstrate replication competent virus in the absence of any detectable plasma viremia, even using highly sensitive assays for HIV RNA and DNA. Mouse models are likely to play a key role in this scientific agenda, moving forward.

Dr. Larisa Poluektova has been working in this field for many years, and we have been working together since 2006. Originally focused on neuropathogenesis work, more recently our collaborative activities have been in the development of nanoformulated antiretroviral therapy (ART) under the direction of Dr Howard Gendelman [1–3]. Nanomedicines contain crystalline drug particles of small diameter, coated with low-molecular-weight excipients to produce specific sizes, charges, and shapes that optimize cell and tissue penetrance. We have been working on nanoformulations of existing antiretroviral agents, and humanized mouse work has been pivotal. Building on what we have learned from the mouse experiments, we have moved into studies in nonhuman primate and hope to advance to clinical trials in humans. This emerging area of discovery has potential to make enormous changes in the field and advance treatment. While highly successful if taken correctly by infected patients, current ART is limited by the need for lifelong daily therapy, by poor tissue penetration, and by adverse effects. Suboptimal adherence to therapy may promote the development of virologic resistance and treatment failure. Nanoformulated ART may be able to be administered intermittently, and thereby improve medication adherence, and also has potential for decreased adverse effects and improved tissue penetrance. Investigations of long-acting formulations are also underway for HIV prevention.

"Humanized Mice for HIV Research" covers all these topics, and more. From an in depth review of the genetic background of mice and tips for humanization through understanding of human immune cells, the book moves on to HIV biology and pathogenesis and how humanized mice can advance the field. With discussion of specific cellular and humoral immune responses, the book includes reviews of development of conventional and novel therapeutics for HIV treatment and prevention. Finally, other human-specific or selective pathogens are presented including dengue, tuberculosis, and malaria, all causes of enormous amounts of human disease. The last section moves to new horizons and exciting prospects for the future from experts in the field.

This is an essential book for scientists and their students and will provide them with comprehensive and up-to-date information about the role of humanized mice in HIV research. Despite a wealth of scholarly articles on this topic, including many from the authors in the book, there are very few comprehensive textbooks about humanized mice in HIV research—a gap that has now been filled very nicely.

Omaha, NE
2013

Susan Swindells (M.B.B.S.)
Terry K. Watanabe Chair for HIV/AIDS
Research and Care, and Professor of
Internal Medicine and Infectious Diseases

References

1. Dash PK, Gendelman HE, Roy U, Balkundi S, Alnouti Y, Mosley RL, Gelbard HA, McMillan J, Gorantla S, Poluektova LY. Long-acting nanoformulated antiretroviral therapy elicits potent antiretroviral and neuroprotective responses in HIV-1-infected humanized mice. AIDS. 2012;26(17):2135–44. doi:10.1097/QAD.0b013e328357f5ad.
2. Gorantla S, Poluektova L, Gendelman HE. Rodent models for HIV-associated neurocognitive disorders. Trends Neurosci. 2012;35(3):197–208.
3. Roy U, McMillan J, Alnouti Y, Gautum N, Smith N, Balkundi S, Dash P, Gorantla S, Martinez-Skinner A, Meza J, Kanmogne G, Swindells S, Cohen SM, Mosley RL, Poluektova L, Gendelman HE. Pharmacodynamic and antiretroviral activities of combination nanoformulated antiretrovirals in HIV-1-infected human peripheral blood lymphocyte-reconstituted mice. J Infect Dis. 2012;206(10):1577–88.

Preface

In 2012, our international editorial team, whose members are listed below, implemented work on a *comprehensive* textbook, or collection, entitled "Humanized Mice for HIV Research." In its current completed form, this detailed document is intended to serve as a scientific guide for graduate students, fellows, and investigators in bench science, academicians (e.g., hematologists, immunologists, virologists), clinicians (e.g., infectious disease specialists), and persons in the pharmaceutical industry (e.g., drug developers, vaccine developers, and pharmacologists/toxicologists) in the field of HIV and beyond. Importantly, humanized mice are the only animals, aside from chimpanzees, that are susceptible to HIV infection. Thus, humanized mice are an ideal platform for the study of HIV.

HIV has been, and still is, intensively investigated. However, the lack of robust small animal models has hindered progress in the basic understanding of HIV infection and pathogenesis. This lack also poses a considerable challenge for preclinical testing and the prioritization of new drug and vaccine candidates.

Stable, multilineage human hematopoietic engraftment can now be routinely achieved in immunodeficient mice. Surveillance of the development of human hematopoietic and lymphoid tissues in the mouse environment by researchers with different expertise provides valuable information. This book provides information on a wide range of different approaches, applications, ideas, observations, hypotheses, and insights. We expect this exchange of information to help facilitate exploration of HIV pathogenesis, and the development of new treatments and preventative approaches that will accelerate progress toward the eradication of this disease.

We sincerely appreciate the great efforts of all of our contributors, and apologize to anyone we may have left out with important new findings, observations, developments, or ideas to share. With the help of humanized mouse models, we hope to progress to an HIV/AIDS-free world. We expect that efforts to control other human-specific infections will also benefit from broadening the application of humanized mice to biomedical research.

Warm regards,

Professor, Department of Pharmacology and Experimental Neuroscience, University of Nebraska Medical Center, Omaha, NE, USA — Larisa Y. Poluektova, MD, PhD

Professor of Medicine, The University of North Carolina at Chapel Hill, Center for AIDS Research, Chapel Hill, NC, USA — J. Victor Garcia, PhD

Professor, Center for Human Retrovirus Research, Laboratory of Viral Pathogenesis, Institute for Virus Research, Kyoto University, Kyoto, Japan — Yoshio Koyanagi, MD, PhD

Professor and Chair Division of Hematology, University Hospital Zürich, Zürich, Switzerland — Markus G. Manz, MD

Associate Professor of Medicine, Center for Immunology and Inflammatory Diseases, Massachusetts General Hospital and, Ragon Institute of MGH, MIT and Harvard, Charlestown, MA, USA — Andrew M. Tager, M.D.

Contents

Part II Understanding of Human Immune Cells Development and Function in Mouse Environment

Part III Humanized Mice for HIV-1 Virus Biology and Pathogenesis. Is It Possible to Address?

Contributors

Ramesh Akkina Department of Microbiology, Immunology and Pathology, Colorado State University, Fort Collins, CO, USA

Marcus Altfeld Ragon Institute of Massachusetts General Hospital, Massachusetts Institute of Technology and Harvard, Harvard Medical School, Boston, MA, USA

Heinrich-Pette-Institute, Leibniz Institute for Experimental Virology, Hamburg, Germany

Maya Caroline André Department of Hematology and Oncology, University Children's Hospital of Tuebingen, Tuebingen, Germany

University Children's Hospital Basel (UKBB), Basel, Switzerland

B. Berkhout Laboratory of Experimental Virology, Department of Medical Microbiology, Center for Infection and Immunity Amsterdam (CINIMA), Academic Medical Center, University of Amsterdam (AMC-UvA), Amsterdam, The Netherlands

Kathleen A. Burke University of Southern California, Los Angeles, CA, USA

Dennis R. Burton Department of Immunology and Microbial Science, International AIDS Vaccine Initiative Neutralizing Antibody Center, and Center for HIV/AIDS Vaccine Immunology and Immunogen Discovery, The Scripps Research Institute, La Jolla, CA, USA

Ragon Institute of Massachusetts General Hospital, Massachusetts Institute of Technology, and Harvard University, Boston, MA, USA

Paula M. Cannon Department of Molecular Microbiology & Immunology, Keck School of Medicine, University of Southern California, Los Angeles, CA, USA

Mireille Centlivre Centre d'Immunologie et des Maladies Infectieuses-Paris (CIMI-Paris), Sorbonnes Universités, UPMC University Paris, France

Laboratory of Immunity and Infection, Institut National de la Santé et de la Recherche Médicale, Université Pierre et Marie Curie, Paris, France

J. Judy Chang Ragon Institute of Massachusetts General Hospital, Massachusetts Institute of Technology and Harvard, Harvard Medical School, Boston, MA, USA

Department of Infectious Diseases, Monash University, Melbourne, VIC, Australia

Morgan L. Chateau Division of Infectious Diseases, Department of Medicine, UNC Center for AIDS Research, University of North Carolina School of Medicine, Chapel Hill, NC, USA

Shailesh K. Choudhary Department of Medicine, Gene Therapy Center, University of North Carolina at Chapel Hill, Chapel Hill, NC, USA

Paul Curley Department of Molecular and Clinical Pharmacology, University of Liverpool, Liverpool, UK

Paul W. Denton Division of Infectious Diseases, Department of Medicine, UNC Center for AIDS Research, University of North Carolina School of Medicine, Chapel Hill, NC, USA

Division of Infectious Diseases, UNC Center for AIDS Research, University of North Carolina School of Medicine, Chapel Hill, NC, USA

James P. Di Santo Innate Immunity Unit, Institut Pasteur, Paris, France

Inserm U668, Paris, France

Madeleine Duc Dodon Laboratoire de Biologie Moléculaire de la Cellule, Unité Mixte de Recherche 5239, Centre National de la Recherche Scientifique, Ecole Normale Supérieure de Lyon, Lyon Cedex 7, France

SFR UMS3444 BioSciences Lyon-Gerland-Lyon Sud (UMS3444), Lyon Cedex 7, France

Erica Eggers School of Nursing, University of California, Los Angeles, CA, USA

Adrian A. Epstein Department of Pharmacology & Experimental Neuroscience, University of Nebraska Medical Center, Omaha, NE, USA

Colin M. Exline University of Southern California, Los Angeles, CA, USA

Gerold Feuer Department of Microbiology and Immunology, SUNY Upstate Medical University, Syracuse, NY, USA

Humurine Technologies, LaVerne, CA, USA

Richard A. Flavell Department of Immunobiology, Yale University, New Haven, CT, USA

Howard Hughes Medical Institute, Yale University, New Haven, CT, USA

Shigeyoshi Fujiwara Department of Infectious Diseases, National Research Institute for Child Health and Development, Tokyo, Japan

J. Victor Garcia Division of Infectious Diseases, Department of Medicine, UNC Center for AIDS Research, University of North Carolina School of Medicine, Chapel Hill, NC, USA

Louis Gazzolo Laboratoire de Biologie Moléculaire de la Cellule, Unité Mixte de Recherche 5239, Centre National de la Recherche Scientifique, Ecole Normale Supérieure de Lyon, Lyon Cedex 7, France

SFR UMS3444 BioSciences Lyon-Gerland-Lyon Sud (UMS3444), Lyon Cedex 7, France

Howard E. Gendelman Department of Pharmacology and Experimental Neuroscience, University of Nebraska Medical Center, Omaha, NE, USA

Stanton L. Gerson Case Comprehensive Cancer Center, National Center for Regenerative Medicine, Seidman Cancer Center, University Hospitals Case Medical center and Case Western Reserve University, Cleveland, OH, USA

Santhi Gorantla Department of Pharmacology & Experimental Neuroscience, University of Nebraska Medical Center, Omaha, NE, USA

Udo F. Hartwig Department of Medicine 3—Hematology, Internal Oncology & Pneumology, Building for R&D, University Medical Center of Johannes Gutenberg-University Mainz, Mainz, Germany

Ursula Hofer University of Southern California, Los Angeles, CA, USA

Nathalia G. Holt University of Southern California, Los Angeles, CA, USA

Zheng Hu First Hospital of Jilin University, Changchun, China

Ken-Ichi Imadome Department of Infectious Diseases, National Research Institute for Child Health and Development, Tokyo, Japan

Mamoru Ito Central Institute for Experimental Animals, Kawasaki, Kanagawa, Japan

Ryoji Ito Central Institute for Experimental Animals, Kawasaki, Kanagawa, Japan

Muazzam Jacobs Division of Immunology, Department of Clinical and Laboratory Science and Institute of Infectious Diseases and Molecular Medicine, UCT Medical School, University of Cape Town, Cape Town, Western Cape, South Africa

National Health Laboratory Service, Johannesburg, South Africa

Ikumi Katano Central Institute for Experimental Animals, Kawasaki, Kanagawa, Japan

Scott G. Kitchen Division of Hematology-Oncology, The David Geffen School of Medicine at UCLA, The UCLA AIDS Institute, Los Angeles, CA, USA

Larisa V. Kovtonyuk Division of Hematology, University Hospital Zürich, Zürich, Switzerland

Yoshio Koyanagi Institute for Virus Research, Kyoto University, Kyoto, Japan

Antoinette Labuschagné Division of Immunology, Department of Clinical and Laboratory Science and Institute of Infectious Diseases and Molecular Medicine, UCT Medical School, University of Cape Town, Cape Town, Western Cape, South Africa

Julie Lang Department of Immunology and Microbiology, University of Colorado Denver School of Medicine, Aurora, CO, USA

Nicolas Legrand Department of Cell Biology and Histology, Center for Immunology Amsterdam (CIA), Academic Medical Center, University of Amsterdam (AMC-UvA), Amsterdam, The Netherlands

AXENIS, Paris, France

Hao Wei Li Columbia Center for Translational Immunology, Columbia University College of Physicians and Surgeons, New York, NY, USA

Qingsheng Li Nebraska Center for Virology, School of Biological Sciences, University of Nebraska-Lincoln, Lincoln, NE, USA

Yan Li Innate Immunity Unit, Institut Pasteur, Paris, France

Inserm U668, Paris, France

Silvia Lopez-Lastra Innate Immunity Unit, Institut Pasteur, Paris, France

Inserm U668, Paris, France

Markus G. Manz Division of Hematology, University Hospital Zürich, Zürich, Switzerland

Guillemette X. Masse Innate Immunity Unit, Institut Pasteur, Paris, France

Inserm U668, Paris, France

Anuja Mathew Division of Infectious Diseases and Immunology, Department of Medicine, University of Massachusetts Medical School, Worcester, MA, USA

Go Matsuda Department of Infectious Diseases, National Research Institute for Child Health and Development, Tokyo, Japan

Joseph M. McCune Divison of Experimental Medicine, Department of Medicine, University of California, San Francisco, CA, USA

San Francisco, CA, USA

JoEllyn M. McMillan Department of Pharmacology and Experimental Neuroscience, University of Nebraska Medical Center, Omaha, NE, USA

Sonja Meixlsperger Institute of Experimental Immunology, University of Zürich, Zürich, Switzerland

Brian Moldt Department of Immunology and Microbial Science, International AIDS Vaccine Initiative Neutralizing Antibody Center, and Center for HIV/AIDS Vaccine Immunology and Immunogen Discovery, The Scripps Research Institute, La Jolla, CA, USA

Orla Mulhern University of Southern California, Los Angeles, CA, USA

Christian Münz Institute for Experimental Immunology, University of Zürich, Zürich, Switzerland

Charles Preston Neff Department of Medicine, Division of Allergy and Clinical Immunology Aurora, University of Colorado, Denver, Aurora, CO, USA

Tomonori Nochi Laboratory of Functional Morphology, Tohoku University Graduate School of Agricultural Science, Sendai, Miyagi, Japan

Division of Infectious Diseases, Department of Medicine, UNC Center for AIDS Research, University of North Carolina School of Medicine, Chapel Hill, NC, USA

Jun-ichi Nunoya Lineberger Comprehensive Cancer Center, The University of North Carolina at Chapel Hill, Chapel Hill, NC, USA

Department of Microbiology and Immunology, The University of North Carolina at Chapel Hill, Chapel Hill, NC, USA

Department of Microbiology, Dokkyo Medical University, Tochigi, Japan

Jill E. Oldenburg University of Southern California, Los Angeles, CA, USA

Andrew Owen Department of Molecular and Clinical Pharmacology, University of Liverpool, Liverpool, UK

Roberta Pelanda Department of Immunology and Microbiology, University of Colorado Denver School of Medicine, Aurora, CO, USA

Alexander Ploss Department of Molecular Biology, Princeton University, Princeton, NJ, USA

Larisa Y. Poluektova Department of Pharmacology & Experimental Neuroscience, University of Nebraska Medical Center, Omaha, NE, USA

Yulan Qing Case Comprehensive Cancer Center, National Center for Regenerative Medicine, Seidman Cancer Center, University Hospitals Case Medical center and Case Western Reserve University, Cleveland, OH, USA

Anthony Rongvaux Department of Immunobiology, Yale University, New Haven, CT, USA

Kei Sato Laboratory of Viral Pathogenesis, Institute for Virus Research, Kyoto University, Kyoto, Japan

Jean-Pierre Yves Scheerlinck Centre for Animal Biotechnology, Faculty of Veterinary and Agricultural Sciences, The University of Melbourne, Victoria, Australia

Edward Seung Center for Immunology and Inflammatory Diseases, Massachusetts General Hospital, Harvard Medical School, Charlestown, MA, USA

Saki Shimizu The UCLA AIDS Institute, Los Angeles, USA

Leonard D. Shultz Jackson Laboratory, Bar Harbor, ME, USA

Roberto F. Speck Division of Infectious Diseases and Hospital Epidemiology, University Hospital of Zürich, University of Zürich, Zürich, Switzerland

Hergen Spits Tytgat Institute for Liver and Intestinal Research, AMC-UvA, Amsterdam, The Netherlands

Helene Strick-Marchand Unité Immunité Innée, Institut Pasteur, Inserm U668, Paris, France

Till Strowig Helmholtz Center for Infection Research, Braunschweig, Germany

Lishan Su Lineberger Comprehensive Cancer Center, The University of North Carolina at Chapel Hill, Chapel Hill, NC, USA

Department of Microbiology and Immunology, The University of North Carolina at Chapel Hill, Chapel Hill, NC, USA

Dong Sung An Division of Hematology-Oncology, The David Geffen School of Medicine at University of California, Los Angeles, USA

Megan Sykes Columbia Center for Translational Immunology, Columbia University College of Physicians and Surgeons, New York, NY, USA

Andrew M. Tager Center for Immunology and Inflammatory Diseases, Massachusetts General Hospital, Harvard Medical School, Charlestown, MA, USA

Pulmonary and Critical Care Unit, Massachusetts General Hospital, Harvard Medical School, Charlestown, MA, USA

Takeshi Takahashi Central Institute for Experimental Animals, kawasaki-ku, Kawasaki, Japan

Hitoshi Takizawa Division of Hematology, University Hospital Zürich, Zürich, Switzerland

International Research Center for Medical Sciences, Kumamoto University, Kumamoto, Japan

Julien Villaudy Laboratoire de Biologie Moléculaire de la Cellule, Unité Mixte de Recherche 5239, Centre National de la Recherche Scientifique, Ecole Normale Supérieure de Lyon, Lyon Cedex 7, France

SFR UMS3444 BioSciences Lyon-Gerland-Lyon Sud (UMS3444), Lyon Cedex 7, France

Angela Wahl Division of Infectious Diseases, Center for AIDS Research, University of North Carolina School of Medicine, Chapel Hill, NC, USA

Charles Wood Nebraska Center for Virology, School of Biological Sciences, University of Nebraska-Lincoln, Lincoln, NE, USA

Yong-Guang Yang First Hospital of Jilin University, Changchun, China

Columbia Center for Translational Immunology, Columbia University College of Physicians and Surgeons, New York, NY, USA

Jerome A. Zack Division of Hematology-Oncology, The David Geffen School of Medicine at UCLA, The UCLA AIDS Institute, Los Angeles, CA, USA

Chapter 1
Mamoru Ito's Vision for the Future of Humanized Mouse Models

Mamoru Ito

Since development of NOD-*scid* IL-2Rg^{null} (NOG, NSG) and BALB/cA-Rag2^{null} IL-2Rg^{null} (BRG) at the start of the twenty-first century, given the high capacity of human cells and tissues to engraft and differentiate in these models, studies using humanized mice have universally attracted researchers' attention. This chapter describes past, present, and future xenotransplantation mouse models, with a particular focus on developments in Japan and the future technological progress needed for the use of humanized mouse models in translational research fields like HIV-1 infection.

1.1 History of the Development of Immunodeficient Mice in Japan

Centuries before xenotransplantation studies began in Japan and the USA, scientists in Europe were conducting cross-species transplantation studies using newborn animals dating as far back as the sixteenth century [1]. Importantly, failures and limitations associated with these transplants have helped improve understanding of basic immune mechanisms that control tissue compatibility. Finally, the discovery of immunodeficient nude mice dramatically increased the performance of xenotransplantation studies and opened a new door for performing xenotransplantation experiments on small laboratory animals [2]. While there are volumes of information regarding these developments for Europe and North America, little has been published regarding progress in this field of research in Asia. Thus, this chapter briefly describes the history of immunodeficient mice in Japan. Such developments

M. Ito (✉)
Central Institute for Experimental Animals, 3-25-12 Tonomachi, Kawasaki-ku, Kawasaki, Kanagawa 210-0821, Japan
e-mail: mito@ciea.or.jp

L. Y. Poluektova et al. (eds.), *Humanized Mice for HIV Research*,
DOI 10.1007/978-1-4939-1655-9_1

began in Japan in 1973 when Dr. Tatsuji Nomura imported the nude mice from Dr. Friis in Denmark. Dr. Nomura was a Japanese pioneer in this field and helped found the International Nude Mice Workshop, referred to as the International Workshop of Immunodeficient Mice following the fifth workshop onward, with Drs. Rygaard and Povlsen of Denmark. This workshop created a foundation to study how nude mice could be used in biomedical fields which was held nine times between 1972 and 1997 at different locations around the world.

Dr. Nomura actively expanded the initial mouse colony from Dr. Friis, and 30,000 nude mice were produced over the following 3 years. In parallel, he formed a consortium with public institutes and pharmaceutical companies to perform cancer research using nude mice with support from the Japanese Ministry of Public Welfare. In conjunction with his work on expanding the initial colony of nude mice, Dr. Nomura continued the development of new immunodeficient mice as described later. After successful development of NOG mice as a results of his effort, in 2006, Dr. Nomura hosted the International Workshop of Humanized Mice in Tokyo, Japan, which has been attended by researchers every 3 years [3]. With these contributions, Dr. Nomura has helped to progress the field of immunodeficient mice in Japan and throughout the world.

Using Dr. Nomura's work as a foundation, in the early 1980s the Central Institute for Experimental Animals (CIEA) attempted to improve the recipient for xenotransplantation by crossing Dr. Nomura's nude mice with X-linked immunodeficient mice (XID) and beige mice, which were one of the few immunodeficient mouse models available in Japan at the time. Unfortunately, these initial attempts were unsuccessful. In 1985, the CIEA introduced a new immunodeficient mouse model, severe combined immunodeficiency (SCID). SCID mice lack T and B cells [4] and were discovered in the USA by Dr. Melvin Bosma in 1983.

In subsequent work by Dr. Joseph McCune and colleagues in the USA using SCID mice, human T and B cells were successfully generated following transplantation of human fetal liver and thymus into these mutant mice in 1988. These humanized mice, termed *SCID-hu*, are able to maintain human T cells and have been of major interest to researchers, particularly in the study of HIV-1 infection [5]. However, *SCID-hu* mice cannot be used in Japan due to bioethical concerns about the use of human fetal organs. Thus, nonhuman SCID mice have been mainly used for basic immunology and cancer studies in Japan.

Human peripheral blood mononucleated cell (PBMC), as well as fetal organs, can be engrafted into SCID mice [6]. In turn, such mice have been used in Japan for studies involving HIV-1 infection. The *scid* gene (formally, $Prkdc^{scid}$) was introduced into the NOD mouse inbred strain to generate NOD-*scid* mice showing ability to support high levels of HIV-1 viremia after transplantation of human cells [7–9]. Then, international groups of collaborating scientists reported that human hematopoietic stem cells (HSC) differentiate when transplanted into NOD-*scid* mice [10, 11]. Therefore, these mice have been used extensively in stem cell biology for more than a decade, until the development of NOG and BRG mice.

Until the early 1990s, immunodeficient mice had only been obtained accidentally, following a spontaneous mutation. Targeting technology using embryonic stem

(ES) cells, established in 1989 by Italian-born American molecular geneticist and Nobel prize recipient Dr. Mario Capecchi, helped to pave the way for artificially developing numerous immunodeficient mouse models [12]. In 2002, artificially generated IL-2Rg knockout mice were crossed with NOD-*scid* mice to create NOG mice, which can inactivate the gene encoding IL-2Rg [13]. Around this same time, our lab also developed BRG mice by inactivating the gene encoding IL-2Rg from BALB/cA-Rag2^{null} mice. In 2004, Dr. Marcus Manz and colleagues at the University Hospital Zurich reported a humanized mouse model using these BRG mice [14]. In 2005, Dr. Leonard Shultz of the Jackson Laboratory in the USA generated NSG mice, which are similar to NOG mice [15]. In 2010, Dr. Takiguchi and colleagues at Kumamoto University in Japan generated NOD-*scid-Jak3^{null}* mice, which have the same immunodeficiency as NOG and NSG mice [16]. These immunodeficient mouse models have been critical in the recent progress in normal and diseased human cell/tissue transplantation for regenerative medicine, cancer and therapeutics development.

1.2 Currently Available Humanized Mouse Models Generated Using NOG, NSG, and BRG Mice and Their Limitations

In general, the following three strains of immunodeficient mice are currently used to generate humanized mouse models: NOG [13], NSG [15], and BRG [14]. The common characteristics of immunodeficient mice are that they are deficient in T, B and NK cells, due to SCID/RAG2^{null} and inactivation of IL-2Rg. Inactivation of IL-2Rg allows for a high level of engraftment and differentiation of human cells into NOD-*scid* and BALB/cA-Rag2^{null} mice. Still, the reason why inactivation of IL-2Rg supports engraftment and differentiation of xenografts is unclear. It is quite possible that inactivation of IL-2Rg is linked to the dysfunction of cytokines responsible for T, B, and natural killer (NK) cell proliferation and differentiation. Our team recently demonstrated a crucial role of interferon gamma (IFNg)-producing CD11c+B220+CD122+ cells in xenograft rejection. IFNg-producing cells constitute a subpopulation of plasmacytoid dendritic cells and are absent in NOG mice [17]. Production of IFNg is impaired in IL-2Rg-deficient mice [18], which suggests that IFNg has an important role in xenograft rejection. The genetic backgrounds of NOG/NSG and BRG mice are the NOD and BALB/cA inbred strains, respectively. The engraftment rate of human cells is generally considered to increase in NOG/NSG mice compared with BRG mice. This is thought to be because the NOD strain has SIRPa polymorphism similar to human and reduced innate immunity, whereas the BALB/cA strain does not [9, 19].

Engraftment of xenografts, including human cells and tissues, is extremely effective in NOG, NSG, and BRG mice compared with conventional immunodeficient mice like NOD-*scid* and C.B-17-*scid*. The high engraftment capacity of these mice enables improved humanized mouse models to be generated. In general, two techniques are used to generate humanized mice. One involves the transfer of mature human PBMC, and the other involves the transfer of HSCs isolated from human cord blood, bone marrow, or fetal liver (e.g., BLT). In the PBMC technique, transferred mature lymphocytes, CD3+ cells in particular, essentially attack the mouse, resulting in early death due to severe graft versus host disease (GVHD). For instance, NOG mice die at 2 weeks after intravenous transfer of 1×10^7 PBMC. In contrast, severe GVHD does not occur in NOD mice as seen in NOG mice. NOD mice survive more than 2 months, and the GVHD occurs only by intraperitoneal transfer of 1×10^7 PBMC [20].

Human CD3+ cells infiltrate the organs of the NOD mice. These proliferating cells are considered to be xenoreactive and can secrete various cytokines in response to the mouse cells. This secretion of cytokines results in further proliferation and activation of the human cells in a paracrine manner. Severe GVHD does not occur when human PBMC are transferred into NOG mice that have been depleted of major histocompatibility complex. Instead, human T cells proliferate less in these mice than in NOG mice with normal levels of the major histocompatibility complex (unpublished data). It is speculated that humanized mice can be generated by the transfer of particular cells, such as NK cells, that are purified from human PBMC. On the other hand, the long-term maintenance of these cells in NOG mice is expected to be difficult. For example, when human NK cells isolated from PBMC are transferred, the cells only survive in mouse peripheral blood for approximately a week (unpublished data). However, depending on the study, such humanized mice could be used for short-term experiments [21, 22].

In contrast to PBMC transfer, various hematopoietic cells differentiate from HSC in NOG/NSG/BRG mice, and such humanized mice have been of particular interest to researchers. When HSC are transferred into NOG mice, myeloid cells typically develop after 3–4 weeks; B cells typically develop after 6–8 weeks, and T cells typically develop after 10–12 weeks. T cells differentiate into CD4+ or CD8+ cells in NOG mice, whereas T-cell differentiation rarely occurs in NOD-*scid* mice. Conversely, certain cell lineages, such as erythrocytes and granulocytes, rarely develop at all, even in NOG, NSG, and BRG mice. The reasons for this phenomenon are beginning to be understood, and it appears that mouse factors are unable to compensate for the absence of human factors responsible for the differentiation of these cells.

T and B cells that differentiate from HSC in NOG mice can be maintained for as long as 1 year without GVHD. At one time, it was expected that such mice could be used to develop hematolymphoid humanized mice with a complete immune system. However, humanized NOG mice do not produce antigen-specific human immunoglobulin (Ig) G antibodies, even when they are challenged with antigens. Antigen-specific cytotoxic T lymphocytes (CTL) were not also induced in NOG mice. This lack of responsiveness may be because human T/B cells and antigen-presenting cells do not interact in humanized NOG mice when human T cells are educated in the mouse thymus. By using NOG/NSG mice that express class I or II human leukocyte antigens (HLA), antigen-specific human IgG antibodies and CTL can be induced following transfer of HLA-matched HSC [23–27]. Thus, human immune responses can be partially elicited in immunodeficient mice that express HLA.

Humanized mice with engrafted human T cells following transfer of PBMC and HSC can be used to evaluate anti-HIV-1 drugs. Still, such mouse models cannot be used to research immunological responses to HIV-1 infection or the development of an HIV-1 vaccine due to functional deficiency of human T and B cells resulting in the lack of robust adaptive immune responses. NOD-*scid* mice that have received human fetal bone marrow, liver, and thymus (i.e., BLT mice), have a working human immune system and can be used as a HIV-1 infection model [28]. However, there are bioethical concerns about the use of such mice, and consequently, they cannot be used in Japan.

1.3 Novel Humanized Mouse Models Generated Using Improved Immunodeficient Mice

Recently, to overcome the disadvantages of conventional NOG, NSG, and BRG mice, several improved immunodeficient mouse models have been developed primarily through the introduction of various human genes [23–34]. Our team has developed and improved several immunodeficient mouse models (http://www.ciea.or.jp/kiban-s/index.html). Our new models were primarily established through the introduction of human cytokine genes and mutated mouse genes (unpublished data). Mice that have been modified to have genes encoding HLA are of interest because they exhibit human immune response following transfer of human haplotype-matched HSC.

Here, we briefly describe the characteristics of the humanized mouse models of particular interest developed by CIEA. In NOG mice expressing human interleukin (IL)-2 following HSC transfer, human NK cells generally developed 4 weeks before T and B cells. These NK cells accounted for 80–90 % of human cells in NOG-hIL-2 mice, and the NK cells effectively suppressed the growth of NK-sensitive K562 leukemia cells in vivo (paper submitted). Myeloid lineage cells, including granulocytes and monocytes, successfully formed in NOG mice that expressed human granulocyte macrophage colony-stimulating factor (GM-CSF) and IL-3, whereas they rarely form in conventional NOG mice. Additionally, a passive cutaneous anaphylactic reaction was successfully elicited in these humanized NOG mice following intracutaneous inoculation of sera from pollenosis patients followed by intravenous inoculation of pollen antigen and Evans Blue dye [35]. HSC did not differentiate when transferred into NOG mice that express human IL-4, but the reason for this remains unclear. When these mice were transplanted with HSC, they showed mild GVHD, and human cells could be maintained for a longer time period due to the shift of T cells to Th2 cells.

These improved humanized mouse models can be used to study human diseases. The evaluation of the use of such models in this context will be left to experts in different areas of biomedical disciplines, such as regeneration, development, infectious diseases and vaccines, and cancer and therapeutics.

1.4 Generation of Humanized Mice in the Future

As outlined earlier, the CIEA research group has developed various improved humanized NOG mouse models, and this work is ongoing. Here, based on our experience, we describe how improved humanized mouse models can be generated in the future (see Fig. 1.1).

As part of our vision for improving immunodeficient mouse models, the following items are needed: the creation of recipients for xenotransplants, the development of techniques to create artificial organs, and the development of methods to create a plentiful and stable supply of stem cells. Here, we focus on humanized mouse models with functioning human immune systems, given that such models are needed to research mechanisms underlying human immune disorders and to develop drugs to treat these disorders.

1.4.1 Improvements in Immunodeficient Mouse Models

Presently, NOG and NSG mice are the best-suited recipients for the generation of humanized mouse models. However, these mice still have cells responsible for innate immunity, such as granulocytes, macrophages, and dendritic cells. It is possible that elimination of innate immunity cells will generate more appropriate recipients.

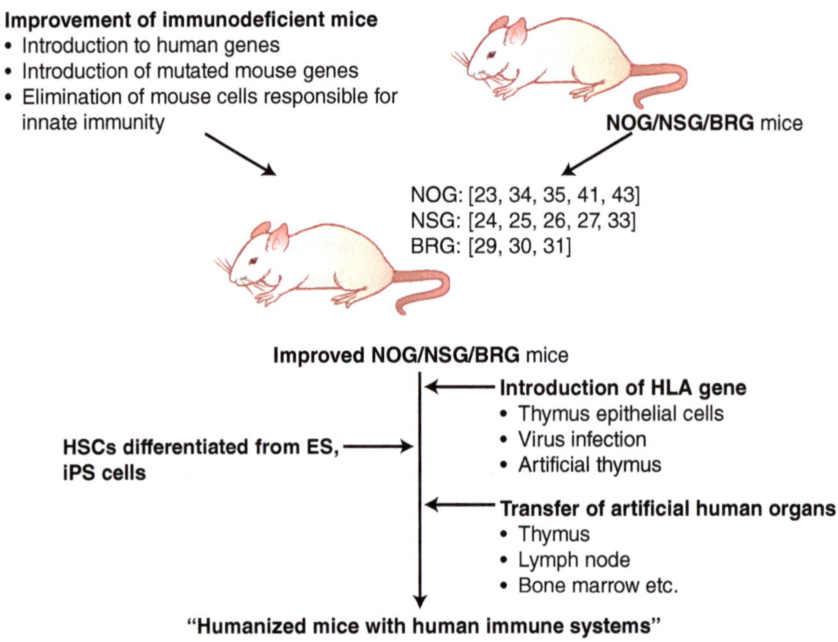

Fig. 1.1 Schematic diagram illustrating the possible future generation of humanized mice

Another possible improvement is to introduce human genes into immunodeficient mice that support the human cell growth and differentiation, as mouse factors cannot always compensate for the absence of such human factors. Indeed, we introduced human cytokine genes to generate improved humanized mouse models. Further crossing of these mouse models is expected to allow the creation of additional useful models. Nonetheless, it is time-consuming and technically difficult to cross three or more transgenic mouse strains to generate mice that express multiple transgenes. Furthermore, in mice with multiple transgenes, the transgenes can influence each other, and consequently, mice might not exhibit the expected characteristics. Knock-in technology using ES cells is an alternative strategy to generate improved humanized mouse models [29, 31, 32]. Recently, we successfully developed ES cells derived from NOG mice embryos. By performing homologous recombination using these cells, we can easily generate knock-in NOG mice, in which mouse genes are replaced with human genes. Remarkable progress in transgenic technology was recently made, that is a genome editing using CRISPR/CAS, TALENs and zinc-finger nucleases to generate targeting animals without using ES cells [36-38]. These techniques are expected to be a useful future tool for more quickly generating improved humanized NOG mice models.

1.4.2 Artificial Organ Formation

Humanized mouse models with human organs are uncommon. Nevertheless, human ovary and endometrium were successfully generated using such models [39, 40]. Furthermore, a human liver has been formed after implantation of human liver cells into urokinase-type plasminogen activator transgenic NOG mice (uPA-NOG), fumarylacetoacetate hydrolase KO BRG mice (BRG-FAH), herpes simplex virus type 1 thymidine kinase transgenic NOG mice (TK-NOG), and FK506-binding domain-fused caspase 8 transgenic BRG mice (AFC8-BRG), in which mouse liver cells were spontaneously or artificially destroyed [41–43]. These mice also appear to be useful as models for hepatitis C and B infection and for drug metabolism research [44].

Implanting human cells into NOG mice is considered the preferred method to generate humanized mice with artificial organs. To generate xeno-organs in mammals, the laboratory of Dr. Nakauchi at Tokyo University recently developed a novel technique termed "blastocyst complementation." In this technique, normal ES cells are injected into blastocysts of organ-deficient mammals. This technique was successfully used to develop a rat pancreas in a mouse [45], which grew to the same size as a normal mouse pancreas. By injecting human ES or induced pluripotent (iPS) cells, rather than rat ES cells, it may be possible to develop mice with human organs. For example, by injecting human ES or iPS cells into blastocysts of thymus-deficient BRG-nu mice, which we have already developed, there is potential to develop mice with a human thymus. However, the use of this technique is currently limited in Japan due to bioethical concerns. In another attempt, there is an interest report on artificial lymph nodes by the research group of Dr. Takeshi Watanabe at Kyoto University [46].

1.4.3 A Plentiful and Stable Supply of Stem Cells is Needed to Generate Humanized Mice

Stem cells, such as HSC, are needed to generate humanized mice, but they are relatively difficult to obtain and are heterogeneous. These difficulties are particularly pronounced when haplotype-matched HSC are used in NOG mice that express HLA. To overcome these issues, techniques are needed that allow HSC to be generated from ES or iPS cells. There are several reports on the generation of CD34+ cells from ES cells in vitro [47, 48]. Still, HSC generated from ES/iPS cells have not been successfully developed into multilineage hematopoietic cells in vivo. A plentiful and stable supply of HSC is expected to be useful for future routine generation of humanized mouse models.

1.4.4 Mice with Human Immune Systems

Mice with human immune systems are needed for vaccine development and to evaluate immunological responses to newly developed drugs. The development of mice with humanized thymus and lymph nodes is considered key to the generation of mice with human immune systems. Following transfer of haplotype-matched HSC, antigen-specific IgG antibodies are produced in HLA-DR Tg NOG/NSG mice, and CTL are induced in HLA-A2 NSG mice. Results indicate that expression of HLA assists in the generation of mice with human immune systems. Hence, an immediate focus should be placed on the generation of HLA-NOG mice that express a dominant HLA haplotype, such as D4 or D9 of HLA class I and A2 or A24 of HLA class II. By crossing mice that express HLA-DR or HLA-A, there is the potential to generate improved humanized mouse models in which human antibodies are produced, and human CTL are induced. Another approach is to inject multiple DNA isolated from persons with different HLA haplotypes into mouse embryos in order to generate mice that express multiple HLA haplotypes. Using this technique, mice can be generated that simultaneously express two or three types of HLA class I/II. Still, the expression levels of these antigens in such mice remain obscure.

There are many HLA haplotypes, which have practical applications in personalized medicine. For example, an iPS cell bank is being established and promoted specifically for the Japanese population to provide cell transplantation therapy. This resource is located at the Center of iPS Cell Research and Application (CiRA) at Kyoto University in Japan. CiRA has estimated that to cover 90 % of the Japanese population, approximately 140 iPS cell lines homozygous for HLA-A, HLA-B, or HLA-DR are needed. It is not possible to generate HLA-NOG mice that cover all potential haplotypes of Japanese individuals. Still, the number of HLA haplotypes required could be less than 140. In future work, several techniques could be used to overcome this challenge. First, it may be possible to generate human thymus epithelial cell lines of various HLA haplotypes. These cells could be implanted into mouse thymus and educate human T cells to interact with B cells and antigen-presenting cells. Unfortunately, no such cell lines are currently available. Second, it may be possible to express HLA either specifically in mouse thymus or systemically by infecting mice with a virus vector. Nonetheless, virus vectors that can potentially be used to infect the thymus have only recently been developed. In 2010, Chu et al. [49] reported that the adeno-associated virus may be a candidate, but its efficacy remains uncertain. Third, it may be possible to generate an artificial thymus from human ES or iPS cells. This would be a major advancement in the generation of humanized mice with human immune systems. However, a technical advance is required in order to generate an artificial thymus that covers all potential HLA haplotypes.

Human immune responses occur in HLA-NOG mice, but they appear to be weaker than true human immune responses. The reason for this may be that NOG mice have severe leukopenia and marked atrophy of the thymus and lymph nodes. Dr. Paul Denton et al. [50] reported that following transplantation of human fetal BLT, NOD-*scid* mice develop intestinal human T cells throughout the small and large intestine, whereas NSG and BRG mice do not. This failure of human cells to develop in the intestines of NSG and BRG mice is due to the inactivation of IL-2Rg in these mice. Inactivation of Il-2Rg is expected to impair the differentiation of lymphoid tissue inducer cells responsible for lymph node development. This can be overcome by specifically expressing IL-2Rg in the lymph nodes of IL-2Rg-deficient mice, including NOG mice.

In conclusion, there are issues associated with currently available humanized mouse models. These issues need to be resolved to generate improved humanized mouse models through which specific human diseases are reproduced. In turn, these models have the potential for use in the study of the mechanisms underlying diseases and in the development of drugs to treat diseases.

Acknowledgments This work was supported by a grant from the Research Foundation for Scientific Research (S) (#22220007) of the Ministry of Education, Culture, Sports, Science, and Technology (MEXT) and a grant from Research on Emerging and Re-emerging Infectious Diseases from the Ministry of Health, Labour, and Welfare, Japan. This review is dedicated to Prof. Tatsuji Nomura, the late director of the Central Institute for Experimental Animals (CIEA), whose forethought helped in guiding our research.

References

1. Deschamps JY, Roux FA, Sai P, Gouin E. History of xenotransplantation. Xenotransplantation. 2005;12(2):91–109.
2. Isaacson JH, Cattanach B. Two new 'hairless' mutants—Sha and Hfh11. Mouse News Lett. 1962;27:31.
3. Manz MG, Di Santo JP. Renaissance for mouse models of human hematopoiesis and immunobiology. Nat Immunol. 2009;10(10):1039–42.
4. Bosma GC, Custer RP, Bosma MJ. A severe combined immunodeficiency mutation in the mouse. Nature. 1983;301(5900):527–30.

5. McCune JM, Namikawa R, Kaneshima H, Shultz LD, Lieberman M, Weissman IL. The SCID-hu mouse: murine model for the analysis of human hematolymphoid differentiation and function. Science. 1988;241(4873):1632–9.
6. Mosier DE, Gulizia RJ, Baird SM, Wilson DB. Transfer of a functional human immune system to mice with severe combined immunodeficiency. Nature. 1988;335(6187):256–9.
7. Koyanagi Y, Tanaka Y, Tanaka R, Misawa N, Kawano Y, Tanaka T, et al. High levels of viremia in hu-PBL-NOD-scid mice with HIV-1 infection. Leukemia. 1997;11(Suppl 3):109–12.
8. Koyanagi Y, Tanaka Y, Kira J, Ito M, Hioki K, Misawa N, et al. Primary human immunodeficiency virus type 1 viremia and central nervous system invasion in a novel hu-PBL-immunodeficient mouse strain. J Virol. 1997;71(3):2417–24.
9. Shultz LD, Schweitzer PA, Christianson SW, Gott B, Schweitzer IB, Tennent B, et al. Multiple defects in innate and adaptive immunologic function in NOD/LtSz-scid mice. J Immunol. 1995;154(1):180–91.
10. Larochelle A, Vormoor J, Hanenberg H, Wang JC, Bhatia M, Lapidot T, et al. Identification of primitive human hematopoietic cells capable of repopulating NOD/SCID mouse bone marrow: implications for gene therapy. Nat Med. 1996;2(12):1329–37.
11. Dick JE, Bhatia M, Gan O, Kapp U, Wang JC. Assay of human stem cells by repopulation of NOD/SCID mice. Stem Cells. 1997;15(Suppl 1):199–203 (discussion 4–7).
12. Capecchi MR. Altering the genome by homologous recombination. Science. 1989;244(4910):1288–92.
13. Ito M, Hiramatsu H, Kobayashi K, Suzue K, Kawahata M, Hioki K, et al. NOD/SCID/gamma(c)(null) mouse: an excellent recipient mouse model for engraftment of human cells. Blood. 2002;100(9):3175–82.
14. Traggiai E, Chicha L, Mazzucchelli L, Bronz L, Piffaretti JC, Lanzavecchia A, et al. Development of a human adaptive immune system in cord blood cell-transplanted mice. Science. 2004;304(5667):104–7.
15. Shultz LD, Lyons BL, Burzenski LM, Gott B, Chen X, Chaleff S, et al. Human lymphoid and myeloid cell development in NOD/LtSz-scid IL2R gamma null mice engrafted with mobilized human hemopoietic stem cells. J Immunol. 2005;174(10):6477–89.
16. Sato Y, Takata H, Kobayashi N, Nagata S, Nakagata N, Ueno T, et al. Failure of effector function of human CD8+ T Cells in NOD/SCID/JAK3(-)/(-) immunodeficient mice transplanted with human CD34+ hematopoietic stem cells. PloS ONE. 2010;5(10):e13109.
17. Ito R, Katano I, Ida-Tanaka M, Kamisako T, Kawai K, Suemizu H, et al. Efficient xenoengraftment in severe immunodeficient NOD/Shi-scid IL2rgammanull mice is attributed to a lack of CD11c+B220+CD122+ cells. J Immunol. 2012;189(9):4313–20.
18. Ohteki T, Fukao T, Suzue K, Maki C, Ito M, Nakamura M, et al. Interleukin 12-dependent interferon gamma production by CD8alpha+ lymphoid dendritic cells. J Exp Med. 1999;189(12):1981–6.
19. Takenaka K, Prasolava TK, Wang JC, Mortin-Toth SM, Khalouei S, Gan OI, et al. Polymorphism in Sirpa modulates engraftment of human hematopoietic stem cells. Nat Immunol. 2007;8(12):1313–23.
20. Ito R, Katano I, Kawai K, Hirata H, Ogura T, Kamisako T, et al. Highly sensitive model for xenogenic GVHD using severe immunodeficient NOG mice. Transplantation. 2009;87(11):1654–8.
21. Shiokawa M, Takahashi T, Murakami A, Kita S, Ito M, Sugamura K, et al. In vivo assay of human NK-dependent ADCC using NOD/SCID/gammac(null) (NOG) mice. Biochem Biophys Res Commun. 2010;399(4):733–7.
22. Ito A, Ishida T, Yano H, Inagaki A, Suzuki S, Sato F, et al. Defucosylated anti-CCR4 monoclonal antibody exercises potent ADCC-mediated antitumor effect in the novel tumor-bearing humanized NOD/Shi-scid, IL-2Rgamma(null) mouse model. Cancer Immunol Immunother. 2009;58(8):1195–206.
23. Suzuki M, Takahashi T, Katano I, Ito R, Ito M, Harigae H, et al. Induction of human humoral immune responses in a novel HLA-DR-expressing transgenic NOD/Shi-scid/gammacnull mouse. Int Immunol. 2012;24(4):243–52.

24. Strowig T, Gurer C, Ploss A, Liu YF, Arrey F, Sashihara J, et al. Priming of protective T cell responses against virus-induced tumors in mice with human immune system components. J Exp Med. 2009;206(6):1423–34.
25. Shultz LD, Saito Y, Najima Y, Tanaka S, Ochi T, Tomizawa M, et al. Generation of functional human T-cell subsets with HLA-restricted immune responses in HLA class I expressing NOD/SCID/IL2r gamma(null) humanized mice. Proc Natl Acad Sci U S A. 2010;107(29):13022–7.
26. Sato Y, Nagata S, Takiguchi M. Effective elicitation of human effector CD8+ T Cells in HLA-B*51:01 transgenic humanized mice after infection with HIV-1. PloS ONE. 2012;7(8):e42776.
27. Danner R, Chaudhari SN, Rosenberger J, Surls J, Richie TL, Brumeanu TD, et al. Expression of HLA class II molecules in humanized NOD. Rag1KO.IL2RgcKO mice is critical for development and function of human T and B cells. PloS ONE. 2011;6(5):e19826.
28. Denton PW, Garcia JV. Novel humanized murine models for HIV research. Curr HIV/AIDS Rep. 2009;6(1):13–9.
29. Willinger T, Rongvaux A, Takizawa H, Yancopoulos GD, Valenzuela DM, Murphy AJ, et al. Human IL-3/GM-CSF knock-in mice support human alveolar macrophage development and human immune responses in the lung. Proc Natl Acad Sci U S A. 2011;108(6):2390–5.
30. Strowig T, Rongvaux A, Rathinam C, Takizawa H, Borsotti C, Philbrick W, et al. Transgenic expression of human signal regulatory protein alpha in Rag2-/-gamma(c)-/- mice improves engraftment of human hematopoietic cells in humanized mice. Proc Natl Acad Sci U S A. 2011;108(32):13218–23.
31. Rongvaux A, Willinger T, Takizawa H, Rathinam C, Auerbach W, Murphy AJ, et al. Human thrombopoietin knockin mice efficiently support human hematopoiesis in vivo. Proc Natl Acad Sci U S A 2011;108(6):2378–83.
32. Rathinam C, Poueymirou WT, Rojas J, Murphy AJ, Valenzuela DM, Yancopoulos GD, et al. Efficient differentiation and function of human macrophages in humanized CSF-1 mice. Blood. 2011;118(11):3119–28.
33. King MA, Covassin L, Brehm MA, Racki W, Pearson T, Leif J, et al. Human peripheral blood leucocyte non-obese diabetic-severe combined immunodeficiency interleukin-2 receptor gamma chain gene mouse model of xenogeneic graft-versus-host-like disease and the role of host major histocompatibility complex. Clin Exp Immunol. 2009;157(1):104–18.
34. Ito R, Negishi N, Irie N, Matsuo K, Suzuki D, Katano I, et al. Osteosclerosis and inhibition of human hematopoiesis in NOG mice expressing human Delta-like 1 in osteoblasts. Exp Hematol. 2012;40(11):953–63 (e3).
35. Ito R, Takahashi T, Katano I, Kawai K, Kamisako T, Ogura T, Ida-Tanaka M, Suemizu H, Nunomura S, Ra C, Mori A, Aiso S and Ito M. Establishment of a human allergy model using human IL-3/GM-CSF-transgenic NOG mice. *J Immunol*. 2013;191:2890–99.
36. Cong L, Ran FA, Cox D, Lin S, Barretto R, Habib N, Hsu PD, Wu X, Jiang W, Marraffini LA and Zhang F. Multiplex genome engineering using CRISPR/Cas systems. *Science*. 2013;339:819–23.
37. Tong C, Huang G, Ashton C, Wu H, Yan H and Ying QL. Rapid and cost-effective gene targeting in rat embryonic stem cells by TALENs. *J Genet Genomics*. 2012;39:275–80.
38. Geurts AM, Cost GJ, Freyvert Y, Zeitler B, Miller JC, Choi VM, Jenkins SS, Wood A, Cui X, Meng X, Vincent A, Lam S, Michalkiewicz M, Schilling R, Foeckler J, Kalloway S, Weiler H, Menoret S, Anegon I, Davis GD, Zhang L, Rebar EJ, Gregory PD, Urnov FD, Jacob HJ and Buelow R. Knockout rats via embryo microinjection of zinc-finger nucleases. *Science*. 2009;325:433.
39. Terada Y, Terunuma-Sato Y, Kakoi-Yoshimoto T, Hasegawa H, Ugajin T, Koyanagi Y, et al. Development of human Graafian follicles following transplantation of human ovarian tissue into NOD/SCID/gammac null mice. Am J Reprod Immunol. 2008;60(6):534–40.
40. Matsuura-Sawada R, Murakami T, Ozawa Y, Nabeshima H, Akahira J, Sato Y, et al. Reproduction of menstrual changes in transplanted human endometrial tissue in immunodeficient mice. Hum Reprod. 2005;20(6):1477–84.

41. Suemizu H, Hasegawa M, Kawai K, Taniguchi K, Monnai M, Wakui M, et al. Establishment of a humanized model of liver using NOD/Shi-scid IL2Rgnull mice. Biochem Biophys Res Commun. 2008;377(1):248–52.
42. He Z, Zhang H, Zhang X, Xie D, Chen Y, Wangensteen KJ, et al. Liver xeno-repopulation with human hepatocytes in Fah-/-Rag2-/- mice after pharmacological immunosuppression. Am J Pathol. 2010;177(3):1311–9.
43. Hasegawa M, Kawai K, Mitsui T, Taniguchi K, Monnai M, Wakui M, et al. The reconstituted 'humanized liver' in TK-NOG mice is mature and functional. Biochem Biophys Res Commun. 2011;405(3):405–10.
44. Bility MT, Zhang L, Washburn ML, Curtis TA, Kovalev GI, Su L. Generation of a humanized mouse model with both human immune system and liver cells to model hepatitis C virus infection and liver immunopathogenesis. Nat Protoc. 2012;7(9):1608–17.
45. Kobayashi T, Yamaguchi T, Hamanaka S, Kato-Itoh M, Yamazaki Y, Ibata M, et al. Generation of rat pancreas in mouse by interspecific blastocyst injection of pluripotent stem cells. Cell. 2010;142(5):787–99.
46. Kobayashi Y, Kato K, Watanabe T. Synthesis of functional artificial lymphoid tissues. Discov Med. 2011;12(65):351–62.
47. Vodyanik MA, Bork JA, Thomson JA, Slukvin, II. Human embryonic stem cell-derived CD34+ cells: efficient production in the coculture with OP9 stromal cells and analysis of lymphohematopoietic potential. Blood. 2005;105(2):617–26.
48. Srivastava AS, Nedelcu E, Esmaeli-Azad B, Mishra R, Carrier E. Thrombopoietin enhances generation of CD34+ cells from human embryonic stem cells. Stem Cells. 2007;25(6):1456–61.
49. Chu Q, Moreland RJ, Gao L, Taylor KM, Meyers E, Cheng SH, et al. Induction of immune tolerance to a therapeutic protein by intrathymic gene delivery. Mol Ther. 2010;18(12):2146–54.
50. Denton PW, Nochi T, Lim A, Krisko JF, Martinez-Torres F, Choudhary SK, et al. IL-2 receptor gamma-chain molecule is critical for intestinal T-cell reconstitution in humanized mice. Mucosal Immunol. 2012;5(5):555–66.

Part I
Mouse Genetic Background and Human Hematopoietic Stem Cells Biology

Chapter 2
Humanized Mice as Models for Human Disease

Joseph M. McCune and Leonard D. Shultz

2.1 Introduction

It has been over a quarter century since "humanized mouse" models were first introduced. As before, such models facilitate the preclinical analysis of therapeutic compounds and vaccines, and otherwise assist in decision-making processes as ideas move from the bench to the clinic. At best, these models even provide a more complete understanding of pathologic mechanisms in vivo, enabling the focused discovery of better interventions for use in treating human disease. Conversely, and as has been evident throughout the history of humanized mice, it is important that their limitations be acknowledged, especially by investigators close to the field, lest inappropriate experiments be designed and/or inaccurate predictions be made.

This book is devoted to state-of-the-art descriptions of the potential utility of these models, and this introductory chapter frames that potential in terms of ultimate goals and hurdles yet to cross. In doing this, two general messages emerge: first, great strides have been made in the development and optimization of these models in recent years and, secondly, more work needs to be done before their full potential and practical applications can be fully realized.

J. M. McCune (✉)
Divison of Experimental Medicine, Department of Medicine, University of California, UCSF Box 1234, San Francisco, CA 94143-1234, USA
e-mail: MMcCune@medsfgh.ucsf.edu

1001 Potrero Ave. Building 3, 6th Floor, Room 601, San Francisco, CA 94110, USA

L. D. Shultz
Jackson Laboratory, Bar Harbor, ME, USA

L. Y. Poluektova et al. (eds.), *Humanized Mice for HIV Research*,
DOI 10.1007/978-1-4939-1655-9_2

2.2 The Rationale Design of Animal Models

By definition, a model is something that is designed to mimic something else. In the case of an animal model of a human disease, the immediate contradiction arises: since no two humans are alike, who should the animal "model"? To circumvent this issue, it is perhaps more pragmatic to instead pose the question: what problem is being addressed? The answer to this question dictates the design and development of the model, not to mention the need for its creation at all.

Likely as not, different mouse models will be used to address dissimilar questions and some questions will not be answerable through the use of any existing mouse model. By example, the human fetal and adult immune systems are distinct from one another in function if not phenotype, emanating in the fetus from a multilineage hematopoietic stem cell (HSC) in the liver and bone marrow that gives rise to tolerogenic (e.g., FoxP3+ Treg) T cells, and from another HSC population in the adult bone marrow that gives rise to immunoreactive (e.g., Th1 and Th2) T cells [1, 2]. It follows that the human fetal immune response would best be studied in a mouse model constructed with human fetal tissues whereas adult-type immune responses would be more appropriately studied in a model constructed with adult bone marrow-derived HSCs. As another example, it is clear that the lymphoid architecture of immunodeficient mice (including *Prkdc^{scid}/Prkdc^{scid}*, hereafter abbreviated as *scid* mice, and especially *scid IL2rg^{null}* mice) is underdeveloped relative to that of wild-type mice or, for that matter, normal humans, making it unlikely that it can support physiologic trafficking and differentiation of human lymphoid cells following human HSC engraftment [3, 4]. Until and unless this problem is solved, it is difficult to imagine how "normal" human immune function can be optimally studied using such mice.

Once a question has been posed and a model designed, it would seem best to satisfy certain other criteria:

1. *The model must be shown to mimic that aspect of human physiology and/or pathophysiology that it is designed to "model."* This demonstration is the type of iterative work that forms the foundation of translational research, and there is no more important venue for it to occur than in the design and use of animal models. Thus, careful clinical observation should frame the design, development, and optimization of a given model. To the extent that similar (and, hopefully, identical) observations can be made in the model, it is more likely that new data obtained from it will be pertinent to the human case.
2. *The model must be constructed in a way that provides for reproducible results.* In general, this means that there should be openly available and standardized procedures that lead to definable mouse xenogeneic chimeras with measurable parameters of human hematolymphoid engraftment and immune function, ones that meet the usual level of reproducibility demanded of any experimental endeavor. External validation of novel findings is otherwise difficult. This criterion prompts careful consideration of two linked issues. First, since there is broad interindividual variation in almost every feature of human biology (including, for example,

the capacity of cells or hematolymphoid organs to engraft in a mouse and/or to sustain HIV replication thereafter), [5] it follows that reproducible results will be most likely obtained within a given cohort of mice if all members of the cohort are prepared with cells and tissues from the same human donor. Since data from humanized mouse models would ideally be applicable to more than a single individual, it follows also that it would be important to have the ability to repeat experiments in the context of multiple cohorts of humanized mice, bearing cells and tissues from multiple human donors.

3. *In most instances, the model should be capable of relatively high throughput analysis of interventions* such as drugs and vaccines, amenable to the analysis of different congeners in varying doses and by different routes, with appropriate positive and negative controls, and with a sufficient number of nearly identical animals in each subgroup to enable statistical analysis of the data. Thus, while isolated observations made in one or two humans have occasionally offered profound insights into normal and abnormal human biology (consider, for instance, the first vaccinations against smallpox by Jenner or the first use of insulin by Banting and Best), [6] it seems far less likely that statistically nonverifiable results from small numbers of humanized mice will have lasting clinical impact.

The bottom line: in the idealized case, it would be possible to make multiple cohorts of many mice, with each cohort using tissue from the same human donor, each with roughly similar quantitative and qualitative levels of engraftment, and all accurately mimicking a given aspect of human physiology or pathophysiology that is of interest to study and otherwise not approachable in vitro.

2.3 A Case in Point: The SCID-hu Thy/Liv Model

When first confronted with the above issues some 25 years ago, there was interest in designing an animal model that would enable the preclinical evaluation of antiviral compounds and vaccines against HIV. With this goal in mind, it was important that the animal be small (so that issues of cost and number could be minimized), that it harbor human hematolymphoid organs (e.g., thymus, lymph node, spleen, and bone marrow) and constituent primary cells that are normally infected by HIV in people, and that it be permissive for infection with primary isolates of HIV. Given these constraints, an attempt was made to engraft interactive human hematolymphoid organs into a mouse in such a way that they would not be rejected. Thus was born the SCID-hu Thy/Liv mouse and its various derivatives, e.g., those engrafted with bone, lymph node, bone–thymus–spleen (the "BTS" mouse), or all of the above (the "full house mouse") [7–11].

The decision to implant human fetal tissues and human HSC into the immunodeficient CB17 *scid/scid* (SCID) mouse was prompted by the following logic: (1) it seemed less likely that (human versus mouse) graft-versus-host disease (GVHD) would occur if human fetal HSC were allowed to differentiate into T and B cells in the mouse environment and, hence, come to see the mouse as "self"; [12] (2) it seemed more likely that physiologic human immune cell maturation would occur within implanted human, as opposed to endogenous murine, parenchymal microenvironments; and, not least (3) successful engraftment of a functional human immune system could be easily ascertained: CB17 *scid/scid* mice, but not other stocks of immunodeficient mice available at the time, were known to otherwise succumb to *Pneumocystis carinii* pneumonia [13].

Initial experiments using the SCID-hu model showed robust engraftment, vascularization, and growth of the conjoint Thy/Liv organ that was formed by coimplantation of human fetal liver and human fetal thymus under the kidney capsule [7, 8]; no signs or symptoms of xenogeneic GVHD (due to the unanticipated movement of murine myeloid cells into the human Thy/Liv organ, promoting negative selection of developing human T cells that recognized the H-2^d background of the CB17-*scid/scid* mouse) [7, 14–15]; detectable replication and spread of primary isolates of HIV [16]; human antibody production and class switching in engrafted human fetal lymph nodes [7, 17]; and, remarkably, the absence of *Pneumocystis carinii* pneumonia in all engrafted mice [7]. Given these results, countless attractive applications presented themselves for further consideration: from preclinical analysis of antiviral compounds and vaccines against a wide variety of human pathogens to isolation of human hematopoietic progenitor cells to dissection of the rules of tolerance induction in humans to creation of an expanded battery of humanized mouse models for the analysis of other human organ systems (e.g., the central nervous system).

Over the next decade, with funding from the National Institutes of Health (NIH) as well as from biotech/pharma, and with the hard work and good thought of many, these notions were put to test. Not surprisingly, some worked but most did not:

1. The SCID-hu Thy/Liv model could be optimized and standardized, and created in cohorts of 50 or more mice per human tissue donor for the preclinical analysis of antiviral compounds against HIV [18, 19]. With the help of a relatively large and committed staff of highly trained and expert researchers, it has been possible to create as many as 40 such cohorts on an annual basis and to infect, dose, and analyze the results of antiviral drugs within them in a reproducible manner. Given the data now on hand, the predictive powers of this model have proven to be considerable: antiretroviral compounds found to be active in it have also been found to be active in humans; those not active in it are also inactive in humans. In addition, use of the model has provided unique information about the bioavailability and mechanism of action of antiviral compounds in vivo.
2. The Thy/Liv model and its derivatives (including the SCID-hu "Bone" and "BTS" models) enabled the discovery and definition of human HSC for clinical use [20–22], including application of gene-modified HSCs for the treatment of HIV disease [23, 24]. This application and recent modifications continue to be in play, enabling, by example, the recent discrimination between fetal and adult human HSCs [1].

3. The Thy/Liv implant has been found to faithfully mimic the structure and physiology of the normal human thymus [7, 8, 25–29]. It accordingly represents a convenient model to study many aspects of human T cell differentiation and function that could not be studied in vitro and that are only studied with difficulty, if at all, in humans or in nonhuman primates.
4. The SCID-hu Bone model can be used to study the effects of irradiation [30] and exogenously provided cytokines [31] on human HSCs and for the analysis of species- and organ-specific metastasis of human malignancies [32, 33].

On the other hand, and after considerable effort, many other "potential applications" failed. Thus, peripheral human immune responses are observed but found to be neither useful nor relevant, in part because the level of human cell engraftment in the periphery is low but also because important cues provided by the murine lymphoid architecture (e.g., endothelial adhesion molecules, cytokines, chemokines, and the like) are incompatible with (i.e., do not bind to and/or appropriately activate) human cells; it is logistically difficult to carry out longitudinal studies in SCID-hu mice, e.g., sampling peripheral blood or implanted organs over time; and it is difficult to reproducibly engraft other important components of the human immune system, e.g., lymph node, spleen, liver, and gut associated lymphoid tissue. It was these deficiencies, indeed, that sparked interest in the development of secondary generations of humanized mouse models.

2.4 The Next Wave of Humanized Mice

Humanized mouse models developed to address the deficiencies of the SCID-hu models have been well summarized in recent reviews [3, 34–40] and extensively discussed in the remaining chapters of this book. An important distinguishing characteristic is their higher level of multilineage peripheral engraftment, manifest even in the peripheral blood. This attribute is obtained upon the use of mouse stocks that are even more immunodeficient than the CB17-*scid/scid* strain [e.g., NOD-*scid/scid Il2rg*^{null} (NSG), BALB/c-*Rag2*^{null}*Il2rg*^{null} (BRG), etc.], younger (e.g., engrafted with human cells shortly after birth), and/or irradiated [3]. As a consequence, it is now possible to sample human cells and to quantitate signs of HIV replication by drawing peripheral blood. This feature greatly simplifies the use of the models and enables experimental designs (e.g., longitudinal analyses) that cannot be carried out in the SCID-hu Thy/Liv mouse.

These advances not only open the door to a wide range of experimental possibilities, they also make the use of the models more widely accessible (provided that newly developed stocks of humanized mouse models are made readily available to all through a public repository). It is accordingly not surprising that they have been embraced with enthusiasm.

2.5 Barriers to Cross

Imagination being the powerful force that it is, investigators who are outside the field of humanized mouse technology will be motivated to pursue important applications that may or may not be feasible. It will be incumbent upon those who are working in the field to clearly discuss what can and cannot be done using currently available models. Working thereafter from a platform allowing continuous standardization, optimization, and iterative tests of relevance, the ultimate practical reality of the models will become evident over time. Likely as not, and as in the case of the SCID-hu models, some applications will be forthcoming and others will not.

At this juncture, the current battery of humanized mouse model presents three general problems, ones that should either be solved or accepted as insurmountable barriers to future use:

1. *The presence of clinical and subclinical GVHD:* In the same way that the "hu-PBL-SCID" mouse (in which adult human peripheral blood mononuclear cells were injected intraperitoneally into SCID mice) [41] developed high levels of GVHD [42, 43], many if not all of the current "humanized mouse" models are similarly affected. The frequency and manifestations of clinically apparent disease have been outlined in a number of recent reports [44, 45] and, until proven otherwise, it is reasonable to assume that subclinical GVHD may occur as well. If so, the associated levels of immune activation may make it easier to detect certain endpoints (e.g., viremia after infection with HIV), but it is not clear if a model harboring background levels of GVHD is relevant to the analysis of human diseases in which GVHD is not normally present. Furthermore, ongoing efforts to document and use primary human immune responses in these models should acknowledge that they are occurring in the context of GVHD. In future iterations of the current humanized mouse models, it will be important to understand why such GVHD occurs and how to prevent it from happening (e.g., by creating strains of SCID mice lacking murine MHC Class I and Class II molecules) [46]. Should that not be possible, the use of these models would most appropriately be focused on the analysis of xeno-GVHD.
2. *The absence of normal lymphoid structure and function:* In a normal immune response in mice or in humans, antigen presenting cells may move from distant sites through afferent lymphatics into draining lymph nodes, interacting thereafter with T and B cell subpopulations to prompt antigen-specific proliferation and differentiation of cells with cognate receptors; these then traffic out of the node through efferent lymphatics and into distal effector sites. Unfortunately, most of the infrastructure underlying such physiologic responses is not present in any of the humanized mouse models. In the *IL2rg*^{null} mouse models, for instance, the endogenous murine lymph nodes are poorly developed and unlikely to sustain normal levels of human immune cell trafficking and differentiation [3, 4]. Even in SCID-hu mice engrafted with human fetal lymph nodes, it is not clear whether the supporting vascular and lymphatic endothelium is human and/or mouse, and whether such endothelial structures support the physiologic trafficking of human

immune cells. Should humanized mice be developed for the analysis of human immune responses, e.g., to vaccines, this limitation should be addressed.

3. *The inability to create large cohorts of animals from single human donors:* To date, most if not all studies using the newer versions of humanized mice have been carried out with an unspecified number of human donors contributing to each cohort in a given experiment and, in some cases, as few as three mice in a given group. Alternatively, some studies have detailed the use of cohorts comprised of animals created from multiple human donors. Each approach is problematic. Use of a small number of test animals severely limits the number of conditions (e.g., of doses, controls, etc.) that might be tested in parallel in a single experiment and the number of animals that can be included in each subgroup. Such restrictions are even more compromising if there is a broad range of interindividual variation in engraftment, a degree of inter-assay variation in the measurement of endpoints, and/or attrition of animals during the course of the experiment. In future iterations of these humanized models, it will be important to develop and to optimize procedures and practices that allow for the practical and reproducible creation of larger cohorts. For instance, advances in the derivation of functional human hematopoietic stem cells and thymic epithelial cells from iPS cells may ultimately provide virtually unlimited numbers of cells for construction of humanized mouse models. In the meantime, it is important for investigators to state what criteria were used to decide which animals were selected for study (e.g., viral load, level of engraftment including *absolute* cell numbers, occurrence of GVHD, and criteria for removing animals from experiments).

The above issues highlight two general goals: first, it is important to show that a given humanized mouse model can be created in sufficient numbers to carry out experiments that are standardized, reproducible, and statistically verifiable; second, the events that are then documented in the model must be shown to have correlates in humans. It is only by achieving the first goal that the second can be addressed.

The steps and iterations required to optimize and to standardize these models will be numerous, time-consuming, and costly. Several approaches are possible. As occurred in the optimization and standardization of the SCID-hu Thy/Liv model, there can be an intensive, focused, stable, multimillion dollar investment made by the NIH and by private funders. Alternatively, collaborative data sharing amongst multiple academic investigators should provide an equally robust effort of similar strength. The current book provides a venue for such a collective effort, one that might lead to continued improvement of humanized mouse models and an ever greater capacity to apply them to the solution of unsolved problems of medical importance in humans.

At the end of the day, humanized mouse models will hopefully provide important input into the rapid advance of basic science into the clinic. They should support hypothesis-driven research that can inform meaningful and well-informed decisions along the course of this pathway. Even if such input is to show that a given intervention does not work, that is an incredibly useful (and time- and resource-saving) input to the multiyear, multimillion dollar pathway of drug development. To the extent that the humanized mouse models actually model humans, it may be possible to reach this dream.

Acknowledgements We would like to thank Drs. Sandra Bridges, Cheryl Stoddart, and Jerry Zack for their careful reading of this manuscript as well as for their many contributions to humanized mouse technology.

References

1. Mold JE, Venkatasubrahmanyam S, Burt TD, Michaelsson J, Rivera JM, Galkina SA, Weinberg K, Stoddart CA, McCune JM. Fetal and adult hematopoietic stem cells give rise to distinct T cell lineages in humans. Science. 2010;330(6011):1695–9.
2. Mold JE, McCune JM. Immunological tolerance during fetal development: from mouse to man. Adv Immunol. 2012;115:73–111.
3. Shultz LD, Brehm MA, Garcia-Martinez JV, Greiner DL. Humanized mice for immune system investigation: progress, promise and challenges. Nat Rev Immunol. 2012;12(11):786–98.
4. Denton PW, Nochi T, Lim A, Krisko JF, Martinez-Torres F, Choudhary SK, Wahl A, Olesen R, Zou W, Di Santo JP, Margolis DM et al. Il-2 receptor gamma-chain molecule is critical for intestinal T-cell reconstitution in humanized mice. Mucosal Immunol. 2012;5(5):555–66.
5. Long BR, Stoddart CA. Alpha interferon and HIV infection cause activation of human T cells in NSG-BLT mice. J Virol. 2012 86(6):3327–36.
6. Bacchetti P, Deeks SG, McCune JM. Breaking free of sample size dogma to perform innovative translational research. Sci Transl Med. 2011;3(87):87ps24.
7. McCune JM, Namikawa R, Kaneshima H, Shultz LD, Lieberman M, Weissman IL. The scid-hu mouse: Murine model for the analysis of human hematolymphoid differentiation and function. Science. 1988;241(4873):1632–9.
8. Namikawa R, Weilbaecher KN, Kaneshima H, Yee EJ, McCune JM. Long-term human hematopoiesis in the SCID-hu mouse. J Exp Med. 1990;172(4):1055–63.
9. Shih CC, Kaneshima H, Rabin L, Namikawa R, Sager P, McGowan J, McCune JM. Postexposure prophylaxis with zidovudine suppresses human immunodeficiency virus type 1 infection in SCID-hu mice in a time-dependent manner. J Infect Dis. 1991;163(3):625–7.
10. Kyoizumi S, Baum CM, Kaneshima H, McCune JM, Yee EJ, Namikawa R. Implantation and maintenance of functional human bone marrow in SCID-hu mice. Blood. 1992;79(7):1704–11.
11. Fraser CC, Kaneshima H, Hansteen G, Kilpatrick M, Hoffman R, Chen BP. Human allogeneic stem cell maintenance and differentiation in a long-term multilineage SCID-hu graft. Blood. 1995;86(5):1680–93.
12. Billingham RE, Brent L, Medawar PB. Actively acquired tolerance of foreign cells. Nature. 1953;172(4379):603–6.
13. Walzer PD, Kim CK, Linke MJ, Pogue CL, Huerkamp MJ, Chrisp CE, Lerro AV, Wixson SK, Hall E, Shultz LD. Outbreaks of *Pneumocystis carinii* pneumonia in colonies of immunodeficient mice. Infect Immun. 1989;57(1):62–70.
14. McCune J, Kaneshima H, Krowka J, Namikawa R, Outzen H, Peault B, Rabin L, Shih CC, Yee E, Lieberman M, et al. The SCID-hu mouse: a small animal model for HIV infection and pathogenesis. Annu Rev Immunol. 1991;9:399–429.
15. Peault B, Namikawa R, Krowka J, McCune J. Experimental human hematopoiesis in immunodeficient scid mice engrafted with fetal blood-forming organs. In: Edwards RG Editor. Fetal tissue transplants in medicine. Cambridge: Cambridge University Press; 1992. p. 77–94.
16. Namikawa R, Kaneshima H, Lieberman M, Weissman IL, McCune JM. Infection of the SCID-hu mouse by HIV-1. Science. 1988;242(4886):1684–6.
17. Vandekerckhove BA, Jones D, Punnonen J, Schols D, Lin HC, Duncan B, Bacchetta R, de Vries JE, Roncarolo MG. Human Ig production and isotype switching in severe combined immunodeficient-human mice. J Immunol. 1993;151(1):128–37.

18. Stoddart CA, Bales CA, Bare JC, Chkhenkeli G, Galkina SA, Kinkade AN, Moreno ME, Rivera JM, Ronquillo RE, Sloan B, Black PL. Validation of the SCID-hu Thy/Liv mouse model with four classes of licensed antiretrovirals. PLoS ONE. 2007;2(7):e655.
19. Rabin L, Hincenbergs M, Moreno MB, Warren S, Linquist V, Datema R, Charpiot B, Seifert J, Kaneshima H, McCune JM. Use of standardized SCID-hu Thy/Liv mouse model for preclinical efficacy testing of anti-human immunodeficiency virus type 1 compounds. Antimicrob Agents Chemother. 1996;40(3):755–62.
20. Baum CM, Weissman IL, Tsukamoto AS, Buckle AM, Peault B. Isolation of a candidate human hematopoietic stem-cell population. Proc Natl Acad Sci U S A. 1992;89(7):2804–8.
21. Peault B, Weissman I, Baum C. Analysis of candidate human blood stem cells in "humanized" immune-deficiency SCID mice. Leukemia. 1993;7(Suppl 2):98–101.
22. Peault B, Weissman IL, Baum C, McCune JM, Tsukamoto A. Lymphoid reconstitution of the human fetal thymus in scid mice with $CD34^+$ precursor cells. J Exp Med. 1991;174(5):1283–86.
23. Bonyhadi ML, Moss K, Voytovich A, Auten J, Kalfoglou C, Plavec I, Forestell S, Su L, Bohnlein E, Kaneshima H. RevM10-expressing T cells derived in vivo from transduced human hematopoietic stem-progenitor cells inhibit human immunodeficiency virus replication. J Virol. 1997;71(6):4707–16.
24. Su L, Lee R, Bonyhadi M, Matsuzaki H, Forestell S, Escaich S, Bohnlein E, Kaneshima H. Hematopoietic stem cell-based gene therapy for acquired immunodeficiency syndrome: efficient transduction and expression of RevM10 in myeloid cells in vivo and in vitro. Blood. 1997;89(7):2283–90.
25. Vandekerckhove BA, Krowka JF, McCune JM, de Vries JE, Spits H, Roncarolo MG. Clonal analysis of the peripheral T cell compartment of the SCID-hu mouse. J Immunol. 1991;146(12):4173–9.
26. Vandekerckhove BA, Baccala R, Jones D, Kono DH, Theofilopoulos AN, Roncarolo MG. Thymic selection of the human T cell receptor v beta repertoire in SCID-hu mice. J Exp Med. 1992;176(6):1619–24.
27. Vandekerckhove BA, Namikawa R, Bacchetta R, Roncarolo MG. Human hematopoietic cells and thymic epithelial cells induce tolerance via different mechanisms in the SCID-hu mouse thymus. J Exp Med. 1992;175(4):1033–43.
28. Roncarolo MG, Vandekerckhove B. Scid-hu mice as a model to study tolerance after fetal stem cell transplantation. Bone Marrow Transplant. 1992;9(Suppl 1):83–4.
29. Baccala R, Vandekerckhove BA, Jones D, Kono DH, Roncarolo MG, Theofilopoulos AN. Bacterial superantigens mediate T cell deletions in the mouse severe combined immunodeficiency-human liver/thymus model. J Exp Med. 1993;177(5):1481–5.
30. Kyoizumi S, McCune JM, Namikawa R. Direct evaluation of radiation damage in human hematopoietic progenitor cells in vivo. Radiat Res. 1994;137(1):76–83.
31. Kyoizumi S, Murray LJ, Namikawa R. Preclinical analysis of cytokine therapy in the SCID-hu mouse. Blood. 1993;81(6):1479–88.
32. Namikawa R, Ueda R, Kyoizumi S. Growth of human myeloid leukemias in the human marrow environment of SCID-hu mice. Blood. 1993;82(8):2526–36.
33. Shtivelman E, Namikawa R. Species-specific metastasis of human tumor cells in the severe combined immunodeficiency mouse engrafted with human tissue. Proc Natl Acad Sci U S A. 1995;92(10):4661–5.
34. Akkina R. New generation humanized mice for virus research: comparative aspects and future prospects. Virology. 2013;435(1):14–28.
35. Garcia S, Freitas AA. Humanized mice: current states and perspectives. Immunol Lett. 2012;146(1–2):1–7.
36. Ito R, Takahashi T, Katano I, Ito M. Current advances in humanized mouse models. Cell Mol Immunol. 2012;9(3):208–14.
37. Denton PW, Garcia JV. Humanized mouse models of HIV infection. AIDS Rev. 2011;13(3):135–48.

38. Legrand N, Ploss A, Balling R, Becker PD, Borsotti C, Brezillon N, Debarry J, de Jong Y, Deng H, Di Santo JP, Eisenbarth S, et al. Humanized mice for modeling human infectious disease: challenges, progress, and outlook. Cell Host Microbe. 2009;6(1):5–9.
39. Nischang M, Gers-Huber G, Audige A, Akkina R, Speck RF. Modeling HIV infection and therapies in humanized mice. Swiss Med Wkly. 2012;142:w13618.
40. Drake AC, Chen Q, Chen J. Engineering humanized mice for improved hematopoietic reconstitution. Cell Mol Immunol. 2012;9(3):215–24.
41. Mosier DE, Gulizia RJ, Baird SM, Wilson DB. Transfer of a functional human immune system to mice with severe combined immunodeficiency. Nature. 1988;335(6187):256–9.
42. Williams SS, Umemoto T, Kida H, Repasky EA, Bankert RB. Engraftment of human peripheral blood leukocytes into severe combined immunodeficient mice results in the long term and dynamic production of human xenoreactive antibodies. J Immunol. 1992;149(8):2830–6.
43. Duchosal MA, Eming SA, McConahey PJ, Dixon FJ. Characterization of hu-pbl-scid mice with high human immunoglobulin serum levels and graft-versus-host disease. Am J Pathol. 1992;141(5):1097–1113.
44. Greenblatt MB, Vbranac V, Tivey T, Tsang K, Tager AM, Aliprantis AO. Graft versus host disease in the bone marrow, liver and thymus humanized mouse model. PLoS ONE. 2012;7(9):e44664.
45. Ali N, Flutter B, Sanchez Rodriguez R, Sharif-Paghaleh E, Barber LD, Lombardi G, Nestle FO. Xenogeneic graft-versus-host-disease in NOD-SCID Il-2rgamma null mice display a T-effector memory phenotype. PLoS ONE. 2012;7(8):e44219.
46. King MA, Covassin L, Brehm MA, Racki W, Pearson T, Leif J, Laning J, Fodor W, Foreman O, Burzenski L, Chase TH et al. Human peripheral blood leucocyte non-obese diabetic-severe combined immunodeficiency interleukin-2 receptor gamma chain gene mouse model of xenogeneic graft-versus-host-like disease and the role of host major histocompatibility complex. Clin Exp Immunol. 2009;157(1):104–18.

Chapter 3
Role of Mouse Innate Immunity in Immunodeficient Mice for Xenotransplantation

Ryoji Ito, Ikumi Katano and Mamoru Ito

Abbreviations

BRG	BALB/c $Rag2^{null}IL2r\gamma^{null}$
NOD	Nonobese diabetic
NOG, NSG	NOD-*scid* $IL2r\gamma^{null}$
SCID	Severe combined immunodeficiency
HSC	Hematopoietic stem cell

3.1 Natural Killer (NK) Cells

Many studies have suggested that NK cells play a crucial role in xenograft rejection. In vivo depletion of NK cells in C.B-17-*scid* or NOD-*scid* mice by treatment using NK cell-specific antibodies (e.g., anti-asialo-GM1 and anti-TM-β1) resulted in significantly higher engraftment rates of reconstituted human cells [1]. NOD-based immunodeficient mice are better recipients for xenoengraftment than BALB/c, C57BL/6, or other strains, due to reduced NK cell activity and additional deficiencies in innate immunity [2, 3]. Shultz et al. demonstrated that NK cell activity was markedly lower in NOD-*scid* mice than in C.B-17-*scid* mice, and the engraftment rate of human lymphoblastoid T cells was higher in NOD-*scid* mice than in C.B-17-*scid* mice [2]. From these findings, researchers have attempted to establish novel immunodeficient mouse strains with NK cell deficiency for further acceleration of xenoengraftment. Higher success rates for xenoengraftment of human hematopoietic stem cells (HSCs) have been observed in NOD-*scid* $\beta 2m^{null}$ and

R. Ito (✉) · I. Katano · M. Ito
Central Institute for Experimental Animals, 3-25-12 Tonomachi, Kawasaki-ku, Kawasaki, Kanagawa 210-0821, Japan
e-mail: rito@ciea.or.jp

L. Y. Poluektova et al. (eds.), *Humanized Mice for HIV Research*,
DOI 10.1007/978-1-4939-1655-9_3

NOD-Rag1^{null}Prf1^{null} mice, which lack NK activity [4, 5]. Currently, immunodeficient mice with a disrupted IL-2 receptor common gamma chain (γc) gene are a mainstream tool for xenograft animal models such as NOG, NSG, and BRG mice [6–8]. These strains lack NK cells and produce better xenoengraftment results than do other strains, particularly in the differentiation of mature leukocytes from transplanted human HSCs. The improvement in engraftment is attributed to the inactivation of the γc gene, which is shared by cytokines such as IL-2, IL-4, IL-9, IL-7, IL-15, and IL-21 and is important for the differentiation and function of not only T, B, and NK cells but also innate immune cells such as macrophages, dendritic cells (DCs), and others [9–13]. In particular, IL-2, IL-7, and IL-15 play critical roles in NK cell differentiation and functional maturation (e.g., cytotoxicity and production of inflammatory cytokines) [14].

Xenograft rejection by human NK cells has been investigated extensively by using pig-to-primate mixed chimeras [15]. NK cells exert significant cytotoxicity against xenografts compared with allografts because the level of inhibitory signals from xenograft MHC molecules associated with NK cell receptors is lower than that from allogenic MHC molecules. In fact, HLA-E gene-expressed porcine cells, but not normal porcine cells, were found to be protected from human NK cytotoxicity [16]. Human NK cytotoxicity against porcine cells is being evaluated for application to transplant therapy using xenograft tissues.

3.2 Macrophages

To improve engraftment of human hematopoietic cells, depletion of host macrophages in immunodeficient mice is effective [17–19]. Verstegen et al. reported that human leukocytes were efficiently engrafted in HSC-transplanted *scid* mice after macrophage depletion by administration of dichloromethylene diphosphonate (CL2MDP) liposomes [18]. Moreover, Hu et al. demonstrated that host macrophages inhibited human thrombopoiesis and restored human platelets by macrophage depletion [19].

Several cytokines promote the maturation and activation of macrophages. An IL-12 cytokine is a prerequisite for the activation of innate immunity as well as induction of Th1 responses. Because macrophages produce IL-12, a positive feedback pathway through IL-12 is activated and exerts immune barriers by producing IFNγ and several chemokines (MCP-1, MIP-1α, and MIP-1β) Ohteki et al. demonstrated that this activation pathway is impaired in RAG2$^{-/-}$IL-2Rβ$^{-/-}$ or RAG2$^{-/-}$γc$^{-/-}$ mice, which lack IL-15 signaling because receptors for IL-15 use both the IL-2Rβ and γc subunits [10]. They also indicated that IL-15 is an upstream activator for induction of IL-12, because the production of IL-12 and IFNγ by macrophages is also severely impaired in IL-15$^{-/-}$ mice. These results suggest that IL-15 signaling is critical for the functional maturation of macrophages, and the lack of IL-15 receptors in immunodeficient mice may be important for successful xenoengraftment.

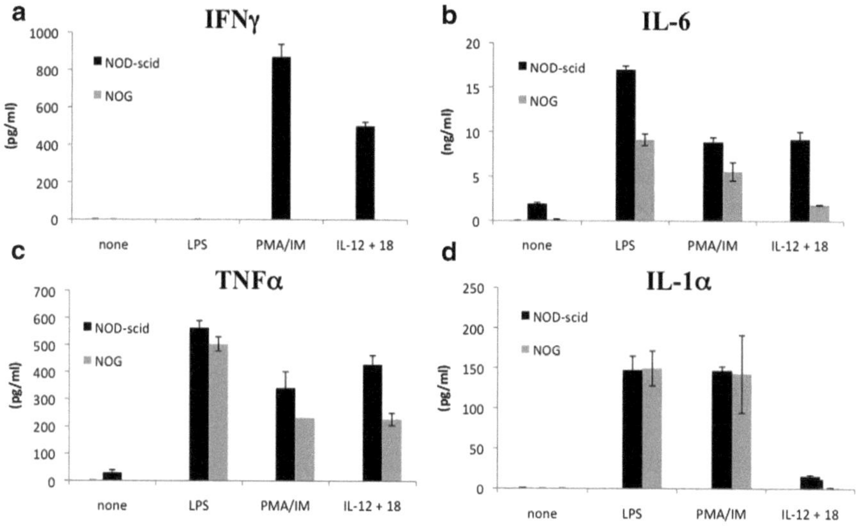

Fig. 3.1 Comparison of cytokine production by macrophages from NOD-*scid* and NOG mice. Peritoneal macrophages were isolated from 8- to 9-week-old NOD-*scid* and NOG mice, and then stimulated with LPS (final concentration, 10 μg/ml), PMA (10 ng/ml)/ionomycin (1 μg/ml), and IL-12 (10 ng/ml)/IL-18 (40 ng/ml) in vitro. After 24 h, the levels of cytokines IFNγ (**a**), IL-6 (**b**), TNFα (**c**), and IL-1α (**d**) in culture supernatants were determined by ELISA

Recently, we compared the production of inflammatory cytokines such as IFNγ, TNFα, IL-1α, and IL-6 from macrophages of NOD-*scid* or NOG mice (unpublished data). The production of IFNγ in the supernatant of cultured macrophages was completely absent in NOG mice when they were stimulated with PMA/ionomycin and IL-12/IL-18 (Fig. 3.1a). Likewise, the IL-6 level was lower in NOG mice than in NOD-*scid* mice following stimulation with LPS, PMA/ionomycin, and IL-12/IL-18 (Fig. 3.1b). Conversely, no significant difference between NOD-*scid* and NOG mice was evident in the production of TNFα and IL-1α (Fig. 3.1c, 3.1d). These results suggest that cytokine production is partially reduced in macrophages of NOG mice compared with NOD-*scid* mice, which may be due to the absence of γc-signaling-dependent production.

Mice with an NOD genetic background in combination with the *scid* mutation or RAG deficiency provide better results in xenoengraftment than do those with a BALB/c or C57BL/6 background. NOD-*scid* mice exhibit a decreased number of macrophages and functional impairment of them, where LPS-induced IL-1 production was dramatically reduced in macrophages differentiated by stimulation with IFNγ and M-CSF [2, 20].

A novel mechanism of xenograft recognition by host macrophages was reported recently. Signal regulatory protein-α (sirpα), which is a critical immune inhibitory receptor on macrophages, interacts with the CD47 ligand on the xenograft to prevent phagocytosis [21–24]. This interaction displays species specificity, thereby critically regulating graft rejection in the xeno-environment. Strowig et al. generated human sirpα–expressing BRG mice. These mice showed significantly increased engraftment levels of human cells compared with non-transgenic BRG mice [24]. The sirpα polymorphism in the NOD genetic background (which is similar to human sirpα) leads to enhanced binding to human CD47, which may activate CD47-induced signaling pathways to support xenoengraftment [23, 25, 26]. Legrand et al. demonstrated that BRG mice harboring NOD-sirpα showed increased engraftment ratios of human T and NK cells compared to conventional BRG mice [23]. Overall, the CD47-sirpα interaction leads to improved xenoengraftment as a result of phagocyte tolerance in host macrophages.

3.3 Complement Systems

Complement systems work through three different pathways, namely the classic, alternative, and lectin pathways. Complement factors comprise a membrane attack complex that damages target cells and contributes to adaptive immunity to eliminate foreign materials.

Previously, several C5-deficient strains have been reported [27], and sera from homozygous C5-deficient strains lack complement hemolytic activity. All C5-deficient strains, including NOD mice [28], contain an identical 2-bp depletion in an exon near the 5′ end of the C5 gene. The sera from NOD-*scid* mice, but not C.B-17-*scid* mice, have been shown to lack hemolytic complement activity against sheep red blood cells [2]. The C5 deficiency in the NOD strain may result in the higher engraftment rate of transplanted human cells. Recently, complement-dependent injury against human cells was demonstrated by transplantation of human mesenchymal stem cells (MSCs) into C3-deficient mice [29]. Adoptive transfer of MSCs into C3-deficient mice resulted in reduced cellular injury in vivo compared with transfer into wild-type mice. The authors postulate that complement may be involved in recognizing and injuring transplanted MSCs.

An initial immunological obstacle by complement systems, also called natural antibody-mediated hyperacute rejection (HAR), is observed in pig-to-primate xenotransplantation. This rejection is initiated by the binding of anti-galactose-α1,3-galactose (Gal) natural antibodies produced from primate cells to Gal on the pig vascular endothelium and sequential activation of the complement cascade [30]. Several genetically modified pigs [31–33], such as Gal-transferase-knockout pigs, have demonstrated partial success in the induction of immunological tolerance by regulating complement-dependent HAR. Taken together, these studies suggest that inhibition of complement activation could be a novel therapeutic strategy for xenotransplantation.

3.4 Dendritic Cells: CD11c⁺B220⁺CD122⁺ Cells

Our previous studies suggested that DCs play a potentially pivotal role in xenograft rejection, compared with NK cells. The production levels of inflammatory cytokines, such as IFNγ and IL-6, were markedly reduced in NOG mice and CD11c⁺ DC-depleted NOD-*scid* mice spleen cells but not in NK cell-depleted NOD-*scid* mice spleen cells [6]. Based on these findings, we speculated that the subpopulation of CD11c⁺ DCs has a strong potential for xenograft rejection in mice.

Generally, DCs are fractionated into several subpopulations based on surface markers and functional properties [34]. Conventional DCs (cDCs) are characterized by high expression of CD11c and act as professional antigen presenting cells (APCs) for stimulation of naive T cells [35]. Plasmacytoid DCs (pDCs) are characterized by expression of CD11c and B220, which play crucial roles in producing type-I IFNs against viruses *via* Toll-like receptors (TLRs) [36]. We recently demonstrated that CD122-expressing cells in the pDC fraction (CD11c⁺B220⁺CD122⁺ cells), which are lacking in NOG mice, have a greater potential to induce xenograft rejection than NK cells (Fig. 3.2a, 3.2b; [37]). CD11c⁺B220⁺CD122⁺ cells have been reported to be IFNs-producing killer dendritic cells (IKDCs), which share characteristics of DC and NK cells and produce larger amounts of IFNγ than NK cells [38, 39].

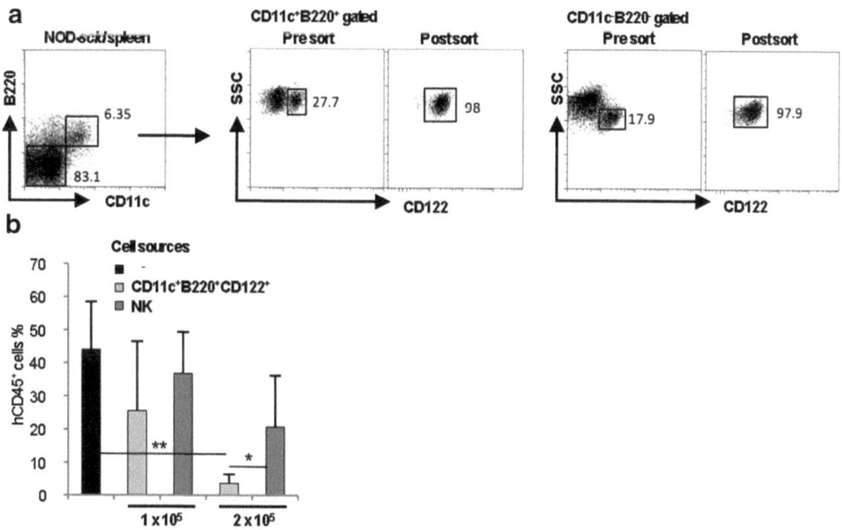

Fig. 3.2 Suppressive effects of CD11c⁺B220⁺CD122⁺ and NK cells on xenoengraftment. **a** CD11c⁺B220⁺CD122⁺ cells or NK cells were isolated from the CD11c⁺B220⁺or CD11c⁻B220⁻ fraction of NOD-*scid* mouse spleen cells, and 1×10^5 or 2×10^5 cells were transplanted into NOG mice before hPBMC transplantation. **b** At 4 weeks posttransplantation, red blood cell-lysed PBMCs were collected from the transplanted NOG mice, and the efficacy of human cell engraftment was assessed by flow cytometry. From [37]. Copyright 2012. The American Association of Immunologists, Inc

Although the functional properties of $CD11c^{+}B220^{+}CD122^{+}$ cells resemble those of NK cells, the phenotypic and morphologic features of $CD11c^{+}B220^{+}CD122^{+}$ cells are distinct from those of NK cells. IFNγ produced by $CD11c^{+}B220^{+}CD122^{+}$ cells may be important in the rejection of xenografts because $CD11c^{+}B220^{+}CD122^{+}$ cells from IFNγ-deficient mice did not suppress xenoengraftment. Consistent with our results, Lin et al. transplanted pig cells into T-cell-depleted IFNγ-deficient mice and found that engraftment was significantly enhanced in the IFNγ-deficient mice [40]. Considering the role of IFNγ in xenograft rejection, we speculate that macrophages can be activated by IFNγ and recruited to the graft site through secretion of chemokines. In addition, cytotoxicity mediated by perforin/granzyme, FasL, and TRAIL of $CD11c^{+}B220^{+}CD122^{+}$ cells and/or NK cells may be upregulated in an autocrine manner by IFNγ production. These emerging lines of evidence suggest that the absence of $CD11c^{+}B220^{+}CD122^{+}$ cells in immunodeficient mice may be critical for the engraftment of xenotransplants.

In conclusion, multiple dysfunctions in innate immunity, such as those associated with NK cells, macrophages, complement systems, and DCs in NOG or NSG mice, may result in successful xenoengraftment and lead to remarkable advances in humanized mouse xenograft models.

Acknowledgement This work was supported by a grant from the Research Foundation for Pharmaceutical Sciences, and by Grants-in-Aid for Young Scientists (B) (#22700458) and Scientific Research (S) (#18100005) from the Ministry of Education, Culture, Sports, Science, and Technology (MEXT) of Japan.

References

1. Tsuchida M, Brown SA, Tutt LM, Tan J, Seehafer DL, Harris JP, et al. A model of human anti-T-cell monoclonal antibody therapy in SCID mice engrafted with human peripheral blood lymphocytes. Clin Transplant. 1997; 11(5 Pt 2):522–8.
2. Shultz LD, Schweitzer PA, Christianson SW, Gott B, Schweitzer IB, Tennent B, et al. Multiple defects in innate and adaptive immunologic function in NOD/LtSz-scid mice. J Immunol. 1995;154(1):180–91.
3. Greiner DL, Shultz LD, Yates J, Appel MC, Perdrizet G, Hesselton RM, et al. Improved engraftment of human spleen cells in NOD/LtSz-scid/scid mice as compared with C.B-17-scid/scid mice. Am J Pathol. 1995;146(4):888–902.
4. Kollet O, Peled A, Byk T, Ben-Hur H, Greiner D, Shultz L, et al. beta2 microglobulin-deficient (B2m(null)) NOD/SCID mice are excellent recipients for studying human stem cell function. Blood. 2000;95(10):3102–5.
5. Shultz LD, Banuelos S, Lyons B, Samuels R, Burzenski L, Gott B, et al. NOD/LtSz-Rag1nullPfpnull mice: a new model system with increased levels of human peripheral leukocyte and hematopoietic stem-cell engraftment. Transplantation. 2003;76(7):1036–42.
6. Ito M, Hiramatsu H, Kobayashi K, Suzue K, Kawahata M, Hioki K, et al. NOD/SCID/gamma(c)(null) mouse: an excellent recipient mouse model for engraftment of human cells. Blood. 2002;100(9):3175–82.
7. Shultz LD, Lyons BL, Burzenski LM, Gott B, Chen X, Chaleff S, et al. Human lymphoid and myeloid cell development in NOD/LtSz-scid IL2R gamma null mice engrafted with mobilized human hemopoietic stem cells. J Immunol. 2005;174(10):6477–89.

8. Traggiai E, Chicha L, Mazzucchelli L, Bronz L, Piffaretti JC, Lanzavecchia A, et al. Development of a human adaptive immune system in cord blood cell-transplanted mice. Science. 2004;304(5667):104–7.
9. Sugamura K, Asao H, Kondo M, Tanaka N, Ishii N, Ohbo K, et al. The interleukin-2 receptor gamma chain: its role in the multiple cytokine receptor complexes and T cell development in XSCID. Annu Rev Immunol. 1996;14:179–205.
10. Ohteki T, Suzue K, Maki C, Ota T, Koyasu S. Critical role of IL-15-IL-15R for antigen-presenting cell functions in the innate immune response. Nat Immunol. 2001;2(12):1138–43.
11. Ohteki T, Fukao T, Suzue K, Maki C, Ito M, Nakamura M, et al. Interleukin 12-dependent interferon gamma production by CD8alpha + lymphoid dendritic cells. J Exp Med. 1999;189(12):1981–6.
12. Kovanen PE, Leonard WJ. Cytokines and immunodeficiency diseases: critical roles of the gamma(c)-dependent cytokines interleukins 2, 4, 7, 9, 15, and 21, and their signaling pathways. Immunol Rev. 2004;202:67–83.
13. Leonard WJ. Cytokines and immunodeficiency diseases. Nat Rev Immunol. 2001;1(3):200–8.
14. Di Santo JP. Natural killer cell developmental pathways: a question of balance. Annu Rev Immunol. 2006;24:257–86.
15. Forte P, Lilienfeld BG, Baumann BC, Seebach JD. Human NK cytotoxicity against porcine cells is triggered by NKp44 and NKG2D. J Immunol. 2005;175(8):5463–70.
16. Lilienfeld BG, Crew MD, Forte P, Baumann BC, Seebach JD. Transgenic expression of HLA-E single chain trimer protects porcine endothelial cells against human natural killer cell-mediated cytotoxicity. Xenotransplantation. 2007;14(2):126–34.
17. Hu Z, Van Rooijen N, Yang YG. Macrophages prevent human red blood cell reconstitution in immunodeficient mice. Blood. 2011;118(22):5938–46.
18. Verstegen MM, van Hennik PB, Terpstra W, van den Bos C, Wielenga JJ, van Rooijen N, et al. Transplantation of human umbilical cord blood cells in macrophage-depleted SCID mice: evidence for accessory cell involvement in expansion of immature $CD34^{+}CD38$- cells. Blood. 1998;91(6):1966–76.
19. Hu Z, Yang YG. Full reconstitution of human platelets in humanized mice after macrophage depletion. Blood. 2012;120(8):1713–6.
20. Serreze DV, Gaedeke JW, Leiter EH. Hematopoietic stem-cell defects underlying abnormal macrophage development and maturation in NOD/Lt mice: defective regulation of cytokine receptors and protein kinase C. Proc Natl Acad Sci U S A. 1993;90(20):9625–9.
21. Ide K, Wang H, Tahara H, Liu J, Wang X, Asahara T, et al. Role for CD47-SIRPalpha signaling in xenograft rejection by macrophages. Proc Natl Acad Sci U S A. 2007;104(12):5062–6.
22. Wang H, VerHalen J, Madariaga ML, Xiang S, Wang S, Lan P, et al. Attenuation of phagocytosis of xenogeneic cells by manipulating CD47. Blood. 2007;109(2):836–42.
23. Legrand N, Huntington ND, Nagasawa M, Bakker AQ, Schotte R, Strick-Marchand H, et al. Functional CD47/signal regulatory protein alpha (SIRP(alpha)) interaction is required for optimal human T- and natural killer- (NK) cell homeostasis in vivo. Proc Natl Acad Sci U S A. 2011;108(32):13224–9.
24. Strowig T, Rongvaux A, Rathinam C, Takizawa H, Borsotti C, Philbrick W, et al. Transgenic expression of human signal regulatory protein alpha in $Rag2^{-}/^{-}gamma(c)^{-}/^{-}$ mice improves engraftment of human hematopoietic cells in humanized mice. Proc Natl Acad Sci U S A. 2011;108(32):13218–23.
25. Takenaka K, Prasolava TK, Wang JC, Mortin-Toth SM, Khalouei S, Gan OI, et al. Polymorphism in Sirpa modulates engraftment of human hematopoietic stem cells. Nat Immunol. 2007;8(12):1313–23.
26. Yamauchi T, Takenaka K, Urata S, Shima T, Kikushige Y, Tokuyama T, et al. Polymorphic Sirpa is the genetic determinant for NOD-based mouse lines to achieve efficient human cell engraftment. Blood. 2013;121(8):1316–25.
27. Wetsel RA, Fleischer DT, Haviland DL. Deficiency of the murine fifth complement component (C5). A 2-base pair gene deletion in a 5′-exon. J Biol Chem. 1990;265(5):2435–40.

28. Baxter AG, Cooke A. Complement lytic activity has no role in the pathogenesis of autoimmune diabetes in NOD mice. Diabetes. 1993;42(11):1574–8.
29. Li Y, Lin F. Mesenchymal stem cells are injured by complement after their contact with serum. Blood. 2012;120(17):3436–43.
30. Ezzelarab M, Ayares D, Cooper DK. Carbohydrates in xenotransplantation. Immunol Cell Biol. 2005;83(4):396–404.
31. Phelps CJ, Koike C, Vaught TD, Boone J, Wells KD, Chen SH, et al. Production of alpha 1,3-galactosyltransferase-deficient pigs. Science. 2003;299(5605):411–4.
32. Kuwaki K, Tseng YL, Dor FJ, Shimizu A, Houser SL, Sanderson TM, et al. Heart transplantation in baboons using alpha1,3-galactosyltransferase gene-knockout pigs as donors: initial experience. Nat Med. 2005;11(1):29–31.
33. Menoret S, Plat M, Blancho G, Martinat-Botte F, Bernard P, Karam G, et al. Characterization of human CD55 and CD59 transgenic pigs and kidney xenotransplantation in the pig-to-baboon combination. Transplantation. 2004;77(9):1468–71.
34. Shortman K, Naik SH. Steady-state and inflammatory dendritic-cell development. Nat Rev Immunol. 2007;7(1):19–30.
35. Banchereau J, Steinman RM. Dendritic cells and the control of immunity. Nature. 1998;392(6673):245–52.
36. Kadowaki N, Ho S, Antonenko S, Malefyt RW, Kastelein RA, Bazan F, et al. Subsets of human dendritic cell precursors express different toll-like receptors and respond to different microbial antigens. J Exp Med. 2001;194(6):863–9.
37. Ito R, Katano I, Ida-Tanaka M, Kamisako T, Kawai K, Suemizu H, et al. Efficient xenoengraftment in severe immunodeficient NOD/Shi-scid IL2rgammanull mice is attributed to a lack of $CD11c^{+}B220^{+}CD122^{+}$ cells. J Immunol. 2012;189(9):4313–20.
38. Chan CW, Crafton E, Fan HN, Flook J, Yoshimura K, Skarica M, et al. Interferon-producing killer dendritic cells provide a link between innate and adaptive immunity. Nat Med. 2006;12(2):207–13.
39. Taieb J, Chaput N, Menard C, Apetoh L, Ullrich E, Bonmort M, et al. A novel dendritic cell subset involved in tumor immunosurveillance. Nat Med. 2006;12(2):214–9.
40. Lin ML, Zhan Y, Nutt SL, Brady J, Wojtasiak M, Brooks AG, et al. NK cells promote peritoneal xenograft rejection through an IFN-gamma-dependent mechanism. Xenotransplantation. 2006;13(6):536–46.

Chapter 4
Mouse Genetic Background and Human Hematopoietic Stem Cells Biology; Tips for Humanization

Larisa V. Kovtonyuk and Hitoshi Takizawa

4.1 Genetic Backgrounds and Hematopoietic Cytokines for Better Function of Human HSC and Immune System

Xenotransplantation is defined as transplantation of living cells, tissues, or organs from one to another species. Major barrier for successful xenotransplant engraftment is bidirectional immune response between host and donor, triggered by recognition of "non-self" antigens and eventually resulting in deconstruction of host tissues or donor xenograft. Degree of this xenogenic immune reactions depends on what host immuno-components are activated by donor xenografts [1]: (i) humoral adoptive immune response mediated by presence of host natural antibodies that reacts to xenograft-derived molecules, and subsequent activation of complement system causing graft rejection; (ii) T cell-mediated cellular immunity as blockage of T-cell response significantly suppressed xenograft rejection; (iii) innate immune cells such as natural killer (NK) cells, macrophages, dendritic cells (DCs), neutrophils that detect exogenous molecules with species-specific modification or polymorphism via cell surface receptors, e.g., pattern recognition receptors. Ligation of the receptors on innate immune cells leads to cell activation and immediate rejection of xenografts via engulfment, complement system- or antibody-mediated mechanism.

Minimum requirement for xenotransplantation model to avoid the abovementioned xenogeneic immune reactions is the use of immunodeficient animals that lack acquired and ideally innate immunity, and thus permit sustained engraftment of xenotransplant. Among laboratory animals, small rodents, such as mice, have been utilized for the last several decades to develop immunodeficient strain by

H. Takizawa (✉) · L. V. Kovtonyuk
Division of Hematology, University Hospital Zürich, CH-8091 Zürich, Switzerland
e-mail: hitoshi.takizawa@usz.ch

H. Takizawa
International Research Center for Medical Sciences, Kumamoto University, 860-0811 Kumamoto, Japan

L. Y. Poluektova et al. (eds.), *Humanized Mice for HIV Research*,
DOI 10.1007/978-1-4939-1655-9_4

spontaneous gene mutation or artificial gene targeting, and engraft with human cells or tissues to reconstitute human hemato-lymphoid system (HHLS) for understanding human physiology and pathophysiology [2, 3]. Since history of development of immunodeficient mice has been well described in many other reviews [4, 5], major breakthroughs are discussed here briefly: (1) Severe combined immunodeficiency (SCID) mice was found on C.B17 background, which lack functional B and T cells because of impaired V(D)J recombination [6–8]; (2) Non-obese diabetic (NOD) mice crossed with SCID mice (NOD/SCID) have additional functional defects in innate immunity (macrophages, NK cells, and complement system), allowing better engraftment of xenotransplant than C.B17-SCID [9]; (3) Mice with homozygous deletion of interleukin-2 receptor gamma chain gene (*Il2rg*$^{-/-}$), a common cytokine-receptor gamma chain shared by other cytokine receptors (IL2-, IL4-, IL7-, IL9-, IL15- and IL21-R), was genetically engineered to have NK cell deficiency [10, 11], and subsequently crossed with Balb/c mice deficient for *Rag2* gene encoding a V(D)J recombination enzyme [12] (Balb/c Rag2$^{-/-}$IL2r$\gamma^{-/-}$, termed as BRG hereafter), or with NOD/SCID (termed as NSG or NOG) [13, 14]. Not only gene mutation but also mouse genetic background *per se* seems to be a determinant for high human cell engraftment: SCID or Rag2$^{-/-}$Il2r$\gamma^{-/-}$ mice on Balb/c genetic background is more permissive for human cell engraftment compared to equivalent mutants on C57BL/6 background [15, 16]. NOD background is superior to other strains, e.g., C.B-17, Balb/c or C57BL/6 [17] because NOD mice have lower levels of NK cell activity [9, 14] and a polymorphism in signal regulatory protein alpha (*Sirpa*) gene that cross-reacts with human ligands and suppresses phagocytosis of human cells by mouse macrophages ([18, 19] and in detail, discussed below).

Hematopoiesis is one of the somatic organ systems with high cell turnover and highly-organized developmental hierarchy, in which self-renewing hematopoietic stem cells (HSCs) in the bone marrow (BM) produce all hematopoietic and immune cells to maintain functional hemato-immune system throughout lifetime of an individual. Differentiation, maturation, and cell death of blood and immune cells are tightly controlled by many growth factors, chemokines, cytokines, and adhesion molecules through receptor-ligand interaction, and the subsequent intracellular signal activation. Some of these soluble factors are secreted by hematopoietic cells, while others are produced by non-hematopoietic cells in the BM, e.g., mesenchymal stroma, osteoblasts, endothelial cells [20]. In HHLS models, the factors that are produced only by non-hematopoietic mouse cells, need to be cross-reactive to respective human receptors to support human cell engraftment and reconstitution [21]. Therefore, non-cross-reactive human factors have been supplemented in HHLS mice by in vivo administration of recombinant protein, lentiviral or plasmid vector coding human proteins, or genetic replacement of mouse factors with human counterparts, and shown to enhance human hematopoiesis and immune function [22–26].

4.2 Human HSC Biology in Mouse Environment

Human HSC has been immunophenotyped by prospective isolation of living subpopulation with antibodies against cell surface markers, and subsequent functional test of the isolated cells. In vitro culture systems have been developed to presumably test multilineage and self-renewal potential of the isolated HSC candidate, including methylcellulose-based colony forming-unit (CFU) assay or long-term culture-initiating cells (LTC-IC) on stromal cells. However, as these *in vitro* assays often use hyper/hypo-physiologic concentration of cytokines and limited to maintain HSC self-renewal, they do not reflect *in vivo* situations and distinguish self-renewing HSCs from other non-self-renewing progenitors. Thus, transplantation of human HSC candidates into HHLS mice has been employed as so far "a gold standard" surrogate assay to test in vivo function of authentic HSCs. Human HSCs capable of repopulating HHLS mice in vivo, so-called SCID repopulating cells (SRCs), have been enriched in lineage marker negative (Lin^-), $CD34^+$, $CD90^+$ [27, 28] in human fetal liver (FL)/ bone marrow (FBM), umbilical cord blood (UCB), and adult BM (ABM) [29], and further subdivided as $CD38^-$ [30, 31], $CD45RA^-$ [32], Rhadamin-123^- [33], and $CD49f^+$ [34]. However, serial transplanting ability of the human HSCs defined above are limited in a mouse, unlike mouse HSCs that are transplantable over a couple of rounds of transplantation [35], suggesting that mouse environment *per se* does not support human HSC self-renewal.

HSCs are localized in a specialized BM microenvironment, referred to as "niche," which provides necessary and sufficient factors to maintain HSC quiescence, self-renewal and multilineage differentiation capacity [20]. Cellular niche component consists of various cell types including osteoblasts/osteoclasts [36–38], endothelial cells [39], adipocytes [40], Schwann cells [41], macrophages [42, 43], CXCL12 abundant reticular (CAR) cells [44], and leptin receptor$^+$ [45], nestin$^+$ [46], or platelet-derived growth factor receptor (PDGFR)-α^+ [47] mesenchymal stromal cells (MSC) (Fig. 4.1). These niche cells in different location of BM produce a variety of molecular niche factors to control steady state HSC homeostasis through ligand/cell surface receptor interactions: Angiopoietin-1 (ANG-1)/Tie2 [48], CXCL12/CXCR4 [39, 49, 50], Stem cell factor (SCF)/c-Kit [45], Thrombopoietin (THPO)/c-Mpl [51, 52], Tumor growth factor (TGF)-β/TGF receptor-2 [41], Wnt/Frizzle [53, 54], N-cadherin [36, 54, 55], E-selectin ([56]; Fig. 4.1).

The knowledge on cellular and molecular BM niche components as mentioned above has been obtained from the studies with genetically modified mice, whereas human counterpart niche components remain largely unknown. Given that LTC-IC represents human HSC potential *in vitro*, stromal cells that maintain LTC-ICs presumably recapitulate human BM microenvironment and serve as HSC niche cells. Similarly, it has been demonstrated that $CD146^+$ MSCs that are found in human BM can self-renew and form hematopoietic microenvironment *in vivo* in a heterotopic site with expression of ANG-1 [57]. Since little is known about other niche cells and factors for human HSC maintenance, and cross-reactivity of mouse factors to human HSCs, establishment of *in vivo* system to engineer human BM microenvironment might help to better support human HSCs in a mouse environment.

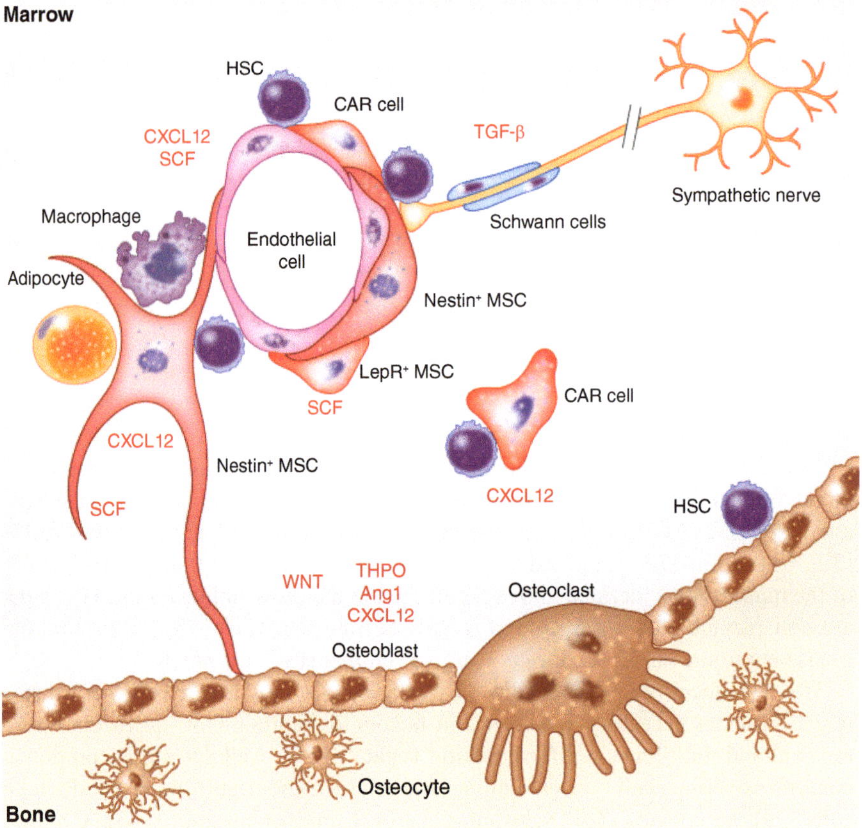

Fig. 4.1 Possible cellular and molecular HSC niche components in mouse BM. Niche cells (*black* instead of black) and factors (*red*) previously reported are depicted. HSC hematopoietic stem cell, CAR cell, CXCL12 abundant reticular cell, MSC, mesenchymal stromal cell, LepR, leptin receptor, ANG-1, angiopoietin-1, SCF, stem cell factor, THPO, thrombopoietin, TGF-β, tumor growth factor-β

4.3 Mouse Hematopoietic Stem Cell or Niche Defect in SCID BM

First, HHLS model was developed by two groups, by engrafting CB17-SCID mice with human tissues [58] or peripheral blood leukocytes [59], and observed human B/T differentiation as well as human antibody secretion in the peripheral blood. C.B17-SCID mice have a spontaneous point mutation in *Prkdc* gene (protein kinase, DNA activated, catalytic polypeptide) involving in V(D)J recombination [7, 8], and therefore, no functional B and T cells but intact innate immunity and NK function, leading to low level of human cell engraftment.

Prkdc also regulates non-homologous end-joining (NHEJ) pathway, a process to repair double stranded breaks upon DNA damage [60]. It was found that DNA-damaging agents restore T cell receptor rearrangement, thereby resulting in normal

T cell development, but also induce lymphoma with high frequency [61]. Furthermore, C.B17-SCID show "leaky" phenotype: with age, normal functional B and T cell are spontaneously generated and serum immunogloblin level is increased accordingly [62–64].

It was recently demonstrated that CB17-scid mice have impaired self-renewal capacity of HSC in lethal irradiation transplantation, and upon non-conditioned transplantation, are more permissive for exogenous mouse HSC engraftment than control recipient, allowing 8–10% of donor hematopoietic stem/progenitor cell (HSPC) in BM [65]. As $Rag1^{-/-}$ mice, another immunodeficient strain with defective V(D)J recombination, did not show similar result, this defect is not due to immunodeficiency but impaired competitive repopulating ability of SCID mutant HSCs leading to niche occupancy defect. In addition, genetically modified *Prkdc* mutants show HSPC loss caused by DNA damage accumulation during embryonic development and start to die of pancytopenia shortly after the birth [66]. Since similar phenotype was observed in HSCs lacking NHEJ-related gene [67, 68], functional impairment of SCID HSCs is dependent on NHEJ machinery rather than V(D)J recombination, as in part supported by the recent finding that HSCs have active NHEJ machinery to be resistance to DNA damage [69].

4.4 Role of Mouse Innate Immunity in the Engraftment of Human Cells: NOG versus NSG

Innate immune cells such as NK cells, DCs, or macrophages play an important role for xenograft rejection. Human cell engraftment in HHLS mice was significantly improved when C.B17-SCID mice were crossed to NOD strain that lacks complement system, NK cells, and macrophages [70, 71]. NK activity in NOD/SCID was significantly reduced compared to C.B17-SCID but not completely abrogated [64]. Further elimination of NK cells has been achieved by anti-CD122 antibody that targets both NK and myeloid cells or genetically engineered mutation in *Il2rg* gene [13, 14], and lead to facilitate engraftment of human SRCs [72, 73]. Two types of mutations in IL2Rγ have been introduced into NOD/SCID independently: one mutant with complete null mutation resulting in no cell surface expression of IL2Rγ, and no binding to ligands at all (NSG, NOD/LtSz-scid $Il2r\gamma^{-/-}$ [14]), and another with a truncated intracellular domain of IL2Rγ that still binds ligands but does not transduce downstream signals (NOG, NOD/Shi-scid$Il2r\gamma^{-/-}$ [13]). Recent comparative analysis, where limited number of human HSCs were transplanted into NSG, NOG, and anti-CD122 treated NOD/SCID, demonstrated that NOG and NSG mice show significant improvement in human SRC detection over NOD/SCID model [74]. In addition, the higher sensitivity to detect SRCs in NSG than NOG suggests that the remaining IL2Rγ chain extracellular domain in NOG might leave HSCs undifferentiated in the BM and thus result in less repopulation and ignorance of SRCs.

Besides NK cell-mediated innate immune barrier, another key genetic determinant for xenograftment has been identified in *Sirpa* gene responsible for high human cell engraftment in NOD strain [18]. SIRPα is mainly expressed on macrophages, granulocytes, and DCs, and upon binding CD47 transduces a negative "don't eat me" signal in macrophages preventing phagocytic activity [75, 76]. It was shown that NOD-type SIRPα with polymorphism in immunoglobulin like domain V (IgV) can cross-react with human CD47 unlike SIRPα from other strains, and therefore, block mouse macrophage engulfment against engrafted human cells, leading to efficient homing and engraftment [18]. This finding was further proved by significant improvement of human cell engraftment in HHLS mice with overexpression of mouse CD47 in transplanted-human $CD34^+$ cells, genetic introduction of human SIRPα in BRG strain, or crossing of C57/BL6.$Rag2^{-/-}Il2r\gamma^{-/-}$ with C57/BL6.NOD harboring NOD-type SIRPα [19, 77, 78]. Consistently, disruption of SIRPα–CD47 signal axis in macrophages by neutralizing antibody or chemical-mediated depletion of macrophages enhances human HSC engraftment [79, 80].

4.5 Lymphoid Tissue Development: NOD/SCID versus NOG versus NSG Mice

Introduction of SCID mutation into NOD background reduces activity of mouse innate immunity against engrafted human cells, and also rescues NOD mice from development of T cell-mediated autoimmune insulin-dependent diabetes mellitus [81]. Despite improved human cell engraftment, NOD/SCID mice develop a spontaneous thymic lymphoma with high incidence, in 70 % of mice by the age of 40 weeks [64, 82]. A mean lifespan of NOD/SCID mice is only 8.5 months [64] and thus severely impedes long-term study of human immunity and hematopoiesis. Genetic crosses of NOD/SCID onto $IL2R\gamma^{-/-}$ mice extend their lifespan up to on average 22 months with free of thymoma [14].

Engraftment of human $CD45^+$ cells in thymus was significantly higher in NOG or NSG than NOD/SCID [12, 13, 64, 71, 74] (Table 4.1). The reconstituted thymi of NOG or NSG contain human $CD45^+CD3^+$ cells, majority of which are CD4 and CD8 double positive T cells [74], and these human T cells display polyclonal T cell receptor (TCR)-αβ [83] and functional antigen response [84, 85]. Similarly, spleen has been better reconstituted with $CD19^+IgM^+$ B, $CD3^+$ T, and $CD56^+$ NK cells in NOD or NSG than other strains. Functional analysis revealed that human B cells developed in mouse spleen resemble the characteristics of B-1 B cells that represent a minor population in humans and express auto/polyreactive antibody [86]. Although the human antibody repertoire in HHLS mice is similar to that observed in humans [87, 88], class switching from IgM to IgG and affinity maturation of human antibody occur poorly in current HHLS models [2, 64]. Development of peripheral lymph nodes (pLNs), e.g., axillary, inguinal LNs is also limited in any of IL2Rγ-deficient immunocompromised mice including BRG, NOD, NSG, while mesenteric LN is usually formed [5, 89]. Given the fact that transfer of relatively high number of T cells develops pLNs in BRG model [90, 91], it will be possible to overcome this limitation upon robust T cell reconstitution by long-term sustained HSC engraftment.

Table 4.1 Representative humanization protocol in HHLS models

Mouse strain abbreviation	Mouse age	Conditioning	Human cells	Cell dose	Injection route	Analysis (wkpt)	% hCD45$^+$				% hCD34$^+$ in BM CD45$^+$	Refs.
							PB	Thy	SPL	BM		
BRG	newborn	400 cGy IR	CB CD34$^+$	$3.8–12 \times 10^4$	i.h.	4–26	n.d.	>70	10–80	10–70	n.d.	[89]
	newborn	350 cGy IR	FL CD34$^+$	$2–20 \times 10^5$	i.p.	6–11	>50	45–99	>50	>50	n.d.	[5, 97]
BLT (NOD/SCID)	6–10 week	Implantation of FL and FT and 200–300 cGy IR	FL CD34$^+$	$1–5 \times 10^5$	i.t.v.	18–25	40	n.d.	50	50	n.d.	[126]
	6–8 week	Implantation of FL and FT and 325 cGy IR	FL CD34$^+$	$2–25 \times 10^5$	i.t.v.	26	49	99	42	48	n.d.	[116]
NOG	8–13 week	225 cGy IR	CB Lin$^-$	$2.3–3.0 \times 10^5$	i.f.	11–13	n.d.	8	4	16–22	6	[74]
	n.d.	240 cGy IR	CB CD34$^+$	0.4×10^5	i.t.v.	20	30	n.d.	70	35	3	[13]
NSG	newborn	100 cGy IR	CB Lin$^-$CD34$^+$	10^5	i.f.v.	12	69	78–96	55	73	n.d.	[85]
	newborn	100 cGy IR	CB CD34$^+$	3×10^4	i.f.v.	12	50	n.d.	n.d.	n.d.	n.d.	[127]
	newborn	150 cGy IR	CB CD34$^+$	1×10^6	i.f.v.	8–12	>50	>90	55	64	n.d.	[96]
	8–13 week	225 cGy IR	CB Lin$^-$	$2.3–3.0 \times 10^5$	i.f.	11–13	n.d.	6	4	43	6	[74]
	n.d.	325 cGy IR	mbPB CD34$^+$	7.0×10^5	i.t.v.	8–10	6	37	60	35	n.d.	[14]
	7–10 week	2 × 25 mg/kg BSU i.p.	CB CD34$^+$	20×10^5	i.t.v.	24	60–65	n.d.	n.d.	5	n.d.	[104]
	3–4week	2 × 25 mg/kg BSU i.p.	CB CD34$^+$	2×10^5	i.t.v.	22	41	99	74	83	n.d.	[105]
BRG-hTHPO^{KI}	newborn	300 cGy IR	CB/FL CD34$^+$	$1.0–2.0 \times 10^5$	i.h.	12–16	10	n.d.	27	80	12	[22]

Table 4.1 (continued)

Mouse strain abbreviation	Mouse age	Conditioning	Human cells	Cell dose	Injection route	Analysis (wkpt)	% $hCD45^+$				% $hCD34^+$ in BM $CD45^+$	Refs.
							PB	Thy	SPL	BM		
BRG-hIL3/ GM-CSF^{KI}	newborn	400 cGy IR	CB/FL $CD34^+$	1.0–2.0×10^5	i.h.	8–12	18	n.d.	20	50	15	[23]
BRG-hM-CSF^{KI}	newborn	300 cGy IR	FL $CD34^+$	1.0–2.0×10^5	i.h.	12	35	n.d.	25	42	n.d.	[24]
BRG-$hSirp^{Tg}$	newborn	360 cGy IR	FL $CD34^+$	1.0–2.0×10^5	i.h.	10–12	23	n.d.	70	66	n.d.	[78]
NSG-hIL-3/ GM-SF/ SCF^{Tg}	newborn	100 cGy IR	FL $CD34^+$	1.5–2.0×10^5	i.h.	8–12	50	n.d.	n.d.	n.d.	0.2 $CD34^+$ $CD38^-$	[128]
NSG $mbhSCF^{Tg}$	newborn	150 cGy IR	CB Lin^-CD34^+ $CD38^-$	0.5–5.3×10^4	i.f.v.	8–35	83.1	n.d.	94.5	97.1	n.d.	[125]
	newborn	No IR	CB $CD34^+$	3.0×10^4	i.t.v.	12	13	46	70	57	0.27$CD34^+$ $CD38^-$	[121]
NSG-$\beta2m^{-/-}$ HLA-$A2^{Tg}$	newborn	150 cGy IR	CB Lin^-CD34^+ $CD38^-$	0.5–3.0×10^4	i.f.v.	16–32	n.d.	n.d.	87	71	n.d.	[124]
NSG-HLA-$A2^{Tg}$	newborn	100 cGy IR	FL $CD34^+$	1.0–3.0×10^5	i.h.	12	40	n.d.	>60	28	n.d.	[123]
NSG-HLA-$DR4^{Tg}$	n.d.	350 cGy IR	CB $CD34^+$	8.0×10^4	i.t.v.	24	n.d.	84	82	69	7	[122]
NOG-I-$A^{-/-}$HLA-$DR4^{Tg}$	6 week	120 cGy IR	CB $CD34^+$	1.0×10^5	n.d.	12–20	n.d.	n.d.	20–80	70	n.d.	[129]

Table 4.1 (continued)

Mouse strain abbreviation	Mouse age	Conditioning	Human cells	Cell dose	Injection route	Analysis (wkpt)	% hCD45$^+$				% hCD34$^+$ in BM CD45$^+$	Refs.
							PB	Thy	SPL	BM		

BRG Balb/c Rag2$^{-/-}$ IL2rγ$^{-/-}$, *NSG* NOD/SCID IL2rγ$^{-/-}$, *NOG* NOD/SCID IL2rγ$^{-/-}$, *NRG* NOD Rag2$^{-/-}$ IL2rγ$^{-/-}$, *BLT* bone marrow-liver-thymus, *IL* interleukin, *THPO* thrombopoietin, *GM-CSF* granulocyte macrophage colony-stimulating factor, *mbhSCF* membrane-bound human stem cell factor, *HLA* human leukocyte antigen, *KI* knock-in, *Tg* transgenic, *IR* irradiation, *BSU* busulfan, *CB* cord blood, *FL* fetal liver, *FT* fetal thymus, *i.f.v* intrafacial vein, *i.h.* intrahepatic, *i.p.* intraperitoneal, *i.f.* intrafemoral, *i.t.v.* intratail vein, *PB* peripheral blood, *Spl* spleen, *Thy* thymus, *BM* bone marrow, *wk* week old, *wkpt* weeks post transplantation, *n.d.* not discribed or determined

4.6 Parameters in Donor and Recipient that Influence Humanization Efficiency

Main source of human HSCs to engraft HHLS models is FL, UCB, ABM, and pharmacologically mobilized peripheral blood (mbPB). Frequency of $CD34^+$ cells containing human HSPCs is around 0.5–5 % in FL, UCB, and ABM, and 0.04–0.4 % in mobilized PB [29, 92]. Since $CD34^+$ cells are functionally heterogeneous, in vivo functional studies have been performed by transplanting limiting dose of cells into HHLS mice to determine the frequency of SRCs, and found the highest in FL and the lowest in mbPB (SRC frequency: FL≥CB>BM≥mbPB) [22, 23, 93, 94].

Choice of age and sex of recipient animals seems to matter: injection of CB-derived $CD34^+$ cells into 6–8 week BRG mice results in $20 \pm 16\,\%$ engraftment in BM and partial thymic repopulation [95], whereas newborn of BRG shows 10–60 % human chimerism in BM and robust development of functional T cells [89, 96, 97] (Table. 4.1). This suggests more efficient de novo human cell repopulation in newborn than in young adult; with sex of recipient mouse, limiting dilution and serial transplantation of the defined human HSCs into NSG male or female demonstrated that NSG females support higher engraftment in BM and spleen, and more robust self-renewal of human HSC than do males, suggesting that sex-specific mechanism is crucial for human HSC engraftment and maintenance [98]. It is, however, of note that a range of $1\text{–}2 \times 10^5$ $CD34^+$ cells does not seem to make a significant difference in human cell engraftment [98].

A major hurdle for successful HSC engraftment during transplantation is BM homing, a highly coordinated multistep process in which injected HSCs circulate in blood, cross BM endothelium barrier, and finally settle down to an emptied BM niche to self-renewal and differentiate into multilineage blood cells [99]. In xenotransplantation, this process is further complicated by other parameters as cross-reactivity of homing, migration and growth factors, availability of HSC niche space in host BM, growth competition between mouse and human HSCs. One strategy to avoid homing-associated complications is to inject cells directly into the hematopoietic site, either bone of adult mice (intra-femoral injection) to escape from host immune surveillance [100, 101], or liver of newborn (intra-hepatic injection) to recapitulate natural homing process [89]. Intraosseous injection allows detecting human SRCs with 15-fold higher sensitivity compared to intravenous injection [73, 100], although it might stress and disturb BM microstructure [102]. Other injection technique, intrafacial or intraperitoneal injection seems to work at least in newborn as comparable to or better than intrahepatic injection [85, 97].

To achieve high and sustained human HSC engraftment, niche space in host needs to be created by depletion of mouse HSCs and hematopoietic cells. To this end, most of the studies with HHLS models employ sublethal dose of irradiation. Since each immunodeficient strain has different irradiation sensitivity [74], the irradiation dose should be optimized considering both animal mortality and human engraftment. It will be interesting to test efficacy of human HSC engraftment with other myeloablative regimens than irradiation, such as cyclophosphamide [103], busulfan [104, 105], or neutralizing antibody against mouse c-Kit [106]. Representative humanization protocols with HHLS models are shown in Table 4.1.

4.7 How to Improve "Humanization": Multilineage Cell Engraftment and Maintenance

For the past decades, a lot of efforts have been made to develop different HHLS models with genetic crossing and gene targeting, and to study physiology of human hematopoiesis and immunity, as well as pathogenesis of human diseases including autoimmunity, infection, and cancer (reviewed in [2, 3]). Despite remarkable achievements, there are some limitations that remain to be overcome to have robust HHLS models: (i) unstable human cell reconstitution and maintenance; (ii) limited functionality of human innate and acquired immunity; (iii) inefficient platelet and red blood cell production. This topic will not be discussed here, as it is not highly relevant to HIV infection and also covered in other review [2].

First, long-term stabilization of human cell engraftment over experiments and different donor cells requires more number of human HSCs to be transplanted and better maintenance of self-renewal capacity of transplanted HSCs *in vivo*. However, human samples available for transplantation are usually limited. One approach to improve this is to develop an *in vitro* culture system that allows expansion of functional HSCs. This has been already considered and up to 17-fold expansion has been achieved at the best as determined by in vivo NOD/SCID repopulating assay (reviewed in [107]). Another way to supply human HSCs is to generate human HSCs from human embryonic stem cells (ESCs) or induced pluripotent stem cells (iPSCs) that can be maintained and expanded *in vitro*. However, the process of HSC generation from pluripotent stem cells would be highly complicated by the evidences that many extrinsic and intrinsic factors are temporarily and spatially involved in the HSC emergence during hematopoietic ontogeny (reviewed in [108, 109]). Many studies have attempted to mimic *in vivo* early HSC development by in vitro culture of PSCs with supplementation of fluid factors or coculture with feeder cells, e.g., fibroblasts or stromal cells [110, 111]. Yet, *in vitro* HSC development from iPSCs is largely limited to production of *in vivo* transplantable HSCs. To bypass exogenous factor-mediated HSC specification, a recent work has demonstrated that human iPSC-derived teratoma formed in NSG mice contain human HSCs that are transplantable into another recipient [112]. Introduction of a certain sets of transcription factors in mouse or human fibroblasts, or in human iPSCs have shown to activate hemogenic program that generates in vitro hematopoietic colony forming cells or *in vivo* blood repopulating cells with limited self-renewal [113–115]. Although these studies have significantly advanced our knowledge to expand or develop HSC *ex vivo*, unlimited supply of transplantable human HSCs for preclinical and clinical application remains challenging.

To improve *in vivo* engraftment or maintenance of human HSCs, human *SIRPA* and *THPO* gene have been introduced into BRG mice by genetic engineering

(SIRPα[Tg] BRG and THPO[KI] BRG, respectively), and these models showed high engraftment level and enhanced self-renewal activity of human HSC in BM with no obvious impact on quality of mouse life [22, 78] (Table 4.1). An alternative strategy to support human HSC function is providing mice with human microenvironments by cotransplantation of human tissues or human niche cells. Some groups have developed a model where human FL-derived CD34^{+} cells are transplanted into sublethally irradiated NOD/SCID or NSG previously implanted with autologous human fetal thymus and liver, so-called BM-liver-thymus (BLT) model, and achieved high human engraftment and robust immune response [116, 117]. Yet, a broad application of BLT model is limited by necessity of easy access to human fetal tissues. In contrast, implantation of human MSCs is relatively easy, as they can grow from their self-renewing progenitors in *in vitro* culture of human BM efficiently [57]. *In vivo* implantation of MSC progenitor or MSC-derived ossicle develop heterotopic bone organ that recruits functional HSC to the site [118, 119]. This system might serve as “stem cell trap” to provide more optimal microenvironment for robust human HSC maintenance.

Second, because of low or little cross-reactivity of some of the mouse-soluble factors to human receptor [21], mouse environment provides suboptimal conditions for human HSPCs to fully differentiate to the mature cells that in turn migrate to an appropriate tissue and exert their function upon activation. Various strategies have been taken to improve differentiation and maturation of human immune cells as follows: *in vivo* injection of recombinant human cytokines, human cytokine-coding plasmid or virus to enhance erythropoiesis, myelopoiesis, thymopoiesis, and NK cell development (reviewed in [120]); genetic replacement of myeloid-supporting mouse cytokines with human counterparts to promote myeloid cell differentiation [22–25]; transgene expression of human SIRPα, membrane-bound SCF, MHC I and II to augment human cell engraftment, mast cell development, and functional T cell development [78, 121–125]. More new animals or novel methodology are expected to be developed and revolutionize HHLS models.

References

1. Yang Y-G, Sykes M. Xenotransplantation: current status and a perspective on the future. Nat Rev Immunol. 2007;7:519–31.
2. Rongvaux A, Takizawa H, Strowig T, Willinger T, Eynon EE, Flavell RA, Manz MG. Human hemato-lymphoid system mice: current use and future potential for medicine. Annu Rev Immunol. 2013;31:635–74.
3. Shultz LD, Brehm MA, Garcia-Martinez JV, Greiner DL. Humanized mice for immune system investigation: progress, promise and challenges. Nat Rev Immunol. 2012;12:786–98.
4. Shultz LD, Ishikawa F, Greiner DL. Humanized mice in translational biomedical research. Nat Rev Immunol. 2007;7:118–30.
5. Legrand N, Weijer K, Spits H. Experimental models to study development and function of the human immune system in vivo. J Immunol. 2006;176:2053–8.
6. Bosma GC, Custer RP, Bosma MJ. A severe combined immunodeficiency mutation in the mouse. Nature. 1983;301:527–30.

7. Schuler W, Weiler IJ, Schuler A, Phillips RA, Rosenberg N, Mak TW, Kearney JF, Perry RP, Bosma MJ. Rearrangement of antigen receptor genes is defective in mice with severe combined immune deficiency. Cell. 1986;46:963–72.
8. Malynn BA, Blackwell TK, Fulop GM, Rathbun GA, Furley AJ, Ferrier P, Heinke LB, Phillips RA, Yancopoulos GD, Alt FW. The Scid defect affects the final step of the immunoglobulin VDJ recombinase mechanism. Cell. 1988;54:453–60.
9. Hesselton RM, Greiner DL, Mordes JP, Rajan TV, Sullivan JL, Shultz LD. High levels of human peripheral blood mononuclear cell engraftment and enhanced susceptibility to human immunodeficiency virus type 1 infection in NOD/LtSz-Scid/Scid mice. J Infect Dis. 1995;172:974–82.
10. DiSanto JP, Müller W, Guy-Grand D, Fischer A, Rajewsky K. Lymphoid development in mice with a targeted deletion of the interleukin 2 receptor gamma chain. Proc Natl Acad Sci U S A. 1995;92:377–81.
11. Cao X, Shores EW, Hu-Li J, Anver MR, Kelsall BL, Russell SM, Drago J, Noguchi M, Grinberg A, Bloom ET. Defective lymphoid development in mice lacking expression of the common cytokine receptor gamma chain. Immunity. 1995;2:223–38.
12. Shinkai Y, Rathbun G, Lam KP, Oltz EM, Stewart V, Mendelsohn M, Charron J, Datta M, Young F, Stall AM. RAG-2-deficient mice lack mature lymphocytes owing to inability to initiate V(D)J rearrangement. Cell. 1992;68:855–67.
13. Ito M, Hiramatsu H, Kobayashi K, et al. NOD/SCID/gamma(c)(null) mouse: an excellent recipient mouse model for engraftment of human cells. Blood. 2002;100:3175–82.
14. Shultz LD, Lyons BL, Burzenski LM, et al. Human lymphoid and myeloid cell development in NOD/LtSz-scid IL2R gamma null mice engrafted with mobilized human hemopoietic stem cells. J Immunol. 2005;174:6477–89.
15. Mazurier F, Fontanellas A, Salesse S, Taine L, Landriau S, Moreau-Gaudry F, Reiffers J, Peault B, Di Santo JP, de Verneuil H. A novel immunodeficient mouse model–RAG2 x common cytokine receptor gamma chain double mutants–requiring exogenous cytokine administration for human hematopoietic stem cell engraftment. J Interferon Cytokine Res. 1999;19:533–41.
16. Greiner DL, Hesselton RA, Shultz LD. SCID mouse models of human stem cell engraftment. Stem Cells. 1998;16:166–77.
17. Brehm MA, Cuthbert A, Yang C, et al. Parameters for establishing humanized mouse models to study human immunity: analysis of human hematopoietic stem cell engraftment in three immunodeficient strains of mice bearing the IL2rgamma(null) mutation. Clin Immunol. 2010;135:84–98.
18. Takenaka K, Prasolava TK, Wang JCY, Mortin-Toth SM, Khalouei S, Gan OI, Dick JE, Danska JS. Polymorphism in Sirpa modulates engraftment of human hematopoietic stem cells. Nat Immunol. 2007;8:1313–23.
19. Yamauchi T, Takenaka K, Urata S, et al. Polymorphic Sirpa is the genetic determinant for NOD-based mouse lines to achieve efficient human cell engraftment. Blood. 2013;121:1316–25.
20. Mercier FE, Ragu C, Scadden DT. The bone marrow at the crossroads of blood and immunity. Nat Rev Immunol. 2012;12:49–60.
21. Manz MG. Human-hemato-lymphoid-system mice: opportunities and challenges. Immunity. 2007;26:537–41.
22. Rongvaux A, Willinger T, Takizawa H, et al. Human thrombopoietin knockin mice efficiently support human hematopoiesis in vivo. Proc Natl Acad Sci U S A. 2011;108:2378–83.
23. Willinger T, Rongvaux A, Takizawa H, et al. Human IL-3/GM-CSF knock-in mice support human alveolar macrophage development and human immune responses in the lung. Proc Natl Acad Sci U S A. 2011;108:2390–5.
24. Rathinam C, Poueymirou WT, Rojas J, Murphy AJ, Valenzuela DM, Yancopoulos GD, Rongvaux A, Eynon EE, Manz MG, Flavell RA. Efficient differentiation and function of human macrophages in humanized CSF-1 mice. Blood. 2011;118:3119–28.
25. Rongvaux A, Willinger T, Martinek J, et al. Development and function of human innate immune cells in a humanized mouse model. Nat Biotechnol. 2014;32:364–372.

26. Strowig T, Rongvaux A, Rathinam C, Takizawa H, Borsotti C, Philbrick W, Eynon EE, Manz MG, Flavell RA. Transgenic expression of human signal regulatory protein alpha in Rag2-/-gamma(c)-/- mice improves engraftment of human hematopoietic cells in humanized mice. Proc Natl Acad Sci USA. 2011;108:13218–13223.
27. Baum CM, Weissman IL, Tsukamoto AS, Buckle AM, Peault B. Isolation of a candidate human hematopoietic stem-cell population. Proc Natl Acad Sci U S A. 1992;89:2804–8.
28. Craig W, Kay R, Cutler RL, Lansdorp PM. Expression of Thy-1 on human hematopoietic progenitor cells. J Exp Med. 1993;177:1331–42.
29. Kondo M, Wagers AJ, Manz MG, Prohaska SS, Scherer DC, Beilhack GF, Shizuru JA, Weissman IL. Biology of hematopoietic stem cells and progenitors: implications for clinical application. Annu Rev Immunol. 2003;21:759–806.
30. Larochelle A, Vormoor J, Hanenberg H, et al. Identification of primitive human hematopoietic cells capable of repopulating NOD/SCID mouse bone marrow: implications for gene therapy. Nat Med. 1996;2:1329–37.
31. Bhatia M, Wang JC, Kapp U, Bonnet D, Dick JE. Purification of primitive human hematopoietic cells capable of repopulating immune-deficient mice. Proc Natl Acad Sci U S A. 1997;94:5320–5.
32. Majeti R, Park CY, Weissman IL. Identification of a hierarchy of multipotent hematopoietic progenitors in human cord blood. Cell Stem Cell. 2007;1:635–45.
33. McKenzie JL, Takenaka K, Gan OI, Doedens M, Dick JE. Low rhodamine 123 retention identifies long-term human hematopoietic stem cells within the Lin-CD34+CD38− population. Blood. 2007;109:543–5.
34. Notta F, Doulatov S, Laurenti E, Poeppl A, Jurisica I, Dick JE. Isolation of single human hematopoietic stem cells capable of long-term multilineage engraftment. Science. 2011;333:218–21.
35. Allsopp RC, Morin GB, DePinho R, Harley CB, Weissman IL. Telomerase is required to slow telomere shortening and extend replicative lifespan of HSCs during serial transplantation. Blood. 2003;102:517–20.
36. Zhang J, Niu C, Ye L, et al. Identification of the haematopoietic stem cell niche and control of the niche size. Nature. 2003;425:836–41.
37. Calvi LM, Adams GB, Weibrecht KW, et al. Osteoblastic cells regulate the haematopoietic stem cell niche. Nature. 2003;425:841–6.
38. Mansour A, Abou-Ezzi G, Sitnicka E, Jacobsen SEW, Wakkach A, Blin-Wakkach C. Osteoclasts promote the formation of hematopoietic stem cell niches in the bone marrow. J Exp Med. 2012;209:537–49.
39. Ding L, Morrison SJ. Haematopoietic stem cells and early lymphoid progenitors occupy distinct bone marrow niches. Nature. 2013;495:231–5.
40. Naveiras O, Nardi V, Wenzel PL, Hauschka PV, Fahey F, Daley GQ. Bone-marrow adipocytes as negative regulators of the haematopoietic microenvironment. Nature. 2009;460:259–63.
41. Yamazaki S, Ema H, Karlsson G, Yamaguchi T, Miyoshi H, Shioda S, Taketo MM, Karlsson S, Iwama A, Nakauchi H. Nonmyelinating Schwann cells maintain hematopoietic stem cell hibernation in the bone marrow niche. Cell. 2011;147:1146–58.
42. Chow A, Lucas D, Hidalgo A, et al. Bone marrow CD169+ macrophages promote the retention of hematopoietic stem and progenitor cells in the mesenchymal stem cell niche. J Exp Med. 2011;208:261–71.
43. Winkler IG, Sims NA, Pettit AR, et al. Bone marrow macrophages maintain hematopoietic stem cell (HSC) niches and their depletion mobilizes HSCs. Blood. 2010;116:4815–28.
44. Omatsu Y, Sugiyama T, Kohara H, Kondoh G, Fujii N, Kohno K, Nagasawa T. The essential functions of adipo-osteogenic progenitors as the hematopoietic stem and progenitor cell niche. Immunity. 2010;33:387–99.
45. Ding L, Saunders TL, Enikolopov G, Morrison SJ. Endothelial and perivascular cells maintain haematopoietic stem cells. Nature. 2012;481:457–62.

46. Méndez-Ferrer S, Michurina TV, Ferraro F, Mazloom AR, Macarthur BD, Lira SA, Scadden DT, Ma'ayan A, Enikolopov GN, Frenette PS. Mesenchymal and haematopoietic stem cells form a unique bone marrow niche. Nature. 2010;466:829–34.
47. Morikawa S, Mabuchi Y, Kubota Y, et al. Prospective identification, isolation, and systemic transplantation of multipotent mesenchymal stem cells in murine bone marrow. J Exp Med. 2009;206:2483–96.
48. Arai F, Hirao A, Ohmura M, Sato H, Matsuoka S, Takubo K, Ito K, Koh GY, Suda T. Tie2/angiopoietin-1 signaling regulates hematopoietic stem cell quiescence in the bone marrow niche. Cell. 2004;118:149–61.
49. Sugiyama T, Kohara H, Noda M, Nagasawa T. Maintenance of the hematopoietic stem cell pool by CXCL12-CXCR4 chemokine signaling in bone marrow stromal cell niches. Immunity. 2006;25:977–88.
50. Greenbaum A, Hsu Y-MS, Day RB, Schuettpelz LG, Christopher MJ, Borgerding JN, Nagasawa T, Link DC. CXCL12 in early mesenchymal progenitors is required for haematopoietic stem-cell maintenance. Nature. 2013;495:227–30.
51. Yoshihara H, Arai F, Hosokawa K, et al. Thrombopoietin/MPL signaling regulates hematopoietic stem cell quiescence and interaction with the osteoblastic niche. Cell Stem Cell. 2007;1:685–97.
52. Qian H, Buza-Vidas N, Hyland CD, Jensen CT, Antonchuk J, Månsson R, Thoren LA, Ekblom M, Alexander WS, Jacobsen SEW. Critical role of thrombopoietin in maintaining adult quiescent hematopoietic stem cells. Cell Stem Cell. 2007;1:671–84.
53. Butler JM, Nolan DJ, Vertes EL, et al. Endothelial cells are essential for the self-renewal and repopulation of Notch-dependent hematopoietic stem cells. Cell Stem Cell. 2010;6:251–64.
54. Sugimura R, He XC, Venkatraman A, et al. Noncanonical Wnt signaling maintains hematopoietic stem cells in the niche. Cell. 2012;150:351–65.
55. Hosokawa K, Arai F, Yoshihara H, et al. Cadherin-based adhesion is a potential target for niche manipulation to protect hematopoietic stem cells in adult bone marrow. Cell Stem Cell. 2010;6:194–8.
56. Winkler IG, Barbier V, Nowlan B, Jacobsen RN, Forristal CE, Patton JT, Magnani JL, Lévesque J-P. Vascular niche E-selectin regulates hematopoietic stem cell dormancy, self renewal and chemoresistance. Nat Med. 2012;18:1651–7.
57. Sacchetti B, Funari A, Michienzi S, et al. Self-renewing osteoprogenitors in bone marrow sinusoids can organize a hematopoietic microenvironment. Cell. 2007;131:324–36.
58. McCune JM, Namikawa R, Kaneshima H, Shultz LD, Lieberman M, Weissman IL. The SCID-hu mouse: murine model for the analysis of human hematolymphoid differentiation and function. Science. 1988;241:1632–9.
59. Mosier DE, Gulizia RJ, Baird SM, Wilson DB. Transfer of a functional human immune system to mice with severe combined immunodeficiency. Nature. 1988;335:256–9.
60. Fulop GM, Phillips RA. The scid mutation in mice causes a general defect in DNA repair. Nature. 1990;347:479–82.
61. Danska JS, Pflumio F, Williams CJ, Huner O, Dick JE, Guidos CJ. Rescue of T cell-specific V(D)J recombination in SCID mice by DNA-damaging agents. Science. 1994;266:450–55.
62. Bosma GC, Fried M, Custer RP, Carroll A, Gibson DM, Bosma MJ. Evidence of functional lymphocytes in some (leaky) Scid mice. J Exp Med. 1988;167:1016–33.
63. Nonoyama S, Smith FO, Ochs HD. Specific antibody production to a recall or a neoantigen by SCID mice reconstituted with human peripheral blood lymphocytes. J Immunol. 1993;151:3894–901.
64. Shultz LD, Schweitzer PA, Christianson SW, Gott B, Schweitzer IB, Tennent B, McKenna S, Mobraaten L, Rajan TV, Greiner DL. Multiple defects in innate and adaptive immunologic function in NOD/LtSz-scid mice. J Immunol. 1995;154:180–91.
65. Qing Y, Lin Y, Gerson SL. An intrinsic BM hematopoietic niche occupancy defect of HSC in scid mice facilitates exogenous HSC engraftment. Blood. 2012;119:1768–71.
66. Zhang S, Yajima H, Huynh H, et al. Congenital bone marrow failure in DNA-PKcs mutant mice associated with deficiencies in DNA repair. J Cell Biol. 2011;193:295–305.

67. Rossi DJ, Bryder D, Seita J, Nussenzweig A, Hoeijmakers J, Weissman IL. Deficiencies in DNA damage repair limit the function of haematopoietic stem cells with age. Nature. 2007;447:725–9.
68. Nijnik A, Woodbine L, Marchetti C, et al. DNA repair is limiting for haematopoietic stem cells during ageing. Nature. 2007;447:686–90.
69. Mohrin M, Bourke E, Alexander D, Warr MR, Barry-Holson K, Le Beau MM, Morrison CG, Passegué E. Hematopoietic stem cell quiescence promotes error-prone DNA repair and mutagenesis. Cell Stem Cell. 2010;7:174–85.
70. Serreze DV, Leiter EH. Defective activation of T suppressor cell function in nonobese diabetic mice. Potential relation to cytokine deficiencies. J Immunol. 1988;140:3801–7.
71. Baxter AG, Cooke A. Complement lytic activity has no role in the pathogenesis of autoimmune diabetes in NOD mice. Diabetes. 1993;42:1574–8.
72. Tanaka T, Kitamura F, Nagasaka Y, Kuida K, Suwa H, Miyasaka M. Selective long-term elimination of natural killer cells in vivo by an anti-interleukin 2 receptor beta chain monoclonal antibody in mice. J Exp Med. 1993;178:1103–7.
73. McKenzie JL, Gan OI, Doedens M, Dick JE. Human short-term repopulating stem cells are efficiently detected following intrafemoral transplantation into NOD/SCID recipients depleted of CD122+ cells. Blood. 2005;106:1259–61.
74. McDermott SP, Eppert K, Lechman ER, Doedens M, Dick JE. Comparison of human cord blood engraftment between immunocompromised mouse strains. Blood. 2010;116:193–200.
75. Barclay AN, Brown MH. The SIRP family of receptors and immune regulation. Nat Rev Immunol. 2006;6:457–64.
76. Matozaki T, Murata Y, Okazawa H, Ohnishi H. Functions and molecular mechanisms of the CD47-SIRPalpha signalling pathway. Trends Cell Biol. 2009;19:72–80.
77. Legrand N, Huntington ND, Nagasawa M, et al. Functional CD47/signal regulatory protein alpha (SIRP(alpha)) interaction is required for optimal human T- and natural killer- (NK) cell homeostasis in vivo. Proc Natl Acad Sci USA. 2011;108:13224–9.
78. Strowig T, Rongvaux A, Rathinam C, Takizawa H, Borsotti C, Philbrick W, Eynon EE, Manz MG, Flavell RA. Transgenic expression of human signal regulatory protein alpha in Rag2–/– gamma(c)–/– mice improves engraftment of human hematopoietic cells in humanized mice. Proc Natl Acad Sci U S A. 2011;108:13218–23.
79. Terpstra W, Leenen PJ, van den Bos C, et al. Facilitated engraftment of human hematopoietic cells in severe combined immunodeficient mice following a single injection of Cl2MDP liposomes. Leukemia. 1997;11:1049–54.
80. Rozemuller H, Knaän-Shanzer S, Hagenbeek A, van Bloois L, Storm G, Martens ACM. Enhanced engraftment of human cells in RAG2/gammac double-knockout mice after treatment with CL2MDP liposomes. Exp Hematol. 2004;32:1118–25.
81. Christianson SW, Shultz LD, Leiter EH. Adoptive transfer of diabetes into immunodeficient NOD-scid/scid mice. Relative contributions of CD4+ and CD8+ T-cells from diabetic versus prediabetic NOD.NON-Thy-1a donors. Diabetes. 1993;42:44–55.
82. Prochazka M, Gaskins HR, Shultz LD, Leiter EH. The nonobese diabetic scid mouse: model for spontaneous thymomagenesis associated with immunodeficiency. Proc Natl Acad Sci U S A. 1992;89:3290–94.
83. Yahata T, Ando K, Nakamura Y, Ueyama Y, Shimamura K, Tamaoki N, Kato S, Hotta T. Functional human T lymphocyte development from cord blood CD34+ cells in nonobese diabetic/Shi-scid, IL-2 receptor gamma null mice. J Immunol. 2002;169:204–9.
84. Hiramatsu H, Nishikomori R, Heike T, Ito M, Kobayashi K, Katamura K, Nakahata T. Complete reconstitution of human lymphocytes from cord blood CD34+ cells using the NOD/SCID/gammacnull mice model. Blood. 2003;102:873–80.
85. Ishikawa F, Yasukawa M, Lyons B, Yoshida S, Miyamoto T, Yoshimoto G, Watanabe T, Akashi K, Shultz LD, Harada M. Development of functional human blood and immune systems in NOD/SCID/IL2 receptor {gamma} chain(null) mice. Blood. 2005;106:1565–73.
86. Berland R, Wortis HH. Origins and functions of B-1 cells with notes on the role of CD5. Annu Rev Immunol. 2002;20:253–300.

87. Becker PD, Legrand N, van Geelen CMM, et al. Generation of human antigen-specific monoclonal IgM antibodies using vaccinated "human immune system" mice. PLoS ONE. 2010;5(10):e13137. doi:10.1371/journal.pone.0013137.
88. Ippolito GC, Hoi KH, Reddy ST, et al. Antibody repertoires in humanized NOD-scid-IL2Rγ(null) mice and human B cells reveals human-like diversification and tolerance checkpoints in the mouse. PLoS ONE. 2012;7:e35497.
89. Traggiai E, Chicha L, Mazzucchelli L, Bronz L, Piffaretti J-C, Lanzavecchia A, Manz MG. Development of a human adaptive immune system in cord blood cell-transplanted mice. Science. 2004;304:104–7.
90. van Rijn RS, Simonetti.ER, Hagenbeek A, et al. A new xenograft model for graft-versus-host disease by intravenous transfer of human peripheral blood mononuclear cells in RAG2−/− gammac−/− double-mutant mice. Blood. 2003;102:2522–31.
91. Neagu MR, Ziegler P, Pertel T, Strambio-De-Castillia C, Grütter C, Martinetti G, Mazzucchelli L, Grütter M, Manz MG, Luban J. Potent inhibition of HIV-1 by TRIM5-cyclophilin fusion proteins engineered from human components. J Clin Invest. 2009;119:3035–47.
92. Tjønnfjord GE, Steen R, Veiby OP, Friedrich W, Egeland T. Evidence for engraftment of donor-type multipotent CD34+ cells in a patient with selective T-lymphocyte reconstitution after bone marrow transplantation for B-SCID. Blood. 1994;84:3584–89.
93. Coulombel L. Identification of hematopoietic stem/progenitor cells: strength and drawbacks of functional assays. Oncogene. 2004;23:7210–22.
94. Matsumura T, Kametani Y, Ando K, et al. Functional CD5+ B cells develop predominantly in the spleen of NOD/SCID/gammac(null) (NOG) mice transplanted either with human umbilical cord blood, bone marrow, or mobilized peripheral blood CD34+ cells. Exp Hematol. 2003;31:789–97.
95. Kerre TCC, De Smet G, De Smedt M, Zippelius A, Pittet MJ, Langerak AW, De Bosscher J, Offner F, Vandekerckhove B, Plum J. Adapted NOD/SCID model supports development of phenotypically and functionally mature T cells from human umbilical cord blood CD34(+) cells. Blood. 2002;99:1620–26.
96. Holt N, Wang J, Kim K, et al. Human hematopoietic stem/progenitor cells modified by zinc-finger nucleases targeted to CCR5 control HIV-1 in vivo. Nat Biotechnol. 2010;28:839–47.
97. Gimeno R, Weijer K, Voordouw A, Uittenbogaart CH, Legrand N, Alves NL, Wijnands E, Blom B, Spits H. Monitoring the effect of gene silencing by RNA interference in human CD34+ cells injected into newborn RAG2−/− gammac−/− mice: functional inactivation of p53 in developing T cells. Blood. 2004;104:3886–93.
98. Notta F, Doulatov S, Dick JE. Engraftment of human hematopoietic stem cells is more efficient in female NOD/SCID/IL-2Rgc-null recipients. Blood. 2010;115:3704–07.
99. Tavor S, Petit I, Porozov S, Goichberg P, Avigdor A, Sagiv S, Nagler A, Naparstek E, Lapidot T. Motility, proliferation, and egress to the circulation of human AML cells are elastase dependent in NOD/SCID chimeric mice. Blood. 2005;106:2120–27.
100. Yahata T, Ando K, Sato T, Miyatake H, Nakamura Y, Muguruma Y, Kato S, Hotta T. A highly sensitive strategy for SCID-repopulating cell assay by direct injection of primitive human hematopoietic cells into NOD/SCID mice bone marrow. Blood. 2003;101:2905–13.
101. Mazurier F, Doedens M, Gan OI, Dick JE. Rapid myeloerythroid repopulation after intrafemoral transplantation of NOD-SCID mice reveals a new class of human stem cells. Nat Med. 2003;9:959–63.
102. Chabner KT, Adams GB, Qiu J, Moskowitz M, Marsters ES, Topulos GP, Scadden DT. Direct vascular delivery of primitive hematopoietic cells to bone marrow improves localization but not engraftment. Blood. 2004;103:4685–86.
103. Volk A, Hartmann S, Muik A, Geiss Y, Königs C, Dietrich U, Laer von D, Kimpel J. Comparison of three humanized mouse models for adoptive T cell transfer. J Gene Med. 2012;14:540–48.
104. Hayakawa J, Hsieh MM, Uchida N, Phang O, Tisdale JF. Busulfan produces efficient human cell engraftment in NOD/LtSz-Scid IL2Rgamma(null) mice. Stem Cells. 2009;27:175–82.

105. Singh M, Singh P, Gaudray G, et al. An improved protocol for efficient engraftment in NOD/LTSZ-SCIDIL-2Rγnull mice allows HIV replication and development of anti-HIV immune responses. PLoS ONE. 2012;7:e38491.
106. Czechowicz A, Kraft D, Weissman IL, Bhattacharya D. Efficient transplantation via antibody-based clearance of hematopoietic stem cell niches. Science. 2007;318:1296–99.
107. Takizawa H, Schanz U, Manz MG. Ex vivo expansion of hematopoietic stem cells: mission accomplished? Swiss Med Wkly. 2011;141:w13316.
108. Dzierzak E, Speck NA. Of lineage and legacy: the development of mammalian hematopoietic stem cells. Nat Immunol. 2008;9:129–36.
109. Cumano A, Godin I. Ontogeny of the hematopoietic system. Annu Rev Immunol. 2007;25:745–85.
110. Kennedy M, Awong G, Sturgeon CM, Ditadi A, LaMotte-Mohs R, Zúñiga-Pflücker JC, Keller G. T lymphocyte potential marks the emergence of definitive hematopoietic progenitors in human pluripotent stem cell differentiation cultures. Cell Rep. 2012;2:1722–35.
111. Choi K-D, Vodyanik MA, Togarrati PP, et al. Identification of the hemogenic endothelial progenitor and its direct precursor in human pluripotent stem cell differentiation cultures. Cell Rep. 2012;2:553–67.
112. Amabile G, Welner RS, Nombela-Arrieta C, et al. In vivo generation of transplantable human hematopoietic cells from induced pluripotent stem cells. Blood. 2013;121:1255–64.
113. Pereira C-F, Chang B, Qiu J, et al. Induction of a hemogenic program in mouse fibroblasts. Cell Stem Cell. 2013;13:205–18.
114. Szabo E, Rampalli S, Risueño RM, Schnerch A, Mitchell R, Fiebig-Comyn A, Levadoux-Martin M, Bhatia M. Direct conversion of human fibroblasts to multilineage blood progenitors. Nature. 2010;468:521–26.
115. Doulatov S, Vo LT, Chou SS, et al. Induction of multipotential hematopoietic progenitors from human pluripotent stem cells via respecification of lineage-restricted precursors. Cell Stem Cell. 2013;13:459–70.
116. Melkus MW, Estes JD, Padgett-Thomas A, Gatlin J, Denton PW, Othieno FA, Wege AK, Haase AT, Garcia JV. Humanized mice mount specific adaptive and innate immune responses to EBV and TSST-1. Nat Med. 2006;12:1316–22.
117. Akkina R. New generation humanized mice for virus research: comparative aspects and future prospects. Virology. 2013;435:14–28.
118. Lee J, Li M, Milwid J, et al. Implantable microenvironments to attract hematopoietic stem/cancer cells. Proc Natl Acad Sci U S A. 2012;109:19638–43.
119. Scotti C, Piccinini E, Takizawa H, Todorov A, Bourgine P, Papadimitropoulos A, Barbero A, Manz MG, Martin I. Engineering of a functional bone organ through endochondral ossification. Proc Natl Acad Sci U S A. 2013;110:3997–4002.
120. Drake AC, Chen Q, Chen J. Engineering humanized mice for improved hematopoietic reconstitution. Cell Mol Immunol. 2012;9:215–24.
121. Brehm MA, Racki WJ, Leif J, et al. Engraftment of human HSCs in nonirradiated newborn NOD-scid IL2rγ null mice is enhanced by transgenic expression of membrane-bound human SCF. Blood. 2012;119:2778–88.
122. Danner R, Chaudhari SN, Rosenberger J, Surls J, Richie TL, Brumeanu T-D, Casares S. Expression of HLA class II molecules in humanized NOD.Rag1KO.IL2RgcKO mice is critical for development and function of human T and B cells. PLoS ONE. 2011;6:e19826.
123. Strowig T, Gurer C, Ploss A, et al. Priming of protective T cell responses against virus-induced tumors in mice with human immune system components. J Exp Med. 2009;206:1423–34.
124. Shultz LD, Saito Y, Najima Y, et al. Generation of functional human T-cell subsets with HLA-restricted immune responses in HLA class I expressing NOD/SCID/IL2r gamma(null) humanized mice. Proc Natl Acad Sci U S A. 2010;107:13022–27.
125. Takagi S, Saito Y, Hijikata A, et al. Membrane-bound human SCF/KL promotes in vivo human hematopoietic engraftment and myeloid differentiation. Blood. 2012;119:2768–77.

126. Lan P, Tonomura N, Shimizu A, Wang S, Yang Y-G. Reconstitution of a functional human immune system in immunodeficient mice through combined human fetal thymus/liver and CD34+ cell transplantation. Blood. 2006;108:487–92.
127. Kumar P, Ban H-S, Kim S-S, et al. T cell-specific siRNA delivery suppresses HIV-1 infection in humanized mice. Cell. 2008;134:577–86.
128. Billerbeck E, Barry WT, Mu K, Dorner M, Rice CM, Ploss A. Development of human CD4+ FoxP3+ regulatory T cells in human stem cell factor-, granulocyte-macrophage colony-stimulating factor-, and interleukin-3-expressing NOD-SCID IL2Rγ(null) humanized mice. Blood. 2011;117:3076–86.
129. Suzuki M, Takahashi T, Katano I, Ito R, Ito M, Harigae H, Ishii N, Sugamura K. Induction of human humoral immune responses in a novel HLA-DR-expressing transgenic NOD/Shi-scid/γcnull mouse. Int Immunol. 2012;24:243–52.

Chapter 5
Biology of Human Hematopoietic Stem Cell Xenotransplantation in Mice

Zheng Hu and Yong-Guang Yang

Abbreviations

HSC	Hematopoietic stem cell
HHLS	Human hematopoietic and lymphoid systems
SCID	Severe combined immune deficiency
Prkdc	Protein kinase, DNA-activated, catalytic polypeptide
RAG	Recombination activating gene
SIRPα	Signal regulatory protein alpha
EPO	Erythropoietin
GM-CSF	Granulocyte-macrophage colony-stimulating factor
M-CSF	Macrophage colony-stimulating factor
NOD	Nonobese diabetic
NSG	NOD/SCID IL2 receptor gamma chain knockout
TCD	T cell-depleted
RBC	Red blood cell
TPO	Thrombopoietin
WBC	White blood cell

Y.-G. Yang (✉)
Columbia Center for Translational Immunology, Columbia University College of Physicians and Surgeons, New York, NY, USA
e-mail: yy2324@columbia.edu

Z. Hu · Y.-G. Yang
First Hospital of Jilin University, Changchun, China

L. Y. Poluektova et al. (eds.), *Humanized Mice for HIV Research*,
DOI 10.1007/978-1-4939-1655-9_5

5.1 Introduction

Humanized mice with functional human hematopoietic and lymphoid systems (referred to as HHLS mice [1]) are particularly useful in the study of human hematopoiesis and immune function under physiological and pathological conditions. Human hematopoietic stem cell (HSC) xenotransplantation is required for creating HHLS mice [2–4]. Although currently human HSC engraftment and differentiation can be reproducibly established in immunodeficient nonobese diabetic (NOD)/severe combined immune deficiency (SCID) mice or their derivatives, the creation of HHLS mice with all lineages of human blood cells remains a challenge. The barriers to achieve human HSC xenotransplantation in immunodeficient mice can be broadly classified as immunological and nonimmunological (i.e., unfavorable or disadvantageous hematopoietic microenvironment).

5.2 Immunological Barriers to Human HSC Transplantation in Mice

In general, transplants across species (i.e., xenotransplantation) are subject to vigorous immunologic rejection. In addition to adaptive immunity, the extensive molecular incompatibilities between the donor and the host induce robust innate immune responses to xenografts [5]. Innate immune responses include humoral and cellular responses, both of which can be elicited by xenoreactive antibodies following xenotransplantation [6]. Cellular innate xenoresponses can also be induced by xenografts in the absence of antibodies. Antibody-independent innate immune responses are the main immunological barriers in successfully making HHLS mouse models using immunodeficient mice, such as those with SCID mutation (i.e., a mutation in the *Prkdc* gene) or targeted disruption of recombination activating gene (RAG). Innate immune cells include phagocytic cells (monocytes/macrophages and neutrophils), natural killer (NK) cells, and cells producing inflammatory mediators (basophils, eosinophils, and mast cells). Activation of innate immunity is controlled by the balance between activating and inhibitory receptors. As such, the combinatorial effect of xenoantigen recognition by the activating receptors and the incompatibility of inhibitory receptor-ligand interactions leads to the rapid and robust responses against xenografts [5]. The common cytokine γ-chain (also known as IL2Rγ-chain) is a critical component of the high-affinity receptors shared by multiple cytokines (interleukin (IL)-2, IL-4, IL-7, IL-9, IL-15, and IL-21) [7]. The availability of immunodeficient mice with targeted disruption of the common cytokine receptor γ-chain gene makes it possible to prevent human hematopoietic cells from rejection by NK cells [8, 9]. The inability of donor CD47 to interact with recipient signal regulatory protein alpha (SIRPα) has been found to be a major mechanism triggering macrophage-mediated rejection of xenogeneic hematopoietic cells (see discussion below). Since the NOD mouse SIRPα interacts with human CD47 better than SIRPα on other mouse backgrounds [10], the use of immunodeficient mice on the NOD background helps to attenuate macrophage-mediated rejection of human hematopoietic cells. Thus, among the currently available immunodeficient mice, NOD/SCID IL2Rγ-chain knockout (NSG) mice are most permissive for human hematopoietic reconstitution via HSC transplantation.

5.3 Nonimmunological Barriers to Human HSC Transplantation in Mice

Hematopoiesis is a complex process of blood cell development and differentiation from HSCs, regulated by hematopoietic cytokines and adhesion molecules [11]. For example, erythropoietin (EPO) and IL-3 are required for erythrocyte development [12], IL-15 is required for NK cell differentiation and survival [13], granulocyte-macrophage colony-stimulating factor (GM-CSF) is important for dendritic cell differentiation [14], and macrophage colony-stimulating factor (M-CSF; also known as CSF-1) is essential to the development of monocytes/macrophages [15]. Because a number of mouse cytokines such as mouse IL-15, GM-CSF, IL-4, M-CSF, and IL-3, have no function on human cells [16–20], it is impossible to establish perfectly normal human hematopoiesis in conventional mice, even in the complete absence of immunological rejection. Nonetheless, human $CD34^+$ cell transplantation can lead to sustained and high-level human hematopoietic reconstitution in immunodeficient mice. However, these humanized mice exhibit clear abnormalities in cell lineage proportions, such as poor development of human NK cells and cells of the monocytic lineage. Human NK cell development and function can be significantly improved by the presence of human IL-15 and IL-15/IL-15Rα complex (IL-15 coupled to IL-15Rα) [21]. Similarly, significantly improved reconstitution with human monocytic lineage cells can be achieved in knock-in mice expressing human IL-3 and GM-CSF [22] or expressing CSF-1 [23]. Since no improvement was seen after depletion of mouse phagocytic cells, the only important immune cell population present in NSG mice [24], the unfavorable hematopoietic environment, but not immunological rejection, is predominantly attributed to the poor reconstitution with human NK and monocytic lineage cells in HHLS mice.

An earlier study suggests that recipient HSCs have a competitive advantage that limits the repopulation of the xenogeneic donor hematopoietic cells in a lineage-specific manner [25]. C.B-17 SCID mice receiving SCID and T cell-depleted (TCD) rat bone marrow cells showed a marked reduction in rat myeloid repopulation without affecting rat lymphoid repopulation when compared with the C.B-17 SCID mice receiving TCD rat bone marrow cells alone. However, repopulation of both myeloid and lymphoid cells was inhibited in C.B-17 SCID mice receiving TCD C.B-17 mouse and rat bone marrow cells [25]. A recent study, showing that different stem and progenitor cells reside in distinct cellular niches in bone marrow [26], may support the possibility that a lineage-specific competition might contribute to the poor reconstitution of human myeloid cells in immunodeficient mice lacking T and B cells.

5.4 Limited Improvement of Human Red Blood Cell and Platelet Reconstitution by Providing Human Hematopoietic Cytokines

Despite high levels of human $CD45^+$ nucleated blood cell reconstitution, the levels of human red blood cells (RBCs) and platelets are extremely low in blood circulation in HHLS mice [24, 27]. RBCs are derived from HSCs in bone marrow and subsequently migrate into the blood stream after nuclear extrusion [28]. The presence of a large number of human $CD235a^+CD45^-$ erythroid cells, which are at a stage of differentiation beyond erythroblasts (i.e., normoblasts or later), in the bone marrow from HHLS mice indicates that the lack of human RBCs in blood could be due to either the failure of human erythroid precursors to fully mature, or to poor survival or rejection of mature human RBCs in these mice. Since human cells may not react or react inefficiently to mouse hematopoietic cytokines, unfavorable hematopoietic environment has been considered the major cause for the absence of mature human RBCs in HHLS mice. However, unlike the reconstitution of human NK and monocytic lineage cells which is highly responsive to treatment with human cytokines [21, 29], presence of human EPO and IL-3 does not significantly improve human RBC reconstitution in HHLS mice [29].

Similar to mature RBCs, immunodeficient mice grafted with human $CD34^+$ cells also show extremely poor reconstitution with human platelets [27]. Megakaryocytes account for a very small fraction (less than 1 %) of bone marrow cells [30]. Platelets are generated from megakaryocytes in the bone marrow and are subsequently released into the bloodstream. The presence of human $CD61^+$, morphologically mature megakaryocytes in the bone marrow of HHLS mice indicates that (1) the mouse environment is capable of supporting efficient megakaryocyte differentiation, and (2) the poor human platelet reconstitution is unlikely, or at least not exclusively caused by inefficient human megakaryocyte differentiation in mice. This is in accordance with the study [31] showing that immunodeficient knock-in mice expressing human thrombopoietin (TPO), a major growth factor in megakaryocyte development, have no significant advantage in human platelet reconstitution after $CD34^+$ cell transplantation.

5.5 Macrophage-Mediated Xenorejection Is a Major Limiting Factor Towards the Successful Reconstitution of Human RBCs and Platelets in HHLS Mice

Failure to significantly increase human RBCs or platelets by human cytokines suggests that xenorejection is likely to be an important factor limiting human RBC and platelet reconstitution in HHLS mice. This possibility was confirmed by the report of rapid clearance of normal human RBCs and platelets in NOD/SCID and NSG mice [24]. The absence of human RBCs and platelets in human $CD34^+$ cell-grafted NSG mice rules out the possibility of xenorejection by T, B or NK cells. Significant increase in human RBCs and platelets in HHLS mice after phagocytic cell depletion by clodronate-liposomes demonstrates that phagocytic cells are effectors which eliminate human RBCs and platelets in HHLS mice [24, 27].

CD47 is a ligand of SIRPα, an immune inhibitory receptor on macrophages [32, 33]. CD47-SIRPα pathway prevents normal hematopoietic cells and platelets from phagocytosis [32–34]. It has been reported that the lack of efficient cross-reaction between donor CD47 and recipient SIRPα causes rejection of xenogeneic hematopoietic cells by macrophages [34, 35]. However, given the fact that human CD47 is capable of interacting with NOD mouse SIRPα [10], destruction of human RBCs and platelets by mouse phagocytic cells in NOD/SCID and NSG mice is unlikely to be caused by a failure of CD47-SIRPα interaction. In support of this possibility, studies have shown that in NOD/SCID mice rejection of human RBCs and platelets occurs significantly more rapidly when compared with RBCs and platelets from the CD47-deficient mice, respectively [24, 27]. Further understanding of the mechanisms triggering vigorous rejection of human RBCs and platelets by mouse phagocytic cells may help construct HHLS mice with all lineages of human blood cells.

It is not clear why mouse phagocytic cells destruct only human RBCs and platelets, but not WBCs. Similar phenomenon was observed in $CD47^{+/+}$ mouse-to-$CD47^{-/-}$ mouse bone marrow chimeras, in which CD47-deficient RBCs, but not WBCs, were destroyed by $CD47^{+/+}$ phagocytic cells [36]. In HHLS mice, macrophage depletion can induce full reconstitution of human platelets in periphery [27]; however, the level of human RBCs remains much lower compared to human lymphocytes [24]. This suggests that factors such as unfavorable hematopoietic microenvironment may contribute to the poor human RBC reconstitution in HHLS mice; therefore, treatment with human cytokines may potentially improve human RBC reconstitution in phagocytic cell-depleted HHLS mice [24]. On the other hand, since macrophages play an important role in erythroid cell differentiation, depletion of bone marrow macrophages may also affect erythropoiesis. A recent study shows that bone marrow $CD169^+$ macrophages can provide a niche, promoting erythropoiesis under homeostasis and stress [37]. In agreement with that, treatment with clodronate-liposomes has shown to delay reticulocytosis and hematocrit recovery in mice with phenylhydrazine-induced anemia [37]. Selected depletion of mouse macrophages may prevent destruction of human RBCs while sparing human erythropoiesis, if the macrophage-mediated erythroid cell differentiation occurs in a species-specific manner. Furthermore, given the inefficient development of human monocytic cells in mice [29], strategies towards optimization of human monocyte-macrophage repopulation may also promote human erythroid cell differentiation in HHLS mice.

5.6 Summary

To date, the construction of HHLS mice with perfectly normal human hematopoiesis remains a challenge. Despite these limitations, the currently available HHLS mice have been proved to be a highly valuable and irreplaceable in vivo model for biomedical and translational research. A better understanding of the biology of human HSC xenotransplantation in mice will help us to optimally humanize the mouse hematopoietic environment and improve currently available HHLS mouse models. Moreover, strategies to prevent mouse macrophage-mediated xenograft destruction will be needed for developing HHLS mice with human RBCs and platelets.

Acknowledgments The work from the authors' laboratories discussed in this chapter was supported by grants from the National Institutes of Health (NIH) (R01 AI064569 and P01AI045897), Chinese Ministry of Science and Technology (2015CB964400), Chinese Ministry of Education (IRT1133), and NSFC (81273334 and 81200397).

References

1. Rongvaux A, Takizawa H, Strowig T, Willinger T, Eynon EE, Flavell RA, et al. Human hemato-lymphoid system mice: current use and future potential for medicine. Ann Rev Immunol. 2013;31:635–74. (PubMed PMID: 23330956).
2. Lan P, Tonomura N, Shimizu A, Wang S, Yang YG. Reconstitution of a functional human immune system in immunodeficient mice through combined human fetal thymus/liver and CD34+ cell transplantation. Blood. 2006;108(2):487–92.
3. Lan P, Wang L, Diouf B, Eguchi H, Su H, Bronson R, et al. Induction of human T-cell tolerance to porcine xenoantigens through mixed hematopoietic chimerism. Blood. 2004;103(10):3964–9.
4. Traggiai E, Chicha L, Mazzucchelli L, Bronz L, Piffaretti JC, Lanzavecchia A, et al. Development of a human adaptive immune system in cord blood cell-transplanted mice. Science. 2004;304(5667):104–7. (PubMed PMID: 15064419. Epub 2004/04/06. eng).
5. Yang YG, Sykes M. Xenotransplantation: current status and a perspective on the future. Nat Rev Immunol. 2007;7(7):519–31. (PubMed PMID: 17571072. Epub 2007/06/16. eng).
6. Wang H, Yang YG. Innate cellular immunity and xenotransplantation. Curr Opin Organ Transplant. 2012;17(2):162–7. (PubMed PMID: 22262106. Pubmed Central PMCID: 3324331. Epub 2012/01/21. eng).
7. Sugamura K, Asao H, Kondo M, Tanaka N, Ishii N, Ohbo K, et al. The interleukin-2 receptor gamma chain: its role in the multiple cytokine receptor complexes and T cell development in XSCID. Annu Rev Immunol. 1996;14:179–205. (PubMed PMID: 8717512. Epub 1996/01/01. eng).
8. Cao X, Shores EW, Hu-Li J, Anver MR, Kelsall BL, Russell SM, et al. Defective lymphoid development in mice lacking expression of the common cytokine receptor gamma chain. Immunity. 1995;2(3):223–38. (PubMed PMID: 7697543. Epub 1995/03/01. eng).
9. Ito M, Hiramatsu H, Kobayashi K, Suzue K, Kawahata M, Hioki K, et al. NOD/SCID/gamma(c)(null) mouse: an excellent recipient mouse model for engraftment of human cells. Blood. 2002;100(9):3175–82. (PubMed PMID: 12384415. Epub 2002/10/18. eng).
10. Takenaka K, Prasolava TK, Wang JC, Mortin-Toth SM, Khalouei S, Gan OI, et al. Polymorphism in Sirpa modulates engraftment of human hematopoietic stem cells. Nat Immunol. 2007;8(12):1313–23. (PubMed PMID: 17982459. Epub 2007/11/06. eng).

11. Orkin SH, Zon LI. Hematopoiesis: an evolving paradigm for stem cell biology. Cell. 2008;132(4):631–44. (PubMed PMID: 18295580. Pubmed Central PMCID: 2628169. Epub 2008/02/26. eng).
12. Giarratana MC, Kobari L, Lapillonne H, Chalmers D, Kiger L, Cynober T, et al. Ex vivo generation of fully mature human red blood cells from hematopoietic stem cells. Nat Biotechnol. 2005;23(1):69–74. (PubMed PMID: 15619619. Epub 2004/12/28. eng).
13. Mrozek E, Anderson P, Caligiuri MA. Role of interleukin-15 in the development of human CD56+ natural killer cells from CD34+ hematopoietic progenitor cells. Blood. 1996;87(7):2632–40. (PubMed PMID: 8639878. Epub 1996/04/01. eng).
14. Rosenzwajg M, Canque B, Gluckman JC. Human dendritic cell differentiation pathway from CD34+ hematopoietic precursor cells. Blood. 1996;87(2):535–44. (PubMed PMID: 8555475. Epub 1996/01/15. eng).
15. Stec M, Weglarczyk K, Baran J, Zuba E, Mytar B, Pryjma J, et al. Expansion and differentiation of CD14+ CD16(-) and CD14++ CD16+ human monocyte subsets from cord blood CD34+ hematopoietic progenitors. J Leukoc Biol. 2007;82(3):594–602.(PubMed PMID: 17595380. Epub 2007/06/28. eng).
16. Eisenman J, Ahdieh M, Beers C, Brasel K, Kennedy MK, Le T, et al. Interleukin-15 interactions with interleukin-15 receptor complexes: characterization and species specificity. Cytokine. 2002;20(3):121–9. (PubMed PMID: 12453470. Epub 2002/11/28. eng).
17. Metcalf D. The molecular biology and functions of the granulocyte-macrophage colony-stimulating factors. Blood. 1986;67(2):257–67. (PubMed PMID: 3002522. Epub 1986/02/01. eng).
18. Mosmann TR, Yokota T, Kastelein R, Zurawski SM, Arai N, Takebe Y. Species-specificity of T cell stimulating activities of IL 2 and BSF-1 (IL 4): comparison of normal and recombinant, mouse and human IL 2 and BSF-1 (IL 4). J immunol. 1987;138(6):1813–6. (PubMed PMID: 3493289. Epub 1987/03/15. eng).
19. Fixe P, Praloran V. Macrophage colony-stimulating-factor (M-CSF or CSF-1) and its receptor: structure-function relationships. Eur Cytokine Netw. 1997;8(2):125–36. (PubMed PMID: 9262961. Epub 1997/06/01. eng).
20. Stevenson LM, Jones DG. Cross-reactivity amongst recombinant haematopoietic cytokines from different species for sheep bone-marrow eosinophils. J Comp Pathol. 1994;111(1):99–106. (PubMed PMID: 7962731. Epub 1994/07/01. eng).
21. Huntington ND, Legrand N, Alves NL, Jaron B, Weijer K, Plet A, et al. IL-15 trans-presentation promotes human NK cell development and differentiation in vivo. J Exp Med. 2009;206(1):25–34. (PubMed PMID: 19103877. Pubmed Central PMCID: 2626663. Epub 2008/12/24. eng).
22. Willinger T, Rongvaux A, Takizawa H, Yancopoulos GD, Valenzuela DM, Murphy AJ, et al. Human IL-3/GM-CSF knock-in mice support human alveolar macrophage development and human immune responses in the lung. Proc Natl Acad Sci U S A. 2011;108(6):2390–5. (PubMed PMID: 21262803. Pubmed Central PMCID: 3038773. Epub 2011/01/26. eng).
23. Rathinam C, Poueymirou WT, Rojas J, Murphy AJ, Valenzuela DM, Yancopoulos GD, et al. Efficient differentiation and function of human macrophages in humanized CSF-1 mice. Blood. 2011;118(11):3119–28. (PubMed PMID: 21791433. Epub 2011/07/28. eng).
24. Hu Z, Van Rooijen N, Yang YG. Macrophages prevent human red blood cell reconstitution in immunodeficient mice. Blood. 2011;118(22):5938–46. (PubMed PMID: 21926352. Pubmed Central PMCID: 3228505).
25. Gritsch HA, Sykes M. Hematopoietic competition limits xenogenic myeloid reconstitution in SCID mice. Transplant Proc. 1996;28(2):708. (PubMed PMID: 8623357. Epub 1996/04/01. eng).
26. Ding L, Morrison SJ. Haematopoietic stem cells and early lymphoid progenitors occupy distinct bone marrow niches. Nature. 2013;495(7440):231–5. (PubMed PMID: 23434755. Pubmed Central PMCID: 3600153. Epub 2013/02/26. eng).
27. Hu Z, Yang YG. Full reconstitution of human platelets in humanized mice after macrophage depletion. Blood. 2012;120(8):1713–6. (PubMed PMID: 22773384. Pubmed Central PMCID: 3429310).

28. Qiu LB, Dickson H, Hajibagheri N, Crocker PR. Extruded erythroblast nuclei are bound and phagocytosed by a novel macrophage receptor. Blood. 1995;85(6):1630–9. (PubMed PMID: 7888682. Epub 1995/03/15. eng).
29. Chen Q, Khoury M, Chen J. Expression of human cytokines dramatically improves reconstitution of specific human-blood lineage cells in humanized mice. Proc Natl Acad Sci U S A. 2009;106(51):21783–8. (PubMed PMID: 19966223. Pubmed Central PMCID: 2789167. Epub 2009/12/08. eng).
30. Patel SR, Hartwig JH, Italiano JE Jr. The biogenesis of platelets from megakaryocyte proplatelets. J Clin Invest. 2005;115(12):3348–54. (PubMed PMID: 16322779. Pubmed Central PMCID: 1297261. Epub 2005/12/03. eng).
31. Rongvaux A, Willinger T, Takizawa H, Rathinam C, Auerbach W, Murphy AJ, et al. Human thrombopoietin knockin mice efficiently support human hematopoiesis in vivo. Proc Natl Acad Sci U S A. 2011;108(6):2378–83. (PubMed PMID: 21262827. Pubmed Central PMCID: 3038726. Epub 2011/01/26. eng).
32. Oldenborg PA, Zheleznyak A, Fang YF, Lagenaur CF, Gresham HD, Lindberg FP. Role of CD47 as a marker of self on red blood cells. Science. 2000;288(5473):2051–4.
33. Jiang P, Lagenaur CF, Narayanan V. Integrin-associated protein is a ligand for the P84 neural adhesion molecule. J Biol Chem. 1999;274(2):559–62. (PubMed PMID: 9872987. Epub 1999/01/05. eng).
34. Wang H, VerHalen J, Madariaga ML, Xiang S, Wang S, Lan P, et al. Attenuation of phagocytosis of xenogeneic cells by manipulating CD47. Blood. 2007;109(2):836–42.
35. Ide K, Wang H, Tahara H, Liu J, Wang X, Asahara T, et al. Role for CD47-SIRPalpha signaling in xenograft rejection by macrophages. Proc Natl Acad Sci U S A. 2007;104(12):5062–6. (PubMed PMID: 17360380. Pubmed Central PMCID: 1829264. Epub 2007/03/16. eng).
36. Wang H, Madariaga ML, Wang S, Van Rooijen N, Oldenborg PA, Yang YG. Lack of CD47 on nonhematopoietic cells induces split macrophage tolerance to CD47null cells. Proc Natl Acad Sci U S A. 2007;104(34):13744–9. (PubMed PMID: 17699632. Pubmed Central PMCID: 1949343. Epub 2007/08/19. eng).
37. Chow A, Huggins M, Ahmed J, Hashimoto D, Lucas D, Kunisaki Y, et al. CD169(+) macrophages provide a niche promoting erythropoiesis under homeostasis and stress. Nat Med. 2013;19(4):429–36. (PubMed PMID: 23502962. Epub 2013/03/19. eng).

Chapter 6
Impact of the Mouse IL-2Rγ Chain on Lymphoid Tissue Development and Human Reconstitution in Immunodeficient Mice

Paul W. Denton, Tomonori Nochi and J. Victor Garcia

6.1 Introduction

The development of humanized mouse models has dramatically expanded the research landscape by facilitating in vivo modeling of many human diseases [1–23]. Different humanized mouse models have been bioengineered for these studies. Differences between humanized mouse models can include the immunodeficient mouse strain manipulated and whether the human hematopoietic stem cell (hHSC) transplant only or the bone marrow-liver-thymus (BLT) humanization strategy was utilized. To achieve the best experimental outcomes possible, it is critical that key attributes of different pairings of a given humanization protocol (e.g., hHSC only or BLT) with specific strains of immunodeficient mice be considered. As an aid in this process, this review compares lymphoid tissue development and human reconstitution among different human reconstitution strategies performed in mouse strains with or without a mouse interleukin (IL)-2Rγ chain. To best characterize the role of the mouse IL-2Rγ chain in lymphoid tissue development and human reconstitution, particular emphasis has been placed on reviewing manuscripts that include direct comparisons between humanized mice generated in immunodeficient mouse strains with or without a mouse IL-2Rγ chain.

P. W. Denton (✉) · J. V. Garcia · T. Nochi
Division of Infectious Diseases, Department of Medicine, UNC Center for AIDS Research, University of North Carolina School of Medicine, Genetic Medicine Building 2046, Chapel Hill, NC 27599-7042, USA
e-mail: pauden@rm.dk

T. Nochi
(current address) Laboratory of Functional Morphology, Tohoku University Graduate School of Agricultural Science, Sendai, Miyagi, Japan

L. Y. Poluektova et al. (eds.), *Humanized Mice for HIV Research*,
DOI 10.1007/978-1-4939-1655-9_6

6.2 Immunodeficient Mouse Strains and Humanization Strategies

A key consideration in the generation of humanized mice is the choice of immunodeficient mouse strain. In the mid-1990s, nonobese diabetic (NOD) mice were crossed with severe combined immunodeficient (SCID) mice to develop NOD/SCID mouse strains [24, 25]. Bone marrow engraftment of NOD/SCID mice with human hematopoietic stem cells (NOD/SCID-hHSC) results in a self-renewing source of human B cells, monocyte/ macrophages, dendritic cells, and natural killer (NK) cells that systemically reconstitute the transplanted mouse [26–28]. When planning experiments with NOD/SCID mice, it is important to consider the facts that the endogenous mouse thymus in these animals is vestigial such that it cannot produce human T cells following hHSC engraftment and that the lifespan of these animals is limited due to their predisposition to develop thymic lymphomas [25, 29]. Despite these constraints, the high and sustained levels of human myeloid, dendritic and B cells in NOD/SCID-hHSC have made this model useful for a variety of applications [6, 17, 28]. The lack of human T cells in NOD/SCID-hHSC mice led to the development of a combination approach, which would permit the development of a more complete human immune system (including human T cells) in humanized mice. BLT humanized mice are bioengineered by a bone marrow transplant of hHSC that is paired with the implantation under the renal capsule of autologous human liver and thymus [30, 31]. In BLT mice, the implanted human thymic tissue develops into a human thymus (also referred to as a thymic organoid) with normal human thymic structure and function [30–33]. T cell progenitors migrate to this thymus, mature into human leukocyte antigen (HLA)-restricted human T cells, exit the thymus, systemically populate the BLT mouse and perform effector functions such as HLA-restricted responses to Epstein–Barr virus (EBV) and HIV [31, 34].

The mouse IL-2Rγ chain is an essential component in the receptors for six different cytokines (IL-2, IL-4, IL-7, IL-9, IL-15, IL-21) [35]. Additional immunodeficient mouse strains have been developed that exhibit improved hHSC engraftment by modifying NOD/SCID [36, 37] and Rag2 knockout [38, 39] mice to prevent mouse IL-2Rγ chain function. Two different NOD/SCID IL-2Rγ chain mice have been developed and are designated as NOG [36] and NSG [37] mice. When compared to NOD/SCID mice, NOG and NSG mice are slightly more radiosensitive, do not develop thymic lymphomas, and consequently have a longer lifespan [37]. Both neonate and adult mice lacking the mouse IL-2Rγ chain have been transplanted with hHSC leading to humanization [40–42]. One of the most important differences between mice lacking a functional mouse IL-2Rγ chain and NOD/SCID mice is that the endogenous mouse thymus in NOG, NSG and Rag2γc mice supports human T cell production leading to systemic reconstitution of these mice with human T cells together with human B cells, human monocyte/macrophages, human dendritic cells, and human NK cells [41–43].

6.3 Primary and Secondary Lymphoid Tissues

Several reports have discussed the human reconstitution of the bone marrow, thymus, and spleen in humanized mice [23, 44, 45]. However, only four published studies report direct comparisons in the human reconstitution of these tissues between mice with and without a functional IL-2Rγ chain [36, 41, 46, 47]. The approach taken for the comparisons in three of these reports was to transplant similar numbers of hHSC into NOG or NSG mice and mice from the isogenic parent NOD/SCID stain [36, 41, 47]. Ito et al., Hiramatsu et al., and McDermott et al. each used NOD/SCID-hHSC and NOG-hHSC mice for their comparison [36, 41, 47]. McDermott et al. also used NSG-hHSC mice [47]. Their consensus findings were that transplantation of equal numbers of hHSC into NOG and NSG mice yielded significantly higher human reconstitution levels in the peripheral blood, bone marrow and spleen relative to NOD/SCID mice [36, 41, 47]. In addition, the production of human T cells by the mouse thymus was confirmed in the NOG-hHSC mice [41, 47] and NSG-hHSC mice [47].

In the fourth report [46], each humanized mouse was generated utilizing an optimized number of hHSC which is very different from the strategy tested in the other three reports. Specifically, NOD/SCID-BLT mice were transplanted with approximately 10 times more hHSC compared to NSG-BLT, NSG-hHSC, and Rag2γc-hHSC. Under these optimized conditions, the absolute numbers of splenic human T cells were not significantly different between these four types of humanized mice and limited differences were observed in the numbers of human T cell in bone marrow [46]. This paper did report that the human thymus present in BLT mice harbors a greater number of human thymocytes compared to a mouse thymus in hHSC transplant only mice [46]. A key point for consideration in the design and interpretation of experiments utilizing NSG-BLT mice is that human thymopoiesis in these animals occurs in both the implanted human thymus and the endogenous mouse thymus with the majority of human thymocytes present in the human thymus [46]. The fact that NSG-BLT mice have two separate locations with ongoing human thymopoiesis suggests that T cells in these mice could be heterogeneous in their major histocompatibility complex (MHC) restrictions which could complicate some experimental analyses.

Together, the available literature supports the conclusion that the presence or absence of a mouse IL-2Rγ chain does not impact humanization of bone marrow or spleen regardless of whether the hHSC transplant only or BLT method of humanization is utilized (Fig. 6.1). Regarding the endogenous mouse thymus in these strains, the production of human T cells from a mouse thymus only occurs in mice without a functional mouse IL-2Rγ chain [26–28, 40–43]. The development of a human thymus does not depend on the presence or absence of a mouse IL-2Rγ chain. Rather transplant of human tissues, as in BLT mice, is required for development of a human thymus (Fig. 6.1).

The mouse IL-2Rγ chain is a component of the IL-7 receptor [35]. This molecule is important, but dispensable, for lymph node organogenesis [48]. Since many humanized mice are bioengineered in mice lacking an IL-2Rγ chain, we examined

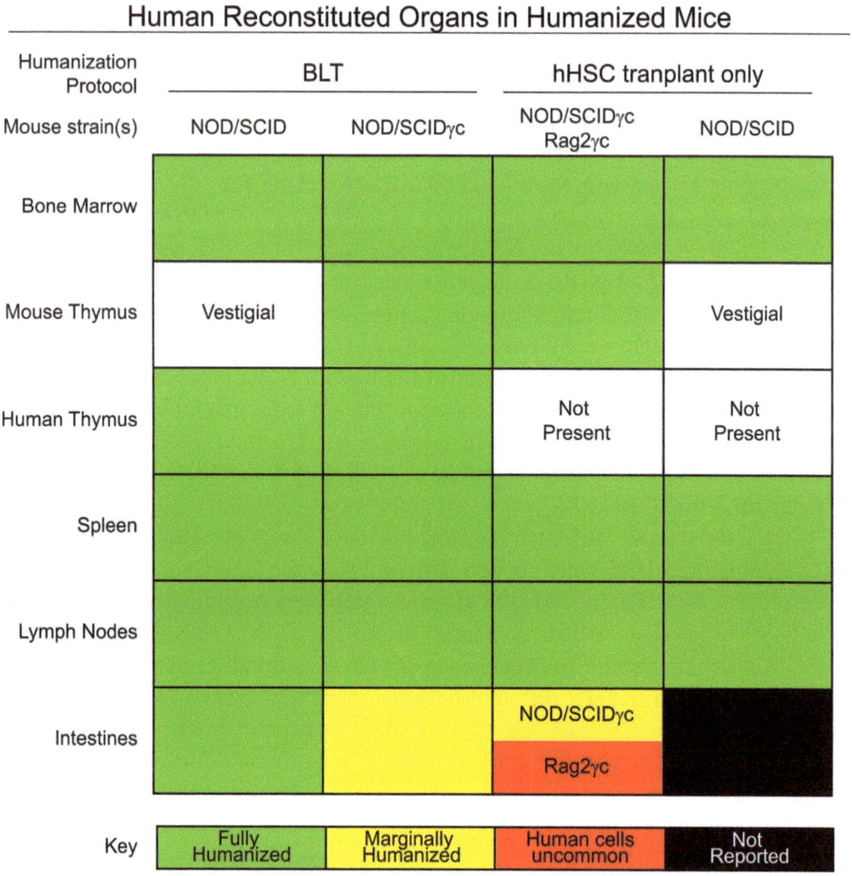

Fig. 6.1 Relative human reconstitution in the lymphoid organs of different humanized mice. At the *bottom* of the figure is a key describing the relative level of human reconstitution assigned to each color. In the mice not implanted with human thymus tissue, that organ is listed as not present. NOD/SCIDγc=NOD/SCID IL2R gamma chain deficient mice, Rag2γc=Rag2^{null} IL2R gamma chain^{null} mice. *NOD* nonobese diabetic, *SCID* severe combined immune deficiency, *BLT* bone marrow-liver-thymus, *hHSC* human hematopoietic stem cell

the literature and found that the presence of lymph nodes in humanized mice has been reported in at least 44 manuscripts [3, 9, 10, 16, 19, 30, 31, 34, 37, 41–43, 46, 49–79]. Specific lymph nodes identified include: cervical [9, 30, 37, 74, 75, 77], mediastenal [75], axillary [9, 30, 37, 76], brachial [30, 37, 76], renal [30]; iliac [3], inguinal [30, 37, 76], mesenteric [3, 9, 19, 30, 41–43, 46, 65–71, 76, 78], and popliteal [79]. The most commonly reported lymph nodes in humanized mice are the mesenteric, cervical, axillary, and brachial. Why lymph nodes in the caudal region of humanized mice are less common has not been determined. The fact that 39 of these 44 manuscripts utilized mouse IL-2Rγ chain deficient mouse strains and still were able to detect lymph nodes [3, 9, 10, 16, 19, 30, 34, 37, 41–43, 46, 49, 50, 52–54, 56–72, 74–78] confirms that a functional mouse IL-2Rγ chain is dispensable for lymph node organogenesis [48] in humanized mice.

One critical function of secondary lymphoid tissues (i.e., spleen and lymph nodes) is to facilitate the development of humoral immune responses. The broad B cell repertoires [80–83] and antigen-specific human antibody responses [9, 34, 54, 55, 73, 84, 85] that have been reported in different types of humanized mice indicates that these animals have some degree of spleen and lymph node function in this regard, although T cell help has clearly not been optimized in these models [57, 81, 85, 86]. Shultz et al. suggested that human T cells in humanized mice may require thymic selection with HLA in order to provide adequate T cell help to B cells for T-dependent antibody production [87], yet T-dependent antibody production has been reported in humanized mice lacking human thymic stroma for T cell education [9, 88]. Furthermore, there is a question of whether lymphoid follicles and germinal centers, anatomical regions that facilitate T cell help, are present within the secondary lymphoid tissues of humanized mice. Lymphoid follicles and germinal center-like structures have been reported to be both present [16, 30, 31, 41, 43, 89] and absent [8, 37, 82] in humanized mice. Beyond direct observation with immunohistochemistry approaches, B cell development and function has been used as a surrogate marker for the presence of lymphoid follicles and germinal centers in humanized mice. A post-germinal center surface phenotype on B cells in humanized mice has been reported [9, 88] while others have identified B cells as maturing independent of a germinal center [68, 85, 90]. In summary, further work is required to reach a consensus regarding the development of lymphoid follicles and germinal centers in humanized mice and to define any role of the mouse IL-2Rγ chain in these processes.

6.4 Intestinal Lymphoid Tissues

At least 15 manuscripts have examined the presence of human immune cells within intestines of humanized mice [34, 43, 46, 59, 61, 70, 73, 77, 91–97]. For hHSC transplant only mice, the results reported lack both quantitation and a broad analysis of the different human hematopoietic lineages present in the mouse intestines [43, 70, 91–94, 97]. In contrast, several groups have reported that multiple lineages of human immune cells (including T cells, B cells, macrophages, NK cells, and dendritic cells) are distributed through the intestines of BLT mice [34, 73, 77, 95]. To fully characterize the differential intestinal humanization between various humanize mice, one manuscript reported direct, quantitative comparisons of intestinal humanization between NOD/SCID-BLT, NSG-BLT, NSG-hHSC and Rag2γc-hHSC mice [46]. Quantified intestinal cell populations in this report included the lamina propria lymphocytes (LPL) and intraepithelial lymphocytes (IEL) from both the small intestine (SI) and large intestine (LI): SI LPL; SI IEL; LI LPL; and LI IEL.

NOD/SCID BLT mice exhibited the highest numbers of human T cells in each of these four intestinal populations [46]. Regardless of the human reconstitution method (i.e., –BLT or –hHSC transplant only), intestinal T cell levels were consistently lower when the immunodeficient mouse strain used lacked a functional IL-2Rγ chain (Fig. 6.1) [46]. These data indicate that the presence of a functional mouse IL-2Rγ chain is critical for robust intestinal human T cell engraftment.

The intestinal immune system is composed of gut-associated lymphoid tissue (GALT) (e.g., Peyer's patches and isolated lymphoid follicles) and intestinal lamina propria. Many humanized mice are bioengineered in mice lacking an IL-2Rγ chain (a component of the IL-7 receptor) and IL-7 signaling in lymphoid tissue organizer cells and lymphoid tissue inducer cells is indispensible for GALT organogenesis [98–104]. Therefore, we examined the literature for reports describing the presence of GALT in humanized mice. One manuscript reported undefined lymphoid aggregates (e.g., isolated lymphoid follicles) in the intestines of NOD/SCID-BLT mice [73]. These aggregates contained $CD4^+$ T cells, macrophages, and dendritic cells and were susceptible to HIV-induced pathology [73]. Given that the humanized mouse research field is so heavily reliant upon immunodeficient mice without a functional mouse IL-2Rγ chain and there are translational research questions regarding human GALT pathologies that could be studied if humanized mice exhibit GALT structures, defining the presence or absence of GALT in different types of humanized mice is a very important area of future study.

6.5 Conclusions

Lymphocyte migration into and egress from lymphoid tissues is the result of a series of poorly understood but highly complex species-specific interactions between cell adhesion molecules, integrins, chemokine receptors, and chemokines. Here we discussed the remarkable observation that human lymphocytes traffic in humanized mice such that lymphoid tissues develop and are populated with human cells in these mice. For the bone marrow, thymus, spleen, and to a lesser extent lymph nodes, the presence of a functional mouse IL-2Rγ chain is dispensable for lymphoid tissue development and human reconstitution. However, utilization of the BLT humanization protocol on mice with and without a functional mouse IL-2Rγ chain has provided compelling evidence that this molecule is important for intestinal human reconstitution. Specifically, NOD/SCID-BLT mice have the highest levels of intestinal humanization relative to NSG-BLT mice, as well as to NSG-hHSC and Rag2γc-hHSC mice. Thus, immunodeficient mouse strains with improved hHSC engraftment may exhibit deficiencies in human reconstitution in certain tissues (e.g., intestines). Understanding these distinctions increases our ability to accurately interpret previously published humanized mouse data and to properly plan future experiments.

6.6 Future Directions

There are two major research areas for future investigation related to human lymphoid tissue development in humanized mice. First, a clear understanding of the cross-species activity for individual cytokines, chemokines, and adhesion molecules would offer critical insights into the optimal combination of human transgenes for appropriate human lymphoid tissue development. There are ongoing efforts by multiple research groups to apply human growth factors exogenously [65, 105–107] or develop mice that transgenically expressing human development and growth factors [108–111]. Yet a comprehensive description of cross-species activity is still lacking [112]. Second, when humanized mice are created by a transplant of hHSC only, all of the human cells present in these mice are hematopoietic in origin. The mice still lack nonhematopoietic cells (e.g., epithelial cells) that are involved in human immune development and function. Pairing current human reconstitution techniques with previously developed strategies for the implantation into mice of human tissues beyond the human thymus in BLT mice (e.g., human bone marrow and human intestines) could introduce additional nonhematopoietic cells into the humanize mouse models [113, 114]. Successful implementation of such strategies would provide the best possible environment for proper lymphoid tissue development in humanized mice.

Acknowledgments This work was supported in part by the University of North Carolina (UNC) Center for AIDS Research grant P30 AI50410; by National Institute of Health grants AI096113 and AI08263 (J.V.G.); and by a Research Fellowship of the Japan Society for the Promotion of Science (T.N.).

References

1. King M, Pearson T, Rossini AA, Shultz LD, Greiner DL. Humanized mice for the study of type 1 diabetes and beta cell function. Ann N Y Acad Sci. 2008;1150:46–53.
2. Chang NH, Inman RD, Dick JE, Wither JE. Bone marrow-derived human hematopoietic stem cells engraft NOD/SCID mice and traffic appropriately to an inflammatory stimulus in the joint. J Rheumatol. 2010;37(3):496–502.
3. Kwant-Mitchell A, Ashkar AA, Rosenthal KL. Mucosal innate and adaptive immune responses against herpes simplex virus type 2 in a humanized mouse model. J Virol. 2009;83(20):10664–76.
4. Pedroza-Gonzalez A, Xu K, Wu T-C, Aspord C, Tindle S, Marches F, et al. Thymic stromal lymphopoietin fosters human breast tumor growth by promoting type 2 inflammation. J Exp Med. 2011;208(3):479–90.
5. Sellier-Leclerc AL, Duval A, Riveron S, Macher MA, Deschenes G, Loirat C, et al. A humanized mouse model of idiopathic nephrotic syndrome suggests a pathogenic role for immature cells. J Am Soc Nephrol. 2007;18(10):2732–9.
6. Islas-Ohlmayer M, Padgett-Thomas A, Domiati-Saad R, Melkus MW, Cravens PD, Martin Mdel P, et al. Experimental infection of NOD/SCID mice reconstituted with human CD34+ cells with Epstein–Barr virus. J Virol. 2004;78(24):13891–900.

7. Ma SD, Hegde S, Young KH, Sullivan R, Rajesh D, Zhou Y, et al. A new model of Epstein–Barr virus infection reveals an important role for early lytic viral protein expression in the development of lymphomas. J Virol. 2011;85(1):165–77.
8. Strowig T, Gurer C, Ploss A, Liu YF, Arrey F, Sashihara J, et al. Priming of protective T cell responses against virus-induced tumors in mice with human immune system components. J Exp Med. 2009;206(6):1423–34.
9. Yajima M, Imadome K, Nakagawa A, Watanabe S, Terashima K, Nakamura H, et al. A new humanized mouse model of Epstein–Barr virus infection that reproduces persistent infection, lymphoproliferative disorder, and cell-mediated and humoral immune responses. J Infect Dis. 2008;198(5):673–82.
10. Washburn ML, Bility MT, Zhang L, Kovalev GI, Buntzman A, Frelinger JA, et al. A humanized mouse model to study Hepatitis C virus infection, immune response, and liver disease. Gastroenterology. 2011;140(4):1334–44.
11. Smith MS, Goldman DC, Bailey AS, Pfaffle DL, Kreklywich CN, Spencer DB, et al. Granulocyte-colony stimulating factor reactivates human cytomegalovirus in a latently infected humanized mouse model. Cell Host Microbe. 2010;8(3):284–91.
12. Yu CI, Gallegos M, Marches F, Zurawski G, Ramilo O, Garcia-Sastre A, et al. Broad influenza-specific CD8+ T-cell responses in humanized mice vaccinated with influenza virus vaccines. Blood. 2008;112(9):3671–8.
13. Libby SJ, Brehm MA, Greiner DL, Shultz LD, McClelland M, Smith KD, et al. Humanized nonobese diabetic-scid IL2rgammanull mice are susceptible to lethal *Salmonella typhi* infection. Proc Natl Acad Sci U S A. 2010;107(35):15589–94.
14. Song J, Willinger T, Rongvaux A, Eynon EE, Stevens S, Manz MG, et al. A mouse model for the human pathogen *Salmonella typhi*. Cell Host Microbe. 2010;8(4):369–76.
15. Firoz Mian M, Pek EA, Chenoweth MJ, Ashkar AA. Humanized mice are susceptible to *Salmonella typhi* infection. Cell Mol Immunol. 2011;8(1):83–7.
16. Unsinger J, McDonough JS, Shultz LD, Ferguson TA, Hotchkiss RS. Sepsis-induced human lymphocyte apoptosis and cytokine production in "humanized" mice. J Leukoc Biol. 2009;86(2):219–27.
17. Bente DA, Melkus MW, Garcia JV, Rico-Hesse R. Dengue fever in humanized NOD/SCID mice. J Virol. 2005;79(21):13797–9.
18. Jaiswal S, Pearson T, Friberg H, Shultz LD, Greiner DL, Rothman AL, et al. Dengue virus infection and virus-specific HLA-A2 restricted immune responses in humanized NOD-scid IL2rgammanull mice. PLoS ONE. 2009;4(10):e7251.
19. Kuruvilla JG, Troyer RM, Devi S, Akkina R. Dengue virus infection and immune response in humanized RAG2(-/-)gamma(c)(-/-) (RAG-hu) mice. Virology. 2007;369(1):143–52.
20. Mota J, Rico-Hesse R. Humanized mice show clinical signs of dengue fever according to infecting virus genotype. J Virol. 2009;83(17):8638–45.
21. Subramanya S, Kim SS, Abraham S, Yao J, Kumar M, Kumar P, et al. Targeted delivery of small interfering RNA to human dendritic cells to suppress dengue virus infection and associated proinflammatory cytokine production. J Virol. 2010;84(5):2490–501.
22. Vasilakis N, Cardosa J, Diallo M, Sall AA, Holmes EC, Hanley KA, et al. Sylvatic dengue viruses share the pathogenic potential of urban/endemic dengue viruses. J Virol. 2010;84(7):3726-7 (author reply 7–8).
23. Denton PW, Garcia JV. Humanized mouse models of HIV infection. AIDS Rev. 2011;13(3):135–48.
24. Koyanagi Y, Tanaka Y, Tanaka R, Misawa N, Kawano Y, Tanaka T, et al. High levels of viremia in hu-PBL-NOD-scid mice with HIV-1 infection. Leukemia. 1997;11(Suppl 3):109–12.
25. Shultz LD, Schweitzer PA, Christianson SW, Gott B, Schweitzer IB, Tennent B, et al. Multiple defects in innate and adaptive immunologic function in NOD/LtSz-scid mice. J Immunol. 1995;154(1):180–91.
26. Lowry PA, Shultz LD, Greiner DL, Hesselton RM, Kittler EL, Tiarks CY, et al. Improved engraftment of human cord blood stem cells in NOD/LtSz-scid/scid mice after irradiation or multiple-day injections into unirradiated recipients. Biol Blood Marrow Transplant. 1996 Feb;2(1):15–23.

27. Larochelle A, Vormoor J, Hanenberg H, Wang JC, Bhatia M, Lapidot T, et al. Identification of primitive human hematopoietic cells capable of repopulating NOD/SCID mouse bone marrow: implications for gene therapy. Nat Med. 1996;2(12):1329–37.
28. Cravens PD, Melkus MW, Padgett-Thomas A, Islas-Ohlmayer M, Del P MM, Garcia JV. Development and activation of human dendritic cells in vivo in a xenograft model of human hematopoiesis. Stem Cells. 2005;23(2):264–78.
29. Greiner DL, Hesselton RA, Shultz LD. SCID Mouse models of human stem cell engraftment. Stem Cells. 1998;16(3):166–77.
30. Lan P, Tonomura N, Shimizu A, Wang S, Yang YG. Reconstitution of a functional human immune system in immunodeficient mice through combined human fetal thymus/liver and CD34+ cell transplantation. Blood. 2006;108(2):487–92.
31. Melkus MW, Estes JD, Padgett-Thomas A, Gatlin J, Denton PW, Othieno FA, et al. Humanized mice mount specific adaptive and innate immune responses to EBV and TSST-1. Nat Med. 2006;12(11):1316–22.
32. McCune JM, Namikawa R, Kaneshima H, Shultz LD, Lieberman M, Weissman IL. The SCID-hu mouse: murine model for the analysis of human hematolymphoid differentiation and function. Science. 1988;241(4873):1632–9.
33. Namikawa R, Weilbaecher KN, Kaneshima H, Yee EJ, McCune JM. Long-term human hematopoiesis in the SCID-hu mouse. J Exp Med. 1990;172(4):1055–63.
34. Brainard DM, Seung E, Frahm N, Cariappa A, Bailey CC, Hart WK, et al. Induction of robust cellular and humoral virus-specific adaptive immune responses in human immunodeficiency virus-infected humanized BLT mice. J Virol. 2009;83(14):7305–21.
35. Leonard WJ. Cytokines and immunodeficiency diseases. Nat Rev Immunol. 2001;1(3):200–8.
36. Ito M, Hiramatsu H, Kobayashi K, Suzue K, Kawahata M, Hioki K, et al. NOD/SCID/gamma(c)(null) mouse: an excellent recipient mouse model for engraftment of human cells. Blood. 2002;100(9):3175–82.
37. Shultz LD, Lyons BL, Burzenski LM, Gott B, Chen X, Chaleff S, et al. Human lymphoid and myeloid cell development in NOD/LtSz-scid IL2R gamma null mice engrafted with mobilized human hemopoietic stem cells. J Immunol. 2005;174(10):6477–89.
38. Goldman JP, Blundell MP, Lopes L, Kinnon C, Di Santo JP, Thrasher AJ. Enhanced human cell engraftment in mice deficient in RAG2 and the common cytokine receptor gamma chain. Br J Haematol. 1998;103(2):335–42 (130(5):662–70).
39. Mazurier F, Fontanellas A, Salesse S, Taine L, Landriau S, Moreau-Gaudry F, et al. A novel immunodeficient mouse model-RAG2 x common cytokine receptor gamma chain double mutants-requiring exogenous cytokine administration for human hematopoietic stem cell engraftment. J Interferon Cytokine Res. 1999;19(5):533–41.
40. Gimeno R, Weijer K, Voordouw A, Uittenbogaart CH, Legrand N, Alves NL, et al. Monitoring the effect of gene silencing by RNA interference in human CD34+ cells injected into newborn RAG2-/- gammac-/- mice: functional inactivation of p53 in developing T cells. Blood. 2004;104(13):3886–93.
41. Hiramatsu H, Nishikomori R, Heike T, Ito M, Kobayashi K, Katamura K, et al. Complete reconstitution of human lymphocytes from cord blood CD34+ cells using the NOD/SCID/gammacnull mice model. Blood. 2003;102(3):873–80.
42. Traggiai E, Chicha L, Mazzucchelli L, Bronz L, Piffaretti JC, Lanzavecchia A, et al. Development of a human adaptive immune system in cord blood cell-transplanted mice. Science. 2004;304(5667):104–7.
43. Ishikawa F, Yasukawa M, Lyons B, Yoshida S, Miyamoto T, Yoshimoto G, et al. Development of functional human blood and immune systems in NOD/SCID/IL2 receptor {gamma} chain(null) mice. Blood. 2005;106:1565–73.
44. Akkina R. New generation humanized mice for virus research: comparative aspects and future prospects. Virology. 2013;435(1):14–28.

45. Shultz LD, Brehm MA, Garcia-Martinez JV, Greiner DL. Humanized mice for immune system investigation: progress, promise and challenges. Nat Rev Immunol. 2012;12(11):786–98.
46. Denton PW, Nochi T, Lim A, Krisko JF, Martinez-Torres F, Choudhary SK, et al. IL-2 receptor gamma-chain molecule is critical for intestinal T-cell reconstitution in humanized mice. Mucosal Immunol. 2012;5(5):555–66.
47. McDermott SP, Eppert K, Lechman ER, Doedens M, Dick JE. Comparison of human cord blood engraftment between immunocompromised mouse strains. Blood. 2010;116(2):193–200.
48. Cao X, Shores EW, Hu-Li J, Anver MR, Kelsall BL, Russell SM, et al. Defective lymphoid development in mice lacking expression of the common cytokine receptor gamma chain. Immunity. 1995;2(3):223–38.
49. Chateau ML, Denton PW, Swanson MD, McGowan I, Garcia JV. Rectal transmission of Transmitted/Founder HIV-1 is efficiently prevented by topical 1 % tenofovir in BLT humanized mice. PLoS ONE. 2013;8(3):e60024.
50. Sato K, Nie C, Misawa N, Tanaka Y, Ito M, Koyanagi Y. Dynamics of memory and naive CD8+ T lymphocytes in humanized NOD/SCID/IL-2Rgammanull mice infected with CCR5-tropic HIV-1. Vaccine. 2010;28(Suppl 2):B32–7.
51. Onoe T, Kalscheuer H, Danzl N, Chittenden M, Zhao G, Yang YG, et al. Human natural regulatory T cell development, suppressive function, and postthymic maturation in a humanized mouse model. J Immunol. 2011;187(7):3895–903.
52. Dash PK, Gendelman HE, Roy U, Balkundi S, Alnouti Y, Mosley RL, et al. Long-acting NanoART elicits potent antiretroviral and neuroprotective responses in HIV-1 infected humanized mice. AIDS. 2012;26(17):2135-44
53. Watanabe S, Terashima K, Ohta S, Horibata S, Yajima M, Shiozawa Y, et al. Hematopoietic stem cell-engrafted NOD/SCID/IL2Rgamma null mice develop human lymphoid systems and induce long-lasting HIV-1 infection with specific humoral immune responses. Blood. 2007;109(1):212–8.
54. Sango K, Joseph A, Patel M, Osiecki K, Dutta M, Goldstein H. Highly active antiretroviral therapy potently suppresses HIV infection in humanized Rag2-/-gammac-/- mice. AIDS Res Hum Retroviruses. 2010;26(7):735–46.
55. Tonomura N, Habiro K, Shimizu A, Sykes M, Yang YG. Antigen-specific human T-cell responses and T cell-dependent production of human antibodies in a humanized mouse model. Blood. 2008;111(8):4293–6.
56. Watanabe S, Ohta S, Yajima M, Terashima K, Ito M, Mugishima H, et al. Humanized NOD/SCID/IL2R {gamma}null Mice Transplanted with hematopoietic stem cells under nonmyeloablative conditions show prolonged life spans and allow detailed analysis of human immunodeficiency virus type 1 pathogenesis. J Virol. 2007;81(23):13259–64.
57. Lang J, Kelly M, Freed BM, McCarter MD, Kedl RM, Torres RM, et al. Studies of lymphocyte reconstitution in a humanized mouse model reveal a requirement of T cells for human B cell maturation. J Immunol. 2013;190(5):2090–101.
58. Joo SY, Chung YS, Choi B, Kim M, Kim JH, Jun TG, et al. Systemic human T cell developmental processes in humanized mice cotransplanted with human fetal thymus/liver tissue and hematopoietic stem cells. Transplantation. 2012;94(11):1095–102.
59. Hur EM, Patel SN, Shimizu S, Rao DS, Gnanapragasam PN, An DS, et al. Inhibitory effect of HIV-specific neutralizing IgA on mucosal transmission of HIV in humanized mice. Blood. 2012;120(23):4571–82.
60. Zou W, Denton PW, Watkins RL, Krisko JF, Nochi T, Foster JL, et al. Nef functions in BLT mice to enhance HIV-1 replication and deplete CD4+ CD8+ thymocytes. Retrovirology. 2012;9(1):44.
61. Denton PW, Estes JD, Sun Z, Othieno FA, Wei BL, Wege AK, et al. Antiretroviral pre-exposure prophylaxis prevents vaginal transmission of HIV-1 in humanized BLT mice. PLoS Med. 2008;5(1):e16.

62. Denton PW, Olesen R, Choudhary SK, Archin NM, Wahl A, Swanson MD, et al. Generation of HIV latency in BLT humanized mice. J Virol. 2012;86(1):630–4.
63. Denton PW, Othieno F, Martinez-Torres F, Zou W, Krisko JF, Fleming E, et al. One percent tenofovir applied topically to humanized BLT mice and used according to the CAPRISA 004 experimental design demonstrates partial protection from vaginal HIV infection, validating the BLT model for evaluation of new microbicide candidates. J Virol. 2011;85(15):7582–93.
64. Chateau M, Swanson MD, Garcia JV. Inefficient vaginal transmission of tenofovir resistant HIV-1. J Virol. 2013;87(2):1274–7.
65. Pek EA, Chan T, Reid S, Ashkar AA. Characterization and IL-15 dependence of NK cells in humanized mice. Immunobiology. 2010;216(1–2):218–24.
66. Ma SD, Yu X, Mertz JE, Gumperz JE, Reinheim E, Zhou Y, et al. An Epstein–Barr virus (EBV) mutant with enhanced BZLF1 expression causes lymphomas with abortive lytic EBV infection in a humanized mouse model. J Virol. 2012;86(15):7976–87.
67. Heuts F, Gavier-Widen D, Carow B, Juarez J, Wigzell H, Rottenberg ME. CD4+ cell-dependent granuloma formation in humanized mice infected with mycobacteria. Proc Natl Acad Sci U S A. 2013;110(16):6482–7.
68. Scheeren FA, Nagasawa M, Weijer K, Cupedo T, Kirberg J, Legrand N, et al. T cell-independent development and induction of somatic hypermutation in human IgM+ IgD+ CD27+ B cells. J Exp Med. 2008;205(9):2033–42.
69. Berges BK, Wheat WH, Palmer BE, Connick E, Akkina R. HIV-1 infection and CD4 T cell depletion in the humanized Rag2-/-gamma c-/- (RAG-hu) mouse model. Retrovirology. 2006;3:76.
70. Choudhary SK, Archin NM, Cheema M, Dahl NP, Garcia JV, Margolis DM. Latent HIV-1 infection of resting CD4(+) T cells in the humanized Rag2(-)/(-) gammac(-)/(-) mouse. J Virol. 2012;86(1):114–20.
71. Berges BK, Akkina SR, Remling L, Akkina R. Humanized Rag2(-/-) gammac(-/-) (RAG-hu) mice can sustain long-term chronic HIV-1 infection lasting more than a year. Virology. 2010;397(1):100–3.
72. Kalscheuer H, Danzl N, Onoe T, Faust T, Winchester R, Goland R, et al. A model for personalized in vivo analysis of human immune responsiveness. Sci Transl Med. 2012;4(125):125ra30.
73. Sun Z, Denton PW, Estes JD, Othieno FA, Wei BL, Wege AK, et al. Intrarectal transmission, systemic infection, and CD4+ T cell depletion in humanized mice infected with HIV-1. J Exp Med. 2007;204(4):705–14.
74. Gorantla S, Makarov E, Finke-Dwyer J, Gebhart CL, Domm W, Dewhurst S, et al. CD8+ cell depletion accelerates HIV-1 immunopathology in humanized mice. J Immunol. 2010;184(12):7082–91.
75. Gorantla S, Sneller H, Walters L, Sharp JG, Pirruccello SJ, West JT, et al. Human immunodeficiency virus type 1 pathobiology studied in humanized BALB/c-Rag2-/-gammac-/- mice. J Virol. 2007;81(6):2700–12.
76. Legrand N, Huntington ND, Nagasawa M, Bakker AQ, Schotte R, Strick-Marchand H, et al. Functional CD47/signal regulatory protein alpha (SIRP(alpha)) interaction is required for optimal human T- and natural killer- (NK) cell homeostasis in vivo. Proc Natl Acad Sci U S A. 2011;108(32):13224–9.
77. Wahl A, Swanson MD, Nochi T, Olesen R, Denton PW, Chateau M, et al. Human breast milk and antiretrovirals dramatically reduce oral HIV-1 transmission in BLT humanized mice. PLoS Pathog. 2012;8(6):e1002732.
78. Denton PW, Krisko JF, Powell DA, Mathias M, Kwak YT, Martinez-Torres F, et al. Systemic administration of antiretrovirals prior to exposure prevents rectal and intravenous HIV-1 transmission in humanized BLT mice. PLoS ONE. 2010;5(1):e8829.
79. Murooka TT, Deruaz M, Marangoni F, Vrbanac VD, Seung E, von Andrian UH, et al. HIV-infected T cells are migratory vehicles for viral dissemination. Nature. 2012;490(7419):283–7.
80. Ippolito GC, Hoi KH, Reddy ST, Carroll SM, Ge X, Rogosch T, et al. Antibody repertoires in humanized NOD-scid-IL2Rgamma(null) mice and human B cells reveals human-like diversification and tolerance checkpoints in the mouse. PLoS ONE. 2012;7(4):e35497.

81. Chang H, Biswas S, Tallarico AS, Sarkis PT, Geng S, Panditrao MM, et al. Human B-cell ontogeny in humanized NOD/SCID gammac(null) mice generates a diverse yet auto/poly- and HIV-1-reactive antibody repertoire. Genes Immun. 2012;13(5):399–410.
82. Becker PD, Legrand N, van Geelen CM, Noerder M, Huntington ND, Lim A, et al. Generation of human antigen-specific monoclonal IgM antibodies using vaccinated "Human Immune System" mice. PLoS ONE. 2010;5(10):e13137.
83. Marodon G, Desjardins D, Mercey L, Baillou C, Parent P, Manuel M, et al. High diversity of the immune repertoire in humanized NOD.SCID.gamma c-/- mice. Eur J Immunol. 2009;39(8):2136–45.
84. Jaiswal S, Pazoles P, Woda M, Shultz LD, Greiner DL, Brehm MA, et al. Enhanced humoral and HLA-A2-restricted dengue virus-specific T-cell responses in humanized BLT NSG mice. Immunology. 2012;136(3):334–43.
85. Biswas S, Chang H, Sarkis PT, Fikrig E, Zhu Q, Marasco WA. Humoral immune responses in humanized BLT mice immunized with West Nile virus and HIV-1 envelope proteins are largely mediated via human CD5(+) B cells. Immunology. 2011;134(4):419–33.
86. Watanabe Y, Takahashi T, Okajima A, Shiokawa M, Ishii N, Katano I, et al. The analysis of the functions of human B and T cells in humanized NOD/shi-scid/gammac(null) (NOG) mice (hu-HSC NOG mice). Int Immunol. 2009;21(7):843–58.
87. Shultz LD, Saito Y, Najima Y, Tanaka S, Ochi T, Tomizawa M, et al. Generation of functional human T-cell subsets with HLA-restricted immune responses in HLA class I expressing NOD/SCID/IL2r gamma(null) humanized mice. Proc Natl Acad Sci U S A. 2010;107(29):13022–7.
88. Choi B, Chun E, Kim M, Kim ST, Yoon K, Lee KY, et al. Human B cell development and antibody production in humanized NOD/SCID/IL-2Rgamma(null) (NSG) mice conditioned by busulfan. J Clin Immunol. 2011;31(2):253–64.
89. Pearson T, Shultz LD, Miller D, King M, Laning J, Fodor W, et al. Non-obese diabetic-recombination activating gene-1 (NOD-Rag1 null) interleukin (IL)-2 receptor common gamma chain (IL2r gamma null) null mice: a radioresistant model for human lymphohaematopoietic engraftment. Clin Exp Immunol. 2008;154(2):270–84.
90. White RE, Ramer PC, Naresh KN, Meixlsperger S, Pinaud L, Rooney C, et al. EBNA3B-deficient EBV promotes B cell lymphomagenesis in humanized mice and is found in human tumors. J Clin Invest. 2012;122(4):1487–502.
91. Waldron-Lynch F, Henegariu O, Deng S, Preston-Hurlburt P, Tooley J, Flavell R, et al. Teplizumab induces human gut-tropic regulatory cells in humanized mice and patients. Sci Transl Med. 2012;4(118):118ra12.
92. Hofer U, Baenziger S, Heikenwalder M, Schlaepfer E, Gehre N, Regenass S, et al. RAG2-/- gamma(c)-/- mice transplanted with CD34+cells from human cord blood show low levels of intestinal engraftment and are resistant to rectal transmission of human immunodeficiency virus. J Virol. 2008;82(24):12145–53.
93. Holt N, Wang J, Kim K, Friedman G, Wang X, Taupin V, et al. Human hematopoietic stem/progenitor cells modified by zinc-finger nucleases targeted to CCR5 control HIV-1 in vivo. Nat Biotechnol. 2010;28(8):839–47.
94. Berges BK, Akkina SR, Folkvord JM, Connick E, Akkina R. Mucosal transmission of R5 and X4 tropic HIV-1 via vaginal and rectal routes in humanized Rag2-/- gammac -/- (RAG-hu) mice. Virology. 2008;373(2):342–51.
95. Ringpis GE, Shimizu S, Arokium H, Camba-Colon J, Carroll MV, Cortado R, et al. Engineering HIV-1-resistant T-cells from short-hairpin RNA-expressing hematopoietic stem/progenitor cells in humanized BLT mice. PLoS ONE. 2012;7(12):e53492.
96. Shimizu S, Hong P, Arumugam B, Pokomo L, Boyer J, Koizumi N, et al. A highly efficient short hairpin RNA potently down-regulates CCR5 expression in systemic lymphoid organs in the hu-BLT mouse model. Blood. 2010;115(8):1534–44.
97. Cimbro R, Vassena L, Arthos J, Cicala C, Kehrl JH, Park C, et al. IL-7 induces expression and activation of integrin alpha4beta7 promoting naive T-cell homing to the intestinal mucosa. Blood. 2012;120(13):2610–9.
98. Mebius RE. Organogenesis of lymphoid tissues. Nat Rev Immunol. 2003;3(4):292–303.

99. van de Pavert SA, Mebius RE. New insights into the development of lymphoid tissues. Nat Rev Immunol. 2010;10(9):664–74.
100. Eberl G, Littman DR. Thymic origin of intestinal alphabeta T cells revealed by fate mapping of RORgammat+cells. Science. 2004;305(5681):248–51.
101. Eberl G. Inducible lymphoid tissues in the adult gut: recapitulation of a fetal developmental pathway?. Nat Rev Immunol. 2005;5(5):413–20.
102. Lorenz RG, Chaplin DD, McDonald KG, McDonough JS, Newberry RD. Isolated lymphoid follicle formation is inducible and dependent upon lymphotoxin-sufficient B lymphocytes, lymphotoxin beta receptor, and TNF receptor I function. J Immunol. 2003;170(11):5475–82.
103. Ivanov II, Diehl GE, Littman DR. Lymphoid tissue inducer cells in intestinal immunity. Curr Top Microbiol Immunol. 2006;308:59–82.
104. Hamada H, Hiroi T, Nishiyama Y, Takahashi H, Masunaga Y, Hachimura S, et al. Identification of multiple isolated lymphoid follicles on the antimesenteric wall of the mouse small intestine. J Immunol. 2002;168(1):57–64.
105. Chen Q, Khoury M, Chen J. Expression of human cytokines dramatically improves reconstitution of specific human-blood lineage cells in humanized mice. Proc Natl Acad Sci U S A. 2009;106(51):21783–8.
106. Huntington ND, Legrand N, Alves NL, Jaron B, Weijer K, Plet A, et al. IL-15 trans-presentation promotes human NK cell development and differentiation in vivo. J Exp Med. 2009;206(1):25–34.
107. van Lent AU, Dontje W, Pouw SM, Nagasawa M, Siamari R, Bakker AQ, Pouw SM, et al. IL-7 enhances thymic human T cell development in "human immune system" Rag2-/-IL-2Rgammac-/- mice without affecting peripheral T cell homeostasis. J Immunol. 2009;183(12):7645–55.
108. Willinger T, Rongvaux A, Strowig T, Manz MG, Flavell RA. Improving human hemato-lymphoid-system mice by cytokine knock-in gene replacement. Trends Immunol. 2011;32(7):321–7.
109. Ito R, Takahashi T, Katano I, Ito M. Current advances in humanized mouse models. Cell Mol Immunol. 2012;9(3):208–14.
110. Rongvaux A, Willinger T, Takizawa H, Rathinam C, Auerbach W, Murphy AJ, et al. Human thrombopoietin knockin mice efficiently support human hematopoiesis in vivo. Proc Natl Acad Sci U S A. 2011;108(6):2378–83.
111. Billerbeck E, Barry WT, Mu K, Dorner M, Rice CM, Ploss A. Development of human CD4+ FoxP3+ regulatory T cells in human stem cell factor, GM-CSF and interleukin 3 expressing NOD SCID IL2R{gamma}NULL humanized mice. Blood. 2011;117(11):3076–86.
112. Manz MG. Human-hemato-lymphoid-system mice: opportunities and challenges. Immunity. 2007;26(5):537–41.
113. Kyoizumi S, Baum CM, Kaneshima H, McCune JM, Yee EJ, Namikawa R. Implantation and maintenance of functional human bone marrow in SCID-hu mice. Blood. 1992;79(7):1704–11.
114. Zhang Z, Jin L, Champion G, Seydel KB, Stanley SL Jr. Shigella infection in a SCID mouse-human intestinal xenograft model: role for neutrophils in containing bacterial dissemination in human intestine. Infect Immun. 2001;69(5):3240–7.

Chapter 7
BM Hematopoietic Niche Occupancy Defect of HSC in *Scid* Mice

Yulan Qing and Stanton L. Gerson

7.1 Introduction

Hematopoietic reconstitution is a prerequisite for the studies of human hematology and immunology. In order for the hematopoietic cell to engraft, differentiate, and function, some basic requirements need to be met: Cells must not be rejected, need to locate appropriately, and be supported by the host environment [1]. In this chapter, we are going to focus on the proper localization of HSCs.

7.2 HSC Niche and Long-Term Hematopoiesis

HSC function is closely coupled to the cells' unique cell-cycle kinetics, in which adult HSCs are predominantly quiescent, undergoing proliferation in response to stress [2]. Loss of quiescence or increased proliferation of HSCs results in loss of self-renewal or stem cell exhaustion [3]. The balance between HSC quiescence and proliferation is delicately regulated by both by intrinsic and extrinsic factors.

The intrinsic factors include cell cycle regulators, signaling molecules and transcription factors [4, 5]. Several cell cycle regulators have been shown to play critical regulatory roles in hematopoietic stem/progenitor cell (HSC/HPC) proliferation, including p57, p27, Rb family, and the D-type cyclins and their catalytic partners Cdk4/6. Studies from mouse models have shown that inactivation of these genes in HSCs results in increased proliferation and loss of quiescence associated with loss of self-renewal ability. Intracellular signaling pathways are also essential for HSC cell cycle regulation. For example, activation of PI3K cascade by deletion of

S. L. Gerson (✉) · Y. Qing
Case Comprehensive Cancer Center, National Center for Regenerative Medicine, Seidman Cancer Center, University Hospitals Case Medical center and Case Western Reserve University, Cleveland, OH 44106 USA
e-mail: slg5@case.edu

L. Y. Poluektova et al. (eds.), *Humanized Mice for HIV Research*,
DOI 10.1007/978-1-4939-1655-9_7

Phosphatase and tensin homolog (PTEN), a negative regulator of PI3K, attenuated the signaling promoted HSC quiescence, increased HSC cycling [6, 7]. PI3K signaling suppresses the expression of transcription factors FoxOs. Accordingly, a loss of FoxO3a alone or triple mutants for FoxOs1, 3, and 4 lead to increased proliferation and apoptosis of HSC, and in turn self-renewal activity [8–10].

The extracellular signals are from surrounding cells found within the highly specialized bone marrow (BM) microenvironment termed the HSC niche [11–13]. Adult HSCs reside in the niche, which plays a pivotal role in regulating the survival and self-renewal ability of HSCs, protecting them from exhaustion, while preventing their excessive proliferation. Two major HSC niches are currently proposed to exist in BM: the endosteal osteoblastic niche and the perivascular endothelial niche. Though it is unclear if both niches fulfill a similar function, or provide distinct properties, or work synergistically on HSC function, studies have agreed that niche occupancy is essential for HSC function, as HSCs outside the niche do not self-renew and commence the process of differentiation to ultimately produce mature blood cells [14].

The niche regulates HSC quiescence and function in several ways including cell-to-cell interaction, production of growth factors, cytokines, and extracellular matrix proteins. BM niche produces angiopoietin (Ang-1), which bind to cell surface receptor Tie2 on HSCs. The Ang-1/Tie2 interaction retained HSC in the niche, enhanced HSC quiescence, and thus sustained long-term hematopoiesis [15]. Chemokines and their receptors regulate HSC migration, homing, and mobilization from BM. Stromal-derived factor-1 (SDF-1) is secreted by a variety of cells within the BM niche, while its receptor C-X-C chemokine receptor type 4 (CXCR4) is highly expressed on the surface of HSC. SDF-1 is a potent chemoattractant for HSC [16, 17], deletion of CXCR4 from adult BM HSCs results in loss of HSC quiescence, and impaired HSC homing [18, 19]. Stem cell factor (SCF)/c-Kit pathway is also crucial for long-term maintenance and self-renewal of adult HSCs [20, 21]. Studies in mice with a loss of function *Kit* mutations with different severity have revealed severe defects in overall HSC function [22].

Sustained long-term HSC engraftment requires the accessibility of the recipient's BM niche. Under steady state conditions, a few sites are open to HSC lodging despite the presence of homing signals. There are several approaches to open the niche for transplanted HSCs. Conditioning recipients with myeloablative agents, such as radiation and cytotoxic chemotherapy, is usually efficient at opening the niche, both by killing of HSC and more commonly, by inducing cell death among progenitors, leading to increased proliferation and differentiation signals for HSC to leave the niche. While immune cells appear to contribute to graft maintenance, immunodeficiency of the host can only marginally improve donor HSC engraftment [23]. Cytokine signalings also contribute to stable HSC niche occupancy. Administration of ACK2, an antibody that blocks c-Kit function, led to the transient removal of >98% of endogenous HSCs in mice. Subsequent transplantation of these mice with donor HSCs led to chimerism levels of up to 90% [24, 25]. In addition, recipients with genetically 'weak' HSC compartments, such as mice bearing mutations in Kit ($Kit^{W/Wv}$), can accept HSC, allowing long-term engraftment without conditioning [26]. The W allele encodes a truncated c-Kit protein that lacks kinase activity; the Wv allele contains a C to T point mutation at position 2007 of the c-Kit sequence, which results in the change of the theronine at position 660 to methionine and significantly reduced kinase activity. $Kit^{W/Wv}$ mice display a severe macrocytic anemia characterized by an underlying stem cell defect [27, 28].

7.3 BM Niche Occupancy Is a Characteristic of HSC

Interaction between HSC and BM niche is essential for HSC to sustain long-term hematopoiesis at steady state and in transplantation settings. HSC occupancy in the BM niche can be measured by two ways, one is to measure the location of HSC in the BM by microscopy—a time point assessment. Interaction between HSC and BM endosteal niche is required for maintenance of HSC quiescence, and the distance between HSC and BM endosteal niche correlates with the primitiveness of HSC- physical niche occupancy and, indirectly with steady state quiescence. Briefly, purified HSCs are labeled with fluorescence dyes and transplanted into irradiated recipients, hours later, femur or calvarium BM from the recipients are subjected to microscopy analysis [29, 30]. With a combination of high-resolution confocal microscopy and two-photon imaging, the localization of HSCs to the niche can be determined. Due to the nature of quick alteration of primitive HSC and short life of fluorescence dyes, this approach is powerful to analyze HSCs lodging to the niche transiently short after the transplantation.

A second measure of BM niche occupancy is long-term hematopoiesis. At steady state, quiescent HSCs reside in the niche, the available niches are rare, exogenous HSCs are not able to undergo significant engraftment. Only under a certain circumstance, for example, a mutant host with "weak HSC", can wild type (WT) HSCs outcompete the endogenous mutant HSCs for the niche, engraft, and contribute to long-term hematopoiesis. In short, 5×10^6 BM cells are transplanted into recipients without any preconditioning. The chimeric state of donor contribution over 16 weeks in peripheral blood and BM HSCs of recipient mice will be measured to assess the ability of infused cells to compete for and occupy the niche (Fig. 7.1). By measuring the contribution of donor-derived hematopoiesis, the HSC niche occupancy can be determined. This is a powerful and sensitive assay, which measures the functional outcome of BM niche occupancy, and we term it HSC hematopoietic BM niche occupancy assay [31].

7.4 BM Hematopoietic Niche Occupancy Defect of Scid HSCs

Scid mice are "permissive" to exogenous HSC transplantation, suggesting that scid HSCs are disadvantaged against WT HSCs in competing for the BM niche to sustain long-term hematopoiesis. To assess the BM hematopoietic niche occupancy

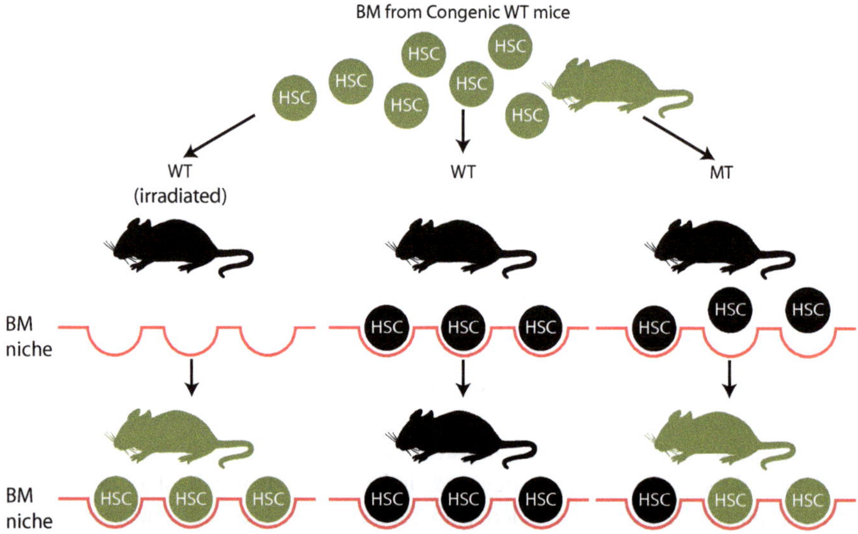

Fig. 7.1 Scheme of hematopoietic stem cell (HSC) bone marrow (BM) hematopoietic niche occupancy assay. 5×10^6 congenic WT BM (normally B6.SJL (CD45.1, Boy J), since most of the laboratory genetic altered mice are C57BL/6 background) are transplanted into nonablated recipient mice. If the resident cells have strong niche occupancy function, the recipients will retain their own hematopoiesis; or if the resident cells lack niche occupant competency mutant type (MT), donor HSCs will engraft and displace the recipient hematopoiesis, donor-derived hematopoiesis will be detected. HSC BM niche occupancy is analyzed as the chimeric state of donor contribution over 16 weeks in peripheral blood and BM HSCs of recipient mice. When the WT recipients were irradiated before transplantation, BM niches in the recipient became available, and WT donor HSCs can engraft and reconstitute the hematopoiesis in the recipients

of scid HSC, 5×10^6 WT BM cells were transplanted into scid and WT recipients without conditioning. As expected, little if any measurable stem cell engraftment occurred in WT recipients. In contrast, transplanted WT BM made a long-term multilineage contribution to hematopoiesis in *scid* recipients. By 16 weeks after transplantation, 12–18 % of myeloid cells in blood and 8 to 10 % of lineage-negative (Lin)- stem-cell antigen 1 $(SCA1)^+KIT^+$ (LSK) cells in BM were WT donor derived (Fig. 7.2a–c). In contrast, HSCs in *Rag1*$^{-/-}$ mice have a minimal BM hematopoietic niche occupancy defect, allowing very slight (< 1 %) engraftment of WT HSCs under the conditions used (Fig. 7.2a–c). Similarly, marginal HSC engraftment was detected in *Rag2*$^{-/-}$ mice without conditioning [23]. These results indicate that HSCs in scid mice have an intrinsic defect in BM hematopoietic niche occupancy, and importantly, this niche occupancy defect is independent of the immunodeficiency.

Scid BM also had a significant competitive repopulation disadvantage. When mixed 1:1 with WT BM and transplanted into lethally irradiated WT recipients, *scid* BM yields < 1 % overall chimerism in peripheral blood of recipients. Because *scid* BM cannot generate T and B lymphocytes, myeloid $(Mac1^+)$ chimerism in the recipient peripheral blood would closely represent donor contributions. *Scid* BM

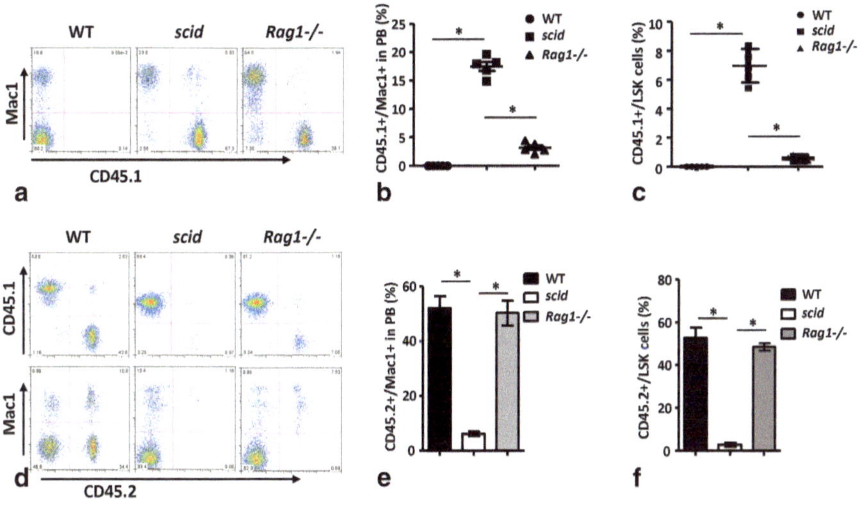

Fig. 7.2 *Scid* HSCs are defective in BM hematopoietic niche occupancy and competitive repopulation. **a–c** Engraftment of WT HSC into unconditioned *scid* and $Rag1^{-/-}$ mice. WT BM cells (5×10^6; CD45.1) were transplanted into unconditioned WT, *scid,* and $Rag1^{-/-}$ mice (CD45.2). Sixteen weeks after transplantation, donor chimerisms of multilineages in peripheral blood of recipients were analyzed by Fluorescence-activated cell sorting (FACS), and representative results are shown in part **a**. Myeloid chimerisms in the peripheral blood (PB) were quantitated in part **b,** donor chimerism of myeloid cells is the $CD45.1^+Mac1^+$ portion of the total $Mac1^+$ cells. Engraftment of transplanted WT BM HSCs (LSK) in the recipients were analyzed in part **c,** donor chimerism of HSC is the $CD45.1^+$LSK portion of the total LSK cells. **d–f** Competitive repopulation assay of *scid* and $Rag1^{-/-}$ HSCs. A total of 2×10^6 BM from 8-week-old healthy WT, *scid,* or $Rag1^{-/-}$ mice (CD45.2) were mixed with age-matched WT competitors (CD45.1) at 1:1 ratio and transplanted into lethally irradiated WT recipients (CD45.1). Sixteen weeks after transplantation, donor chimerisms of myeloid cells ($Mac1^+$) in the peripheral blood (shown in **d**) and HSCs (LSK) in the BM were analyzed and quantitated in parts **e** and **f**. Donor chimerisms of $Mac1^+$ cells in the peripheral blood are calculated as the $CD45.2^+Mac1^+$ portion of the total $Mac1^+$ cells. Donor chimerisms of LSK in the BM are calculated as the $CD45.2^+$LSK portion of the total LSK. Error bars indicate SD; significance was determined by a student 2-tailed t test. $^*P < 0.005$. (Republished with permission of the American Society of Hematology, from "An intrinsic BM hematopoietic niche occupancy defect of HSC in *scid* mice facilitates exogenous HSC engraftment." 2012 Feb 16; 119 [7]

was impaired in myeloid generation, contributing only approximately 7 % of peripheral blood myeloid chimerism (Fig. 7.2d–e). Consistently, HSCs derived from *scid* BM in the transplant recipients' BM comprised <5 % of the total HSC population (Fig. 7.2f). $Rag1^{-/-}$ mice was evaluated in the competitive repopulation assay. Although the overall chimerism derived from $Rag1^{-/-}$ BM in the peripheral blood of recipients was significantly decreased from WT BM because of a lack of lymphocytes production (Fig. 7.2d), $Rag1^{-/-}$ BM did not exhibit a competitive repopulation disadvantage in terms of myeloid chimerism, contributing almost 50 % of the myeloid cells in the peripheral blood and approximately 50 % of HSC in the BM of the recipients (Fig. 7.2d–f).

HSCs in *scid* mice have intrinsic defects in competitive repopulation and BM hematopoietic niche occupancy [31]. The HSC defect in *scid* but not $Rag1^{-/-}$ mice is due to the underlying genetic defect of these mice rather than differences in immunodeficiency. The *scid* mutation is a spontaneous nonsense mutation in protein kinase, DNA-activated catalytic polypeptide (Prdkc), which is the catalytic subunit of DNA-dependent kinase (DNA PKcs) in the nonhomologous end-joining (NHEJ) pathway. Although the NHEJ pathway is involved in both DNA repair and V(D)J recombination [32], these functions are distinct from the vantage point of hematopoiesis. Since HSCs in $Rag1^{-/-}$ mice and $Rag2^{-/-}$ mice retain functional, it is the deficiency in NHEJ DNA repair rather than the inability of V(D)J recombination, or immunodeficiency, that results in the HSC defect in *scid* mice. We speculate that the *scid* defect leads to the inability of the HSC population to repair spontaneous DNA double strand breaks while quiescent in the BM niche, forcing the cells into cell cycle where they rely on homologous recombination, and as a result, chronically have far fewer cells in the quiescent state occupying the niche. As a direct consequence of less quiescent cells in the niche, they are more prone to spontaneous competitive engraftment by exogenous donor HSC. This is not the case for $Rag1^{-/-}$ HSC which can still repair these breaks, therefore, maintain quiescence and are resistance to exogenous HSC engraftment.

BM hematopoietic niche occupancy capacity and HSC competitive repopulation activity are separable. Heterozygous Kit mutant mice exhibit a large HSC self-renewal defect [22], HSCs from heterozygous *c-Kit*^{Wv} mice have been examined for both activities. In competitive repopulation assay, when mixed with WT BM at 1:1 ratio, *c-Kit*^{Wv} BM displayed significantly impaired repopulation activity, contributing <15% chimerism in BM HSC population (Fig. 7.3a and [22]. When 5×10^6 WT BM cells were transplanted into *c-Kit*^{Wv} mice without preconditioning, similar with WT mice, little if any measurable donor HSC engraftment was observed in the recipients (Fig. 7.3b), even 10^7 WT BM donor cells did not yield detectable engraftment [33], indicating that *c-Kit*^{Wv} HSCs were able to lodge to the niche, did not allow exogenous HSC engraftment. These results also suggest that BM hematopoietic niche occupancy capacity is a distinct characteristic of HSC.

7.5 Summary

Both competitive repopulation and BM hematopoietic niche occupancy defects of scid HSCs facilitate the engraftment of exogenous HSCs, including human HSCs. Exogenous HSCs can access the niches in *scid* mice, start self-renew, outcompete *scid* HSCs in repopulation, and contribute to long-term hematopoiesis. This finding is important to the interpretation of hematopoietic xenotransplantation in the scid model system and suggests that scid conditions may not mimic human/human transplantation and may allow both preferential engraftment and survival of donor HSC populations.

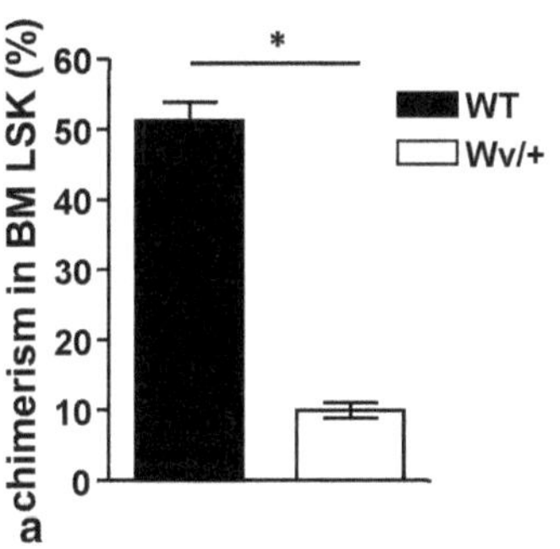

a

Recipients and no. of cells transplanted	Mice with Donor-derived hematopoiesis no./total 8 wk	16 wk
WT		
5x10^6	0/10	0/10
c-Kit$^{Wv/+}$		
5x10^6	0/10	0/10

b

Fig. 7.3 *c-Kit*$^{Wv/+}$ HSCs are defective in competitive repopulation, but not BM hematopoietic niche occupancy. **a** Competitive repopulation assay of *c-Kit*$^{Wv/+}$ HSCs. 2×10^6 BM from 8-week-old healthy WT, or *c-Kit*$^{Wv/+}$ mice (CD45.2) were mixed with age-matched WT competitors (CD45.1) at 1:1 ratio and transplanted into lethally irradiated WT recipients (CD45.1). Sixteen weeks after transplantation, donor chimerisms of HSCs (LSK) in the BM were analyzed and quantitated. Error bars indicate SD; significance was determined by a student 2-tailed t test. *$P < 0.005$. **b** Engraftment of WT HSC into unconditioned *c-Kit*$^{Wv/+}$ mice. WT BM cells (5×10^6; CD45.1) were transplanted into unconditioned WT or *c-Kit*$^{Wv/+}$ mice (CD45.2). Eight and 16 weeks after transplantation, donor chimerisms in peripheral blood and BM LSK fraction of recipients were analyzed

References

1. Manz MG. Human-hemato-lymphoid-system mice: opportunities and challenges. Immunity. 2007;26(5):537–41.
2. Wilson A, Laurenti E, Trumpp A. Balancing dormant and self-renewing hematopoietic stem cells. Curr Opin Genet Dev. 2009;19(5):461–8.
3. Orford KW, Scadden DT. Deconstructing stem cell self-renewal: genetic insights into cell-cycle regulation. Nat Rev Genet. 2008;9(2):115–28.
4. Pietras EM, Warr MR, Passegue E. Cell cycle regulation in hematopoietic stem cells. J Cell Biol. 2011;195(5):709–20.
5. Rossi L, Lin KK, Boles NC, Yang L, King KY, Jeong M, et al. Less is more: unveiling the functional core of hematopoietic stem cells through knockout mice. Cell Stem Cell. 2012;11(3):302–17.
6. Yilmaz OH, Valdez R, Theisen BK, Guo W, Ferguson DO, Wu H, et al. Pten dependence distinguishes haematopoietic stem cells from leukaemia-initiating cells. Nature. 2006;441(7092):475–82.
7. Yilmaz OH, Morrison SJ. The PI-3kinase pathway in hematopoietic stem cells and leukemia-initiating cells: a mechanistic difference between normal and cancer stem cells. Blood Cells Mol Dis. 2008;41(1):73–6.
8. Miyamoto K, Araki KY, Naka K, Arai F, Takubo K, Yamazaki S, et al. Foxo3a is essential for maintenance of the hematopoietic stem cell pool. Cell Stem Cell. 2007;1(1):101–12.
9. Yalcin S, Zhang X, Luciano JP, Mungamuri SK, Marinkovic D, Vercherat C, et al. Foxo3 is essential for the regulation of ataxia telangiectasia mutated and oxidative stress-mediated homeostasis of hematopoietic stem cells. J Biol Chem. 2008;283(37):25692–705.

10. Tothova Z, Kollipara R, Huntly BJ, Lee BH, Castrillon DH, Cullen DE, et al. FoxOs are critical mediators of hematopoietic stem cell resistance to physiologic oxidative stress. Cell. 2007;128(2):325–39.
11. Papayannopoulou T, Scadden DT. Stem-cell ecology and stem cells in motion. Blood. 2008;111(8):3923–30.
12. Wilson A, Trumpp A. Bone-marrow haematopoietic-stem-cell niches. Nat Rev Immunol. 2006;6(2):93–106.
13. Ema H, Suda T. Two anatomically distinct niches regulate stem cell activity. Blood. 2012;120(11):2174–81.
14. Lo Celso C, Scadden DT. The haematopoietic stem cell niche at a glance. J Cell Sci. 2011;124(Pt 21):3529–35.
15. Arai F, Hirao A, Ohmura M, Sato H, Matsuoka S, Takubo K, et al. Tie2/angiopoietin-1 signaling regulates hematopoietic stem cell quiescence in the bone marrow niche. Cell. 2004;118(2):149–61.
16. Broxmeyer HE. Chemokines in hematopoiesis. Curr Opin Hematol. 2008;15(1):49–58.
17. Lapidot T, Dar A, Kollet O. How do stem cells find their way home? Blood. 2005;106(6):1901–10.
18. Nie Y, Han YC, Zou YR. CXCR4 is required for the quiescence of primitive hematopoietic cells. J Exp Med. 2008;205(4):777–83.
19. Sugiyama T, Kohara H, Noda M, Nagasawa T. Maintenance of the hematopoietic stem cell pool by CXCL12-CXCR4 chemokine signaling in bone marrow stromal cell niches. Immunity. 2006;25(6):977–88.
20. Thoren LA, Liuba K, Bryder D, Nygren JM, Jensen CT, Qian H, et al. Kit regulates maintenance of quiescent hematopoietic stem cells. J Immunol. 2008;180(4):2045–53.
21. Geissler EN, Russell ES. Analysis of the hematopoietic effects of new dominant spotting (W) mutations of the mouse. I. Influence upon hematopoietic stem cells. Exp Hematol. 1983;11(6):452–60.
22. Sharma Y, Astle CM, Harrison DE. Heterozygous kit mutants with little or no apparent anemia exhibit large defects in overall hematopoietic stem cell function. Exp Hematol. 2007;35(2):214–20.
23. Bhattacharya D, Rossi DJ, Bryder D, Weissman IL. Purified hematopoietic stem cell engraftment of rare niches corrects severe lymphoid deficiencies without host conditioning. J Exp Med. 2006;203(1):73–85.
24. Czechowicz A, Kraft D, Weissman IL, Bhattacharya D. Efficient transplantation via antibody-based clearance of hematopoietic stem cell niches. Science. 2007;318(5854):1296–9.
25. Xue X, Pech NK, Shelley WC, Srour EF, Yoder MC, Dinauer MC. Antibody targeting KIT as pretransplantation conditioning in immunocompetent mice. Blood. 2010;116(24):5419–22.
26. Migliaccio AR, Carta C, Migliaccio G. In vivo expansion of purified hematopoietic stem cells transplanted in nonablated W/Wv mice. Exp Hematol. 1999;27(11):1655–66.
27. Fleischman RA. From white spots to stem cells: the role of the Kit receptor in mammalian development. Trends Genet. 1993;9(8):285–90.
28. Russell ES. Hereditary anemias of the mouse: a review for geneticists. Adv Genet. 1979;20:357–459.
29. Lo Celso C, Fleming HE, Wu JW, Zhao CX, Miake-Lye S, Fujisaki J, et al. Live-animal tracking of individual haematopoietic stem/progenitor cells in their niche. Nature. 2009;457(7225):92–6.
30. Xie Y, Yin T, Wiegraebe W, He XC, Miller D, Stark D, et al. Detection of functional haematopoietic stem cell niche using real-time imaging. Nature. 2009;457(7225):97–101.
31. Qing Y, Lin Y, Gerson SL. An intrinsic BM hematopoietic niche occupancy defect of HSC in scid mice facilitates exogenous HSC engraftment. Blood. 2012;119(7):1768–71.
32. Lieber MR. The mechanism of double-strand DNA break repair by the nonhomologous DNA end-joining pathway. Annu Rev Biochem. 2010;79:181–211.
33. Antonchuk J, Hyland CD, Hilton DJ, Alexander WS. Synergistic effects on erythropoiesis, thrombopoiesis, and stem cell competitiveness in mice deficient in thrombopoietin and steel factor receptors. Blood. 2004;104(5):1306–13.

Chapter 8
Improvement of Human Multilineage Hematopoietic Engraftment by Cytokine Knock-in Replacement in Human-Hemato-Lymphoid System Mice

Anthony Rongvaux, Markus G. Manz and Richard A. Flavell

A primary aim of biomedical research is to gain a better understanding of human physiology and to use this knowledge to prevent or cure human diseases. Due to practical and ethical barriers to the experimentation on human subjects, many studies are conducted on small animal models such as the mouse. However, mice are not men and the knowledge gained from animal experimentation is not always applicable to humans [1, 2]. In this context, mice repopulated with a human hemato-lymphoid system (HHLS) represent a useful small animal model for the study of human hematopoiesis and immune function in vivo [2, 3].

HHLS mice are generated by transplantation of human hematopoietic stem and progenitor cells (HSPCs) and/or human fetal tissues into recipient mice deficient in the innate and adaptive arms of the immune response [2, 3]. The first models of HHLS mice were developed in the late 1980s [4–6] and have been undergoing successive improvements since then [7, 8]. The strains of mice currently used as recipients for human hematopoietic engraftment share three characteristics. First, they lack B and T cells due to the severe combined immune deficiency (*scid*) mutation in the gene encoding the DNA-activated protein kinase, DNA activated, catalytic polypeptide (PRKDC) protein [4, 5], or due to deletion of one of the two *Rag* genes [9–11]. Second, deletion of the *Il2rg* gene that encodes the common gamma chain (γ_c) of cytokine receptors abolishes interleukin (IL)-15 (and IL-2, -4, -7, -9, -21) signaling and results in the absence of natural killer (NK) cells [9, 11–13]. Third, the interaction between the signaling regulatory protein alpha (SIRPα) receptor

R. A. Flavell (✉) · A. Rongvaux
Department of Immunobiology, Yale University, New Haven, CT 06520, USA
e-mail: richard.flavell@yale.edu

Howard Hughes Medical Institute, Yale University, New Haven, CT 06520, USA

M. G. Manz
Division of Hematology, University Hospital Zürich, 8091 Zurich, Switzerland

L. Y. Poluektova et al. (eds.), *Humanized Mice for HIV Research*,
DOI 10.1007/978-1-4939-1655-9_8

expressed on mouse macrophages and the CD47 ligand on human cells provides an inhibitory signal to mouse macrophages and confers phagocytic tolerance for the human xenograft [14, 15].

Cross-species interaction between SIRPα expressed on mouse cells and CD47 on human cells can be achieved using several approaches. Mice of the nonobese diabetic (NOD) genetic background contain a strain-specific polymorphism in the *Sirpa* gene, which affects the glycosylation of the mouse SIRPα protein and renders it cross-reactive with human CD47 [14, 16]. This polymorphism in *Sirpa* is most likely the main determinant of the capacity of the NOD background to support high levels of human cell engraftment, because backcrossing of the NOD-*Sirpa* allele onto the Balb/c RAG2$^{-/-}$ $\gamma_c^{-/-}$ genetic background is sufficient to support engraftment levels comparable to those observed in NOD-based recipients [16]. An alternative approach to induce SIRPα/CD47 cross-reactivity consists in expressing, by lentiviral transduction, the mouse *Cd47* gene in human HSPCs prior to the transplantation into recipient mice [16]. Finally, our labs have generated a human *SIRPA* bacterial artificial chromosome (BAC)-transgenic mouse, in which the entire gene encoding human SIRPα is inserted in the mouse genome [17]. This transgene induces mouse-to-human phagocytic tolerance, as demonstrated by the delayed clearance of human red blood cells injected into those mice, compared to nontransgenic mice. Upon transplantation of human HSPCs, the human *SIRPA* transgene resulted in significantly increased engraftment and maintenance of human hematopoietic cells, as well as improved immune function such as the production of antigen-specific immunoglobulin M (IgM) and immunoglobin G (IgG) in response to immunization [17].

High levels of human hematopoietic cell engraftment, upon transplantation of human HSPCs, are achieved when using NOD *scid* $\gamma_c^{-/-}$ (NOG (12) or NSG (13)) or hSIRPα^{tg} RAG2$^{-/-}$ $\gamma_c^{-/-}$ (SRG (17)) mice as recipients. Furthermore, human multilineage hematopoietic development is observed, with the presence of most human cell types present including T and B lymphocytes, NK cells, monocytes/macrophages, dendritic cells and sometimes low levels of erythrocytes and platelets. However, the terminal differentiation, homeostasis, physiologic relative cellular composition, and/or effector function of most of these human cells is suboptimal. It has been hypothesized that this condition is due to reduced or absent cross-reactivity between cytokines secreted by mouse tissues and the human receptors expressed on hematopoietic cells as well as nonphysiologic education of human adaptive immune cells in the xenogeneic environment [2, 18, 19]. To circumvent this limitation, several strategies have been developed to deliver human cytokines in the mouse host.

The injection of recombinant cytokines into recipient mice is the most straightforward approach to deliver human factors, and it has been used since the earliest models of HHLS mice [20–23]. This method has the disadvantage of being costly and labor-intensive as cytokines have short half lives in vivo and commonly daily injections are needed for prolonged periods of time. More cost effective methods consist in using lentiviral delivery of cytokine-encoding cDNA that results in constitutive synthesis and secretion of the cytokine in vivo [24], or hydrodynamic injection of plasmid DNA that leads to transient high-level secretion in circulation of cytokines, mostly by liver cells [25]. These methods can prove useful to boost the development of specific lineages of human cells, or for proof-of-concept experiments to test the efficacy of a candidate cytokine to support human hematopoietic development and function. However, the results of experiments performed using any of these protocols of cytokine delivery should be interpreted with caution, because they generally result in overexpression resulting in high concentrations of cytokines that may not be representative of physiological conditions and may induce artefactual effects on human cell development and function. Moreover expression is commonly systemic rather than physiologically normal local delivery which is the hallmark of cytokine biology.

To circumvent the need for exogenous administration of cytokines or cytokine-encoding vectors, genetically modified mice expressing human cytokines have been developed [26–28]. The transgenic (over)-expression of a cytokine-encoding cDNA under the control of a constitutive promoter, such as the cytomegalovirus (CMV) promoter, presents the same limitations as the previously described protocols of delivery (i.e., nonphysiological regulation of gene expression). In order to achieve more physiological expression of the human gene, therefore, BAC-transgenesis, in which the entire human gene (including the promoter and all regulatory sequences) is inserted in the mouse genome, should be favored over artificial ectopic expression in the wrong tissues by the use of constitutive promoters, cDNA constructs and the like.

Finally, our laboratories have been developing a method of gene humanization by knock-in replacement, mostly based on the velocigene technology of Regeneron Inc. [19, 29]. This method consists in replacing a portion of the mouse genome (from the initiation codon to the stop codon of a gene of interest) by its human counterpart. As most of the regulatory sequences (including promoter, 5' and 3' untranslated regions (UTR)) are of mouse origin, the transcription of the "humanized" gene is achieved in the most physiological conditions. The genes selected for such gene replacement have to meet at least two main criteria [18, 19]. First, the candidate genes have to encode cytokines that play the major roles in hematopoiesis; these cytokines should not be (fully) cross-reactive from mouse to human and mostly nonhematopoietic cells should produce them. Cytokines made by hematopoietic cells will be produced by the engrafted human cells and need not be provided endogenously. The second criteria relates to the cross-reactivity of the human cytokine on mouse cells. Indeed, in the case of homozygous replacement of a gene, defects in mouse hematopoietic cell development could ensue if the human cytokine is not fully cross-reactive on human cells as the mouse copies are eliminated in this way. On the one hand, this defect in mouse cells can be an advantage because it opens a niche that can be colonized by human cells upon engraftment of HSPCs. On the other hand, defects in these mouse cells could result in health defects or even lethality of the mouse. Therefore, only genes dispensable for health and survival of the mouse can be effectively humanized by knock-in replacement. We have recently reported the generation of three strains of mice with knock-in replacement of cytokine-encoding genes [30–32], all in the $RAG2^{-/-}\ \gamma_c^{-/-}$ genetic background, and we describe these mice in the following sections.

8.1 Thrombopoietin

Hematopoietic stem cells (HSCs) are rare cells that have the unique properties of self-renewal capacity and the potential to generate all cell types of the hematopoietic system, for the entire life of the organism [33, 34]. Several lines of evidence suggest that functional human HSCs are not efficiently maintained in classical models of HHLS mice: Large numbers of $CD34^+$cells (which contain all HSCs) need to be injected in order to achieve successful engraftment; engraftment levels decline with time, and the serial transplantation of human $CD34^+$cells from a primary recipient mouse to a secondary recipient is highly inefficient, indicating a loss of human HSC (or *scid* repopulating cells, SRC) over time.

Thrombopoietin (TPO) is one of the critical factors secreted in the bone marrow niche and required for the maintenance of functional, quiescent, and self-renewing HSCs [35–38]. The homozygous humanization of the gene encoding TPO by gene replacement [30] ($TPO^{h/h}$) resulted in a reduction in the mouse $Lin^-Sca1^+cKit^+$ (LSK) cell population, which is known to contain mouse HSCs, leaving an open niche for human HSCs. After transplantation of human HSPCs ($CD34^+$cells) into $RAG2^{-/-}$ $\gamma_c^{-/-}$ $TPO^{h/h}$, we observed increased numbers and frequencies of the human cell population known to be enriched in HSCs ($CD34^+CD38^{low}CD90^+CD45RA^+$cells) compared to control ($TPO^{m/m}$) recipient mice. Furthermore, human $CD34^+$cells isolated from the bone marrow of $TPO^{h/h}$ mice had a higher capacity to repopulate secondary recipient mice, showing that genetic humanization of TPO favors the maintenance of more functional and self-renewing human HSCs in the mouse. Phenotypically, this improved function of human HSCs resulted in increased engraftment levels of human hematopoietic cells in the bone marrow, that were maintained without decline for up to 6–7 months [30]. Despite this significant improvement through humanization of TPO, the maintenance and function of human HSCs remains suboptimal in HHLS mice. Additional factors or other creative approaches will be beneficial to reconstruct a more human bone marrow niche in the mouse [39].

8.2 Interleukin-3 and GM-CSF (CSF2)

The genes encoding IL-3 and granuocyte-macrophage colony stimulating factor (GM-CSF) are closely linked in the genome (less than 10 kb apart), and therefore we humanized both genes with a single knock-in replacement event, thus, generating $RAG2^{-/-}$ $\gamma_c^{-/-}$ IL-3/GM-$CSF^{h/h}$ mice [31]. Both cytokines play important roles in the development and maturation of myeloid cells, but we have so far characterized in detail and reported only the role of GM-CSF in this model.

GM-CSF is critically required for the terminal differentiation and function of lung alveolar macrophages [40, 41]. As human GM-CSF is not cross-reactive on mouse cells, nonengrafted GM-$CSF^{h/h}$ mice phenocopied GM-CSF-deficient mice and developed pulmonary alveolar proteinosis due to functional defects in mouse alveolar macrophages [31]. Upon transplantation of human $CD34^+$ cells, human alveolar macrophages developed in IL-3/GM-$CSF^{h/h}$ recipient mice, thereby replacing the defective mouse alveolar macrophages, but they were barely detectable in IL-3/GM-$CSF^{m/m}$ control recipients. Interestingly, in engrafted IL-3/GM-$CSF^{h/h}$ mice, human alveolar macrophages were able to partially rescue the pulmonary alveolar proteinosis phenotype observed in nonengrafted mice of the same strain. Besides the maintenance of lung homeostasis, alveolar macrophages play an important role in the immune response to mucosal infections by production of proinflammatory and antiviral cytokines, such as IL-6 or type I interferons. Accordingly, we observed robust expression of these human cytokines in the lung of engrafted IL-3/GM-$CSF^{h/h}$ mice in response to intranasal infection with influenza virus [31]. This model should be useful for studies of the response to lung mucosal infections or other lung pathologies.

8.3 M-CSF (CSF-1)

Macrophage colony stimulating factor (M-CSF) is another cytokine required for myelopoiesis, particularly for the development of monocytic/macrophage cells. The knock-in replacement of M-CSF in $RAG2^{-/-}$ $\gamma_c^{-/-}$ M-$CSF^{h/h}$ mice did not have any detectable phenotypic effect on mouse myeloid cell populations, suggesting that human M-CSF is at least partially cross-reactive on mouse cells [32]. After transplantation of human $CD34^+$ cells, the percentage of human myeloid cells ($CD33^+CD14^+$) in lymphoid tissues was increased from less than 5 % in M-$CSF^{m/m}$ recipients to up to 20–30 % in M-$CSF^{h/h}$ mice. These cells were also present in nonlymphoid tissues of M-$CSF^{h/h}$ mice, such as the liver, lungs and peritoneum. Finally, in response to lipopolysaccharide (LPS) stimulation in vivo, M-CSF gene humanization resulted in a 2–3-fold increase in the serum concentrations of human cytokines such as tumor necrosis factor alpha (TNFα) and IL-6 [32]. Major phenotypic and functional differences in monocyte subpopulations exist between the mouse and human species, and models to study human monocytes in vivo are currently lacking [42, 43]. M-CSF-humanized mice should prove a useful tool to study the development, activation, migration, and differentiation of human monocytes subsets in vivo, in homeostatic and diverse pathological conditions.

These three examples illustrate how the humanization by knock-in replacement of cytokine-encoding genes can impact favorably the development, homeostasis and function of human cells in HHLS mice. The same approach can be used to humanize additional cytokines and improve the development and function of other human hematopoietic cell types. Furthermore, we anticipate that the combination of multiple human cytokines in a single recipient mouse will have additive or synergistic effects on human hematopoiesis. This will result in HHLS mice with more complete and functional human innate and adaptive immune responses. Such an improved model will provide a much-needed tool for predictive preclinical human research in vivo.

References

1. Mestas J, Hughes CC. Of mice and not men: differences between mouse and human immunology. J Immunol (Baltimore, Md: 1950). 2004;172(5):2731–8. PubMed PMID: 14978070. Epub 2004/02/24. eng.
2. Rongvaux A, Takizawa H, Strowig T, Willinger T, Eynon EE, Flavell RA, et al. Human hemato-lymphoid system mice: current use and future potential for medicine. Annu Rev Immunol. 2013;31:635–74. PubMed PMID: 23330956. Epub 2013/01/22. eng.
3. Shultz LD, Brehm MA, Garcia-Martinez JV, Greiner DL. Humanized mice for immune system investigation: progress, promise and challenges. Nat Rev Immunol. 2012;12(11):786–98. PubMed PMID: 23059428. Epub 2012/10/13. eng.
4. Mosier DE, Gulizia RJ, Baird SM, Wilson DB. Transfer of a functional human immune system to mice with severe combined immunodeficiency. Nature. 1988;335(6187):256–9. PubMed PMID: 2970594. Epub 1988/09/15. eng.
5. McCune JM, Namikawa R, Kaneshima H, Shultz LD, Lieberman M, Weissman IL. The SCID-hu mouse: murine model for the analysis of human hematolymphoid differentiation and function. Science (New York, NY). 1988;241(4873):1632–9. PubMed PMID: 2971269. Epub 1988/09/23. eng.
6. Kamel-Reid S, Dick JE. Engraftment of immune-deficient mice with human hematopoietic stem cells. Science (New York, NY). 1988;242(4886):1706–9. PubMed PMID: 2904703. Epub 1988/12/23. eng.
7. Legrand N, Weijer K, Spits H. Experimental models to study development and function of the human immune system in vivo. J Immunol (Baltimore, Md: 1950). 2006;176(4):2053–8. PubMed PMID: 16455958. Epub 2006/02/04. eng.
8. Shultz LD, Ishikawa F, Greiner DL. Humanized mice in translational biomedical research. Nat Rev Immunol. 2007;7(2):118–30. PubMed PMID: 17259968. Epub 2007/01/30. eng.
9. Mazurier F, Fontanellas A, Salesse S, Taine L, Landriau S, Moreau-Gaudry F, et al. A novel immunodeficient mouse model–RAG2 x common cytokine receptor gamma chain double mutants–requiring exogenous cytokine administration for human hematopoietic stem cell engraftment. J Interferon Cytokine Res. 1999;19(5):533–41. PubMed PMID: 10386866. Epub 1999/07/01. eng.
10. Shultz LD, Lang PA, Christianson SW, Gott B, Lyons B, Umeda S, et al. NOD/LtSz-Rag1null mice: an immunodeficient and radioresistant model for engraftment of human hematolymphoid cells, HIV infection, and adoptive transfer of NOD mouse diabetogenic T cells. J Immunol (Baltimore, Md: 1950). 2000;164(5):2496–507. PubMed PMID: 10679087. Epub 2000/02/29. eng.
11. Traggiai E, Chicha L, Mazzucchelli L, Bronz L, Piffaretti JC, Lanzavecchia A, et al. Development of a human adaptive immune system in cord blood cell-transplanted mice. Science (New York, NY). 2004;304(5667):104–7. PubMed PMID: 15064419. Epub 2004/04/06. eng.
12. Ito M, Hiramatsu H, Kobayashi K, Suzue K, Kawahata M, Hioki K, et al. NOD/SCID/gamma(c)(null) mouse: an excellent recipient mouse model for engraftment of human cells. Blood. 2002;100(9):3175–82. PubMed PMID: 12384415. Epub 2002/10/18. eng.
13. Ishikawa F, Yasukawa M, Lyons B, Yoshida S, Miyamoto T, Yoshimoto G, et al. Development of functional human blood and immune systems in NOD/SCID/IL2 receptor {gamma} chain(null) mice. Blood. 2005;106(5):1565–73. PubMed PMID: 15920010. Pubmed Central PMCID: 1895228. Epub 2005/05/28. eng.
14. Takenaka K, Prasolava TK, Wang JC, Mortin-Toth SM, Khalouei S, Gan OI, et al. Polymorphism in Sirpa modulates engraftment of human hematopoietic stem cells. Nat Immunol. 2007;8(12):1313–23. PubMed PMID: 17982459. Epub 2007/11/06. eng.
15. Takizawa H, Manz MG. Macrophage tolerance: CD47-SIRP-alpha-mediated signals matter. Nat Immunol. 2007;8(12):1287–9. PubMed PMID: 18026079. Epub 2007/11/21. eng.
16. Legrand N, Huntington ND, Nagasawa M, Bakker AQ, Schotte R, Strick-Marchand H, et al. Functional CD47/signal regulatory protein alpha (SIRP(alpha)) interaction is required for

optimal human T- and natural killer- (NK) cell homeostasis *in vivo*. Proc Natl Acad Sci U S A. 2011;108(32):13224–9. PubMed PMID: 21788504. Pubmed Central PMCID: 3156191. Epub 2011/07/27. eng.
17. Strowig T, Rongvaux A, Rathinam C, Takizawa H, Borsotti C, Philbrick W, et al. Transgenic expression of human signal regulatory protein alpha in Rag2$^{-/-}$gamma(c)$^{-/-}$ mice improves engraftment of human hematopoietic cells in humanized mice. Proc Natl Acad Sci U S A. 2011;108(32):13218–23. PubMed PMID: 21788509. Pubmed Central PMCID: 3156175. Epub 2011/07/27. eng.
18. Manz MG. Human-hemato-lymphoid-system mice: opportunities and challenges. Immunity. 2007;26(5):537–41. PubMed PMID: 17521579. Epub 2007/05/25. eng.
19. Willinger T, Rongvaux A, Strowig T, Manz MG, Flavell RA. Improving human hemato-lymphoid-system mice by cytokine knock-in gene replacement. Trends Immunol. 2011;32(7):321–7. PubMed PMID: 21697012. Epub 2011/06/24. eng.
20. Lapidot T, Pflumio F, Doedens M, Murdoch B, Williams DE, Dick JE. Cytokine stimulation of multilineage hematopoiesis from immature human cells engrafted in SCID mice. Science (New York, NY). 1992;255(5048):1137–41. PubMed PMID: 1372131. Epub 1992/02/28. eng.
21. van Lent AU, Dontje W, Nagasawa M, Siamari R, Bakker AQ, Pouw SM, et al. IL-7 enhances thymic human T cell development in "human immune system" Rag2$^{-/-}$IL-2Rgammac$^{-/-}$ mice without affecting peripheral T cell homeostasis. J Immunol (Baltimore, Md: 1950). 2009;183(12):7645–55. PubMed PMID: 19923447. Epub 2009/11/20. eng.
22. Huntington ND, Legrand N, Alves NL, Jaron B, Weijer K, Plet A, et al. IL-15 trans-presentation promotes human NK cell development and differentiation *in vivo*. J Exp Med. 2009;206(1):25–34. PubMed PMID: 19103877. Pubmed Central PMCID: 2626663. Epub 2008/12/24. eng.
23. Hu Z, Van Rooijen N, Yang YG. Macrophages prevent human red blood cell reconstitution in immunodeficient mice. Blood. 2011;118(22):5938–46. PubMed PMID: 21926352. Pubmed Central PMCID: 3228505. Epub 2011/09/20. eng.
24. O'Connell RM, Balazs AB, Rao DS, Kivork C, Yang L, Baltimore D. Lentiviral vector delivery of human interleukin-7 (hIL-7) to human immune system (HIS) mice expands T lymphocyte populations. PloS One. 2010;5(8):e12009. PubMed PMID: 20700454. Pubmed Central PMCID: 2917362. Epub 2010/08/12. eng.
25. Chen Q, Khoury M, Chen J. Expression of human cytokines dramatically improves reconstitution of specific human-blood lineage cells in humanized mice. Proc Natl Acad Sci U S A. 2009;106(51):21783–8. PubMed PMID: 19966223. Pubmed Central PMCID: 2789167. Epub 2009/12/08. eng.
26. Nicolini FE, Cashman JD, Hogge DE, Humphries RK, Eaves CJ. NOD/SCID mice engineered to express human IL-3, GM-CSF and Steel factor constitutively mobilize engrafted human progenitors and compromise human stem cell regeneration. Leukemia. 2004;18(2):341–7. PubMed PMID: 14628073. Epub 2003/11/25. eng.
27. Brehm MA, Racki WJ, Leif J, Burzenski L, Hosur V, Wetmore A, et al. Engraftment of human HSCs in nonirradiated newborn NOD-scid IL2rgamma null mice is enhanced by transgenic expression of membrane-bound human SCF. Blood. 2012;119(12):2778–88. PubMed PMID: 22246028. Pubmed Central PMCID: 3327456. Epub 2012/01/17. eng.
28. Takagi S, Saito Y, Hijikata A, Tanaka S, Watanabe T, Hasegawa T, et al. Membrane-bound human SCF/KL promotes *in vivo* human hematopoietic engraftment and myeloid differentiation. Blood. 2012;119(12):2768–77. PubMed PMID: 22279057. Pubmed Central PMCID: 3327455. Epub 2012/01/27. eng.
29. Valenzuela DM, Murphy AJ, Frendewey D, Gale NW, Economides AN, Auerbach W, et al. High-throughput engineering of the mouse genome coupled with high-resolution expression analysis. Nat Biotechnol. 2003;21(6):652–9. PubMed PMID: 12730667. Epub 2003/05/06. eng.
30. Rongvaux A, Willinger T, Takizawa H, Rathinam C, Auerbach W, Murphy AJ, et al. Human thrombopoietin knockin mice efficiently support human hematopoiesis *in vivo*. Proc Natl

Acad Sci U S A. 2011;108(6):2378–83. PubMed PMID: 21262827. Pubmed Central PMCID: 3038726. Epub 2011/01/26. eng.
31. Willinger T, Rongvaux A, Takizawa H, Yancopoulos GD, Valenzuela DM, Murphy AJ, et al. Human IL-3/GM-CSF knock-in mice support human alveolar macrophage development and human immune responses in the lung. Proc Natl Acad Sci U S A. 2011;108(6):2390–5. PubMed PMID: 21262803. Pubmed Central PMCID: 3038773. Epub 2011/01/26. eng.
32. Rathinam C, Poueymirou WT, Rojas J, Murphy AJ, Valenzuela DM, Yancopoulos GD, et al. Efficient differentiation and function of human macrophages in humanized CSF-1 mice. Blood. 2011;118(11):3119–28. PubMed PMID: 21791433. Epub 2011/07/28. eng.
33. Kondo M, Wagers AJ, Manz MG, Prohaska SS, Scherer DC, Beilhack GF, et al. Biology of hematopoietic stem cells and progenitors: implications for clinical application. Annu Rev Immunol. 2003;21:759–806. PubMed PMID: 12615892. Epub 2003/03/05. eng.
34. Doulatov S, Notta F, Laurenti E, Dick JE. Hematopoiesis: a human perspective. Cell Stem Cell. 2012;10(2):120–36. PubMed PMID: 22305562. Epub 2012/02/07. eng.
35. Fox N, Priestley G, Papayannopoulou T, Kaushansky K. Thrombopoietin expands hematopoietic stem cells after transplantation. J Clin Invest. 2002;110(3):389–94. PubMed PMID: 12163458. Pubmed Central PMCID: 151089. Epub 2002/08/07. eng.
36. Kirito K, Fox N, Kaushansky K. Thrombopoietin stimulates Hoxb4 expression: an explanation for the favorable effects of TPO on hematopoietic stem cells. Blood. 2003;102(9):3172–8. PubMed PMID: 12855555. Epub 2003/07/12. eng.
37. Qian H, Buza-Vidas N, Hyland CD, Jensen CT, Antonchuk J, Mansson R, et al. Critical role of thrombopoietin in maintaining adult quiescent hematopoietic stem cells. Cell Stem Cell. 2007;1(6):671–84. PubMed PMID: 18371408. Epub 2008/03/29. eng.
38. Yoshihara H, Arai F, Hosokawa K, Hagiwara T, Takubo K, Nakamura Y, et al. Thrombopoietin/MPL signaling regulates hematopoietic stem cell quiescence and interaction with the osteoblastic niche. Cell Stem Cell. 2007;1(6):685–97. PubMed PMID: 18371409. Epub 2008/03/29. eng.
39. Scotti C, Piccinini E, Takizawa H, Todorov A, Bourgine P, Papadimitropoulos A, et al. Engineering of a functional bone organ through endochondral ossification. Proc Natl Acad Sci U S A. 2013;110(10):3997–4002. PubMed PMID: 23401508. Pubmed Central PMCID: PMC3593845. Epub 2013/02/13. eng.
40. Dranoff G, Crawford AD, Sadelain M, Ream B, Rashid A, Bronson RT, et al. Involvement of granulocyte-macrophage colony-stimulating factor in pulmonary homeostasis. Science (New York, NY). 1994;264(5159):713–6. PubMed PMID: 8171324. Epub 1994/04/29. eng.
41. Stanley E, Lieschke GJ, Grail D, Metcalf D, Hodgson G, Gall JA, et al. Granulocyte/macrophage colony-stimulating factor-deficient mice show no major perturbation of hematopoiesis but develop a characteristic pulmonary pathology. Proc Natl Acad Sci U S A. 1994;91(12):5592–6. PubMed PMID: 8202532. Pubmed Central PMCID: 44042. Epub 1994/06/07. eng.
42. Auffray C, Sieweke MH, Geissmann F. Blood monocytes: development, heterogeneity, and relationship with dendritic cells. Annu Rev Immunol. 2009;27:669–92. PubMed PMID: 19132917. Epub 2009/01/10. eng.
43. Chow A, Brown BD, Merad M. Studying the mononuclear phagocyte system in the molecular age. Nat Rev Immunol. 2011;11(11):788–98. PubMed PMID: 22025056. Epub 2011/10/26. eng.

Part II
Understanding of Human Immune Cells Development and Function in Mouse Environment

Chapter 9
Cytokine Species-Specificity and Humanized Mice

Jean-Pierre Yves Scheerlinck

9.1 Introduction

Cytokines can be defined as small, almost always soluble (although some can also be membrane-bound) proteins, produced to regulate immune responses and body functions often associated with immunity. They often, but not always, are synthetized by immune cells, act at very low concentrations due to their high affinity for their receptor(s) and exert their influence mostly at short range. Importantly, some of them can also be made, under specific circumstances, by cells other than leukocytes. As such they can be compared to hormones, with the main difference being that the cells producing cytokines are generally not part of specific, organized organs or glands, but have a tendency to be disseminated throughout the body.

Since cytokines play a key role in the regulation of immune responses and the interactions between the immune system and other organs (for example the neural system or the liver), it is critical that they be carefully considered in the development of animal models for any disease involving the immune system in general and HIV infection in particular. While in most cases the cytokines' general structure, function and receptors are maintained between humans and mice, there are some exceptions. For example, mice have a specific IL3R-β chain ($\beta_{IL\text{-}3}$), in addition to the common IL-3R-β chain (β_c), which is shared between IL-3, IL-5 and GM-CSF and is the only β-chain in humans [1]. Despite this general conservation between humans and mice, one can expect that in some cases mouse cytokines will not activate human cells and vice versa (i.e., there is not cross-reaction between cytokines of the two species; in other words, the cytokines are species specific). It should be noted that species specificity does not need to be absolute but could be quantitative in that a much higher concentration of the cytokine is required when from a different species. For example, about a thousand times more murine IL-15 is required to activate human cells compared to human IL-15 [2].

J.-P. Yves Scheerlinck (✉)
Centre for Animal Biotechnology, Faculty of Veterinary and Agricultural Sciences,
The University of Melbourne, Victoria 3010, Australia
e-mail: j.scheerlinck@unimelb.edu.au

L. Y. Poluektova et al. (eds.), *Humanized Mice for HIV Research*,
DOI 10.1007/978-1-4939-1655-9_9

When the cells expressing the cytokine and the cells reacting to the cytokine (i.e., expressing the receptor for the cytokine) have the same origin in a humanized mouse, this does not matter a great deal. However, if the cells expressing the receptor for the cytokine and the cells producing the cytokine are of different origins in the humanized mouse, this becomes very important. While the limitations resulting from the lack of cytokine cross-reactivity between species have been identified as a shortcoming in humanized mice [3], the full implications have not been explored systematically. Here, we address this issue and suggest a way to better identify cytokines more likely to be species specific: based on amino acid identity between mice and humans.

9.2 Humanized Mice

In order to understand the implications of cytokine species specificity in the context of mouse models for HIV, it is essential to briefly consider the basic properties and ways in which mice can be 'humanized'. A basic prerequisite for a mouse model for HIV research is the ability of HIV to infect cells present in the animal. In addition, in as far as possible, a functional human-like immune system must be present in the mouse model. This can, to various degrees, be achieved by creating 'humanized mice', which were defined by Shultz et al as 'immuno-deficient mice that have been engrafted with human primary haematopoietic cells and tissues that generate a functional human immune system' [4]. Basic requirements for the generation of humanized mice are that the transferred human cells must not be rejected, must be located in their normal environment and supported by that environment [5]. We will consider both the source of human immune cells and different types of immunosuppressed mice in the context of species-specific cytokines.

9.2.1 Sources of Human Immune Cells

In their simplest form, human immune cells can be provided by peripheral blood lymphocytes (PBL) [4, 5]. However, the transferred lymphocytes will unleash an immune response towards the mouse host cells, resulting in a xenogeneic graft-versus-host response, which will interfere with subsequent studies. In addition, because the mouse cells do not express the human leukocyte antigen (HLA) molecules, these cells cannot present antigen to the mature human T cells. To address these issues it is possible to repopulate the mouse immune system with human cells by transferring human hematopoietic stem cells, bone marrow cells or peripheral blood following G-CSF treatment. As a result, human precursor cells, including precursor T cells, will develop. These human precursor T cells will migrate to the mouse thymus to complete their differentiation. Hence, they will encounter mouse major histocompatibility complex (MHC) within the thymus. Therefore, the human

T cells are educated to the mouse MHC background and not the HLA background. Indeed, positive and negative selection within the thymus will result in these human T cells only recognizing antigen presented in the context of the mouse MHC. As a result, only mouse cells can now present antigen to the human T cells. Thus in such a model, mouse cytokines produced by the mouse antigen-presenting cells (APC) will need to cross-react with human cells as only mouse cells will be able to activate the human T cells. Conversely, human cytokines produced by the T cells will need to cross-react with mouse APC to allow the activation of the mouse APC by cytokines produced by the human T cells.

To resolve this issue, coimplantation of human fetal liver and thymus under the kidney capsule can be performed. This transplantation results in the generation of HLA-restricted human T cells. However, there is still a relatively poor human immune system, primarily because only low levels of human immune cells are present. To circumvent this problem, human bone marrow can be transferred, in addition to liver and thymus resulting in the bone marrow, liver, thymus (BLT) model [4]. While the human immune system is much more functional in the BLT model, vaccination studies are still limited because IgM antibodies are primarily produced, with only limited class switching occurring, suggesting that immunoregulatory mechanisms are incomplete. In such a model, cross-reactivity between mouse and human cytokines is particularly important for the correct development of the human cells within the 'mouse organ scaffold' (except for the human thymus).

9.2.2 *Sources of Immunosuppressed Mice*

The recipient mice need to be severely immunosuppressed in order to accept the xenogeneic graft. This immunosuppressed state can be achieved in several different ways. Initially mice were discovered with a mutation in the protein-kinase, DNA-activated, catalytic polypeptide (Prkdc), which resulted in severe combined immunodeficiency (SCID). While these mice were immunodeficient because of their inability to recombine the T cell receptor and immunoglobulin genes, later in life some recombination event could occur, resulting in leakage [6]. Subsequently, a similar phenotype was obtained by targeted mutation of the recombination-activating gene 1 (Rag1) and Rag2 loci, circumventing the issue of leakiness. However, these mice still have a mouse innate immune system including mouse natural killer (NK) T cells, mouse macrophages, which limit the engraftment of human cells. As such, the mouse innate immune cells will produce cytokines, which depending on the type, will or will not activate human cells. To overcome this, SCID mice can be crossed with nonobese diabetic (NOD) mice resulting in NOD–SCID mice with additional, but not absolute, deficiencies in innate immune cell types [6]. By crossing these mice with mice deficient for the IL-2 receptor γ-chain (IL-2Rγ) further improvements in acceptance of human cells were obtained. The IL-2Rγ is needed for signal transduction through the high-affinity receptors for IL-2, IL-4, IL-7, IL-9, IL-15 and IL-21. Hence, targeted mutations at the IL2rg locus coding for the

IL-2Rγ resulted in complete NK T cells impairment [4, 6]. Recent additional improvements include the transgenic expression of HLA genes and knockout of the mouse MHC genes. The idea is that in these mice the human T cells are educated in the mouse thymus but against HLA molecules expressed by the transgene. The approach would result in a more effective education of human T cell in the mouse without requiring a more complicated transplantation of human thymus cells. Indeed theoretically in these mice, T cells will be educated by the mouse thymus, hence using mouse cytokines.

9.2.3 Expressing Human Cytokines in Immuno-Deficient Mice

In an attempt to promote the development of the transferred human immune system, human cytokines have been overexpressed in the mouse under the form of 'knock-ins'. This is particularly important when the mouse cytokine cannot activate human cells. For example, IL-3 and GM-CSF do not cross-react between mouse and human [5]. It is therefore not surprising that knocking-in the genes for human IL-3 and human GM-CSF into immunodeficient mice increases the number of functional human macrophages in these humanized mice [4]. Another example consists of transiently expressing human cytokines (IL-15 and Flt-3/Flk-2 ligand) in the humanized mouse in order to improve the development of human NK cell populations [7]. While IL-15 has been reported to cross-react between humans and mice [5], mouse IL-15 is about three orders of magnitude less effective on human cells compared to human IL-15 [2] providing an explanation for this result. The case of thrombopoietin is interesting because this cytokine is cross-reactive between mice and humans [5]. As a result, one would expect that the mouse cytokine would be able to sustain the human cells. Nevertheless, here again knocking-in the gene for thrombopoetin into immunodeficient mice prior to reconstitution with human cells results in increased human myelopoiesis and ability for human stem cell renewal while decreasing lymphopoeisis. At the same time, it also results in thrombocytopenia and decreased hematopoietic stem cells in the mouse host [4, 8]. Thus, while this approach leads to some improvements, at the same time it creates a new set of issues and complications. Also, this approach will always be limited to a relatively small number of knocked-in human cytokines. Therefore, despite substantial progress it remains that humanized mice do not possess a complete and fully functional human immune system.

Even in the best-case scenario if a complete and functional human immune system can be created in a mouse, the resulting mouse model will still have a human immune system while at the same time retaining mouse tissues for other organs. Therefore species-specific cytokines produced by these transferred human cells and their (differentiated) progenitors will be able to affect only these human cells and not the mouse cells from any other organ. Conversely, species-specific cytokines made by mouse cells in the majority of mouse organs cannot affect the transferred human cells. Hence, the implications of this coexistence of two parallel organ systems need to be considered in detail.

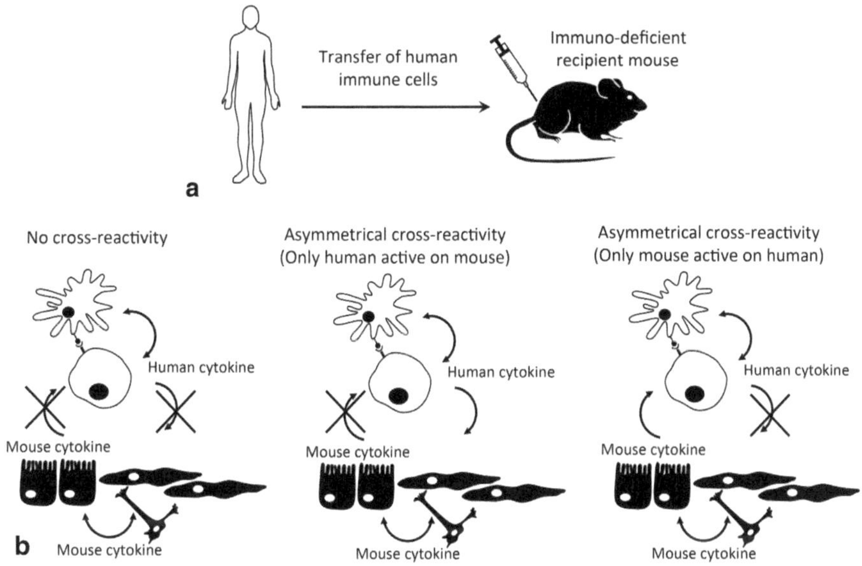

Fig. 9.1 Schematic representation of the effect of species-specific cytokines produced by human leukocytes and mouse nonleukocytes in humanized mice. **a** Humanized mice contain human immune cells that have been transferred into an immunodeficient mouse. While in many cases the human immune system is not completely functional, even a functional immune system will be deficient in its interactions with the other organs of the host mouse. **b** For species-specific cytokines, the human immune system will not be able to interact with a wide variety of cells of the mouse host and vice versa. Asymmetrical cross-reactivity in which the human cytokine will activate mouse cells (but not the reverse) will allow for the human immune system to influence the mouse organs. However, mouse cytokines will not be in a position to influence the human immune system. This often results in lack of complete establishment of a functional human immune system in the mouse host. Asymmetrical cross-reactivity whereby the mouse cytokine can activate the human immune system but not reverse will lead to the lack of key physiological effects of the immune system on the affected organs of the mouse.

9.3 Cytokines as Messengers Between the Immune System and Other Organs

Cytokines are produced mainly by leukocytes and by and large act on leukocytes. Thus in the ideal situation, when all the leukocytes in a mouse model of HIV are replaced by human leukocytes (Fig. 9.1a), correct communication between these cells should occur, irrespective of cytokine species specificity (Fig. 9.1b). However, cytokines can also affect nonleukocytes and regulate other body functions, such as for example, the effects they have on the neural system or on hepatocytes during acute phase responses. In a similar way, but outside of the scope of this review, leukocytes also interact with, for example, endothelial cells during transmigration and species specificity of the ligands and receptors of the human cells migrating through mouse tissues will be important in the way immune cells are able to reach their destination.

If the human cytokines produced by the transferred human leukocytes are species-specific, these cytokines will not perform their nonimmune function (Fig. 9.1b). For example, it is well documented that tumor necrosis factor (TNF) and interferon (IFN)-γ both induce fever and cause pain. However, human IFN-γ is unable to bind the murine receptor [9], while human TNF-α can activate mouse cells [10] possibly resulting in very different outcomes in a humanized mouse depending on which cytokine is responsible for the physiological observation. Thus, due to redundancy in the cytokine network and cross-reactivity of certain but not all cytokines, it is possible that a function is observed but that the underlying mechanism and possibly also the associated feedback loops might be different in humanized mice compared to humans. Conversely, if mouse-derived cells produce species-specific cytokines, these will not be able to affect the human leukocytes, so that mouse organs are impaired in conveying a signal back to the human immune system (Fig. 9.1b).

Thus, for species-specific cytokines we can expect that there would be a disconnect between the leukocyte-mediated functions of the cytokines and the functions of the cytokine mediated by other cells. The same argument can be put forward when nonimmune cells produce the cytokine. These cytokines are produced by host mouse cells and will affect other mouse cells more than they would affect the transferred human leukocytes. Thus here also, there would be a discontinuity between the origin of the cytokine and its function on different cell types (leukocytes vs. all other cell types).

9.3.1 *Cytokines Produced by Multiple Cell Types and Interacting with (Multiple) Receptors on Different Cells*

In addition, an interesting but complicated and artificial situation arises when a cytokine is produced by more than one cell type and some of these cell types happen to be part of the transferred human cells (and their progenitors), while others are of mouse origin. Indeed, for species-specific cytokines, this situation would result in two different sources of the cytokine, each being able to only affect cells of their own species.

The situation can become even more artificial if there is asymmetrical cross-reactivity (i.e. if the cytokine from one species, say humans, can cross-react on mouse cells, but the mouse cytokine cannot activate human cells). Indeed, under these conditions there are two sources of cytokines with different effects. The human cytokine becomes the 'super cytokine', able to activate all cells, while the homologous mouse cytokine is the 'poor relative', only able to affect the mouse and not the human cells (Fig. 9.1b). One such complex example is IL-6, which is produced by a wide range of cells including not only leukocytes such as macrophages, T cells, B cells and granulocytes; but also by smooth muscle cells, chondroblasts, osteoblasts and fibroblasts. In a humanized mouse, the IL-6 produced by white blood cells is human and since human IL-6 activates both human and mouse cells [5], it can be considered a 'super cytokine'. However, at the same time the mouse

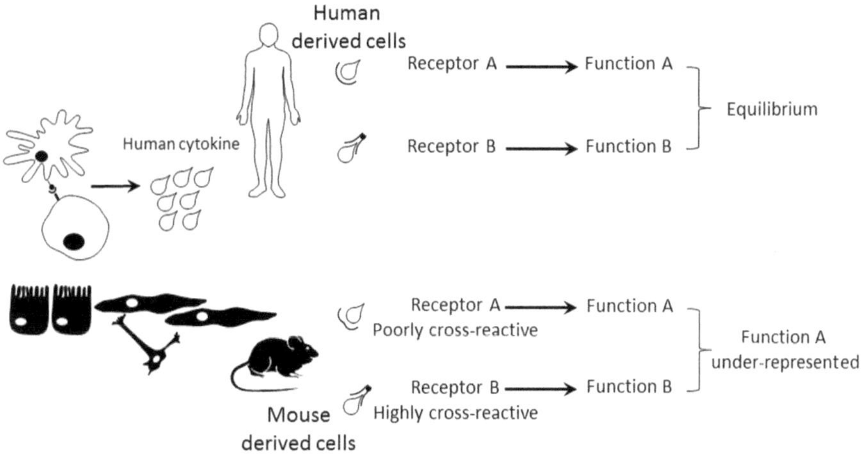

Fig. 9.2 Schematic representation of the effect of one cytokine interacting with several receptors, each having a different degree of cross-reactivity for the cytokine. The human cytokine is produced and will interact optimally with each of its receptors on human cells. When interacting with the mouse cytokine receptors it is likely that the degree of cross-reactivity will differ for each receptor. This could lead to the functions associated with the receptor that has a high degree of cross-reactivity to be overrepresented compared to the function mediated by the poorly cross-reactive receptor.

smooth muscle, chondroblasts, osteoblasts and fibroblasts will produce 'poor relative' mouse IL-6, which can only activate mouse cells but not human cells [5]. As if the situation is not complicated enough, the IL-6 receptor is expressed not only by many leukocytes (T cells, activated B cells, monocytes), but also by nonleukocytes such as hepatocytes, where it plays a role in induction of acute-phase proteins. In some humanized mice (namely BLT mice [4]) human liver cells will be present. These cells could potentially be activated by human IL-6 but not by mouse IL-6, while the mouse liver cells can be activated by both mouse and human IL-6. In other humanized mice models, no human liver cells are present, potentially giving rise to very different outcomes.

Another complex situation arises when a cytokine has the ability to interact with more than one receptor (Fig. 9.2). For example, according to the KEGG cytokine–cytokine receptor interaction pathway (http://www.genome.jp/kegg/pathway/hsa/hsa04060.html), TNF superfamily member 10 (TNFSF10) also known as TNF-related apoptosis inducing ligand (TRAIL) or APO-2 ligand can bind five different receptors (TRAILR1-4 and TR1). In another example, the IL-15 receptor can consist of different subunits, which combine to form different receptors with different affinities for both the human and mouse IL-15 giving rise to complex titration curves, which differ on different cells [2]. Thus, it is possible that the degree of cross-reactivity of the human and mouse cytokine will differ between each of these receptors expressed on different cells and/or associated with different functions. This creates an imbalance between these functions that are mediated by cross-reactive versus species-specific receptors or receptor complexes in the presence of the cytokine from the nonmatching species (Fig. 9.2).

9.3.2 *Multi-Subunit Cytokines*

The case of subunit cytokines is also interesting, as species-specificity reactivity does not necessarily needs to apply equally to both subunits giving rise to some interesting possibilities. For example, IL-12 is composed of a p35 and a p40 subunit with the heterodimer being active the active cytokine [11]. While murine IL-12 activates human cells, human IL-12 does not activate mouse cells. Studies using interspecies heterodimers demonstrated that the p40 subunit was cross-reactive and that the p35 subunit was responsible for the observed asymmetrically species specificity [12]. Interestingly, while p35 is expressed in both lymphoid tissue (e.g. spleen and thymus) and nonlymphoid tissue (e.g. lung and brain), the p40 subunit is expressed only in lymphoid cells [12]. In addition, the p40 homodimer is a competitive inhibitor of the heterodimer in that it binds the receptor yet does not activate the cell. Presumably, due to the cross-reactivity of the p40 in the heterodimer, the p40 homodimer will also be cross-reactive between human and mouse and inhibit both cell types. To complicate matters further, the cross-reactive p40 chain can also associate with another polypeptide, p19 to form the cytokine IL-23, whereas the species-specific p35 chain can associate with the Ebi3 chain to form IL-35 [13].

Hence under circumstances when species-specific cytokines are: (i) made by cells of both mouse and human origin, (ii) have receptors that are expressed on cells from both mouse and human origin, (iii) have more than one receptors mediating different functions and/or (iv) composed of multiple subunit; extreme care needs to be taken when interpreting findings in humanized mice.

9.4 Cross-Reactivity and Amino Acid Identity of Humans and Mice Cytokines

Arguments put forward in the previous section suggest that cytokine species specificity is highly relevant to humanized mice. Hence, it is worthwhile investigating what is known about cross-reactivity of cytokines between mice and humans. Obviously, species specificity of cytokines, can best be determined experimentally using bioassays or receptor binding studies. However, this is time consuming and in many cases the results of these experiments, especially when negative, are likely to be underreported. Across a broad range of species the ability of cytokines to cross-react between species is linked to their percentage amino acid identity [14]. In general terms, the higher the amino acid identity between equivalent cytokines in different species, the more the three-dimensional, physical and chemical structures of the proteins will be similar and therefore the higher the probability that cross-reaction will occur. Therefore, in the absence of experimental data, an informed guess can be made about the likely cross-reactivity, based on amino acid identity between the homologous cytokines in human and mice. This is obviously only a very rough tool and should be considered in terms of statistical probability of cross-reaction rather than as a replacement of experimental determination of cross-reaction.

In Tables 9.1, 9.2 and 9.3, cytokines as defined in the KEGG database (http://www.genome.jp/kegg/brite.html), were ranked by percentage of amino acid identity between human and mouse cytokine homologues. The amino acid identity between mouse and human cytokines varies from 100 to 31 %. At above 80 % amino acid identity (Table 9.1) there is a high probability of the cytokines being cross-reactive between human and mouse and therefore these cytokines particularly in the higher end of the amino acid identity scale, are expected to be of little consequence in humanized mice. In a range between approximately 80 and 60 % amino acid identity between cytokines of different species (Table 9.2), there is a gray zone where cross-reaction does occur but in many cases there is either no cross-reaction or only one-way cross-reaction. Below, 60 % amino acid identity (Table 9.3) it is expected that there will be much less chance of cross-reaction occurring, particularly the two-way cross-reaction that is often important in humanized mouse models.

9.5 Interactions Between the Pathogen and the Humanized Mouse

Many pathogens, especially viruses, have evolved to evade immune responses through a large number of strategies. Some of these strategies involve immunomodulation at the level of cytokines including the use of cytokine analogues, cytokine receptor analogues and molecules interfering with cytokine production or its effect on target cells [19, 20]. In many cases viruses have 'stolen' host cytokine-related genes and used these to subvert immune responses [20]. As a result of selection pressure exerted by the immune system on the pathogen these cytokine-related viral genes are critical to the survival of the virus and are important in the context of vaccine development [21]. The capture of cytokine genes by viruses and their subsequent selection can only occur in the species, which are the natural host to the infective agent. Indeed, there is little opportunity, and no selection pressure, for obtaining immunomodulatory mechanisms from species that are not naturally infected by the pathogen, suggesting that immunomodulatory mechanisms might to some degree be species specific. Therefore, even if, under experimental setting, the pathogen can be made to infect the species used as a model, it is unlikely that all immunomodulatory mechanisms would be active in that experimental model. Thus in these cases, the model would not mimic all immunomodulatory function that the pathogen has in its natural host and at least theoretically, important immune evasion mechanisms might be missed.

Applying this insight to humanized mouse models of HIV is not easy for a number of reasons. First, it is not possible to predict as yet undiscovered immunomodulatory

Table 9.1 Cytokines from the KEGG database likely to (at least one-way) cross-react between mice and humans based on percentage of amino acid identity between humans and mice

KEGG	Cytokine	%	KEGG	Cytokine	%
K04358	Fibroblast growth factor 12	100	K12457	Neurotrophin 4	91
K04358	Fibroblast growth factor 8	100	K05460	Hepatocyte growth factor	91
K04358	Fibroblast growth factor 16	100	K13375	Transforming growth factor beta-1	90
K04358	Fibroblast growth factor 9	100	K05459	Insulin-like growth factor 1	90
K04358	Fibroblast growth factor 18	99	K13376	Transforming growth factor beta-2	90
K05497	Growth differentiation factor 11	99	K04359	Platelet derived growth factor B	90
K04358	Fibroblast growth factor 14	99	K05474	TNF ligand superfamily member 12	89
K04358	Fibroblast growth factor 17	99	K05462	Ephrin-A 2	89
K13377	Transforming growth factor beta-3	98	K05457	Neuregulin 3	88
K05465	Angiopoietin 1	97	K05490	Interleukin 17B	88
K04663	Bone morphogenetic protein 7	97	K04358	Fibroblast growth factor 3	88
K04358	Fibroblast growth factor 11	97	K05417	Interleukin 11	88
K04662	Bone morphogenetic protein 4	97	K05448	Vascular endothelial growth factor B	88
K05463	Ephrin-B	97	K05462	Ephrin-A 5	87
K04355	Brain-derived neurotrophic factor	97	K04664	Growth differentiation factor 6	87
K05421	B-cell stimulating factor 3	97	K05450	Platelet derived growth factor C	87
K04667	Inhibin, beta B	96	K05456	Neuregulin 2	86
K05497	Growth differentiation factor 8	96	K04663	Bone morphogenetic protein 8A	86
K04667	Inhibin, beta A	96	K05466	Angiopoietin 2	86
K05463	Ephrin-B 1	96	K04358	Fibroblast growth factor 5	86
K04356	Neurotrophin 3	96	K02582	Nerve growth factor, beta	85
K05463	Ephrin-B 3	96	K05449	Vascular endothelial growth factor C	85
K04358	Fibroblast growth factor 1	96	K05503	Bone morphogenetic protein 10	85
K05502	Bone morphogenetic protein 1	95	K05492	Interleukin 17D	85
K04358	Fibroblast growth factor 20	95	K05450	Platelet derived growth factor D	85

Table 9.1 (continued)

KEGG	Cytokine	%	KEGG	Cytokine	%
K10033	C-X-C motif chemokine 14	95	K05473	TNF ligand super-family memb. 11	85
K04358	Fibroblast growth factor 2	95	K13769	Insulin-like growth factor 2	84
K05480	Ectodysplasin-A	95	K05449	Vascular endothelial gr. factor C	84
K04358	Fibroblast growth factor 7	94	K05462	Ephrin-A 1	84
K05455	Neuregulin 1	94	K05420	Ciliary neurotrophic factor	83
K04358	Fibroblast growth factor 13	94	K05424	Leptin	83
K10031	C-X-C motif chemokine 12	93	K05496	Bone morphogenetic protein 3B	83
K04358	Fibroblast growth factor 10	93	K05461	KIT ligand[a]	83
K04358	Fibroblast growth factor 6	93	K05462	Ephrin-A 4	83
K05462	Ephrin-A 3	93	K04667	Inhibin, beta E	82
K04663	Bone morphogenetic protein 5	93	K05488	IL-1 family, member 10 (theta)	82
K05452	Glial cell-derived neuro-trophic factor	93	K04668	Left–right determi-nation 2	82
K04359	Platelet derived growth factor A	92	K04668	Left–right determi-nation 1	82
K04662	Bone morphogenetic protein 2	92	K05464	Growth arrest-specific 6	82
K04664	Growth differentiation factor 5	91	K04358	Fibroblast growth factor 22	81
K04663	Bone morphogenetic protein 6	91	K05453	Macrophage CSF 1[b]	81
K05483	IL-1 family, member 5 (delta)	91			

[a] Mouse active on human cells, but human not active on mouse cells (SCF [5])
[b] Human active on mouse cells, but mouse not active on human cells [5]

mechanisms. As such, it is not possible to experimentally determine whether these would affect the humanized mice. Fortunately, most of the known immunomodulatory mechanisms affect cells directly involved in immune regulation [22]. In the case of humanized mice these would be predominantly transferred human cells. Since these cells are derived from the natural host, there should be no problem. However, it is impossible to exclude that some immunomodulatory mechanisms would be initiated or would contain at least one step that involves a nonleukocyte and therefore a mouse cell. For example it is well documented that HIV has the ability to interfere with type I IFNs [23]. In a humanized mouse sources of type I IFNs would include, human leukocytes (B cells, T cells, NK cells, macrophages and dendritic cells) as well as murine endothelial, epithelial cells and some neurons. Since

Table 9.2 Cytokines from the KEGG database that might cross-react (possibly asymmetrically) between mice and humans based on percentage amino acid identity between human and mouse

KEGG	Cytokine	%	KEGG	Cytokine	%
K05448	Vascular endothelial growth factor A	80	K05444	Interleukin 19	70
K12499	C-C motif chemokine 5	80	K05495	Growth differentiation factor 1	70
K05503	Bone morphogenetic protein 9	80	K05407	Platelet factor 4	70
K04358	Fibroblast growth factor 4	80	K05454	Fms-related tyrosine kinase 3 ligand	69
K05496	Bone morphogenetic protein 3	79	K12671	C-X-C motif chemokine 10	68
K05500	Inhibin, alpha	79	K05505	C-X-C motif chemokine 1	68
K04358	Fibroblast growth factor 21	79	K05416	C-X-C motif chemokine 9	68
K03156	TNF superfamily, member 2[a]	79	K12672	C-X-C motif chemokine 11	68
K04666	Nodal	79	K14624	C-C motif chemokine 2	68
K05422	Cardiotrophin 1	79	K04519	Interleukin 1 beta	68
K05437	Erythropoietin	79	K05425	Interleukin 12B[b]	67
K04664	Growth differentiation factor 7	78	K05438	Growth hormone 1	67
K05445	Interleukin 22	78	K05447	Interleukin 28A	67
K05491	Interleukin 17C	78	K05448	Vascular endothelial gr. factor A/B	66
K05512	C-C motif chemokine 19	78	K05444	Interleukin 24	66
K03161	TNF ligand superfamily member 5	77	K05482	Interleukin 18	66
K05477	TNF ligand superfamily member 14	77	K05513	C-C motif chemokine 28	66
K05493	Interleukin 17E	77	K05514	C-C motif chemokine 17	64
K05481	Interleukin 1 receptor antagonist	76	K05508	C-X3-C motif chemokine 1	64
K12964	C-C motif chemokine 4	76	K05414	Interferon alpha 13	64
K05419	Leukemia inhibitory factor	76	K05414	Interferon alpha 8	64
K05478	TNF ligand superfamily member 15	76	K14625	C-C motif chemokine 20	64
K04667	Inhibin, beta C	76	K04357	Epidermal growth factor	64
K05444	Interleukin 20	76	K05447	Interleukin 28B	63
K05458	Neuregulin 4	76	K05498	Growth differentiation factor 9B	63

Table 9.2 (continued)

KEGG	Cytokine	%	KEGG	Cytokine	%
K05408	C-C motif chemokine 3	75	K05511	C-C motif chemokine 15/23	63
K05426	Interleukin 23, alpha subunit p19	75	K05414	Interferon alpha 1	63
K00431	Thyroid peroxidase	74	K05514	C-C motif chemokine 22	63
K04663	Bone morphogenetic protein 8B	74	K05505	C-X-C motif chemokine 2	63
K04389	TNF ligand superfamily member 6	74	K05486	Interleukin 1 family, member 8 (eta)	63
K03157	Lymphotoxin beta	73	K05513	C-C motif chemokine 27	63
K05475	TNF ligand superfamily member 13	73	K05509	C-C motif chemokine 7	62
K04665	anti-Mullerian hormone	73	K05489	Interleukin 17	62
K05443	Interleukin 10	73	K05504	Growth differentiation factor 15	62
K05433	Interleukin 15[b,c]	73	K05431	Interleukin 7[b]	62
K05468	Lymphotoxin alpha	72	K05510	C-C motif chemokine 6/9	61
K05498	Growth differentiation factor 9	72	K04383	Interleukin 1 alpha	61
K05512	C-C motif chemokine 21C	71	K05439	Prolactin	61
K05423	Granulocyte CSF[b]	71	K05414	Interferon alpha 6	61
K05495	Growth differentiation factor 3	71	K05434	Interleukin 21	61
K05512	C-C motif chemokine 21A	71	K05507	C motif chemokine 1	61
K05428	Interleukin 5	71	K05487	Interleukin 1 family, member 9	60
K04358	Fibroblast growth factor	71	K05414	Interferon alpha 5	60
K05471	TNF ligand superfamily member 8	70	K05414	Interferon alpha 7	60

[a] Cross-reaction human cytokine on mouse cells [10]
[b] Cross-reaction between mouse and human [5, 11]
[c] Eisenman et al. [2] suggests large quantitative differences for IL-15

there is no evolutionary pressure for HIV to develop a mechanism to affect mouse cells, the ability of the virus to affect these cells would be fortuitous. In addition, little data is available about the species specificity of type I IFNs (many different proteins all with the potential to show at least some species specificity), further complicating interpretation of results. Hence, studies involving immunomodulatory mechanisms, particularly if they contain at least one interaction with mouse-derived cells, could be problematic.

Table 9.3 Cytokines from the KEGG database unlikely to be cross-react both ways between mice and humans based on percentage amino acid identity between human and mouse

KEGG	Cytokine	%	KEGG	Cytokine	%
K13072	C-C motif chemokine 25	59	K05479	TNF ligand superfamily member 18	52
K05442	Interferon, tau-1	59	K05506	C-X-C motif chemokine 5	51
K05414	Interferon alpha 2	59	K04358	Fibroblast growth factor 15/19	51
K05514	C-C motif chemokine 24	59	K10029	Platelet basic protein	50
K05509	C-C motif chemokine 11	59	K05514	C-C motif chemokine 26	50
K05406	Interleukin 12A[a]	59	K05415	Interferon beta	49
K05435	Interleukin 13	59	K05418	Oncostatin M	49
K04721	TNF ligand superfamily member 10	58	K10035	C-X-C motif chemokine 16	48
K05414	Interferon alpha 4	58	K10032	C-X-C motif chemokine 13	47
K05429	Interleukin 2[b]	58	K05469	TNF ligand superfamily member 4	45
K05414	Interferon alpha 14	57	K05514	C-C motif chemokine 1	42
K05494	Interleukin 17F	57	K05436	Thymic stromal lymphopoietin	42
K05432	Interleukin 9	57	K05430	Interleukin 4[c]	41
K05470	TNF ligand superfamily member 7	56	K04687	Interferon gamma[c]	41
K05476	TNF ligand superfamily member 13B	56	K05405	Interleukin 6[b]	41
K05427	Granulocyte-macrophage CSF[c]	55	K05441	Interferon, kappa	37
K05467	Angiopoietin 4	55	K05472	TNF ligand superfamily member 9	34
K05484	IL-1 family, member 6 (epsilon)	55	K04736	Interleukin 3[c]	31
K05509	C-C motif chemokine 8	54			

[a] Mouse cytokine active on human cells, but human not active on mouse cells [11]
[b] Human cytokine active on mouse cells, but mouse not/less active on human cells [2, 5]
[c] No cross-reaction [5, 9, 15–18]

9.6 Concluding Remarks

When considering animal models involving combining cells from mice and humans it is critical to consider that a proportion of the cytokines will not cross-react between species and that this lack of cross-reaction might well be asymmetrical. This has important implications not only for the development of the transplanted human immune cells but also for the induction of immune responses. In particular, complications can arise at the interface between the transplanted immune system and the host, including, for example, the interactions between the nervous system and the immune system. While some of these issues can be specifically addressed by creating new models in which certain human cytokines are knocked-in and therefore address specific concerns, only a very limited number of the many non-cross-reacting cytokines can be manipulated in that way. As a result when considering animal models it is imperative to focus on certain aspects of the host/pathogen interaction in which these issues are not critical, while acknowledging that the model is unsuitable for the study of other aspects. Ignoring these issues might result in some findings not being able to be translated to the situation in humans hence potentially misleading researchers in their quest to disease understanding and control.

Finally, the species specificity issues highlighted here for cytokines are equally valid for a range of other proteins that are produced by transplanted cells from humans and affect or are affected by proteins from the mouse host. These could include but are not limited to, proteins involved for example in the formation of immune organs or expressed on mouse endothelial cells and involved in the regulation of immune cell trafficking.

Acknowledgement The author is grateful to Dr. Neil Young of The University of Melbourne for his help in analyzing amino acid identity between human and mouse cytokines using the KEGG database.

References

1. Ozaki K, Leonard WJ. Cytokine and cytokine receptor pleiotropy and redundancy. J Biol Chem. 2002;277(33):29355–8. (PubMed PMID: 12072446. Epub 2002/06/20. eng).
2. Eisenman J, Ahdieh M, Beers C, Brasel K, Kennedy MK, Le T, et al. Interleukin-15 interactions with interleukin-15 receptor complexes: characterization and species specificity. Cytokine. 2002;20(3):121–9. (PubMed PMID: 12453470. Epub 2002/11/28. eng).
3. Akkina R. New generation humanized mice for virus research: comparative aspects and future prospects. Virology. 2013;435(1):14–28. (PubMed PMID: 23217612. Epub 2012/12/12. eng).
4. Shultz LD, Brehm MA, Garcia-Martinez JV, Greiner DL. Humanized mice for immune system investigation: progress, promise and challenges. Nat Rev Immunol. 2012;12(11):786–98. (PubMed PMID: 23059428. Epub 2012/10/13. eng).
5. Manz MG. Human-hemato-lymphoid-system mice: opportunities and challenges. Immunity. 2007;26(5):537–41. (PubMed PMID: 17521579. Epub 2007/05/25. eng).
6. Shultz LD, Ishikawa F, Greiner DL. Humanized mice in translational biomedical research. Nat Rev Immunol. 2007;7(2):118–30. (PubMed PMID: 17259968. Epub 2007/01/30. eng).
7. Chen Q, Khoury M, Chen J. Expression of human cytokines dramatically improves reconstitution of specific human-blood lineage cells in humanized mice. Proc Natl Acad Sci U S A. 2009;106(51):21783–8. (PubMed PMID: 19966223. Pubmed Central PMCID: 2789167. Epub 2009/12/08. eng).
8. Rongvaux A, Willinger T, Takizawa H, Rathinam C, Auerbach W, Murphy AJ, et al. Human thrombopoietin knockin mice efficiently support human hematopoiesis in vivo. Proc Natl Acad Sci U S A. 2011;108(6):2378–83. (PubMed PMID: 21262827. Pubmed Central PMCID: 3038726. Epub 2011/01/26. eng).

9. Smith MR, Muegge K, Keller JR, Kung HF, Young HA, Durum SK. Direct evidence for an intracellular role for IFN-gamma. Microinjection of human IFN-gamma induces Ia expression on murine macrophages. J Immunol. 1990;144(5):1777–82. (PubMed PMID: 2106553. Epub 1990/03/01. eng).
10. Eskandari MK, Nguyen DT, Kunkel SL, Remick DG. WEHI 164 subclone 13 assay for TNF: sensitivity, specificity, and reliability. Immunol Invest. 1990;19(1):69–79. (PubMed PMID: 2110931. Epub 1990/02/01. eng).
11. Trinchieri G. Interleukin-12: a proinflammatory cytokine with immunoregulatory functions that bridge innate resistance and antigen-specific adaptive immunity. Annu Rev Immunol. 1995;13:251–76. (PubMed PMID: 7612223. Epub 1995/01/01. eng).
12. Schoenhaut DS, Chua AO, Wolitzky AG, Quinn PM, Dwyer CM, McComas W, et al. Cloning and expression of murine IL-12. J Immunol. 1992;148(11):3433–40.
13. Vignali DA, Kuchroo VK. IL-12 family cytokines: immunological playmakers. Nat Immunol. 2012;13(8):722–8. (PubMed PMID: 22814351. Epub 2012/07/21. eng).
14. Scheerlinck JP. Functional and structural comparison of cytokines in different species. Vet Immunol Immunopathol. 1999;72(1–2):39–44. (PubMed PMID: 20080128).
15. Hemmi S, Merlin G, Aguet M. Functional characterization of a hybrid human-mouse interferon gamma receptor: evidence for species-specific interaction of the extracellular receptor domain with a putative signal transducer. Proc Natl Acad Sci U S A. 1992;89(7):2737–41. (PubMed PMID: 1532657. Pubmed Central PMCID: 48737. Epub 1992/04/01. eng).
16. Hemmi S, Peghini P, Metzler M, Merlin G, Dembic Z, Aguet M. Cloning of murine interferon gamma receptor cDNA: expression in human cells mediates high-affinity binding but is not sufficient to confer sensitivity to murine interferon gamma. Proc Natl Acad Sci U S A. 1989;86(24):9901–5. (PubMed PMID: 2532365. Pubmed Central PMCID: 298610. Epub 1989/12/01. eng).
17. Paul WE. Interleukin-4: a prototypic immunoregulatory lymphokine. Blood. 1991;77(9):1859–70. (PubMed PMID: 2018830. Epub 1991/05/01. eng).
18. Boulay JL, Paul WE. Hematopoietin sub-family classification based on size, gene organization and sequence homology. Curr biol. 1993;3(9):573–81. (PubMed PMID: 15335670. Epub 1993/09/01. Eng).
19. Engel P, Angulo A. Viral immunomodulatory proteins: usurping host genes as a survival strategy. Adv Exp Med Biol. 2012;738:256–76. (PubMed PMID: 22399384. Epub 2012/03/09. eng).
20. Epperson ML, Lee CA, Fremont DH. Subversion of cytokine networks by virally encoded decoy receptors. Immunol Rev. 2012;250(1):199–215. (PubMed PMID: 23046131. Epub 2012/10/11. eng).
21. Scheerlinck JP, Nahar MF, Every A. Immune memory resilience in the face of immunomodulatory pathogens: Implications for vaccine development. Curr Trends Immunol. 2012;13:95–102.
22. Reuter MA, Pombo C, Betts MR. Cytokine production and dysregulation in HIV pathogenesis: lessons for development of therapeutics and vaccines. Cytokine Growth Factor Rev. 2012;23(4–5):181–91. (PubMed PMID: 22743036. Pubmed Central PMCID: 3582023. Epub 2012/06/30. eng).
23. Hughes R, Towers G, Noursadeghi M. Innate immune interferon responses to human immunodeficiency virus-1 infection. Rev Med Virol. 2012;22(4):257–66. (PubMed PMID: 22359246. Epub 2012/02/24. eng).

Chapter 10
Human T-Cell Biology in a Mouse Environment

Nicolas Legrand and Hergen Spits

Abbreviations

BLT	Bone marrow, liver, thymus
BRG	BALB/c Rag2$^{-/-}$ IL-2R$\gamma_c^{-/-}$
BRGS	BALB/c Rag2$^{-/-}$ IL-2R$\gamma_c^{-/-}$ SIRPα^{NOD}
DC	Dendritic cells
FDC	Follicular dendritic cells
FL	Fetal liver
FRC	Fibroblastic reticular cells
GvHD	Graft-versus-host disease
HIS	Human immune system
hHSPC	Human hematopoietic stem and progenitor cells
hTSPC	Human thymus seeding progenitor cells
ISP	Immature single positive
LN	Lymph node
LTi	Lymphoid tissue inducer
NOD	Nonobese diabetic
NOG	NOD/Shi-scid/γc^{null}
NSG	NOD.Cg-Prkdc^{scid} $IL2rg^{tmlWjl}$/Sz
SCID	Severe combined immuno-deficiency
T_{FH}	T follicular helper
UCB	Umbilical cord blood

N. Legrand (✉)
AXENIS, Paris, France
e-mail: nicolas.legrand@pasteur.fr

H. Spits
Tytgat Institute for Liver and Intestinal Research, AMC-UvA, Amsterdam, The Netherlands
e-mail: hergen.spits@amc.uva.nl

L. Y. Poluektova et al. (eds.), *Humanized Mice for HIV Research*,
DOI 10.1007/978-1-4939-1655-9_10

10.1 A Large Collection of Mice Humanized for the Immune System

The experimental, prospective analysis of human hematopoietic development and function is complex, for technical, ethical, and practical reasons. Human immunology mostly relies either on in vitro systems, or on observations in clinical settings. Tackling these limitations, and facing the lack of appropriate models to study HIV, several pioneering groups developed in the late 1980s chimeric mouse models of human immunity based on the transplantation of human hematopoietic cells into various immunodeficient mouse recipients [1–3]. The history of humanized mouse models of hematopoiesis has already been reviewed before, providing an extensive overview of the incremental optimization strategies that have been applied over the past two decades to meet desired features for such "human immune system" (HIS) mice [4–10].

In brief, three main criteria have been demonstrated to strongly impact on the stability of the human hematopoietic xenograft in HIS mice: (i) the degree of immunodeficiency of the mouse recipients (the inactivation of mouse Rag2 and IL-2Rγ_c gene expression being particularly beneficial, as it fully abrogates the mouse T/B and NK cell development, respectively); (ii) their genetic background (the nonobese diabetic (NOD) and BALB/c strains being for instance particularly permissive, in contrast to the C57BL/6 strain); and (iii) the age at which the mice receive the xenotransplant (newborns being more permissive than adult animals). Additional many technical variations (e.g., site of transplantation, origin of the human hematopoietic cells, coengraftment of human tissues, and treatment of the recipient mice with exogenous products) have been applied, each with specific advantages and drawbacks. Among those, HIS mouse models generated through the transplantation of human hematopoietic stem and progenitor cells (hHSPC) are particularly attractive, as they should provide systems supporting multilineage human hematopoietic reconstitution in a durable fashion, allowing long-term studies in a more physiological environment. The past decade has seen the emergence of two major families of HIS mouse models: One based on the transplantation of hHSPC suspensions alone and another on hHSPC transplantation together with human fetal thymic tissues. Both families of models make use of various immunodeficient mouse genetic backgrounds, mainly NOD-based and BALB/c-based.

10.2 Humanized Mouse Generation with Hematopoietic Progenitors Only

NOD background-based mouse strains, either originating from the Jackson Laboratories in the USA (NSG [NOD.Cg-*Prkdc*^{scid} *IL2rg*^{tmlWjl}/Sz] mice) or from the Central Institute for Experimental Animals in Japan (NOG [NOD/Shi-scid/γc^{null}] mice), are widely used to generate HIS mice. Human hematopoietic reconstitution was demonstrated in both newborn and adult NSG/NOG mice, although newborn animals are more permissive to the xenograft [11–15]. A common strategy nowadays involves conditioning of newborn animals by sublethal irradiation followed by a single injection with a suspension of 10^5–10^6 human umbilical cord blood (UCB) or fetal liver (FL) $CD34^+$ cells, a cellular fraction known to be enriched for hHSPC. Transplantation of as few as ten purified $CD34^+CD38^-CD90^+CD45RA^-Lin^-$ hHSPC resulted already in human hematopoietic reconstitution in NSG or NOG mice [16, 17]. The particularly high efficiency of xenotransplantation in the NOD background can be explained by the combination of defective function of mouse complement, mouse plasmacytoid dendritic cells (DC), and/or mouse phagocytes [18–20]. Of note, due to the *scid* mutation, NSG/NOG mice exhibit an ineffective DNA repair machinery and are therefore particularly radio-sensitive, which may explain the relatively short life-span of NSG mice (<6 months) after completion of the humanization procedure [21].

We and others have shown that newborn (≤5-day-old) BALB/c $Rag2^{-/-}$ IL-$2R\gamma_c^{-/-}$ (BRG) mice can also serve as a recipient for a hHPC xenograft, following the aforementioned straight-forward reconstitution strategy [22, 23]. Human hematopoietic reconstitution is directly dependent on the transplantation age, since 1-week and 2-week-old recipient BRG animals exhibit a severely reduced capacity to support the generation of a human hematopoietic ($hCD45^+$) cell graft, which is virtually absent when starting from adult BRG animals [22, 24]. Of note, human hematopoietic reconstitution is inefficient in C57BL/6 $Rag2^{-/-}$ IL-$2R\gamma_c^{-/-}$ recipients, both with newborn or adult animals [24, 25].

10.3 Humanized Mouse Generation with an Additional Human Thymus Organoid

The second major family of HIS mice—known as BLT (for "bone marrow, liver, thymus") mice—is reminiscent of the original SCID-hu Thy/Liv mice [1]. In this model, xenotransplantation of human fetal thymus and fetal liver fragments is performed under the kidney capsule of immune-deficient C.B-17 SCID mice, a natural immune-deficient mouse strain in which T- and B-cell development is inhibited due to a defective V(D)J recombination process—although T and B cells are detectable in some mice, a phenomenon known as "T/B cell leakiness" [26–28]. As a result, a human thymus organoid engrafts and becomes the major site of human thymopoiesis. Still, only a minority of the SCID-hu Thy/Liv animals (~1 out of 3) contain a detectable population of human hematopoietic cells, which all belong to the T cell lineage and only represent a minor fraction (~0.7%) of all leukocytes [29]. Partly because mouse T and B cells may arise and mouse NK cells are normal in the C.B-17 SCID background, the human xenograft is eventually lost over time, probably due to immune rejection. To resolve these issues, a similar technical strategy has later been applied to NOD-based immuno-deficient mouse strains to generate BLT mice [30–32]. Similarly to SCID-hu Thy/Liv mice, the BLT model is intrinsically more challenging to construct due to specific technical constrains—e.g., dependency on human fetal tissues, requirement for delicate surgery, but it exhibits attractive features—in particular a relatively large human peripheral T-cell population which has permitted major achievements in the HIV field [33]. Nevertheless, it was recently reported that BLT mice develop over time a wasting syndrome due to graft-versus-host disease (GvHD), a feature that may limit the use of BLT mice in long-term settings [34–36].

10.4 Human T-Cell Development in HIS Mice

The most recent, aforementioned HIS mouse models (NSG/NOG, BRG, and BLT) all support de novo, multilineage human hematopoiesis, including human T-cell development. BRG-HIS and NSG/NOG-HIS mice rely on the mouse thymus for the generation of human T cells, as illustrated by the absence of human T cells in thymus-deficient nude BRG-*nu/nu* mice after hHSPC inoculation [37]. In contrast, human T-cell development in BLT mice mostly takes place in the ectopic human thymic organoid deriving from the human fetal tissues engrafted under the kidney capsule. This fundamental distinction obviously imposes major differences on their relative capacity to support human T-cell production, as human thymopoiesis is likely to be more efficient in a human microenvironment.

It is currently unknown whether the hHSPC established in HIS mice are able to generate human thymus seeding progenitor cells (hTSPC) as efficiently and using the same differentiation pathway as in human individuals. Furthermore, different hTSPC might colonize a mouse thymus (NSG/NOG; BRG) vs. a human thymic organoid (BLT), an issue that requires further investigations. It has been described that $CD34^{+}CD38^{low}$ human thymocytes contain the earliest immature thymocyte population, since they mostly lack TCR gene rearrangements and contain T cell, DC, and NK cell precursor activities [38–41]. This cell population resembles $CD34^{+}CD45RA^{+}CD7^{+}CD38^{low}$ cells found in human UCB, which are also able to give rise to T cells [42, 43]. Of note, it has been shown that human fetal bone marrow $CD34^{+}CD45RA^{+}CD7^{+}$ cells can be actively recruited to immuno-deficient mouse (NOD.*scid*) thymic lobes, where they commit to the T cell lineage [44].

Overall, one can therefore speculate that hTSPC are generated in HIS mice from the transplanted $CD34^{+}$ hHSPC and are then able to colonize the murine thymus. In BLT mice, hTSPC populate both the murine thymus and the human thymic organoid of BLT mice—potentially resulting in a more effective engraftment. It is likely that different sources of human hHSPC, which significantly differ in their developmental status (e.g., FL, UCB, and adult bone marrow) result in major differences in the degree of hTSPC generation and thymus engraftment. For instance, we have consistently observed that thymus colonization occurs at least two weeks earlier in FL $CD34^{+}$ hHSPC-injected BRG mice, as compared to UCB $CD34^{+}$ hHSPC-injected BRG mice (unpublished observation), and similar kinetic differences have been reported for total human hematopoietic reconstitution [45]. These observations might be due to intrinsic differences when considering the differentiation capacity of each of these hHSPC sources. It is also possible that the frequency of pluripotent $CD34^+CD38^-$ hHSPC present in the $CD34^+$ cell population used as the original inoculum determines the degree of colonization, since this frequency is significantly higher when using FL samples (30–50 % vs. < 1 % in UCB).

Colonization of the mouse thymus in BRG-HIS or NSG/NOG-HIS mice is initiated 2–4 weeks after the hHSPC transplantation, and leads to limited accumulation of human thymocytes in situ. NSG/NOG-HIS mice tend to exhibit lower human thymocyte numbers ($0.1–0.5 \times 10^6$ cells) than BRG-HIS mice ($1–5 \times 10^6$ cells) [11–13, 15, 22, 23, 46, 47]. In contrast, the size of the human thymic organoid in NSG-BLT mice after hHSPC transplantation is much higher ($\sim 70 \times 10^6$ human cells), with high-interindividual variability [48]. Thymic tissues (mouse and human organoid) get organized with cortex- and medulla-like regions, which mostly contain immature, $CD4^+CD8^+$ double-positive thymocytes and mature, $CD4^+$ or $CD8^+$ single-positive thymocytes, respectively—as seen in mouse or human thymus control tissues [11, 14, 23, 49–51].

Altogether, the human thymocyte differentiation pattern in HIS mice is similar to what is reported from normal human thymus samples with respect to the major cell surface markers [52]. As in the human thymus, the rare $CD34^+CD1a^-$ cells observed in the thymus of HIS mice may contain dual T/NK cell precursors, which would correlate with local, relatively high accumulation of discrete human NK cell sub-populations [53, 54]. Acquisition of CD1a expression marks commitment towards the T-cell lineage, and $CD34^+CD1a^+$ cells subsequently differentiate into $CD4^+$ immature single positive (ISP) cells, which contain precursors of both αβ and γδ T cells [52]. The heterogeneous $CD4^+CD8^+$ double-positive immature thymocyte population expresses variable levels (mostly from negative to dim) of CD3ε on cell surface, whereas mature $CD4^+$ or $CD8^+$ thymocytes express high levels of surface CD3ε and have lost the expression of CD1a. Most of the mature thymocytes (~ 99 %) belong to the Tαβ lineage. A fraction (1–5 %) of $CD4^+$ mature thymocytes resemble "natural" regulatory T cells ($CD25^+GITR^+FOXP3^+$), both phenotypically and functionally [55–59].

10.5 Human Peripheral T-Cell Homeostasis in HIS Mice

Once they reach a mature stage, human thymocytes (including γδT cells and T_{reg} cells) colonize lymphoid organs and peripheral tissues of HIS mice, with kinetics that differ from other human hematopoietic lineages. Sequential analysis of human cell content in peripheral organs of BRG-HIS mice shows initial colonization by immature/naive B cells, followed by gradual T-cell accumulation, which correlates with accumulation of mature/memory-like B cells and serum human IgG over time [46, 60]. At early time points after reconstitution, reconstitution of human cells in HIS mice therefore mimic the situation of young infants or lymphopenic patients, e.g., shortly after hHSPC transplantation, who exhibit relatively immature, partially functional immune systems [61].

The relative proportion of human hematopoietic cell subsets in peripheral lymphoid organs of all HIS mouse models is reminiscent of a mouse situation, with high proportion of lymphocytes and low frequency of granulocytes, in contrast with normal human blood. Overall, the number of human leukocytes found in adult BRG-HIS mice is relatively low as compared to what is observed for murine leukocytes in immuno-competent mice (e.g., for BRG-HIS mice: 20–30 % of BALB/c leukocyte numbers in the bone marrow, 1–2 % in the thymus, and 1–5 % in the spleen) [22–24, 46]. Transplantation of hHSPC into NOD-based immuno-deficient mice results in higher numbers of human hematopoietic cells (two to tenfold depending on the secondary lymphoid organ) than in BRG-HIS mice [24, 45]. This difference is apparent only when the actual cell numbers are determined by counting, as similar relative frequencies of human ($hCD45^+$ vs. $mCD45^+$) leukocytes have been reported in several NOD-based and BRG-HIS mice.

Major differences have been reported between BRG, NSG/NOG, and BLT mice regarding the actual frequency and phenotype of human T cells in the secondary lymphoid organs of hHSPC-reconstituted animals. In the spleen, T cells usually represent less than 10 % of human leukocytes in adult (~14-week-old) BRG-HIS mice [22–24, 46], around 10–20 % in NSG/NOG-HIS mice [11–15] and >50 % of human leukocytes in BLT mice [30–32]. Large numbers of naive ($CCR7^+CD45RA^+$ or $CD27^+CD45RO^-$) T cells are observed in BLT and NSG/NOG-HIS mice, whereas BRG-HIS mouse T cells mostly exhibit a phenotype of activated, cycling ($Ki67^+$ or BrdU-incorporating), apoptosis-prone cells [24, 55]. BRG-HIS mice also tend to show the accumulation in peripheral lymphoid organs of $CD4^+CD8^+$ double-positive T cells, which may correspond to a population of human activated T cells usually observed in inflammatory conditions [24, 62].

The major difference in reconstitution of human hematopoietic cells between NOD-based and BRG-HIS mice is likely to be caused by mouse phagocytic activity against human xenografts in the BRG-HIS mouse model [19, 21, 24]. Depletion of mouse phagocytes using clodronate-containing liposomes has been shown to enhance human hematopoietic cell accumulation in various HIS mouse settings [4, 63]. Further evidence for a major role of phagocytic cells in limiting human cell engraftment in BRG mice comes from analysis of the phagocytosis-inhibiting receptor SIRPα. CD47, the ligand for SIRPα, is broadly expressed on hematopoietic and nonhematopoietic cells [64, 65]. Mouse phagocytes in the BRG mice are unable to integrate signals from human CD47, whereas in NOD mice these phagocytes express a human CD47-compatible allele of the phagocyte-inhibiting SIRPα receptor [19].

Full inactivation of antihuman phagocytic activity through optimal hCD47/$SIRP\alpha^{NOD}$ interactions thus explains the relatively high degree of human hematopoietic cell accumulation in NSG/NOG mice. We and others have demonstrated that a similar level of human cell accumulation can be obtained in the BRG recipient mice, either by transgenic expression of human SIRPα [21], enforced expression of mouse CD47 DNA into hHSPC prior to transplantation, or by using congenic BRG $SIRP\alpha^{NOD}$ (BRGS) mice [24]. We have shown that human T cells, NK cells, and hHSPC are particularly sensitive to phagocyte-mediated cell removal and that compatible CD47/SIRPα interactions leads to the selective accumulation of naive, resting T cells in the BRG background and severe reduction of the frequency of activated, inflammation-associated $CD4^+CD8^+$ T cells in secondary lymphoid organs [24]. Of note, similar genetic engineering or congenic approaches also allow for efficient human hematopoietic reconstitution in the "nonpermissive" C57BL/6 mouse background [24, 66].

Both NSG/NOG-HIS mice and BLT mice share a compatible CD47/SIRPα mediated tolerance of mouse phagocytes, and the difference between these two models in terms of peripheral human T cell accumulation should be caused by another feature. Considering the size of the thymopoiesis-supporting organ in BLT mice, it is likely that human mature T cell export to peripheral organs is higher in these mice. It is estimated that ~ 1–2 % of thymocytes are exported to peripheral lymphoid organs each day [67–69], and that reduced thymopoiesis—as seen for instance in BRG-HIS and NSG/NOG-HIS mice—can only be partially compensated by an enhancement of thymus cell export [70]. Furthermore, the human thymic organoid in BLT mice is a nonnegligible source of human IL-7, a central factor in the development and peripheral maintenance of naive T cells that is expressed by bone marrow stromal cells and thymus epithelial cells, as well as various cell types in secondary and tertiary lymphoid tissues [71–73]. Mouse IL-7 interacts with the human IL-7 receptor, but less efficiently than human IL-7. Therefore mouse IL-7 does not fully compensate for human IL-7 absence in HIS mice, and human IL-7 supplementation in humanized mice indeed potentiates human T-cell development and/or peripheral survival [6, 14, 47, 74].

10.6 Human Hematopoietic Cell Colonization of Lymph Nodes in HIS Mice

Human hematopoietic cell reconstitution of mouse lymph nodes (LN) is described as particularly limited in several HIS mouse models making use of the IL-2Rγ_c deficiency. This feature might be due to the inability of human hematopoietic cells (in particular human T cells) to migrate to these tissues, or to defective LN microarchitecture characteristics impairing entry/retention of hematopoietic cells. It is well documented that LN organogenesis requires signaling via the IL-7 receptor—composed of IL-7Rα and IL-2Rγ_c chains—and IL-2Rγ_c-deficient hosts therefore lack LN, with the notable exception of mesenteric LN. Deficiency in IL-7 signaling in IL-2Rγ_c-deficient mice consequently leads to defective generation of bone marrow-derived $CD4^+CD3^-IL\text{-}7R\alpha^+$ lymphoid tissue inducer (LTi) cells [75]. When coinjected to NOD.*scid* BLT mice—i.e., a HIS mouse model without LN organogenesis defect—human and mouse $CD4^+$ T cells can colonize mesenteric LN with similar efficiencies, whereas entry of human $CD4^+$ T cells to peripheral LN is 50 % less efficient [76]. A less efficient entry of human hematopoietic cells into mouse peripheral LN combined to specific microenvironment architecture defects in IL-2Rγ_c-deficient immuno-deficient mouse strains might thus explain the relatively poor accumulation of human immune cells in such tissues. Chemokines promoting recruitment of hematopoietic cells to LN—such as CCL19, CCL20, or CCL21—are mostly expressed by nonhematopoietic tissues and show limited sequence identity (65–75 %) between mouse and human [10, 77]. Such interspecies incompatibilities may limit the efficacy of human hematopoietic cell chemo-attraction to mouse peripheral LN.

Interestingly, studies reporting improved LN colonization by human hematopoietic cells are obtained in HIS mouse models generated with optimized humanization protocols. For instance, more systematic peripheral lymph node reconstitution as been described in BRG-HIS mice cotransplanted with UCB $CD34^+$ cells and autologous $CD2^-CD3^-CD34^-$ "support cells" [46], or in BRGS-HIS mice generated with FL $CD34^+CD38^-$ hHSPC [24]. As the most up-to-date HIS mouse models allow for enhanced accumulation of human hematopoietic cells, the potential emergence of discrete human hematopoiesis-derived subsets, such as human LTi-like cells, is also reported [24, 78]. It is not clear yet whether such human cells could impact on mouse LN organogenesis and/or organization over time, an interesting issue that deserves further investigations. Of note, BRG-HIS mouse LN reconstitution was recently described as a long-term, T-cell-dependent phenomenon, since LN containing only human B cells were never observed in such mice [46]. Overall, it can therefore not be excluded that long-term cross-talk between human hematopoietic cells and mouse stromal/epithelial components leads to beneficial features on LN tissue microarchitecture—a feature that may now be possible to explore in the latest generation of HIS mice.

10.7 Human T-Cell Repertoire and Selection in HIS Mice

The analysis of the T-cell repertoire available in various HIS mouse models has been performed, either by flow cytometry for the relative representation of the various TCR-Vβ families or by TCR-Vβ CDR3 length monitoring. CDR3 length analysis provides a fair indication of TCR-Vβ repertoire diversity. Overall, TCR-Vβ repertoire in the thymus or peripheral lymphoid organs of BLT or NSG/NOG mice was indistinguishable from the repertoire of control, human PBMCs, whereas BRG-HIS mice exhibited a more restricted, oligoclonal repertoire [11, 23, 30, 48, 51, 57, 79–81]. This observation was consistent with the notion that the NOD-based HIS mouse models are more permissive to the accumulation of human hematopoietic cells than the BRG strain, as discussed earlier. Furthermore, oligoclonality of the TCR-Vβ repertoire in BRG-HIS mice is probably further reinforced by the fact that human T cells are cycling in an extensive manner in this model [55], a feature that is corrected in presence of compatible CD47/SIRPα interactions, like in BRGS-HIS mice [24].

The exact nature of MHC restricting elements for HIS mouse human T cells remains a matter of debate. It is usually considered that positive selection of developing thymocytes is mostly mediated by cortical thymic epithelial cells. Still, there are experimental conditions in which alternative positive selection pathways might exist, based on interactions with hematopoiesis-derived cells. For instance, *nu/nu* H-2^b mice receiving a H-2^k thymus graft or a rat thymus xenograft mostly mount a H-2^b-restricted anti-LCMV cytotoxic T cell response [82]. Furthermore, *nu/nu* H-2^b mice receiving a H-2^k thymus graft and partially reconstituted with T cell incompetent (Rag1$^{-/-}$) H-2^k bone marrow progenitor cells were able to mount T cell responses restricted to both host and donor MHC molecules [82].

These results tend to argue in favor of flexibility during the process of T-cell selection, and the mouse/human chimeric composition of the hematopoietic compartment in HIS mice should deliver a mixed H-2/HLA restriction of the human T-cell repertoire. Detection of H-2 vs. HLA-restricted antigen-specific T-cell responses further depends on other aspects, such as the niche in which the antigen is presented (e.g., specific cell subsets supporting the replication of a live pathogen) and the technical tools used to detect a T-cell response (e.g., tetramers, peptide pools), which might be not ideal in some specific HIS mouse situations. HLA-restricted, antigen-specific human CD8$^+$ T-cells responses can be detected in BLT or HLA-expressing HIS mice infected with human-specific, lymphotropic viruses such as HIV, EBV, or dengue virus [31, 83–86]. It remains unclear though whether the HLA molecules expressed by the mouse recipients should match with the HLA haplotype of the hHSPC donor cells for maximized efficacy.

Last, the question of T cell negative selection in HIS mice also remains particularly elusive. In theory, deletion of human T cells reactive to mouse tissue antigens should be occurring in the mouse thymus, based on interactions with AIRE-expressing medullary thymic epithelial cells. Such a feature should not be observed in the human thymic organoid of BLT mice, which would by definition only permit negative selection based on expression of human genes, therefore fitting with the relative dichotomy between BLT and BRG/NSG/NOG HIS mouse models as far as GVHD incidence is concerned [34–36]. Still, functional tolerance could also be imposed through the generation of human regulatory T cells, for instance in the thymus via the interactions with mouse and/or human DC [36, 87, 88], but this point remains to be clarified.

10.8 Improving Human T-Cell Biology in HIS Mice

All current HIS mouse models share similar limitations that provide leads for the identification of the next HIS mouse optimization increment. In steady state conditions, the number of human hematopoietic cells present in central and peripheral organs is relatively low, lymph node organogenesis is limited but appears to be connected to T-cell reconstitution levels. The concentration of plasma immunoglobulins is low, very variable between individual HIS mice and usually represents ~1–10% of normal mouse and human levels. Last, immune responses in HIS mice are most of the time faint, with high-interindividual variability. In particular, B-cell responses are characterized by weak immunoglobulin isotype switch to IgG and low frequency of somatic hypermutations. High inter- and intramodel variability could in part be due to the very diverse reconstitution protocols used, when considering hHSPC origin, potential coinjection of support cell population and cytokine-based precultures, as well as housing conditions for mouse colonies.

In this context, a more careful identification and description of the various T cell subsets present in HIS mice would be of importance to determine for instance to what extent T_{H17} [89] or follicular helper T (T_{FH}) cells are actually present and functional in humanized mice. The T_{FH} subset is of particular interest, as it plays a critical role in the triggering of the germinal center formation, antigen-specific B-cell activation and antigen-specific maintenance [90, 91]. The development of T_{FH} cells is strictly dependent on the expression of the inducible T-cell costimulator (ICOS) and the cytokine IL-21, which expression pattern might be suboptimal in HIS mice. Fine-tuned supplementation of HIS mice with ICOS and IL-21 might be a way to ensure the proper generation of T_{FH} cells around immunization procedures. The T_{FH} cells express the B cell follicle homing receptor CXCR5, and it is also a possibility that the appropriate chemokine ligand CXCL13 (in theory of hematopoietic cell origin) is not properly expressed in HIS mice.

Mice and humans hematopoietic systems differ significantly and exhibit profound interspecies incompatibilities that directly impact on the construction of HIS mouse models [6, 10, 92]. In some cases, specific murine mouse products may limit per se the mouse colonization by human cells and therefore have to be invalidated. For instance, deletion of the mouse Flt3/Flk2 receptor strongly reduces the number of murine DC, which cannot outcompete their human counterparts anymore, and renders human DC selectively reactive to exogenous treatment with the cross-reactive ligand FLT3-L [93].

Supplementation of HIS mice with human gene products might also be particularly helpful to optimize human T-cell development and/or function. Several supplementation strategies have been tested already for a variety of human cytokines, such as using exogenous products for injections [14, 47, 54, 94, 95], genetic engineering of the transplanted hHSPC [47], transgenic expression of human genes [96], replacement of mouse genes by their human equivalent [97–99], or hydrodynamic delivery of human DNA-encoding plasmids [74, 100]. All these approaches are valuable to provide support to human T cells during their development or triggering of an immune response, and could be integrated into vaccination protocols as immuno-stimulatory strategies to improve T-B-DC interplay. This would for instance be the case with human IL-15 agonist [80], human IL-12 [101], or human GM-CSF/IL-4 [102]. On a more general level, optimization of immunization procedures to potentiate cytotoxic and/or helper T-cell responses is required in the field of humanized mice, with a nonexhaustive list of parameters that might be addressed, such as antigen formulation and dose, immunization site, adjuvant type, sequence of successive immunizations, or time frames between immunizations.

In parallel, efforts are required to further improve humanization of the nonhematopoietic compartments in the mouse recipients used to generate HIS mice. This could be achieved via genetic approaches (which may be particularly laborious) or by exogenous supplementation of specific cell subsets using suspensions of purified human cells. Apart from the aforementioned LTi cells, lymphoid microarchitecture limitations might also explain the relatively defective germinal center reaction that can be observed in HIS mice. At least two nonhematopoietic cell populations might have to be supplemented in HIS mice to permit the optimal initiation of human immune responses. First, it is very likely that the nonhematopoietic follicular dendritic cells (FDC) of HIS mice are of murine origin and might therefore be unable to efficiently crosstalk with human leukocytes. Considering the critical role of FDC in the organization of lymphoid microarchitecture and support to B cell memory [103, 104], it seems rather critical to establish a human FDC compartment in HIS mice. Of note, the murine FDC precursor was very recently identified as a perivascular progenitor cell expressing the platelet-derived growth factor receptor β [105], and we expect that the identification and inoculation of human FDC progenitor cells into HIS mice will be a valuable approach to obtain improved human immune cell functions. Last, the fibroblastic reticular cell (FRC) network, which is known to be critical for the migration and maintenance of T cells in lymphoid organs, is probably heavily disturbed in HIS mice, as observed in lymphopenic HIV-infected patients [106]. Treatment of the animals with lymphotoxin-β (LTβ), a key factor for FRC network maintenance, might represent a valuable strategy to positively impact on lymphoid micro-architecture in HIS mice. Alternatively, the in vitro generation of FRC from mesenchymal stem cells [107] could represent a valuable source for in vivo supplementation of HIS mice.

10.9 Concluding Remarks

Development and function of the human hematopoietic xenograft, in particular human T cells, in HIS mouse models have been strongly improved over the past decade. It is now possible to detect antigen-specific T- and B-cell responses using a variety of assays. Nevertheless, the frequency, intensity, and quality of these responses are still very weak when compared to responses in humans and supplementary optimization increments are required to obtain immune responses that are systematic, robust, and accurate. Considering the recent progress made in the field, one can be optimistic about what the coming years will deliver in further refinements and improvements of HIS mouse models to a point that they can serve as a robust and useful preclinical platform to address human unmet medical needs.

Acknowledgments We thank Dr. Mireille Centlivre for valuable suggestions and critical reading of the manuscript.

References

1. McCune JM, Namikawa R, Kaneshima H, Shultz LD, Lieberman M, Weissman IL. The SCID-hu mouse: murine model for the analysis of human hematolymphoid differentiation and function. Science. 1988;241(4873):1632–9.
2. Kamel-Reid S, Dick JE. Engraftment of immune-deficient mice with human hematopoietic stem cells. Science. 1988;242(4886):1706–9.

3. Mosier DE, Gulizia RJ, Baird SM, Wilson DB. Transfer of a functional human immune system to mice with severe combined immunodeficiency. Nature. 1988;335(6187):256–9.
4. Legrand N, Weijer K, Spits H. Experimental models to study development and function of the human immune system in vivo. J Immunol. 2006;176(4):2053–8.
5. Legrand N, Ploss A, Balling R, Becker PD, Borsotti C, Brezillon N, et al. Humanized mice for modeling human infectious disease: challenges, progress, and outlook. Cell Host Microbe. 2009;6(1):5–9.
6. Manz MG. Human-hemato-lymphoid-system mice: opportunities and challenges. Immunity. 2007;26(5):537–41.
7. Manz MG, Di Santo JP. Renaissance for mouse models of human hematopoiesis and immunobiology. Nat Immunol. 2009;10(10):1039–42.
8. Shultz LD, Brehm MA, Garcia-Martinez JV, Greiner DL. Humanized mice for immune system investigation: progress, promise and challenges. Nature reviews Immunology. 2012;12(11):786–98. (Epub 2012/10/13).
9. Ito R, Takahashi T, Katano I, Ito M. Current advances in humanized mouse models. Cell Mol Immunol. 2012;9(3):208–14. (Epub 2012/02/14).
10. Rongvaux A, Takizawa H, Strowig T, Willinger T, Eynon EE, Flavell RA, et al. Human hemato-lymphoid system mice: current use and future potential for medicine. Annu Rev Immunol. 2013;31:635–74. (Epub 2013/01/22).
11. Hiramatsu H, Nishikomori R, Heike T, Ito M, Kobayashi K, Katamura K, et al. Complete reconstitution of human lymphocytes from cord blood $CD34^+$ cells using the NOD/SCID/gammacnull mice model. Blood. 2003;102(3):873–80.
12. Ito M, Hiramatsu H, Kobayashi K, Suzue K, Kawahata M, Hioki K, et al. NOD/SCID/gamma(c)(null) mouse: an excellent recipient mouse model for engraftment of human cells. Blood. 2002;100(9):3175–82.
13. Yahata T, Ando K, Nakamura Y, Ueyama Y, Shimamura K, Tamaoki N, et al. Functional human T lymphocyte development from cord blood $CD34^+$ cells in nonobese diabetic/Shi-scid, IL-2 receptor gamma null mice. J Immunol. 2002;169(1):204–9.
14. Shultz LD, Lyons BL, Burzenski LM, Gott B, Chen X, Chaleff S, et al. Human lymphoid and myeloid cell development in NOD/LtSz-scid IL2R gamma null mice engrafted with mobilized human hemopoietic stem cells. J Immunol. 2005;174(10):6477–89.
15. Ishikawa F, Yasukawa M, Lyons B, Yoshida S, Miyamoto T, Yoshimoto G, et al. Development of functional human blood and immune systems in NOD/SCID/IL2 receptor {gamma} chainnull mice. Blood. 2005;106(5):1565–73.
16. Notta F, Doulatov S, Dick JE. Engraftment of human hematopoietic stem cells is more efficient in female NOD/SCID/IL-2Rgc-null recipients. Blood. 2010;115(18):3704–7. (Epub 2010/03/09).
17. Majeti R, Park CY, Weissman IL. Identification of a hierarchy of multipotent hematopoietic progenitors in human cord blood. Cell Stem Cell. 2007;1(6):635–45.
18. Shultz LD, Schweitzer PA, Christianson SW, Gott B, Schweitzer IB, Tennent B, et al. Multiple defects in innate and adaptive immunologic function in NOD/LtSz-scid mice. J Immunol. 1995;154(1):180–91.
19. Takenaka K, Prasolava TK, Wang JC, Mortin-Toth SM, Khalouei S, Gan OI, et al. Polymorphism in Sirpa modulates engraftment of human hematopoietic stem cells. Nat Immunol. 2007;8(12):1313–23.
20. Ito R, Katano I, Ida-Tanaka M, Kamisako T, Kawai K, Suemizu H, et al. Efficient xenoengraftment in severe immunodeficient NOD/Shi-scid IL2rgammanull mice is attributed to a lack of $CD11c^+B220^+CD122^+$ cells. J Immunol. 2012;189(9):4313–20. (Epub 2012/09/29).
21. Strowig T, Rongvaux A, Rathinam C, Takizawa H, Borsotti C, Philbrick W, et al. Transgenic expression of human signal regulatory protein alpha in $Rag2^{-/-}$gamma(c)$^{-/-}$ mice improves engraftment of human hematopoietic cells in humanized mice. Proc Natl Acad Sci U S A. 2011;108(32):13218–23. (Epub 2011/07/27).
22. Gimeno R, Weijer K, Voordouw A, Uittenbogaart CH, Legrand N, Alves NL, et al. Monitoring the effect of gene silencing by RNA interference in human $CD34^+$ cells injected into newborn $RAG2^{-/-}$ gammac$^{-/-}$ mice: functional inactivation of p53 in developing T cells. Blood. 2004;104(13):3886–93.

23. Traggiai E, Chicha L, Mazzucchelli L, Bronz L, Piffaretti JC, Lanzavecchia A, et al. Development of a human adaptive immune system in cord blood cell-transplanted mice. Science. 2004;304(5667):104–7.
24. Legrand N, Huntington ND, Nagasawa M, Bakker AQ, Schotte R, Strick-Marchand H, et al. Functional CD47/signal regulatory protein alpha (SIRP(alpha)) interaction is required for optimal human T- and natural killer- (NK) cell homeostasis in vivo. Proc Natl Acad Sci U S A. 2011;108(32):13224–9. (Epub 2011/07/27).
25. Mazurier F, Fontanellas A, Salesse S, Taine L, Landriau S, Moreau-Gaudry F, et al. A novel immunodeficient mouse model–RAG2 x common cytokine receptor gamma chain double mutants—requiring exogenous cytokine administration for human hematopoietic stem cell engraftment. J Interferon Cytokine Res. 1999;19(5):533–41.
26. Bosma GC, Custer RP, Bosma MJ. A severe combined immunodeficiency mutation in the mouse. Nature. 1983;301(5900):527–30.
27. Bosma GC, Fried M, Custer RP, Carroll A, Gibson DM, Bosma MJ. Evidence of functional lymphocytes in some (leaky) scid mice. J Exp Med. 1988;167(3):1016–33.
28. Bosma MJ. B and T cell leakiness in the scid mouse mutant. Immunodefic Rev. 1992;3(4):261–76. (Epub 1992/01/01).
29. Krowka JF, Sarin S, Namikawa R, McCune JM, Kaneshima H. Human T cells in the SCID-hu mouse are phenotypically normal and functionally competent. J Immunol. 1991;146(11):3751–6.
30. Melkus MW, Estes JD, Padgett-Thomas A, Gatlin J, Denton PW, Othieno FA, et al. Humanized mice mount specific adaptive and innate immune responses to EBV and TSST-1. Nat Med. 2006;12(11):1316–22.
31. Brainard DM, Seung E, Frahm N, Cariappa A, Bailey CC, Hart WK, et al. Induction of robust cellular and humoral virus-specific adaptive immune responses in human immunodeficiency virus-infected humanized BLT mice. J Virol. 2009;83(14):7305–21.
32. Lan P, Tonomura N, Shimizu A, Wang S, Yang YG. Reconstitution of a functional human immune system in immunodeficient mice through combined human fetal thymus/liver and $CD34^+$ cell transplantation. Blood. 2006;108(2):487–92.
33. Denton PW, Garcia JV. Mucosal HIV-1 transmission and prevention strategies in BLT humanized mice. Trends Microbiol. 2012;20(6):268–74. (Epub 2012/04/17).
34. Greenblatt MB, Vbranac V, Tivey T, Tsang K, Tager AM, Aliprantis AO. Graft versus host disease in the bone marrow, liver and thymus humanized mouse model. PloS One. 2012;7(9):e44664. (Epub 2012/09/08).
35. Covassin L, Jangalwe S, Jouvet N, Laning J, Burzenski L, Shultz LD, et al. Human immune system development and survival of NOD-scid IL2rgamma (NSG) mice engrafted with human thymus and autologous hematopoietic stem cells. Clin Exp Immunol. 2013; 174(3):372–88. (Epub 2013/07/23).
36. Lockridge JL, Zhou Y, Becker YA, Ma S, Kenney SC, Hematti P, et al. Mice engrafted with human fetal thymic tissue and hematopoietic stem cells develop pathology resembling chronic graft-versus-host disease. Biol Blood Marrow Transplant. 2013;19(9):1310–22. (Epub 2013/06/29).
37. Scheeren FA, Nagasawa M, Weijer K, Cupedo T, Kirberg J, Legrand N, et al. T cell-independent development and induction of somatic hypermutation in human $IgM^+IgD^+CD27^+$ B cells. J Exp Med. 2008;205(9):2033–42.
38. Dik WA, Pike-Overzet K, Weerkamp F, de Ridder D, de Haas EF, Baert MR, et al. New insights on human T cell development by quantitative T cell receptor gene rearrangement studies and gene expression profiling. J Exp Med. 2005;201(11):1715–23.
39. Res P, Martinez-Caceres E, Cristina Jaleco A, Staal F, Noteboom E, Weijer K, et al. $CD34^+$CD38dim cells in the human thymus can differentiate into T, natural killer, and dendritic cells but are distinct from pluripotent stem cells. Blood. 1996;87(12):5196–206.
40. Blom B, Spits H. Development of human lymphoid cells. Annu Rev Immunol. 2006;24:287–320.

41. Blom B, Verschuren MC, Heemskerk MH, Bakker AQ, van Gastel-Mol EJ, Wolvers-Tettero IL, et al. TCR gene rearrangements and expression of the pre-T cell receptor complex during human T-cell differentiation. Blood. 1999;93(9):3033–43.
42. Haddad R, Guardiola P, Izac B, Thibault C, Radich J, Delezoide AL, et al. Molecular characterization of early human T/NK and B-lymphoid progenitor cells in umbilical cord blood. Blood. 2004;104(13):3918–26.
43. Hao QL, Zhu J, Price MA, Payne KJ, Barsky LW, Crooks GM. Identification of a novel, human multilymphoid progenitor in cord blood. Blood. 2001;97(12):3683–90. (Epub 2001/06/05).
44. Haddad R, Guimiot F, Six E, Jourquin F, Setterblad N, Kahn E, et al. Dynamics of thymus-colonizing cells during human development. Immunity. 2006;24(2):217–30.
45. Lepus CM, Gibson TF, Gerber SA, Kawikova I, Szczepanik M, Hossain J, et al. Comparison of human fetal liver, umbilical cord blood, and adult blood hematopoietic stem cell engraftment in NOD-scid/gammac$^{-/-}$, Balb/c-Rag1$^{-/-}$gammac$^{-/-}$, and C.B-17-scid/bg immunodeficient mice. Hum Immunol. 2009;70(10):790–802.
46. Lang J, Kelly M, Freed BM, McCarter MD, Kedl RM, Torres RM, et al. Studies of lymphocyte reconstitution in a humanized mouse model reveal a requirement of T cells for human B cell maturation. J Immunol. 2013;190(5):2090–101. (Epub 2013/01/22).
47. van Lent AU, Dontje W, Nagasawa M, Siamari R, Bakker AQ, Pouw SM, et al. IL-7enhances thymic human T cell development in "human immune system" Rag2$^{-/-}$IL-2Rgammac$^{-/-}$ mice without affecting peripheral T cell homeostasis. J Immunol. 2009;183(12):7645–55.
48. Denton PW, Nochi T, Lim A, Krisko JF, Martinez-Torres F, Choudhary SK, et al. IL-2 receptor gamma-chain molecule is critical for intestinal T-cell reconstitution in humanized mice. Mucosal Immunol. 2012;5(5):555–66. (Epub 2012/05/10).
49. Sun Z, Denton PW, Estes JD, Othieno FA, Wei BL, Wege AK, et al. Intrarectal transmission, systemic infection, and CD4^{+} T cell depletion in humanized mice infected with HIV-1. J Exp Med. 2007;204(4):705–14.
50. Joo SY, Chung YS, Choi B, Kim M, Kim JH, Jun TG, et al. Systemic human T cell developmental processes in humanized mice cotransplanted with human fetal thymus/liver tissue and hematopoietic stem cells. Transplantation. 2012;94(11):1095–102. (Epub 2012/12/12).
51. Huntington ND, Alves NL, Legrand N, Lim A, Strick-Marchand H, Plet A, et al. Autonomous and extrinsic regulation of thymopoiesis inhuman immune system (HIS) mice. Eur J Immunol. 2011;41(10):2883–93. (Epub 2011/07/09).
52. Spits H. Development of alphabeta T cells in the human thymus. Nat Rev Immunol. 2002;2(10):760–72.
53. Pek EA, Chan T, Reid S, Ashkar AA. Characterization and IL-15 dependence of NK cells in humanized mice. Immunobiology. 2011;216(1–2):218–24. (Epub 2010/07/16).
54. Huntington ND, Legrand N, Alves NL, Jaron B, Weijer K, Plet A, et al. IL-15 trans-presentation promotes human NK cell development and differentiation in vivo. J Exp Med. 2009;206:25–34.
55. Legrand N, Cupedo T, van Lent AU, Ebeli MJ, Weijer K, Hanke T, et al. Transient accumulation of human mature thymocytes and regulatory T cells with CD28 superagonist in "human immune system" Rag2($^{-/-}$)gammac($^{-/-}$) mice. Blood. 2006;108(1):238–45.
56. Onoe T, Kalscheuer H, Danzl N, Chittenden M, Zhao G, Yang YG, et al. Human natural regulatory T cell development, suppressive function, and postthymic maturation in a humanized mouse model. J Immunol. 2011;187(7):3895–903. (Epub 2011/08/31).
57. Kalscheuer H, Danzl N, Onoe T, Faust T, Winchester R, Goland R, et al. A model for personalized in vivo analysis of human immune responsiveness. Sci Transl Med. 2012;4(125):125ra30. (Epub 2012/03/17).
58. Billerbeck E, Barry WT, Mu K, Dorner M, Rice CM, Ploss A. Development of human CD4^{+}FoxP3^{+} regulatory T cells in human stem cell factor-, granulocyte-macrophage colony-stimulating factor-, and interleukin-3-expressing NOD-SCID IL2Rgamma(null) humanized mice. Blood. 2011;117(11):3076–86. (Epub 2011/01/22).

59. Jiang Q, Zhang L, Wang R, Jeffrey J, Washburn ML, Brouwer D, et al. FoxP3^{+}CD4^{+} regulatory T cells play an important role in acute HIV-1 infection in humanized Rag2$^{-/-}$gammaC$^{-/-}$ mice in vivo. Blood. 2008;112(7):2858–68.
60. Becker PD, Legrand N, van Geelen CM, Noerder M, Huntington ND, Lim A, et al. Generation of human antigen-specific monoclonal IgM antibodies using vaccinated "human immune system" mice. PloS one. 2010;5(10).
61. Vuyyuru R, Patton J, Manser T. Human immune system mice: current potential and limitations for translational research on human antibody responses. Immunol Res. 2011;51(2–3):257–66. (Epub 2011/11/01).
62. Parel Y, Chizzolini C. CD4^{+} CD8^{+} double positive (DP) T cells in health and disease. Autoimmun Rev. 2004;3(3):215–20.
63. Rozemuller H, Knaan-Shanzer S, Hagenbeek A, van Bloois L, Storm G, Martens AC. Enhanced engraftment of human cells in RAG2/gammac double-knockout mice after treatment with CL2MDP liposomes. Exp Hematol. 2004;32(11):1118–25.
64. Brown EJ, Frazier WA. Integrin-associated protein (CD47) and its ligands. Trends Cell Biol. 2001;11(3):130–5.
65. Sarfati M, Fortin G, Raymond M, Susin S. CD47 in the immune response: role of thrombospondin and SIRP-alpha reverse signaling. Curr Drug Targets. 2008;9(10):842–50.
66. Yamauchi T, Takenaka K, Urata S, Shima T, Kikushige Y, Tokuyama T, et al. Polymorphic Sirpa is the genetic determinant for NOD-based mouse lines to achieve efficient human cell engraftment. Blood. 2013;121(8):1316–25. (Epub 2013/01/08).
67. Berzins SP, Godfrey DI, Miller JF, Boyd RL. A central role for thymic emigrants in peripheral T cell homeostasis. Proc Natl Acad Sci U S A. 1999;96(17):9787–91.
68. Gabor MJ, Scollay R, Godfrey DI. Thymic T cell export is not influenced by the peripheral T cell pool. Eur J Immunol. 1997;27(11):2986–93. (Epub 1997/12/12).
69. Scollay RG, Butcher EC, Weissman IL. Thymus cell migration. Quantitative aspects of cellular traffic from the thymus to the periphery in mice. Eur J Immunol. 1980;10(3):210–8. (Epub 1980/03/01).
70. Almeida AR, Borghans JA, Freitas AA. T cell homeostasis: thymus regeneration and peripheral T cell restoration in mice with a reduced fraction of competent precursors. J Exp Med. 2001;194(5):591–9.
71. Alves NL, Huntington ND, Mention JJ, Richard-Le Goff O, Di Santo JP. Cutting edge: a thymocyte-thymic epithelial cell cross-talk dynamically regulates intrathymic IL-7 expression in vivo. J Immunol. 2010;184(11):5949–53.
72. Mazzucchelli R, Durum SK. Interleukin-7 receptor expression: intelligent design. Nat Rev Immunology. 2007;7(2):144–54.
73. Huang HY, Luther SA. Expression and function of interleukin-7 in secondary and tertiary lymphoid organs. Semin Immunol. 2012;24(3):175–89. (Epub 2012/03/27).
74. O'Connell RM, Balazs AB, Rao DS, Kivork C, Yang L, Baltimore D. Lentiviral vector delivery of human interleukin-7 (hIL-7) to human immune system (HIS) mice expands T lymphocyte populations. PloS One. 2010;5(8):e12009.
75. Cupedo T, Mebius RE. Cellular interactions in lymph node development. J Immunol. 2005;174(1):21–5.
76. Murooka TT, Deruaz M, Marangoni F, Vrbanac VD, Seung E, von Andrian UH, et al. HIV-infected T cells are migratory vehicles for viral dissemination. Nature. 2012;490(7419):283–7. (Epub 2012/08/03).
77. Bromley SK, Mempel TR, Luster AD. Orchestrating the orchestrators: chemokines in control of T cell traffic. Nat Immunol. 2008;9(9):970–80. (Epub 2008/08/20).
78. Bernink JH, Peters CP, Munneke M, Te Velde AA, Meijer SL, Weijer K, et al. Human type 1 innate lymphoid cells accumulate in inflamed mucosal tissues. Nat Immunol. 2013;14(3):221–9. (Epub 2013/01/22).
79. Pham HP, Manuel M, Petit N, Klatzmann D, Cohen-Kaminsky S, Six A, et al. Half of the T-cell repertoire combinatorial diversity is genetically determined in humans and humanized mice. Eur J Immunol. 2012;42(3):760–70. (Epub 2011/11/23).

80. Huntington ND, Alves NL, Legrand N, Lim A, Strick-Marchand H, Mention JJ, et al. IL-15 transpresentation promotes both human T-cell reconstitution and T-cell-dependent antibody responses in vivo. Proc Natl Acad Sci U S A. 2011;108(15):6217–22. (Epub 2011/03/30).
81. Marodon G, Desjardins D, Mercey L, Baillou C, Parent P, Manuel M, et al. High diversity of the immune repertoire in humanized NOD.SCID.gamma c−/− mice. Eur J Immunol. 2009;39(8):2136–45. (Epub 2009/07/03).
82. Zinkernagel RM, Althage A. On the role of thymic epithelium vs. bone marrow-derived cells in repertoire selection of T cells. Proc Natl Acad Sci U S A. 1999;96(14):8092–7.
83. Sato Y, Nagata S, Takiguchi M. Effective elicitation of human effector CD8(+) T cells in HLA-B*51:01 transgenic humanized mice after infection with HIV-1. PloS One. 2012;7(8):e42776. (Epub 2012/08/11).
84. Strowig T, Gurer C, Ploss A, Liu YF, Arrey F, Sashihara J, et al. Priming of protective T cell responses against virus-induced tumors in mice with human immune system components. J Exp Med. 2009;206(6):1423–34.
85. Shultz LD, Saito Y, Najima Y, Tanaka S, Ochi T, Tomizawa M, et al. Generation of functional human T-cell subsets with HLA-restricted immune responses in HLA class I expressing NOD/SCID/IL2r gamma(null) humanized mice. Proc Natl Acad Sci U S A. 2010;107(29):13022–7.
86. Jaiswal S, Pearson T, Friberg H, Shultz LD, Greiner DL, Rothman AL, et al. Dengue virus infection and virus-specific HLA-A2 restricted immune responses in humanized NOD-scid IL2rgammanull mice. PloS One. 2009;4(10):e7251. (Epub 2009/10/06).
87. Jung KC, Park CG, Jeon YK, Park HJ, Ban YL, Min HS, et al. In situ induction of dendritic cell-based T cell tolerance in humanized mice and nonhuman primates. J Exp Med. 2011;208(12):2477–88. (Epub 2011/10/26).
88. Watanabe N, Wang YH, Lee HK, Ito T, Cao W, Liu YJ. Hassall's corpuscles instruct dendritic cells to induce $CD4^{+}CD25^{+}$ regulatory T cells in human thymus. Nature. 2005;436(7054):1181–5.
89. Rajesh D, Zhou Y, Jankowska-Gan E, Roenneburg DA, Dart ML, Torrealba J, et al. Th1 and Th17 immunocompetence in humanized NOD/SCID/IL2rgammanull mice. Hum Immunol. 2010;71(6):551–9. (Epub 2010/03/20).
90. King C. New insights into the differentiation and function of T follicular helper cells. Nature reviews Immunology. 2009;9(11):757–66.
91. Rodriguez-Bayona B, Ramos-Amaya A, Bernal J, Campos-Caro A, Brieva JA. Cutting edge: IL-21 derived from human follicular helper T cells acts as a survival factor for secondary lymphoid organ, but not for bone marrow, plasma cells. J Immunol. 2012;188(4):1578–81. (Epub 2012/01/18).
92. Mestas J, Hughes CC. Of mice and not men: differences between mouse and human immunology. J Immunol. 2004;172(5):2731–8.
93. Di Santo JP Mention JJ. Dendritic cell-boosted humanized immune system mice. PCT/US2010/029800. WO 2010/115115 A1. Patent. 2010.
94. Bjorkstrom NK, Riese P, Heuts F, Andersson S, Fauriat C, Ivarsson MA, et al. Expression patterns of NKG2A, KIR, and CD57 define a process of CD56dim NK-cell differentiation uncoupled from NK-cell education. Blood. 2010;116(19):3853–64.
95. Wils EJ, Rombouts EJ, van Mourik I, Spits H, Legrand N, Braakman E, et al. Stem cell factor consistently improves thymopoiesis after experimental transplantation of murine or human hematopoietic stem cells in immunodeficient mice. J Immunol. 2011;187(6):2974–81. (Epub 2011/08/24).
96. Takagi S, Saito Y, Hijikata A, Tanaka S, Watanabe T, Hasegawa T, et al. Membrane-bound human SCF/KL promotes in vivo human hematopoietic engraftment and myeloid differentiation. Blood. 2012;119(12):2768–77. (Epub 2012/01/27).
97. Willinger T, Rongvaux A, Strowig T, Manz MG, Flavell RA. Improving human hemato-lymphoid-system mice by cytokine knock-in gene replacement. Trends Immunol. 2011;32(7):321–7. (Epub 2011/06/24).

98. Willinger T, Rongvaux A, Takizawa H, Yancopoulos GD, Valenzuela DM, Murphy AJ, et al. Human IL-3/GM-CSF knock-in mice support human alveolar macrophage development and human immune responses in the lung. Proc Natl Acad Sci U S A. 2011;108(6):2390–5. (Epub 2011/01/26).
99. Rongvaux A, Willinger T, Takizawa H, Rathinam C, Auerbach W, Murphy AJ, et al. Human thrombopoietin knockin mice efficiently support human hematopoiesis in vivo. Proc Natl Acad Sci U S A. 2011;108(6):2378–83. (Epub 2011/01/26).
100. Chen Q, Khoury M, Chen J. Expression of human cytokines dramatically improves reconstitution of specific human-blood lineage cells in humanized mice. Proc Natl Acad Sci U S A. 2009;106(51):21783–8.
101. De Giovanni C, Nicoletti G, Landuzzi L, Romani F, Croci S, Palladini A, et al. Human responses against HER-2-positive cancer cells in human immune system-engrafted mice. Br J Cancer. 2012;107(8):1302–09. (Epub 2012/08/30).
102. Chen Q, He F, Kwang J, Chan JK, Chen J. GM-CSF and IL-4 stimulate antibody responses in humanized mice by promoting T, B, and dendritic cell maturation. J Immunol. 2012;189(11):5223–9. (Epub 2012/10/24).
103. Vinuesa CG, Linterman MA, Goodnow CC, Randall KL. T cells and follicular dendritic cells in germinal center B-cell formation and selection. Immunol Rev. 2010;237(1):72–89.
104. Cyster JG. B cell follicles and antigen encounters of the third kind. Nat Immunol. 2010;11(11):989–96. (Epub 2010/10/21).
105. Krautler NJ, Kana V, Kranich J, Tian Y, Perera D, Lemm D, et al. Follicular dendritic cells emerge from ubiquitous perivascular precursors. Cell. 2012;150(1):194–206. (Epub 2012/07/10).
106. Zeng M, Paiardini M, Engram JC, Beilman GJ, Chipman JG, Schacker TW, et al. Critical role for CD4 T cells in maintaining lymphoid tissue structure for immune cell homeostasis and reconstitution. Blood. 2012;120:1856–67. (Epub 2012/05/23).
107. Amc-Thomas P, Maby-El Hajjami H, Monvoisin C, Jean R, Monnier D, Caulet-Maugendre S, et al. Human mesenchymal stem cells isolated from bone marrow and lymphoid organs support tumor B-cell growth: role of stromal cells in follicular lymphoma pathogenesis. Blood. 2007;109(2):693–702. (Epub 2006/09/21).

Chapter 11
Thymic Education of Human T Cells and Regulatory T Cell Development in Humanized Mice

Hao Wei Li, Yong-Guang Yang and Megan Sykes

11.1 Overview of T Cell Selection and Maturation

The immune system of mammals is needed to fight infections while maintaining tolerance to the self. T cells play a central role in these processes. The development of T cells in the thymus endows them with the ability to recognize and mount immune responses to millions of pathogen-derived nonself-antigens while still being tolerant to "self" antigens. Mouse studies demonstrate that the thymus can perform three critical functions for development of a functional, self-tolerant T cell repertoire, namely positive selection, negative selection and generation of $CD4^{+}CD25^{+}Foxp3^{+}$ regulatory T cells (Tregs). T cell progenitors migrate to the thymus, where they develop into T cells. Absence of thymus, such as in nude mice and human patients lacking a functional FOXN1 gene, results in the absence of normal functional T cells [1]. Newly generated T cells undergo further post-thymic maturation to become fully functional in the periphery [2]. Intrathymic positive selection of T cells results in a repertoire that most efficiently recognizes exogenous peptide antigens presented by the same "self" MHC molecules. Engagement by T cell receptor (TCR) of naïve T cells of the same MHC/self-peptide complexes expressed on peripheral antigen-presenting cells (APCs), as those on which they are positively selected in the thymus, provides a critical survival signal [3]. These processes sustain a functional T cell pool. Negative selection and generation of Tregs ensure that the majority of strongly autoreactive T cells are purged and those that escape negative selection are kept in check by Tregs in the periphery [4]. Positive and negative

M. Sykes (✉)
Columbia Center for Translational Immunology, Columbia University College of Physicians and Surgeons, 650 West 168th Street, Black Building 1512, New York, NY 10032, USA
e-mail: Megan.Sykes@columbia.edu

H. W. Li · Y.-G. Yang
Columbia Center for Translational Immunology, Columbia University College of Physicians and Surgeons, New York, NY, USA

Y.-G. Yang
First Hospital of Jilin University, Changchun, China

L. Y. Poluektova et al. (eds.), *Humanized Mice for HIV Research*,
DOI 10.1007/978-1-4939-1655-9_11

selection of developing thymocytes is achieved through differential affinities of the TCR on the thymocytes with MHC-peptide complexes on the thymic epithelial cells and intrathymic APCs. Thymocytes expressing TCR with low and intermediate affinities for MHC-peptide complexes expressed on thymic epithelium are positively selected, enabling effective recognition of peptides presented by autologous MHC molecules in the periphery. Positively selected thymocytes expressing TCR with high affinities for self peptides presented by autologous MHC molecules undergo apoptosis during negative selection [5]. This process, also known as central tolerance, ensures the elimination of the strongly self-reactive T cells and is one of the key mechanisms for establishing self-tolerance.

11.2 Humanized Mouse Models Involving Human T Cell Development

Models allowing investigation of human T cell development, such as humanized mouse models, provide a unique opportunity to investigate the human immune system and T cell development. Initially, humanized mice were generated by injecting human mature immune cells, such as peripheral blood mononuclear cells (PBMCs) into severe combined immunodeficiency (SCID) mice [6]. Later, humanized mice were established by injecting human hematopoietic stem cells (HSCs) into irradiated neonatal or adult immunodeficient mice [7, 8]. Engraftment of human HSCs gives rise to human immune cells, including T cells, B cells, monocytes, etc. However, both types of humanized mouse models have limitations for the study of human T cell development. Grafting of human PBMCs includes mature T cells and thus does not allow the investigation of human T cell development. In addition, mature human T cells that are reactive to mouse xenoantigens can dominate and preclude good functional immunity [9, 10]. T cells in humanized mice generated by transplanting human HSCs develop in the mouse thymus rather than the human thymus. Thymopoiesis occurs at only a low level and the small number of T cells generated show functional defects [11, 12] (discussed in detail below). Therefore, these models are not optimal for the investigation of human T cell development.

Early work by McCune et al. demonstrated that implantation of human fetal thymic tissue under the kidney capsule of unconditioned adult SCID mice together with intravenous or intrathymic injection of fetal liver cells led to robust thymopoiesis [13]. However, very few human T cells populated the periphery and these were only detected transiently, from week 4 to 10 postimplantation. Implantation of fetal thymic tissue together with fetal liver led to more persistent T cells in peripheral blood [14]. Despite the presence of human IgG and human HLA Class I^+ cells in peripheral blood in some animals, human hematopoietic reconstitution of non-T cells seemed to be largely confined to the human graft microenvironment [13]. The failure to achieve systemic human hematopoiesis in this model may reflect the susceptibility of human hematopoietic cells to rapid destruction by mouse macrophages in CB.17-SCID mice [15]. Subsequently, we combined i.v. injection of human HSCs and implantation of human thymic tissue under the kidney capsule in nonobese diabetic SCID (NOD/SCID) mice and achieved much more robust human T cell and APC reconstitution. It is likely that the reduced phagocytosis of human cells in mice on the NOD background, reflecting the compatibility of the NOD SIRPα allele with human CD47 [16], explains the improved human hematopoietic reconstitution in our model over the original model described by McCune et al. In our model, humanized mice are generated by transplanting human fetal thymic/liver tissue under the kidney capsule and coinjecting fetal liver-derived $CD34^+$ HSCs from the same donors to irradiated adult NOD/SCID mice [17–20]. The model has been replicated and termed the "BLT" mouse [21]. The transplantation of autologous thymic tissue is an important improvement to the injection of human HSCs. In this model, not only major immune cell populations, such as B cells, dendritic cells and monocytes, but also large numbers of T cells appear in peripheral lymphoid tissues, which develop significant structure by 12 weeks post-transplant [17, 19]. Because human T cells in these mice develop in autologous thymic tissues, this model is highly relevant and valuable for the study of human T cell development and offers advantages compared to other models. First, in sharp contrast to humanized mice generated by grafting human HSCs only [7, 8, 11], thymopoiesis occurs at a high level, with thymocyte numbers ($> 10^8$) that are similar to or even greater than those in a normal, immunocompetent mouse thymus [20]. Large numbers of T cells migrate to the peripheral lymphoid tissues [17–19]. Importantly, robust immune functions are seen, such as spontaneous rejection of xenogeneic skin grafts and class-switched antibody responses following immunization [17, 18, 22]. Secondly, this model not only allows the investigation of human T cell development in physiological conditions, but also enables characterization of human T cell generation in pathological conditions, such as viral infection [21, 23, 24]. Thirdly, as this model allows the de novo development of human T cells in human thymic tissue, it can be utilized to investigate immune abnormalities arising from HSC-intrinsic factors.

We have modified this model by transplanting adult bone marrow-derived $CD34^+$ HSCs and partially HLA-matched, T-cell depleted allogeneic fetal thymic tissue to NOD/SCID/ Common γ-chain-deficient (NSG) mice [25]. These humanized mice, similar to those generated by grafting autologous fetal thymic tissue and fetal liver-derived $CD34^+$ cells, demonstrate polyclonal human T cell reconstitution and robust T cell function [25]. This humanized mouse model allows the study of T cell development and function in any individual human and is thus termed the "Personalized Immune" (PI) mouse model. With this PI mouse model, we are characterizing the development and function of T cells in type I diabetic patients, addressing the hypothesis that T cell abnormalities in these patients are intrinsically determined in their HSCs. In this chapter, we will summarize knowledge on human T cell development learned in humanized mouse models generated by transplantation of human fetal thymic tissue and fetal or adult $CD34^+$ HSCs.

11.3 Positive Selection

Interactions of T cell progenitors entering the thymus with MHC/peptide complexes expressed by thymic stromal cells, including thymic epithelial cells (TECs), trigger positive selection [26, 27]. Studies using transgenic mice exclusively expressing MHC I [28] or II [29] molecules on cortical TECs demonstrated that these cells are required for positive selection of CD8 or CD4 T cells, respectively. During positive selection, thymocytes expressing a TCR that is able to recognize an autologous MHC/self-peptide complex expressed on cortical TECs with at least low to medium affinity survive this process [5, 30]. This low level of self-reactivity is critical for the homeostasis [3] and immune functions of T cells in the periphery [30, 31].

In humanized mice generated by transplanting only human HSCs to immunodeficient mice, T cells develop in the mouse thymus, where human thymocytes are positively selected by mouse MHC molecules expressed on mouse TECs. The ability of murine MHC to positively select human T cells is not unexpected in view of previous work. Ample evidence from our porcine thymus transplant models shows that xenogeneic MHC molecules are able to mediate positive selection. A series of early studies demonstrated that mouse T cells [32–35] or human T cells [36, 37] developed robustly, with diverse repertoires, in porcine thymus. Importantly, porcine MHC molecules were shown to positively select mouse T cells with no contribution from the mouse MHC [35]. Nevertheless, mouse T cells selected by porcine MHC molecules could still respond to cognate antigen stimulation and were able to control infection in vivo, apparently due to cross-reactivity of the diverse repertoire combined with post-thymic selection for T cells that cross-reacted with mouse MHC [32]. The ability of xenogeneic MHC molecules to mediate positive selection is presumably due to the broad MHC cross-reactivity for TCR that is conserved between species.

Although the native thymi of these mice receiving human HSCs alone demonstrate relatively normal histologic structure due to colonization by developing human T cells, the number of thymocytes is low (about 1 million/thymus) compared to the much larger numbers in human thymic grafts (of the order of 100 fold more), and this difference is reflected in lymphoid tissues [11]. Multiple factors may contribute to the low level of thymopoiesis in this model. Failed thymic structural development, as TECs depend on interactions with thymocytes for their own development [38, 39], is partly circumvented by introducing human HSCs close to the time of birth [8]. Incompatibility of adhesion molecules and cytokines between mouse and human may lead to decreased homing of human thymocyte progenitors to and decreased survival of developing thymocytes in the mouse thymus, respectively. Indeed, supplementation of human cytokines, such as IL-7 [40] and IL-15 [41], increased human thymopoiesis. In addition, increased thymopoiesis (but still only of a few million thymocytes) was found in recipient mice expressing a single human HLA Class II molecule (HLA-DR4), showing that reduced ability of mouse MHC molecules to positively select human thymocytes plays a role [11]. However, these issues arising from xenoincompatibility cannot be generalized to all species, as robust human thymopoiesis occurs in pig thymic grafts (again, about 100-fold the level in the native mouse thymus). In fact, humanized mice grafted with fetal pig thymic tissues and human HSCs demonstrate thymopoiesis as robust as that seen in recipients of human thymic tissues and HSCs from the same donor [36, 37]. Thus, the mouse thymus is not an optimal microenvironment for the development of human T cells in humanized mice generated by grafting human HSCs alone. In contrast, in humanized mice generated by transplantation of autologous fetal thymic tissues and fetal liver-derived $CD34^+$ HSCs, T cell development occurs in autologous thymic tissue and the problems described above are thereby circumvented, allowing robust thymopoiesis. Large numbers of thymocytes ($>1\times10^8$/thymic graft) [20] and splenic T cells (10–20 million/spleen) can be recovered 10–12 weeks post transplantation. The developing thymocytes show a normal ratio of $CD4^+$, $CD8^+$, $CD4^+CD8^+$ and $CD4^-CD8^-$ subsets [17, 19].

11.4 Post-thymic Interactions Between T Cells and APCs

T cells that have survived positive and negative selection in the thymus mature and are exported to the periphery, where their interactions with APCs are critical for their survival, further maturation, function, and homeostasis. TCRs interact most efficiently in the periphery with the MHC/peptide complexes on which they are positively selected in the thymus [3, 42], providing critical signals for naïve T cell survival in the periphery [3]. In addition, self peptide-MHC complexes promote the responsiveness of peripheral T cells to their cognate antigens [31, 43] by acting as co-agonists to enhance their functional sensitivity [30, 44].

In mouse models, proliferation of T cells occurs when they are transferred to lymphopenic hosts, including rapid and slow proliferation known as lymphopenia-induced proliferation (LIP) [3, 45]. Recognition of commensal microorganisms causes the rapid proliferation [45], whereas the slower LIP is dependent on interactions between TCR and self MHC/peptide complexes and γc cytokines, such as IL-7 and IL-15 [3, 45]. Although insights into homeostatic and lymphopenia-driven proliferation of T cells have been obtained in mouse models, it is more difficult to investigate these events in humans, underscoring the importance of a suitable humanized mouse model to study homeostatic proliferation of human T cells. We have addressed this issue using humanized mice generated by grafting human fetal thymic tissue and HSCs [19]. Transfer of $CD45RO^-$ naïve T cells isolated from humanized mice, generated with human thymic tissue and intravenously-administered HSCs, to T cell-deficient humanized mice reconstituted with only the HSCs from the same donor, led to two forms of LIP, similar to results in $RAG^{-/-}$ mice receiving naïve mouse T cells [45]. Like mouse T cells, human naïve T cells undergoing rapid proliferation acquired a memory phenotype and production of IFN-γ, while T cells undergoing slow proliferation retained a naïve phenotype and did not produce IFN-γ. Importantly, the recovery of transferred T cells that had undergone LIP was correlated with the level of human APC chimerism in the secondary recipients and no cells were recovered from adoptive recipients lacking any human hematopoiesis. This result suggests that the LIP of transferred T cells was dependent on interactions with autologous APCs in the periphery, probably via interactions between TCR and autologous HLA molecules [19]. Thus, the survival of human T cells, like that of mouse T cells, depends on interactions with autologous MHC in the periphery. This requirement might in part explain the absence of T cells in the peripheral lymphoid tissues in the humanized mouse model established by McCune et al. [13]. Because no human APCs were found in the peripheral lymphoid tissues in that model, T cells egressing the thymus were unable to interact with peripheral APCs to receive a survival signal, resulting in a lack of accumulation of T cells. Consistently, the addition of intravenous CD34 cells to human fetal thymus and liver grafts in NOD/SCID mice led not only to the presence of human APCs in the periphery, but also to markedly increased T cell reconstitution [17, 18]. Moreover, only the T cells in animals that received CD34 cells intravenously were sufficiently functional to reject xenografts spontaneously [17], consistent with the need for tonic interactions with positive selecting MHC/peptide complexes in the periphery to maintain T cell function [30, 31, 43, 44].

Mouse studies demonstrate that T cells termed recent thymic emigrants, which have recently completed intrathymic development and been exported to the periphery, undergo further post-thymic maturation to gain full immune function. This process requires their entry into secondary lymphoid organs to interact with APCs, but may be independent of TCR engagement with self MHC-peptide complexes [2, 46]. Due to the difficulty in investigating this process, it is unclear whether similar phenomena prevail in humans. However, humanized mice generated with autologous thymic tissue and i.v. injection of HSCs can be a useful tool to unravel this process.

Acquisition of MHC preference for peptide antigen recognition occurs through positive selection. This preference is of critical importance in achieving efficient immune responses to antigens. One example is the interactions between CD4 T helper cells and B cells in the production of antigen-specific antibodies. Because human CD4 T cells developing in humanized mice grafted only with human HSCs are selected by and thus preferentially recognize antigens presented by mouse MHC, they are not able to provide efficient help to antigen-specific human B cells via the interactions of T cell receptor and HLA molecules on B cells. This may contribute to the failure to induce antigen-specific IgG production following immunization. Consistent with this notion, transgenic expression of a human HLA class II molecule in recipient mice enhances IgG responses [11]. In contrast, humanized mice generated by grafting autologous thymic tissue and HSCs should have effective interactions between CD4 T cells and B cells, since T cells develop in the thymic tissue from the same donor and are thus positively selected on their own HLA molecules. Consistently, immunization of these mice leads to production of antigen-specific IgG [22]. Another demonstration of the importance of having the positive selecting MHC also present in the periphery is provided by human T cells developing in pig thymic tissue. In humanized mice grafted with pig thymic tissue and human HSCs, robust thymopoiesis and T cell reconstitution is seen. However, T cell responses to tetanus toxoid following immunization were markedly reduced compared to those in mice with an autologous human thymus graft [47]. Presumably T cells developing in the pig thymic tissue and thus positively selected on pig MHC molecules were not able to efficiently recognize antigen presented by human APCs in the periphery.

11.5 Negative Selection

Mouse studies demonstrate that negative selection is critical for the induction of tolerance to self. The majority of autoreactive T cells are purged by this process, resulting in a T cell pool that is tolerant to self antigens [4, 48]. The negative selection process has been demonstrated in several conditions in humanized mice receiving fetal thymic tissue and HSCs. In some mice receiving fresh human thymic tissues, human anti-mouse xenogeneic GVHD can be seen about 12–16 weeks post-transplantation, as reflected by hunched posture, hair loss, and dermatitis. Target organs show typical GVHD histopathologies [49]. In contrast, the incidence of xenogeneic GVHD is clearly reduced by the use of freeze/thawed thymic tissue and administration of T cell-depleting antibody following transplantation, both of which deplete pre-existing thymocytes, before and after they migrate to the periphery respectively (Fig. 11.1, reference 25 and unpublished data). T cells from humanized mice that are grafted with freeze/thawed autologous thymic tissue and HSCs do not mount

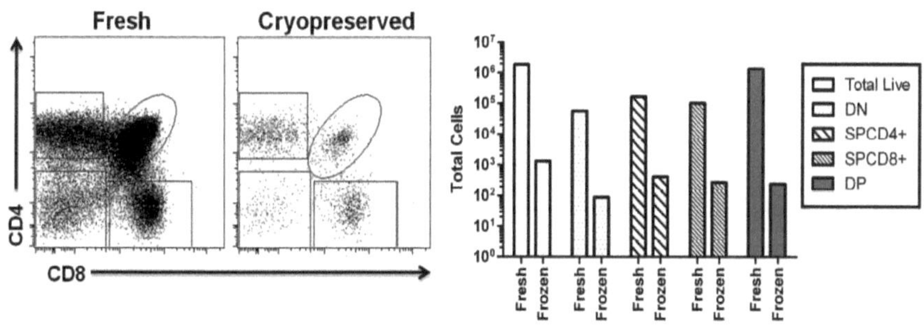

Fig. 11.1 Cryopreservation depletes thymocytes in human fetal graft. *Fresh* and *cryopreserved* thymus (~0.002 g of a single piece of tissue) from the same donor were dissociated and stained for live cells and thymocytes markers *CD4* and *CD8*. 1×10^5 total events were collected and subgated on live ($DAPI^-$) thymocytes. FCM plots are shown on the *left* and total cell number for CD4CD8 double negative, single positive and double positive populations are shown in the graft at the *right*. Cryopreservation decreased single positive CD4 and CD8 cells 510 fold and 454 fold respectively after adjusting for tissue weight difference before freezing (*SPCD4* fresh 1.9×10^5 vs. cryopreserved 425 cells; *SPCD8* fresh 1.1×10^5 vs. cryopreserved 275 cells; from Kalscheuer et al. [25]. Reprinted with permission from AAAS)

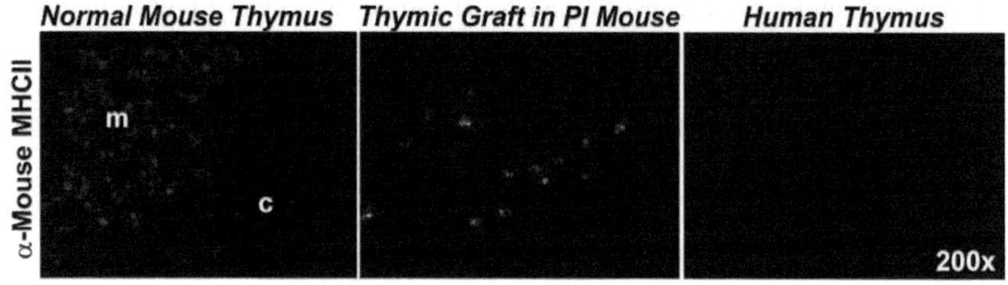

Fig. 11.2 Antigen-presenting cells from the recipient mouse in human thymic graft of PI mouse. The *thymic graft from a PI mouse* reconstituted with adult CD34$^+$ cells (*center*), a thymus from a normal C57BL/6 mouse (*left*) and a thymus from a human (*right*) were sectioned and stained with anti-mouse pan-MHC class II mAb to reveal the presence of mouse-derived MHC class II positive cells. In the *normal mouse thymus*, *m* denotes the medullary and *c* denotes the cortical region. When tested for cross-reactivity, the anti-mouse MHCII antibody did not bind to *human thymus* tissue (*right*). (From Kalscheuer et al. [25]. Reprinted with permission from AAAS)

responses to mouse xenoantigens in in-vitro mixed lymphocyte reaction (MLR) cultures, demonstrating tolerance of human T cells to mouse xenoantigens [47]. Preexisting thymocytes within human fetal thymic grafts can cause xenogeneic GVHD in these humanized mice because they have not undergone negative selection in the presence of mouse APCs. Mouse APCs populate the freeze/thawed human thymic tissue (Fig. 11.2) when grafted to NSG mice receiving T cell depleting mAb [25], resulting in negative selection in the human thymic graft during de novo human T cell development.

We have investigated the induction of tolerance to pig xenoantigens by thymic transplantation. Our early studies demonstrated that transplantation of pig thymic tissue to thymectomized and T and NK cell-depleted mice led to generation of a mouse T cell repertoire that was specifically tolerant to the recipient mouse and donor pig, suggesting that thymic transplantation could be a potential approach to inducing tolerance to pig xenoantigens in humans [34, 50]. Negative selection of mouse developing T cells by the pig thymus was found to be responsible for much of the tolerance in this model [33], but a role for Tregs was also suggested [51]. Later studies showed that cotransplantation of pig thymus prevented rejection of pig kidney grafts in baboons, further indicating the potential of this approach to induce tolerance to pig xenoantigens [52]. Humanized mice have been used in our recent studies to investigate this approach for human T cell tolerance. Transplantation of pig thymic tissue with human HSCs leads to generation of a human T cell repertoire that is specifically tolerant to the donor pig xenoantigens [36, 53]. T cells from spleens of humanized mice transplanted with $SLA^{d/d}$ pig thymic tissues and human HSCs showed strong responses to human alloantigens and to third party $SLA^{c/c}$ pig xenoantigens, with specific unresponsiveness to the pig donor and mouse recipient [47]. Alternatively, induction of pig/human mixed hematopoietic chimerism in humanized mice generated by grafting human fetal thymic tissue and autologous human HSCs with pig bone marrow cells also led to specific unresponsiveness of human T cells to donor pig xenoantigens [18]. Pig MHC Class II$^+$ cells were found in the human thymic grafts in the pig/human mixed chimeric humanized mice, suggesting that negative selection of human T cells by these pig MHC Class II$^+$ APCs led to the deletion of pig-reactive human T cells [18]. These data suggest that human T cells become tolerant to pig xenoantigens by undergoing negative selection in pig thymus or in human thymus containing pig bone marrow-derived APCs. Pig thymus transplantation and induction of mixed porcine hematopoietic chimerism thus are two potential solutions for inducing human T cell tolerance to pig xenoantigens. Induction of tolerance is likely to be essential for successful pig-to-human xenogeneic organ transplantation.

11.6 Treg Generation and Maturation

Ample studies demonstrate that CD4$^+$CD25$^+$Foxp3$^+$ naturally occurring Tregs play a critical role in maintenance of self tolerance [54, 55]. Tregs are a promising candidate for therapy and prophylaxis against multiple diseases [56]. Mouse "natural" Tregs are generated in the thymus [54] in processes involving both positive and negative selection [5, 57]. The process by which Tregs are generated, including the location and molecular signals, has been under active investigation [5, 57, 58]. Naturally occurring Tregs exist in humans [59, 60], but little is known about their development. Thus, humanized mouse models that enable the investigation of human Treg development are needed. Although human Tregs with in vitro suppressive activities are present in humanized mice generated by transplantation of human HSCs alone, these Tregs develop in the mouse thymus and the developmental process may not fully recapitulate that of normal human Tregs [61]. We therefore investigated human Treg development in humanized mice grafted with fetal thymic tissue and HSCs [20]. CD25$^+$CD127^{low} thymocytes were present at similar percentages among CD4 single positive thymocytes in thymic grafts of the humanized mice and in human fetal thymic tissues. This subset of T cells expresses both Foxp3 and Helios, which are typical markers for human natural Tregs. Like the Tregs in fetal thymus, thymus-derived Tregs from humanized mice express HLA-DR and are mainly CD45RA$^-$ and CD45RO$^+$ [20]. These data demonstrate that human Tregs develop normally in the thymic grafts of the humanized mice. Tregs were also detected in multiple tissues of humanized mice, including peripheral blood, spleen and lymph nodes. Despite the similarity of Tregs in the thymus grafts, there are differences between Tregs in the PBMCs of humanized mice and those in adult human PBMCs. A greater percentage of CD4$^+$CD25$^+$Foxp3$^+$ Tregs showed a naïve phenotype (CD45RA$^+$CD45RO$^-$) in the PBMCs of humanized mice compared to human PBMCs and the percentage of HLA-DR$^+$ Tregs was lower in the PBMCs of humanized mice. However, a subset of peripheral Tregs in humanized mice expressed CD45RO and HLA-DR, suggesting that post-thymic encounter with self MHC-peptide complexes on human APCs had taken place. The lower percentage of these "activated" Tregs in humanized mice compared to adult PBMCs might be explained simply by the difference in age of the human immune system in each type of host and/or by the greater exposure of humans to microorganisms compared to the mice. Functional assays demonstrated that Tregs isolated from spleens of humanized mice are as potent on a "per cell" basis as those from human peripheral blood in suppressing anti-CD3-induced proliferation of $CD4^{+}CD25^{-}$ conventional T cells [20].

11.7 Concluding Remarks

In summary, insights into human T cell development have been obtained using humanized mice generated by transplantation of autologous fetal thymic tissue and HSCs. These studies collectively indicate that thymocyte development, including positive selection, negative selection and generation of Tregs occur in human thymus grafts and that humanized mice generated by transplantation of autologous fetal thymic tissue and HSCs (Fig. 11.3) are of considerable utility for the study of human T cell development in both physiological and pathological conditions. In addition to allowing the detailed investigation of normal human T cell development,

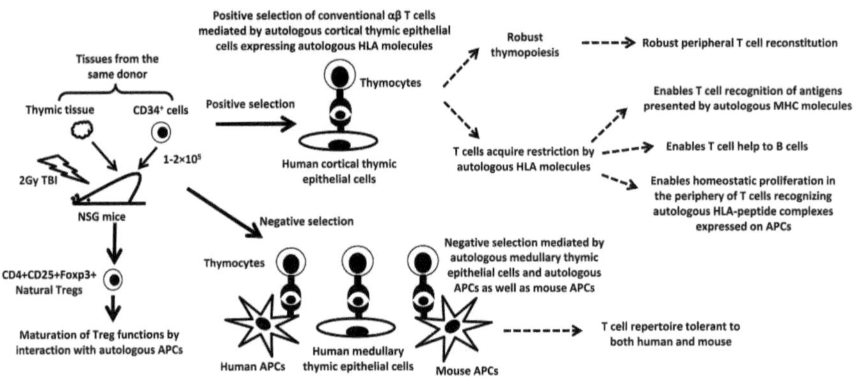

Fig. 11.3 Positive and *negative selection* of T cells and development of *$CD4^{+}CD25^{+}Foxp3$ natural Tregs* in humanized mice generated by grafting autologous human fetal thymic tissue and fetal liver-derived *$CD34^{+}$* cells. NSG mice are conditioned by *2 Gy total body irradiation* (*TBI*) followed by implantation of human fetal *thymic tissue* under the kidney capsule and injection of autologous fetal liver-derived *$CD34^{+}$* cells. Human lymphoid progenitors home to the autologous thymic tissue and undergo selection and maturation, resulting in robust thymopoiesis. *Positive selection*, which is mediated by autologous cortical thymic epithelial cells expressing *HLA molecules*, allow conventional αβ T cells to optimally recognize peptides presented by autologous HLA molecules expressed on peripheral antigen-presenting cells (*APCs*), permitting T cells to respond to antigen, provide help to B cells and undergo homeostatic proliferation. Negative selection in the thymus is mediated by autologous human medullary thymic epithelial cells, autologous *human APCs* in the thymic graft and mouse bone marrow-derived APCs. This negative selection results in a human T cell repertoire that is tolerant to both human tissue donor and the recipient mouse. *$CD4^{+}CD25^{+}Foxp3$ natural Tregs* are generated in human thymic tissue. Their interactions with autologous APCs in the periphery may lead to further maturation

this model has enabled studies of human T cell homeostasis and provides a model for testing therapeutic strategies to induce tolerance for the treatment of human diseases, such as thymus transplantation to induce tolerance to pig xenoantigens for xenogeneic organ transplantation. Moreover, this model can be further optimized by using genetic manipulation. For example, transplantation of thymic tissue and TCR gene-transduced autologous CD34+ HSCs enables the study of the development of antigen-specific T cell clones. While humanized mice generated by grafting of fetal tissues allows the investigation of human T cell development in general, the "PI" humanized mouse model makes it possible to explore the roles of abnormalities of T cell development in the pathogenesis of immune-mediated diseases, such as autoimmune diseases. The "PI" humanized mice thus provide a novel and powerful tool to shed light on human diseases and in which to test immunotherapies in a personalized fashion.

Acknowledgments We thank Shavree Washington for assistance with the manuscript. The work of the authors discussed in this chapter was supported by the following NIH grants: RO1AI084074 (to Sykes), RC1HL100117 and R01 AI064569 (to Yang) and P01AI045897 (to Sykes and Yang) and JDRF grant 1-2007-1057 (to Sykes).

References

1. Mecklenburg L, Tychsen B, Paus R. Learning from nudity: lessons from the nude phenotype. Exp Dermatol. 2005;14(11):797–810.
2. Fink PJ, Hendricks DW. Post-thymic maturation: young T cells assert their individuality. Nat Rev Immunol. 2011;11(8):544–9.
3. Surh CD, Sprent J. Homeostasis of naive and memory T cells. Immunity. 2008;29(6):848–62.
4. Joller N, Peters A, Anderson AC, Kuchroo VK. Immune checkpoints in central nervous system autoimmunity. Immunol Rev. 2012;248(1):122–39.
5. Moran AE, Hogquist KA. T-cell receptor affinity in thymic development. Immunology. 2012;135(4):261–7.
6. Mosier DE, Gulizia RJ, Baird SM, Wilson DB. Transfer of a functional human immune system to mice with severe combined immunodeficiency. Nature. 1988;335(6187):256–9.
7. Ishikawa F, Yasukawa M, Lyons B, Yoshida S, Miyamoto T, Yoshimoto G, et al. Development of functional human blood and immune systems in NOD/SCID/IL2 receptor γ chainnull mice. Blood. 2005;106(5):1565–73.
8. Traggiai E, Chicha L, Mazzucchelli L, Bronz L, Piffaretti J-C, Lanzavecchia A, et al. Development of a human adaptive immune system in cord blood cell-transplanted mice. Science. 2004;304(5667):104–7.
9. Tary-Lehmann M, Lehmann PV, Schols D, Roncarolo MG, Saxon A. Anti-SCID mouse reactivity shapes the human CD4+ T cell repertoire in hu-PBL-SCID chimeras. J Exp Med. 1994;180(5):1817–27.
10. Tary-Lehmann M, Saxon A. Human mature T cells that are anergic in vivo prevail in SCID mice reconstituted with human peripheral blood. J Exp Med. 1992;175(2):503–16.
11. Danner R, Chaudhari SN, Rosenberger J, Surls J, Richie TL, Brumeanu T-D, et al. Expression of HLA class II molecules in humanized NOD. Rag1KO.IL2RgcKO mice is critical for development and function of human T and B cells. PLoS ONE. 2011;6(5):e19826.
12. Watanabe Y, Takahashi T, Okajima A, Shiokawa M, Ishii N, Katano I, et al. The analysis of the functions of human B and T cells in humanized NOD/shi-scid/γcnull (NOG) mice (hu-HSC NOG mice). Int Immunol. 2009;21(7):843–58.

13. McCune J, Namikawa R, Kaneshima H, Shultz L, Lieberman M, Weissman I. The SCID-hu mouse: murine model for the analysis of human hematolymphoid differentiation and function. Science. 1988;241(4873):1632–9.
14. Namikawa R, Weilbaecher KN, Kaneshima H, Yee EJ, McCune JM. Long-term human hematopoiesis in the SCID-hu mouse. J Exp Med. 1990;172(4):1055–63.
15. Wang H, VerHalen J, Madariaga ML, Xiang S, Wang S, Lan P, et al. Attenuation of phagocytosis of xenogeneic cells by manipulating CD47. Blood. 2007;109(2):836–42.
16. Takenaka K, Prasolava TK, Wang JCY, Mortin-Toth SM, Khalouei S, Gan OI, et al. Polymorphism in Sirpa modulates engraftment of human hematopoietic stem cells. Nat Immunol. 2007;8(12):1313–23.
17. Lan P, Tonomura N, Shimizu A, Wang S, Yang Y-G. Reconstitution of a functional human immune system in immunodeficient mice through combined human fetal thymus/liver and $CD34^+$ cell transplantation. Blood. 2006;108(2):487–92.
18. Lan P, Wang L, Diouf B, Eguchi H, Su H, Bronson R, et al. Induction of human T-cell tolerance to porcine xenoantigens through mixed hematopoietic chimerism. Blood. 2004;103(10):3964–9.
19. Onoe T, Kalscheuer H, Chittenden M, Zhao G, Yang Y-G, Sykes M. Homeostatic expansion and phenotypic conversion of human T cells depend on peripheral interactions with APCs. J Immunol. 2010;184(12):6756–65.
20. Onoe T, Kalscheuer H, Danzl N, Chittenden M, Zhao G, Yang Y-G, et al. Human natural regulatory T cell development, suppressive function, and postthymic maturation in a humanized mouse model. J Immunol. 2011;187(7):3895–903.
21. Melkus MW, Estes JD, Padgett-Thomas A, Gatlin J, Denton PW, Othieno FA, et al. Humanized mice mount specific adaptive and innate immune responses to EBV and TSST-1. Nat Med. 2006;12(11):1316–22.
22. Tonomura N, Habiro K, Shimizu A, Sykes M, Yang Y-G. Antigen-specific human T-cell responses and T cell-dependent production of human antibodies in a humanized mouse model. Blood. 2008;111(8):4293–6.
23. Brainard DM, Seung E, Frahm N, Cariappa A, Bailey CC, Hart WK, et al. Induction of robust cellular and humoral virus-specific adaptive immune responses in human immunodeficiency virus-infected humanized BLT mice. J Virol. 2009;83(14):7305–21.
24. Hongo D, Hadidi S, Damrauer S, Garrigue V, Kraft D, Sachs DH, et al. Porcine thymic grafts protect human thymocytes from HIV-1-induced destruction. J Infect Dis. 2007;196(6):900–10.
25. Kalscheuer H, Danzl N, Onoe T, Faust T, Winchester R, Goland R, et al. A model for personalized in vivo analysis of human immune responsiveness. Sci Transl Med. 2012;4(125):125ra30.
26. Anderson G, Jenkinson EJ. Lymphostromal interactions in thymic development and function. Nat Rev Immunol. 2001;1(1):31–40.
27. Starr TK, Jameson SC, Hogquist KA. Positive and negative selection of T cells. Annu Rev Immunol. 2003;21(1):139–76.
28. Capone M, Romagnoli P, Beermann F, MacDonald HR, van Meerwijk JPM. Dissociation of thymic positive and negative selection in transgenic mice expressing major histocompatibility complex class I molecules exclusively on thymic cortical epithelial cells. Blood. 2001;97(5):1336–42.
29. Laufer TM, DeKoning J, Markowitz JS, Lo D, Glimcher LH. Unopposed positive selection and autoreactivity in mice expressing class II MHC only on thymic cortex. Nature. 1996;383(6595):81–5.
30. Lo W-L, Allen P. Self-peptides in TCR repertoire selection and peripheral T cell function. Curr Top Microbiol Immunol. 2014;373:49–67.
31. Krogsgaard M, Juang J, Davis MM. A role for "self" in T-cell activation. Semin Immunol. 2007;19(4):236–44.
32. Zhao Y, Fishman JA, Sergio JJ, Oliveros JL, Pearson DA, Szot GL, et al. Immune restoration by fetal pig thymus grafts in T cell-depleted, thymectomized mice. J Immunol. 1997;158(4):1641–9.

33. Zhao Y, Sergio JJ, Swenson K, Arn JS, Sachs DH, Sykes M. Positive and negative selection of functional mouse CD4 cells by porcine MHC in pig thymus grafts. J Immunol. 1997;159(5):2100–7.
34. Zhao Y, Swenson K, Sergio JJ, Arn JS, Sachs DH, Sykes M. Skin graft tolerance across a discordant xenogeneic barrier. Nat Med. 1996;2(11):1211–6.
35. Zhao Y, Swenson K, Sergio JJ, Sykes M. Pig MHC mediates positive selection of mouse CD4+ T cells with a mouse MHC-restricted TCR in pig thymus grafts. J Immunol. 1998;161(3):1320–6.
36. Nikolic B, Gardner JP, Scadden DT, Arn JS, Sachs DH, Sykes M. Normal development in porcine thymus grafts and specific tolerance of human T cells to porcine donor MHC. J Immunol. 1999;162(6):3402–7.
37. Shimizu I, Fudaba Y, Shimizu A, Yang Y-G, Sykes M. Comparison of human T cell repertoire generated in xenogeneic porcine and human thymus grafts. Transplantation. 2008;86(4):601–10.
38. Alves NL, Huntington ND, Rodewald H-R, Di Santo JP. Thymic epithelial cells: the multitasking framework of the T cell "cradle". Trends Immunol. 2009;30(10):468–74.
39. van Ewijk W, Shores EW, Singer A. Crosstalk in the mouse thymus. Immunol Today. 1994;15(5):214–7.
40. van Lent AU, Dontje W, Nagasawa M, Siamari R, Bakker AQ, Pouw SM, et al. IL-7 enhances thymic human T cell development in "human immune system" Rag2-/-IL-2Rgammac-/- mice without affecting peripheral T cell homeostasis. J Immunol. 2009;183(12):7645–55.
41. Huntington ND, Alves NL, Legrand N, Lim A, Strick-Marchand H, Plet A, et al. Autonomous and extrinsic regulation of thymopoiesis inhuman immune system (HIS) mice. Eur J Immunol. 2011;41(10):2883–93.
42. Surh CD, Sprent J. Homeostatic T cell proliferation: how far can T cells be activated to self-ligands? J Exp Med. 2000;192(4):F9–14.
43. Stefanova I, Dorfman JR, Germain RN. Self-recognition promotes the foreign antigen sensitivity of naive T lymphocytes. Nature. 2002;420(6914):429–34.
44. Lo W-L, Felix NJ, Walters JJ, Rohrs H, Gross ML, Allen PM. An endogenous peptide positively selects and augments the activation and survival of peripheral CD4+ T cells. Nat Immunol. 2009;10(11):1155–61.
45. Kieper WC, Troy A, Burghardt JT, Ramsey C, Lee JY, Jiang H-Q, et al. Cutting edge: Recent immune status determines the source of antigens that drive homeostatic T cell expansion. J Immunol. 2005;174(6):3158–63.
46. Fink PJ. The biology of recent thymic emigrants. Annu Rev Immunol. 2013;31(1):31–50.
47. Kalscheuer H, Onoe T, Dahmani A, Li H-W, Hölzl M, Yamada K, et al. Xenograft tolerance and immune function of human T cells developing in pig thymus xenografts. J Immunol. 2014;192(7):3442–50.
48. Gallegos AM, Bevan MJ. Central tolerance: good but imperfect. Immunol Rev. 2006;209(1):290–6.
49. Greenblatt MB, Vbranac V, Tivey T, Tsang K, Tager AM, Aliprantis AO. Graft versus host disease in the bone marrow, liver and thymus humanized mouse model. PLoS ONE. 2012;7(9):e44664.
50. Lee LA, Gritsch HA, Sergio JJ, Arn JS, Glaser RM, Sablinski T, et al. Specific tolerance across a discordant xenogeneic transplantation barrier. Proc Natl Acad Sci U S A. 1994;91(23):10864–7.
51. Rodriguez-Barbosa JI, Zhao Y, Barth R, Zhao G, Arn JS, Sachs DH, et al. Enhanced CD4 reconstitution by grafting neonatal porcine tissue in alternative locations is associated with donor-specific tolerance and suppression of preexisting xenoreactive T cells. Transplantation. 2001;72(7):1223–31.
52. Yamada K, Yazawa K, Shimizu A, Iwanaga T, Hisashi Y, Nuhn M, et al. Marked prolongation of porcine renal xenograft survival in baboons through the use of [alpha]1,3-galactosyltransferase gene-knockout donors and the cotransplantation of vascularized thymic tissue. Nat Med. 2005;11(1):32–4.

53. Habiro K, Sykes M, Yang YG. Induction of human T-cell tolerance to pig xenoantigens via thymus transplantation in mice with an established human immune system. Am J Transplant. 2009;9(6):1324–9.
54. Shevach EM. Regulatory T cells in autoimmmunity. Annu Rev Immunol. 2000;18(1):423–49.
55. Shevach EM. Mechanisms of foxp3$^+$ T regulatory cell-mediated suppression. Immunity. 2009;30(5):636–45.
56. Riley JL, June CH, Blazar BR. Human T regulatory cell therapy: take a billion or so and call me in the morning. Immunity. 2009;30(5):656–65.
57. Bettini ML, Vignali DAA. Development of thymically derived natural regulatory T cells. Ann N Y Acad Sci. 2010;1183(1):1–12.
58. Yuan X, Malek TR. Cellular and molecular determinants for the development of natural and induced regulatory T cells. Hum Immunol. 2012;73(8):773–82.
59. Baecher-Allan C, Brown JA, Freeman GJ, Hafler DA. CD4$^+$ CD25high regulatory cells in human peripheral blood. J Immunol. 2001;167(3):1245–53.
60. Stephens LA, Mottet C, Mason D, Powrie F. Human CD4$^+$ CD25$^+$ thymocytes and peripheral T cells have immune suppressive activity in vitro. Eur J Immunol. 2001;31(4):1247–54.
61. Billerbeck E, Barry WT, Mu K, Dorner M, Rice CM, Ploss A. Development of human CD4$^+$ FoxP3$^+$ regulatory T cells in human stem cell factor–, granulocyte-macrophage colony-stimulating factor–, and interleukin-3–expressing NOD-SCID IL2Rγ null humanized mice. Blood. 2011;117(11):3076–86.

Chapter 12
Human B-Cell Development in a Mouse Environment

Julie Lang and Roberta Pelanda

12.1 Introduction

B lymphocytes play a versatile role in the immune system's fight against infections. These cells are the source of pathogen-specific and neutralizing antibodies (Abs) as well as a memory storage of previous pathogen encounters. Some B cells are also "innately" capable of quickly activating a response to a pathogen by secreting low-affinity Abs that weaken the infection before more specific responses are achieved. Furthermore, B cells present antigen and produce cytokines and chemokines that regulate T cells and other blood cell types to coordinate the immune response during the early and late stages of infection. Therefore, B cells are an essential component of an effective and sustained immune response against microbes including viruses such as HIV.

Recent HIV studies have demonstrated that human B cells do have the ability to produce HIV-specific Abs that broadly neutralize most viral clades [1–3]. As an effective HIV vaccine remains elusive, studies aimed at investigating the basis of rare protective anti-HIV Ab responses are critical to direct successful immunization strategies. Within this context, hematopoietic humanized mice (hu-mice) represent a relevant experimental tool to explore basic and translational mechanisms of human B-cell development, B-cell activation, and Ab responses. Immunodeficient mice transplanted with human hematopoietic stem cells develop a human immune system that roughly recapitulates the establishment and complexity of the natural system. In this animal model, human B cells, T cells, NK cells, monocytes, and dendritic cells differentiate from the donor hematopoietic stem cells and together coordinate the development and function of the immune system. These chimeric mice are useful,

R. Pelanda (✉) · J. Lang
Department of Immunology and Microbiology, University of Colorado Denver School of Medicine, 12800 East 19th Avenue, Mail Stop 8333, 80045-2508 Aurora, CO, USA
e-mail: Roberta.Pelanda@ucdenver.edu

L. Y. Poluektova et al. (eds.), *Humanized Mice for HIV Research*,
DOI 10.1007/978-1-4939-1655-9_12

therefore, not only to study how human B cells respond to a vaccine or an infectious agent, but also to discover and characterize the immune factors (cellular or otherwise) that influence the development of these responses. Moreover, since each set of hu-mice is generated from distinct human stem cell donors, these mice also provide an experimental tool to study the effects of natural human genetic variation on B cells.

This chapter will review the kinetics, properties, and limitations of human B-cell development in hu-mice and will discuss how the hu-mouse model can be further improved to support enhanced B-cell numbers and function.

12.2 Development of Human B Cells in Mice

The bone marrow is the physiological site of B-cell production in mice and humans. Human B cells similarly develop in the bone marrow of hu-mice with clearly identifiable maturational stages [4–8]. Within human B cells developing in the mouse, Ig gene rearrangements display normal kinetics with the light chain following the heavy chain ([8]; and Lang and Pelanda, unpublished observations). Moreover, these rearrangements result in a peripheral B-cell population with a roughly normal lambda:kappa light chain ratio (about 1:1) ([9]; Lang and Pelanda, unpublished observations).

CD19, a marker restricted to the B-cell lineage, is first expressed on pro-B cells and is retained on the surface of B cells throughout maturation before being downmodulated on plasma cells. The expression of the B-cell receptor (BCR) components and isotypes defines the $CD19^{+}$ bone marrow cells as precursor (pre, H-chain only), immature (H and L chains, IgM), transitional/naive (IgM and IgD), and switched memory (IgA, IgE, or IgG) B cells [10, 11]. Throughout human B-cell development, expression of cell-surface “markers” guides our definition of B-cell stages ([10, 12–17]; Fig. 12.1). For the majority of these markers the absolute expression level varies as the B-cell progresses from an immature through the transitional stage to a mature, naive B cell, with the transitional stage generally presenting an intermediate level of expression [10, 12–14, 18]. The careful selection and measurement of a panel of these markers is required for the proper analysis and definition of B-cell development [13–14, 19]. Using such analyses, we and others have determined that $CD34^{+}CD10^{+}$ lymphocyte precursors in the bone marrow of hu-mice proceed through normal B-cell development with clearly identifiable pro, pre, immature, and transitional stages of maturation and with a modality that is largely similar to B cells developing in human bone marrow [4–8, 20].

12.3 B Cell Tolerance in hu-mice

As in the mouse [21–22], B cells developing in human bone marrow undergo a process of selection (central tolerance) that removes many specificities reacting with local self-antigens [23]. By analyzing the B-cell repertoire of hu-mice, two studies have investigated whether B-cell tolerance operates in the context of this model

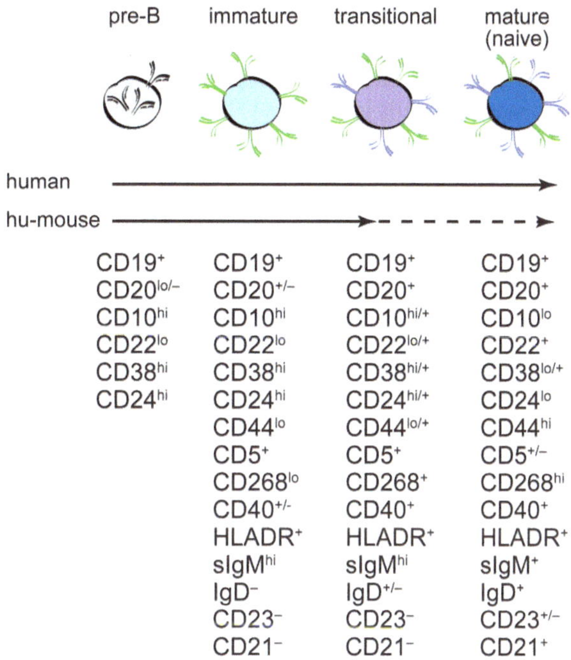

Fig. 12.1 Markers associated with B-cell development stages in humans. The *black*, *green*, and *violet* cell receptors on pre-B cells, immature B cells, and transitional/mature B cells represent the pre-BCR, IgM, and IgD, respectively. The *solid line* illustrates normal developmental stages of human B-cell maturation and the *dashed line* illustrates the defective maturation in the hu-mouse

[24–25]. They found that the prevalence of two autoimmune-associated heavy (VH4-34) and light (Vκ4-1) chains progressively decreases from the immature to the more mature B-cell subsets, suggesting some level of tolerance induction. Inconsistent with the above results though, a Hep-2 antinuclear antibody assay indicated that the frequency of autoreactive B cells does not decrease from the immature to the newly emigrant and mature B-cell stages as it does in humans [24, 26]. This suggests that B-cell tolerance might be defective in hu-mice. Nonetheless, such conclusion requires a more formal demonstration. Our lab (in collaboration with David Nemazee, The Scripps Institute) has been investigating central B-cell tolerance by using hu-mice that ubiquitously express an anti-human kappa chimeric protein as a synthetic neo self-antigen specific for all human $Ig\kappa^{+}$ B cells. Our studies indicate that the developing $Ig\kappa^{+}$ B cells (~50 % of all immature B cells) are for the most part excluded from entering the peripheral B-cell population, a phenotype consistent with the induction of tolerance (Lang and Pelanda, unpublished observations). Thus, future studies are needed to precisely define the prevalence and mechanisms of B-cell tolerance in hu-mice.

12.4 B Cell Maturation Defect in hu-mice

Unlike the normal development of immature B cells in the bone marrow of hu-mice, a defect has been reported in the differentiation of immature/transitional B cells into naive, mature B cells [27]. For instance, while IgM^{+} B cells in the bone marrow of hu-mice express CD40, HLA-DR, CD5, CD44, and CD268 and high levels of CD10, CD24, and CD38, which are markers of immature/transitional B cells, they most often do not express CD21 and CD22, which label late transitional and mature B cells in humans [7]. This phenotype is consistent with a B-cell developmental block at an early transitional stage (Fig. 12.1). These transitional B cells make up the majority of the B-cell population in the spleen of hu-mice, particularly the younger ones (see below). Although this B cell defect has been clearly established in several hu-mouse models, one should also consider that B cell poiesis is ongoing not only in the bone marrow, but also at detectable levels in the spleen of many hu-mice [8, 28]. Therefore, the skewed representation of transitional B cells in the spleen may in part be due to the continual generation of these cells in situ. Many studies have also reported that most peripheral human B cells in hu-mice atypically express CD5 and have suggested that hu-mice select for the development of a B1 B cell subset [9, 29]. However, it is important to note that in humans this antigen marks B-cell maturation stages and not necessarily a functional subset [12, 15]. Therefore, the expression of CD5 on B cells of hu-mice is more likely related to the fact that most of these cells are immature rather than B1.

12.5 Mature B Cells Accumulate in Lymph Nodes (LNs) and Spleens of Older hu-Mice

Unlike mouse bone marrow chimeras that, within a few weeks, achieve a steady-state engraftment that is maintained for months to even years, the engraftment of the human hematopoietic system within the mouse is dynamic: it takes longer to appear and changes continually over time. This dynamic engraftment has unique characteristics in the various lymphatic tissues: it establishes early in the bone marrow and thymus, it is slightly delayed in the blood and spleen, and is greatly delayed in the LNs, where engraftment of mesenteric LNs precedes that of peripheral LNs by weeks (Fig. 12.2, and [7]). Furthermore, chimerism gradually wanes in the bone marrow, and then in blood and spleen, eventually extinguishing the engrafted human immune system.

By studying the kinetics of human hematopoietic engraftment, our group has found that the defect in B-cell maturation in hu-mice is not an absolute inherent

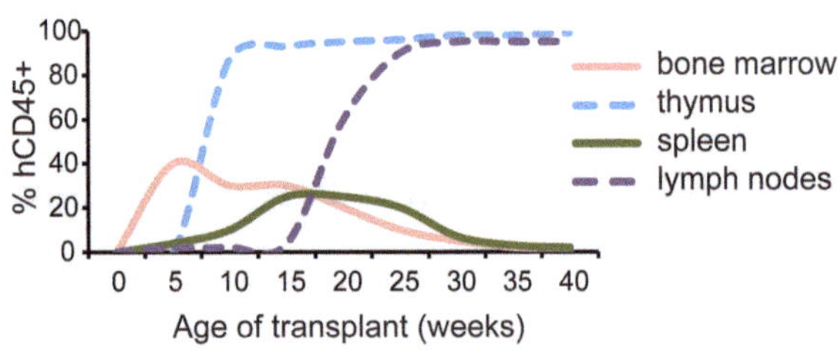

Fig. 12.2 Dynamic engraftment of human hematopoietic cells in hu-mice. The percentage of human CD45⁺ cells relative to total human and mouse CD45⁺ cells is diagrammed over time

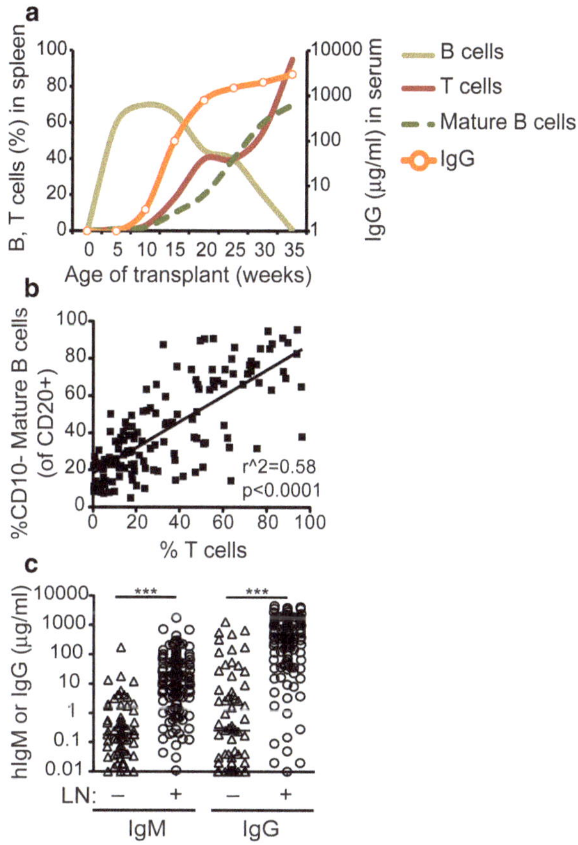

Fig. 12.3 B-cell maturation and function depend on the presence of T cells. **a** Dynamic engraftment of human lymphocytes in mice. The schematic illustrates the frequency of T and B cells in the spleen following stem cell transplantation. The functional engraftment is also portrayed as the presence of mature B cells and human IgG. **b** Correlation of T cells with mature B cells in the spleen of hu-mice. Each symbol represents data from an individual mouse. **c** Human IgM and IgG in sera of individual hu-mice relative to the presence of engrafted LNs in the same mouse. (Panels **b** and **c** courtesy of ref. [7]; pp. 2090–210. Copyright 2013. The American Association of Immunologists)

inability to progress beyond the transitional stage [7]. The B cells that populate the LNs of hu-mice, in fact, are $CD10^{lo}$, $CD21^{hi}$, $CD22^{hi}$, and $CD44^{hi}$, which are characteristics of mature B cells. In addition, the LN B cells can be activated, as evident by the presence of IgG class-switched cells and of B cells that express CD27 and CD11c memory antigens, as well as CD80, CD86, CD25, and CD69 activation markers.

In contrast to the LN, B cells in the spleen of hu-mice display a range of maturational stages from the predominant transitional to the less frequent mature, activated, and memory class-switched types [7]. This range is well demonstrated by the broad distribution of CD10, CD24, and CD38 expression within the splenic B-cell population of an individual hu-mouse and among multiple hu-mice. Importantly, our studies with BALB/c/$Rag2^{null}$/$Il2r\gamma^{null}$ (BRG) hu-mice have demonstrated that the frequency of mature B cells in the spleen increases with time, such that 20 weeks after transplantation more than 75 % of hu-mice have a mature B-cell population (Fig. 12.3a, and [7]). LNs, which harbor mainly mature B cells, are also more frequent in older hu-mice [7] and the increased numbers of mature B cells in these older animals drives higher IgM and IgG concentrations in their sera (Fig. 12.3b). Thus, there exists a "goldilocks" time period for mature B cells in hu-mice, between 16 and 24 weeks after transplant, during which a significant population of mature B cells is present in LNs and spleen, and human hematopoiesis is still ongoing in the bone marrow. Notably, the dynamic characteristic of B-cell maturation appears to be common to hu-mice whether they are generated on a BRG [5, 7, 27] or nonobese diabetic, NOD (NSG or NOG, [4, 8, 24, 28–30]; Lang and Pelanda, unpublished observations) genetic background. However, when compared to age-matched BRG mice, NOD recipients have a slightly higher proportion of mature B cells in the spleen (Lang and Pelanda, unpublished observations), most likely as a consequence of higher human chimerism and, therefore, cell numbers [31–34]. Nonetheless, the B cells in the NOD recipients secrete significantly lower human Ig, particularly IgG, suggesting inferior function relative to B cells in BRG mice ([4, 7, 8, 24, 30, 35]; Lang and Pelanda, unpublished observations).

In spite of their defective maturation, B cells are over-represented relative to other hematopoietic cell types, particularly during the first 16 weeks after transplant. This phenotype is thought to be largely contingent on the inefficient function of mouse cytokines on some human hematopoietic cell types [36–39]. B-cell activating factor (BAFF or BlyS) is considered the most relevant cytokine for B cells [40]. In mice, it not only promotes the survival of mature B cells, but it also contributes to Ig class switch and to the differentiation of immature B cells into transitional and mature B cells [41–42]. Mirroring the stages during which BAFF functions, the expression of CD268 (BAFF-R) on B cells begins at the immature stage and progressively increases throughout differentiation into a mature B cell. Similar to B cells in their native hosts, those that develop in hu-mice begin to express CD268 at the immature cell stage, but the expression remains low on all splenic and LN B cells [7]. Schmidt and colleagues have shown that mouse BAFF is abundant in hu-mice and, although it binds human CD268, it fails to properly signal [43]. The low CD268 expression on human B cells likely represents internalization of the receptor upon binding large amounts of mouse BAFF. We have detected human BAFF mRNA in tissue of hu-mice at levels that positively correlate with numbers of mature splenic B cells (Lang and Pelanda, unpublished observations). In addition, treatment of hu-PBL mice with human BAFF significantly increases the survival of mature B cells [43], and BAFF injections into hu-PBL and hu-mice result in higher human Ig serum concentrations ([35]; and Lang and Pelanda, unpublished observations). Overall, these data suggest that ectopic expression of human BAFF may improve the survival and perhaps the generation of mature B cells in hu-mice. Efforts are ongoing to develop new recipient mouse strains that express human BAFF with the goal of enhancing B-cell maturation and function.

12.6 Necessity of T Cells for the Maturation of Human B Cells

The rise of mature B cells in the spleen of hu-mice over time and their unique presence in LNs suggest that the specific human factors that are responsible for this phenotype are produced in a tissue and time-specific manner. T cells were found to gradually accumulate in peripheral lymphoid tissue of hu-mice starting around 12 weeks after transplant (Fig. 12.3a, [7, 44–45]). These T cells slowly accumulate in the spleen and readily populate the LNs along with mature B cells. Notably, we have observed a highly significant correlation between T-cell frequency and mature B cells in the spleen (Fig. 12.3b). Supporting a model in which T cells mediate B-cell maturation, addition of syngeneic, exogenous T cells to hu-mice expedites B-cell maturation, while in vivo depletion of T cells retards this process [7]. Furthermore, addition of T cells to transitional B-cell cultures also mediates B-cell maturation [15].

While detailed reports of B-cell maturation studies in T-cell-deficient mice and humans are not available to our knowledge, some studies have shown that patients with T-cell immunodeficiency display a transitional (cord blood-like) and/or functionally impaired B-cell phenotype [13, 46–47]. The precise mechanism by which T cells contribute to B-cell maturation is an active area of investigation and studies with hu-mice will be particularly useful in this context. Some data suggest that activation of T cells might play a role in the B-cell maturation process [7, 48]. CD40, MHC, and T-cell cytokines are initial relevant candidates, although it remains possible that the effect of T cells is indirect and B-cell maturation requires another cell population that is directly modulated by T cells.

Presently, the correlation between T cell and mature B-cell numbers has practical implications for the use of hu-mice: the frequency of T cells in the peripheral blood can be used to infer the maturation state of B cells and, thus, help select proper animals for experimentation [7]. These findings, moreover, may have important connotations for cord blood transplantation, which now represents more than 25 % of all human hematopoietic transplants due to their increased availability and reduced HLA-match requirements. The reconstitution kinetics of human cord blood transplantation largely mirrors the development of the human immune system in mice and include the delayed T cell appearance that is likely responsible for the unfortunate high-mortality rate associated with infections. The state of B-cell maturation in these patients has not been studied although the B-cell function is known to be reduced [49]. Moreover, a report of improved early B-cell function when cord blood is not depleted of T cells [50] is consistent with a role of T cells in B-cell maturation. Therefore, the hu-mouse can serve as a model system to characterize and test improvements in cord blood transplantations.

12.7 Human Ab Responses in Mice

Analyses of the Ig heavy- and light-chain usage among the human B cells residing in the mouse have confirmed the presence of a diverse heavy- and light-chain repertoire, a repertoire that is grossly indistinguishable from that of B cells in humans [24]. Nevertheless, Ab responses in hu-mice have been reported to be weak and sporadic [7, 44, 48, 51, 52]. This defect appears to be partly due to the inefficient maturation of B cells because enhanced immunization responses have been observed in older hu-mice, which display higher numbers of mature B cells. This improvement was significant for IgM and IgG responses to a T-independent antigen (e.g., NP-Ficoll) and for IgM responses to a T-dependent antigen (e.g., DTaP) [7, 44]. T-cell-dependent IgG responses were also slightly improved [7], suggesting the presence of both a population of competent B cells capable of Ig class-switch and a productive cognate B–T cell collaboration. However, even in the presence of T cells and mature B cells, the Ab responses to immunogenic challenges in hu-mice remain inferior to those observed in humans and mice. Thus, important factors necessary for normal B-cell function and T–B collaboration are still lacking in hu-mice. Our histological studies noted that although the lymphoid tissue of animals bearing significant numbers of mature B cells displays an increased colocalization of T and B cells, the lymphoid architecture remains abnormal [7], potentially still limiting Ab responses. Another issue relates to the possibility that human thymocytes are educated on mouse MHC instead of human HLA antigens, thus affecting B–T cell cognate interaction. Indeed, supplying a human MHC class II allele [53] or cotransplanting a human thymus [29, 54, 55] enhances T-dependent Ab responses in hu-mice. Nevertheless, these responses remain inferior to those in intact mice and humans leaving this issue only partly resolved.

In humans and mice, preimmune serum is often used as a control for immunization responses, which is appropriate given the steady state of their immune systems. In hu-mice, however, similar to changing B- and T-cell frequencies and function, human Ig levels increase with age and differ greatly among individual chimeras (Fig. 12.3c, and [56]), providing a challenge for the determination of antigen-specific responses. Comparing Ab responses among hu-mice is complicated by two major factors both correlating with sera Ig concentrations: (1) a nonspecific background that is measured even in enzyme-linked immunosorbent assay (ELISA) plates that are not coated with antigen; and (2) a polyreactive response to antigen, most notably of the IgM isotype that is detected in both unimmunized and immunized mice [7]. Thus, the exact measure of Ab responses remains a challenge, emphasizing the need of a well-controlled, standardized assay to allow comparisons of Ab responses in hu-mice among different laboratories.

12.8 Concluding Remarks

The past decade has witnessed an extensive characterization of human B-cell development in immunodeficient BALB/c and NOD mice. In these models, the human B cells are able to mature, class-switch, and produce Ig to both "natural" antigens and immunological challenges. The development of human B cells in the mouse follows a similar progression as is observed in human bone marrow, albeit with reduced efficiency to mature past the transitional B cell stage. This immature state along with defective T-B collaboration and localization within secondary lymphoid organs are likely responsible for the inferior Ab responses of B cells in hu-mice. Nonetheless, there are already many practical applications of the current hu-mice for B cell studies, including those of Epstein-Barr virus (EBV) infection [57–59], HIV infection [60–61], B-cell depleting therapies [62–63], and production of humanized monoclonal Abs [56].

Our current knowledge of human B-cell development and function in the mouse is useful for both experimental strategies in this system and for practical applications in humans. The hu-mouse model is an excellent system to dissect mechanisms of human B-cell development, including selection and tolerance. Hu-mice generated with fetal or cord blood stem cells may be representative of emerging immune systems, such as the case for newborns, cord blood transplantations, and patients recovering from immunodepletion therapies (i.e., rituximab). It is well established that immunizations in these circumstances are not as effective as in a stable, adult immune system. Thus, this model may be useful to help guide childhood vaccination strategies and treat patients with immunodeficiency, autoimmunity, and transplantation.

References

1. Burton DR, Ahmed R, Barouch DH, Butera ST, Crotty S, Godzik A, et al. A blueprint for HIV vaccine discovery. Cell Host Microbe. 2012;12(4):396–407.
2. Kwong PD, Mascola JR. Human antibodies that neutralize HIV-1: identification, structures, and B cell ontogenies. Immunity. 2012;37(3):412–25.
3. Corti D, Lanzavecchia A. Broadly neutralizing antiviral antibodies. Annu Rev Immunol. 2013;31:705–42.
4. Choi B, Chun E, Kim M, Kim ST, Yoon K, Lee KY, et al. Human B cell development and antibody production in humanized NOD/SCID/IL-2Rgamma(null) (NSG) mice conditioned by busulfan. J Clin Immunol. 2011;31(2):253–64.
5. Gimeno R, Weijer K, Voordouw A, Uittenbogaart CH, Legrand N, Alves NL, et al. Monitoring the effect of gene silencing by RNA interference in human $CD34^+$ cells injected into newborn $RAG2^{-/-}$ $gammac^{-/-}$ mice: functional inactivation of p53 in developing T cells. Blood. 2004;104(13):3886–93.
6. Hiramatsu H, Nishikomori R, Heike T, Ito M, Kobayashi K, Katamura K, et al. Complete reconstitution of human lymphocytes from cord blood $CD34^+$ cells using the NOD/SCID/gammacnull mice model. Blood. 2003;102(3):873–80.
7. Lang J, Kelly M, Freed BM, McCarter MD, Kedl RM, Torres RM, et al. Studies of lymphocyte reconstitution in a humanized mouse model reveal a requirement of T cells for human B cell maturation. J Immunol. 2013;190(5):2090–101.
8. Watanabe Y, Takahashi T, Okajima A, Shiokawa M, Ishii N, Katano I, et al. The analysis of the functions of human B and T cells in humanized NOD/shi-scid/gammac(null) (NOG) mice (hu-HSC NOG mice). Int Immunol. 2009;21(7):843–58.
9. Matsumura T, Kametani Y, Ando K, Hirano Y, Katano I, Ito R, et al. Functional $CD5^+$ B cells develop predominantly in the spleen of NOD/SCID/gammac(null) (NOG) mice transplanted either with human umbilical cord blood, bone marrow, or mobilized peripheral blood $CD34^+$ cells. Exp Hematol. 2003;31(9):789–97.
10. Perez-Andres M, Paiva B, Nieto WG, Caraux A, Schmitz A, Almeida J, et al. Human peripheral blood B-cell compartments: a crossroad in B-cell traffic. Cytometry B Clin Cytom. 2010;78(Suppl 1):S47–60.
11. Blom B, Spits H. Development of human lymphoid cells. Annu Rev Immunol. 2006;24:287–320.
12. Carsetti R, Rosado MM, Wardmann H. Peripheral development of B cells in mouse and man. Immunol Rev. 2004;197:179–91.

13. Cuss AK, Avery DT, Cannons JL, Yu LJ, Nichols KE, Shaw PJ, et al. Expansion of functionally immature transitional B cells is associated with human-immunodeficient states characterized by impaired humoral immunity. J Immunol. 2006;176(3):1506–16.
14. Sims GP, Ettinger R, Shirota Y, Yarboro CH, Illei GG, Lipsky PE. Identification and characterization of circulating human transitional B cells. Blood. 2005;105(11):4390–8.
15. Lee J, Kuchen S, Fischer R, Chang S, Lipsky PE. Identification and characterization of a human CD5+ pre-naive B cell population. J Immunol. 2009;182(7):4116–26.
16. Hoffkes HG, Schmidtke G, Schmucker U, Uppenkamp M, Brittinger G. Immunophenotyping of B lymphocytes by multiparametric flow cytometry in bone marrow aspirates of healthy adults. Ann Hematol. 1995;71(3):123–8.
17. Veneri D, Ortolani R, Franchini M, Tridente G, Pizzolo G, Vella A. Expression of CD27 and CD23 on peripheral blood B lymphocytes in humans of different ages. Blood Transfus. 2009;7(1):29–34.
18. Palanichamy A, Barnard J, Zheng B, Owen T, Quach T, Wei C, et al. Novel human transitional B cell populations revealed by B cell depletion therapy. J Immunol. 2009;182(10):5982–93.
19. Caraux A, Klein B, Paiva B, Bret C, Schmitz A, Fuhler GM, et al. Circulating human B and plasma cells. Age-associated changes in counts and detailed characterization of circulating normal CD138- and CD138+ plasma cells. Haematologica. 2010;95(6):1016–20.
20. Lang J, Weiss N, Freed BM, Torres RM, Pelanda R. Generation of hematopoietic humanized mice in the newborn BALB/c-Rag2null Il2rgammanull mouse model: a multivariable optimization approach. Clin Immunol. 2011;140(1):102–16.
21. Pelanda R, Torres RM. Central B-cell tolerance: where selection begins. Cold Spring Harb Perspect Biol. 2012;4(4):a007146.
22. Shlomchik MJ. Activating systemic autoimmunity: B's, T's, and tolls. Curr Opin Immunol. 2009;21(6):626–33.
23. Meffre E, Wardemann H. B-cell tolerance checkpoints in health and autoimmunity. Curr Opin Immunol. 2008;20(6):632–8.
24. Chang H, Biswas S, Tallarico AS, Sarkis PT, Geng S, Panditrao MM, et al. Human B-cell ontogeny in humanized NOD/SCID gammac(null) mice generates a diverse yet auto/poly- and HIV-1-reactive antibody repertoire. Genes Immun. 2012;13(5):399–410.
25. Ippolito GC, Hoi KH, Reddy ST, Carroll SM, Ge X, Rogosch T, et al. Antibody repertoires in humanized NOD-scid-IL2Rgamma(null) mice and human B cells reveals human-like diversification and tolerance checkpoints in the mouse. PLoS ONE. 2012;7(4):e35497.
26. Wardemann H, Yurasov S, Schaefer A, Young JW, Meffre E, Nussenzweig MC. Predominant autoantibody production by early human B cell precursors. Science. 2003;301(5638):1374–7.
27. Vuyyuru R, Patton J, Manser T. Human immune system mice: current potential and limitations for translational research on human antibody responses. Immunol Res. 2011;51(2–3):257–66.
28. Kim M, Choi B, Kim SY, Yang JH, Roh CR, Lee KY, et al. Co-transplantation of fetal bone tissue facilitates the development and reconstitution in human B cells in humanized NOD/SCID/IL-2Rgammanull (NSG) mice. J Clin Immunol. 2011;31(4):699–709.
29. Biswas S, Chang H, Sarkis PT, Fikrig E, Zhu Q, Marasco WA. Humoral immune responses in humanized BLT mice immunized with West Nile virus and HIV-1 envelope proteins are largely mediated via human CD5+ B cells. Immunology. 2011;134(4):419–33.
30. Wang X, Qi Z, Wei H, Tian Z, Sun R. Characterization of human B cells in umbilical cord blood-transplanted NOD/SCID mice. Transpl Immunol. 2012;26(2–3):156–62.
31. Strowig T, Rongvaux A, Rathinam C, Takizawa H, Borsotti C, Philbrick W, et al. Transgenic expression of human signal regulatory protein alpha in Rag2^{-}/$^{-}$gamma(c)$^{-}$/$^{-}$ mice improves engraftment of human hematopoietic cells in humanized mice. Proc Natl Acad Sci U S A. 2011;108(32):13218–23.
32. Brehm MA, Cuthbert A, Yang C, Miller DM, DiIorio P, Laning J, et al. Parameters for establishing humanized mouse models to study human immunity: analysis of human hematopoietic stem cell engraftment in three immunodeficient strains of mice bearing the IL2rgamma(null) mutation. Clin Immunol. 2010;135(1):84–98.

33. Lepus CM, Gibson TF, Gerber SA, Kawikova I, Szczepanik M, Hossain J, et al. Comparison of human fetal liver, umbilical cord blood, and adult blood hematopoietic stem cell engraftment in NOD-scid/gammac^{-}/$^{-}$, Balb/c-Rag1^{-}/$^{-}$gammac^{-}/$^{-}$, and C.B-17-scid/bg immunodeficient mice. Hum Immunol. 2009;70(10):790–802.
34. Ito M, Kobayashi K, Nakahata T. NOD/Shi-scid IL2rgamma(null) (NOG) mice more appropriate for humanized mouse models. Curr Top Microbiol Immunol. 2008;324:53–76.
35. Giassi LJ, Pearson T, Shultz LD, Laning J, Biber K, Kraus M, et al. Expanded CD34+ human umbilical cord blood cells generate multiple lymphohematopoietic lineages in NOD-scid IL2rgamma(null) mice. Exp Biol Med (Maywood). 2008;233(8):997–1012.
36. Rongvaux A, Willinger T, Takizawa H, Rathinam C, Auerbach W, Murphy AJ, et al. Human thrombopoietin knockin mice efficiently support human hematopoiesis in vivo. Proc Natl Acad Sci U S A. 2011;108(6):2378–83.
37. Willinger T, Rongvaux A, Strowig T, Manz MG, Flavell RA. Improving human hemato-lymphoid-system mice by cytokine knock-in gene replacement. Trends Immunol. 2011;32(7):321–7.
38. Rathinam C, Poueymirou WT, Rojas J, Murphy AJ, Valenzuela DM, Yancopoulos GD, et al. Efficient differentiation and function of human macrophages in humanized CSF-1 mice. Blood. 2011;118(11):3119–28.
39. van Lent AU, Dontje W, Nagasawa M, Siamari R, Bakker AQ, Pouw SM, et al. IL-7 enhances thymic human T cell development in "human immune system" Rag2^{-}/$^{-}$IL-2Rgammac^{-}/$^{-}$ mice without affecting peripheral T cell homeostasis. J Immunol. 2009;183(12):7645–55.
40. Mackay F, Figgett WA, Saulep D, Lepage M, Hibbs ML. B-cell stage and context-dependent requirements for survival signals from BAFF and the B-cell receptor. Immunol Rev. 2010;237(1):205–25.
41. Mackay F, Browning JL. BAFF: a fundamental survival factor for B cells. Nat Rev Immunol. 2002;2(7):465–75.
42. Rowland SL, Leahy KF, Halverson R, Torres RM, Pelanda R. BAFF receptor signaling aids the differentiation of immature B cells into transitional B cells following tonic BCR signaling. J Immunol. 2010;185(8):4570–81.
43. Schmidt MR, Appel MC, Giassi LJ, Greiner DL, Shultz LD, Woodland RT. Human BLyS facilitates engraftment of human PBL derived B cells in immunodeficient mice. PLoS ONE. 2008;3(9):e3192.
44. Traggiai E, Chicha L, Mazzucchelli L, Bronz L, Piffaretti JC, Lanzavecchia A, et al. Development of a human adaptive immune system in cord blood cell-transplanted mice. Science. 2004;304(5667):104–7.
45. Gorantla S, Sneller H, Walters L, Sharp JG, Pirruccello SJ, West JT, et al. Human immunodeficiency virus type 1 pathobiology studied in humanized BALB/c-Rag2^{-}/$^{-}$gammac^{-}/$^{-}$ mice. J Virol. 2007;81(6):2700–12.
46. Small TN, Keever C, Collins N, Dupont B, O'Reilly RJ, Flomenberg N. Characterization of B cells in severe combined immunodeficiency disease. Hum Immunol. 1989;25(3):181–93.
47. Roifman CM, Zhang J, Chitayat D, Sharfe N. A partial deficiency of interleukin-7R alpha is sufficient to abrogate T-cell development and cause severe combined immunodeficiency. Blood. 2000;96(8):2803–7.
48. Chen Q, He F, Kwang J, Chan JK, Chen J. GM-CSF and IL-4 stimulate antibody responses in humanized mice by promoting T, B, and dendritic cell maturation. J Immunol. 2012;189(11):5223–9.
49. Bunin N, Small T, Szabolcs P, Baker KS, Pulsipher MA, Torgerson T. NCI, NHLBI/PBMTC first international conference on late effects after pediatric hematopoietic cell transplantation: persistent immune deficiency in pediatric transplant survivors. Biol Blood Marrow Transplant. 2012;18(1):6–15.
50. Chiesa R, Gilmour K, Qasim W, Adams S, Worth AJ, Zhan H, et al. Omission of in vivo T-cell depletion promotes rapid expansion of naive CD4+ cord blood lymphocytes and restores adaptive immunity within 2 months after unrelated cord blood transplant. Br J Haematol. 2012;156(5):656–66.

51. Shultz LD, Ishikawa F, Greiner DL. Humanized mice in translational biomedical research. Nat Rev Immunol. 2007;7(2):118–30.
52. Legrand N, Weijer K, Spits H. Experimental models to study development and function of the human immune system in vivo. J Immunol. 2006;176(4):2053–8.
53. Danner R, Chaudhari SN, Rosenberger J, Surls J, Richie TL, Brumeanu TD, et al. Expression of HLA class II molecules in humanized NOD. Rag1KO.IL2RgcKO mice is critical for development and function of human T and B cells. PLoS ONE. 2011;6(5):e19826.
54. Tonomura N, Habiro K, Shimizu A, Sykes M, Yang YG. Antigen-specific human T-cell responses and T cell-dependent production of human antibodies in a humanized mouse model. Blood. 2008;111(8):4293–6.
55. Kalscheuer H, Danzl N, Onoe T, Faust T, Winchester R, Goland R, et al. A model for personalized in vivo analysis of human immune responsiveness. Sci Transl Med. 2012;4(125):125ra30.
56. Becker PD, Legrand N, van Geelen CM, Noerder M, Huntington ND, Lim A, et al. Generation of human antigen-specific monoclonal IgM antibodies using vaccinated "human immune system" mice. PLoS ONE. 2010;5(10):e13137.
57. Sato K, Misawa N, Nie C, Satou Y, Iwakiri D, Matsuoka M, et al. A novel animal model of Epstein-Barr virus-associated hemophagocytic lymphohistiocytosis in humanized mice. Blood. 2011;117(21):5663–73.
58. Wahl A, Linnstaedt SD, Esoda C, Krisko JF, Martinez-Torres F, Delecluse HJ, et al. A cluster of virus-encoded microRNAs accelerates acute systemic Epstein-Barr virus infection but does not significantly enhance virus-induced oncogenesis in vivo. J Virol. 2013;87(10):5437–46.
59. White RE, Ramer PC, Naresh KN, Meixlsperger S, Pinaud L, Rooney C, et al. EBNA3B-deficient EBV promotes B cell lymphomagenesis in humanized mice and is found in human tumors. J Clin Invest. 2012;122(4):1487–502.
60. Akkina R. Human immune responses and potential for vaccine assessment in humanized mice. Curr Opin Immunol. 2013;25(3):403–9.
61. Hur EM, Patel SN, Shimizu S, Rao DS, Gnanapragasam PN, An DS, et al. Inhibitory effect of HIV-specific neutralizing IgA on mucosal transmission of HIV in humanized mice. Blood. 2012;120(23):4571–82.
62. Glorius P, Baerenwaldt A, Kellner C, Staudinger M, Dechant M, Stauch M, et al. The novel tribody [(CD20)(2)xCD16] efficiently triggers effector cell-mediated lysis of malignant B cells. Leukemia. 2013;27(1):190–201.
63. Heider KH, Kiefer K, Zenz T, Volden M, Stilgenbauer S, Ostermann E, et al. A novel Fc-engineered monoclonal antibody to CD37 with enhanced ADCC and high proapoptotic activity for treatment of B-cell malignancies. Blood. 2011;118(15):4159–68.

Chapter 13
The Analysis of the Functions of Human B and T Cells in Humanized NOG Mice

Takeshi Takahashi

Over the past two decades, reconstitution of human hematopoietic and immune systems in mice has been explored using several severely-immunodeficient mouse models including NOD/scid, NOG, NSG, or BRG [1, 2]. These parental mouse strains have been modified extensively by introducing human genes or replacing mouse genes with the corresponding human genes [3]. These models are currently utilized in many fields of research, and the establishment of such humanized mice has resulted in various advances including engraftment of human hematopoietic stem cells (HSCs) [4], differentiation of multiple lineages of human cells [5, 6], and enhancement of immune responses [7–9]. However, for improved application of this technology, it is important to recognize several immunological features intrinsic to humanized mice and to consider carefully the choice of mouse strains. This chapter describes the properties of human B and T lymphocytes that develop and constitute major subpopulations in NOG mice transplanted with HSC (huHSC-NOG).

B lymphocytes Development of human B lymphocytes in NOG mice can be detected in peripheral blood at ~4–6 weeks after HSC transplantation [10]. In conventional NOG mice, most human $CD45^+$ cells in the peripheral blood are $CD19^+$ B cells, and this population rapidly increases 2–3 months after HSC transplantation. However, as development of human T cells becomes evident after 3 months, the frequency of B cells decreases gradually compared to those observed at early time points.

The phenotype of human B cells in huHSC-NOG mice generally resembles that of B cells from healthy human donors [11]. In spleen, $CD19^+$ cells comprise mainly IgM^+IgD^- and IgM^+IgD^+ cells. The phenotype of the IgM^+IgD^- population is similar to that of immature B cells in bone marrow, whereas the IgM^+IgD^+ B cells seem to represent more mature B cells. Along with T-cell differentiation, the phenotype of

T. Takahashi (✉)
Central Institute for Experimental Animals, 3-25-12 Tono-machi,
kawasaki-ku, Kawasaki 210-0821, Japan
e-mail: takeshi-takahashi@ciea.or.jp

L. Y. Poluektova et al. (eds.), *Humanized Mice for HIV Research*,
DOI 10.1007/978-1-4939-1655-9_13

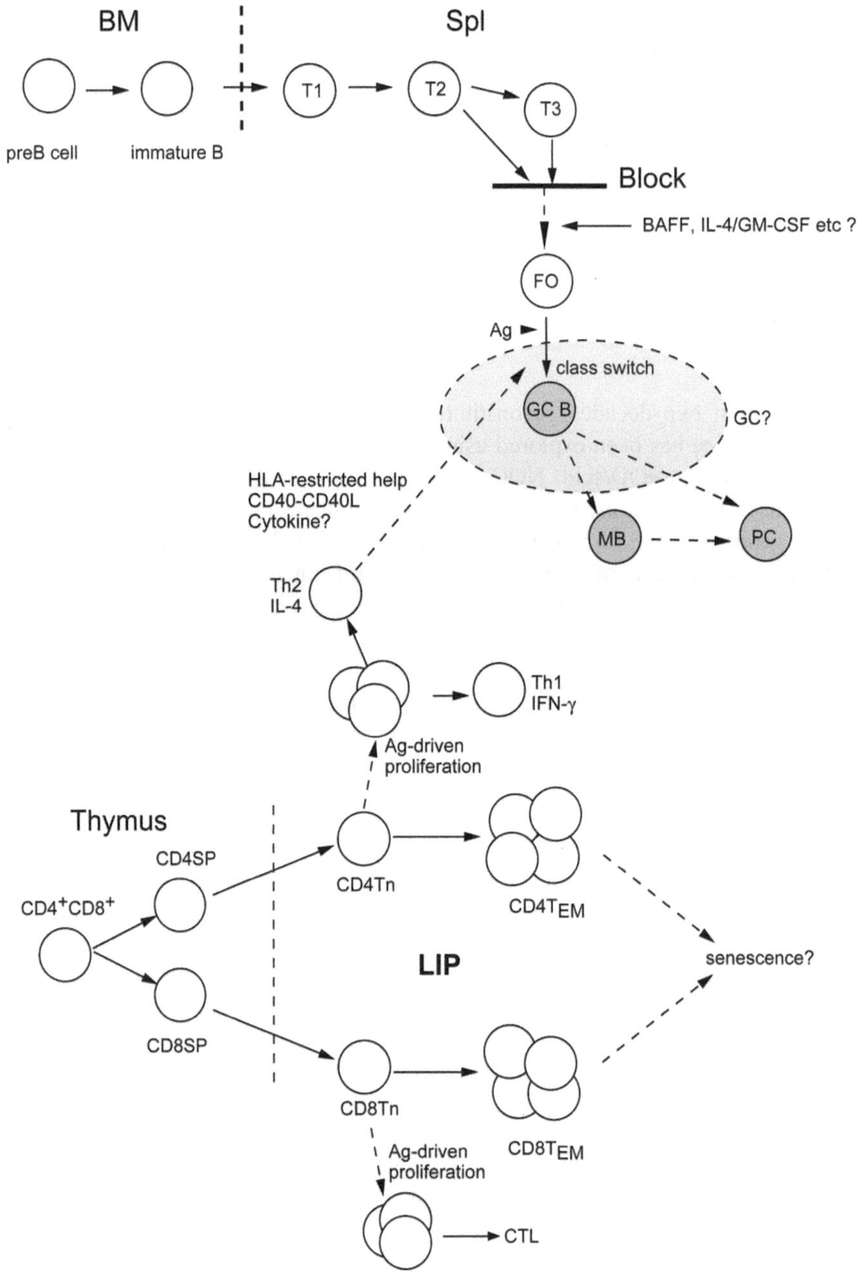

Fig. 13.1 Schematic of immune responses by human B and T cells in humanized mice. Human B cells develop in mouse bone marrow (*BM*). However, differentiation of most B cells is blocked around the transitional stages (T1–T3) before reaching mature follicular cells (FO). T cells are positively selected in the thymus in a process that depends largely on the mouse major histocompatibility complex (MHC). After migrating into the periphery, T cells are subjected to lymphopenia-induced proliferation (*LIP*), which might result in loss of T-cell function. In the case where human leukocyte antigens (HLAs) are

IgM^+IgD^+ B cells becomes more mature, resembling that of follicular B cells from normal humans [12], and a few $CD27^+$ memory B cells appear in the late period after HSC transplantation. However, there are significant differences between human B cells from donors and human B cells expressed in huHSC-NOG mice. The most prominent differences are the high CD5 and low CD21 expression levels in B cells from huHSC-NOG mouse, suggesting incomplete differentiation of human B cells in humanized mice (Fig. 13.1) [13, 14].

Regarding immunological function, human B cells from huHSC-NOG mice produce IgG in vitro in response to mitogen and anti-CD40 antibody stimulation in the presence of IL-21, suggesting that the class-switching molecular machinery is functional [14]. In addition, in vitro stimulation of human B cells from huHSC-NOG mice induces the expression of activation-induced deaminase (AID), a critical molecule for class-switching and somatic hyper mutation (SHM) [14].

B-cell function in hu-HSC NOG mice has been evaluated in a number of studies in vivo, with differing results. Initial experiments demonstrated that B cells could respond to various antigens, and that the specific antibodies produced were IgM-dominant, with rare instances of IgG [15, 16]. Thus, it was assumed that human T cells and B cells did not interact; i.e., most human T cells were positively selected by mouse major histocompatibility complex (MHC) in the mouse thymus and could not recognize peptide-human leukocyte antigen (pHLA) complexes on human B cells. To circumvent the mismatch of MHC-restricted T cells and pHLA on B cells, several HLA-DR transgenic NOG or NSG mice were generated [8, 9]. Reconstitution of these mice with HSC revealed that an IgG response was possible when the HLA-DR haplotype was matched between donor HSC and recipient mice, suggesting that humoral immune responses can be mediated in human-HSC transferred mice, albeit with limited magnitude.

Affinity maturation of antibodies is another important aspect of humoral immunity. However, due to the weak IgG response in huHSC-NOG mice, affinity maturation of antibodies has not been detected in humanized mice. It should be noted that conventional humanized mice have poorly organized follicular structure in the secondary lymphoid organs and germinal centers (GC; Fig. 13.1). Given the key role of GCs in B-cell response, this defect raises the concern of whether naive B cells can differentiate into memory B cells or plasma cells in humanized mice in a manner similar to differentiation in the human lymph node (LN) or spleen. These potential similarities require further clarification.

T lymphocytes Human T cells appear in the peripheral blood in huHSC-NOG mice at about 3 months post-HSC transplantation [10]. However, human T cells do not develop in athymic nude NOG mice (nu/nu NOG) (T.T. unpublished data), suggesting that the thymus in the recipient mouse is required for T-cell development. Most human thymocytes in huHSC-NOG mice are positively selected by mouse MHC,

---

matched between the hematopoietic stem cell (HSC) and recipient mouse, T cells can be activated in an HLA-restricted manner and differentiate into effector cells, which can support class-switching in B cells. No clear evidence has suggested germinal center (*GC*) formation or differentiation of antigen-specific memory B (*MB*) cells and plasma cells (*PC*), which are represented by *shading*

as evidenced by the fact that the number of $CD8^+$ or $CD4^+$ human T cells is reduced in beta-2-microglobulin (β2m)- or I-Aβ-deficient NOG mice [14], respectively. The development of a few human T cells in these MHC-deficient mice suggests that human thymocytes could differentiate into mature T cells in a thymic epithelial cell (TEC)-independent manner. Developing thymocytes in close proximity might provide HLA signals to facilitate maturation [17].

T-cell-mediated immunity in huHSC-NOG mice has been investigated in several reports using virus infection, including immune responses against Epstein-Barr virus (EBV) [15, 18]. Interestingly, mice that harbored solely human B cells developed B-cell lymphoma, whereas animals with both human B and T cells were protected from the development of B-cell lymphoma [15, 19]. In these animals, EBV-specific $CD8^+$ cytotoxic T lymphocytes (CTL) were detected by the specific HLA tetramer [20], and they produced cytokines in response to in vitro exposure to an EBV-transformed autologous B-cell line [18, 19]. These results suggest that human $CD8^+$ T cells developed in humanized mice have sufficient function to eradicate EBV. Considering that positive selection of human thymocytes depends on mouse MHC, it is curious that human $CD8^+$ T cells in huHSC-NOG mice can recognize EBV-derived antigens presented on HLA. One explanation is that a significant number of human $CD8^+$ T cells are positively selected through a TEC-independent mechanism, as described above. Recently, class I-HLA (HLA I)-expressing NSG mice were developed. In these animals transplanted with HLA haplotype-matched HSC, human T cells showed much clearer HLA-restricted immune responses, suggesting functional human $CD8^+$ T cells are maintained in the mouse environment [7, 19].

Regarding $CD4^+$ T cells, initial experiments using conventional NOG mice failed to show evidence of humoral immune responses since antigen-specific IgG is rarely produced by immunization or infections. HLA-DR-expressing transgenic NSG or NOG mice were capable of mounting successful IgG responses with HLA-matched HSC-transplantation, suggesting that human $CD4^+$ T cells in humanized mice can be activated in an antigen-specific manner and are able to exert helper function to B cells through pHLA and TCR interactions [8, 9]. In our hands, T cells from HLA-DR4 transgenic I-$A\beta^{-/-}$ NOG mice differentiated into IFN-γ-producing Th1 cells or IL-4-producing Th2 cells in response to in vitro stimulation (T.T., manuscript in preparation). Thus, human $CD4^+$ T cells in the mouse environment maintain the ability to differentiate into various lineages of effector cells. Likewise, differentiation into other effector lineages such as Th-17 or inducible regulatory T cells (iTreg) would be possible when the appropriate cytokine milieu is provided in mice.

It should be emphasized that T-cell homeostasis in humanized mice is not physiologically similar to that in normal humans. For example, a study using BRG mice showed that the human T cells were quickly labeled by BrdU [21]. In addition, the $CD45RA^+CD62L^{hi}$naive (Tn) T-cell phenotype in huHSC-NOG mice rapidly changed to the $CD45RO^+CD62L^{lo}$ effector/memory (T_{EM}) phenotype to produce abundant IFN-γ[9], and the accumulated T_{EM}-like cells did not proliferate or produce IL-2 in response to in vitro stimulation [14]. These studies suggest that human T cells in humanized mice are under strong pressure of lymphopenia (Fig. 13.1). Under extreme lymphopenic conditions, such as those found in NOG mice, a few human

T cells emigrate from the atrophic thymus and massively proliferate in response to an excessive amount of antigenic and cytokine signals. This lymphopenia-induced proliferation might be responsible for the reported impairment of T-cell function in conventional huHSC-NOG mice. Importantly, the phenotypes of human T cells in humanized mice are similar to those in patients with impaired thymopoiesis who had received cord-blood transplantation [22].

The bone marrow/liver/thymus (BLT) model might provide clues to the necessary components for reconstitution of a naive T-cell pool in huHSC-NOG mice [23]. In this animal, the T-cell phenotype resembles that of normal human T cells; i.e., a relatively high frequency of Tn cells is maintained. One prominent difference between the NSG- or NOG-based model and the BLT mouse is the size of the thymus. As mentioned previously, atrophy of the thymus in NOG mice is so severe that the number of human thymocytes in huHSC-NOG mice does not usually exceed 10^7. In contrast, the thymus/liver (Thy/Liv) organoid transplanted in a kidney capsule contains more than 10^8 cells [24]. Therefore, the enormous supply of T cells from the Thy/Liv organoid might help maintain Tn cells in BLT mice. Another difference between the two models is the presence or absence of LNs. LN development and LN number are significantly impaired in NSG or NOG mice [25], whereas NOD/scid mice, on which BLT mice are generated, have a normal number of LNs. This difference is attributable to the γc-deficiency in the former since the absence of IL-7-signal causes a significant decrease in lymph-tissue inducer cells (LTi) [26]. Considering that LNs produce growth factors including IL-7, the decreased LN number may affect human T-cell homeostasis. Thus, restoring LN development in γc-deficient mice is an intriguing approach to improving the naive T-cell pool.

The quality of quasi-human immune systems in humanized mice has been improved markedly by the introduction of human genes. Regarding acquired immunity, it is noteworthy that the development of human myeloid cells, including dendritic cells or macrophages, was greatly enhanced in human TPO-knock-in (KI) mice [6], human IL-3/GM-CSF KI mice [5] and IL-3/GM-CSF transgenic mice [27]. These myeloid cells will support human lymphocyte function. Additionally, various HLA-transgenic mice are useful for inducing HLA-restricted immune responses in the mouse environment. In the near future, integration of these multiple strains will enable the recapitulation of human immune responses in the mouse environment. Development of these animal models may lead to therapeutic approaches to the treatment of chronic diseases such as HIV-infection or autoimmune diseases.

References

1. McCune JM, Namikawa R, Kaneshima H, Shultz LD, Lieberman M, Weissman IL. The SCID-hu mouse: murine model for the analysis of human hematolymphoid differentiation and function. Science. 1988;241(4873):1632–9.
2. Shultz LD, Schweitzer PA, Christianson SW, Gott B, Schweitzer IB, Tennent B, et al. Multiple defects in innate and adaptive immunologic function in NOD/LtSz-scid mice. J Immunol. 1995;154(1):180–91.

3. Willinger T, Rongvaux A, Strowig T, Manz MG, Flavell RA. Improving human hemato-lymphoid-system mice by cytokine knock-in gene replacement. Trends Immunol. 2011;32(7):321–7.
4. Strowig T, Rongvaux A, Rathinam C, Takizawa H, Borsotti C, Philbrick W, et al. Transgenic expression of human signal regulatory protein alpha in Rag2−/−{gamma}c−/− mice improves engraftment of human hematopoietic cells in humanized mice. Proc Natl Acad Sci U S A. 2011;108(32):13218–23.
5. Willinger T, Rongvaux A, Takizawa H, Yancopoulos GD, Valenzuela DM, Murphy AJ, et al. Human IL-3/GM-CSF knock-in mice support human alveolar macrophage development and human immune responses in the lung. Proc Natl Acad Sci U S A. 2011;108(6):2390–5.
6. Rongvaux A, Willinger T, Takizawa H, Rathinam C, Auerbach W, Murphy AJ, et al. Human thrombopoietin knockin mice efficiently support human hematopoiesis in vivo. Proc Natl Acad Sci U S A. 2011;108(6):2378–83.
7. Shultz LD, Saito Y, Najima Y, Tanaka S, Ochi T, Tomizawa M, et al. Generation of functional human T-cell subsets with HLA-restricted immune responses in HLA class I expressing NOD/SCID/IL2r gamma(null) humanized mice. Proc Natl Acad Sci U S A. 2010;107(29):13022–7.
8. Danner R, Chaudhari SN, Rosenberger J, Surls J, Richie TL, Brumeanu TD, et al. Expression of HLA class II molecules in humanized NOD. Rag1KO.IL2RgcKO mice is critical for development and function of human T and B cells. PLoS ONE. 2011;6(5):e19826.
9. Suzuki M, Takahashi T, Katano I, Ito R, Ito M, Harigae H, et al. Induction of human humoral immune responses in a novel HLA-DR-expressing transgenic NOD/Shi-scid/gammacnull mouse. Int Immunol. 2012;24(4):243–52.
10. Ito M, Hiramatsu H, Kobayashi K, Suzue K, Kawahata M, Hioki K, et al. NOD/SCID/gamma(c)(null) mouse: an excellent recipient mouse model for engraftment of human cells. Blood. 2002;100(9):3175–82.
11. Hiramatsu H, Nishikomori R, Heike T, Ito M, Kobayashi K, Katamura K, et al. Complete reconstitution of human lymphocytes from cord blood CD34+ cells using the NOD/SCID/gammacnull mice model. Blood. 2003;102(3):873–80.
12. Lang J, Kelly M, Freed BM, McCarter MD, Kedl RM, Torres RM, et al. Studies of lymphocyte reconstitution in a humanized mouse model reveal a requirement of T cells for human B cell maturation. J Immunol. 2013;190(5):2090–101.
13. Matsumura T, Kametani Y, Ando K, Hirano Y, Katano I, Ito R, et al. Functional CD5+ B cells develop predominantly in the spleen of NOD/SCID/gammac(null) (NOG) mice transplanted either with human umbilical cord blood, bone marrow, or mobilized peripheral blood CD34+ cells. Exp Hematol. 2003;31(9):789–97.
14. Watanabe Y, Takahashi T, Okajima A, Shiokawa M, Ishii N, Katano I, et al. The analysis of the functions of human B and T cells in humanized NOD/shi-scid/gammac(null) (NOG) mice (hu-HSC NOG mice). Int Immunol. 2009;21(7):843–58.
15. Traggiai E, Chicha L, Mazzucchelli L, Bronz L, Piffaretti JC, Lanzavecchia A, et al. Development of a human adaptive immune system in cord blood cell-transplanted mice. Science. 2004;304(5667):104–7.
16. Baenziger S, Tussiwand R, Schlaepfer E, Mazzucchelli L, Heikenwalder M, Kurrer MO, et al. Disseminated and sustained HIV infection in $CD34^+$ cord blood cell-transplanted Rag2−/−gamma c−/− mice. Proc Natl Acad Sci U S A. 2006;103(43):15951–6.
17. Choi EY, Jung KC, Park HJ, Chung DH, Song JS, Yang SD, et al. Thymocyte-thymocyte interaction for efficient positive selection and maturation of CD4 T cells. Immunity. 2005;23(4):387–96.
18. Yajima M, Imadome K, Nakagawa A, Watanabe S, Terashima K, Nakamura H, et al. A new humanized mouse model of Epstein-Barr virus infection that reproduces persistent infection, lymphoproliferative disorder, and cell-mediated and humoral immune responses. J Infect Dis. 2008;198(5):673–82.
19. Strowig T, Gurer C, Ploss A, Liu YF, Arrey F, Sashihara J, et al. Priming of protective T cell responses against virus-induced tumors in mice with human immune system components. J Exp Med. 2009;206(6):1423–34.

20. Sato K, Misawa N, Nie C, Satou Y, Iwakiri D, Matsuoka M, et al. A novel animal model of Epstein-Barr virus-associated hemophagocytic lymphohistiocytosis in humanized mice. Blood. 2011;117(21):5663–73.
21. Legrand N, Cupedo T, van Lent AU, Ebeli MJ, Weijer K, Hanke T, et al. Transient accumulation of human mature thymocytes and regulatory T cells with CD28 superagonist in "human immune system" Rag2(−/−)gammac(−/−) mice. Blood. 2006;108(1):238–45.
22. Komanduri KV, St John LS, de Lima M, McMannis J, Rosinski S, McNiece I, et al. Delayed immune reconstitution after cord blood transplantation is characterized by impaired thymopoiesis and late memory T-cell skewing. Blood. 2007;110(13):4543–51.
23. Melkus MW, Estes JD, Padgett-Thomas A, Gatlin J, Denton PW, Othieno FA, et al. Humanized mice mount specific adaptive and innate immune responses to EBV and TSST-1. Nat Med. 2006;12(11):1316–22.
24. Denton PW, Nochi T, Lim A, Krisko JF, Martinez-Torres F, Choudhary SK, et al. IL-2 receptor gamma-chain molecule is critical for intestinal T-cell reconstitution in humanized mice. Mucosal Immunol. 2012;5(5):555–66.
25. Cao X, Shores EW, Hu-Li J, Anver MR, Kelsall BL, Russell SM, et al. Defective lymphoid development in mice lacking expression of the common cytokine receptor gamma chain. Immunity. 1995;2(3):223–38.
26. Meier D, Bornmann C, Chappaz S, Schmutz S, Otten LA, Ceredig R, et al. Ectopic lymphoid-organ development occurs through interleukin 7-mediated enhanced survival of lymphoid-tissue-inducer cells. Immunity. 2007;26(5):643–54.
27. Ito R, Takahashi T, Katano I, Kawai K, Kamisako T, Ogura T, et al. Establishment of a human allergy model using human IL-3/GM-CSF-transgenic NOG mice. J Immunol. 2013;191(6):2890–9.

Chapter 14
NK Cell Development in Human Immune System (HIS) Mice and Their Role in HIV Pathogenesis

Yan Li, Silvia Lopez-Lastra, Guillemette X. Masse and James P. Di Santo

14.1 Introduction

Natural killer (NK) cells were initially identified by their ability to spontaneously lyse tumor cells without prior activation, and early efforts focused on understanding the mechanisms controlling NK cell cytotoxic activity. It was later learned that NK cells use germ line-encoded inhibitory and activating NK cell receptors that distinguish self-MHC from "altered," "stressed," or nonself-MHC molecules expressed at the cell surface. Thus, NK cells and adaptive lymphocytes use fundamentally different strategies to detect invading pathogens, with NK cells surveying self MHC for alterations, as opposed to B and T lymphocytes that detect foreign pathogens or pathogen-derived peptides, respectively.

The essential role of NK cells in antiviral defense is demonstrated by the increased susceptibility to infections (particularly herpes and papillomavirus) observed in NK cell-deficient patients [1]. Upon infection, MHC class I expression is downmodulated on the infected cells, and as such, become targets for NK cell immunosurveillance. Some viruses have evolved mechanisms that allow their potential escape from NK cell recognition. These include virus encoded proteins that complex and inhibit ligands for NK cell activating receptors, as well as proteins that interfere with soluble factors required for the inflammatory response. In contrast, cellular transformation is more generally associated with the inducible expression of "stress" proteins that are recognized by activating receptors, such as NKG2D.

Silvia Lopez-Lastra has contributed equally to this chapter.

J. P. Di Santo (✉) · Y. Li · S. Lopez-Lastra · G. X. Masse
Innate Immunity Unit, Institut Pasteur, 25 rue du Docteur Roux, 75724 Paris, France
e-mail: disanto@pasteur.fr

Inserm U668, Paris, France

L. Y. Poluektova et al. (eds.), *Humanized Mice for HIV Research*,
DOI 10.1007/978-1-4939-1655-9_14

Only a fraction of peripheral NK cells (in mice or man) truly act as “natural killers,” while most NK cells possess additional effector molecules that allow them to modify cells within their microenvironment. These include soluble cytokines and chemokines that can amplify immune recognition via upregulation of MHC molecules on target cells (IFN-γ), activate neutrophils (TNF-α), recruit hematopoietic cells (GM-CSF), and potentially suppress the inflammatory response (IL-10). Thus, NK cells are involved in different steps of the immune response (detection, first-line defense, and subsequent amplification), they interface directly with cells involved in the adaptive immune responses (dendritic cell cross-talk), and they are potentially involved in the regulation of autoimmunity.

14.2 NK Cells in Mice and Man

14.2.1 NK Cell Development: A Stepwise Transcriptional Process that Generates Diverse Functional Subsets

The development of fully differentiated NK cells from pluripotent hematopoietic stem cells (HSCs) is controlled by the integrated influence of cytokines and growth-regulatory molecules that promote proliferation together with transcription factors (TFs) that activate or repress lineage-specific gene expression [2]. The result is a stepwise transformation that is accompanied by an evolving pattern of receptor expression whose combined signals control the acquisition of NK cell functional responses [3]. Three main stages of maturing NK cells can be recognized: (1) NK cell precursors (NKP) that are committed to the NK cell lineage, (2) immature NK (iNK) cells that are acquiring phenotypic and functional competence, and (3) mature NK (mNK) cells that have completed the developmental process.

The process by which a hematopoietic stem cell (HSC) becomes a committed NKP is perhaps the most poorly characterized in terms of receptor expression and transcriptional activity. Recently, this stage has been refined with the use of reporter mice and additional cell surface markers that define developmental intermediates between the common lymphoid progenitor (CLP) and a “refined” NKP with potential to give rise to only NK cells [4–6]. In contrast, human pro-NK ($CD34^+CD38^+CD45RA^+CD7^+$) and pre-NK ($CD34^+CD45RA^+CD117^+CD94^-CD122^+$) subsets retain some multi-lineage potential suggesting that these are not homogeneous populations of human NKP [7, 8].

The transcriptional activity that restricts NKP potential to the NK cell lineage is not fully understood. The inhibitor of DNA binding 2 (Id2) is critical for NK development by preventing the transcriptional activity of E box TFs [9]. Still, this is not sufficient to ensure NK cell lineage development, since Id2 is also required to generate distinct innate lymphoid cell (ILC) subsets from CLP [10]. Presumably a combination of TFs are involved that may include Tox (Thymocyte selection-associated HMG TOX protein), T-bet (Tbx21), and Eomesodermin (Eomes) [11, 12]. Recently,

Ets1 has emerged as an early regulator of NK precursor development controlling Id2 and Tbx21 expression [13]. It is perhaps through combinatorial TFs expression that the NK lineage diverges from CLP. This initial TFs profile may also encode cytokine receptors, such as CD122, required for IL15 responsiveness.

The transition from NKP to iNK cells generates cells with partially mature NK phenotypes, that lack cytotoxic and INF-γ secretion capacities. In the mouse, this stage is marked by acquisition of NK1.1 and NKp46, while in human NK cells this involves upregulation of CD56 and NKp46 expression [14]. Several TFs are expressed in iNK cells, including the nuclear factor induced by IL-3 (Nfil3/E4bp4) [15–17] and Tox [18], but little is known about how these TFs control NK cell maturation, surface expression, and functionality. Gata3 plays a role in NK cell differentiation at this stage by regulating cytokine production [19].

The terminal differentiation of mNK cells is marked by the acquisition of DX5/CD49b in mice, whereas in humans, this stage is achieved with the generation of $CD56^{hi}CD16^{-}$ and $CD56^{+}$ $CD16^{+}$ NK cell subsets. In both species, mature NK cells express receptors of the CD94/NKG2C family and also variable and stochastic expression of MHC class I-specific NK cell receptors (KIR in human, Ly49 in mouse). Many of these inhibitory receptors are involved in sensing target cell MHC expression and are critical regulators of NK cell maturation in the bone marrow and activation in the periphery. The mechanism behind the "education" process that generates KIR/Ly49 repertoires remains one of the unsolved enigmas in NK cell biology. Other surface molecules involved in processes such as cell adhesion and activation increase expression on mature NK cells compared to iNK, including CD11a, CD11b, CD16, and CD43. The functional capacity of NK cells also increases as cells progress to mNK: perforin, granzymes, cytokines (IFN-γ, TNFs), concomitantly with an upregulation of Blimp1, Eomes, Tbx21, MITF, MEF, and CEBP-γ. While not all TF/target gene associations are completely understood, both Blimp1 and Tbx21 appear to directly regulate cytokine, perforin, and granzymes B production [11, 20].

14.2.2 Roles for Cytokines in NK Cell Development and Function

NK cell differentiation is regulated by several distinct cytokines among which the common γ-chain-dependent cytokines have a decisive role. In particular, IL-2, IL-7, and IL-15 display specific and overlapping roles in the generation and maintenance of NK cell and may have a differential impact on the development of specific NK subsets [21]. Studies have shown that $CD56^{hi}CD16^{-}$ NK cells in man express the high affinity trimeric IL-2Rαβγ receptor and are exquisitely sensitive to low levels of IL-2, whereas $CD56^{+}$ $CD16^{+}$ NK cells in man and all NK cells in mice express the intermediate affinity IL-2Rβγ complex and can only be triggered by higher (nonphysiological) levels of IL-2 [22–24]. Moreover, mice or humans deficient of IL-2 or IL-2Rα do not have apparent defect in peripheral NK cell reconstitution [25–27]. These studies indicate that IL-2 may have a dose-dependent effect on the maturation and function of peripheral NK subtypes, but does not play a role in mainstream NK cell development in either species. IL-15 exerts its effects via "trans-presentation" by soluble IL-15Rα or through membrane-bound IL-15Rα to cells expressing the IL-2Rβγ complex [28–35]. In contrast to IL-2, mice deficient for IL-15 or IL-15Rα have defects in NK population, which confirms the primary role of IL-15 for the development and homeostasis of NK cells [36, 37]. Abundant granzyme B and perforin can be produced from prestored mRNA in NK cells after stimulation of IL-2 or IL-15 to increase NK cell-mediated cytolytic activity [38, 39].

Unlike IL-2 and IL-15, IL-7Rα is not expressed by most circulating NK cells. However, a subset of NK1.1^{+} cells in murine thymus expresses IL-7Rα as do the human CD56^{hi}CD16^{-} subset [40, 41]. Mouse CD127^{+} NK cells distinguish themselves from classical NK cells by their dependency on IL-7 and GATA-3, and display high cytokine secretion potential but low cytolytic activity. Whether IL-7 can boost the homeostasis of NK cell subsets in vivo remains to be tested.

Apart from common γ-chain-dependent cytokines, many other cytokines can function to regulate mature NK cell responses. For example, IFN-α/β is known to induce cytolytic activity of NK cell during viral infections, IL-12 and IL-18 are potent stimulators of IFN-γ secretion, and IL-27 can promote "regulatory" activity (IL-10) from human NK cells [42]. In addition, negative regulators of NK cell function likely include IL-10, TGF-β and in some cases, type I interferons [43].

14.2.3 Mechanisms of NK-Cell-Mediated Responses

The major effector functions of NK cells (cytotoxicity and cytokine secretion) are controlled through an integration of signals that arise from an array of activating and inhibiting receptors and the presence of pro-inflammatory cytokines. The inhibitory receptors, encoded by the killer Ig-like receptor (KIR) multi-gene family in humans and the C-type lectin-like genes family (Ly49) in mice, are responsible for the MHC "self-tolerance" through recognition of host MHC class I molecules (HLA in humans) [44]. Loss of inhibition allows the subsequent activation of the "educated" NK cells in immune responses ("missing-self hypothesis"). Rarely the situation is "black and white" with complete loss of MHC expression. More often, it is the balance between activation and inhibition that determines the outcome. For example, when a NK cell recognizes a stress- or infection-induced ligand via an activating receptor (such as CD16, NKG2D, or natural cytotoxicity receptors NKp30, NKp44 and NKp46) in a low MHC-I expression setting, the NK cell becomes activated and the "killing machinery" is engaged. This involves the exocytosis of cytotoxic granules containing perforin and granzymes via an immunological synapse. Additionally, small peptides and the pro-apoptotic molecules Fas ligand and TRAIL (TNF-related apoptosis-inducing ligand) may also be delivered. These induce a death-inducing signaling complex that leads to the induction of caspase-mediated apoptosis [45, 46].

NK cells bear the CD16 receptor (Fcγ RIIIa) for IgG and can mediate antibody-dependent cellular cytotoxicity (ADCC) especially against antibody-coated tumor cells. CD16 engagement triggers NK cell activation, cytokine production, and degranulation that lead to target cell lysis. Some HIV-specific peptides have been proposed as possible ADCC targets for a NK cell-based vaccine, as reviewed below, as well as, epitopes unregulated during tumorigenesis [47].

Human and mouse NK cells are responsive to soluble factors present in the environment, such as IL-12 and IL-18, that lead to robust secretion of IFN-γ, TNF-α, granulocyte macrophage-colony-stimulating factor (GM-CSF), IL-10, and IL-13 [48]. Some subsets of NK cells appear more responsive in this respect ($CD56^{hi}CD16^{-}$ human NK cells, $CD127^{+}$ mouse NK cells), but it should be stressed that production of certain cytokines (for example IFN-γ) appears to be a general property of all NK cells. The cytokine-mediated activation of NK cells shapes the subsequent recruitment and function of other hematopoietic cells. As an example, during infection NK cells regulate CD4 T cell-mediated support for the antiviral CD8 T cells as well as recruitment and activation of myeloid cells [49, 50].

Several observations from in vitro experiments and studies with mouse models as well as with tumor and transplantation patients confirmed the regulatory effect of Treg cells on NK cell activity. Treg cells inhibit NK cell proliferation, cytotoxicity and IFN-γ secretion [51] that may be counteracted by the accumulation of IL-15Rα expressing DCs, or by an IL-2-dependent signal [52]. Another possible player in the crosstalk network around NK cells are neutrophils [53], although the extent to which these two cell types cross-regulate activities in vivo remains unknown.

14.3 NK Cells in HIV Infection

14.3.1 NK Cell-Mediated Control of HIV Infection

NK cell-mediated innate immunity appears to play a pivotal role in restraining HIV-1 viral replication at earlier stages of infection as demonstrated by the correlation between the levels of viremia observed in patients and the expression of specific combinations of KIR-MHC class I alleles [54]. Moreover, increased NK-cell mediated responses have been documented in sero-negative exposed individuals compared to healthy and HIV-1-positive patients [55]. In general, an increase in $CD56^{+}$ $CD16^{+}$ subset and a reduction in $CD56^{hi}CD16^{-}$ NK cells are observed early in HIV-1 infection. As the infection becomes chronic, Siglec-7 expression on NK cells increases, followed by the emergence of a population of functionally anergic NK cells that are $CD56^{-}$ and present increased expression of inhibitory receptors [56, 57]. These observations suggest that "exhausted" NK cell phenotypes can result from excessive activation in the context of prolonged virus replication.

The putative mechanisms that NK cells use to restrict HIV replication are not fully elucidated. Both cytolytic and secretory mechanisms are possible, including granule-mediated lysis of virally infected target cells and/or production of soluble factors that limit viral spread. Along the latter, NK cells can produce CC-chemokine ligand 3 (CCL3), CCL4, and CCL5, all of which are ligands for CCR5 co-receptor required for HIV entry [58].

For many years, the focus of research has been on the identification of viral proteins or epitopes that could be targets for elimination of virally infected cells by ADCC. However, the wide variety of viral strains complicates the identification of conserved structures capable of triggering a robust and broadly reactive ADCC response.

Epidemiological studies have delved into the importance of KIR-HLA interactions in the NK cell-mediated control of HIV, which is already extensively appreciated for T-cell responses, however, there is not yet complete consensus regarding the specific allele variants associated with better HIV-1 disease outcomes [59.

14.3.2 Viral Evasion and Modulation of NK Cell Phenotype and Function in HIV-Infected Patients

NK cell function in HIV-infected patients has been shown to be defective in many respects. One direct mechanism by which HIV can directly modulate NK function is through its viral proteins. HIV protein Tat inhibits NK-mediated cytolysis by blocking calcium channel activity. This in turn inhibits the cytolytic capacity of NK cells through two distinct mechanisms: reduced degranulation and inhibition of FasL upregulation [60, 61]. Another example concerns the HIV envelope protein gp120. Binding of gp120 to CXCR4, constitutively expressed on NK cells, can induce autophagy leading to NK cell death [62].

While many viruses downregulate MHC class I molecule during infection, HIV-infected cells retain relatively normal levels of MHC class I molecules. Instead, HIV protein Nef, selectively reduces expression of HLA-A and HLA-B that are mainly used to present viral peptides to cytotoxic T cells. In contrast, Nef does not modify expression of HLA-C and HLA-E molecules that serve as ligands for inhibitory NK cell receptors [63–65]. Nolting et al. reported that chronic HIV-1 infection is associated with a specific defect of NKG2D-mediated NK cell activation, due to elevated levels of the soluble (shed) MICA that are released by HIV-1-infected $CD4^{+}$ T cells into patient sera. Increased serum MICA results in NKG2D downregulation and a profound dysregulation of NK cells effector functions [66]. Hence, by differentially modulating HLA molecules and NKG2D ligands on infected cells, HIV escapes from most antiviral activities mediated by CD8 T cell and NK cells.

HIV infection perturbs the expression of activating and inhibitory receptors on NK cells resulting in the modulation of NK cell function. For example, NK cells from HIV-infected patients have increased expression of receptors like CD94, CD161, and KIR2DL1 [67, 68], and decreased expression of NKp44 and 2B4 [69, 70]. The molecular mechanisms that generate NK receptor expression modulation during HIV infection remain unknown. Possible causes include HCMV coinfection or chronic activation of immune system following loss of containment of gut commensal flora.

NK cell activation can be modulated by a "cross-talk" between NK cells and other immune cells, especially dendritic cells. pDCs are the major type I IFN-producing cells during viral infection, and IFN α/β primes NK cells for cytolytic activity. During the course of disease, HIV-infected pDC are progressively depleted resulting in a decrease of IFN-α levels [71, 72]. This may be one mechanism behind the suppression of NK-mediated lysis of autologous-infected $CD4^+$ T cells in HIV-infected patients. "Classical" cDC from HIV patients produce less IL-12 and subsequently fail to activate NK cells for IFN-γ and TNF-α secretion [73, 74]. Saidi et al. demonstrated that the cytokine high mobility group box protein 1 (HMGB1) plays a role as an immune "alarmin" during HIV infection [75]. HMGB1 is an endogenous danger signal produced by innate effectors such as NK cells, which can be released during states of cellular stress or damage. HMGB1 normally promotes the NK-DC cross talk but, in the context of chronic HIV infection, triggers viral replication in DC and blocks NK-mediated killing of infected DC thus contributing to viral persistence [76]. Increased levels of circulating HMGB1 are detected in HIV-1-infected individuals and are associated with high viral loads [77]. Finally, $FoxP3^+$ Tregs can suppress both T and NK cell responses, and increased numbers of activated Tregs have been observed in HIV patients [78]. Treg suppression of NK cell function may be mediated by TGF-β-dependent NKG2D downregulation [79]. Moreover, circulating levels of other immunosuppressive cytokines (like IL-10) are increased in HIV-infected patients [80]. Collectively, the dysregulated NK cell "cross talk" results in reduced NK cell functions and an inefficient antiviral response during chronic HIV infection.

While it is generally assumed that NK cells play a critical role in the control of HIV infection, in some cases, a paradoxical role for NK cells in promoting HIV replication or pathology has been observed. A conserved peptide in gp41 (called 3S) can induce the expression of NKp44 ligands (NKp44L) on noninfected $CD4^+$ T cells rendering them targets for NK cell elimination [81]. Nef can inhibit NKp44L expression thereby protecting infected $CD4^+$ T cells from NK cell lysis [82]. Blocking NKp44L expression by anti-3S antibodies protects autologous $CD4^+$ T cells in SHIV-infected macaques from NK cell cytotoxic activity, and can preserve CD4 central memory T cells [83]. Viral proteins may, therefore, reconfigure innate immune cell function to promote viral persistence.

14.3.3 Advances in HIV Therapies that Target NK Cells

In the past three decades, tremendous efforts have been devoted to create an efficient vaccine to combat HIV. The discovery of new broadly neutralizing Abs and vector-based vaccines to induce cell-mediated immunity provide hope; however, we still lack an efficient vaccine that can induce HIV-specific adaptive immunity. In recent years, the role for innate immune responses, and especially those mediated by NK cells, in the early resistance to HIV and in the control of persistent infection have been extensively studied. While HAART treatment can provide an important degree of viremic control, NK cell numbers and functions are generally not normalized during treatment suggesting the existence of additional mechanistic deficiencies [80, 84–86].

Several new approaches have been proposed to restore NK cell function in HIV-infected individuals. Expression of inhibitory receptors on NK cells increases in HIV-infected patients, and blocking these receptors with mAbs or small molecule compounds may help increase NK cell activation. Indeed, mAb masking of inhibitory KIRs increases the cytolytic activities against autologous-infected cells [87]. Still, NK cells may be generally activated by this approach, and potential side effects need to be carefully assessed before use in HIV patients.

As cytokine levels are dysregulated in HIV-infected patients, exogenous delivery of IL-2, IL-15, and/or IFN-α/β may help boost the NK cell homeostasis and function. In particular, IL-2 has been approved clinically as a therapy to treat metastatic melanoma and renal cancers [88]. Although the treatment is associated with dose-dependent toxicity, anti-tumor NK and cytotoxic T cell responses are significantly improved. Still, immune cells are activated and Tregs are expanded during IL-2 treatment [89], so this treatment may lead to increased HIV infection and higher viral load. The combination of HAART and low-dose IL-2 treatment may minimize the viral production while enhancing the antiviral activities. IL-15 also holds promise as an adjuvant for HIV therapy. Compared to IL-2, IL-15 is less toxic and has lower mitogenic potential [90]. IL-15 inhibits apoptosis of NK and CD8 T cells from HIV-infected patients by upregulating the expression of Bcl-X_L. However, recent animal study suggests that although transient stimulation of IL-15/IL-15Rα complexes in vivo induces NK cell proliferation and activation, prolonged stimulation may impair NK phenotype and function [91].

14.4 Human Immune System (HIS) Mice: A Tool to Study Human NK Cell Biology In Vivo

14.4.1 Human NK Cell Reconstitution in HIS Mice

Given the central role of NK cells in maintaining immune homeostasis and for immune responses during infection, inflammation, and cancer, an in vivo model that can allow a better understanding of human NK cell development and function has multiple applications. Human immune system (HIS) mice are generated after transfer of human CD34$^+$ cells into suitably conditioned immunodeficient mice; several different HIS mouse models are available and the reader is referred to recent reviews that describe these models in detail [93]. Most HIS mice demonstrate clear evidence of multilineage human hematopoietic reconstitution, with strong development of lymphoid cells and lower levels of myeloid cell reconstitution. Still, within the lymphoid lineage, not all lymphocyte subsets are reconstituted with equal efficiency. Human B and T cells are in general, well represented in HIS mice. In contrast, human NK cell reconstitution, as well as other innate lymphocyte subsets (γδ T cells, NK-T cells) remain relatively low in these models.

Despite the low levels of NK cell reconstitution, several studies managed to characterize the phenotypes and functions of the few NK cells that develop in NOD-based (NOG and NSG) HIS mice. Human NK cells are present in bone marrow and spleen of reconstituted NOG mice, and show cytolytic function against K562 tumor cells at high effector to target (E:T) ratios [94]. Human NK cells from $CD34^+$-transplanted NSG mice and supplemented with human IL-7 were generated that lack the surface expression of inhibitory receptors like KIRs. Most of these human NK cells displayed a $CD56^{hi}CD16^-$ phenotype and appeared functionally inert as they failed to kill K562 cells at low E:T ratios and only poorly produced IFN-γ after stimulation [95]. Another study showed that almost half of $NKp46^+$ NK cells in NSG mice are $CD56^-$ and require in vivo or in vitro activation to become $CD56^+$ [96].

Similarly, systemic NK reconstitution in BALB/c $Rag2^{-/-}Il2rg^{-/-}$ (BRG) recipients is poor in terms of absolute numbers of $CD3^-NKp46^+$ cells [30], [97, 98]. Human NK cells from BRG mice express NKp46, NKG2D, CD122, CD94/NKG2A, and CD161 similar to their native human counterparts; however, the frequency of KIR^+ NK cells is reduced. Despite this difference, human NK cells in BRG mice appear functionally competent and degranulate when cocultured with tumor cell lines [30, 99].

In the NOD–SCID–bone marrow–liver–thymus model (BLT mice), T cell development is prominent and highly efficient due to T cell education in a human thymic microenvironment. This results in a systemic T cell reconstitution (lymphoid and nonlymphoid tissue, including gut, salivary glands, skin) that generates strong adaptive T cell responses. Nevertheless, NK cell reconstitution in BLT mice is still poor in terms of number of $CD3^-CD56^+$ NK cells [100].

14.4.2 Approaches to Improve Human NK Cell Reconstitution and Function in HIS Mice

NK cells represent 10–15 % of lymphocytes in human blood, whereas in the current HIS models $CD56^+$ NK cells rarely exceed 1–3 % of the human $CD45^+$ cell gate. Multiple mechanisms may explain the poor human NK cell reconstitution and function in HIS mouse models. One possibility is that IL-15 availability differs in the different recipient strains. NOD mice have a defective IL-15 allele, which normally contributes to mouse NK cell defect in this strain and might also affect human NK cell development in NSG HIS mice [101]. While the IL-15 allele in the BALB/c background is functional, the levels of human NK cells in BRG HIS mice are not dramatically improved compared to NSG HIS mice. This observation raises the intriguing possibility that mouse IL-15 might not be the driving force behind human NK cell development in HIS mice. While mouse IL-15 can cross-react weakly with human cells [30], human IL-15 is more efficient in stimulating IL-15-reponsive lymphocytes. Interestingly, in vivo neutralization of mouse IL-15 had little effect on human NK cell homeostasis in HIS mice, whereas blocking human IL-15 strongly reduced human NK cell development [30]. Moreover, exogenous treatment of HIS mice with human IL-15 was shown to strongly improve human NK cell development and function in HIS mice [30, 98, 102].

Another difference in HIS models involves the host macrophage response to xenografted cells. SIRPα is an inhibitory receptor expressed by macrophages that recognizes CD47 expressed on most nucleated normal cells [103]. The interaction between SIRPα and CD47 delivers a "Do not eat me" signal to the macrophage and thereby protects the $CD47^{+}$ target cell from phagocytosis. Human CD47 poorly interacts with most mouse SIRPα alleles, although in NOD mice, $SIRP\alpha^{NOD}$ binds human CD47 with high affinity and provides a mechanism to explain the higher reconstitution of human xenografts in NOD-based immunodeficient mice [103]. As such, another possible reason for the poor human NK cell development in BRG HIS mice could be the enhanced phagocytosis of human NK cells due to poor SIRPα–CD47 interactions. This hypothesis has been tested using three different approaches: forced expression of mCD47 on human cells before reconstitution of BRG mice, creation of BRG $SIRP\alpha^{NOD}$ congenic mice and creation of BRG hSIRPα transgenic mice [104, 105]. In all these cases, human NK cell numbers are increased significantly but the frequencies of these cells are still low compared to that observed in human peripheral blood. Clearly additional mechanisms play a role in this process and further work is required in this area.

A plethora of hematopoietic and nonhematopoeitic cells intimately regulate NK cell development and function. NK differentiation involves the interaction of NK cells with stromal cells and DC, pDC, Treg, and myeloid cells [106–108]. However, many of these cell types are not efficiently reconstituted in HIS mice, and stromal cells are by definition of mouse origin in the HIS model. Studies have shown that distinct cytokines can selectively boost the engraftment of DC and Tregs in HIS mice [102, 109, 110]. In the future, it will be interesting to systematically study the effects of diverse human cytokines (alone or in combination) for NK cell homeostasis and function in HIS mouse; for these studies, hydrodynamic injection may offer a flexible approach to rapidly test cytokine effects.

14.4.3 Future Development of HIS Mice to Study NK Cell Responses in HIV Infection

HIV tropism is limited to human and chimpanzees [111]. As many countries either ban or severely restrict the research on great apes, there is a pressing need to identify alternative in vivo models that can recapitulate the kinetics of HIV infection in man, and can be used to develop new therapeutics and screen vaccine candidates. A relatively simple, reproducible, small sized ,and less costly animal model, ideally based in rodents, could provide a means to better understand HIV pathophysiology and test new therapies.

In recent years, HIS mice have been used to model human HIV infections [111–114]. HIS mice have several unique advantages for the study of HIV. First, multiple cohorts of HIS mice can be generated from multiple, genetically distinct donors. This overcomes the intragenetic variables generated from clinical studies on pooled data from patient samples but also allows for analysis of genetic differences. Second, the continuous supply of newly generated immune cells in vivo during the course of HIV infection enables longitudinal studies on HIV viral persistence and evolution. Third, both laboratory-adapted and clinical isolates of HIV virus can be used in HIS mice to study HIV transmission and human immune response. Fourth, the access to all immune tissues, and in particular mucosal sites, will allow in-depth investigation of the biology of HIV, which has been limited in humans because of the difficulty in obtaining biopsies from infected patients. Moreover, multiple routes of HIV transmission have been shown to elicit pathology in HIS mice. Finally, new therapeutic or prophylactic approaches may be tested in HIS mice because of the capacity of these models to recapitulate normal human immune responses.

In recent years, HIS mice have provided a better understanding of the early events of HIV infection and represent a useful platform to test new therapeutic approaches [115–118]. The success in modeling HIV infection in HIS mice has been tempered by the relatively weak anti-HIV adaptive immune responses that can be elicited in this context [112–114]. It is not known whether anti-HIV NK cell responses occur in infected HIS mice, partly due to the inadequate reconstitution of human NK cells in HIS mice in general. Some of the aforementioned strategies to boost NK cell homeostasis in HIS mice should facilitate the study of NK cells during HIV infection.

As explained earlier, it is not fully understood if NK cells are beneficial or detrimental in the control of HIV infection. HIS mice offer a means to dissect the role for NK cells in HIV biology. Whether human NK cells eliminate virally infected T cells and DC or alternatively whether they attack uninfected human CD4 T cells expressing NKp44 ligands induced by gp41-derived peptides can be tested. Once established, blocking antibodies or compounds could be tested to modulate the beneficial detrimental effect from human NK cells on the evolution of HIV infection and disease.

Harnessing the power of NK-mediated ADCC in HIS model may be a new approach to identify novel HIV therapeutics. Previous studies on antibody responses against HIV infection were focused on neutralizing effect to block the entry of HIV viruses. Recent findings indicate that many previously identified nonneutralizing antibodies may actually play a role via ADCC. HIS mice with boosted NK cells could offer an in vivo model to evaluate the ADCC ability of those antibodies. In addition, as activated and proliferating T cells are the major targets for HIV infection, administration of anti-CD25 antibody may specifically deplete those cells by NK-mediated ADCC. This hypothesis can be tested using HIS mice boosted with human NK cells.

The absence of classical and nonclassical HLA class I molecules in most HIS models pose an issue of proper human NK cell education and function. In current HIS models, NK cells express low levels of KIRs, produce less IFN-γ, and display lower cytolytic activity after stimulating with cytokines and tumor cells. Despite that human T and B cells in secondary lymphoid tissues might serve as a source of HLA class I molecules, inefficient "educating" HLA-KIR interactions may also account for these phenotypic and functional NK cell deficiencies observed in HIS mice. The use of HLA class I expressing HIS mice [119, 120] may provide a means to study the impact of MHC-KIR interactions to shape NK-mediated responses during HIV infection.

Vaginal and rectal transmission of HIV accounts for the majority of new transmissions and early virus replication occurs primarily in gastrointestinal tract [111]. As such, NK-mediated innate immunity may be critical for the early control of HIV replication at mucosal sites. However, the reconstitution of human NK cells at these sites is rather poor in most HIS models. Whether this is secondary to poor lymphoid structure development [121], or to defective NK cell maturation and homing, remains unclear. Alternatively, activation of the murine innate system during HIV infection, due to indirect effects of human cytokines that cross-react with mouse hematopoietic and stromal components, may also dampen human innate (and adaptive) immune responses. These are challenging issues for the future use and development of HIS mice to study HIV infection and disease pathogenesis.

14.5 Concluding Remarks

HIS mouse models provide a novel tool to better understand the contribution of NK cells to the immune response against HIV. Functional subsets of human NK cells can develop in HIS mice; these same mice have been shown to be permissive for HIV infection, replication, and its subsequent disease manifestations. HIS mice can therefore provide a basis to experimentally decipher the role for human NK cells in HIV pathophysiology. The ability to modulate NK cell function in HIS mice constitutes a tractable translational tool that could contribute to the development of NK cell-based therapies and vaccines that limit HIV replication or accelerate the immune control of the disease.

References

1. Orange JS. Human natural killer cell deficiencies. Curr Opin Allergy Clin Immunol. 2006;6:399–409.
2. Di Santo JP. Natural killer cell developmental pathways: a question of balance. Annu Rev Immunol. 2006;24:257–86.
3. Lanier LL. Up on the tightrope: natural killer cell activation and inhibition. Nat Immunol. 2008;9:495–502.

4. Fathman JW, Bhattacharya D, Inlay MA, Seita J, Karsunky H, Weissman IL. Identification of the earliest natural killer cell-committed progenitor in murine bone marrow. Blood. 2011;118:5439–47.
5. Carotta S, Pang SH, Nutt SL, Belz GT. Identification of the earliest NK-cell precursor in the mouse BM. Blood. 2011;117:5449–52.
6. Rosmaraki EE, Douagi I, Roth C, Colucci F, Cumano A, Di Santo JP. Identification of committed NK cell progenitors in adult murine bone marrow. Eur J Immunol. 2001;31:1900–9.
7. Freud AG, Yokohama A, Becknell B, Lee MT, Mao HC, Ferketich AK, Caligiuri MA. Evidence for discrete stages of human natural killer cell differentiation in vivo. J Exp Med. 2006;203:1033–43.
8. Freud AG, Caligiuri MA. Human natural killer cell development. Immunol Rev. 2006;214:56–72.
9. Yokota Y, Mansouri A, Mori S, Sugawara S, Adachi S, Nishikawa S, Gruss P. Development of peripheral lymphoid organs and natural killer cells depends on the helix-loop-helix inhibitor Id2. Nature. 1999;397:702–6.
10. Spits H, Di Santo JP. The expanding family of innate lymphoid cells: regulators and effectors of immunity and tissue remodeling. Nat Immunol. 2011;12:21–7.
11. Gordon SM, Chaix J, Rupp LJ, Wu J, Madera S, Sun JC, Lindsten T, Reiner SL. The transcription factors T-bet and Eomes control key checkpoints of natural killer cell maturation. Immunity. 2012;36:55–67.
12. Di Santo JP. The innate side of TOX. Nat Immunol. 2010;11:885–6.
13. Ramirez K, Chandler KJ, Spaulding C, Zandi S, Sigvardsson M, Graves BJ, Kee BL. Gene deregulation and chronic activation in natural killer cells deficient in the transcription factor ETS1. Immunity. 2012;36:921–32.
14. Di Santo JP. Natural killer cells: diversity in search of a niche. Nat Immunol. 2008;9:473–5.
15. Male V, Nisoli I, Gascoyne DM, Brady HJ. E4BP4: an unexpected player in the immune response. Trends Immunol. 2012;33:98–102.
16. Kamizono S, Duncan GS, Seidel MG, Morimoto A, Hamada K, Grosveld G, Akashi K, Lind EF, Haight JP, Ohashi PS, Look AT, Mak TW. Nfil3/E4bp4 is required for the development and maturation of NK cells in vivo. J Exp Med. 2009;206:2977–86.
17. Gascoyne DM, Long E, Veiga-Fernandes H, Boer J de, Williams O, Seddon B, Coles M, Kioussis D, Brady HJ. The basic leucine zipper transcription factor E4BP4 is essential for natural killer cell development. Nat Immunol. 2009;10:1118–24.
18. Aliahmad P, Seksenyan A, Kaye J. The many roles of TOX in the immune system. Curr Opin Immunol. 2012;24:173–7.
19. Samson SI, Richard O, Tavian M, Ranson T, Vosshenrich CA, Colucci F, Buer J, Grosveld F, Godin I, Di Santo JP. GATA-3 promotes maturation, IFN-gamma production, and liver-specific homing of NK cells. Immunity. 2003;19:701–11.
20. Kallies A, Carotta S, Huntington ND, Bernard NJ, Tarlinton DM, Smyth MJ, Nutt SL. A role for Blimp1 in the transcriptional network controlling natural killer cell maturation. Blood. 2011;117:1869–79.
21. Rochman Y, Spolski R, Leonard WJ. New insights into the regulation of T cells by gamma(c) family cytokines. Nat Rev Immunol. 2009;9:480–90.
22. Tsudo M, Goldman CK, Bongiovanni KF, Chan WC, Winton EF, Yagita M, Grimm EA, Waldmann TA. The p75 peptide is the receptor for interleukin 2 expressed on large granular lymphocytes and is responsible for the interleukin 2 activation of these cells. Proc Natl Acad Sci USA. 1987;84:5394–8.
23. Caligiuri MA, Zmuidzinas A, Manley TJ, Levine H, Smith KA, Ritz J. Functional consequences of interleukin 2 receptor expression on resting human lymphocytes. Identification of a novel natural killer cell subset with high affinity receptors. J Exp Med. 1990;171:1509–26.
24. Nagler A, Lanier LL, Phillips JH. Constitutive expression of high affinity interleukin 2 receptors on human CD16-natural killer cells in vivo. J Exp Med. 1990;171:1527–33.
25. Sadlack B, Lohler J, Schorle H, Klebb G, Haber H, Sickel E, Noelle RJ, Horak I. Generalized autoimmune disease in interleukin-2-deficient mice is triggered by an uncontrolled activation and proliferation of CD4+ T cells. Eur J Immunol. 1995;25:3053–9.

26. Willerford DM, Chen J, Ferry JA, Davidson L, Ma A, Alt FW. Interleukin-2 receptor alpha chain regulates the size and content of the peripheral lymphoid compartment. Immunity. 1995;3:521–30.
27. Orange JS. Human natural killer cell deficiencies and susceptibility to infection. Microbes Infect. 2002;4:1545–58.
28. Dubois S, Mariner J, Waldmann TA, Tagaya Y. IL-15Ralpha recycles and presents IL-15 In trans to neighboring cells. Immunity. 2002;17:537–47.
29. Giron-Michel J, Giuliani M, Fogli M, Brouty-Boye D, Ferrini S, Baychelier F, Eid P, Lebousse-Kerdiles C, Durali D, Biassoni R, Charpentier B, Vasquez A, Chouaib S, Caignard A, Moretta L, Azzarone B. Membrane-bound and soluble IL-15/IL-15Ralpha complexes display differential signaling and functions on human hematopoietic progenitors. Blood. 2005;106:2302–10.
30. Huntington ND, Legrand N, Alves NL, Jaron B, Weijer K, Plet A, Corcuff E, Mortier E, Jacques Y, Spits H, Di Santo JP. IL-15 trans-presentation promotes human NK cell development and differentiation in vivo. J Exp Med. 2009;206:25–34.
31. Mortier E, Woo T, Advincula R, Gozalo S, Ma A. IL-15Ralpha chaperones IL-15 to stable dendritic cell membrane complexes that activate NK cells via trans presentation. J Exp Med. 2008;205:1213–25.
32. Olsen SK, Ota N, Kishishita S, Kukimoto-Niino M, Murayama K, Uchiyama H, Toyama M, Terada T, Shirouzu M, Kanagawa O, Yokoyama S. Crystal structure of the interleukin-15. interleukin-15 receptor alpha complex: insights into trans and cis presentation. J Biol Chem. 2007;282:37191–204.
33. Stonier SW, Schluns KS. Trans-presentation: a novel mechanism regulating IL-15 delivery and responses. Immunol Lett. 2010;127:85–92.
34. Mortier E, Bernard J, Plet A, Jacques Y. Natural, proteolytic release of a soluble form of human IL-15 receptor alpha-chain that behaves as a specific, high affinity IL-15 antagonist. J Immunol. 2004;173:1681–8.
35. Mortier E, Quemener A, Vusio P, Lorenzen I, Boublik Y, Grotzinger J, Plet A, Jacques Y. Soluble interleukin-15 receptor alpha (IL-15Ralpha)-sushi as a selective and potent agonist of IL-15 action through IL-15R beta/gamma. Hyperagonist IL-15 x IL-15R alpha fusion proteins. J Biol Chem. 2006;281:1612–9.
36. Kennedy MK, Glaccum M, Brown SN, Butz EA, Viney JL, Embers M, Matsuki N, Charrier K, Sedger L, Willis CR, Brasel K, Morrissey PJ, Stocking K Schuh JC, Joyce S, Peschon JJ. Reversible defects in natural killer and memory CD8 T cell lineages in interleukin 15-deficient mice. J Exp Med. 2000;191:771–80.
37. Lodolce JP, Boone DL, Chai S, Swain RE, Dassopoulos T, Trettin S, Ma A. IL-15 receptor maintains lymphoid homeostasis by supporting lymphocyte homing and proliferation. Immunity. 1998;9:669–76.
38. Fehniger TA, Cai SF, Cao X, Bredemeyer AJ, Presti RM, French AR, Ley TJ. Acquisition of murine NK cell cytotoxicity requires the translation of a pre-existing pool of granzyme B and perforin mRNAs. Immunity. 2007;26:798–811.
39. Gamero AM, Ussery D, Reintgen DS, Puleo CA, Djeu JY. Interleukin 15 induction of lymphokine-activated killer cell function against autologous tumor cells in melanoma patient lymphocytes by a CD18-dependent, perforin-related mechanism. Cancer Res. 1995;55:4988–94.
40. Vosshenrich CA, Garcia-Ojeda ME, Samson-Villeger SI, Pasqualetto V, Enault L, Richard-Le Goff O, Corcuff E, Guy-Grand D, Rocha B, Cumano A, Rogge L, Ezine S, Di Santo JP. A thymic pathway of mouse natural killer cell development characterized by expression of GATA-3 and CD127. Nat Immunol. 2006;7:1217–24.
41. Hanna J, Bechtel P, Zhai Y, Youssef F, McLachlan K, Mandelboim O. Novel insights on human NK cells' immunological modalities revealed by gene expression profiling. J Immunol. 2004;173:6547–63.
42. Laroni A, Gandhi R, Beynon V, Weiner HL. IL-27 imparts immunoregulatory function to human NK cell subsets. PLoS ONE. 2011;6:e26173.

43. Biron CA, Nguyen KB, Pien GC, Cousens LP, Salazar-Mather TP. Natural killer cells in antiviral defense: function and regulation by innate cytokines. Annu Rev Immunol. 1999;17:189–220.
44. Inngjerdingen M, Kveberg L, Naper C, Vaage JT. Natural killer cell subsets in man and rodents. Tissue Antigens. 2011;78:81–8.
45. Ghosh AK, Basu S. Fas-associated factor 1 is a negative regulator in capsaicin induced cancer cell apoptosis. Cancer Lett. 2010;287:142–9.
46. Zamani AG, O. BI, Durakbasi-Dursun G, Ural O, Erdal E, Yildirim MS. Evaluation of death pathway genes FAS and FASL polymorphisms in chronic HBV infection. Int J Immunogenet. 2013;40:482–7
47. Wang SY, Racila E, Taylor RP, Weiner GJ. NK-cell activation and antibody-dependent cellular cytotoxicity induced by rituximab-coated target cells is inhibited by the C3b component of complement. Blood. 2008;111:1456–63.
48. Cooper MA, Fehniger TA, Turner SC, Chen KS, Ghaheri BA, Ghayur T, Carson WE, Caligiuri MA. Human natural killer cells: a unique innate immunoregulatory role for the CD56(bright) subset. Blood. 2001;97:3146–51.
49. Waggoner SN, Cornberg M, Selin LK, Welsh RM. Natural killer cells act as rheostats modulating antiviral T cells. Nature. 2012;481:394–8.
50. Martin-Fontecha A, Thomsen LL, Brett S, Gerard C, Lipp M, Lanzavecchia A, Sallusto F. Induced recruitment of NK cells to lymph nodes provides IFN-gamma for T(H)1 priming. Nat Immunol. 2004;5:1260–5.
51. Laouar Y, Sutterwala FS, Gorelik L, Flavell RA. Transforming growth factor-beta controls T helper type 1 cell development through regulation of natural killer cell interferon-gamma. Nat Immunol. 2005;6:600–7.
52. Gasteiger G, Hemmers S, Firth MA, Le Floc'h A, Huse M, Sun JC, Rudensky AY. IL-2-dependent tuning of NK cell sensitivity for target cells is controlled by regulatory T cells. J Exp Med. 2013;210:1167–78.
53. Costantini C, Micheletti A, Calzetti F, Perbellini O, Pizzolo G, Cassatella MA. Neutrophil activation and survival are modulated by interaction with NK cells. Int Immunol. 2010;22:827–38.
54. Martin MP, Carrington M. Natural killer cells and HIV-1 disease. Curr Opin in HIV AIDS. 2006;1:226–31.
55. Scott-Algara D, Truong LX, Versmisse P, David A, Luong TT, Nguyen NV, Theodorou I, Barre-Sinoussi F, Pancino G. Cutting edge: increased NK cell activity in HIV-1-exposed but uninfected Vietnamese intravascular drug users. J Immunol. 2003;171:5663–7.
56. Brunetta E, Fogli M, Varchetta S, Bozzo L, Hudspeth KL, Marcenaro E, Moretta A, Mavilio D. The decreased expression of Siglec-7 represents an early marker of dysfunctional natural killer-cell subsets associated with high levels of HIV-1 viremia. Blood. 2009;114:3822–30.
57. Mavilio D, Lombardo G, Benjamin J, Kim D, Follman D, Marcenaro E, O'Shea MA, Kinter A, Kovacs C, Moretta A, Fauci AS. Characterization of CD56-/CD16+ natural killer (NK) cells: a highly dysfunctional NK subset expanded in HIV-infected viremic individuals. Proc Natl Acad Sci USA. 2005;102:2886–91.
58. Oliva A, Kinter AL, Vaccarezza M, Rubbert A, Catanzaro A, Moir S, Monaco J, Ehler L, Mizell S, Jackson R, Li Y, Romano JW, Fauci AS. Natural killer cells from human immunodeficiency virus (HIV)-infected individuals are an important source of CC-chemokines and suppress HIV-1 entry and replication in vitro. J Clin Invest. 1998;102:223–31.
59. Jost S, Altfeld M. Control of human viral infections by natural killer cells. Annu Rev Immunol. 2013;31:163–94.
60. Yang Y, Dong B, Mittelstadt PR, Xiao H, Ashwell JD. HIV Tat binds Egr proteins and enhances Egr-dependent transactivation of the Fas ligand promoter. J Biol Chem. 2002;277:19482–7.
61. Zocchi MR, Rubartelli A, Morgavi P, Poggi A. HIV-1 Tat inhibits human natural killer cell function by blocking L-type calcium channels. J Immunol. 1998;161:2938–43.
62. Espert L, Denizot M, Grimaldi M, Robert-Hebmann V, Gay B, Varbanov M, Codogno P, Biard-Piechaczyk M. Autophagy is involved in T cell death after binding of HIV-1 envelope proteins to CXCR4. J Clin Invest. 2006;116:2161–72.

63. Bonaparte MI, Barker E. Killing of human immunodeficiency virus-infected primary T-cell blasts by autologous natural killer cells is dependent on the ability of the virus to alter the expression of major histocompatibility complex class I molecules. Blood. 2004;104:2087–94.
64. Cohen GB, Gandhi RT, Davis DM, Mandelboim O, Chen BK, Strominger JL, Baltimore D. The selective downregulation of class I major histocompatibility complex proteins by HIV-1 protects HIV-infected cells from NK cells. Immunity. 1999;10:661–71.
65. Collins KL, Baltimore D. HIV's evasion of the cellular immune response. Immunol Rev. 1999;168:65–74.
66. Nolting A, Dugast AS, Rihn S, Luteijn R, Carrington MF, Kane K, Jost S, Toth I, Nagami E, Faetkenheuer G, Hartmann P, Altfeld M, Alter G. MHC class I chain-related protein A shedding in chronic HIV-1 infection is associated with profound NK cell dysfunction. Virology. 2010;406:12–20.
67. Galiani MD, Aguado E, Tarazona R, Romero P, Molina I, Santamaria M, Solana R, Pena J. Expression of killer inhibitory receptors on cytotoxic cells from HIV-1-infected individuals. Clin Exp Immunol. 1999;115:472–6.
68. Parato KG, Kumar A, Badley AD, Sanchez-Dardon JL, Chambers KA, Young CD, Lim WT, Kravcik S, Cameron DW, Angel JB. Normalization of natural killer cell function and phenotype with effective anti-HIV therapy and the role of IL-10. AIDS. 2002;16:1251–56.
69. Eger KA, Unutmaz D. Perturbation of natural killer cell function and receptors during HIV infection. Trends Microbiol. 2004;12:301–3.
70. Fogli M, Costa P, Murdaca G, Setti M, Mingari MC, Moretta L, Moretta A, De Maria A. Significant NK cell activation associated with decreased cytolytic function in peripheral blood of HIV-1-infected patients. Eur J Immunol. 2004;34:2313–21.
71. Feldman S, Stein D, Amrute S, Denny T, Garcia Z, Kloser P, Sun Y, Megjugorac N, Fitzgerald-Bocarsly P. Decreased interferon-alpha production in HIV-infected patients correlates with numerical and functional deficiencies in circulating type 2 dendritic cell precursors. Clin Immunol. 2001;101:201–10.
72. Hosmalin A, Lebon P. Type I interferon production in HIV-infected patients. J Leukoc Biol. 2006;80:984–93.
73. Mavilio D, Lombardo G, Kinter A, Fogli M, La Sala A, Ortolano S, Farschi A, Follmann D, Gregg R, Kovacs C, Marcenaro E, Pende D, Moretta A, Fauci AS. Characterization of the defective interaction between a subset of natural killer cells and dendritic cells in HIV-1 infection. J Exp Med. 2006;203:2339–50.
74. Tasca S, Tambussi G, Nozza S, Capiluppi B, Zocchi MR, Soldini L, Veglia F, Poli G, Lazzarin A, Fortis C. Escape of monocyte-derived dendritic cells of HIV-1 infected individuals from natural killer cell-mediated lysis. AIDS. 2003;17:2291–8.
75. Saidi H, Melki MT, Gougeon ML. HMGB1-dependent triggering of HIV-1 replication and persistence in dendritic cells as a consequence of NK-DC cross-talk. PLoS ONE. 2008;3:e3601.
76. Melki MT, Saidi H, Dufour A, Olivo-Marin JC, Gougeon ML. Escape of HIV-1-infected dendritic cells from TRAIL-mediated NK cell cytotoxicity during NK-DC cross-talk-a pivotal role of HMGB1. PLoS Pathog. 2010;6:e1000862.
77. Troseid M, Nowak P, Nystrom J, Lindkvist A, Abdurahman S, Sonnerborg A. Elevated plasma levels of lipopolysaccharide and high mobility group box-1 protein are associated with high viral load in HIV-1 infection: reduction by 2-year antiretroviral therapy. AIDS. 2010;24:1733–7.
78. Chougnet CA, Shearer GM. Regulatory T cells (Treg) and HIV/AIDS: summary of the September 7–8, 2006 workshop. AIDS Res Hum Retroviruses. 2007;23:945–52.
79. Ghiringhelli F, Menard C, Martin F, Zitvogel L. The role of regulatory T cells in the control of natural killer cells: relevance during tumor progression. Immunol Rev. 2006;214:229–38.
80. Iannello A, Debbeche O, Samarani S, Ahmad A. Antiviral NK cell responses in HIV infection: II. viral strategies for evasion and lessons for immunotherapy and vaccination. J Leukoc Biol. 2008;84:27–49.

81. Vieillard V, Strominger JL, Debre P. NK cytotoxicity against CD4+ T cells during HIV-1 infection: a gp41 peptide induces the expression of an NKp44 ligand. Proc Natl Acad Sci USA. 2005;102:10981–6.
82. Fausther-Bovendo H, Sol-Foulon N, Candotti D, Agut H, Schwartz O, Debre P, Vieillard V. HIV escape from natural killer cytotoxicity: nef inhibits NKp44 L expression on CD4+ T cells. AIDS. 2009;23:1077–87.
83. Vieillard V, Crouzet J, Boufassa F, Sennepin A, Ho Tsong Fang R, Debre P, Meyer L. Specific anti-gp41 antibodies predict HIV-1 disease progression. J Acquir Immune Defic Syndr. 2012;61:403–5.
84. Chehimi J, Azzoni L, Farabaugh M, Creer SA, Tomescu C, Hancock A, Mackiewicz A, D'Alessandro L, Ghanekar S, Foulkes AS, Mounzer K, Kostman J, Montaner LJ. Baseline viral load and immune activation determine the extent of reconstitution of innate immune effectors in HIV-1-infected subjects undergoing antiretroviral treatment. J Immunol. 2007;179:2642–50.
85. Goodier MR, Imami N, Moyle G, Gazzard B, Gotch F. Loss of the CD56hiCD16- NK cell subset and NK cell interferon-gamma production during antiretroviral therapy for HIV-1: partial recovery by human growth hormone. Clin Exp Immunol. 2003;134:470–6.
86. Ostrowski SR, Ullum H, Pedersen BK, Gerstoft J, Katzenstein TL. 2B4 expression on natural killer cells increases in HIV-1 infected patients followed prospectively during highly active antiretroviral therapy. Clin Exp Immunol. 2005;141: 526–33.
87. De Maria A, Ferraris A, Guastella M, Pilia S, Cantoni C, Polero L, Mingari MC, Bassetti D, Fauci AS, Moretta L. Expression of HLA class I-specific inhibitory natural killer cell receptors in HIV-specific cytolytic T lymphocytes: impairment of specific cytolytic functions. Proc Natl Acad Sci USA. 1997;94:10285–8.
88. Rosenberg SA, Yang JC, Topalian SL, Schwartzentruber DJ, Weber JS, Parkinson DR, Seipp CA, Einhorn JH, White DE. Treatment of 283 consecutive patients with metastatic melanoma or renal cell cancer using high-dose bolus interleukin 2. JAMA: J Am Med Assoc. 1994;271:907–13.
89. Kottke T, Galivo F, Wongthida P, Diaz RM, Thompson J, Jevremovic D, Barber GN, Hall G, Chester J, Selby P, Harrington K, Melcher A, Vile RG. Treg depletion-enhanced IL-2 treatment facilitates therapy of established tumors using systemically delivered oncolytic virus. Mol Ther. 2008;16:1217–26.
90. Ahmad A, Ahmad R, Iannello A, Toma E, Morisset R, Sindhu ST. IL-15 and HIV infection: lessons for immunotherapy and vaccination. Curr HIV Res. 2005;3:261–70.
91. Elpek KG, Rubinstein MP, Bellemare-Pelletier A, Goldrath AW, Turley SJ. Mature natural killer cells with phenotypic and functional alterations accumulate upon sustained stimulation with IL-15/IL-15Ralpha complexes. Proc Natl Acad Sci USA. 2010;107:21647–52.
92. Azzoni L, Foulkes AS, Papasavvas E, Mexas AM, Lynn KM, Mounzer K, Tebas P, Jacobson JM, Frank I, Busch MP, Deeks SG, Carrington M, O'Doherty U, Kostman J, Montaner LJ. Pegylated Interferon alfa-2a monotherapy results in suppression of HIV type 1 replication and decreased cell-associated HIV DNA integration. J Infec Dis. 2013;207: 213–22.
93. Rongvaux A, Takizawa H, Strowig T, Willinger T, Eynon EE, Flavell RA, Manz MG. Human hemato-lymphoid system mice: current use and future potential for medicine. Annu Rev Immunol. 2013;31:635–74.
94. Hiramatsu H, Nishikomori R, Heike T, Ito M, Kobayashi K, Katamura K, Nakahata T. Complete reconstitution of human lymphocytes from cord blood CD34+ cells using the NOD/SCID/gammacnull mice model. Blood. 2003;102:873–80.
95. Andre MC, Erbacher A, Gille C, Schmauke V, Goecke B, Hohberger A, Mang P, Wilhelm A, Mueller I, Herr W, Lang P, Handgretinger R, Hartwig UF. Long-term human CD34+ stem cell-engrafted nonobese diabetic/SCID/IL-2R gamma(null) mice show impaired CD8+ T cell maintenance and a functional arrest of immature NK cells. J Immunol. 2010;185:2710–20.
96. Strowig T, Chijioke O, Carrega P, Arrey F, Meixlsperger S, Ramer PC, Ferlazzo G, Munz C. Human NK cells of mice with reconstituted human immune system components require preactivation to acquire functional competence. Blood. 2010;116:4158–67.

97. Gimeno R, Weijer K, Voordouw A, Uittenbogaart CH, Legrand N, Alves NL, Wijnands E, Blom B, Spits H. Monitoring the effect of gene silencing by RNA interference in human CD34+ cells injected into newborn RAG2-/- gammac-/- mice: functional inactivation of p53 in developing T cells. Blood. 2004;104:3886–93.
98. Pek EA, Chan T, Reid S, Ashkar AA. Characterization and IL-15 dependence of NK cells in humanized mice. Immunobiol. 2011;216:218–24.
99. Kwant-Mitchell A, Pek EA, Rosenthal KL, Ashkar AA. Development of functional human NK cells in an immunodeficient mouse model with the ability to provide protection against tumor challenge. PLoS ONE. 2009;4:e8379.
100. Calderon VE, Valbuena G, Goez Y, Judy BM, Huante MB, Sutjita P, Johnston RK, Estes DM, Hunter RL, Actor JK, Cirillo JD, Endsley JJ. A humanized mouse model of tuberculosis. PLoS ONE. 2013;8:e63331.
101. Suwanai H, Wilcox MA, Mathis D, Benoist C. A defective Il15 allele underlies the deficiency in natural killer cell activity in nonobese diabetic mice. Proc Natl Acad Sci USA. 2010;107:9305–10.
102. Chen Q, Khoury M, Chen J. Expression of human cytokines dramatically improves reconstitution of specific human-blood lineage cells in humanized mice. Proc Natl Acad Sci USA. 2009;106:21783–8.
103. Takenaka K, Prasolava TK, Wang JC, Mortin-Toth SM, Khalouei S, Gan OI, Dick JE, Danska JS. Polymorphism in Sirpa modulates engraftment of human hematopoietic stem cells. Nat Immunol. 2007;8:1313–23.
104. Legrand N, Huntington ND, Nagasawa M, Bakker AQ, Schotte R, Strick-Marchand H, Geus SJ de, Pouw SM, Bohne M, Voordouw A, Weijer K, Di Santo JP, Spits H. Functional CD47/signal regulatory protein alpha (SIRP(alpha)) interaction is required for optimal human T- and natural killer- (NK) cell homeostasis in vivo. Proc Natl Acad Sci USA. 2011;108:13224–9.
105. Strowig T, Rongvaux A, Rathinam C, Takizawa H, Borsotti C, Philbrick W, Eynon EE, Manz MG, Flavell RA. Transgenic expression of human signal regulatory protein alpha in Rag2-/-gamma(c)-/- mice improves engraftment of human hematopoietic cells in humanized mice. Proc Natl Acad Sci USA. 2011;108:13218–23.
106. Gerosa F, Gobbi A, Zorzi P, Burg S, Briere F, Carra G, Trinchieri G. The reciprocal interaction of NK cells with plasmacytoid or myeloid dendritic cells profoundly affects innate resistance functions. J Immunol. 2005;174:727–34.
107. Pedroza-Pacheco I, Madrigal A, Saudemont A. Interaction between natural killer cells and regulatory T cells: perspectives for immunotherapy. Cell Mol Immunol. 2013;10:222–9.
108. Roth C, Rothlin C, Riou S, Raulet DH, Lemke G. Stromal-cell regulation of natural killer cell differentiation. J Mol Med (Berl). 2007;85:1047–56.
109. Billerbeck E, Barry WT, Mu K, Dorner M, Rice CM, Ploss A. Development of human CD4+ FoxP3+ regulatory T cells in human stem cell factor-, granulocyte-macrophage colony-stimulating factor-, and interleukin-3-expressing NOD-SCID IL2Rgamma(null) humanized mice. Blood. 2011;117:3076–86.
110. Chen Q, He F, Kwang J, Chan JK, Chen J. GM-CSF and IL-4 stimulate antibody responses in humanized mice by promoting T, B, and dendritic cell maturation. J Immunol. 2012;189:5223–9.
111. Denton PW, Garcia JV. Humanized mouse models of HIV infection. AIDS Rev. 2011;13:135–48.
112. Berges BK, Rowan MR. The utility of the new generation of humanized mice to study HIV-1 infection: transmission, prevention, pathogenesis, and treatment. Retrovirology. 2011;8:65.
113. Nischang M, Gers-Huber G, Audige A, Akkina R, Speck RF. Modeling HIV infection and therapies in humanized mice. Swiss Med Wkly. 2012;142:w13618.
114. Zhang L, Meissner E, Chen J, Su L. Current humanized mouse models for studying human immunology and HIV-1 immuno-pathogenesis. Sci China Life Sci. 2010;53:195–203.

115. Klein F, Halper-Stromberg A, Horwitz JA, Gruell H, Scheid JF, Bournazos S, Mouquet H, Spatz LA, Diskin R, Abadir A, Zang T, Dorner M, Billerbeck E, Labitt RN, Gaebler C, Marcovecchio PM, Incesu RB, Eisenreich TR, Bieniasz PD, Seaman MS, Bjorkman PJ, Ravetch JV, Ploss A, Nussenzweig MC. HIV therapy by a combination of broadly neutralizing antibodies in humanized mice. Nature. 2012;492:118–22.
116. Krisko JF, Martinez-Torres F, Foster JL, Garcia JV. HIV restriction by APOBEC3 in humanized mice. PLoS Pathog. 2013;9:e1003242.
117. Murooka TT, Deruaz M, Marangoni F, Vrbanac VD, Seung E, Andrian UH von, Tager AM, Luster AD, Mempel TR. HIV-infected T cells are migratory vehicles for viral dissemination. Nature. 2012;490:283–7.
118. Neagu MR, Ziegler P, Pertel T, Strambio-De-Castillia C, Grutter C, Martinetti G, Mazzucchelli L, Grutter M, Manz MG, Luban J. Potent inhibition of HIV-1 by TRIM5-cyclophilin fusion proteins engineered from human components. J Clin Invest. 2009;119:3035–47.
119. Ru Z, Xiao W, Pajot A, Kou Z, Sun S, Maillere B, Zhao G, Ojcius DM, Lone YC, Zhou Y. Development of a humanized HLA-A2.1/DP4 transgenic mouse model and the use of this model to map HLA-DP4-restricted epitopes of HBV envelope protein. PLoS ONE. 2012;7:e32247.
120. Shultz LD, Saito Y, Najima Y, Tanaka S, Ochi T, Tomizawa M, Doi T, Sone A, Suzuki N, Fujiwara H, Yasukawa M, Ishikawa F. Generation of functional human T-cell subsets with HLA-restricted immune responses in HLA class I expressing NOD/SCID/IL2r gamma(null) humanized mice. Proc Natl Acad Sci USA. 2010;107:13022–7.
121. Denton PW, Nochi T, Lim A, Krisko JF, Martinez-Torres F, Choudhary SK, Wahl A, Olesen R, Zou W, Di Santo JP, Margolis DM, Garcia JV. IL-2 receptor gamma-chain molecule is critical for intestinal T-cell reconstitution in humanized mice. Mucosal Immunol. 2012;5:555–66.

Chapter 15
Maintenance and Function of Human CD8⁺ T Cells and NK Cells in Humanized Mice

Udo F. Hartwig, Maya C. André and Christian Münz

Abbreviations

BM	Bone marrow
HLA	Human leukocyte antigen
IFN	Interferon
HIS	Human immune system
HSC	Hematopoietic stem cell
IL	Interleukin
KIR	Killer immunoglobulin-like receptor
LN	Lymph node
MHC	Major histocompatibility complex
NK	Natural killer
PB	Peripheral blood
RAG	Recombination activating gene
TCR	T cell receptor
TLR	Toll-like receptor

Udo F. Hartwig and Maya C. André share first authorship.

U. F. Hartwig (✉)
Department of Medicine 3—Hematology, Internal Oncology & Pneumology,
Building for R&D, University Medical Center of Johannes Gutenberg-University Mainz,
Obere Zahlbacherstraße 63, 55131 Mainz, Germany
e-mail: uhartwig@uni-mainz.de

M. C. André
University Children's Hospital, Pediatric Hematology/Oncology, Eberhard-Karls University,
Tübingen, Germany

University Children's Hospital Basel (UKBB), Basel, Switzerland

C. Münz
Institute for Experimental Immunology, University of Zürich, Zurich, Switzerland

L. Y. Poluektova et al. (eds.), *Humanized Mice for HIV Research*,
DOI 10.1007/978-1-4939-1655-9_15

15.1 Introduction

The availability of highly immuno-deficient mice that harbor, in addition to a severe combined immunodeficiency (scid) mutation [1] or deficient recombination activating genes (Rag), [2] loss of function mutations in the IL-2 receptor common gamma chain (γc) has improved reconstitution of human immune system (HIS) mice in so far as complex studies on human T and NK cell immunobiology are now realizable. However, to maintain optimal homeostasis and function of $CD8^+$ T lymphocytes and NK cells in the periphery, HIS mice have to provide human leukocyte antigen (HLA) class I restriction elements, permit interactions with "human" dendritic cells (DC), procure physiological concentrations of "human" cytokines such as IL-7 and IL-15, promote migration and homing and facilitate "licensing" of NK cells in the context of enabling the acquisition of a diverse inhibitory receptor repertoire. Keeping the enormous complexity of any such immunological process in mind, a considerable progress has been achieved during the past decade; however, it is still a long way to go in imagining, designing, and composing the ideal "human"-murine chimera.

15.2 Maintenance and Function of $CD8^+$ T Cells in Humanized Mice

15.2.1 Migration and Homing of $CD8^+$ T cells to Murine Lymphoid Organs of HIS Mice

"Human" T cells either successfully generated in HIS mice following engraftment of umbilical cord blood, human fetal liver, or adult bone marrow (BM)-derived HSCs or, alternatively, by adoptive transfer of PBMCs, preferentially home into the spleen and BM as the formation of lymph nodes is only rarely observed or arrest in a structurally primitive organizational stage [3, 4]. In man, particularly CD62L, CCR7, leukocyte functional antigen-1 (LFA-1), P-selectin glycoprotein ligand-1 (PSGL-1), CXCR3, and very late antigen-4 (VLA-4) are known to play a critical role in mediating $CD8^+$ T cell migration and homing, however, it has so far not been formally evaluated whether these molecules are expressed on HIS-derived $CD8^+$ T cells. With regard to their respective ligands, it has been shown for a variety of selectins and chemokines that the murine orthologs share only 50–80 % homology to their respective human counterparts [3]. In support of these findings, comparative adoptive transfer studies performed to examine homing of differently labeled murine and human T cells into the spleens of NSG mice revealed that human T lymphocytes including $CD8^+$ T cells migrated less efficiently into splenic tissue than murine control cells (Hartwig UF, unpublished results). In contrast, integrins are highly conserved across species and human $CD8^+$ T cells were indeed found to efficiently adhere to murine primary endothelial cells and transmigrate through endothelial layers in vitro [5]. However, a certain cross-reactivity of murine molecules may be sufficient to provide retention signals of human cells to murine niches, as shown by the example of improved homing of human HSCs through the interaction of human CXCR4 chemokine receptor with its murine CXCL12 (SDF-1) ligand [6]. In summary, migration and homing properties of HIS-derived $CD8^+$ T cells have so far been incompletely characterized but the varying tissue distribution strongly argues for a directed migration and homing process occurring in vivo.

15.2.2 Homeostatic Proliferation and Maintenance of CD8⁺ T Cells in Lymphoid Tissues of HIS Mice

Studies in immuno-competent mice have provided strong evidence that continuous T cell receptor (TCR) engagement with self-ligands in the presence of intermittent IL-7 [7] and IL-15 enables survival and homeostasis of naive T lymphocytes [8–11]. In line with this, it was recently demonstrated that TCR contact with self-MHC molecules sensitizes naive $CD8^+$ T cells to IL-7 and IL-15 and thus maintains survival, whereas the presence of IL-2 is crucial for induction and optimization of $CD8^+$ T cell-mediated immune responses [12, 13]. As reviewed in detail in Chap. 10, the development of human $\alpha\beta$-TCR expressing $CD8^+$ T cells and their homing to secondary lymphoid tissues is achieved in HIS mice for time periods extending 6 months. However, consistent with the data obtained in ordinary mice, recent studies in HSC-engrafted, HLA-A2 transgenic NSG (NSG-A2) mice demonstrate that development of functional human $CD8^+$ T cells is clearly improved in the presence of HLA-A2 molecules [4, 14, 15]. Moreover, $CD8^+$ T cells developing in the NOD/scid and NSG bone-liver-thymus (BLT) model undergo efficient thymic selection and contain a highly diverse, HLA class I-restricted functional repertoire [16, 17]. Together, these results suggest that expression of human MHC class I clearly improves the generation of high numbers of HLA-restricted fully functional $CD8^+$ T cells in HIS mice. In addition to the role of TCR-MHC interactions, repetitive injections of human IL-7 [18–20], or lentiviral-mediated IL-7 expression in vivo [21] may facilitate thymic lymphopoiesis and long-term engraftment in HIS mice. However, in line with the recent data suggesting that continuous IL-7 exposure will induce IFN-γ-triggered cell death of $CD8^+$ T cells [7], we observed a nonphysiological, artificial phenotypic skewing of T cell subpopulations during long-term IL-7 supplementation with a selective expansion of memory $CD4^+$ T cells and ultimately vanishing $CD8^+$ T cells [20]. Apart from exogenous administration of IL-7, administration of IL-15 linked to its IL-15Rα "sushi" domain distinctly increased $CD8^+$ T cell expansion in HIS mice demonstrating the central role of IL-15 in maintaining development, homeostatic proliferation, and survival of "human" T cells in HIS mice [22, 23]. Moreover, the importance of IL-15 in the maintenance of T cell memory in HIS mice has recently been shown for central memory CMV-reactive $CD8^+$ T cells following adoptive transfer into NSG mice [24]. Interestingly, HSC engraftment into human stem cell factor-, granulocyte-macrophage colony-stimulating factor-, and IL-3 expressing NSG-SGM3 mice revealed not only elevated myeloid cell frequencies, i.e., particularly DCs, but also increased levels of CD4$^+$ FoxP3$^+$ regulatory T cells and to a lesser extent CD8$^+$ T cells [25]. These data support earlier findings, suggesting that, in addition to orthotopic HLA expression and appropriate homeostatic cytokine levels, sufficient numbers of human APCs that provide HLA-restricted and costimulatory signals will promote homeostatic CD8$^+$ T cell expansion and functionality [26]. In summary, expansion and maintenance of human CD8$^+$ T cells in HIS mice is a multifactorial process that can be achieved in current models but will substantially advance in γc-deficient mice that express HLA class I restriction elements in combination with cytokines that promote myeloid and lymphoid expansion and differentiation [27].

15.2.3 CD8$^+$ T Cell Function in HIS Mice

Comparative studies show that HIS-derived peripheral CD8$^+$ T cells can be unspecifically stimulated in vitro, however, the responses are modest when compared to human controls [28]. In contrast, huNSG-derived CD3$^+$ T cells that were reconstituted in the presence of IL-7 exhibited a potent in vitro alloreactivity against human HLA-mismatched DCs, indicating that IL-7 supplementation to HIS mice supports function [20]. Accordingly, efficient CD8$^+$ T cell-mediated immune responses have been reported in various HIS mouse models of viral diseases (HIV, EBV, Dengue virus), intracellular bacteria (*M. tuberculosis*) or of human cancer [3]. Thus, we will here exemplarily summarize results from experiments on EBV and leukemia models obtained in our own laboratories. Following engraftment of HLA-A2$^+$ HSCs into NSG or NSG-A2 recipients, mature T cells could be effectively primed to elicit protective T cell responses against EBV-induced lymphomas [4]. Although T cells developed in both groups of mice with similar frequencies, more lytic EBV (BRLF1) peptide-specific CD8$^+$ T cells were found in huNSG-A2 mice [4]. As patient-derived xenotransplantation (PDX) models composed of tumor specimens and HSCs from the identical patient are difficult to achieve, newer approaches in immunotherapy have been pursued in NSG mice engrafted, e.g., with patient-derived acute myeloid leukemia (AML) blasts [29] and engraftment of HLA-matched CD8$^+$ T cells from healthy donors. Adoptive transfer of in vitro generated AML-reactive or reprogrammed AML (HMMR/Rhamm)-specific CD8$^+$ T cells into these mice results in strong reduction or even prevention of leukemic growth being indicative of intact cytotoxic functions of CD8$^+$ T cells after transfer into the murine milieu [30, 31]. Given the fact that CD8$^+$ T cells require appropriate TCR-MHC interactions ideally located in the thymic tissue, interactions with DCs that would ideally occur in preformed lymph node structures, the presence of sufficiently cross-reactive homing molecules and physiological concentrations of supportive cytokines to be able to exert a complex functionality, further studies and potential improvements of CD8$^+$ T cell physiology in future generations of HIS mice would be desirable.

15.3 Maintenance and Function of NK Cells in Humanized Mice

15.3.1 Migration, Homing, and Tissue Distribution of HIS-Derived NK Cells

Although NK cells express a variety of chemotactic receptors, little is known on recirculation and trafficking of human NK cells in man. Even less is known about the expression of integrins, chemokines, and selectins in HIS mice that might mediate adhesion, activation, migration, and homing of "human" NK cells. Analysis of NK cell distribution in man has often been erroneous as classical NK cells had not been distinguished from the growing population of retinoic acid receptor gamma$^+$ (RORγt$^+$) innate lymphoid cells that share common NK cell markers such as NKp46 [32]. Keeping this potential pitfall in mind, "human" NKp46$^+$CD56$^+$ cells can be identified in PB, BM, spleen, lung, and liver of both of huBRG [22] and huNSG mice [20, 33], although at low frequencies around 0.5–2% of "human" lymphocytes. In the context of trafficking and migration of NK cells in man, the tissue-homing molecules CCR5, CCR6, CXCR3, and CXCR4 and the lymph node-homing markers CCR7 and CD62L have been identified to be of unique importance, however, the expression of these molecules has not been formally evaluated in huBRG or huNSG-derived NK cells. With respect to their corresponding ligands, the protein identities of human chemokines, selectins, and integrins were compared to their respective murine counterparts to facilitate extrapolation to the conditions prevailing in HIS mice [3]. Although many important integrins appear highly conserved between mouse and man, the amino acid sequences of numerous selectins or chemokines, respectively, exhibit little homology between mouse and man indicating that homing and trafficking rather than migration might cause particular difficulties in the currently available immune-permissive mouse strains.

15.3.2 Cytokine Requirements of HIS-Derived NK Cells

A large body of evidence has been compiled that emphasizes the importance of the cytokine IL-15 in promoting differentiation, survival, and the acquisition of functional competence of NK cells [34]. IL-15 signals are transmitted through the IL2/15R beta and common γ chain, however, it is the delivery of IL-15 to these signaling molecules that is unique in so far as IL-15 has to be presented with its respective receptor *in trans* to induce efficient signal transduction [35]. Keeping in mind that this trans-presentation of IL-15 has to be provided either by "human" dendritic cells (DCs) that are expressed only in modest numbers in HIS mice (see Chap. 16) or "human" epithelial cells that are constitutively absent in humanized mice, developing NK cells in HIS mice are likely to face suboptimal concentrations of IL-15. In line with the two observations that murine IL-15 is only weakly cross-reactive to human cells and that a human but not a murine IL-15-blocking mAb ablated NK cell emergence during reconstitution of huBRG mice, [22] the presence of "human" NK cells should presumably be attributed to marginable but sufficient "human" IL-15 concentrations. Indeed, treatment of huBRG mice with human IL-15/IL-15Rα complexes resulted in significantly higher numbers of NK cells together with a distinctly matured phenotype and functionality (see below) [22, 36]. Applying the method of hydrodynamic injection of plasmid DNA encoding for human IL-15 and Flt-3/Flk-2 ligand in huNSG mice, Chen and colleagues were able to enhance NK cell numbers and to prime phenotype and functionality (as specified below) [37]. Of note, in C57/BL6 mice prolonged in vivo exposure to IL-15/IL-15Rα complexes led to a marked accumulation of killer cell lectin-like receptor subfamily G member-1^{+} (KLRG-1) mature NK cells with a considerably impaired activation status, impairments in cytotoxicity, a reduced proliferative activity and nonphysiologic alterations in the balance between activating and inhibitory receptors [38]. Thus, given the evidence that IL-15 is critical for NK cells the transient and limited application of IL-15/IL-15Rα complexes to HIS mice will sustain and improve both NK cell homeostasis and function, enabling translational researchers to optimize their humanized immune system model to study anti-inflammatory or anti-tumor responses of NK cells.

15.3.3 Homeostatic Proliferation and Maintenance of HIS-Derived NK Cells

In contrast to T cells that survive and proliferate in response to antigen recognition by rearranged receptors, NK cell homeostasis is maintained by other factors, particularly IL-15. The molecular basis for the survival-promoting effects of IL-15 was characterized by demonstrating that NK cell apoptosis is prevented by IL-15-induced restriction of the proapoptotic BH3-only protein Bim and enhancement of the prosurvival protein Mcl-1 [39]. Studies in huBRG mice demonstrated that treatment with IL-15/IL-15Rα complexes results in increased NK cell lymphopoiesis that was attributed to improved proliferation and survival as both the number of Ki67^{+} and Bcl-xL^{+} cells increased [22]. Irrespective of IL-15 supplementation, data on NK cell numbers in huBRG [22] and huNSG mice [20, 33] indicate that "human" NK cells accumulate in higher levels in NSG mice. In line with the observation that NSG mice obviously exhibit a lower level of murine xenoreactivity towards human cells than BRG mice, the extent of affinity of the murine (NOD) SIRP-α towards human CD47 was shown to be higher in NSG mice [40, 41]. A functional CD47/signal regulatory protein alpha (SIRP-α) interaction is a major determinant in the escape from phagocyte-mediated cell clearance of xenogeneic cells [41, 42]. Thus, ligation of SIRP-α molecules expressed on phagocytes with their respective CD47 molecules expressed on hematopoietic cells will result in improved engraftment when transplantation is enforced over xenogeneic barriers. Accordingly, it was recently shown that forced expression of mouse CD47 by human stem cells will improve engraftment in huBRG mice and promote optimal human T and NK cell homeostasis in vivo [42]. As HIS mice constitutively lack so-called lymphoid tissue inducer cells and consequently lymph node organogenesis (see also Chap. 6), NK cell homeostasis is impaired both at the level of development and survival. To compensate for this lack, NK cells in HIS mice will benefit from exogenous IL-15 supplementation and potentially improvement of the CD47/SIRP-α interaction.

15.3.4 Inhibitory KIR Expression and "Licensing" of HIS-Derived NK Cells

As NK cell functionality importantly corresponds to the NK cell phenotype, we will briefly discuss the emergence of phenotypically distinct NK cell subpopulations and the acquisition of a diverse inhibitory killer immunoglobulin-like receptor (KIR) repertoire here. For further details on the phenotypic differentiation of HIS-derived NK cells the reader should refer Chap. 14. During NK cell ontogeny, a linear differentiation model has been postulated for humans [43] that is also demonstrable in HIS mice and encompasses the distinct stages of immature CD94$^+$CD56$^+$KIR$^-$ precursors, CD56^{br}CD16$^-$KIR$^-$ NK cells that mature to CD56^{dim}CD16$^+$KIR$^-$ and ultimately CD56^{dim}CD16$^+$KIR$^+$ NK cells followed by the emergence of functionally mature CD56^{dim}CD16$^+$KIR$^+$CD57$^+$ NK cells [22, 36]. Own studies on Nk cell generation in huNSG mice supported with Fc-IL7 but not IL-15, revealed a functional arrest of constitutively activated CD56^{bright}CD16$^-$KIR$^-$ NK cells that expressed high levels of the NKp44 receptor, which is normally expressed only on activated NK cells, but simultaneously lacked any IFN-γ secretion upon unspecific stimulation or cytotoxicity against the MHC class I-deficient cell line K562 [22]. In line with this, expression of inhibitory KIRs such as KIR2DL1, 2DL2/DL3, and 3DL1 was absent in NK cells from huNSG mice while the respective KIR-specific mRNA was present for all three KIRs. Interestingly, this predominance of a CD56^{bright}CD16$^-$KIR$^-$ cell population exhibiting little cytotoxicity has been also observed in humans in the early post allogeneic bone marrow transplantation period [44, 45]. Although the cytokine IL-7 has so far not been linked to the maturational conversion of CD-56^{br}CD16$^-$ to CD56^{dim}CD16$^+$ NK cells, it is noteworthy that the absence of the CD-56^{dim}CD16$^+$ NK cell subpopulation was not demonstrable when reconstitution was performed without the exogenous support of IL-7 in both huBRG [22] and huNSG mice [33]. When interpreting this obvious functional arrest of huNSG-derived NK cells in a developmentally immature state, it is important to note that acquisition of functionality, i.e., cytotoxicity, importantly involves the acquisition of inhibitory KIRs. This so-called "licensing of NK cells" comprises a process in which only NK cells that express at least one inhibitory KIR for self-MHC class I molecules are rendered functionally competent [46]. This involvement of KIR-MHC class I interactions in the "fine-tuning" of NK-cell-effector capacities implies that the generation of NK cells in a murine tissue lacking human MHC class I molecules might be difficult, or in other words the licensing of HIS-derived NK cells will rely on the recognition of sufficient numbers of self-MHC molecules expressed on reconstituting "human" cells in the immediate vicinity. However, normal MHC class I expression itself is not required for KIR expression as patients with a TAP I deficiency and absence of MHC class I expression will still express a certain extent of KIR^+ NK cells [47]. Thus, licensing of NK cells in HIS mice has either to occur via inhibitory NK cell receptor engagement by MHC class I molecules on adjacent "human" cells or partially unlicensed NK cells will have to gain functionality when preactivated in vivo or during infections [48].

15.3.5 Acquisition of Functional Maturity of HIS-Derived NK Cells

This in vivo preactivation and maturation of HIS-NK cells can be performed with IL-15/IL-15Rα complexes [22], hydrodynamic tail vein injection of IL-15/Flt-3/Flk-2 ligand [37] or via bystander cell maturation with the toll-like receptor (TLR) 3 agonist poly I:C. [33]. Upon IL-15/IL-15Rα treatment, "human" NK cells exhibit a phenotypic skewing to the more differentiated $CD56^{dim}CD16^+KIR^+$ subpopulation in both huBRG mice [22] and huNSG mice (Andre MC, unpublished observation). Interestingly, this IL-15-induced increase of KIR^+ NK cells did not reflect the expansion of one distinct NK cell clone, as 1–5 different KIR members were present in the KIR^+ NK cell fraction [22]. Functionally, preactivated HIS-NK cells expressed high levels of granzyme B and IFN-γ, degranulated in response to cocultivation with NK cell-susceptible targets such as K562 cells [22, 33], and were able to respond robustly to adenovirus infection in vivo [37]. Collectively, these data suggest that a foremost resting, slightly immature NK cell compartment is induced in HIS mice that can be efficiently rendered functional to enable studies on NK cell-mediated innate immune responses.

15.3.6 Perspectives

Currently used models of HIS mice support the maintenance of $CD8^+$ T cells and NK cells and thus enable complex studies on the functionality of both cell populations in the context of infectious or malignant disease. Although the recent progress in providing ubiquitous transgenic expression of human MHC molecules or human cytokines has been significant, some issues have so far remained unresolved that would considerably improve $CD8^+$ T cell and NK cell physiology in HIS mice (Table 15.1). As such, the tissue-restricted expression of human MHC molecules in the thymic stroma, a refinement of lymph node formation, the expression of human selectins and chemokines to promote directed homing and migration, and the potential expression of tissue-specific, physiological concentrations of growth-promoting cytokines would be desirable when envisaging the ideal human-murine chimera.

Table 15.1 Summary of the current status and future perspectives of CD8+ T cell and NK cell biology in HIS mice

	Immunodeficient mice	Current status of HIS-mice	Potential improvements on hu-mice	References
Migration, homing, and retention signals	Physiological interaction of murine integrins and selectins with cognate ligands, e.g., CD54/102—CD11a/18, CD62E/P—CD62L, CD106—CD49d	Expression of human SIRPa to interact with huCD47 on, e.g., T and NK cells	Orthotopic expression of preferentially human migration and homing molecules with low homology to murine counterparts such as, e.g., CD54/102	[13], [5]
HLA-restriction	MHC I: K, D, L alleles of H-$2^{b,d}$ haplotypes	MHC I: HLA-A2 MHC II: HLA-DR4 MHC II: HLA-DR1	Expression of broad HLA class I and II haplotypes under murine control elements	[4], [15],
Chemokines	Expression of murine chemokines with their cognate receptors on murine granulocytes and cells of myeloid origin	Murine CXCR4 that can interact with human SDF-1	Orthotopic expression of human chemokines such as CCL-19 and CCL-21 under murine control elements to promote trafficking and chemotaxis of T and NK cells	[6]
Cytokines	Expression of murine cytokines with different degrees of homology to human orthologs	Transgenic expression of human IL-3, SCF, and GM-CSF to improve lymphohematopoiesis. Exogenous and ectopic expression of IL-7 and IL-15	Intermittent expression of hu-IL-7 and hu-IL-15 under murine control elements to provide homeostatic cytokines at physiological conditions and kinetics	[19], [21], [22], [23], [27], [25]

Acknowledgements This work has been funded by Deutsche Forschungsgemeinschaft (DFG) (TP6 KFO 183), Stiftung Innovation Rheinland-Pfalz, MAIFOR Program of University Mainz, Excellence Cluster Program "Invidualized Immune Intervention (CI3)" of German Ministry for Education and Research to UFH and grants from the DFG (KFO 183, TP 4), and the Jose Carreras Stiftung (SP 12/03) to MCA. The research of CM was supported by grants from the National Cancer Institute (R01CA108609), the Sassella Foundation (10/02, 11/02 and 12/02), Cancer Research Switzerland (KFS-02652-08-2010), the Association for International Cancer Research (11-0516), KFSP^{MS}, and KFSP^{HLD} of the University of Zurich, the Vontobel, Baugarten, EMDO, and Sobek Foundations, the Fondation Acteria, Novartis, and the Swiss National Science Foundation (310030_143979 and CRSII3_136241).

References

1. Bosma GC, Custer RP, Bosma MJ. A severe combined immunodeficiency mutation in the mouse. Nature. 1983;301:527–30.
2. Mombaerts P, Iacomini J, Johnson RS, Herrup K, Tonegawa S, Papaioannou VE. RAG-1-deficient mice have no mature B and T lymphocytes. Cell. 1992;68:869–77.
3. Rongvaux A, Takizawa H, Strowig T, Willinger T, Eynon EE, Flavell RA, Manz MG. Human hemato-lymphoid system mice: current use and future potential for medicine. Ann Rev Immunol. 2013;31:635–74.
4. Strowig T, Gurer C, Ploss A, Liu YF, Arrey F, Sashihara J, et al. Priming of protective T cell responses against virus-induced tumors in mice with human immune system components. J Exp Med. 2009;206:1423–34.
5. Brunk A, Nonn M, Lang V, Henschler R, Herr W, Hartwig UF. Functional analyses of human T cell extravasation in a humanized NOD/SCID/IL2rcgnull transplantation model. Blood. Annual Meeting of American Society of Hematology; 2009 Dec 5–8, New Orleans, LA, USA. 2009. p. 966.
6. Lapidot T, Kollet O. The essential roles of the chemokine SDF-1 and its receptor CXCR4 in human stem cell homing and repopulation of transplanted immune-deficient NOD/SCID and NOD/SCID/B2mnull mice. Leukemia. 2002;16:1992–2003.
7. Kimura MY, Pobezinsky LA, Guinter TI, Thomas J, Adams A, Park JH, Tai X, Singer A. IL-7 signaling must be intermittent, not continuous, during CD8+ T cell homeostasis to promote cell survival instead of cell death. Nat Immunol. 2013;14(2):143–51.
8. Tanchot C, Lemonnier FA, Pérarnau B, Freias AA, Rocha B. Differential requirements of CD8 naïve or memory T cells. Science. 1997; 276:2057–61.
9. Ma A, Koka R, Burkett P. Diverse functions of IL-2, IL-15, and IL-7 in lymphoid homeostasis. Ann Rev Immunol. 2006;24:657–79.
10. Surh CD, Sprent J. Homeostasis of naïve and memory T cells. Immunity. 2008;29:848–62.
11. Takada K, Jameson SC. Naive T cell homeostasis: from awareness of space to a sense of place. Nat Rev Immunol. 2009;9:823–32.
12. Cho J-H, Kim H-O, Surh CD, Sprent J. T Cell receptor-dependent regulation of lipid rafts controls naive $CD8^+$ T Cell homeostasis. Immunity. 2010;32:214–26.
13. Boyman O, Sprent J. The role of interleukin-2 during homeostasis and activation of the immune system. Nat Rev Immunol. 2012;12:180–90.
14. Jaiswal S, Pearson T, Friberg H, Shultz LD, Greiner DL, Rothman AL, Mathew A. Dengue virus infection and virus-specific HLA-A2 restricted immune responses in humanized NOD-scid IL2rγnull mice. PLoS ONE. 2009;4(10):e7251. doi:10.1371/journal.pone.0007251.
15. Shultz LD, Saito Y, Najima Y, Tanaka S, Ochi T, Tomizawa M, Doi T, Sone A, Suzuki N, Fujiwara H, Yasukawa M, Ishikawa F. Generation of functional human T cell subsets with HLA-restricted immune responses in HLA class I expressing NOD/SCID/IL2rγnull humanized mice. Proc Natl Acad Sci U S A. 2010;107:13022–27.
16. Melkus MW, Estes JD, Padgett-Thomas A, Gatlin J, Denton PW, Othieno FA, Wege AK, Haase AT, Garcia JV. Humanized mice mount specific adaptive and innate immune responses to EBV and TSST-1. Nat Med. 2006;12:1316–22.
17. Tonomura N, Habiro K, Shimizu A, Sykes M, Yang YG. Antigen-specific human T-cell responses and T cell-dependent production of human antibodies in a humanized mouse model. Blood. 2008;111:4293–96.
18. Shultz LD, Lyons BL, Burzenski LM, Gott B, Chen X, Chaleff S, Kotb M, Gillies SD, King M, Mangada J, Greiner DL, Handgretinger R. Human lymphoid and myeloid cell development in NOD/LtSz-scid IL2Rγnull mice engrafted with mobilized human hemopoietic stem cells. J Immunol. 2005;174:6477–89.
19. van Lent AU, Dontje W, Nagasawa M, Siamari R, Bakker AQ, Pouw SM, Maijoor KA, Weijer K, Cornelissen JJ, Blom B, Di Santo JP, Spits H, Legrand N. IL-7 enhances thymic human T cell development in "human immune system" $Rag2^{-/-}$IL-$2Rgamma_c^{-/-}$ mice without affecting peripheral T cell homeostasis. J Immunol. 2009;183:7645–55.

20. André MC, Erbacher A, Gille C, Schmauke V, Goecke B, Hohberger A, Mang P, Wilhelm A, Mueller I, Herr W, Lang P, Handgretinger R, Hartwig UF. Long-term human CD34+ stem cell-engrafted nonobese diabetic/SCID/IL-2R gamma(null) mice show impaired CD8+ T cell maintenance and a functional arrest of immature NK cells. J Immunol. 2010;185(5):2710–20.
21. O'Connell RM, Balazs AB, Rao DS, Kivork C, Yang L, Baltimore D. Lentiviral vector delivery of human interleukin-7 (hIL-7) to human immune system (HIS) mice expands T lymphocyte populations. PLoS ONE. 2010;5:e12009. doi: 10.1371/journal.pone.0012009.
22. Huntington ND, Legrand N, Alves NL, Jaron B, Weijer K, Plet A, Corcuff E, Mortier E, Jacques Y, Spits H, Di Santo JP. IL-15 trans-presentation promotes human NK cell development and differentiation in vivo. J Exp Med. 2009;206:25–34.
23. Huntington ND, Alves NL, Legrand N, Lim A, Strick-Marchand H, Mention JJ, et al. IL-15 trans-presentation promotes both human T-cell reconstitution and T-cell-dependent antibody responses in vivo. Proc Natl Acad Sci U S A. 2011;108:6217–22.
24. Wang X, Berger C, Wong CW, Forman SJ, Riddell SR, Jensen MC. Engraftment of human central memory-derived effector CD8+ T cells in immunodeficient mice. Blood. 2011;117(6):1888–98.
25. Billerbeck E, Barry WT, Mu K, Dorner M, Rice CM, Ploss A. Development of human $CD4^+FoxP3^+$ regulatory T cells in human stem cell factor-, granulocyte-macrophage colony-stimulating factor-, and interleukin-3-expressing NOD-SCID $IL2Rgamma^{null}$ humanized mice. Blood. 2011;7:3076–3086.
26. Onoe T, Kalscheuer H, Chittenden M, Zhao G, Yang Y-G, Sykes M. Homeostatic expansion and phenotypiv conversion of human T cells depend on peripheral interactions with APC. J Immunol. 2010;184:6756–65.
27. Willinger T, Rongvaux A, Strowig T, Manz MG, Flavell RA. Improving human hemato-lymphoid-system mice by cytokine knock-in gene replacement. Trends Immunol. 2011;32:321–27.
28. Watanabe Y, Takahashi T, Okajima A, Shiokawa M, Ishii N, Katano I, Ito R, Ito M, Minegishi M, Minegishi N, Tsuchiya S, Sugamura K. The analysis of the functions of human B and T cells humanized NOD/shi-scid/gc^{null} (NOG) mice (hu-HSC NOG mice). Int Immunol. 2009;21:843–59.
29. Woiterski J, Ebinger M, Witte KE, Goecke B, Heininger V, Philippek M, Bonin M, Schrauder A, Röttgers S, Herr W, Lang P, Handgretinger R, Hartwig UF, André MC. Engraftment of low numbers of pediatric acute lymphoid and myeloid leukemias into NOD/SCID/IL2Rcγnull mice reflects individual leukemogenecity and highly correlates with clinical outcome. Int J Cancer. 2013;133(7):1547–56.
30. Distler E, Wölfel C, Köhler S, Nonn M, Kaus N, Schnürer E, Meyer RG, Wehler TC, Huber C, Wölfel T, Hartwig UF, Herr W. Acute myeloid leukemia (AML)-reactive cytotoxic T lymphocyte clones rapidly expanded from CD8(+) CD62L((high)+) T cells of healthy donors prevent AML engraftment in NOD/SCID IL2Rgamma(null) mice. Exp Hematol. 2008;36(4):451–63.
31. Spranger S, Jeremias I, Wilde S, Leisegang M, Stärck L, Mosetter B, Uckert W, Heemskerk MH, Schendel DJ, Frankenberger B. TCR-transgenic lymphocytes specific for HMMR/Rhamm limit tumor outgrowth in vivo. Blood. 2012;119:3440–9.
32. Spits H, Di Santo JP. The expanding family of innate lymphoid cells: regulators and effectors of immunity and tissue remodeling. Nat Immunol. 2011;12(1):21–7.
33. Strowig T, Chijioke O, Carrega P, Arrey F, Meixlsperger S, Rämer PC, Ferlazzo G, Münz C. Human NK cells of mice with reconstituted human immune system components require preactivation to acquire functional competence. Blood. 2010;116:4158–67.
34. Colucci F, Caligiuri MA, Di Santo JP. What does it take to make a natural killer? Nat Rev Immunol. 2003;3(5):413–25.
35. Sandau MM, Schluns KS, Lefrancois L, Jameson SC. Cutting edge: transpresentation of IL-15 by bone marrow-derived cells necessitates expression of IL-15 and IL-15R alpha by the same cells. J Immunol. 2004;173(11):6537–41.
36. Björkström NK, Riese P, Heuts F, Andersson S, Fauriat C, Ivarsson MA, Björklund AT, Flodström-Tullberg M, Michaëlsson J, Rottenberg ME, Guzmán CA, Ljunggren HG, Malmberg

KJ. Expression patterns of NKG2A, KIR, and CD57 define a process of CD56dim NK-cell differentiation uncoupled from NK-cell education. Blood. 2010;116(19):3853–64.
37. Chen Q, Khoury M, Chen J. Expression of human cytokines dramatically improves reconstitution of specific human-blood lineage cells in humanized mice. Proc Natl Acad Sci U S A. 2009;106:21783–8.
38. Elpek KG, Rubinstein MP, Bellemare-Pelletier A, Goldrath AW, Turley SJ. Mature natural killer cells with phenotypic and functional alterations accumulate upon sustained stimulation with IL-15/IL-15Ralpha complexes. Proc Natl Acad Sci U S A. 2010;107(50):21647–52.
39. Huntington ND, Puthalakath H, Gunn P, Naik E, Michalak EM, Smyth MJ, Tabarias H, Degli-Esposti MA, Dewson G, Willis SN, Motoyama N, Huang DC, Nutt SL, Tarlinton DM, Strasser A. Interleukin 15-mediated survival of natural killer cells is determined by interactions among Bim, Noxa and Mcl-1. Nat Immunol. 2007;8(8):856–63.
40. Takenaka K, Prasolava TK, Wang JC, Mortin-Toth SM, Khalouei S, Gan OI, Dick JE, Danska JS. Polymorphism in Sirpa modulates engraftment of human hematopoietic stem cells. Nat Immunol. 2007;8(12):1313–23.
41. Strowig T, Rongvaux A, Rathinam C, Takizawa H, Borsotti C, Philbrick W, Eynon EE, Manz MG, Flavell RA. Transgenic expression of human signal regulatory protein alpha in Rag2-/-gamma(c)-/- mice improves engraftment of human hematopoietic cells in humanized mice. Proc Natl Acad Sci U S A. 2011;108(32):13218–23.
42. Legrand N, Huntington ND, Nagasawa M, Bakker AQ, Schotte R, Strick-Marchand H, de Geus SJ, Pouw SM, Böhne M, Voordouw A, Weijer K, Di Santo JP, Spits H. Functional CD47/signal regulatory protein alpha (SIRP(alpha)) interaction is required for optimal human T- and natural killer- (NK) cell homeostasis in vivo. Proc Natl Acad Sci U S A. 2011;108(32):13224–9.
43. Freud AG, Yokohama A, Becknell B, Lee MT, Mao HC, Ferketich AK, Caligiuri MA. Evidence for discrete stages of human natural killer cell differentiation in vivo. J Exp Med. 2006;203(4):1033–43.
44. Gottschalk LR, Bray RA, Kaizer H, Gebel HM. Two populations of CD56 (Leu-19)+/CD16+ cells in bone marrow transplant recipients. Bone Marrow Transpl. 1990;5(4):259–64.
45. Jacobs R, Stoll M, Stratmann G, Leo R, Link H, Schmidt RE. CD16- CD56+ natural killer cells after bone marrow transplantation. Blood. 1992;79(12):3239–44.
46. Anfossi N, André P, Guia S, Falk CS, Roetynck S, Stewart CA, Breso V, Frassati C, Reviron D, Middleton D, Romagné F, Ugolini S, Vivier E. Human NK cell education by inhibitory receptors for MHC class. Immunity. 2006;25(2):331–42.
47. de la Salle H, Hanau D, Fricker D, Urlacher A, Kelly A, Salamero J, Powis SH, Donato L, Bausinger H, Laforet M, Jeras M, Spehner D, Bieber T, Falkenrodt A, Cazenave J-P, Trowsdale J, Tongio M-M. Homozygous human TAP peptide transporter mutation in HLA class I deficiency. Science. 1994;265(5169):237–41.
48. Orr MT, Murphy WJ, Lanier LL. 'Unlicensed' natural killer cells dominate the response to cytomegalovirus infection. Nat Immunol. 2010;11(4):321–7.

Chapter 16
Phenotypical and Functional Properties of Antigen-Presenting Cells Derived from Humanized Mice

Maya Caroline André, Sonja Meixlsperger and Christian Münz

Abbreviations

APC	Antigen-presenting cells
BM	Bone marrow
cDC	Conventional DC
moDC	Monocyte-derived DC
pDC	Plasmacytoid DC
Flt3L	Flt3 ligand
G-CSF, M-CSF, GM-CSF	Granulocyte, macrophage, granulocyte-macrophage colony stimulating factor
IFN	Interferon
IL	Interleukin
KI	Knock-in
LPS	Lipopolysaccharide
MCP-1	Monocyte chemoattractant protein-1
MHC	Major histocompatibility complex
PAMP	Pathogen-associated molecular pattern
PB	Peripheral blood
TLR	Toll-like receptor
TNF	Tumor necrosis factor

M. C. André (✉)
Department of Hematology and Oncology, University Children's Hospital of Tuebingen,
Hoppe-Seyler-Str. 1, 72076 Tuebingen, Germany
e-mail: maya.andre@med.uni-tuebingen.de

University Children's Hospital, Basel, Switzerland

S. Meixlsperger · C. Münz
Institute of Experimental Immunology, University of Zurich, Zurich, Switzerland

L. Y. Poluektova et al. (eds.), *Humanized Mice for HIV Research*,
DOI 10.1007/978-1-4939-1655-9_16

16.1 Introduction

Comparatively little data exist on the in vivo development of "human" myeloid cell subsets in humanized mice. Early observations have shown that the "human" surrogate hematopoiesis generated in humanized mice exhibits a distinct bias toward lymphoid development, whereas the myeloid lineage is relatively underrepresented. Facing the importance of monocytes and dendritic cells (DCs) in exerting innate-immune functions and in initiating adaptive-immune responses, the generation of myelopoiesis and mucosal immunity in humanized mice has lately come into the focus of translational scientists. This chapter summarizes the currently available data on the emergence, differentiation, maintenance, and functionality of myeloid antigen-presenting cells (APCs) in humanized mice.

16.2 Generation of Monocytes/Macrophages in Humanized Mice

16.2.1 Cross-Reactivity of Murine to Human Myelopoiesis-Supporting Cytokines

In contrast to lymphoid cell populations that are constitutively absent in NOG, NSG, and BRG substrains, the murine myeloid compartment is in principle preserved albeit functionally deficient [1]. Thus, emerging "human" myeloid cells have to compete with the murine myelopoesis and cytokines in the near vicinity. The proliferation and differentiation from multipotent-hematopoetic stem cells to monocytes is controlled by the concerted action of stem cell factor (SCF), Flt3 ligand (Flt3L), thrombopoetin, granulocyte-macrophage colony stimulating factor (GM-CSF), and interleukin-3 (IL3). Many of them exhibit considerable functional redundancies, however, only the murine homologs of SCF, Flt3 ligand, and thrombopoetin may cross-react with their respective human receptor counterparts on stem and myeloid progenitor cells [1, 2]. Following the initial murine cytokine boost, myeloid progenitor cells derived from human stem cells will then synthesize and secrete "human" GM-CSF and IL3, enabling the subsequent differentiation and maintenance of "human" monocytes.

16.2.2 Emergence, Differentiation, and Maintenance of Monocytes

Subclasses of monocytes have been defined accounting for their differential expression of CD14 and CD16, however the emergence of these developmentally diverse subpopulations has not been formally addressed in humanized mice. It has been described that humanization of NSG mice (huNSG mice) induces myeloid CD33^{+} precursors which subsequently differentiate into CD14^{+} monocytes with a phenotypically normal expression of costimulatory factors such as CD86 and the T cell antigen-presenting molecule HLA-DR [2, 3]. In both huNSG and huBRG mice, CD14^{+} CD33^{+} monocytes/macrophages may decline after 16 weeks posttransplantation depending on the quality of the utilized SC preparations [3, 4] indicating that, next to the assumed fading functionality of human hematopoietic stem cells, the concentrations of essential "human" cytokines may be insufficient to promote long-term myelopoiesis.

16.2.3 Tissue Distribution

The tissue distribution of monocytes has not been studied systematically. In principle, monocytes/macrophages have been detected in the bone marrow (BM), spleen, peripheral blood (PB), lungs, and liver of humanized mice although monocytic-lineage differentiation preferentially takes place in the BM of humanized mice [2–4] Although not formally proven, the varying organ reconstitution and tissue distribution of monocytes/macrophages argues for a directed migration and homing to selected hematopoietic niches. In this context, the role of "human" chemokines or adhesion molecules that are required for this complex process has so far been unaddressed and might be a worthwhile subject for future studies.

16.2.4 Efficiency for Migration/Homing and the Generation of Mucosal Immunity

Migration and homing represent important aspects of monocyte biology, however little is known on the expression of chemokines, selectins, and integrins in humanized mice. In a recently published excellent review, the protein identities of human migration and homing molecules were compared to their respective murine counterparts [5]. In contrast to integrins that are apparently highly conserved between the mouse and human species, the amino acid sequences of selectins and chemokines differ considerably, indicating that homing to murine tissues might cause difficulties in the currently available immune-permissive mouse strains. However, it has been described that the lungs of recipient humanized mice were repopulated with "human" alveolar CD33^{+} CD14^{+} HLA-DR^{+} monocytes/macrophages [2]. In vivo models ideally employing infection with human pulmonary bacterial or viral pathogens would be necessary to verify an existing functionally intact pulmonary mucosal immunity and immune defence.

16.2.5 Secretion of Inflammatory Cytokines by Monocytes/ Macrophages Derived from Humanized Mice

Studies on the secretion of inflammatory cytokines in humanized mice have largely been restricted to the context of bacterial infections or in vivo treatment with lipopolysaccharide (LPS). As cytokine concentrations have mostly been quantified from serum or plasma of infected humanized mice, the evidence for monocytes as the source of inflammatory cytokine generation is rather an indirect one. Using a model of intraperitoneal injection of *Salmonella typhi*, the Fang group [6] recorded elevated concentrations of IL6, monocyte chemoattractant protein-1 (MCP-1), and tumor necrosis factor (TNF). In a caecal ligation and puncture model, important features of human septic diseases were recapitulated in humanized mice as demonstrated amongst others by a marked elevation of both pro- (TNF, IL1b, IL6, and IL8) and anti-inflammatory (IL10) cytokines [7]. Further evidence for the functional competence of macrophages came from the observation that gene silencing of high-mobility group protein B1 was able to ameliorate the sepsis-induced cytokine storm [8]. Of note, the in vivo response to LPS will closely depend on the extent of engraftment or the number of "human" target cells, respectively, as the application of moderate doses of LPS (10 µg/mouse) to mice engrafted with low levels of "human" cells resulted in the secretion of insufficient amounts of proinflammatory cytokines [3]. Collectively, monocytes derived from humanized mice are capable of secreting a comprehensive and diversified array of inflammatory cytokines upon bacterial stimulation, thus enabling the fostering of complex systemic inflammatory responses.

16.2.6 Responsiveness to Inflammatory Cytokines or Toll-Like Receptor Adjuvants

One prerequisite for a functionally competent monocyte is the ability for synthesis and secretion of inflammatory cytokines upon sensing of conserved pathogen-associated molecular patterns (PAMP) by toll-like receptors (TLRs). Unfortunately, it is only partially known whether monocytes from humanized mice express the respective receptors for inflammatory cytokines and/or PAMPs. It was shown that huNSG-derived $CD33^+$ cells express comparable levels of IFN-γR, G-CSFR, GM-CSFR, and M-CSFR when compared to the respective donor cells [2]. In line with this, human myeloid lineage cells generated in huNSG mice exhibited intact functional responses to human cytokines as determined by increased JAK/STAT signaling upon exposure to recombinant human IFN-γ and G-CSF [2] or upregulation of the costimulatory molecule CD86 upon IFN-γ stimulation [3]. In addition, huNSG-derived monocytes expressed normal levels of TLR2 and 4 and responded to in vivo LPS stimulation with adequate synthesis of TNF, IL-6, and IL-8 [2]. Although TLR3 expression was not formally quantified, huNSG-derived monocytes exhibited qualitative and functional responses upon activation with the synthetic analogue of dsRNA polyI:C and increased their synthesis of interferon (IFN)-α and β mRNA [2]. As time kinetics of cytokine secretion resembled human data, these observations indicate that huNSG-derived monocytes may indeed be used to study bacterially-induced inflammatory responses.

16.2.7 Robustness of Phagocytic Function/Ability for Migration

Based on the limited data available, monocytes derived from humanized mice exhibit intact functional responses with regard to their ability for phagocytosis. Upon in vitro culture, huNSG-derived monocytes incorporated fluorescent beads [2] and digested live GFP-expressing *Escherichia coli* [3] or *S. typhimurium* [2] to a comparable extent as human controls. Interestingly, in vivo engulfment of *S. typhi* bacteria by huNSG mice was accompanied with granulomatous inflammation resulting in the emergence of large-palisading epitheloid macrophages and formation of multinucleated giant cells [6]. As the Langhans giant cell formation requires the complex interaction of phagocytosing monocytes and T cells, activation of the CD40–CD40L axis and the presence of IFN-γ [9], this observation indicates that essential pathophysiological mechanisms characteristic for phagocytosis remain intact in humanized mice.

16.2.8 Induction of T Cell Stimulation and Proliferation

In contrast to phagocytic functions, the ability for the induction of T cell stimulation and proliferation seems to be distinctly altered in APCs derived from humanized mice. T cell blast formation and induction of T cell proliferation upon monoclonal antibody-mediated ligation of the T cell receptor CD3 domain with OKT3 was distinctly reduced in huNSG mice when compared to the respective human controls [3]. However, in the face of the immanent deficiencies in T cell function of humanized mice [10], studies requiring the interaction of monocyte-derived macrophages and T cells are difficult to interpret. Ideally, functional studies should be performed ex vivo with autologous human T cells or in vivo soon after additional adoptive transfer of fully functional T cells.

16.2.9 Source of Hematopoetic Stem Cells for Generation of Humanized Mice

The majority of research groups use the transplantation of neonatal recipients with cord blood-derived stem cells. However, preliminary work of our own group demonstrated in a direct comparison that the phenotypical and functional properties of monocytes derived from neonatal huNSG mice differ substantially from those derived from adult recipients [3]. In detail, the myelomonocytic lineage-differentiation in neonatal huNSG mice is delayed and monocytes are phenotypically immature with respect to HLA-DR expression and the emergence of $CD80^+$ $CD86^+$ monocytes. Functionally, neonatal huNSG-derived monocytes are less sensitive towards IFN-γ stimulation and exhibit a reduced T cell stimulating capacity, whereas the phagocytic activity and the ability for cytokine secretion are mature. In contrast, in vivo development of mature and functionally intact human myeloid subsets in neonatal NSG recipients transplanted with cord blood-derived stem cells was reported to be feasible [2]. Although the generation of neonatal huNSG mice is without doubt a useful model to study bacterial neonatal inflammatory responses [3, 11], comprehensive comparative studies including both neonatal and adult recipients together with the respective human controls are urgently warranted to allow the selection of a meaningful preclinical model for trials on human sepsis.

16.2.10 Role of Supportive Human Cytokines

Facing the limited cross-reactivity of many murine cytokines and the previously described limitations in the generation of a "human" myelopoesis, a variety of methods for providing adequate support of human cytokines have been developed. In detail, the injection of G-CSF resulted in recruitment of CD15-$CD33^+$ human myeloid cells into the recipient circulation [2] and hydrodynamic tail vein injection of a DNA vector encoding for human M-CSF significantly enhanced the expression of monocytes/macrophages in PB, BM, lung, liver, and spleen [12]. Extensive approaches have been undertaken to provide ample cytokine concentrations using "knock-in humanization" with replacement of either human CSF-1 [4] or human IL3/GM-CSF [13]. Analysis of huBRG CSF-1 knock-in (KI) mice revealed that the frequency of $CD14^+$ $CD33^+$ monocytes is increased and important costimulatory molecules such as CD40, CD80, CD86, and HLA-DR are upregulated [4]. Functional analysis demonstrated that huBRG CSF-1-derived monocytes exhibit an augmented ability for phagocytosis, display an enhanced capability for migration and responded to in vivo LPS injection with synthesis of proinflammatory cytokines. Transgenic expression of IL-3 and GM-CSF resulted in improved myeloid immune cell reconstitution in the lung and supported the rescue of a preexistent pulmonary alveolar proteinosis syndrome that was induced by the relative lack of murine cytokines [13]. In addition, macrophages of IL3/GM-CSF KI huBRG mice were able to mount antiinfluenza virus immune responses indicating that this model might be ideally suited for future studies on mucosal innate-immune responses. Interestingly, two further attempts to improve human myelopoesis either with forced transgenic expression of membrane-bound SCF [14] or of human thrombopoetin [15] resulted in mice with increased numbers of total $CD33^+$ myeloid cells but distinctly reduced or at best unchanged numbers of APCs. In summary, the generation of myelopoiesis in humanized mice distinctly benefits from induced expression of human cytokines in the murine recipient, which enables the differentiation and functional maturation of monocytes together with the development of a robust mucosal immunity.

16.3 Generation of DCs in Humanized Mice

16.3.1 Characteristics of Human DCs

DCs are very potent APCs that are generated in the BM and efficiently take-up, process, and present antigen. Upon sensing of PAMPs or other danger signals they mature to an activated, mature state in which they upregulate costimulatory molecules and major histocompatibility (MHC) molecules as well as secrete cytokines. DCs are grouped into subsets based on their function, morphology, and the expression or lack of surface markers. In man, the main blood DC subsets in the steady state are conventional (cDCs) and plasmacytoid DCs (pDCs). Both lack markers of other lineages and are high in MHC class-II molecule (i.e. HLA-DR) expression. cDCs are $CD11c^+$ and either highly express CD1c (BDCA1) or CD141 (BDCA3) [16]. pDCs are $CD11c^-$ and positive for CD123, CD303 (BDCA2), and CD304 (BDCA4) [17]. As DCs are a very rare lymphocyte population, earlier studies on DC function were often based on the analysis of in vitro generated monocyte-derived DCs (moDCs). However, recent publications highlight the phenotypic and functional differences between moDCs and primary DC subsets. Therefore, humanized mice present an invaluable tool to study the different DC subsets, especially in lymphoid and peripheral organs that are not easily accessible in the human setting.

16.3.2 cDCs of the Blood and Lymphoid Organs in Humanized Mice

Several reports have documented the presence of "human" DCs in humanized mice. Like human cDCs, cDCs from humanized mice express CD11c, are lineage-negative and $HLA\text{-}DR^+$. They reside in the BM and spleen of huNSG mice and show typical DC morphology [18, 19]. cDCs can also be found in the BM [20], lymph nodes, and liver of huBRG mice [21]. Of note, they also reconstitute in blood and spleen of BLT mice [22, 23]. More recent reports subdivide cDCs into $CD1c^+$ or $CD141^+$ cells and the proportion of DC subsets from huNSG BM is similar to the one seen in human BM [2, 24]. huNSG mice can reconstitute around 1 % of $CD1c^+$ (200,000 per spleen) and 0.4 % $CD141^+$ (80,000 per spleen) cDCs [24]. $CD141^+$ cDCs, the possibly best DC subset for cross-presentation [25], express like their human counterparts mRNA for Clec9A (DNGR-1), Necl2, IRF8, BATF3, and TLR3 [26]. Thus, reconstitution of all conventional DC subsets in huNSG, huNRG and huBRG seems feasible.

16.3.3 pDCs of the Blood and Lymphoid Organs in Humanized Mice

Like cDCs, "human" pDCs can be found in all major humanized mouse strains [18, 20, 23, 27]. In huBRG mice, pDCs were shown to develop either in the periphery or in a transplanted thymus graft [28]. pDCs from huNSG [24] or huBRG [29] mice express CD303 and CD304. The highest percentage of pDCs resides in the BM at about 2 % of "human" CD45$^+$ cells (100,000 per femur), while the spleens of reconstituted mice contain around 1 % (200,000 per spleen) [24]. At lower levels, pDCs also reside in peripheral lymphoid organs. Like pDCs from humans, CD303$^+$ pDCs isolated from the spleen of huNSG mice express mRNA for IRF8, IRF4, DAP12, and TLR7 and 9 [26]. Thus, "human" pDCs reconstitute in severely immune-compromised mice from CD34$^+$ hematopoietic stem cells.

16.3.4 DC Subsets in Non-Lymphoid Tissues of Humanized Mice

Reports describing DCs in non-lymphoid tissues of humanized mice are scarce. Of importance for HIV infection, CD11c$^+$ cDCs are present in the female reproductive [30] and gastro-intestinal tract [31] of BLT mice. Possibly, "human" DCs can even infiltrate the brains of huNSG mice which could play a role when studying the development of NeuroAIDS, however this needs confirmation on the basis of other DC markers than HLA-DR alone [32]. In the future, it will be interesting to analyze the reconstitution of "human" DCs in other organs to better understand and manipulate the induction of an immune response at the site of physiological antigen exposure. For example, detection and possibly augmentation of DC infiltration of the skin in humanized mice could be important as dermal CD14$^+$ DCs are potent inducers of antibody-responses and Langerhans cells efficiently activate cytotoxic T cell responses [33]. However, the occupation of these respective tissues by mouse DCs might in principle limit "human" DC reconstitution at these sites.

16.3.5 Function of DCs in Humanized Mice

In contrast to merely defining "human" DCs in the new humanized mouse models, fewer studies report on their functional maturation. DCs isolated from huBRG mice were shown to exhibit typical DC-like morphology upon LPS and GM-CSF stimulation, stimulate allogeneic T cell proliferation and produce high amounts of IFN-α upon viral infection in vitro [21]. In addition, huNSG and huBRG-derived cDCs or pDCs, respectively, responded to in vitro stimulation of TLR3, 7, 8, and/or 9 with enhanced TNF, IL6, and IFN-α production [26, 27, 29]. More importantly, in vivo injection of TLR agonists into huNSG mice lead to phenotypic maturation of cDCs and pDCs and raised IL-12 and IFN-α levels [24]. Interestingly, intravenous injection of HIV-1 into huBRG mice resulted in productively infected pDCs with distinctly impaired functionality as secondary ex vivo infection with HSV lead to reduced IFN-α production [29]. With respect to bacterial infection, huBRG mice as well as huNSG mice mounted innate immune responses upon *S. typhi* infection as determined by elevated serum cytokine and chemokine levels [6, 34] arguing for in vivo activation of monocytes and/or DCs in humanized mice. Moreover, studies on immune responses to other pathogens or pathogen-derived antigens as well as delayed-type hypersensitivity and superantigen-stimulated T cell responses in humanized mice suggest that the DC compartment in those mice is functional [35, 36]. Of interest, DC function can be manipulated in vivo in humanized mice. For example, treatment of huNOG mice with an α-ICAM-1 antibody prior to LPS stimulation impaired cDC maturation as indicated by reduced costimulatory molecule expression. Blocking DCs in this semi-mature state resulted in antigen-specific T cell tolerance in a huNOG-based diabetes model [37]. Furthermore, injection of Clec9-specific antibodies transiently and partially depleted $CD141^+$ DCs in huNSG mice and abolished cytokine production from this subset in response to TLR agonists in vivo [24]. Taken together, the major constitutive "human" DC subsets of humanized mice seem to be functional based on phenotypic maturation and cytokine secretion.

16.3.6 Boosting DC Numbers in Humanized Mice with Human Cytokines

One major limitation in the generation of the DC compartment in humanized mice is the lack of sufficient concentrations of cytokines, particularly of Flt3L GM-CSF [38]. Even though murine Flt3L is cross-reactive to human cells [1], the concentration seems to be limiting as injection of Flt3L induced higher numbers of DCs. Murine GM-CSF and IL-4, on the other hand, are not cross-reactive to human cells [1]. Combined hydrodynamic-tail-vein injection of DNA encoding GM-CSF, IL-4, and Flt3L boosted $CD11c^+$ $CD209^+$ cDC numbers in lymphoid and non-lymphoid organs [12]. While GM-CSF was sufficient to increase $CD11c^+$ cDC numbers in the spleen, $CD209^+$ cDCs required additional IL-4 supplementation [39]. Those cells were functional as they upregulated maturation markers after LPS stimulation in vivo and produced IL-12 in vitro. Transgenic overexpression of SCF, GM-CSF, and IL-3 in huNSG mice boosted the percentage of $CD1c^+$ cDC in the BM, but did not alter the percentage of $CD304^+$ pDCs [40]. In contrast, injection or lentiviral expression of IL-7 resulted in an increase in pDCs in huBRG mice [41]. In conclusion, DC numbers can be boosted by raising human cytokine levels in vivo, however, cytokine stimulation might result in objectionable and unforeseeable effects. For example, boosting DC numbers with GM-CSF, SCF, and IL-3 unexpectedly expanded functional regulatory T cells in huNSG mice [40]. Thus, due to the pleiotropic effects of cytokines, it will be difficult to selectively expand the DC compartment with systemic cytokine injections into mice with human immune system compartments.

16.4 Conclusions

In summary, the generation of monocytes/macrophages and DCs in humanized mice is feasible but a more comprehensive phenotypical and functional characterization of these subpopulations would be desirable. Given the limited number of data available, the characterization of the expression of "human" chemokines, integrins, and adhesion molecules would improve our understanding of innate-immune responses and would allow improved experimental modulation of APC function. Furthermore, humanized mice present a useful tool to test whether targeting certain DC subsets in vivo in the context of vaccination against pathogens or tumors might be advantageous. Also, a systematic evaluation of the cross-reactivity of murine myeloid-supporting cytokines together with a systematic characterization of later emerging "human" cytokine concentrations would help to define the exact type and cocktail of additional growth hormones/cytokines that humanized mice should be provided with. In this context, the VelociGene strategy comprising the replacement of murine growth hormone/cytokine regions with the respective human counterpart under the control of the murine promoter and the diverse regulatory elements results has brought a substantial progress that will allow further refinement of immuno-permissive mouse strains [42, 43]. Complemented with the appropriate in vitro surrogate assays, the generation of APCs in humanized mice will thus allow researchers in the area of infectious or tumor diseases to gain deeper insights into pathophysiologic complex processes of innate immunity and to achieve meaningful preclinical data for targeted immune-modulation or drug-testing.

Acknowledgments This work was supported by grants from the Deutsche Forschungsgemeinschaft (KFO 183, TP 4) and the Jose Carreras Stiftung (SP 12/03) to MCA. The research of SM and CM was supported by grants from the National Cancer Institute (R01CA108609), the Sassella Foundation (10/02, 11/02, and 12/02), Cancer Research Switzerland (KFS-02652-08-2010), the Association for International Cancer Research (11-0516), $KFSP^{MS}$ and $KFSP^{HLD}$ of the University of Zurich, the Vontobel, Baugarten, EMDO and Sobek Foundations, the Fondation Acteria, Novartis, and the Swiss National Science Foundation (310030_143979 and CRSII3_136241).

References

1. Manz MG. Human-hemato-lymphoid-system mice: opportunities and challenges. Immunity. 2007;26:537–41.
2. Tanaka S, Saito Y, Kunisawa J, Kurashima Y, Wake T, Suzuki N, Shultz LD, Kiyono H, Ishikawa F. Development of mature and functional human myeloid subsets in hematopoietic stem cell-engrafted NOD/SCID/IL2rgammaKO mice. J Immunol. 2012;188:6145–55.
3. Gille C, Orlikowsky TW, Spring B, Hartwig UF, Wilhelm A, Wirth A, Goecke B, Handgretinger R, Poets CF, André MC. Monocytes derived from humanized neonatal NOD/SCID/IL2Rγ (null) mice are phenotypically immature and exhibit functional impairments. Hum Immunol. 2012;73(4):346–54.
4. Rathinam C, Poueymirou WT, Rojas J, Murphy AJ, Valenzuela DM, Yancopoulos GD, Rongvaux A, Eynon EE, Manz MG, Flavell RA. Efficient differentiation and function of human macrophages in humanized CSF-1 mice. Blood. 2011;118(11):3119–28.

5. Rongvaux A, Takizawa H, Strowig T, Willinger T, Eynon EE, Flavell RA, Manz MG. Human hemato-lymphoid system mice: current use and future potential for medicine. Annu Rev Immunol. 2013;31:635–74.
6. Libby SJ, Brehm MA, Greiner DL, Shultz LD, McClelland M, Smith KD, Cookson BT, Karlinsey JE, Kinkel TL, Porwollik S, Canals R, Cummings LA, Fang FC. Humanized nonobese diabetic-scid IL2rgammanull mice are susceptible to lethal Salmonella Typhi infection. Proc Natl Acad Sci U S A. 2010;107:15589–94.
7. Unsinger J, McDonough JS, Shultz LD, Ferguson TA, Hotchkiss RS. Sepsis-induced human lymphocyte apoptosis and cytokine production in "humanized" mice. J Leukoc Biol. 2009;86(2):219–27.
8. Ye C, Choi JG, Abraham S, Wu H, Diaz D, Terreros D, Shankar P, Manjunath N. Human macrophage and dendritic cell-specific silencing of high-mobility group protein B1 ameliorates sepsis in a humanized mouse model. Proc Natl Acad Sci U S A. 2012;109(51):21052–7.
9. Sakai H, Okafuji I, Nishikomori R, Abe J, Izawa K, Kambe N, Yasumi T, Nakahata T, Heike T. The CD40-CD40L axis and IFN-γ play critical roles in Langhans giant cell formation. Int Immunol. 2012;24(1):5–15.
10. André MC, Erbacher A, Gille C, Schmauke V, Goecke B, Hohberger A, Mang P, Wilhelm A, Mueller I, Herr W, Lang P, Handgretinger R, Hartwig UF. Long-term human CD34 + stem cell-engrafted nonobese diabetic/SCID/IL-2R gamma(null) mice show impaired CD8 + T cell maintenance and a functional arrest of immature NK cells. J Immunol. 2010;185(5):2710–20.
11. Ernst W, Zimara N, Hanses F, Männel DN, Seelbach-Göbel B, Wege AK. Humanized mice-a new model to study the influence of drug treatment on neonatal sepsis. Infect Immun. 2013;81(5):1520–31.
12. Chen Q, Khoury M, Chen J. Expression of human cytokines dramatically improves reconstitution of specific human-blood lineage cells in humanized mice. Proc Natl Acad Sci U S A. 2009;106:21783–88.
13. Willinger T, Rongvaux A, Takizawa H, Yancopoulos GD, Valenzuela DM, Murphy AJ, Auerbach W, Eynon EE, Stevens S, Manz MG, Flavell RA. Human IL-3/GM-CSF knock-in mice support human alveolar macrophage development and human immune responses in the lung. Proc Natl Acad Sci U S A. 2011;108(6):2390–5.
14. Takagi S, Saito Y, Hijikata A, Tanaka S, Watanabe T, Hasegawa T, Mochizuki S, Kunisawa J, Kiyono H, Koseki H, Ohara O, Saito T, Taniguchi S, Shultz LD, Ishikawa F. Membrane-bound human SCF/KL promotes in vivo human hematopoietic engraftment and myeloid differentiation. Blood. 2012;119(12):2768–77.
15. Rongvaux A, Willinger T, Takizawa H, Rathinam C, Auerbach W, Murphy AJ, Valenzuela DM, Yancopoulos GD, Eynon EE, Stevens S, Manz MG, Flavell RA. Human thrombopoietin knockin mice efficiently support human hematopoiesis in vivo. Proc Natl Acad Sci U S A. 2011;108(6):2378–83.
16. MacDonald KP, Munster DJ, Clark GJ, Dzionek A, Schmitz J, Hart DN. Characterization of human blood dendritic cell subsets. Blood. 2002;100:4512–20.
17. Dzionek A, Fuchs A, Schmidt P, Cremer S, Zysk M, Miltenyi S, Buck DW, Schmitz J. BDCA-2, BDCA-3, and BDCA-4: three markers for distinct subsets of dendritic cells in human peripheral blood. J Immunol. 2000;165:6037–6046.
18. Ishikawa F, Yasukawa M, Lyons B, Yoshida S, Miyamoto T, Yoshimoto G, Watanabe T, Akashi K, Shultz LD, Harada M. Development of functional human blood and immune systems in NOD/SCID/IL2 receptor gamma chain^{null} mice. Blood. 2005;106:1565–73.
19. Strowig T, Chijioke O, Carrega P, Arrey F, Meixlsperger S, Rämer PC, Ferlazzo G, Münz C. Human NK cells of mice with reconstituted human immune system components require preactivation to acquire functional competence. Blood. 2010;116:4158–67.
20. Kuruvilla JG, Troyer RM, Devi S, Akkina R. Dengue virus infection and immune response in humanized RAG2$^{-/-}$gamma$_c^{-/-}$ (RAG-hu) mice. Virology. 2007;369:143–52.
21. Traggiai E, Chicha L, Mazzucchelli L, Bronz L, Piffaretti JC, Lanzavecchia A, Manz MG. Development of a human adaptive immune system in cord blood cell-transplanted mice. Science. 2004;304:104–7.

22. Brainard DM, Seung E, Frahm N, Cariappa A, Bailey CC, Hart WK, Shin HS, Brooks SF, Knight HL, Eichbaum Q, Yang YG, Sykes M, Walker BD, Freeman GJ, Pillai S, Westmoreland SV, Brander C, Luster AD, Tager AM. Induction of robust cellular and humoral virus-specific adaptive immune responses in human immunodeficiency virus-infected humanized BLT mice. J Virol. 2009;83(14):7305–21.
23. Melkus MW, Estes JD, Padgett-Thomas A, Gatlin J, Denton PW, Othieno FA, Wege AK, Haase AT, Garcia JV. Humanized mice mount specific adaptive and innate immune responses to EBV and TSST-1. Nat Med. 2006;12:1316–22.
24. Meixlsperger S, Leung CS, Rämer PC, Pack M, Vanoaica LD, Breton G, Pascolo S, Salazar AM, Dzionek A, Schmitz J, Steinman RM, Münz C. CD141 + dendritic cells produce prominent amounts of IFN-α after dsRNA recognition and can be targeted via DEC-205 in humanized mice. Blood. 2013;121(25):5034–44.
25. Jongbloed SL, Kassianos AJ, McDonald KJ, Clark GJ, Ju X, Angel CE, Chen CJ, Dunbar PR, Wadley RB, Jeet V, Vulink AJ, Hart DN, Radford KJ. Human CD141^{+} (BDCA-3)$^{+}$ dendritic cells (DCs) represent a unique myeloid DC subset that cross-presents necrotic cell antigens. J Ex Med. 2010;207:1247–60.
26. Poulin LF, Salio M, Griessinger E, Anjos-Afonso F, Craciun L, Chen JL, Keller AM, Joffre O, Zelenay S, Nye E, Le Moine A, Faure F, Donckier V, Sancho D, Cerundolo V, Bonnet D, Reis e Sousa C. Characterization of human DNGR-1 + BDCA3 + leukocytes as putative equivalents of mouse CD8alpha+ dendritic cells. J Exp Med. 2010;207(6):1261–71.
27. Ishikawa F, Niiro H, Iino T, Yoshida S, Saito N, Onohara S, Miyamoto T, Minagawa H, Fujii S, Shultz LD, Harada M, Akashi K. The developmental program of human dendritic cells is operated independently of conventional myeloid and lymphoid pathways. Blood. 2007;110(10):3591–660.
28. Weijer K, Uittenbogaart CH, Voordouw A, Couwenberg F, Seppen J, Blom B, Vyth-Dreese FA, Spits H. Intrathymic and extrathymic development of human plasmacytoid dendritic cell precursors in vivo. Blood 2002;99:2752–59.
29. Zhang L, Jiang Q, Li G, Jeffrey J, Kovalev GI, Su L. Efficient infection, activation, and impairment of pDCs in the BM and peripheral lymphoid organs during early HIV-1 infection in humanized rag2$^{-/-}$gamma C$^{-/-}$ mice in vivo. Blood. 2011;117:6184–92.
30. Denton PW, Estes JD, Sun Z, Othieno FA, Wei BL, Wege AK, Powell DA, Payne D, Haase AT, Garcia JV. Antiretroviral pre-exposure prophylaxis prevents vaginal transmission of HIV-1 in humanized BLT mice. PLoS Med. 2008;5:e16.
31. Sun Z, Denton PW, Estes JD, Othieno FA, Wei BL, Wege AK, Melkus MW, Padgett-Thomas A, Zupancic M, Haase AT, Garcia JV. Intrarectal transmission, systemic infection, and CD4^{+} T cell depletion in humanized mice infected with HIV-1. J Exp Med. 2007;204:705–14.
32. Gorantla S, Makarov E, Finke-Dwyer J, Castanedo A, Holguin A, Gebhart CL, Gendelman HE, Poluektova L. Links between progressive HIV-1 infection of humanized mice and viral neuropathogenesis. Am J Pathol. 2010;177:2938–49.
33. Ueno H, Schmitt N, Klechevsky E, Pedroza-Gonzalez A, Matsui T, Zurawski G, Oh S, Fay J, Pascual V, Banchereau J, Palucka K. Harnessing human dendritic cell subsets for medicine. Immunol Rev. 2010;234:199–212.
34. Song J, Willinger T, Rongvaux A, Eynon EE, Stevens S, Manz MG, Flavell RA, Galan JE. A mouse model for the human pathogen Salmonella typhi. Cell Host Microbe. 2010;8:369–76.
35. Akkina R. New generation humanized mice for virus research: comparative aspects and future prospects. Virology. 2013;435:14–28.
36. Rämer PC, Chijioke O, Meixlsperger S, Leung CS, Münz C. Mice with human immune system components as in vivo models for infections with human pathogens. Immunol Cell Biol. 2011;89:408–16.
37. Jung KC, Park CG, Jeon YK, Park HJ, Ban YL, Min HS, Kim EJ, Kim JH, Kang BH, Park SP, Bae Y, Yoon IH, Kim YH, Lee JI, Kim JS, Shin JS, Yang J, Kim SJ, Rostlund E, Muller WA, Park SH. In situ induction of dendritic cell-based T cell tolerance in humanized mice and nonhuman primates. J Exp Med. 2011;208(12):2477–88.

38. Kingston D, Schmid MA, Onai N, Obata-Onai A, Baumjohann D, Manz MG. The concerted action of GM-CSF and Flt3-ligand on in vivo dendritic cell homeostasis. Blood. 2009;114:835–43.
39. Chen Q, He F, Kwang J, Chan JK, Chen J. GM-CSF and IL-4 stimulate antibody responses in humanized mice by promoting T, B, and dendritic cell maturation. J Immunol. 2012;189:5223–9.
40. Billerbeck E, Barry WT, Mu K, Dorner M, Rice CM, Ploss A. Development of human $CD4^+$ $FoxP3^+$ regulatory T cells in human stem cell factor-, granulocyte-macrophage colony-stimulating factor-, and interleukin-3-expressing NOD-SCID $IL2Rgamma^{null}$ humanized mice. Blood. 2011;117:3076–86.
41. van Lent AU, Dontje W, Nagasawa M, Siamari R, Bakker AQ, Pouw SM, Maijoor KA, Weijer K, Cornelissen JJ, Blom B, Di Santo JP, Spits H, Legrand N. IL-7 enhances thymic human T cell development in "human immune system" $Rag2^{-/-}IL\text{-}2Rgamma_c^{-/-}$ mice without affecting peripheral T cell homeostasis. J Immunol. 2009;183:7645–55.
42. Valenzuela DM, Murphy AJ, Frendewey D, Gale NW, Economides AN, Auerbach W, Poueymirou WT, Adams NC, Rojas J, Yasenchak J, Chernomorsky R, Boucher M, Elsasser AL, Esau L, Zheng J, Griffiths JA, Wang X, Su H, Xue Y, Dominguez MG, Noguera I, Torres R, Macdonald LE, Stewart AF, DeChiara TM, Yancopoulos GD. High-throughput engineering of the mouse genome coupled with high-resolution expression analysis. Nat Biotechnol. 2003;21(6):652–9.
43. Shultz LD, Brehm MA, Garcia-Martinez JV, Greiner DL. Humanized mice for immune system investigation: progress, promise and challenges. Nat Rev Immunol. 2012;12:786–798.

Part III
Humanized Mice for HIV-1 Virus Biology and Pathogenesis. Is It Possible to Address?

Chapter 17
Humanized Mouse Versus Non-human Primate Models of HIV-1 Infection

Qingsheng Li and Charles Wood

17.1 Introduction

Animal models are critical for biomedical research, in particular for human immunodeficiency virus type one (HIV-1), since HIV-1 infection still remains a major burden to global public health. As for 2011, it is estimated that 34 million people are living with HIV-1, 2.5 million people are newly infected, and only a small portion (8 million) of infected people are currently receiving the combined antiretroviral therapy, which is expensive and has to be lifelong [1] (https://www.unaids.org/en/resources/publications/2013/). This dire reality further highlights the importance of animal models in developing vaccine to prevent HIV infection and testing new approaches to purge latently infected reservoir in order to cure HIV infection. An ideal animal model should be able to model HIV-1 transmission, pathogenesis, evaluate the efficacy of antiretroviral agents, immune modulators, and vaccines in preventing, treating, and curing HIV-1 infection in humans. Unfortunately, the universal and ideal model does not exist. Instead, different animal models are often used independently or in combination, of which non-human primates (NHPs) and humanized mice (hu-mice) are the two available models.

In this chapter, we: (1) compare and contrast the pros and cons of NHP and hu-mouse models of HIV-1 infection of humans in general; (2) discuss in detail which model is more relevant in studying HIV-1 transmission and vaccine; and (3) discuss what aspects of these models need to be further improved in order to meet the HIV-1 research need. Since there are many different variables in both models, such as different types of macaques and hu-mice, different types of simian immunodeficiency

Q. Li (✉) · C. Wood
Nebraska Center for Virology, School of Biological Sciences, University of Nebraska-Lincoln, 4240 Fair Street, Lincoln, NE 68583-0900, USA
e-mail: qli4@unl.edu

C. Wood
e-mail: cwood1@unl.edu

L. Y. Poluektova et al. (eds.), *Humanized Mice for HIV Research*,
DOI 10.1007/978-1-4939-1655-9_17

viruses (SIVs) and HIV-1 viruses, different routes and dose of virus infection, we can only compare the best available representatives of NHP and hu-mouse models.

17.2 The Current Status of NHP Models of HIV-1 Infection

More than 30 African NHP species are naturally infected with more than 40 different strains of SIVs [2, 3]; African NHPs have coexisted with SIVs for more than 32,000 years [4] and host the immediate ancestral virus of HIV-1 [5–7] and HIV-2 [8], but infected animals generally do not develop the AIDS-like disease even in the face of a chronic infection with high level of replicating virus [9].

The common chimpanzees in West Central Africa (*Pan troglodytes troglodytes*) are endemically infected with SIVcpzPtt and are the zoonotic source of pandemic HIV-1 group M and non-pandemic group N; Eastern chimpanzees in East Africa (*Pan troglodytes schweinfurthii*) are infected with SIVcpzPts, but this virus has not yet been found in humans [5, 7, 10–13]. Gorillas (*Gorilla gorilla gorilla*) are infected with gorilla SIV (SIVgor) and are the zoonotic source of HIV-1 group P [14, 15]. The simian zoonotic source of HIV-1 group O remains to be identified [15]. Although new data indicate that SIVcpz infections of chimpanzees had negative effects on their health, reproduction, and lifespan [16], the clinical course is still different from HIV-1 infection of humans. For ethical reasons and none/low pathogenic infection of SIV, the endangered species of chimpanzees are not feasible to be used as a model for HIV-1 research [17].

Sooty Mangabeys (*Cercocebus atys*) of African origin are the primate reservoir for HIV-2 [8] and the immediately ancestral virus of SIVmac transmitted to Asian macaques in captivity [18]. Sooty Mangabeys and African green monkeys (genus *Chlorocebus*) do not develop disease with high levels of SIV replication and are mainly used to study the mechanisms of non-pathogenic SIV infection [9, 19–22].

Asian NHPs of macaques, including rhesus macaques (*Macaca mulatta*), cynomolgus macaques (*M. fascicularis*), and pigtailed macaques (*M. nemestrina*) are non-natural hosts to SIVs and develop AIDS-like diseases after infection, of which rhesus macaque has been most widely used in HIV-1 research. Asian NHPs of macaques are regarded as a good model of HIV-1 infection of humans because of the following characteristics: (1) the proximity of macaques to humans, genetically, anatomically, and physiologically [23]; (2) the clinical manifestations and pathogenesis of macaques infected with SIV are similar to humans' infection by HIV-1 [18, 22, 24]; (3) the innate and adaptive immune responses ($CD8^+$ T [25, 26–28] and B cells [29–31]) of macaques to SIV infection are similar to humans' responses to HIV-1. Hence, this model has been widely used for transmission, immunopathogenesis, immune correlates of protection and vaccine efficacy studies, and has gained tremendous insights into the mechanisms of transmission and pathogenesis, and immune correlates of protection. However, this model also has several limitations: (1) macaques are not susceptible to HIV-1 infection; instead only to

SIV or related chimeric virus, expressing HIV-1 envelope (Env-SHIV or SHIV) or reverse-transcriptase (RT-SHIV) in SIV backbone. Although recently, it was reported that pig-tail macaques can support simian-tropic HIV-1 strains that encode only SIV vif protein (stHIV-1) replication [32], however, the virus replication lasted only for several months and its biological relevance to HIV-1 infection remained to be determined; (2) SIV viruses are naturally resistant to many FDA approved anti-HIV-1 drugs, including non-nucleoside reverse transcriptase inhibitors (NNRTIs), some entry inhibitors, and some proteinase inhibitors [33–35]. Although RT-SHIV and Env-SHIV can partly offset this drawback, many preventive/therapeutic regimens used in clinic cannot be studied in this model and vice versa; (3) SIV differs from HIV-1 genotypically and phenotypically, the vaccines designed and tested in this model using SIV or SHIV cannot be directly applied into human clinical trial; (4) only a limited number of SIV viruses are available for macaque studies. Which SIV challenge virus should be used in vaccine protective studies is still being debated [36], since many commonly used SIV challenge viruses in vaccine protective studies have different sensitivity to antibody neutralization and cytotoxic T lymphocytes (CTL)-mediated control. For example, SIVmnE660 can be neutralized more easily than SIVmac251 or SIVmac239 [30, 37], and SHIV89.6P can be controlled by CTL more easily [38]. Thus, the results with uncertain challenge viruses could be either underestimating or exaggerating the protective effect [38], and there are renewed efforts generating better challenge viruses [39, 40]; and (5) macaques and humans are genetically different, especially in major histocompatibility complex (MHC) and T cell receptors (TCR) which are more complex in the macaque species [41–43]. Thus, alternative models are sought to overcome the limitations of the macaque/SIV model. The hu-mice , especially the new generation of hu-mice, has emerged as a good alternative system to study HIV in addition to NHP (Fig. 17.1).

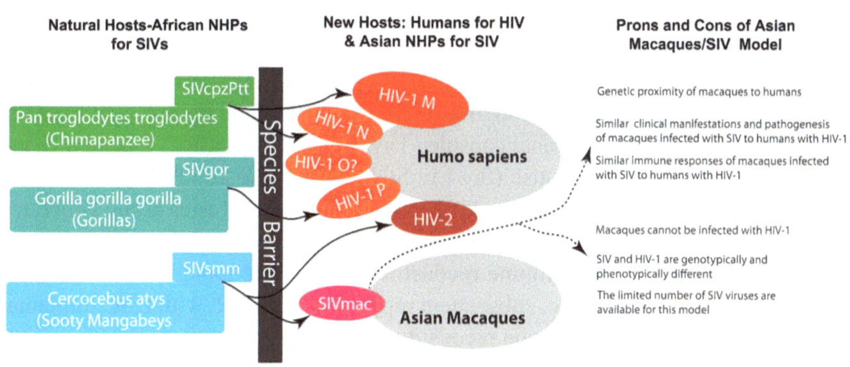

Fig. 17.1 The pros and cons of SIV/macaque model of HIV-1 infection. African NHPs are the natural host of SIVs and generally do not develop AIDS-like disease. The pandemic HIV-1 group M (HIV-1 M) and non-pandemic group N (HIV-1 N) are originated from SIVcpzPtt in Chimpanzees, and non-pandemic HIV-1 P is originated from SIVgov in Gorillas. HIV-2 and SIVmac originated from SIVsmm in sooty Mangabeys. SIVmac infects Asian macaques and cause simian AIDS

17.3 The Current Status of hu-mouse Models of HIV-1 Infection

The major driving force for developing hu-mice is to interpose an in vivo model between in vitro and clinical trials for studying human diseases, since the findings of in vitro experiments cannot be directly tested in human clinical trials due to ethical reasons. Furthermore, macaques are not susceptible to HIV-1 infection, therefore, SIV/macaque always requires a two-stage design and testing in order to move into clinical trials. For example, vectors and immunogen of SIV vaccines tested in the macaque model have to be redesigned into the human version for clinical trials. The hu-mouse model has a potential to serve as an alternative to complement the SIV/macaque model for vaccine studies.

The hu-mice are a heterochimera of the human immune system in the murine body in a delicate balance to avoid human graft versus murine host disease (GVHD) and murine host versus human graft disease (HVGD) while reconstituting the human system. In the past 25 years, this model has gone through several rounds of revolution primarily through two approaches. First, by genetically modifying the mouse to further eliminate murine immune cells and their functions in order to prevent HVGD; and second, by refining the procedures of implantation of human tissues and/or hematopoietic stem cells (HSC) in order to prevent GVHD, to attain a new level of human immune reconstitution in the lymphatic and non-lymphatic tissues, including mucosa. The new generation of hu-mice has drastically expanded its utility and has great potential in studying HIV mucosal transmission, pathogenesis, latency, pre-exposure prophylaxis (PrEP), treatment, and vaccine.

The history of hu-mouse model has been extensively reviewed elsewhere [44–46]; here, we will only highlight the major events in order to better compare NHP and hu-mouse models. The initial two independent groups conducted proof-of-concept experiments in 1988 generating hu-mice by two different approaches [47, 48] based on SCID mice [49]. The seminal paper by McCune [47] laid the conceptual and technical foundation for subsequent stable long-term reconstitution of multi-lineage human immune cells through implanting human fetal thymus and liver tissue fragments under mouse renal capsule (Thy/Liv SCID-hu mice) [50, 51]. Using this first generation of hu-mouse (thy/Liv SCID-hu), some key HIV pathogenesis and treatment questions were studied [52–55]. Meanwhile, Mosier group generated hu-PBL-SCID mice by transferring peripheral blood leukocyte (PBL) to SCID mice [48]. Although the human immune reconstitution is limited and unstable [50, 56, 57] in the hu-PBL-SCID mice, subsequent replacement of PBL with HSC implantation improved the human immune reconstitution [58, 59]

To further eliminate murine NK cells and reduce the "leakiness" of murine functional lymphocytes in some SCID mice, NOD/SCID mice were generated in 1995 [60] by backcrossing SCID and NOD mice, since NOD mice have defects in NK cells, myeloid development and function, and complement pathways [61, 62]. The engraftment of human $CD45^+$ cells in NOD/SCID mice has dramatically increased as compared to the SCID mouse recipients [60, 63]. In addition, to further improve the SCID mouse, RAG-1 [64] and RAG-2 (recombination-activating proteins) [65] deficient mice with no mature T and B cells were generated in 1992. Additionally, the mice with homozygous cytokine common receptor gamma chain mutant, a component of receptors for cytokine IL-2, IL-4, IL-7, IL-9, IL-15, and IL-21 [66–70], were generated (cγ−/−) [68, 69] in 1995. These mice have defects in T and B cells and absence of natural killer cell (NK) activity. With the crossbreeding of different immune deficiency mice above, more severely combined immune deficiency (current generation) mice were generated. These include the BRG (BALB/c_ RAG2−/− cγ−/−) [71, 72] and B6RG (C57BL6_ RAG2−/− cγ−/−) [73], NOG (NOD/Shi-scid *I* cγ−/−) [74], NSG (NOD/LtSzscid *I* cγ−/−) [75], and NOD-RAG1−/− cγ−/− mice [76].

Based on the current generation of immune deficient mice, two general types of hu-mice are commonly generated for biomedical research. First is hu-BLT (bone marrow, liver, and thymus) mice [77] generated through sub-lethal irradiation, implantation of human fetal liver and thymus pieces into the adult mouse left renal capsule, and injection of autologous CD34$^+$ HSC intravenously [77, 78]. Hu-BLT mice are a new generation of hu-mice with a long-term and multi-lineage reconstitution of human hematopoietic system (T, B, NK, DC, and Macrophages) in both lymphatic and mucosal tissues, and can elicit antigen-specific T cell and humoral responses [77, 79–83]. The hu-BLT mouse became the best hu-mouse model for studying HIV-1 mucosal transmission and its prevention, because there is a good immune reconstitution in mucosa, and the T cells can be educated in autologous human thymic tissues [77, 79–81]. Second is the hu-HSC mice generated by sub-lethal irradiation and injection of human CD34$^+$ HSC isolated from fetal liver, umbilical cord blood, or mobilized peripheral blood leukocytes with granulocyte colony stimulating factor (G-CSF) into the new generation of immune deficiency mice [74, 75, 84–87]. It is apparent that injection (intra-hepatic or intra-cardiac) of CD34$^+$ HSC into neonates of the current generation of immune deficient mice leads to much better de novo development of adaptive immune system (B, T, DC, and structured lymphatic organ) as compared with adult recipients [72, 85].

To further improve the human immune responses of the current generation of hu-mice, the human cytokines and growth factors cytokines (GM-CSF [88–90], IL-3 [88, 90], IL4 [88, 89], and IL-15 [91]) and MHC class I (HLA-A*0201 [92–94], HLA-B*51:01 [95]) and II (HLA-DRA and HLA-DRB1:0405 [96], HLA-DR4 [97], HLA-DR1 [93]) or in combination [90] were provided by transgenic, knock-in, vector expression, or hydrodynamic injection (Fig. 17.2). The current generation of hu-mice has increasingly been used in HIV-1 research, because of the following reasons: (1) besides chimpanzees, it is the only model that can directly study HIV-1 infection; (2) it reconstitutes most of human immune system functionally and structurally, thus it can recapitulate many aspects of HIV–host interaction, including CD4$^+$ T-cell depletion, increased CD4$^+$ and CD8$^+$ T-cell turnover, and immune activation [82, 98]; (3) it can be used to study HIV-1 mucosal transmission [79, 99], pathogenesis [88, 98, 100–102], prevention [103, 104], treatment [105–107], and latency [108]; and (4) it is much cheaper than NHP macaque. However, there are several limitations as well: (1) hu-mice are a chimera of human and mouse cells and tissues. Although human immune system is partly reconstituted, the non-lymphatic cells and tissues remain as murine; (2) there is a delay of humoral (3 month PI) and

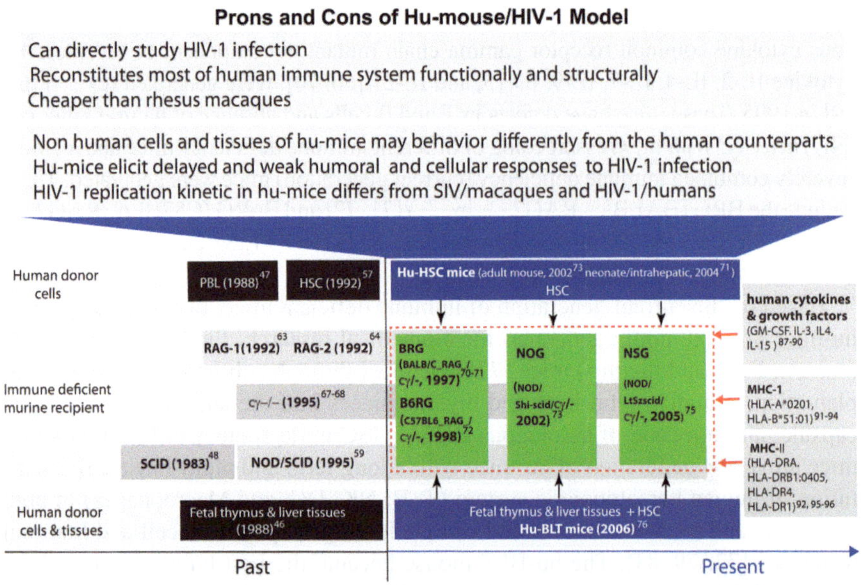

Fig. 17.2 The pros and cons of current hu-mouse model of HIV-1 infection. The two types of new generation of hu-mice: hu-BLT and hu-HSC are developed by improving the implantation method of human cell and tissues and genetically refining the immune deficiency of recipient mouse. Recently, human cytokines, growth factors, and MHC class I and II transgenic mice have further improved the human immune function of current hu-mice

cellular responses (9 weeks PI) in HIV-1-infected hu-BLT mice as compared with SIV/macaque and humans (2 weeks PI) [82]. The adaptive immune responses in hu-mice, even hu-BLT mice, therefore need to be further improved, especially IgG response [82, 87, 109]; (3) the HIV-1 replication kinetics in hu-mice is different from SIV/macaques and HIV-1/humans. In the hu-mice, the virus peaks around 2–3 weeks post infection, but is maintained for several weeks before declining [82], reflecting that there is a delay of the host control of HIV-1 replication.

17.4 Mucosal Transmission of HIV-1 and Its Prevention

HIV-1 is mainly transmitted through mucosal surfaces, such as cervicovagina, foreskin, and anorectum. Better understanding of the early events of HIV-1 mucosal transmission and their underlying mechanisms holds the keys to the better designed microbicide and vaccine. The key body of knowledge on the early events in mucosal transmission of HIV-1 was mainly acquired from the macaque/SIV model, of which atraumatic high-dose or repeated low-dose inoculations of cell-free viruses are often used. For example, in the early vaginal transmission, there is a small infected founder cell population at the portal of entry before systemic virus dissemination [110, 111]; there is a genetic bottleneck as revealed by using single genome amplification in vaginal [112], rectal [113, 114], and penile [115] transmission. Only recently, the infections of macaques, vaginally [116] and rectally, [117] with cell-associated SIV were reported; surprisingly, cell-associated virus that transmits infection across the mucosa was found to be more efficient than cell-free virus [117]. Of cervicovaginal, foreskin, and anorectal routes in SIV/macaque model, anorectal mucosa is the easiest route for transmission, followed by vaginal and penile [115, 118]. Although macaque penile transmission was reported previously, this route of transmission model has been used only very recently [115, 119–121].

In contrast to the long history of the use of macaque/SIV as a model for studying mucosal transmission of HIV-1, hu-mice have been used only recently, since the current generation hu-mice were available, specifically after the hu-BLT mice were developed. However, due to their advantages in being susceptible to HIV-1 infection, cheaper, and easier to manipulate than macaque, the current generation hu-mice are increasingly used in mucosal transmission and prevention studies. This model is especially useful to test microbicide in preventing mucosal transmission of HIV-1 [80, 104, 122]. Except for penile transmission, vaginal [80, 123, 124] and rectal [79, 122, 124] transmission of HIV-1 have both been reported.

17.5 Vaccine

The goal of vaccine development is to elicit protective memory immunity against infection, disease, and death [125, 126]. Macaque-SIV/SHIV model is still the best available model to identify the immune correlates of protection and evaluate vaccine efficacy, since hu-mice have delayed adaptive immune responses, especially very limited IgG response [82, 87, 109]. Currently, human cytokines, growth factors, and MHC class I and II transgenic NSG or NOG mice are generated which may improve this model for vaccine study. Conversely, the current generation of hu-mice is exceptionally useful in testing new preventive and therapeutic strategies, such as human broadly neutralizing antibodies [105], antibody-expressing vector [127], and engineering HIV-1 resistant cells [106, 128]. Its usefulness as a model for testing of HIV-1 vaccines remains to be determined.

17.6 Summary and Prospective

SIV/macaques model has been widely used for HIV-1 research since the middle 1980s and has provided critical insights into the HIV-1 transmission, pathogenesis, treatment, latency, microbicide, and vaccine. However, macaques are genetically distinct from humans, especially in MHC class I and TCR, and are not susceptible to HIV-1 infection. Thus, results derived from this model may not be directly translatable into human clinical trials; for example, vaccines designed and tested in this model using SIV or SHIV have to be redesigned in order to be tested in human clinical trial. The new generation of hu-HSC and hu-BLT mice, especially the hu-BLT mice with transgenic expression of human cytokines, growth factors, and MHC class I and II, offers a new opportunity to study HIV-1 infection using HIV-1 directly. Although there is still room to improve the humoral and cellular immune responses of hu-mice to HIV-1 infection [44, 46, 100, 129, 130], this model already recapitulates many key aspects of mucosal transmission [79, 99], prevention [103, 104], immunopathogenesis [88, 98, 100–102], treatment [105–107], and latency [108]. The new generation of hu-mouse and SIV/macaque models are complementary and together they will overcome the idea that "mice lie and monkeys exaggerate" [131].

References

1. UNAIDS. UNAIDS World AIDS Day Report. 2012.
2. VandeWoude S, Apetrei C. Going wild: lessons from naturally occurring T-lymphotropic lentiviruses. Clin Microbiol Rev. 2006;19(4):728–62.
3. Bibollet-Ruche F, et al., New simian immunodeficiency virus infecting De Brazza's monkeys (*Cercopithecus neglectus*): evidence for a *Cercopithecus* monkey virus clade. J Virol. 2004;78(14):7748–62.
4. Worobey M, et al., Island biogeography reveals the deep history of SIV. Science. 2010;329(5998):1487.
5. Peeters, M., et al. Isolation and partial characterization of an HIV-related virus occurring naturally in chimpanzees in Gabon. AIDS. 1989;3(10):625–30.
6. Gao F, et al. Human infection by genetically diverse SIVSM-related HIV-2 in West Africa. Nature. 1992;358(6386):495–9.
7. Gao F, et al. Origin of HIV-1 in the chimpanzee Pan troglodytes troglodytes. Nature. 1999;397(6718):436–41.
8. Hirsch, VM, et al. An African primate lentivirus (SIVsm) closely related to HIV-2. Nature. 1989;339(6223):389–92.
9. Chahroudi A, et al. Natural SIV hosts: showing AIDS the door. Science. 2012;335(6073):1188–93.
10. Huet T, et al. Genetic organization of a chimpanzee lentivirus related to HIV-1. Nature. 1990;345(6273):356–9.
11. Keele BF, et al. Chimpanzee reservoirs of pandemic and nonpandemic HIV-1. Science. 2006;313(5786):523–6.
12. Sharp PM, Hahn BH. Origins of HIV and the AIDS pandemic. Cold Spring Harb Perspect Med. 2011;1(1):a006841.
13. Li Y, et al. Eastern Chimpanzees, but not bonobos, represent a simian immunodeficiency virus reservoir. J Virol. 2012;86(19):10776–91.
14. Plantier J-C, et al. A new human immunodeficiency virus derived from gorillas. Nat Med. 2009;15(8):871–2.
15. Mourez T, Simon F, Plantier J-C. Non-M variants of human immunodeficiency virus type 1. Clin Microbiol Rev. 2013;26(3):448–61.
16. Keele BF, et al. Increased mortality and AIDS-like immunopathology in wild chimpanzees infected with SIVcpz. Nature. 2009;460(7254):515–9.
17. Great ape debate. Nature. 2011;474(7351):252.

18. Daniel M, et al. Isolation of T-cell tropic HTLV-III-like retrovirus from macaques. Science. 1985;228(4704):1201–4.
19. Silvestri G, et al. Nonpathogenic SIV infection of sooty mangabeys is characterized by limited bystander immunopathology despite chronic high-level viremia. Immunity. 2003;18(3):441–52.
20. Sodora DL, et al. Toward an AIDS vaccine: lessons from natural simian immunodeficiency virus infections of African nonhuman primate hosts. Nat Med. 2009;15(8):861–5.
21. Bosinger SE, et al. Global genomic analysis reveals rapid control of a robust innate response in SIV-infected sooty mangabeys. J Clin Invest. 2009;119(12):3556–72.
22. Van Rompay KK. The use of nonhuman primate models of HIV infection for the evaluation of antiviral strategies. AIDS Res Hum Retroviruses. 2012;28(1):16–35.
23. Sequencing RMG, et al. Evolutionary and biomedical insights from the rhesus macaque genome. Science. 2007;316(5822):222–34.
24. Letvin N, et al. Induction of AIDS-like disease in macaque monkeys with T-cell tropic retrovirus STLV-III. Science. 1985;230(4721):71–3.
25. Yasutomi Y, et al. Simian immunodeficiency virus-specific $CD8^+$ lymphocyte response in acutely infected rhesus monkeys. J Virol. 1993;67(3):1707–11.
26. Allen TM, et al. Tat-specific cytotoxic T lymphocytes select for SIV escape variants during resolution of primary viraemia. Nature. 2000;407(6802):386–90.
27. Jin, X., et al., Dramatic rise in plasma viremia after CD8(+) T cell depletion in simian immunodeficiency virus-infected macaques. J Exp Med. 1999;189(6):991–8.
28. Matano T, et al. Administration of an anti-CD8 monoclonal antibody interferes with the clearance of chimeric simian/human immunodeficiency virus during primary infections of rhesus macaques. J Virol. 1998;72(1):164–9.
29. Schmitz JE, et al. Effect of humoral immune responses on controlling viremia during primary infection of rhesus monkeys with simian immunodeficiency virus. J Virol. 2003;77(3):2165–73.
30. Wu F, et al. Sequential evolution and escape from neutralization of simian immunodeficiency virus SIVsmE660 clones in rhesus macaques. J Virol. 2012;86(16):8835–47.
31. Miller CJ, et al. Antiviral antibodies are necessary for control of simian immunodeficiency virus replication. J Virol. 2007;81(10):5024–35.
32. Hatziioannou T, et al. A macaque model of HIV-1 infection. Proc Natl Acad Sci U S A. 2009;106(11):4425–29.
33. Witvrouw M, et al. Susceptibility of HIV-2, SIV and SHIV to various anti-HIV-1 compounds: implications for treatment and postexposure prophylaxis. Antivir Ther. 2004;9(1):57–65.
34. Parkin NT, Schapiro JM. Antiretroviral drug resistance in non-subtype B HIV-1, HIV-2 and SIV: Antivir Ther. 2004;9(1):3–12.
35. Giuffre AC, et al. Susceptibilities of simian immunodeficiency virus to protease inhibitors. Antimicrob Agents Chemother. 2003;47(5):1756–9.
36. Vlasak J, Ruprecht RM. AIDS vaccine development and challenge viruses: getting real. AIDS. 2006;20(17):2135–40.
37. Lopker M, et al. Heterogeneity in neutralization sensitivities of viruses comprising the simian immunodeficiency virus SIVsmE660 isolate and vaccine challenge stock. J Virol. 2013;87(10):5477–92.
38. Watkins DI, et al. Nonhuman primate models and the failure of the Merck HIV-1 vaccine in humans. Nat Med. 2008;14(6):617–21.
39. Shakirzyanova M, et al. Pathogenic consequences of vaginal infection with CCR5-tropic simian-human immunodeficiency virus SHIVSF162P3N. J Virol. 2012;86(17):9432–42.
40. Ren W, et al. Generation of lineage-related, mucosally transmissible subtype C R5 simian-human immunodeficiency viruses capable of aids development, induction of neurological disease, and coreceptor switching in rhesus macaques. J Virol. 2013;87(11):6137–49.
41. Daza-Vamenta R, et al. Genetic divergence of the rhesus macaque major histocompatibility complex. Genome Res. 2004;14(8):1501–15.

42. Ouyang D, et al. Identification of major histocompatibility complex class I alleles in Chinese rhesus macaques. Acta Biochim Biophys Sin. 2008;40(11):919–27.
43. Otting N, et al. Unparalleled complexity of the MHC class I region in rhesus macaques. Proc Natl Acad Sci U S A. 2005;102(5):1626–31.
44. Shultz LD, et al. Humanized mice for immune system investigation: progress, promise and challenges. Nat Rev Immunol. 2012;12(11):786–98.
45. Shultz LD, Ishikawa F, Greiner DL. Humanized mice in translational biomedical research. Nat Rev Immunol. 2007;7(2):118–30.
46. Ito R, et al. Current advances in humanized mouse models. Cell Mol Immunol. 2012;9(3):208–14.
47. McCune J, et al. The SCID-hu mouse: murine model for the analysis of human hematolymphoid differentiation and function. Science. 1988;241:1632–9.
48. Mosier DE, et al. Transfer of a functional human immune system to mice with severe combined immunodeficiency. Nature. 1988;335(6187):256–9.
49. Bosma GC, Custer RP, Bosma MJ. A severe combined immunodeficiency mutation in the mouse. Nature. 1983;301(5900):527–30.
50. McCune J, et al. The SCID-hu mouse: a small animal model for HIV infection and pathogenesis. Ann Rev Immunol. 1991;9(1):399–429.
51. Namikawa R, et al. Long-term human hematopoiesis in the SCID-hu mouse. J Exp Med. 1990;172(4):1055–63.
52. McCune J, et al. Suppression of HIV infection in AZT-treated SCID-hu mice. Science. 1990;247:564–6.
53. Aldrovandi G, et al. The SCID-hu mouse as a model for HIV-1 infection. Nature. 1993;363:732–6.
54. Bonyhadi M, et al. HIV induces thymus depletion in vivo. Nature. 1993;363:728–32.
55. Stanley S, et al. Human immunodeficiency virus infection of the human thymus and disruption of the thymic microenvironment in the SCID-hu mouse. J Exp Med. 1993;178:1151–63.
56. Mosier D, et al. Human immunodeficiency virus infection of human-PBL-SCID mice. Science. 1991;251: 791–4.
57. Mosier DE, Gulizia RJ, Baird S, Wilson DB. On the SCIDs? Nature. 1989;338:211.
58. Peault B, et al. Lymphoid reconstitution of the human fetal thymus in SCID mice with $CD34^+$ precursor cells. J Exp Med. 1991;174(5):1283–6.
59. Lapidot T, et al. Cytokine stimulation of multilineage hematopoiesis from immature human cells engrafted in SCID mice. Science. 1992;255(5048):1137–41.
60. Shultz LD, et al. Multiple defects in innate and adaptive immunologic function in NOD/LtSz-scid mice. J Immunol. 1995;154(1):180–91.
61. Kataoka S, et al. Immunologic aspects of the nonobese diabetic (NOD) mouse. Abnormalities of cellular immunity. Diabetes. 1983;32(3):247–53.
62. Greiner DL, Hesselton RA, Shultz LD. SCID mouse models of human stem cell engraftment. Stem Cells. 1998;16(3):166–77.
63. Hesselton RM, et al. High levels of human peripheral blood mononuclear cell engraftment and enhanced susceptibility to human immunodeficiency virus type 1 infection in NOD/LtSz-scid/scid mice. J Infect Dis. 1995;172(4):974–82.
64. Mombaerts P, et al. RAG-1-deficient mice have no mature B and T lymphocytes. Cell. 1992;68(5):869–77.
65. Shinkai Y, et al. RAG-2-deficient mice lack mature lymphocytes owing to inability to initiate V(D)J rearrangement. Cell. 1992;68(5):855–67.
66. Noguchi M, et al. Interleukin-2 receptor gamma chain: a functional component of the interleukin-7 receptor. Science. 1993;262(5141):1877–80.
67. Russell S, et al. Interleukin-2 receptor gamma chain: a functional component of the interleukin-4 receptor. Science. 1993;262(5141):1880–3.
68. DiSanto JP, et al. Lymphoid development in mice with a targeted deletion of the interleukin 2 receptor gamma chain. Proc Natl Acad Sci U S A. 1995;92(2):377–81.
69. Cao X, et al. Defective lymphoid development in mice lacking expression of the common cytokine receptor γ chain. Immunity. 1995;2(3):223–38.

70. Asao H, et al. Cutting edge: the common γ-chain is an indispensable subunit of the IL-21 receptor complex. J Immunol. 2001;167(1):1–5.
71. Kirberg J, Berns A, von Boehmer H. Peripheral T cell survival requires continual ligation of the T cell receptor to major histocompatibility complex-encoded molecules. J Exp Med. 1997;186:1269–75.
72. Traggiai E, et al. Development of a human adaptive immune system in cord blood cell-transplanted mice. Science. 2004;304(5667):104–7.
73. Goldman J, et al. Enhanced human cell engraftment in mice deficient in RAG2 and the common cytokine receptor gamma chain. Br J Haematol. 1998;103:335–42.
74. Ito M, et al. NOD/SCID/γ mouse: an excellent recipient mouse model for engraftment of human cells. Blood. 2002;100(9):3175–82.
75. Shultz L, et al. Human lymphoid and myeloid cell development in NOD/LtSz-scid IL2R gamma null mice engrafted with mobilized human hemopoietic stem cells. J Immunol. 2005;174:6477–89.
76. Pearson T, et al. Non-obese diabetic–recombination activating gene-1 (NOD–Rag 1 null) interleukin (IL)-2 receptor common gamma chain (IL 2 rγnull) null mice: a radioresistant model for human lymphohaematopoietic engraftment. Clin Exp Immunol. 2008;154(2):270–84.
77. Melkus MW, et al. Humanized mice mount specific adaptive and innate immune responses to EBV and TSST-1. Nat Med. 2006;12(11):1316–22.
78. Lan P, et al. Reconstitution of a functional human immune system in immunodeficient mice through combined human fetal thymus/liver and $CD34^+$ cell transplantation. Blood. 2006;108(2):487–92.
79. Sun Z, et al. Intrarectal transmission, systemic infection, and $CD4^+$ T cell depletion in humanized mice infected with HIV-1. J Exp Med. 2007;204(4):705–14.
80. Denton PW, et al. Antiretroviral pre-exposure prophylaxis prevents vaginal transmission of HIV-1 in humanized BLT mice. PLoS Med. 2008;5(1):e16.
81. Wheeler LA, et al. Inhibition of HIV transmission in human cervicovaginal explants and humanized mice using CD4 aptamer-siRNA chimeras. J Clin Invest. 2011;121(6):2401–12.
82. Brainard DM, et al. Induction of robust cellular and humoral virus-specific adaptive immune responses in human immunodeficiency virus-infected humanized BLT mice. J Virol. 2009;83(14):7305–21.
83. Dudek TE, et al. Rapid evolution of HIV-1 to functional $CD8^+$ T cell responses in humanized BLT mice. Sci Transl Med. 2012;4(143):143ra98.
84. Hiramatsu H, et al. Complete reconstitution of human lymphocytes from cord blood $CD34^+$ cells using the NOD/SCID/gammacnull mice model. Blood. 2003;102(3):873–80.
85. Brehm MA, et al. Parameters for establishing humanized mouse models to study human immunity: analysis of human hematopoietic stem cell engraftment in three immunodeficient strains of mice bearing the IL2rγnull mutation. Clin Immunol. 2010;135(1):84–98.
86. Traggiai E. An efficient method to make human monoclonal antibodies from memory B cells: potent neutralization of SARS coronavirus. Nat Med. 2004;10:871–5.
87. Watanabe Y, et al. The analysis of the functions of human B and T cells in humanized NOD/shi-scid/gammac(null) (NOG) mice (hu-HSC NOG mice). Int Immunol. 2009;21(7):843–58.
88. Billerbeck E, et al. Development of human $CD4^+FoxP3^+$ regulatory T cells in human stem cell factor, GM-CSF and interleukin 3 expressing NOD SCID IL2RγNULL humanized mice. Blood. 2011;117(11):3076–86.
89. Chen Q, et al. GM-CSF and IL-4 stimulate antibody responses in humanized mice by promoting T, B, and dendritic cell maturation. J Immunol. 2012;189(11):5223–9.
90. Miller PH, et al. Enhanced normal short-term human myelopoiesis in mice engineered to express human-specific myeloid growth factors. Blood. 2013;121(5):e1–4.
91. Chen Q, Khoury M, Chen J. Expression of human cytokines dramatically improves reconstitution of specific human-blood lineage cells in humanized mice. Proc Natl Acad Sci U S A. 2009;106(51):21783–8.

92. Shultz LD, et al. Generation of functional human T-cell subsets with HLA-restricted immune responses in HLA class I expressing NOD/SCID/IL2r gamma(null) humanized mice. Proc Natl Acad Sci U S A. 2010;107(29):13022–7.
93. Billerbeck E, et al. Characterization of human antiviral adaptive immune responses during hepatotropic virus infection in HLA-transgenic human immune system mice. J Immunol. 2013;191(4):1753–64.
94. Strowig T, et al. Priming of protective T cell responses against virus-induced tumors in mice with human immune system components. J Exp Med. 2009;206(6):1423–34.
95. Sato Y, Nagata S, Takiguchi M. Effective elicitation of human effector CD8 ⁺ T cells in HLA-B*51:01 transgenic humanized mice after infection with HIV-1. PLoS One. 2012;7(8):e42776.
96. Suzuki M, et al. Induction of human humoral immune responses in a novel HLA-DR-expressing transgenic NOD/Shi-scid/γcnull mouse. Int Immunol. 2012;24(4):243–52.
97. Danner R, et al. Expression of HLA class II molecules in humanized NOD. Rag1KO.IL-2RgcKO mice is critical for development and function of human T and B cells. PLoS One. 2011;6(5):e19826.
98. Long BR, Stoddart CA. Alpha interferon and HIV infection cause activation of human T cells in NSG-BLT mice. J Virol. 2012;86(6):3327–36.
99. Wahl A, et al. Human breast milk and antiretrovirals dramatically reduce oral HIV-1 transmission in BLT humanized mice. PLoS Pathog. 2012;8(6):e1002732.
100. Dudek TE, et al. Rapid evolution of HIV-1 to functional CD8(+) T cell responses in humanized BLT mice. Sci Transl Med. 2012;4(143):3003984.
101. Palmer BE, et al. In Vivo blockade of the PD-1 receptor suppresses HIV-1 viral Loads and improves $CD4^{+}$ T cell levels in humanized mice. J Immunol. 2013;190(1):211–9.
102. Berges B, et al. HIV-1 infection and CD4 T cell depletion in the humanized Rag2−/−gammac−/− (RAG-hu) mouse model. Retrovirology. 2006;3(1):76.
103. Wainberg M, et al. Systemic administration of antiretrovirals prior to exposure prevents rectal and intravenous HIV-1 transmission in humanized BLT mice. PLoS One. 2010;5(1):e8829.
104. Neff CP, et al. Oral pre-exposure prophylaxis by anti-retrovirals raltegravir and maraviroc protects against HIV-1 vaginal transmission in a humanized mouse model. PLoS One. 2010;5(12):e15257.
105. Klein F, et al. HIV therapy by a combination of broadly neutralizing antibodies in humanized mice. Nature. 2012;492(7427):118–22.
106. Shimizu S, et al. A highly efficient short hairpin RNA potently down-regulates CCR5 expression in systemic lymphoid organs in the hu-BLT mouse model. Blood. 2010;115(8):1534–44.
107. Joseph A, et al. Inhibition of In Vivo HIV infection in humanized mice by gene therapy of human hematopoietic stem cells with a lentiviral vector encoding a broadly neutralizing anti-HIV antibody. J Virol. 2010;84(13):6645–53.
108. Denton PW, et al. Generation of HIV latency in humanized BLT mice. J Virol. 2012;86(1):630–4.
109. Lang J, et al. Studies of lymphocyte reconstitution in a humanized mouse model reveal a requirement of T cells for human B cell maturation. J Immunol. 2013;190(5):2090–101.
110. Li Q, et al. Glycerol monolaurate prevents mucosal SIV transmission. Nature. 2009;458(7241):1034–8.
111. Miller CJ, et al. Propagation and dissemination of infection after vaginal transmission of simian immunodeficiency virus. J. Virol. 2005;79(14):9217–27.
112. Stone M, et al. A limited number of simian immunodeficiency virus (SIV) env variants are transmitted to rhesus macaques vaginally inoculated with SIVmac251. J Virol. 2010;84(14):7083–95.
113. Keele BF, et al. Low-dose rectal inoculation of rhesus macaques by SIVsmE660 or SIVmac251 recapitulates human mucosal infection by HIV-1. J Exp Med. 2009;206(5):1117–34.

114. Liu J, et al. Low-dose mucosal simian immunodeficiency virus infection restricts early replication kinetics and transmitted virus variants in rhesus monkeys. J Virol. 2010;84(19):10406–12.
115. Ma ZM, et al. SIVmac251 is inefficiently transmitted to rhesus macaques by penile inoculation with a single SIVenv variant found in ramp-up phase plasma. AIDS Res Hum Retroviruses. 2011;27(12):1259–69.
116. Sallé B, et al. Infection of macaques after vaginal exposure to cell-associated simian immunodeficiency virus. J Infect Dis. 2010;202(3):337–44.
117. Kolodkin-Gal D, et al. Efficiency of cell-free and cell-associated virus in mucosal transmission of HIV-1/SIV. J Virol. 2013;87(24):13589–97.
118. Miller CJ, et al. Genital mucosal transmission of simian immunodeficiency virus: animal model for heterosexual transmission of human immunodeficiency virus. J Virol. 1989;63(10):4277–84.
119. Qureshi H, et al. Low-dose penile SIVmac251 exposure of rhesus macaques infected with adenovirus type 5 (Ad5) and then immunized with a replication-defective Ad5-based SIV gag/pol/nef vaccine recapitulates the results of the phase IIb step trial of a similar HIV-1 vaccine. J Virol. 2012;86(4):2239–50.
120. Rothaeusler K, et al. Antiviral antibodies and T cells are present in the foreskin of simian immunodeficiency virus-infected rhesus macaques. J Virol. 2012;86(13):7098–106.
121. Yeh WW, et al. The TRIM5 gene modulates penile mucosal acquisition of simian immunodeficiency virus in rhesus monkeys. J Virol. 2011;85(19):10389–98.
122. Denton PW, et al. Systemic administration of antiretrovirals prior to exposure prevents rectal and intravenous HIV-1 transmission in humanized BLT mice. PLoS One. 2010;5(1):e8829.
123. Stoddart CA, et al. Superior human leukocyte reconstitution and susceptibility to vaginal HIV transmission in humanized NOD-scid IL-2Rγ−/− (NSG) BLT mice. Virology. 2011;417(1):154–60.
124. Berges BK, et al. Mucosal transmission of R5 and X4 tropic HIV-1 via vaginal and rectal routes in humanized Rag2−/−γc−/− (RAG-hu) mice. Virology. 2008;373(2):342–51.
125. Plotkin SA. Correlates of protection induced by vaccination. Clin Vaccine Immunol. 2010;17(7):1055–65.
126. Plotkin SA, Gilbert PB. Nomenclature for immune correlates of protection after vaccination. Clin Infect Dis. 2012;54(11):1615–7.
127. Balazs AB, et al. Antibody-based protection against HIV infection by vectored immunoprophylaxis. Nature. 2012;481(7379):81–4.
128. Wilen CB, et al. Engineering HIV-resistant human $CD4^+$ T cells with CXCR4-specific zinc-finger nucleases. PLoS Pathog. 2011;7(4):e1002020.
129. Akkina R. Human immune responses and potential for vaccine assessment in humanized mice. Curr Opin Immunol. 2013;25(3):403–9.
130. Chang H, et al. Human B-cell ontogeny in humanized NOD/SCID [gamma]cnull mice generates a diverse yet auto/poly- and HIV-1-reactive antibody repertoire. Genes Immun. 2012;13(5):399–410.
131. Girard MP, Plotkin SA. HIV vaccine development at the turn of the 21st century. Curr Opin HIV AIDS. 2012;7(1):4–9.

Chapter 18
Host Factor-Mediated Resistance to HIV-1 Infection

Kei Sato

Abbreviations

A	Alanine
AIDS	Acquired immunodeficiency syndrome
APOBEC3	Apolipoprotein B mRNA editing enzyme, catalytic polypeptide-like 3
BLT	Bone marrow/liver/thymus
BST2	Bone marrow stromal antigen 2
C	Cytosine
CBFβ	Core-binding factor, beta subunit
CHMP	Chromatin-modifying protein/charged multivesicular body protein
CTL	$CD8^+$ cytotoxic T lymphocyte
ESCRT	Endosomal sorting complex required for transport
G	Guanine
HIV-1	Human immunodeficiency virus type 1
LEDGF	Lens epithelium-derived growth factor
Nef	Negative factor
NOG	NOD.Cg-*Prkdc*^{scid} *Il2rg*^{tm1Sug}/Jic
NSG	NOD.Cg-*Prkdc*^{scid} *Il2rg*^{tm1Wjl}/SzJ
SIV	Simian immunodeficiency virus
Tat	Transactivator of transcription;
TSG101	Tumor susceptibility gene 101
U	Uracil
Vif	Viral infectivity factor
VPS4	Vacuolar protein sorting 4
Vpu	Viral protein U

K. Sato (✉)
Laboratory of Viral Pathogenesis, Institute for Virus Research, Kyoto University, 53 Shogoinkawara-cho, Sakyo-ku, Kyoto 6068507, Japan
e-mail: ksato@virus.kyoto-u.ac.jp

L. Y. Poluektova et al. (eds.), *Humanized Mice for HIV Research*,
DOI 10.1007/978-1-4939-1655-9_18

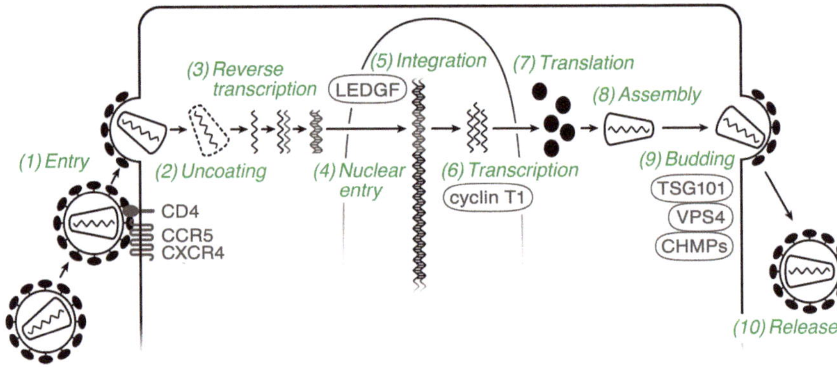

Fig. 18.1 HIV-1 life cycle. HIV-1 particle binds to the receptors, *CD4*, *CCR5*, and/or *CXCR4* molecules, on the surface of a target cell through viral envelope proteins and fuses with the cell (*step 1, entry*). The HIV-1 core containing viral proteins (e.g., reverse transcriptase and integrase) and viral RNA genome (reverse transcription complex; RTC) is released into the cytoplasm and dissociates (*step 2, uncoating*). RTC proceeds to reverse transcription (*step 3, reverse transcription*) and the complex of integrase and viral DNA (preintegration complex; PIC) is imported into the nuclear (*step 4, nuclear entry*). PIC associates with chromosome and integrates viral DNA into cellular genomic DNA (*step 5, integration*). A host factor, *LEDGF*, is crucial for HIV-1 integration. A viral regulatory protein, Tat, initiates and augments HIV-1 transcription by recruiting cellular proteins such as *cyclin T1* (*step 6, transcription*). After translation of viral proteins (*step 7, translation*), viral proteins and genomic RNA are united and assembled (*step 8, assembly*). At the plasma membrane of infected cell, the assembled virus core is wrapped with a lipid bilayer membrane containing HIV-1 envelope proteins and is pinched off by cellular ESCRT machinery, which consists of *TSG101*, *VPS4*, *CHMP* proteins, and others (*step 9, budding*). The nascent budding HIV-1 particle is released (*step 10, release*) and binds to the receptor molecules expressed on other cells, which leads to cell-free infection (i.e., back to step 1). Generally, steps 1–5 are considered as the "early phase" of HIV-1 replication, while steps 6–10 are considered to be the "late phase." Representative "host factors" positively associate with HIV-1 replication are shown in *gray*

18.1 Host Factors

Extensive in vitro studies using cell culture systems have revealed that certain human cellular proteins, called "host factors", positively or negatively regulate HIV-1 infection and replication [1]. For example, cell surface receptors such as CD4 [2, 3], CCR5 [4–6], and/or CXCR4 [7, 8], are utilized by HIV-1 for viral entry into target cells and, thus, serve as positive regulators (Fig. 18.1). In addition, LEDGF is a prerequisite for provirus integration into the human genome mediated by HIV-1 integrase [9], and cyclin T1 is important for efficient viral transcription mediated by Tat, an HIV-1-encoding regulatory protein [10]. Moreover, nascent HIV-1 virions are budded from infected cells by hijacking ESCRT machinery, which consist of TSG101 [11–13], VPS4, CHMP family proteins, and others [14]. In contrast, the human genome encodes at least two proteins capable of negatively regulating HIV-1 replication. These two cellular proteins, APOBEC3G [15] and BST2 (also known

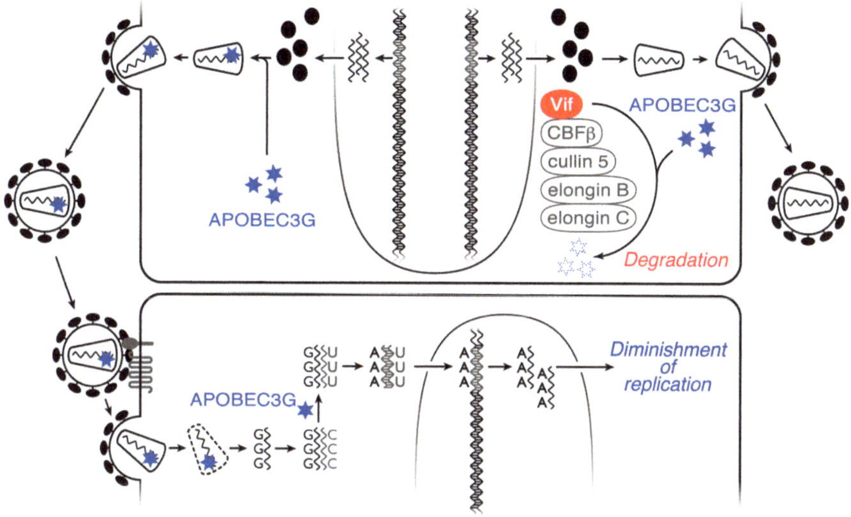

Fig. 18.2 The interplay of *APOBEC3G* and *Vif*. (*Top left* and *bottom*, shaded) In the absence of Vif, APOBEC3G proteins expressed in HIV-1-infected cell are incorporated into the nascent viral particles. In newly infected cell, incoming APOBEC3G induces C-to-U mutation in negative-stranded viral DNA, which is attributed to G-to-A mutation in positive-stranded viral DNA. The G-to-A mutations result in termination mutation (i.e., amino acid mutation to stop codon) and/or nonsynonymous mutation leading to the diminishment of HIV-1 replication. (*Top right*) In the presence of Vif, APOBEC3G proteins are ubiquitinated by Vif-cullin 5-elongin B/C-CBFβ complex and are degraded through the ubiquitin/proteasome-dependent pathway. Because APOBEC3G incorporation into nascent HIV-1 is impaired by the mechanisms of action of Vif, the infectivity of the released viral particle is retained

as tetherin, CD317, and HM1.24) [16, 17] are considered as "restriction factors" and "intrinsic immunity" against HIV-1 infection.

APOBEC3G is one of the APOBEC3 cytidine deaminase family proteins (APOBEC3A, B, C, D, F, G, and H; APOBEC3s), and can be packaged into nascent HIV-1 particles [18, 19]. During reverse transcription, in newly infected cells, hitchhiking APOBEC3G proteins enzymatically remove an amine group (–NH_2) from a cytosine base in the reverse-transcribed viral DNA (i.e., negative-stranded DNA) resulting in a uracil substitution (C-to-U mutation). Since these C-to-U mutations by APOBEC3G are responsible for corresponding G-to-A mutations in positive-stranded proviral DNA, it is widely accepted that APOBEC3G induces G-to-A mutations in HIV-1 genome. The APOBEC3G-mediated G-to-A mutations result in the induction of premature stop codons and/or nonsynonymous mutations in viral proteins, which leads to the diminishment of viral replication (Fig. 18.2).

BST2 is a type II transmembrane glycoprotein and is ubiquitously expressed on the surface of human immune cells [16, 17]. In addition, its expression is augmented by type I interferon treatment [16, 17]. On the surface of HIV-1-infected cells,

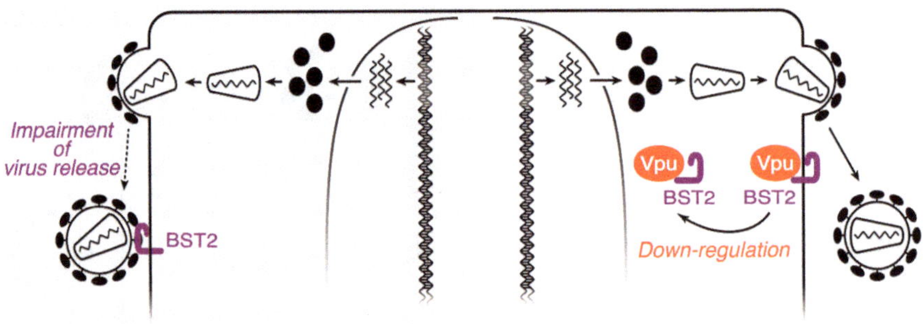

Fig. 18.3 (*Left, shaded*) In the absence of Vpu, BST2 proteins expressed on the surface of an HIV-1-infected cell tether newly budded viral particles and impair their release. (*Right*) In the presence of Vpu, BST2 is downregulated from the surface of HIV-1-infected cell. Because BST2 is sequestered from the budding site by the mechanisms of action of Vpu, nascent HIV-1 particles are efficiently released

BST2 tethers nascent viral particles and impairs their release (Fig. 18.3). Although the anti-viral activity of APOBEC3G is mainly confined to viruses of the *Retroviridae* family, including HIV-1, BST2 can impede the release of various enveloped viruses belonging to not only *Retroviridae* (e.g., HIV-1) [20–26] but also *Arenaviridae* (e.g., Lassa fever virus) [27, 28], *Filoviridae* (e.g., Ebola virus and Marburg virus) [22, 27, 28], *Herpesviridae* (e.g., Kaposi's sarcoma-associated herpesvirus) [29, 30], *Orthomyxoviridae* (e.g., influenza A virus) [31], *Paramyxoviridae* (e.g., Nipah virus) [32], and *Rhabdoviridae* (e.g., vesicular stomatitis virus) [33].

18.2 Viral Factors

In spite of the fact that human cells intrinsically possess mechanisms to diminish HIV-1 replication, HIV-1 infection in humans ultimately causes AIDS. How does HIV-1 circumvent these hurdles?

To counteract the actions of APOBEC3G and BST2, HIV-1 is armed with its own weapons, Vif and Vpu [34]. These two viral proteins act against intrinsic immunity and are known as "accessory proteins" and "viral factors". Vif interacts with APOBEC3G and recruits a cellular E3 ubiquitin ligase complex, which consists of cullin 5, elongin B, and elongin C, thereby leading to the degradation of APOBEC3G in a ubiquitin/proteasome-dependent pathway [18, 19]. In addition, it has been recently reported that Vif hijacks a cellular cofactor, CBFβ, to degrade APOBEC3G [35, 36]. This Vif-mediated APOBEC3G degradation impairs its incorporation into nascent HIV-1 particles (Fig. 18.2). On the other hand, Vpu is a type I transmembrane protein and directly interacts with BST2 through their transmembrane domains [37, 38]. Vpu downregulates BST2 from the surface of HIV-1-infected cells and antagonizes the anti-viral activity of BST2 (Fig. 18.3).

So far, in vitro studies using cell culture systems have been extensively performed and have elucidated the detailed mechanisms of the interaction between "host factors" (APOBEC3G and BST2) and "viral factors" (Vif and Vpu). In other words, the battle between humans and HIV-1 can be interpreted as the interplay between "host factors" and "viral factors", which has been a hot topic in the field of HIV-1 virology. However, because HIV-1 is unable to infect and induce disorders in animals other than humans, investigating the physiological relevance of the interplay between "host factors" and "viral factors" in vivo has been challenging.

18.3 Animal Models for Investigating Host-Virus Protein Interaction

To investigate the dynamics of human-specific virus infections, a human hematopoietic stem cell-transplanted mouse model, called a "humanized mouse model," is one of the best tools to date [39]. It is worth noting that humanized mouse models are able to maintain human leukocytes in a physiological condition for a long period of time (ca. 1 year) [40, 41]. For instance, the expression levels of endogenous *APOBEC3*s in human $CD4^+$ T lymphocytes in the spleen of humanized mice were comparable to those in human peripheral blood [42]. Therefore, by using humanized mouse models, we are able to address the significance of the interplay between "host factors" and "viral factors" in vivo.

18.4 APOBEC3G Versus Vif

A strong preference for G-to-A mutations in HIV-1 sequences has been observed in the specimens of infected patients [43–46]. However, because the sequences of transmitted/founder HIV-1 in clinically infected individuals are undeterminable, previous findings on HIV-1 mutations in patients were based on comparisons to certain HIV-1 laboratory clones. To directly address the significance of the interplay between APOBEC3G and Vif in vivo, either *vif*-deficient or wild-type (i.e., *vif*-proficient) CCR5-tropic HIV-1 molecular clone (strain JR-CSF) [47] was inoculated into humanized NOG mice [41, 42] and humanized NSG BLT mice [48]. Wild-type HIV-1 efficiently replicated in these humanized mice models, whereas *vif*-deficient HIV-1 did not show any viremia [42, 48]. In addition, the depletion of peripheral human $CD4^+$ T lymphocytes, which is the hallmark of HIV-1 infection, was not observed in *vif*-deficient HIV-1-infected humanized mice [42], indicating that Vif is a prerequisite for CCR5-tropic HIV-1 infection and replication in vivo. Moreover, a prominent G-to-A hypermutation was observed in proviral DNA of wild-type (i.e., *vif*-proficient) HIV-1-infected humanized NOG mice [42] and *vif*-deficient HIV-1-infected humanized NSG BLT mice [48]. These findings directly demonstrate that endogenous APOBEC3G induces G-to-A hypermutations in proviral DNA and abrogates *vif*-deficient CCR5-tropic HIV-1 replication in vivo.

Based on the coreceptor usage, HIV-1 can be classified into two types: CCR5-tropic HIV-1 and CXCR4-tropic HIV-1. The former is the major clinical isolates from patients along with transmitted/founder viruses, whereas the latter occasionally emerges during the onset of AIDS [49]. Although *vif*-deficient CCR5-tropic HIV-1 was unable to replicate in vivo [42, 48], a study using humanized NSG BLT mouse model revealed that *vif*-deficient CXCR4-tropic HIV-1 (strain LAI) [50] can display viremia when the virus solution is directly injected into the human thymic implant or is intravenously inoculated [48]. CXCR4 is highly expressed on human $CD4^+$ thymocytes, while CCR5 is less expressed [51, 52]. In addition, it was of particular importance that human thymocytes express lower *APOBEC3G* than peripheral human $CD4^+$ T lymphocytes [48]. Therefore, these observations suggest that CXCR4-tropic HIV-1 is able to sustain its replication without *vif* by hijacking a cell population, which has reduced *APOBEC3G* expression, such as human $CD4^+$ thymocytes.

From in vitro studies, it is known that certain APOBEC3 family proteins other than APOBEC3G, particularly APOBEC3F, can be incorporated into nascent HIV-1 and insert G-to-A mutations in the HIV-1 genome [53, 54]. Similarly to APOBEC3G, Vif degrades APOBEC3F via a ubiquitin/proteasome-dependent pathway. Interestingly, although APOBEC3G-mediated mutations exhibit a dinucleotide signature, GG-to-AG, APOBEC3F has a GA-to-AA preference [18, 19]. Although the ectopic expression of APOBEC3F strongly attenuates HIV-1 infectivity in in vitro cell culture systems [53, 54], it seems controversial whether endogenous APOBEC3F can exert anti-HIV-1 activity in vitro and in vivo [55–57]. However, detailed analyses on proviral DNA sequences in the wild-type (i.e., *vif*-proficient) HIV-1-infected humanized mice have revealed that a mixture of GG-to-AG and GA-to-AA hypermutations in a single amplicon was not observed, and that G-to-A hypermutations in each amplicon can be classified into either GG-to-AG or GA-to-AA [42]. These observations suggest that individual APOBEC3 family proteins, at least APOBEC3G and APOBEC3F, independently participate in G-to-A HIV-1 hypermutation in vivo. Further investigations will reveal the significance of APOBEC3G and APOBEC3F on HIV-1 replication in vivo.

Altogether, Vif is essential for HIV-1 replication, particularly for CCR5-tropic HIV-1, while endogenous APOBEC3 is able to intrinsically diminish HIV-1 replication in vivo. In addition, it has been reported that APOBEC3-mediated mutations can improve CTL recognition and further induce adaptive immune system [58]. On the other hand, an HIV-1-infected humanized NSG BLT mouse model has shown that APOBEC3-mediated mutations can associate with escaping CTL epitopes during acute infection [59]. Humanized mouse models are suitable to investigate the dynamics of viral mutations in vivo [42, 48, 59, 60] and may serve as ideal tools to elucidate the cost and benefit of APOBEC3-mediated editing of HIV-1.

18.5 BST2 Versus Vpu

In vitro experiments have revealed that the potency of HIV-1-infected cells for the release of *vpu*-deficient viral particles is dependent on BST2 expression. For instance, HEK293 cells and Cos-7 cells, which do not express endogenous BST2, are able to produce nascent *vpu*-deficient HIV-1 particles, whereas HeLa cells and Jurkat cells, which express endogenous BST2, are not [16, 17]. Importantly, endogenous BST2 is expressed in primary human $CD4^+$ T cells [16, 61], which are the primary targets for HIV-1 infection in vivo, suggesting that BST2 has the robust potential to regulate HIV-1 replication in vivo. To elucidate the significance of the interplay between BST2 and Vpu in vivo, either *vpu*-deficient or wild-type (i.e., *vpu*-proficient) CCR5-tropic HIV-1 molecular clone (strain AD8) [62–64] was inoculated into a humanized NOG mouse model [41, 61]. Although the growth kinetics of *vpu*-deficient HIV-1 in the humanized mouse model was lower than wild-type HIV-1, *vpu*-deficient HIV-1 efficiently replicated in vivo [61]. In addition, Vpu did not affect the viral cytopathic effect, target cell preference, and the level of viral protein expression [61]. By comparing these findings with those in *vif*-deficient HIV-1-infected humanized mice [42], Vpu appears to be dispensable for viral replication in vivo and had modest effects on viral pathogenesis. However, it was of particular importance that the expression level of BST2 on the surface of HIV-1-infected cells in wild-type HIV-1-inoculated humanized mice was significantly lower than that in *vpu*-deficient HIV-1-inoculated mice, and that the amount of cell-free virions in wild-type HIV-1-infected mice was profoundly higher than that in *vpu*-deficient HIV-1-infected mice [61]. These observations strongly suggest that Vpu augments the release of viral particles by downregulating surface BST2 on HIV-1-infected cells even in vivo.

So, why is BST2 unable to “restrict” HIV-1 replication in vivo despite its potent ability to inhibit viral release? One plausible explanation may be the mode of HIV-1 infection. HIV-1 can be propagated by at least two modes: cell-free virus-mediated infection and cell-to-cell infection [65, 66]. Although it is still controversial [67, 68], certain reports have demonstrated that cell-to-cell HIV-1 infection overcomes BST2-mediated restriction [61, 69]. More importantly, cell-to-cell HIV-1 infection is more efficient than cell-free infection in in vitro cell culture systems [70, 71], and it has been recently reported that cell-to-cell HIV-1 spread efficiently occurs in vivo [72, 73]. Therefore, it seems convincing that BST2 restricts the release of HIV-1 for cell-free infection but does not impair the cell-to-cell spread of HIV-1, which is a major route of viral dissemination in vivo. These findings further provide a caveat that the observations from in vitro cell culture systems do not always reflect the events that occur in vivo and should be readdressed and reevaluated by utilizing animal models such as humanized mouse models.

18.6 Conclusion

Here I summarized the significance of the interplay between "host factors" (APOBEC3G and BST2) and "viral factors" (Vif and Vpu) in HIV-1 replication by utilizing humanized mouse models. To date, rhesus macaque monkeys infected with SIVmac, a primate lentivirus, have been widely utilized as the animal model for HIV-1 infection in humans and have provided novel insights [74]. However, because the structure and function of the genes encoded by SIVmac are different from those encoded by HIV-1 [75], the SIVmac-infected macaque model is unable to recapitulate the interplay between human and HIV-1 proteins on a molecular level. For instance, SIVmac does not encode *vpu* and antagonizes rhesus macaque BST2 by Nef, another viral accessory protein [21, 24, 26, 76]. Therefore, to elucidate the dynamics of viral and host factors and how they affect HIV-1 replication in vivo, it would be important to combine the knowledge provided from SIVmac-infected rhesus macaque models to novel findings brought from HIV-1-infected humanized mouse models. In summary, humanized mouse models are not only able to investigate the dynamics of human-pathogen interaction in vivo but also bridge the knowledge obtained from molecular and cellular biology to clinical science.

References

1. Freed EO. HIV-1 and the host cell: an intimate association. Trends Microbiol. 2004;12(4):170–7.
2. Dalgleish AG, Beverley PC, Clapham PR, Crawford DH, Greaves MF, Weiss RA. The CD4 (T4) antigen is an essential component of the receptor for the AIDS retrovirus. Nature. 1984;312(5996):763–7.
3. Klatzmann D, Champagne E, Chamaret S, Gruest J, Guetard D, Hercend T, Gluckman JC, Montagnier L. T-lymphocyte T4 molecule behaves as the receptor for human retrovirus LAV. Nature. 1984;312(5996):767–8.
4. Deng H, Liu R, Ellmeier W, Choe S, Unutmaz D, Burkhart M, Di Marzio P, Marmon S, Sutton RE, Hill CM, Davis CB, Peiper SC, Schall TJ, Littman DR, Landau NR. Identification of a major co-receptor for primary isolates of HIV-1. Nature. 1996;381(6584):661–6.
5. Dragic T, Litwin V, Allaway GP, Martin SR, Huang Y, Nagashima KA, Cayanan C, Maddon PJ, Koup RA, Moore JP, Paxton WA. HIV-1 entry into $CD4^+$ cells is mediated by the chemokine receptor CC-CKR-5. Nature. 1996;381(6584):667–73.
6. Choe H, Farzan M, Sun Y, Sullivan N, Rollins B, Ponath PD, Wu L, Mackay CR, LaRosa G, Newman W, Gerard N, Gerard C, Sodroski J. The β-chemokine receptors CCR3 and CCR5 facilitate infection by primary HIV-1 isolates. Cell. 1996;85(7):1135–48.
7. Feng Y, Broder CC, Kennedy PE, Berger EA. HIV-1 entry cofactor: functional cDNA cloning of a seven-transmembrane, G protein-coupled receptor. Science. 1996;272(5263):872–7.
8. Doranz BJ, Rucker J, Yi Y, Smyth RJ, Samson M, Peiper SC, Parmentier M, Collman RG, Doms RW. A dual-tropic primary HIV-1 isolate that uses fusin and the beta-chemokine receptors CKR-5, CKR-3, and CKR-2b as fusion cofactors. Cell. 1996;85(7):1149–58.
9. Cherepanov P, Maertens G, Proost P, Devreese B, Van Beeumen J, Engelborghs Y, De Clercq E, Debyser Z. HIV-1 integrase forms stable tetramers and associates with LEDGF/p75 protein in human cells. J Biol Chem. 2003;278(1):372–81.

10. Wei P, Garber ME, Fang SM, Fischer WH, Jones KA. A novel CDK9-associated C-type cyclin interacts directly with HIV-1 Tat and mediates its high-affinity, loop-specific binding to TAR RNA. Cell. 1998;92(4):451–62.
11. VerPlank L, Bouamr F, LaGrassa TJ, Agresta B, Kikonyogo A, Leis J, Carter CA. Tsg101, a homologue of ubiquitin-conjugating (E2) enzymes, binds the L domain in HIV type 1 Pr55^{Gag}. Proc Natl Acad Sci U S A. 2001;98(14):7724–9.
12. Martin-Serrano J, Zang T, Bieniasz PD. HIV-1 and Ebola virus encode small peptide motifs that recruit Tsg101 to sites of particle assembly to facilitate egress. Nat Med. 2001;7(12):1313–9.
13. Garrus JE, von Schwedler UK, Pornillos OW, Morham SG, Zavitz KH, Wang HE, Wettstein DA, Stray KM, Cote M, Rich RL, Myszka DG, Sundquist WI. Tsg101 and the vacuolar protein sorting pathway are essential for HIV-1 budding. Cell. 2001;107(1):55–65.
14. Morita E, Sundquist WI. Retrovirus budding. Annu Rev Cell Dev Biol. 2004;20:395–425.
15. Sheehy AM, Gaddis NC, Choi JD, Malim MH. Isolation of a human gene that inhibits HIV-1 infection and is suppressed by the viral Vif protein. Nature. 2002;418(6898):646–50.
16. Neil SJ, Zang T, Bieniasz PD. Tetherin inhibits retrovirus release and is antagonized by HIV-1 Vpu. Nature. 2008;451(7177):425–30.
17. Van Damme N, Goff D, Katsura C, Jorgenson RL, Mitchell R, Johnson MC, Stephens EB, Guatelli J. The interferon-induced protein BST-2 restricts HIV-1 release and is downregulated from the cell surface by the viral Vpu protein. Cell Host Microbe. 2008;3(4):245–52.
18. Harris RS, Liddament MT. Retroviral restriction by APOBEC proteins. Nat Rev Immunol. 2004;4(11):868–77.
19. Izumi T, Shirakawa K, Takaori-Kondo A. Cytidine deaminases as a weapon against retroviruses and a new target for antiviral therapy. Mini Rev Med Chem. 2008;8(3):231–8.
20. Douglas JL, Viswanathan K, McCarroll MN, Gustin JK, Fruh K, Moses AV. Vpu directs the degradation of the human immunodeficiency virus restriction factor BST-2/Tetherin via a bTrCP-dependent mechanism. J Virol. 2009;83(16):7931–47.
21. Jia B, Serra-Moreno R, Neidermyer W, Rahmberg A, Mackey J, Fofana IB, Johnson WE, Westmoreland S, Evans DT. Species-specific activity of SIV Nef and HIV-1 Vpu in overcoming restriction by tetherin/BST2. PLoS Pathog. 2009;5(5):e1000429.
22. Jouvenet N, Neil SJ, Zhadina M, Zang T, Kratovac Z, Lee Y, McNatt M, Hatziioannou T, Bieniasz PD. Broad-spectrum inhibition of retroviral and filoviral particle release by tetherin. J Virol. 2009;83(4):1837–44.
23. Le Tortorec A, Neil SJ. Antagonism to and intracellular sequestration of human tetherin by the human immunodeficiency virus type 2 envelope glycoprotein. J Virol. 2009;83(22):11966–78.
24. Sauter D, Schindler M, Specht A, Landford WN, Munch J, Kim KA, Votteler J, Schubert U, Bibollet-Ruche F, Keele BF, Takehisa J, Ogando Y, Ochsenbauer C, Kappes JC, Ayouba A, Peeters M, Learn GH, Shaw G, Sharp PM, Bieniasz P, Hahn BH, Hatziioannou T, Kirchhoff F. Tetherin-driven adaptation of Vpu and Nef function and the evolution of pandemic and nonpandemic HIV-1 strains. Cell Host Microbe. 2009;6(5):409–21.
25. Yang SJ, Lopez LA, Hauser H, Exline CM, Haworth KG, Cannon PM. Anti-tetherin activities in Vpu-expressing primate lentiviruses. Retrovirology. 2010;7:13.
26. Zhang F, Wilson SJ, Landford WC, Virgen B, Gregory D, Johnson MC, Munch J, Kirchhoff F, Bieniasz PD, Hatziioannou T. Nef proteins from simian immunodeficiency viruses are tetherin antagonists. Cell Host Microbe. 2009;6(1):54–67.
27. Kaletsky RL, Francica JR, Agrawal-Gamse C, Bates P. Tetherin-mediated restriction of filovirus budding is antagonized by the Ebola glycoprotein. Proc Natl Acad Sci U S A. 2009;106(8):2886–91.
28. Sakuma T, Noda T, Urata S, Kawaoka Y, Yasuda J. Inhibition of Lassa and Marburg virus production by tetherin. J Virol. 2009;83(5):2382–5.
29. Mansouri M, Viswanathan K, Douglas JL, Hines J, Gustin J, Moses AV, Fruh K. Molecular mechanism of BST2/tetherin downregulation by K5/MIR2 of Kaposi's sarcoma-associated herpesvirus. J Virol. 2009;83(19):9672–81.

30. Pardieu C, Vigan R, Wilson SJ, Calvi A, Zang T, Bieniasz P, Kellam P, Towers GJ, Neil SJ. The RING-CH ligase K5 antagonizes restriction of KSHV and HIV-1 particle release by mediating ubiquitin-dependent endosomal degradation of tetherin. PLoS Pathog. 2010;6(4):e1000843.
31. Watanabe R, Leser GP, Lamb RA. Influenza virus is not restricted by tetherin whereas influenza VLP production is restricted by tetherin. Virology. 2011;417(1):50–6.
32. Radoshitzky SR, Dong L, Chi X, Clester JC, Retterer C, Spurgers K, Kuhn JH, Sandwick S, Ruthel G, Kota K, Boltz D, Warren T, Kranzusch PJ, Whelan SP, Bavari S. Infectious Lassa virus, but not filoviruses, is restricted by BST-2/tetherin. J Virol. 2010;84(20):10569–80.
33. Weidner JM, Jiang D, Pan XB, Chang J, Block TM, Guo JT. Interferon-induced cell membrane proteins, IFITM3 and tetherin, inhibit vesicular stomatitis virus infection via distinct mechanisms. J Virol. 2010;84(24):12646–57.
34. Freed EO, Martin MA. HIVs and their replication. In: Knipe DM, Howley PM, editors. Fields virology. 5th ed. Philadelphia: Lippincott; 2007.
35. Jager S, Kim DY, Hultquist JF, Shindo K, LaRue RS, Kwon E, Li M, Anderson BD, Yen L, Stanley D, Mahon C, Kane J, Franks-Skiba K, Cimermancic P, Burlingame A, Sali A, Craik CS, Harris RS, Gross JD, Krogan NJ. Vif hijacks CBF-β to degrade APOBEC3G and promote HIV-1 infection. Nature. 2012;481(7381):371–5.
36. Zhang W, Du J, Evans SL, Yu Y, Yu XF. T-cell differentiation factor CBF-β regulates HIV-1 Vif-mediated evasion of host restriction. Nature. 2012;481(7381):376–9.
37. Douglas JL, Gustin JK, Viswanathan K, Mansouri M, Moses AV, Fruh K. The great escape: viral strategies to counter BST-2/tetherin. PLoS Pathog. 2010;6(5):e1000913.
38. Kirchhoff F. Immune evasion and counteraction of restriction factors by HIV-1 and other primate lentiviruses. Cell Host Microbe. 2010;8(1):55–67.
39. Shultz LD, Brehm MA, Garcia-Martinez JV, Greiner DL. Humanized mice for immune system investigation: progress, promise and challenges. Nat Rev Immunol. 2012;12(11):786–98.
40. Sato K, Koyanagi Y. The mouse is out of the bag: insights and perspectives on HIV-1-infected humanized mouse models. Exp Biol Med (Maywood). 2011;236(8):977–85.
41. Ito M, Hiramatsu H, Kobayashi K, Suzue K, Kawahata M, Hioki K, Ueyama Y, Koyanagi Y, Sugamura K, Tsuji K, Heike T, Nakahata T. NOD/SCID/γ_c^{null} mouse: an excellent recipient mouse model for engraftment of human cells. Blood. 2002;100(9):3175–82.
42. Sato K, Izumi T, Misawa N, Kobayashi T, Yamashita Y, Ohmichi M, Ito M, Takaori-Kondo A, Koyanagi Y. Remarkable lethal G-to-A mutations in vif-proficient HIV-1 provirus by individual APOBEC3 proteins in humanized mice. J Virol. 2010;84(18):9546–56.
43. Fitzgibbon JE, Mazar S, Dubin DT. A new type of G → A hypermutation affecting human immunodeficiency virus. AIDS Res Hum Retroviruses. 1993;9(9):833–8.
44. Janini M, Rogers M, Birx DR, McCutchan FE. Human immunodeficiency virus type 1 DNA sequences genetically damaged by hypermutation are often abundant in patient peripheral blood mononuclear cells and may be generated during near-simultaneous infection and activation of $CD4^+$ T cells. J Virol. 2001;75(17):7973–86.
45. Kieffer TL, Kwon P, Nettles RE, Han Y, Ray SC, Siliciano RF. G → A hypermutation in protease and reverse transcriptase regions of human immunodeficiency virus type 1 residing in resting $CD4^+$ T cells in vivo. J Virol. 2005;79(3):1975–80.
46. Langlois MA, Beale RC, Conticello SG, Neuberger MS. Mutational comparison of the single-domained APOBEC3C and double-domained APOBEC3F/G anti-retroviral cytidine deaminases provides insight into their DNA target site specificities. Nucleic Acids Res. 2005;33(6):1913–23.
47. Kawano Y, Tanaka Y, Misawa N, Tanaka R, Kira JI, Kimura T, Fukushi M, Sano K, Goto T, Nakai M, Kobayashi T, Yamamoto N, Koyanagi Y. Mutational analysis of human immunodeficiency virus type 1 (HIV-1) accessory genes: requirement of a site in the nef gene for HIV-1 replication in activated $CD4^+$ T cells in vitro and in vivo. J Virol. 1997;71(11):8456–66.
48. Krisko JF, Martinez-Torres F, Foster JL, Garcia JV. HIV restriction by APOBEC3 in humanized mice. PLoS Pathog. 2013;9(3):e1003242.

49. Koot M, van't Wout AB, Kootstra NA, de Goede RE, Tersmette M, Schuitemaker H. Relation between changes in cellular load, evolution of viral phenotype, and the clonal composition of virus populations in the course of human immunodeficiency virus type 1 infection. J Infect Dis. 1996;173(2):349–54.
50. Peden K, Emerman M, Montagnier L. Changes in growth properties on passage in tissue culture of viruses derived from infectious molecular clones of HIV-1_{LAI}, HIV-1_{MAL}, and HIV-1_{ELI}. Virology. 1991;185(2):661–72.
51. Kitchen SG, Zack JA. Distribution of the human immunodeficiency virus coreceptors CXCR4 and CCR5 in fetal lymphoid organs: implications for pathogenesis in utero. AIDS Res Hum Retroviruses. 1999;15(2):143–8.
52. Nie C, Sato K, Misawa N, Kitayama H, Fujino H, Hiramatsu H, Heike T, Nakahata T, Tanaka Y, Ito M, Koyanagi Y. Selective infection of $CD4^+$ effector memory T lymphocytes leads to preferential depletion of memory T lymphocytes in R5 HIV-1-infected humanized NOD/SCID/IL-2$R\gamma^{null}$ mice. Virology. 2009;394(1):64–72.
53. Liddament MT, Brown WL, Schumacher AJ, Harris RS. APOBEC3F properties and hypermutation preferences indicate activity against HIV-1 in vivo. Curr Biol. 2004;14(15):1385–91.
54. Wiegand HL, Doehle BP, Bogerd HP, Cullen BR. A second human antiretroviral factor, APOBEC3F, is suppressed by the HIV-1 and HIV-2 Vif proteins. EMBO J. 2004;23(12):2451–8.
55. Chaipan C, Smith JL, Hu WS, Pathak VK. APOBEC3G restricts HIV-1 to a greater extent than APOBEC3F and APOBEC3DE in human primary $CD4^+$ T cells and macrophages. J Virol. 2013;87(1):444–53.
56. Mulder LC, Ooms M, Majdak S, Smedresman J, Linscheid C, Harari A, Kunz A, Simon V. Moderate influence of human APOBEC3F on HIV-1 replication in primary lymphocytes. J Virol. 2010;84(18):9613–7.
57. Miyagi E, Brown CR, Opi S, Khan M, Goila-Gaur R, Kao S, Walker RC Jr, Hirsch V, Strebel K. Stably expressed APOBEC3F has negligible antiviral activity. J Virol. 2010;84(21):11067–75.
58. Casartelli N, Guivel-Benhassine F, Bouziat R, Brandler S, Schwartz O, Moris A. The antiviral factor APOBEC3G improves CTL recognition of cultured HIV-infected T cells. J Exp Med. 2010;207(1):39–49.
59. Dudek TE, No DC, Seung E, Vrbanac VD, Fadda L, Bhoumik P, Boutwell CL, Power KA, Gladden AD, Battis L, Mellors EF, Tivey TR, Gao X, Altfeld M, Luster AD, Tager AM, Allen TM. Rapid evolution of HIV-1 to functional $CD8^+$ T cell responses in humanized BLT mice. Sci Transl Med. 2012;4(143):143ra98.
60. Ince WL, Zhang L, Jiang Q, Arrildt K, Su L, Swanstrom R. Evolution of the HIV-1 env gene in the $Rag2^{-/-}$ $\gamma C^{-/-}$ humanized mouse model. J Virol. 2010;84(6):2740–52.
61. Sato K, Misawa N, Fukuhara M, Iwami S, An DS, Ito M, Koyanagi Y. Vpu augments the initial burst phase of HIV-1 propagation and downregulates BST2 and CD4 in humanized mice. J Virol. 2012;86(9):5000–13.
62. Schubert U, Bour S, Willey RL, Strebel K. Regulation of virus release by the macrophage-tropic human immunodeficiency virus type 1 AD8 isolate is redundant and can be controlled by either Vpu or Env. J Virol. 1999;73(2):887–96.
63. Schubert U, Clouse KA, Strebel K. Augmentation of virus secretion by the human immunodeficiency virus type 1 Vpu protein is cell type independent and occurs in cultured human primary macrophages and lymphocytes. J Virol. 1995;69(12):7699–711.
64. Theodore TS, Englund G, Buckler-White A, Buckler CE, Martin MA, Peden KW. Construction and characterization of a stable full-length macrophage-tropic HIV type 1 molecular clone that directs the production of high titers of progeny virions. AIDS Res Hum Retroviruses. 1996;12(3):191–4.
65. Martin N, Sattentau Q. Cell-to-cell HIV-1 spread and its implications for immune evasion. Curr Opin HIV AIDS. 2009;4(2):143–9.
66. Sattentau Q. Avoiding the void: cell-to-cell spread of human viruses. Nat Rev Microbiol. 2008;6(11):815–26.

67. Casartelli N, Sourisseau M, Feldmann J, Guivel-Benhassine F, Mallet A, Marcelin AG, Guatelli J, Schwartz O. Tetherin restricts productive HIV-1 cell-to-cell transmission. PLoS Pathog. 2010;6(6):e1000955.
68. Kuhl BD, Sloan RD, Donahue DA, Bar-Magen T, Liang C, Wainberg MA. Tetherin restricts direct cell-to-cell infection of HIV-1. Retrovirology. 2010;7:115.
69. Jolly C, Booth NJ, Neil SJ. Cell-cell spread of human immunodeficiency virus type 1 overcomes tetherin/BST-2-mediated restriction in T cells. J Virol. 2010;84(23):12185–99.
70. Carr JM, Hocking H, Li P, Burrell CJ. Rapid and efficient cell-to-cell transmission of human immunodeficiency virus infection from monocyte-derived macrophages to peripheral blood lymphocytes. Virology. 1999;265(2):319–29.
71. Sato H, Orenstein J, Dimitrov D, Martin M. Cell-to-cell spread of HIV-1 occurs within minutes and may not involve the participation of virus particles. Virology. 1992;186(2):712–24.
72. Sigal A, Kim JT, Balazs AB, Dekel E, Mayo A, Milo R, Baltimore D. Cell-to-cell spread of HIV permits ongoing replication despite antiretroviral therapy. Nature. 2011;477(7362):95–8.
73. Murooka TT, Deruaz M, Marangoni F, Vrbanac VD, Seung E, von Andrian UH, Tager AM, Luster AD, Mempel TR. HIV-infected T cells are migratory vehicles for viral dissemination. Nature. 2012;490(7419):283–7.
74. Hatziioannou T, Evans DT. Animal models for HIV/AIDS research. Nat Rev Microbiol. 2012;10(12):852–67.
75. Kirchhoff F. Is the high virulence of HIV-1 an unfortunate coincidence of primate lentiviral evolution? Nat Rev Microbiol. 2009;7(6):467–76.
76. Lim ES, Malik HS, Emerman M. Ancient adaptive evolution of tetherin shaped the functions of Vpu and Nef in human immunodeficiency virus and primate lentiviruses. J Virol. 2010;84(14):7124–34.

Chapter 19
Vaginal and Rectal HIV Transmission in Humanized Mice

Paul W. Denton, Morgan L. Chateau and J. Victor Garcia

19.1 Introduction

Vaginal and rectal HIV exposures account for the vast majority of the 2.5 million new HIV transmissions that occur annually worldwide [1]. A primary goal in the fight against HIV-AIDS is to prevent these new infections, yet the development of interventions designed to prevent mucosal HIV transmissions has been hindered by the lack of adequate models that recapitulate the events taking place during human infection. The highly restrictive species tropism of HIV is the primary reason that there are few models where potential clinical interventions against HIV transmission can be evaluated [2]. Mice are easily manipulated experimentally and would be ideal in vivo experimental platforms for research into mucosal HIV transmission, but unfortunately mouse cells are refractory to HIV infection and when infected cannot support HIV replication [3, 4]. Humanized mice bring the benefits of mice as small-animal models together with the human cells necessary for HIV transmission and replication [5–13].

Reviewed here are immunodeficient mice given a bone marrow transplant of human $CD34^+$ hematopoietic stem cells (hHSC) and then used for vaginal or rectal HIV transmission studies. To date, four different immunodeficient mouse strains have been utilized for this purpose. These strains include: (i) $Rag2^{-/-}$ IL2R $\gamma C^{-/-}$ (commonly referred to as double knockout or DKO; [14]); (ii) $Rag1^{-/-}$ IL2R $\gamma C^{-/-}$ (Rag1; [15]); (iii) nonobese diabetic/SCID (NOD/SCID or N/S; [16–19]); and (iv) N/S IL2R $\gamma C^{-/-}$ (NSG; [20]). DKO, Rag1, and NSG mice humanized with a bone marrow transplantation of hHSC are referred to hereafter as "mouse strain"-hHSC mice. When hHSC bone marrow transplantation is combined with implantation of

P. W. Denton (✉) · M. L. Chateau · J. V. Garcia
Division of Infectious Diseases, Department of Medicine, UNC Center for AIDS Research, University of North Carolina School of Medicine, Genetic Medicine Building 2046, Chapel Hill, NC 27599-7042, USA
e-mail: pauden@rm.dk

L. Y. Poluektova et al. (eds.), *Humanized Mice for HIV Research*,
DOI 10.1007/978-1-4939-1655-9_19

human liver and thymic tissue under the mouse kidney capsule, the resultant chimeric animals are referred to as BLT (bone marrow-liver-thymus) mice [16, 17]. Both BLT and hHSC transplant only humanized mice have been shown to harbor human T cells, B cells, and antigen presenting cells distributed throughout their organ systems (e.g., bone marrow, thymus, spleen, and lymph nodes; [14–17, 20]). However, it is the combination of bone marrow transplantation of hHSC coupled with the implantation of thymic/liver tissue in BLT mice that leads to optimum reconstitution of both the female reproductive tract (FRT) and the gastrointestinal tract of humanized mice with all the human cells that have been identified as important for HIV transmission [17, 21–28]. Another key distinction of BLT mice is that the implanted thymic tissue in BLT mice develops into a *bona fide* human thymus that produces human T cells expressing a broad TCR Vβ repertoire capable of HLA-restricted responses [16, 17, 26, 29].

The presence of HIV target cells in the FRT and intestines of humanized mice suggests that these models could be suitable for the study of vaginal and rectal HIV transmission. When humanized mice were exposed to HIV vaginally, the exposures were performed atraumatically [15, 22, 24, 25, 30–36]. Even though hormonal preconditioning in these models is not necessary for vaginal HIV transmission, two studies included hormonal treatment prior to vaginal exposure for unspecified reasons ([33, 34; Table 19.1). When humanized mice were exposed to HIV rectally, the exposures were either atraumatic [30, 37, 38] or followed simulated receptive anal intercourse ([27, 39;] Table 19.2).

19.2 Vaginal HIV Transmission

In BLT mice, human CD4⁺ T cells, CD68⁺ monocyte/macrophage cells, and clusters of CD11c⁺ dendritic cells are present throughout the FRT including the vagina, ectocervix, endocervix, and uterus [22, 24, 25]. The resultant susceptibility of BLT mice to vaginal HIV infection has been extensively documented by multiple research groups [22, 24, 25, 33, 34, 36]. Four of these studies utilized NSG-BLT mice [24, 33, 34, 36] and two utilized both NSG-BLT and N/S-BLT mice [22, 25]. In the four studies examining vaginal HIV transmission exclusively in NSG-BLT mice, 60 of 70 exposed mice became viremic (86 % transmission; [24, 33, 34, 36]). In one study that utilized both types of BLT mice, a clear relationship was noted between the levels of human reconstitution and vaginal transmission regardless of the strain of mice used [22]. In the other study, NSG-BLT and N/S-BLT mice with similar levels of human reconstitution were shown to also be readily susceptible to vaginal HIV infection (82 % transmission; [25]). The available data makes it is clear that these mice are well-suited for vaginal HIV transmission experiments.

Similarly, DKO-hHSC mice harbor human CD4⁺ cells and CD68⁺ monocyte/macrophage cells in their vagina [30]. Two research groups have reported vaginal HIV infection in DKO-hHSC or Rag1-hHSC mice [15, 30–32, 34, 35]. Data from four of these reports indicates that under their experimental conditions 36 of 37 DKO-hHSC mice became viremic following vaginal exposure to laboratory strains of HIV (97% transmission; [30–32, 35]). It should be noted that two of the manuscripts did not report the number of animals exposed [15, 34]. When all of these studies are examined together, the conclusion is that vaginal HIV transmission is both efficient and reproducible in multiple types of humanized mice.

19.2.1 Prevention of Vaginal HIV Transmission

Of the 11 manuscripts describing vaginal HIV transmission in humanized mice, six used these models to examine HIV prevention strategies (Table 19.1; [24, 25, 31–33, 35]). Preexposure prophylaxis (PrEP) interventions were tested to determine whether infection could be prevented in humanized mice treated either systemically or topically with antiviral drugs, peptides, aptamers, or antibodies. The first study to evaluate an HIV prevention strategy in humanized mice found that systemically applied tenofovir disoproxil fumarate and emtricitabine (reverse transcriptase inhibitors) prevented transmission in 5 of 5 treated NSG-BLT mice (100% protection; [24]). Another study reported that systemically applied raltegravir (an integrase inhibitor) prevented transmission in 6 of 6 treated DKO-hHSC mice (100% protection; [31]). In this study, it was also reported that systemically applied maraviroc (a CCR5 coreceptor inhibitor) prevented transmission in 6 of 6 treated DKO-hHSC mice (100% protection; [31]). The remaining four studies examined topical PrEP strategies for preventing vaginal HIV transmission [25, 32, 33, 35]. In one study, seven different topical interventions were assessed for efficacy in BLT mice [25]. These interventions are listed here along with the numbers of protected vs. treated mice for each agent: a C-peptide fusion inhibitor (C52L; 7 of 7, 100% protection), a membrane-disrupting amphipathic peptide inhibitor (C5A; 8 of 8, 100% protection), a trimeric D-peptide fusion inhibitor (PIE12-Trimer; 5 of 5, 100% protection), a combination of reverse transcriptase inhibitors (tenofovir disoproxil fumarate and emtricitabine; 8 of 9, 89% protection), a single reverse transcriptase inhibitor (tenofovir; 7 of 8, 88% protection), a thioester zinc finger inhibitor that primarily affects virion maturation (TC247; 4 of 7, 57% protection), and a small-molecule Rac inhibitor that affects viral entry from within the target cell (NSC23766; 0 of 4, 0% protection; [25]). In a second study, topically applied maraviroc prevented transmission in 7 of 7 treated DKO-hHSC or Rag1-hHSC mice (100% protection; [32]). In another study, three distinct $CD4^+$ cell targeting aptamer-siRNA chimeras carrying siRNA targeted to CCR5, *gag* or *vif* protected 2 of 4 treated NSG-BLT mice (50% protection; [33]). In the fourth study, anti-HIV neutralizing antibodies were topically applied in DKO-hHSC or Rag1-hHSC mice [35]. VRC01 alone prevented transmission in 7 of 9 treated mice (78% protection) and a pool of four neutralizing antibodies (b12, 4E10, 2F5, and 2G12) prevented transmission in 5 of 5 treated mice (100% protection; [35]). The breadth of the interventions tested in humanized mice highlights the utility of these models for testing different HIV prevention interventions.

Table 19.1. Vaginal HIV transmission in humanized mice

Year/First author (ref.)	Humanized mice	Hormone pretreatment	HIV isolate(s)	Topical PrEP intervention	Systemic PrEP intervention
2008/Denton [24]	NSG-BLT	–	JR-CSF	–	FTC/TDF
2008/Berges [30]	DKO-hHSC	–	BaL NL4-3	–	–
2010/Neff [31]	DKO-hHSC	–	BaL	–	Raltegravir Maraviroc
2011/Akkina [15]	Rag1-hHSC	–	BaL	–	–
2011/Denton [25]	N/S-BLT NSG-BLT	–	JR-CSF	Tenofovir C52L C5A PIE12-Trimer FTC/TDF TC247 NSC23766	–
2011/Neff [32]	DKO-hHSC Rag1-hHSC	–	BaL	Maraviroc	–
2011/Stoddart [22]	N/S-BLT NSG-BLT	–	JR-CSF 81A-G	–	–
2011/Wheeler [33]	N/S-BLT NSG-BLT	Yes	JR-CSF	CD4 aptamer-siRNA chimeras	–
2012/Hur [34]	NSG-BLT DKO-hHSC	Yes	JR-CSF	–	–
2012/Veselinovic [35]	DKO-hHSC Rag1-hHSC	–	BaL	VRCO1 nAb 4 pooled nAb	–
2013/Chateau [36]	NSG-BLT	–	JRSCF JR-CSF K65R[a]	–	–

FTC emtricitabine, *hHSC* human hematopoietic stem cell, *nAb* HIV neutralizing antibody, *TDF* tenofovir disoproxil fumarate

[a] JR-CSF with a K65R mutation in reverse transcriptase

19.2.2 Vaginal Transmission of Antiviral Drug-Resistant HIV

Drug-resistant HIV can be transmitted in humans [40–43] causing treatment failure [44–50]. Since HIV develops resistance to virtually all antiviral drugs currently available, there is great interest in the mucosal transmission of antiviral resistant HIV [51–55]. NSG-BLT mice were vaginally challenged with either wild type HIV-1 or an isogenic HIV containing a K65R mutation in reverse transcriptase that offers resistance to tenofovir [55]. When compared to transmission by the wild-type HIV-1 parent viral isolate (4 of 4 exposed mice became viremic; 100 % transmission), there was a 75 % reduction ($p=0.01$) in vaginal transmission by the K65R isogenic mutant virus (3 of 12 exposed mice became viremic; 25 % transmission; [36]). Furthermore, in infected mice, the drug resistant virus reverted to a wild type (drug sensitive) genotype in plasma and vaginal fluid [36]. These data demonstrate the utility of BLT mice for these types of studies and suggest that drug resistance mutations with predicted fitness cost in vitro (e.g., K65R [56, 57]), also might have a mucosal transmission deficiency that can be characterized in this in vivo model.

19.3 Rectal HIV Transmission

In BLT mice, human $CD4^+$ T cells, $CD68^+$ monocyte/macrophage cells, and $CD11c^+$ dendritic cells are present throughout their small and large intestines, including the rectum [24, 26, 27]. Importantly, the human lymphocytes in the intestines of BLT mice are distributed similarly to those found in human gut [26, 27, 58, 59]. Furthermore, rectal HIV infection in BLT mice has been demonstrated in three different studies. Rectal HIV exposure led to infection in 6 of 7 [27], 12 of 19 [39], and 13 of 25 [38] animals for an average transmission rate of 61 % in BLT mice. Most of these challenges were performed using the primary isolate HIV-$1_{JR\text{-}CSF}$ [60]. Recently it has been demonstrated that in humans, mucosal exposure results in the transmission of one or a few viruses designated as transmitted founder viruses [61–63]. Due to the high relevance of transmitted founder viruses to the human condition, a rectally acquired transmitted founder virus (HIV-1_{THRO} [62]) was evaluated for transmissibility in BLT mice. When BLT mice were exposed rectally to HIV-1_{THRO}, 5 of 8 animals became viremic [38]. Demonstration of the susceptibility of BLT mice to a virus responsible for an actual human transmission event is a significant advancement that will allow for further analysis of the role of transmitted founder viruses in mucosal HIV transmission.

DKO-hHSC mice also harbor human $CD4^+$ cells and $CD68^+$ monocyte/macrophage cells in their large intestines, including the rectum [30], although HIV target cells are uncommon in the large intestine of these mice [26, 37]. Two research groups have examined rectal HIV infection in DKO-hHSC mice [30, 37]. Berges et al. reported that 11 of 14 rectally exposed DKO-hHSC mice became viremic (79 % transmission) after a single exposure to laboratory strains of HIV [30]. In contrast, Hofer et al. reported that 1 of 23 rectally exposed DKO-hHSC mice became viremic (4 % transmission) also after exposure to a laboratory strain of HIV [37]. In the 5 years after these two discordant results were reported, no follow-up manuscripts reporting rectal transmission in DKO-hHSC mice have been published.

19.3.1 Prevention of Rectal HIV Transmission

Of the five manuscripts describing rectal HIV transmission in humanized mice, two used BLT mice to examine HIV prevention by antiviral drug(s) (Table 19.2; [38, 39]). In one study, topically applied tenofovir prevented transmission of HIV-$1_{JR\text{-}CSF}$ in 11 of 12 treated NSG-BLT mice (92 % protection; [38]). This study also reported

Table 19.2. *Rectal* HIV transmission in humanized mice

Year/First author (ref.)	Human-ized mice	sRAI pretreatment	HIV isolate(s)	Topical PrEP intervention	Systemic PrEP intervention
2007/Sun [27]	N/S-BLT	Yes	LAI	–	–
2008/Berges [30]	DKO-hHSC	–	BaL NL4-3	–	–
2008/Hofer [37]	DKO-hHSC	–	YU-2 JR-CSF NL4-3 89.6	–	–
2010/Denton [39]	N/S-BLT NSG-BLT	Yes	JR-CSF	–	FTC/TDF
2013/Chateau [38]	NSG-BLT	–	JR-CSF THRO	Tenofovir	–

FTC emtricitabine, *hHSC* human hematopoietic stem cell, *sRAI* simulated receptive anal intercourse, *TDF* tenofovir disoproxil fumarate

that topically applied tenofovir prevented transmission of the transmitted founder HIV-1 isolate, HIV_{THRO}, in 6 of 6 treated NSG-BLT mice (100% protection; [38]). The other study reported that systemically applied tenofovir disoproxil fumarate and emtricitabine prevented rectal transmission in 9 of 9 treated BLT mice (100% protection; [39]). Thus, both topical and systemic PrEP for the prevention of rectal HIV transmission have been successfully evaluated in humanized mice.

19.4 Validation of BLT Mice for Preclinical Efficacy Testing of HIV Prevention Interventions

One rationale for generating BLT humanized mice as HIV transmission models was to test prevention strategies and a key step in the validation of humanized mice for the testing of these strategies is to compare results obtained using humanized mice to available clinical trial data. Given that some of the same topical and systemic PrEP strategies have been tested in both BLT mice and humans, such comparisons have been made [23]. Briefly, topical tenofovir to prevent vaginal HIV transmission has been tested in BLT mice (88% efficacy; [25]) and in two clinical trials (CAPRISA 004, 39% efficacy and VOICE, discontinued; [64, 65]). Systemic PrEP with tenofovir disoproxil fumarate and emtricitabine for the prevention of vaginal and rectal HIV transmission was tested in BLT mice (100% efficacy each for vaginal and rectal prevention; [24, 39]). In humans, this combination of antiretrovirals has been evaluated in: men that have sex with men (iPrEx, 44% efficacy), sexually active men and women (TDF2, 63% efficacy), heterosexual couples (partners PrEP, 73% efficacy), and sexually active women (FEM-PrEP, discontinued; [66, 67]). Based on the available data, in July 2012 the US Food and Drug Administration approved systemic PrEP with oral Truvada (coformulated tenofovir disoproxil fumarate and emtricitabine) in combination with safer sex practices to reduce the risk of sexually acquired HIV-infection [68].

19.5 Conclusions and Future Directions

Humanized mice have been validated as in vivo models of vaginal and rectal HIV transmission. These successes open avenues for new investigations using humanized mice including: defining the molecular determinants of mucosal HIV transmission; defining the parameters of HIV dissemination following mucosal infection, perhaps using intravital microscopy [69]; and characterizing HIV shedding in the genital tract of infected humanized mice. The results of such studies will provide key insights into host/pathogen interactions in mucosal tissues at the time of and following HIV exposure. In conclusion, humanized mice are reproducible, relevant, and reliable experimental platforms for gaining novel insights into mucosal HIV transmission.

Acknowledgment This work was supported in part by National Institutes of Health grants AI73146 and AI96138 (JVG).

References

1. WHO-UNAIDS. UNAIDS Report on the global AIDS epidemic. Geneva, Switzerland. 2012. http://www.unaids.org/en/media/unaids/contentassets/documents/epidemiology/2012/gr2012/20121120_UNAIDS_Global_Report_2012_with_annexes_en.pdf2012. Accessed 26 April 2013.
2. Hatziioannou T, Ambrose Z, Chung NP, Piatak M Jr., Yuan F, Trubey CM, et al. A macaque model of HIV-1 infection. Proc Natl Acad Sci U S A. 2009;106(11):4425–9.
3. Mariani R, Rutter G, Harris ME, Hope TJ, Krausslich H-G, Landau NR. A block to human immunodeficiency virus type 1 assembly in murine cells. J Virol. 2000;74(8):3859–70.
4. Bieniasz PD, Cullen BR. Multiple blocks to human immunodeficiency virus type 1 replication in rodent cells. J Virol. 2000;74(21):9868–77.
5. Denton PW, Garcia JV. Novel humanized murine models for HIV research. Curr HIV/AIDS Rep. 2009;6(1):13–9.
6. Denton PW, Garcia JV. Humanized mouse models of HIV infection. AIDS Rev. 2011;13(3):135–48.
7. Shultz LD, Brehm MA, Garcia-Martinez JV, Greiner DL. Humanized mice for immune system investigation: progress, promise and challenges. Nat Rev Immunol. 2012;12(11):786–98.
8. Shultz LD, Ishikawa F, Greiner DL. Humanized mice in translational biomedical research. Nat Rev Immunol. 2007;7(2):118–30.
9. Akkina R. New generation humanized mice for virus research: comparative aspects and future prospects. Virology. 2013;435(1):14–28.
10. Berges BK, Rowan MR. The utility of the new generation of humanized mice to study HIV-1 infection: transmission, prevention, pathogenesis, and treatment. Retrovirology. 2011;8:65.
11. Su L. Studying human immunology and immunopathology in humanized mice transplanted with human lymphoid tissues and immune cells. Cell Mol Immunol. 2012;9(3):191–2.

12. Nischang M, Gers-Huber G, Audiga A, Akkina R, Speck R. Modeling HIV infection and therapies in humanized mice. Swiss Med Wkly. 2012;142:w13618.
13. Ito R, Takahashi T, Katano I, Ito M. Current advances in humanized mouse models. Cell Mol Immunol. 2012;9(3):208–14.
14. Traggiai E, Chicha L, Mazzucchelli L, Bronz L, Piffaretti JC, Lanzavecchia A, et al. Development of a human adaptive immune system in cord blood cell-transplanted mice. Science. 2004;304(5667):104–7.
15. Akkina R, Berges BK, Palmer BE, Remling L, Neff CP, Kuruvilla J, et al. Humanized Rag1-/- gammac-/- mice support multilineage hematopoiesis and are susceptible to HIV-1 infection via systemic and vaginal routes. PLoS ONE. 2011;6(6):e20169. doi:10.1371/journal.pone.0020169.
16. Lan P, Tonomura N, Shimizu A, Wang S, Yang YG. Reconstitution of a functional human immune system in immunodeficient mice through combined human fetal thymus/liver and CD34+ cell transplantation. Blood. 2006;108(2):487–92.
17. Melkus MW, Estes JD, Padgett-Thomas A, Gatlin J, Denton PW, Othieno FA, et al. Humanized mice mount specific adaptive and innate immune responses to EBV and TSST-1. Nat Med. 2006;12(11):1316–22.
18. Greiner DL, Shultz LD, Yates J, Appel MC, Perdrizet G, Hesselton RM, et al. Improved engraftment of human spleen cells in NOD/LtSz-scid/scid mice as compared with C.B-17-scid/scid mice. Am J Pathol. 1995;146(4):888–902.
19. Ueda T, Yoshino H, Kobayashi K, Kawahata M, Ebihara Y, Ito M, et al. Hematopoietic repopulating ability of cord blood CD34(+) cells in NOD/Shi-scid mice. Stem Cells. 2000;18(3):204–13.
20. Shultz LD, Lyons BL, Burzenski LM, Gott B, Chen X, Chaleff S, et al. Human lymphoid and myeloid cell development in NOD/LtSz-scid IL2R gamma null mice engrafted with mobilized human hemopoietic stem cells. J Immunol. 2005;174(10):6477–89.
21. Haase AT. Targeting early infection to prevent HIV-1 mucosal transmission. Nature. 2010;464(7286):217–23.
22. Stoddart CA, Maidji E, Galkina SA, Kosikova G, Rivera JM, Moreno ME, et al. Superior human leukocyte reconstitution and susceptibility to vaginal HIV transmission in humanized NOD-scid IL-2Rgamma(-/-) (NSG) BLT mice. Virology. 2011;417(1):154–60.
23. Denton PW, Garcia JV. Mucosal HIV-1 transmission and prevention strategies in BLT humanized mice. Trends Microbiol. 2012;20(6):268–74.
24. Denton PW, Estes JD, Sun Z, Othieno FA, Wei BL, Wege AK, et al. Antiretroviral pre-exposure prophylaxis prevents vaginal transmission of HIV-1 in humanized BLT mice. PLoS Med. 2008;5(1):e16.
25. Denton PW, Othieno F, Martinez-Torres F, Zou W, Krisko JF, Fleming E, et al. One percent tenofovir applied topically to humanized BLT mice and used according to the CAPRISA 004 experimental design demonstrates partial protection from vaginal HIV infection, validating the BLT model for evaluation of new microbicide candidates. J Virol. 2011;85(15):7582–93.
26. Denton PW, Nochi T, Lim A, Krisko JF, Martinez-Torres F, Choudhary SK, et al. IL-2 receptor gamma-chain molecule is critical for intestinal T-cell reconstitution in humanized mice. Mucosal Immunol. 2012;5(5):555–66.
27. Sun Z, Denton PW, Estes JD, Othieno FA, Wei BL, Wege AK, et al. Intrarectal transmission, systemic infection, and CD4+ T cell depletion in humanized mice infected with HIV-1. J Exp Med. 2007;204(4):705–14.
28. Olesen R, Wahl A, Denton PW, Victor Garcia J. Immune reconstitution of the female reproductive tract of humanized BLT mice and their susceptibility to human immunodeficiency virus infection. J Reprod Immunol. 2011;88(2):195–203.
29. Brainard DM, Seung E, Frahm N, Cariappa A, Bailey CC, Hart WK, et al. Induction of robust cellular and humoral virus-specific adaptive immune responses in human immunodeficiency virus-infected humanized BLT mice. J Virol. 2009;83(14):7305–21.

30. Berges BK, Akkina SR, Folkvord JM, Connick E, Akkina R. Mucosal transmission of R5 and X4 tropic HIV-1 via vaginal and rectal routes in humanized Rag2-/- gammac -/- (RAG-hu) mice. Virology. 2008;373(2):342–51.
31. Neff CP, Ndolo T, Tandon A, Habu Y, Akkina R. Oral pre-exposure prophylaxis by anti-retrovirals raltegravir and maraviroc protects against HIV-1 vaginal transmission in a humanized mouse model. PLoS ONE. 2010;5(12):e15257 doi:10.1371/journal.pone.0015257.
32. Neff CP, Kurisu T, Ndolo T, Fox K, Akkina R. A topical microbicide gel formulation of CCR5 antagonist maraviroc prevents HIV-1 vaginal transmission in humanized RAG-hu mice. PLoS ONE. 2011;6(6):e20209. doi:10.1371/journal.pone.0020209.
33. Wheeler LA, Trifonova R, Vrbanac V, Basar E, McKernan S, Xu Z, et al. Inhibition of HIV transmission in human cervicovaginal explants and humanized mice using CD4 aptamer-siRNA chimeras. J Clin Invest. 2011;121(6):2401–12.
34. Hur EM, Patel SN, Shimizu S, Rao DS, Gnanapragasam PN, An DS, et al. Inhibitory effect of HIV-specific neutralizing IgA on mucosal transmission of HIV in humanized mice. Blood. 2012;120(23):4571–82.
35. Veselinovic M, Neff CP, Mulder LR, Akkina R. Topical gel formulation of broadly neutralizing anti-HIV-1 monoclonal antibody VRC01 confers protection against HIV-1 vaginal challenge in a humanized mouse model. Virology. 2012;432(2):505–10.
36. Chateau M, Swanson MD, Garcia JV. Inefficient vaginal transmission of tenofovir resistant HIV-1. J Virol. 2013;87(2):1274–7.
37. Hofer U, Baenziger S, Heikenwalder M, Schlaepfer E, Gehre N, Regenass S, et al. RAG2-/- gamma(c)-/- mice transplanted with CD34+ cells from human cord blood show low levels of intestinal engraftment and are resistant to rectal transmission of human immunodeficiency virus. J Virol. 2008;82(24):12145–53.
38. Chateau ML, Denton PW, Swanson MD, McGowan I, Garcia JV. Rectal transmission of transmitted/founder HIV-1 is efficiently prevented by topical 1% tenofovir in BLT humanized mice. PLoS ONE. 2013;8(3):e60024.
39. Denton PW, Krisko JF, Powell DA, Mathias M, Kwak YT, Martinez-Torres F, et al. Systemic administration of antiretrovirals prior to exposure prevents rectal and intravenous HIV-1 transmission in humanized BLT mice. PLoS ONE. 2010;5(1):e8829.
40. Metzner KJ, Bonhoeffer S, Fischer M, Karanicolas R, Allers K, Joos B, et al. Emergence of minor populations of human immunodeficiency virus type 1 carrying the M184V and L90M mutations in subjects undergoing structured treatment interruptions. J Infect Dis. 2003;188(10):1433–43.
41. Brenner BG, Roger M, Moisi DD, Oliveira M, Hardy I, Turgel R, et al. Transmission networks of drug resistance acquired in primary/early stage HIV infection. AIDS. 2008;22(18):2509–15.
42. Wainberg MA, Moisi D, Oliveira M, Toni TD, Brenner BG. Transmission dynamics of the M184V drug resistance mutation in primary HIV infection. J Antimicrob Chemother. 2011;66(10):2346–9.
43. Hightow-Weidman LB, Hurt CB, Phillips G 2., Jones K, Magnus M, Giordano TP, et al. Transmitted HIV-1 drug resistance among young men of color who have sex with men: a multicenter cohort analysis. J Adolesc Health. 2011;48(1):94–9.
44. Li JF, Lipscomb JT, Wei X, Martinson NA, Morris L, Heneine W, et al. Detection of low-level K65R variants in nucleoside reverse transcriptase inhibitor-naive chronic and acute HIV-1 subtype C infections. J Infect Dis. 2011;203(6):798–802.
45. Little SJ, Holte S, Routy JP, Daar ES, Markowitz M, Collier AC, et al. Antiretroviral-drug resistance among patients recently infected with HIV. N Engl J Med. 2002;347(6):385–94.
46. Borroto-Esoda K, Waters JM, Bae AS, Harris JL, Hinkle JE, Quinn JB, et al. Baseline genotype as a predictor of virological failure to emtricitabine or stavudine in combination with didanosine and efavirenz. AIDS Res Hum Retroviruses. 2007;23(8):988–95.
47. Frentz D, Boucher CA, van de Vijver DA. Temporal changes in the epidemiology of transmission of drug-resistant HIV-1 across the world. AIDS Rev. 2012 ;14(1):17–27.

48. Huang HY, Daar ES, Sax PE, Young B, Cook P, Benson P, et al. The prevalence of transmitted antiretroviral drug resistance in treatment-naive patients and factors influencing first-line treatment regimen selection. HIV Med. 2008;9(5):285–93.
49. Kuritzkes DR, Lalama CM, Ribaudo HJ, Marcial M, Meyer WA 3., Shikuma C, et al. Pre-existing resistance to nonnucleoside reverse-transcriptase inhibitors predicts virologic failure of an efavirenz-based regimen in treatment-naive HIV-1-infected subjects. J Infect Dis. 2008;197(6):867–70.
50. Bansal V, Metzner KJ, Niederost B, Leemann C, Boni J, Gunthard HF, et al. Minority K65R variants and early failure of antiretroviral therapy in HIV-1-infected Eritrean immigrant. Emerg Infect Dis. 2011;17(10):1966–8.
51. Bennett DE, Camacho RJ, Otelea D, Kuritzkes DR, Fleury H, Kiuchi M, et al. Drug resistance mutations for surveillance of transmitted HIV-1 drug-resistance: 2009 update. PLoS ONE. 2009;4(3):e4724.
52. Dumond JB, Yeh RF, Patterson KB, Corbett AH, Jung BH, Rezk NL, et al. Antiretroviral drug exposure in the female genital tract: implications for oral pre- and post-exposure prophylaxis. AIDS. 2007;21(14):1899–907.
53. Kashuba AD, Patterson KB, Dumond JB, Cohen MS. Pre-exposure prophylaxis for HIV prevention: how to predict success. Lancet. 2011. doi:10.1016/S0140-6736(11)61852-7.
54. Patterson KB, Prince HA, Kraft E, Jenkins AJ, Shaheen NJ, Rooney JF, et al. Penetration of tenofovir and emtricitabine in mucosal tissues: implications for prevention of HIV-1 transmission. Sci Transl Med. 2011;3(112):112re4.
55. Johnson VA, Calvez V, Gunthard HF, Paredes R, Pillay D, Shafer R, et al. Update of the drug resistance mutations in HIV-1. Top Antivir Med. 2011;19(4):156–64.
56. Wainberg MA, Miller MD, Quan Y, Salomon H, Mulato AS, Lamy PD, et al. In vitro selection and characterization of HIV-1 with reduced susceptibility to PMPA. Antivir Ther. 1999;4(2):87–94.
57. Wagner BG, Garcia-Lerma JG, Blower S. Factors limiting the transmission of HIV mutations conferring drug resistance: fitness costs and genetic bottlenecks. Sci Rep. 2012;2:320.
58. Abuzakouk M, Carton J, Feighery C, O'Donoghue DP, Weir DG, O'Farrelly C. CD4+ CD8+ and CD8alpha+ beta- T lymphocytes in human small intestinal lamina propria. Eur J Gastroenterol Hepatol. 1998;10(4):325–9.
59. Carton J, Byrne B, Madrigal-Estebas L, O'Donoghue DP, O'Farrelly C. CD4+CD8+ human small intestinal T cells are decreased in coeliac patients, with CD8 expression downregulated on intra-epithelial T cells in the active disease. Eur J Gastroenterol Hepatol. 2004;16(10):961–8.
60. Koyanagi Y, Miles S, Mitsuyasu RT, Merrill JE, Vinters HV, Chen IS. Dual infection of the central nervous system by AIDS viruses with distinct cellular tropisms. Science. 1987;236(4803):819–22.
61. Keele BF, Giorgi EE, Salazar-Gonzalez JF, Decker JM, Pham KT, Salazar MG, et al. Identification and characterization of transmitted and early founder virus envelopes in primary HIV-1 infection. Proc Natl Acad Sci U S A. 2008;105(21):7552–7.
62. Ochsenbauer C, Edmonds TG, Ding H, Keele BF, Decker J, Salazar MG, et al. Generation of transmitted/founder HIV-1 infectious molecular clones and characterization of their replication capacity in CD4 T lymphocytes and monocyte-derived macrophages. J Virol. 2012;86(5):2715–28.
63. Salazar-Gonzalez JF, Salazar MG, Keele BF, Learn GH, Giorgi EE, Li H, et al. Genetic identity, biological phenotype, and evolutionary pathways of transmitted/founder viruses in acute and early HIV-1 infection. J Exp Med. 2009;206(6):1273–89.
64. Abdool Karim Q, Abdool Karim SS, Frohlich JA, Grobler AC, Baxter C, Mansoor LE, et al. Effectiveness and safety of tenofovir gel, an antiretroviral microbicide, for the prevention of HIV infection in women. Science. 2010;329(5996):1168–74.
65. Microbicide Trials Network. MTN statement on decision to discontinue use of Tenofovir gel in VOICE, a major HIV prevention study in women. 2011. http://www.mtnstopshiv.org/node/3909. Accessed: 29 Nov 2011.

66. Grant RM, Lama JR, Anderson PL, McMahan V, Liu AY, Vargas L, et al. Preexposure chemoprophylaxis for HIV prevention in men who have sex with men. N Engl J Med. 2010;363(27):2587–99.
67. Hurt CB, Eron JJ, Jr., Cohen MS. Pre-exposure prophylaxis and antiretroviral resistance: HIV prevention at a cost? Clin Infect Dis. 2011;53(12):1265–70.
68. US-FDA. Truvada for PrEP fact sheet: ensuring safe and proper use. 2012. http://www.fda.gov/downloads/NewsEvents/Newsroom/FactSheets/UCM312279.pdf2012. Accessed 26 April 2013.
69. Murooka TT, Deruaz M, Marangoni F, Vrbanac VD, Seung E, von Andrian UH, et al. HIV-infected T cells are migratory vehicles for viral dissemination. Nature. 2012;490(7419):283–7.

Chapter 20
Oral HIV-1 Transmission in BLT Humanized Mice

Angela Wahl and J. Victor Garcia

Abbreviations

AIDS	Acquired immunodeficiency syndrome
ART	Antiretroviral therapy
BLT	Bone marrow/liver/thymus
CLN	Cervical lymph node
FTC	Emtricitabine
GI	Gastrointestinal
GE	Gastroesophogeal
HIV	Human immunodeficiency virus
IHC	Immunohistochemistry
LCPUFA	Long-chain polyunsaturated fatty acid
NALT	Nasal-associated lymphoid tissue
NOD/SCID	Non-obese diabetic/severe combined immunodeficiency
PBL	Peripheral blood leukocyte
PrEP	Pre-exposure prophylaxis
RNA	Ribonucleic acid
SLPI	Secretory leukocyte protease inhibitor
TDF	Tenofovir

A. Wahl (✉) · J. V. Garcia
Division of Infectious Diseases, Department of Medicine, UNC Center for AIDS Research, University of North Carolina School of Medicine, Genetic Medicine Building 2046, Chapel Hill, NC 27599-7042, USA
e-mail: angela_wahl@med.unc.edu

L. Y. Poluektova et al. (eds.), *Humanized Mice for HIV Research*,
DOI 10.1007/978-1-4939-1655-9_20

20.1 Introduction

The impetus for this research was the fact that the development of effective strategies to prevent human immunodeficiency virus (HIV) acquisition during breastfeeding would be significantly enhanced by a small animal model of oral HIV infection that could be utilized to study the mechanisms for oral transmission of HIV. In addition, a suitable animal model could be used to test the efficacy of novel approaches to prevent oral HIV transmission. Thus, we developed an oral HIV transmission model based on Bone marrow/liver/thymus (BLT) humanized mice [1]. This model has been used to address issues of fundamental importance to the field including: (1) the role of cell-free and cell-associated virus, (2) viral dissemination following exposure, (3) prevention of HIV transmission with antiretrovirals and (4) the effect of human breast milk on HIV transmission.

BLT mice are created by transplanting autologous fetal liver-derived human $CD34^+$ hematopoietic progenitor cells into mice previously implanted with a piece of human fetal liver sandwiched between two small pieces of human fetal thymus [1]. BLT mice show robust reconstitution with virtually all human hematopoietic cell types that are present in primary, secondary and effector immune organs. The immune cells present in BLT mice can mount both humoral and cellular immune responses to model antigens and viruses [1, 2]. Our laboratory has further established that systemic reconstitution with human hematopoietic cells renders BLT mice susceptible to parenteral, rectal and vaginal HIV infection [2–7]. More recently, it was also demonstrated that BLT mice can be utilized to assess the efficacy of pre- and post-exposure antiretroviral prophylactic strategies to prevent mucosal and parenteral HIV infection [3–6].

20.2 Reconstitution of the Oral Cavity and the Upper GI Tract in BLT Humanized Mice

The oral cavity is the first site exposed to HIV that is transmitted during breastfeeding. Therefore, examination of the oral mucosa of BLT mice for the presence of human hematopoietic cells was done using immunohistochemistry (IHC) analysis. It was found that in the oral mucosa of BLT mice, human macrophages are present in the epithelial layer, near the basement membrane, and in the lamina propria [8]. Human dendritic cells are predominantly present in the lamina propria [8]. Human $CD4^+$ and $CD8^+$ T cells are present in the epithelial layer, appearing as a band at the basement membrane, and in the underlying lamina propria [8]. This distribution of cells is very similar to that observed in the oral mucosa of infants [9].

Since mice do not have tonsils, the salivary glands and lymphoid tissues proximal to the oral mucosa of BLT mice were examined for the presence of human cells. The salivary glands of BLT mice were reconstituted with human macrophages, dendritic cells and $CD4^+$ T cells, as wells as human $CD8^+$ T and B cells [8]. In addition, the cervical lymph nodes (CLN), which drain the oral mucosa, and the nasal-associated lymphoid tissue (NALT) of BLT mice were also found to contain human macrophages, dendritic cells and T and B cells [8].

In humans, HIV transmission may occur in the oral cavity and/or in the upper GI tract following an oral exposure. IHC demonstrated that human dendritic cells, macrophages and $CD4^+$ T cells are all present in the esophagus, stomach and upper small intestine of BLT mice as well as at the gastroesophageal (GE) junction, where the esophagus joins the stomach [8]. In regard to the relative distribution of human cells in the upper GI tract of BLT mice, in the esophagus, human macrophages, dendritic cells and $CD4^+$ T cells are located in the basal layer of the epithelium and in the lamina propria [8]. Human dendritic cells, macrophages and $CD4^+$ T cells are found in the epithelium and lamina propria of the stomach as well as the lamina propria of the duodenum [7, 10]. In all these aspects, the upper GI tract of BLT mice resembles that of healthy adults [11–16].

20.3 Efficient Oral HIV Transmission in BLT Humanized Mice

Exposure of BLT mice to a single dose of cell-free HIV-1, administered directly into the oral cavity, resulted in systemic infection and gradual $CD4^+$ T-cell depletion in peripheral blood [8]. Consistent with the presence of HIV target cells in the oral mucosa and upper GI tract of BLT mice, reproducible oral HIV transmission was noted [8]. Specifically, viral RNA was readily detected in the plasma of all BLT mice exposed orally to HIV-1 demonstrating efficient transmission in this model [8]. One important observation made during the course of these experiments was that viral RNA was present in the saliva of the majority of infected BLT mice. As is observed in humans, the viral load in saliva was consistently lower than the viral load in plasma [8, 17, 18].

Currently, following an oral exposure, HIV transmission to neonates is believed to occur in the oral cavity and/or upper GI tract. In order to determine if HIV transmission can occur distal to the oral cavity, HIV-1 was administered directly into the stomach of BLT mice by oral gavage. After exposure by gavage, viral RNA could be readily detected in the plasma of all BLT mice by 2 weeks post-exposure indicating that the mucosal surfaces of BLT mice distal to the oral cavity can be directly infected with HIV [8].

20.4 Oral Transmission of Cell-Associated HIV in BLT Humanized Mice

Biological fluids like blood, semen and cervicovaginal secretions contain both cell-free and cell-associated HIV. Similarly, both forms of the virus are present in the breast milk of HIV-infected women. Currently, it is unclear whether one or both are responsible for the transmission events resulting in infection [19–22]. In this regard, hu-mice have been helpful to determine if cell-free and cell-associated HIV are transmissible following an oral exposure. Once oral transmission of cell-free HIV had been demonstrated, the oral transmission of cell-associated HIV was evaluated. BLT mice were exposed orally to a single dose of HIV-infected cells and monitored for infection by determining the presence of viral RNA in peripheral blood plasma. Remarkably, transmission of cell-associated HIV was readily and reproducibly observed [8]. Similar results were obtained when BLT mice were exposed to cell-associated HIV by gavage, bypassing the oral cavity [8].

Since the vast majority oral HIV transmission events occur in the context of human breast milk, it was important to investigate the impact of breast milk on oral transmission. Previous reports have demonstrated that breast milk has an inhibitory effect on virus infectivity when evaluated with in vitro assays [23–31]. The availability of an in vivo system for oral HIV transmission permitted the first evaluation of the role of human breast milk in oral HIV transmission. Whereas in the absence of breast milk, HIV-RNA was detected in the peripheral blood plasma of all control mice exposed orally to HIV-1, no transmission events were reported when cell-free virus was orally administered in the presence of whole breast milk obtained from HIV-negative women [8]. Together, the in vitro and in vivo results of these experiments highlight the potent HIV inhibitory activity of normal human breast milk and demonstrate that the in vitro HIV inhibitory activity of human breast milk is also capable of efficiently preventing oral transmission of cell-free HIV.

In stark contrast to the robust transmission of cell-associated HIV after oral exposure in the absence of breast milk, cell-associated HIV was not capable of establishing a productive infection when oral exposures occurred in the presence of breast milk [8]. Collectively, these data demonstrate that BLT mice are susceptible to oral transmission of HIV and that human breast milk possesses innate factors that can potently inhibit oral transmission of both cell-free and cell-associated HIV.

20.5 Pre-exposure Administration of Antiretrovirals Prevents Oral HIV Transmission in BLT Humanized Mice

Hu-mice have been extensively used to investigate different aspects of HIV prevention. Therefore, it was pertinent to determine if BLT mice could also be used to study the prevention of oral HIV transmission. Systemic pre-exposure prophylaxis (PrEP) with emtricitabine (FTC) and tenofovir (TDF) has been shown to effectively prevent vaginal, rectal and intravenous HIV transmission [3, 4, 6]. This same combination of antiretrovirals was also highly effective at preventing oral HIV transmission [8]. These experiments serve as a proof-of-principle for the future evaluation of the efficacy of novel and existing antiretroviral therapy (ART) strategies.

20.6 Conclusions and Future Directions

It has been clearly established that human hematopoietic cells generated in situ are capable of repopulating the oral cavity and upper GI tract of BLT humanized mice [8]. The presence of these human cells renders BLT mice susceptible to oral transmission of cell-free and cell-associated HIV. Infection of the oral cavity results in virus shedding into saliva, recapitulating the human condition. Human breast milk can inhibit oral transmission of cell-free and cell-associated HIV and oral transmission of HIV can be prevented with systemic FTC/TDF PrEP. However, it should be noted that similar results have not been seen in other models. Specifically, NOD/SCID and NOD/SCID/$\beta_2m^{-/-}$ mice reconstituted with human peripheral blood leukocytes (hu-PBL mice) failed to demonstrate oral transmission of cell-free HIV following an oral exposure to CXCR4 and CCR5 tropic strains, including HIV-$1_{JR\text{-}CSF}$ [32]. Therefore, BLT mice represent a significant advancement since they are the first highly reproducible model for oral HIV transmission.

Several areas of future investigation and subsequent in vivo models of HIV transmission can be contemplated for this. For example, recent studies of mother–infant transmission pairs indicate that the replicative fitness as well as the length of the variable loop and number of n-linked glycosylation sites in the envelope protein may influence which maternal variant(s) are preferentially transmitted [33–35]. BLT mice could be used to study the molecular and biological properties of HIV strains that provide a selective advantage for oral transmission during breastfeeding. A comparison of viruses transmitted at early, mid and late stages of lactation would indicate if the composition of breast milk, which changes during lactation, influences which viruses are preferentially transmitted. Studies illustrating that ART administered to HIV-infected mothers during pregnancy or postpartum significantly decreases the amount of cell-free but not cell-associated HIV in breast milk [36, 37], combined with reports demonstrating that maternal ART significantly decreases but does not eliminate HIV transmission during breastfeeding, suggest that both cell-free and cell-associated HIV may be transmitted [38, 39]. As the importance of cell-associated HIV in breast milk transmission becomes increasingly more appreciated, antiretroviral drugs and preventative strategies may be needed that reduce the burden of HIV-infected cells in the breast milk of HIV-infected women and/or directly inhibit transmission of cell-associated HIV in infants during breastfeeding.

Inhibitory factors present in breast milk may differ in their ability to inhibit cell-free versus cell-associated HIV infection. While the skim fraction of human breast milk possesses proteins with HIV-inhibitory activity (i.e. mucin, lactoferrin, bile salt-stimulated lipase and secretory leukocyte protease inhibitor (SLPI)) [23–31], the lipid fraction may contain additional factors that can inhibit transmission of cell-free and/or cell-associated HIV. For example, increased concentrations of certain long-chain polyunsaturated fatty acids (LCPUFAs) in breast milk are associated with a decreased risk of HIV breastfeeding transmission [40]. It has been hypothesized that LCPUFAs may inhibit HIV infection by inactivating the virus' envelope, suppressing the release of HIV virions from the host cell membrane and/or enhancing the viability of infected CD4+ T cells [41]. Finally, the availability of a robust and reproducible model of oral HIV transmission will permit the molecular analysis of the early events occurring at the mucosal sites of entry. Such detailed understanding will help the development and implementation of novel preventive strategies that could reduce the number of children acquiring HIV infection during breastfeeding.

Acknowledgements This work was supported in part by National Institutes of Health grants AI071940 and AI073146 (J.V.G.), the University of North Carolina Center for AIDS Research grant P30 AI50410 and the National Institute of Allergy and Infectious Disease Institutional Training grant (5T32AI007273-27) (A.W.). The authors also thank current and past members of the laboratory for their contributions to this work.

References

1. Melkus MW, Estes JD, Padgett-Thomas A, Gatlin J, Denton PW, Othieno FA, Wege AK, Haase AT, Garcia JV. Humanized mice mount specific adaptive and innate immune responses to EBV and TSST-1. Nat Med. 2006;12(11):1316–22.
2. Brainard DM, Seung E, Frahm N, Cariappa A, Bailey CC, Hart WK, Shin HS, Brooks SF, Knight HL, Eichbaum Q, Yang YG, Sykes M, Walker BD, Freeman GJ, Pillai S, Westmoreland SV, Brander C, Luster AD, Tager AM. Induction of robust cellular and humoral virus-specific adaptive immune responses in human immunodeficiency virus-infected humanized BLT mice. J Virol. 2009;83(14):7305–21.
3. Denton PW, Estes JD, Sun Z, Othieno FA, Wei BL, Wege AK, Powell DA, Payne D, Haase AT, Garcia JV. Antiretroviral pre-exposure prophylaxis prevents vaginal transmission of HIV-1 in humanized BLT mice. PLoS Med. 2008;5(1):e16.
4. Denton PW, Krisko JF, Powell DA, Mathias M, Kwak YT, Martinez-Torres F, Zou W, Payne DA, Estes JD, Garcia JV. Systemic administration of antiretrovirals prior to exposure prevents rectal and intravenous HIV-1 transmission in humanized BLT mice. PLoS One. 2010;5(1):e8829.
5. Denton PW, Olesen R, Choudhary SK, Archin NM, Wahl A, Swanson MD, Chateau M, Nochi T, Krisko JF, Spagnuolo RA, Margolis DM, Garcia JV. Generation of HIV latency in humanized BLT mice. J Virol. 2012;86(1):630–4.
6. Denton PW, Othieno F, Martinez-Torres F, Zou W, Krisko JF, Fleming E, Zein S, Powell DA, Wahl A, Kwak YT, Welch BD, Kay MS, Payne DA, Gallay P, Appella E, Estes JD, Lu M, Garcia JV. One percent tenofovir applied topically to humanized BLT mice and used according to the CAPRISA 004 experimental design demonstrates partial protection from vaginal HIV infection, validating the BLT model for evaluation of new microbicide candidates. J Virol. 2011;85(15):7582–93.
7. Sun Z, Denton PW, Estes JD, Othieno FA, Wei BL, Wege AK, Melkus MW, Padgett-Thomas A, Zupancic M, Haase AT, Garcia JV. Intrarectal transmission, systemic infection, and CD4+ T cell depletion in humanized mice infected with HIV-1. J Exp Med. 2007;204(4):705–14.
8. Wahl A, Swanson MD, Nochi T, Olesen R, Denton PW, Chateau M, Garcia JV. Human breast milk and antiretrovirals dramatically reduce oral HIV-1 transmission in BLT humanized mice. PLoS Pathog. 2012;8(6):e1002732.
9. Tugizov SM, Herrera R, Veluppillai P, Greenspan D, Soros V, Greene WC, Levy JA, Palefsky JM. Differential transmission of HIV traversing fetal oral/intestinal epithelia and adult oral epithelia. J Virol. 2012;86(5):2556–70.
10. Denton PWNT, Lim A, Krisko JF, Martinez-Torres F, Choudhary SK, Wahl A, Olesen R, Zou W, Di Santo JP, Margolis DM, Garcia JV. IL2 receptor gamma-chain molecule is critical for intestinal T cell reconstitution in humanized mice. Mucosal Immunol. 2012;5(5):555–66.

11. Geboes K, De Wolf-Peeters C, Rutgeerts P, Janssens J, Vantrappen G, Desmet V. Lymphocytes and Langerhans cells in the human oesophageal epithelium. Virchows Archiv A, Pathol Anat Histopathol. 1983;401(1):45–55.
12. Macdonald TT, Monteleone G. Immunity, inflammation, and allergy in the gut. Science. 2005;307(5717):1920–5.
13. Maynard CL, Weaver CT. Intestinal effector T cells in health and disease. Immunity. 2009;31(3):389–400.
14. Nelson AM, Auerbach A, Man YG. Failure to detect active virus replication in mast cells at various tissue sites of HIV patients by immunohistochemistry. Int J Biol Sci. 2009;5(6):603–10.
15. Rescigno M, Di Sabatino A. Dendritic cells in intestinal homeostasis and disease. J Clin Invest. 2009;119(9):2441–50.
16. Nagai K, noguchi T, Fujiwara S, Kawahara K, Shimada T.Distribution of Langerhans cells in the human esophagus, as revealed by immunohistochemistry. Acta Histochem Cytochem. 2005;38(2):115–9.
17. Balamane M, Winters MA, Dalai SC, Freeman AH, Traves MW, Israelski DM, Katzenstein DA, Klausner JD. Detection of HIV-1 in saliva: implications for case-identification, clinical monitoring and surveillance for drug resistance. Open Virol J. 2010;4:88–93.
18. Liuzzi G, Chirianni A, Clementi M, Bagnarelli P, Valenza A, Cataldo PT, Piazza M.Analysis of HIV-1 load in blood, semen and saliva: evidence for different viral compartments in a cross-sectional and longitudinal study. AIDS. 1996;10(14):F51–6.
19. Hoffman IF, Martinson FE, Stewart PW, Chilongozi DA, Leu SY, Kazembe PN, Banda T, Dzinyemba W, Joshi P, Cohen MS, Fiscus SA. Human immunodeficiency virus type 1 RNA in breast-milk components. J Infect Dis. 2003;188(8):1209–12.
20. Koulinska IN, Villamor E, Chaplin B, Msamanga G, Fawzi W, Renjifo B, Essex M. Transmission of cell-free and cell-associated HIV-1 through breast-feeding. J Acquir Immune Defic Syndr. 2006;41(1):93–9.
21. Lewis P, Nduati R, Kreiss JK, John GC, Richardson BA, Mbori-Ngacha D, Ndinya-Achola J, Overbaugh J. Cell-free human immunodeficiency virus type 1 in breast milk. J Infect Dis. 1998;177(1):34–9.
22. Nduati RW, John GC, Richardson BA, Overbaugh J, Welch M, Ndinya-Achola J, Moses S, Holmes K, Onyango F, Kreiss JK. Human immunodeficiency virus type 1-infected cells in breast milk: association with immunosuppression and vitamin A deficiency. J Infect Dis. 1995;172(6):1461–8.
23. Groot F, Geijtenbeek TB, Sanders RW, Baldwin CE, Sanchez-Hernandez M, Floris R, van Kooyk Y, de Jong EC, Berkhout B. Lactoferrin prevents dendritic cell-mediated human immunodeficiency virus type 1 transmission by blocking the DC-SIGN-gp120 interaction. J Virol. 2005;79(5):3009–15.
24. Habte HH, de Beer C, Lotz ZE, Tyler MG, Kahn D, Mall AS. Inhibition of human immunodeficiency virus type 1 activity by purified human breast milk mucin (MUC1) in an inhibition assay. Neonatology. 2008;93(3):162–70.
25. Harmsen MC, Swart PJ, de Bethune MP, Pauwels R, De Clercq E, The TH, Meijer DK. Antiviral effects of plasma and milk proteins: lactoferrin shows potent activity against both human immunodeficiency virus and human cytomegalovirus replication in vitro. J Infect Dis. 1995;172(2):380–8.
26. McNeely TB, Dealy M, Dripps DJ,, Orenstein JM, Eisenberg SP, Wahl SM. Secretory leukocyte protease inhibitor: a human saliva protein exhibiting anti-human immunodeficiency virus 1 activity in vitro. J Clin Invest. 1995;96(1):456–64.
27. Naarding MA, Dirac AM, Ludwig IS, Speijer D, Lindquist S, Vestman EL, Stax MJ, Geijtenbeek TB, Pollakis G, Hernell O, Paxton WA. Bile salt-stimulated lipase from human milk binds DC-SIGN and inhibits human immunodeficiency virus type 1 transfer to $CD4^+$ T cells. Antimicrob Agents Chemother. 2006;50(10):3367–74.
28. Saeland E, de Jong MA, Nabatov AA, Kalay H, Geijtenbeek TB, van Kooyk Y. MUC1 in human milk blocks transmission of human immunodeficiency virus from dendritic cells to T cells. Mol Immunol. 2009;46(11–12):2309–16.

29. Shugars DC. Endogenous mucosal antiviral factors of the oral cavity. J Infect Dis. 1999;179(Suppl 3):S431–5.
30. Stax MJ, Naarding MA, Tanck MW, Lindquist S, Hernell O, Lyle R, Brandtzaeg P, Eggesbo M, Pollakis G, Paxton WA. Binding of human milk to pathogen receptor DC-SIGN varies with bile salt-stimulated lipase (BSSL) gene polymorphism. PLoS One. 2011;6(2):e17316.
31. Swart PJ, Kuipers ME, Smit C, Pauwels R, deBethune MP, de Clercq E, Meijer DK, Huisman JG. Antiviral effects of milk proteins: acylation results in polyanionic compounds with potent activity against human immunodeficiency virus types 1 and 2 in vitro. AIDS Res Hum Retroviruses. 1996;12(9):769–75.
32. Nakao R, Hanada N, Asano T, Hara T, Abdus Salam M, Matin K, Shimazu Y, Nakasone T, Horibata S, Aoba T, Honda M, Amagasa T, Senpuku H. Assessment of oral transmission using cell-free human immunodeficiency virus-1 in mice reconstituted with human peripheral blood leucocyte. Immunology. 2003;109(2):271–82.
33. Kong X, West JT, Zhang H, Shea DM, M'Soka TJ, Wood C. The human immunodeficiency virus type 1 envelope confers higher rates of replicative fitness to perinatally transmitted viruses than to nontransmitted viruses. J Virol. 2008;82(23):11609–18.
34. Zhang H, Rola M, West JT, Tully DC, Kubis P, He J, Kankasa C, Wood C. Functional properties of the HIV-1 subtype C envelope glycoprotein associated with mother-to-child transmission. Virology. 2010;400(2):164–74.
35. Russell ES, Kwiek JJ, Keys J, Barton K, Mwapasa V, Montefiori DC, Meshnick SR, Swanstrom R. The genetic bottleneck in vertical transmission of subtype C HIV-1 is not driven by selection of especially neutralization-resistant virus from the maternal viral population. J Virol. 2011;85(16):8253–62.
36. Lehman DA, Chung MH, John-Stewart GC, Richardson BA, Kiarie J, Kinuthia J, Overbaugh J. HIV-1 persists in breast milk cells despite antiretroviral treatment to prevent mother-to-child transmission. AIDS. 2008;22(12):1475–85.
37. Shapiro RL, Ndung'u T, Lockman S, Smeaton LM, Thior I, Wester C, Stevens L, Sebetso G, Gaseitsiwe S, Peter T, Essex M. Highly active antiretroviral therapy started during pregnancy or postpartum suppresses HIV-1 RNA, but not DNA, in breast milk. J Infect Dis. 2005;192(5):713–9.
38. Chasela CS, Hudgens MG, Jamieson DJ, Kayira D, Hosseinipour MC, Kourtis AP, Martinson F, Tegha G, Knight RJ, Ahmed YI, Kamwendo DD, Hoffman IF, Ellington SR, Kacheche Z, Soko A, Wiener JB, Fiscus SA, Kazembe P, Mofolo IA, Chigwenembe M, Sichali DS, van der Horst CM, Group BANS. Maternal or infant antiretroviral drugs to reduce HIV-1 transmission. N Engl J Med. 2010;362(24):2271–81.
39. Kilewo C, Karlsson K, Ngarina M, Massawe A, Lyamuya E, Swai A, Lipyoga R, Mhalu F, Biberfeld G. Mitra Plus Study T. Prevention of mother-to-child transmission of HIV-1 through breastfeeding by treating mothers with triple antiretroviral therapy in Dar es Salaam, Tanzania: the Mitra Plus study. J Acquir Immune Defic Syndr. 2009;52(3):406–16.
40. Villamor E, Koulinska IN, Furtado J, Baylin A, Aboud S, Manji K, Campos H, Fawzi WW. Long-chain n-6 polyunsaturated fatty acids in breast milk decrease the risk of HIV transmission through breastfeeding. Am J Clin Nutr. 2007;86(3):682–9.
41. Hsu K. Long-chain polyunsaturated fatty acids as anti-HIV supplementation during breastfeeding. Taiwan J Obstet Gynecol. 2009;48(1):65–8.

Chapter 21
Selective Infection of $CD4^+$ Memory T Cells

Yoshio Koyanagi

Abbreviations

AIDS	Acquired immunodeficiency syndrome
ART	Antiretroviral therapy
CCR5	CC chemokine receptor 5
CTL	Cytotoxic T lymphocytes
HIV-1	Human immunodeficiency virus type 1
HLA	Human leukocyte antigen
LTR	Long terminal repeat
NOD	Nonobese diabetes
SCID	Severe combined immunodeficiency

21.1 Introduction

While it is evident that HIV-1 causes AIDS in humans, the exact mechanism of how HIV-1 targets and depletes $CD4^+$ T cells in vivo remains unclear. Valuable information about the viral target cells and dynamics of viral particles has been obtained from extensive investigations in HIV-1-infected individuals and SIV-infected rhesus macaques. However, in-depth studies including tissue distribution and the homeostasis of $CD4^+$ T cells in HIV-1-infected humanized mice will reveal novel details of HIV pathogenesis.

Y. Koyanagi (✉)
Institute for Virus Research, Kyoto University, 53 Shogoinkawara-cho, Sakyo-ku, Kyoto 606-8507, Japan

L. Y. Poluektova et al. (eds.), *Humanized Mice for HIV Research*,
DOI 10.1007/978-1-4939-1655-9_21

21.2 HIV-1 Infection in Humanized Mice

Two types of humanized mouse models have recently been frequently used to study human virus infection in vivo. To generate these mice, NOD SCID or RAG-2 knockout mice are first backcrossed with common γ-chain (*Il2rγ*) knockout mice [1, 2, 3], which results in severely immunocompromised mice that lack functional lymphocytes and NK cells, and are unable to mount an adaptive immune response. Then, the immunodeficient mice are transplanted with hematopoietic stem cells (HSC) isolated from umbilical cord blood or fetal liver to generate HSC-SCID mice, or with HSC, fetal liver and thymus to make BLT mice [4, 5]. Long-term human hematopoiesis is reconstituted in these mice, which are susceptible to infection by human viruses such as HIV-1 and Epstein-Barr virus (EBV). During virus infection, low levels of induced HIV-1 antibodies have been reported after viral inoculation in BLT mice [6]. Furthermore, HLA-restricted HIV-1 specific $CD4^+$ and $CD8^+$ T cell responses were also induced in BLT mice following HIV-1 infection [6, 7]. These examples help to demonstrate that humanized mice are useful tools for investigating features of HIV-1 infection and pathogenesis in vivo such as HIV-1 viral load (VL) and viral-induced depletion of $CD4^+$ T cells [8–11].

Humanized mice are useful especially for investigating the early stages of HIV-1 infection. Viremia first appears and increases rapidly within 1–3 weeks after the initial inoculation with virus until the VL plateaus at 10^5–10^6 RNA copies per milliliter [12, 13]. Then, a gradual loss of $CD4^+$ T cells follows for 4–6 weeks after CCR5 tropic (R5) HIV-1 infection and 3–4 weeks after CXCR4 tropic (X4) HIV-1 infection [14].

It is well known that HIV-1 pathogenesis depends on the X4 or R5 co-receptor usage. X4 HIV-1 is generally observed only during chronic infection while R5 HIV-1 is the predominant type of HIV-1 found in infected individuals during both acute and chronic phases. One study conducted in rhesus macaques showed that a CXCR4 envelope-inserted SIV, X4 SHIV, caused rapid and complete depletion of most subsets of $CD4^+$ T cells and resulted in mortality due to acute immunodeficiency [15]. On the other hand, R5 SIV infection in rhesus macaques resembled clinical manifestations of HIV-1 infection in humans [16], where the acute onset of immunodeficiency was rare [17, 18], and $CD4^+$ T cells in peripheral blood deceased slowly in number, eventually leading to immunodeficiency. Thus, R5 HIV-1 is an appropriate HIV-1 model. When comparing the rhesus macaque model with humanized mice, a decrease and the stabilization of VL following the initial high level of viremia, known as the "VL set point," was found in HIV-1 infection of the small subset of BLT mice whose human cells expressed an HLA-B57 allele [7], but not in the large group of BLT mice whose human cells expressed more common HLA alleles, and not in HSC-SCID mice. Although CTL responses against HIV-1 antigens in HLA-B57 BLT mice with viral infection appear to control the level of VL set point [7], other immune responses, such as humoral responses are not high [6, 19]. This absence of full-scale human adaptive immune responses against viral proteins may be a large hurdle for humanized mice to serve as a complete model of HIV-1 pathogenesis.

21.3 Memory CD4+ T Cells

The human body is composed of many tissues and circulating cells from which viruses select their preferred targets. The largest population of target cells for HIV is in the CD4+ T cell compartment [20]. T cells are generated in the thymus through negative and positive selection processes with thymic epithelial cells. Naïve CD4+ T (T_N) cells exit from the thymus and circulate in the blood and lymphatic system. They then receive primary immune signals from antigen peptide-MHC II (pMHC II) complexes on antigen presenting cells (APC) and proliferate in lymphoid tissues [21]. Although many of the stimulated T cells disappear after antigen removal, a minority subpopulation of cells, which express the same TCR repertoire and maintain a nondividing state by augmentation of anti-apoptotic protein BCL2 expression, survive as long-lasting memory cells [22]. Memory populations of CD4+ T cells are heterogeneous and include proliferating and resting cells. Some memory CD4+ T cells are released into the blood circulation and others reside in tissues. From the expression profile of specific surface molecules, T_N exclusively express CD45RA, whereas memory/proliferated CD4+ T cells express CD45RO [22, 23].

The mechanisms by which memory cells are maintained are not yet completely defined, but survival and self-renewing proliferation of memory T cells are likely to be controlled by cytokines such as IL-7 and IL-15. Rates of turnover seem to vary in these cells, and the depletion of CD4+ T cells results in compensatory homeostatic proliferation [22, 23]. IL-15 was shown to have a less prominent role for memory CD4+ cells than for NK and memory CD8+ cells in one humanized mouse study, suggesting that IL-15 induced human NK cell proliferation [24]. This difference in IL-15 dependence also correlates closely with the expression levels of the IL-15 receptor, CD122, which is displayed at much lower levels on memory CD4+ cells than on NK and memory CD8+ cells [25]. For this reason, memory CD4+ cells compete less effectively for IL-15 than NK and memory CD8+ cells.

In addition to pMHC II-primed CD4+ T cell populations, memory-phenotype (MP) CD4+ T cells, which arise from T_N cells without intentional immunization, display the same homeostatic requirements [23, 26]. However, because activation by immunization of CD4+ T cells is generally less effective in humanized mice, especially in HSC-SCID mice [27], many of the CD45RO+ CD4+ cells in these animals may in fact be MP CD4+ cells and not pMHC II-primed memory cells. The physiological role of MP CD4+ cells, supposedly made up of self antigen-stimulated CD4+ T cells, is unknown [23, 26].

Memory CD4+ T cells are composed of effector memory T (T_{EM}) cells, central memory T (T_{CM}) cells, and transient effector memory T (T_{TM}) cells, which have a function in between T_{CM} and T_{EM} cells [22]. T_{EM} cells express homing receptors that facilitate migration to nonlymphoid sites of inflammation and produce a variety of cytokines, including IFN-γ, IL-4, and IL-5. In contrast, T_{CM} cells express molecules that facilitate migration to lymphoid tissues (e.g., CCR7 and CD62 L) and do not produce any of the prototypic cytokines of the effector cell lineage after TCR stimulation, but they do secrete IL-2 and proliferate extensively, and potentially differentiate into T_{EM} cells [22, 28]. Since T_{CM} cells survive for years, they are likely long-lived memory cells.

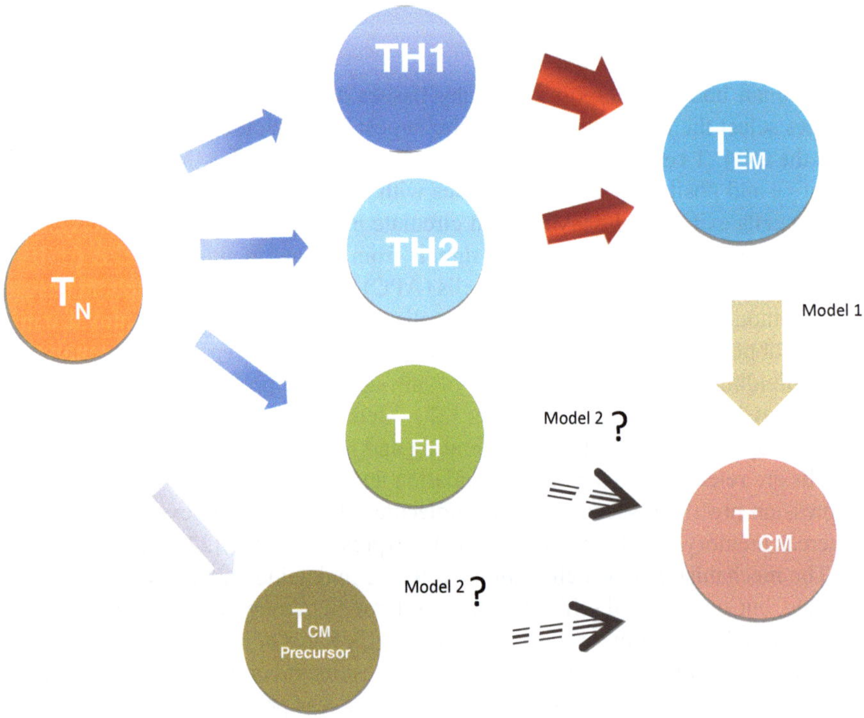

Fig. 21.1 Generation of T_{EM} and T_{CM} cells. T_{EM} and T_{CM} cells are produced from T_N cells. In the first model (pathway I), T_{EM} cells become T_{CM} cells. In the second model (pathway II), T_{CM} precursor cells are generated from T_N cells and become T_{CM} cells. In the third model (pathway III), T_{FH} effector cells become T_{CM} cells [28–30]

Although it remains unknown exactly how T_{EM} and T_{CM} cells originate, pMHC II-primed effector Th1 (IFN-γ-producing) and Th2 (IL-4-producing) $CD4^+$ T cells are known to turn into T_{EM} cells, which are generated from T_{CM} cells (Fig. 21.1, pathway I). This linear differentiation pathway of T_{EM} and T_{CM} cells was demonstrated by experiments performed using pMHC II-primed $CD4^+$ T cells from a TCR-transgenic mouse in naive recipient mice [29]. However, in a study of mice infected with *Listeria monocytogenes*, endogenous T_{CM} cells, and T_{CM} precursor cells were detected immediately after pMHC II-priming (contraction phase) by pMHC II-tetramer staining analysis, and the phenotype and function of T_{CM} cells were maintained for almost a year [30]. These data suggest the existence of a nonlinear differentiation pathway of T_{CM} cell generation. As shown in Fig. 21.1 pathway II, T_{CM} cells appear to be independently generated from T_{EM} cells. However, the question of how HIV infects T_{CM} cells arises because these cells are commonly negative for CCR5 [22]. In germinal centers of lymph nodes, follicular helper T (T_{FH}) cells, which are $CXCR4^{high}$ $CCR7^{low}$ PD-1^+ $ICOS^+$ $CD4^+$ T cells, are generated as one of the effector cells of $CD4^+$ T cells from T_N cells by stimulating pMHC II on B cells. It was proposed that some T_{CM} cells are differentiated from T_{FH} cells after losing the expression of PD-1 in the mouse system (Fig. 21.1, pathway III) [28].

21.4 HIV Targets Memory CD4+ T Cells

T_{CM} cells appear to be a major long-lasting HIV-1 reservoir where replication-competent HIV-1 resides in a latent or nonreplicating state in infected individuals under ART [31]. The level of integrated proviral DNA in T_{CM} cells is significantly higher than in other cells, including T_{EM}, T_N, and T_{TM} cells [31]. In addition, T_{TM} cells have also been reported as a secondary HIV-1 reservoir that harbors distinct proviral DNA in infected individuals with low $CD4^+$ counts, the majority of whom show persistent immune activation [31]. On the other hand, selective HIV-1 antigen expression, detected by a HIV-1 p24 specific antibody, was found in T_{EM} cells of humanized mice [14]. This discrepancy could be due to the phase of HIV-1 infection and detection method used, because samples from chronically infected HIV-1 individuals were analyzed by quantitative PCR, while samples from acutely infected humanized mice were analyzed by flow cytometry. A profound depletion of T_{EM} and T_{TM} cells by SIV has been shown in extra-lymphoid effector sites of acutely infected rhesus macaques [32]. Within several weeks following infection, the population of extra-lymphoid $CCR5^+$ T_{EM} cells, especially in the gastrointestinal (GI) tract, was depleted up to 90 %, while T_{CM} cells were relatively resistant to this initial destruction [33]. In the chronic phase of infection in the SIV model, T_{EM} and T_{TM} cell regeneration levels progressively declined, leading to the development of immunodeficiency in infected rhesus macaques. Interestingly, rhesus macaques that were T_N-depleted by thymectomy showed a similar level of viral replication with SIV infection and memory $CD4^+$ T cell decline compared to those with sham surgery, even though there were impairments in the development of $CD8^+$ T cells and viral-specific antibody responses in the former group [34]. Thus, homeostatic failure to repopulate the T_{CM} cell population, not the T_N cell population, is closely associated with disease progression in the chronic phase of the SIV model. In addition, T_{FH} cells appear to be a major HIV target in lymphoid tissues of chronically infected individuals [35]. These findings suggest that HIV preferentially infects T_{FH} cells and then hides its proviral DNA in the host genome of T_{CM} cells derived from T_{FH} cells. Distribution of $CD4^+$ memory T cells and preferential HIV-1 infection were found in lymphoid and extra-lymphoid tissues of BLT mice [13, 36]. How does HIV-1 target reside in subpopulations of memory $CD4^+$ T cells? Lineage tracing of subpopulations of memory $CD4^+$ T cells will be important for revealing the cell targets that HIV prefers and their dynamics in memory cell populations.

21.5 Homeostasis of CD4+ T Cells in Acute and Chronic Phase of HIV-1 Infection

Depletion of memory $CD4^+$ T cells in lymphoid tissues and peripheral blood is a key pathogenic property of HIV-1. There are still many unanswered questions about how HIV depletes $CD4^+$ T cells by direct or other indirect mechanisms. The balance of destruction and generation is crucial for maintenance of a stable number of T cells in the peripheral blood. The total absolute number of memory CD4$^+$ T cells diminishes from continuous destruction of mature CD4$^+$ T cells in lymphoid organs and peripheral blood with HIV infection during the course of infection. Although many mechanisms have been proposed to explain HIV-mediated depletion of CD4$^+$ T cells using data from in vitro culture experiments and observations of clinical samples from HIV-1-infected individuals, it is important to take into account the stage of infection and main distribution of CD4$^+$ T cells in peripheral blood and tissues.

In the process of crossing the epithelium of the anogenital tract, where Langerhans cells (LC) are localized, HIV targets LC, CD4-, and CCR5-positive dendritic cells, either directly or by capturing viral particles that productively infect CD4$^+$ T cells in local lymphoid tissues [37]. Productive infection results in massive HIV replication and a high level of viremia. In this initial phase, HIV-1 is likely to severely deplete CD4$^+$ T cells in the absence of viral-specific immune responses. Using BLT mice, systemic HIV-1 infection and subsequent CD4$^+$ T cell decline was reproduced by intrarectal transmission [13]. However, the rate of HIV infection in HIV-specific CD4$^+$ T cells in infected humanized mice is unclear. Following this huge systemic assault, untreated individuals enter a chronic phase in which remaining CD4$^+$ T cell compartments are slowly eroded over a period of several years.

Throughout acute and chronic phases of infection, continuous and widespread destruction of memory CD4$^+$ T cells take place in lymphoid tissues and peripheral blood. In addition, the fraction of circulating cells may decrease due to an accumulation of CD4$^+$ T cells in lymphoid tissues since high levels of viral-specific immune responses are induced and many CD4$^+$ T cells become trapped. Apoptosis-related destruction of CD4$^+$ T cells and direct destruction by HIV also occurs [38, 39, 40]. On the other hand, newly generated cells should replenish the loss of mature T cells. Thus, CD4$^+$ T cell depletion in peripheral blood becomes an issue only if cells lost in the periphery cannot be replaced. The most devastating feature of HIV-1 infection is that the virus can have direct pathogenic effects on both mature CD4$^+$ T and on progenitor cells from which they arise. In late phases of infection, the number of CD4$^+$ T cells is significantly decreased from impaired CD4$^+$ T cell regeneration.

A mathematical model of CD4 T depletion in the course of HIV-1 infection developed by Ho et al. suggests that the rate of CD4$^+$ T cell destruction with HIV-1 during chronic infection is too fast to be counteracted by normal CD4$^+$ T cell generation [41]. The majority of apoptotic CD4$^+$ T cells are bystander cells rather than productively infected cells [42, 43]. During the chronic phase of infection, CD4$^+$ T cells in individuals show characteristics of massively and chronically activated phenotypes. This activation is reduced to near-normal levels by treatment with ART [44]. Interestingly, the level of CD4$^+$ T cell activation, mainly in T_{EM} cells, correlates with the viral load better than the degree of CD4$^+$ T cell reduction in both HIV-1-and HIV-2-infected individuals [45]. Studies of HIV-infected individuals and SIV-infected rhesus macaques strongly suggest that massive and chronic CD4$^+$ T cell activation induced by viral infection in the GI tracts plays a role in CD4$^+$ T cell depletion [46]. However, several important questions remain unanswered. In particular, how does T cell activation lead to $CD4^+$ T cell depletion? What signals are acting on $CD4^+$ T cells at the set point of HIV-1 infection (i) to induce such a substantial increase in activated $CD4^+$ T cells, and (ii) to provoke $CD4^+$ T cell depletion? Answering these questions will be crucial to understand $CD4^+$ T cell homeostasis during HIV-1 infection, especially when looking at the memory cell compartments.

Continuous HIV replication occurs in both HSC-SCID mice and BLT mice with HIV-1 infection. Thus, factors affecting the memory cell compartment, such as the duration of HIV-1 infection of $CD4^+$ T cells and other cells, may be revealed from investigating persistent HIV-1 infection in humanized mice. It has been proposed that the virus has evolved strategies to increase the availability of suitable target cells by activating $CD4^+$ T cells. Increasing the availability of target cells would be like "fueling the fire," resulting in further infection and a runaway depletion of $CD4^+$ T cells. However, one group showed from mathematical analyses that the runaway theory of $CD4^+$ T cell depletion may not be explained by activating $CD4^+$ T cells [47]. It is tempting to investigate the dynamics of HIV-1 and infected cells in humanized mice by applying mathematical tools.

21.6 Latent Infection of HIV-1

It has been thought that activated $CD4^+$ T cells become infected and then survive long enough to revert back to a resting state. Because HIV-1 LTR-driven expression is dependent on inducible host transcription factors that are only transiently activated after exposure to antigens, HIV-1 expression may cease as the cells revert to a resting memory state [48]. The result is a stably integrated but transcriptionally silent form of the virus in a cell whose function is to survive for long periods of time. In addition, epigenetic mechanisms, as discussed in Chap. 18, and latent HIV-1 infection in memory $CD4^+$ T cells greatly complicate curing AIDS. Although VL can be suppressed by ART, there is no decline of latently infected HIV-1 positive cells in resting $CD4^+$ memory cells.

21.7 Concluding Remarks

The fact that HIV-1 strategically targets $CD4^+$ memory T cells appears to be a survival tactic for its evolution. These cells exist in heterogeneous populations, either in a proliferating or resting state, and are distributed in multiple tissues in addition to the peripheral blood. Moreover, some memory cells have self-renewal potential. HIV-1 exploits these characteristics during the infection of memory $CD4^+$ T cells to hide and replicate, making eradication of viruses from the body highly challenging, even under intensive anti-HIV treatment. Given the critical role of memory $CD4^+$ T cells during HIV-1 infection, humanized mice will be a powerful tool to investigate these cells in vivo and help in the struggle to cure AIDS in the future.

Acknowledgments I thank Dr. Peter Gee (Kyoto University) for his critical review of the manuscript. This work was partially funded by a Grant-in-Aid for Scientific Research on Innovative Areas (24115008) and Grants-in-Aid for Scientific Research (B24390112) from the Ministry of Education, Culture, Sports, Science and Technology and Research on HIV/AIDS from the Ministry of Health, Labor and Welfare of Japan.

References

1. Ito M, Hiramatsu H, Kobayashi K, Suzue K, Kawahata M, Hioki K, et al. NOD/SCID/gamma(c) (null) mouse: an excellent recipient mouse model for engraftment of human cells. Blood. 2002;100(9):3175–82.
2. Traggiai E, Chicha L, Mazzucchelli L, Bronz L, Piffaretti JC, Lanzavecchia A, et al. Development of a human adaptive immune system in cord blood cell-transplanted mice. Science. 2004;304(5667):104–7.
3. Shultz LD, Schweitzer PA, Christianson SW, Gott B, Schweitzer IB, Tennent B, et al. Human lymphoid and myeloid cell development in NOD/LtSz-scid IL2R γ null mice engrafted with mobilized human hemopoietic stem cells. J Immunol. 2005;174(10):6477–89.
4. Melkus MW, Estes JD, Padgett-Thomas A, Gatlin J, Denton PW, Othieno FA, et al. Humanized mice mount specific adaptive and innate immune responses to EBV and TSST-1. Nat Med. 2006;12(11):1316–22.
5. Lan P, Tonomura N, Shimizu A, Wang S, Yang YG. Reconstitution of a functional human immune system in immunodeficient mice through combined human fetal thymus/liver and $CD34^+$ cell transplantation. Blood. 2006;108(2):487–92.
6. Brainard DM, Seung E, Frahm N, Cariappa A, Bailey CC, Hart WK et al. Induction of robust cellular and humoral virus-specific adaptive immune responses in human immunodeficiency virus-infected humanized BLT mice. J Virol. 2009;83(14):7305–21.
7. Dudek TE, No DC, Seung E, Vrbanac VD, Fadda L, Bhoumik P, et al. Rapid evolution of HIV-1 to functional CD8â° T cell responses in humanized BLT mice. Sci Transl Med. 2012;4(143):143ra98.
8. Denton PW, Garcia JV. Novel humanized murine models for HIV research. Curr. HIV/AIDS Rep. 2009;6(1):13–9.
9. Sato K, Koyanagi Y. The mouse is out of the bag: insights and perspectives on HIV-1 infected humanized mouse models. Exp Biol Med. 2011;236(8):977–85.
10. Shultz LD, Brehm MA, Garcia-Martinez JV, Greiner DL. Humanized mice for immune system investigation: progress, promise and challenges. Nat Rev Immunol. 2012;12(11):786–98.
11. Akkina R. New generation humanized mice for virus research: comparative aspects and future prospects. Virology. 2013;435(1):14–28.
12. Sato K, Misawa N, Fukuhara M, Iwami S, An DS, Ito M et al. VPU augments the initial burst phase of HIV-1 propagation and downregulates BST2 and CD4 in humanized mice. J Virol. 2012;86(9):5000–13.
13. Sun Z, Denton PW, Estes JD, Othieno FA, Wei BL, Wege AK et al. Intrarectal transmission, systemic infection, and $CD4^+$ T cell depletion in humanized mice infected with HIV-1. J Exp Med. 2007;204(4):705–14.
14. Nie C, Sato K, Misawa N, Kitayama H, Fujino H, Hiramatsu H, et al. Selective infection of $CD4^+$ effector memory T lymphocytes leads to preferential depletion of memory T lymphocytes in R5 HIV-1 infected humanized NOD/SCID/IL-2Rgammanull mice. Virology. 2009;394(1):64–72.
15. Nishimura Y, Igarashi T, Donau OK, Buckler-White A, Buckler C, Lafont BA et al. Highly pathogenic SHIVs and SIVs target different $CD4^+$ T cell subsets in rhesus monkeys, explaining their divergent clinical courses. Proc Natl Acad Sci U S A. 2004;101(33):12324–9.

16. Berger EA, Murphy PM, Farber JM. Chemokine receptors as HIV-1 coreceptors: roles in viral entry, tropism, and disease. Annu Rev Immunol. 1999;17:657–700.
17. Ambrose Z, KewalRamani VN, Bieniasz PD, Hatziioannou T. HIV/AIDS: in search of an animal model. Trends Biotechnol. 2007;25(8):333–7.
18. McCune JM. The dynamics of CD4+ T-cell depletion in HIV disease. Nature. 2001;410(6831):974–9.
19. Sato K, Nie C, Misawa N, Tanaka Y, Ito M, Koyanagi Y. Dynamics of memory and naïve CD8+ T lymphocytes in humanized NOD/SCID/IL-2Rgammanull mice infected with CCR5-tropic HIV-1. Vaccine. 2010; 28(Suppl 2):B32–7.
20. Ho DD. Perspectives series: host/pathogen interactions. Dynamics of HIV-1 replication *in vivo*. J Clin Invest. 1997;99(11):2565–720. (Davis, MM. T cell receptor gene diversity and selection. Annu Rev Biochem. 1990;59:475–96)
21. Jenkins, MK et al. *In vivo* activation of antigen-specific CD4 T cells. Annu Rev Immunol. 2001;19:23–45.
22. Sallusto F, Geginat J, Lanzavecchia A. Central memory and effector memory T cell subsets: function, generation, and maintenance. Annu Rev Immunol. 2004;22:745–63.
23. van Leeuwen EM, Sprent J, Surh CD. Generation and maintenance of memory CD4(+) T cells. Curr Opin Immunol. 2009;21(2):167–72.
24. Huntington ND, Legrand N, Alves NL, Jaron B, Weijer K, Plet A, et al. IL-15 trans-presentation promotes human NK cell development and differentiation *in vivo*. J Exp Med. 2009;206(1):25–34.
25. Purton JF, Tan JT, Rubinstein MP, Kim DM, Sprent J, Surh CD. Antiviral CD4+ memory T cells are IL-15 dependent. J Exp Med 2007;204(4):951–61.
26. Younes SA, Punkosdy G, Caucheteux S, Chen T, Grossman Z, Paul WE. Memory phenotype CD4 T cells undergoing rapid, nonburst-like, cytokine-driven proliferation can be distinguished from antigen-experienced memory cells. PLoS Biol. 2011;9(10):e1001171.
27. Watanabe Y, Takahashi T, Okajima A, Shiokawa M, Ishii N, Katano I et al. The analysis of the functions of human B and T cells in humanized NOD/shi-scid/gammac(null) (NOG) mice (hu-HSC NOG mice). Int Immunol. 2009;21(7):843–58.
28. Pepper M, Jenkins MK. Origins of CD4 (+) effector and central memory T cells. Nat Immunol. 2011;12(6):467–71.
29. Löhning M, Hegazy AN, Pinschewer DD, Busse D, Lang KS, Höfer T, et al. Long-lived virus-reactive memory T cells generated from purified cytokine-secreting T helper type 1 and type 2 effectors. J Exp Med. 2008;205(1):53–61.
30. Pepper M, Linehan JL, Pagán AJ, Zell T, Dileepan T, Cleary PP, et al. Different routes of bacterial infection induce long-lived TH1 memory cells and short-lived TH17 cells. Nat Immunol. 2010;11(1):83–9.
31. Chomont N, El-Far M, Ancuta P, Trautmann L, Procopio FA, Yassine-Diab B et al. HIV reservoir size and persistence are driven by T cell survival and homeostatic proliferation. Nat Med. 2009;15(8):893–900.
32. Mattapallil JJ, Douek DC, Hill B, Nishimura Y, Martin M, Roederer M. Massive infection and loss of memory CD4+ T cells in multiple tissues during acute SIV infection. Nature. 2005;434(7037):1093–7.
33. Okoye A, Meier-Schellersheim M, Brenchley JM, Hagen SI, Walker JM, Rohankhedkar M, et al. Progressive CD4+ central memory T cell decline results in CD4+ effector memory insufficiency and overt disease in chronic SIV infection. J Exp Med. 2007;204(9):2171–85.
34. Okoye AA, Rohankhedkar M, Abana C, Pattenn A, Reyes M, Pexton C, et al. Naive T cells are dispensable for memory CD4+ T cell homeostasis in progressive simian immunodeficiency virus infection. J Exp Med. 2012;209(4):641–51.
35. Perreau M, Savoye AL, De Crignis E, Corpataux JM, Cubas R, Haddad EK, et al. Follicular helper T cells serve as the major CD4 T cell compartment for HIV-1 infection, replication, and production. J Exp Med. 2013;210(1):143–56.

36. Wahl A, Swanson MD, Nochi T, Olesen R, Denton PW, et al. Human breast milk and antiretrovirals dramatically reduce oral HIV-1 transmission in BLT humanized mice. PLoS Pathog. 2012;8(6):e1002732.
37. Kawamura T, Kurtz SE, Blauvelt A, Shimada S. The role of Langerhans cells in the sexual transmission of HIV. J Dermatol Sci. 2005;40(3):147–55.
38. Miura Y, Misawa N, Maeda N, Inagaki Y, Tanaka Y, Ito M, et al. Critical contribution of tumor necrosis factor-related apoptosis-inducing ligand (TRAIL) to apoptosis of human $CD4^+$ T cells in HIV-1 infected hu-PBL-NOD-SCID mice. J Exp Med. 2001;193(5):651–60.
39. Holm GH, Gabuzda D. Distinct mechanisms of $CD4^+$ and $CD8^+$ T-cell activation and bystander apoptosis induced by human immunodeficiency virus type 1 virions. J Virol. 2005;79(10):6299–311.
40. Douek DC, Brenchley JM, Betts MR, Ambrozak DR, Hill BJ, Okamoto Y, et al. HIV preferentially infects HIV-specific $CD4^+$ T cells. Nature. 2002;417(6884):95–8.
41. Perelson AS, Essunger P, Cao Y, Vesanen M, Hurley A, Saksela K, et al. Decay characteristics of HIV-1-infected compartments during combination therapy. Nature. 1997;387(6629):188–91.
42. Muro-Cacho CA, Pantaleo G, Fauci AS. Analysis of apoptosis in lymph nodes of HIV-infected persons. Intensity of apoptosis correlates with the general state of activation of the lymphoid tissue and not with stage of disease or viral burden. J Immunol. 1995;154(10):5555–66.
43. Finkel TH, Tudor-Williams G, Banda NK, Cotton MF, Curiel T, Monks C, et al. Apoptosis occurs predominantly in bystander cells and not in productively infected cells of HIV- and SIV-infected lymph nodes. Nat Med. 1995;1(2):129–34.
44. Sloand EM, Kumar PN, Kim S, Chaudhuri A, Weichold FF, Young NS. Human immunodeficiency virus type 1 protease inhibitor modulates activation of peripheral blood CD4 (+) T cells and decreases their susceptibility to apoptosis *in vitro* and *in vivo*. Blood. 1999;94(3):1021–7.
45. Leng Q, Borkow G, Weisman Z, Stein M, Kalinkovich A, Bentwich Z. Immune activation correlates better than HIV plasma viral load with CD4 T-cell decline during HIV infection. J Acquir Immune Defic Syndr. 2001;27(4):389–97.
46. Brenchley JM, Price DA, Schacker TW, Asher TE, Silvestri G, Rao S, et al. Microbial translocation is a cause of systemic immune activation in chronic HIV infection. Nat Med. 2006;12(12):1365–71.
47. Yates A, Stark J, Klein N, Antia R, Callard R. Understanding the slow depletion of memory $CD4^+$ T cells in HIV infection. PLoS Med. 2007;4(5):e177.
48. Eisele E, Siliciano RF. Redefining the viral reservoirs that prevent HIV-1 eradication. Immunity. 2012;37(3):377–88.

Chapter 22
Development and Function of Human $CD4^{+}CD25^{+}FOXP3^{+}$ Regulatory T Cells in Humanized Mouse and HIV-1 Infection

Jun-ichi Nunoya and Lishan Su

Abbreviations

AIDS	Acquired immunodeficiency syndrome
cAMP	Cyclic adenosine monophosphate
FOXP3	Forkhead box P3
HIV-1	Human immunodeficiency virus type 1
HSC	Hematopoietic stem cell
ONTAK	Denileukin difftitox
Tc	Conventional T cell
TCR	T cell receptor
TGF-β	Transforming growth factor-beta
Treg cell	Regulatory T cell
tTreg cell	Thymus-derived Treg cell
pTreg cell	Peripherally-derived Treg cell

J.-i. Nunoya (✉) · L. Su
Lineberger Comprehensive Cancer Center, The University of North Carolina at Chapel Hill, Chapel Hill, NC 27599-7295, USA
e-mail: nunoya@dokkyomed.ac.jp

Department of Microbiology, Dokkyo Medical University, Tochigi, Japan

J.-i. Nunoya · L. Su
Department of Microbiology and Immunology, The University of North Carolina at Chapel Hill, Chapel Hill, NC 27599–7295, USA

L. Y. Poluektova et al. (eds.), *Humanized Mice for HIV Research*,
DOI 10.1007/978-1-4939-1655-9_22

22.1 CD4$^+$CD25$^+$FoxP3$^+$ Regulatory T Cells

Regulatory T (Treg) cells are widely recognized as a suppressor subset in the immune system. Most of the mouse and human Treg cells constitutively express the CD25 molecule (Interleukin (IL)-2 receptor alpha chain) [1]. They also specifically express a transcription factor FOXP3 as a master regulator in their development and function [2]. Stable and high FOXP3 expression is required for suppressive function in human CD4$^+$ T cells, and loss of FOXP3 expression (e.g., long-term culture) decreases their suppression activity [3, 4]. Recent papers also have shown that FOXP3 expression and suppressive activities are enriched in CD4$^+$ CD127$^-$ T cells [5, 6]. Thus, CD25 and CD127 are widely used for identification and purification of Treg cell populations.

Treg cells are continuously produced from the thymus as a functionally distinct and mature T cell subpopulation, and represent about 5 % of CD4$^+$ T lymphocytes in peripheral blood (PB). Thymic development of Treg cells requires high affinity interactions between their T cell receptor (TCR) and self peptide-major histocompatibility complexes presented on thymic stromal cells (Fig. 22.1). These stromal cells also provide costimulatory signals through CD40 and/or CD28 which are necessary for Treg cell development. In addition, IL-2 is required for the in vivo and in vitro activation of Treg cells and sustaining their CD25 expression [7, 8]. Thus, IL-2 is a vital cytokine for the thymic generation and peripheral maintenance of thymus-derived Treg (tTreg) cells. Treg cells can also be induced from conventional T (Tc) cells. Such peripherally-derived Treg (pTreg) cells seem to be functionally different in

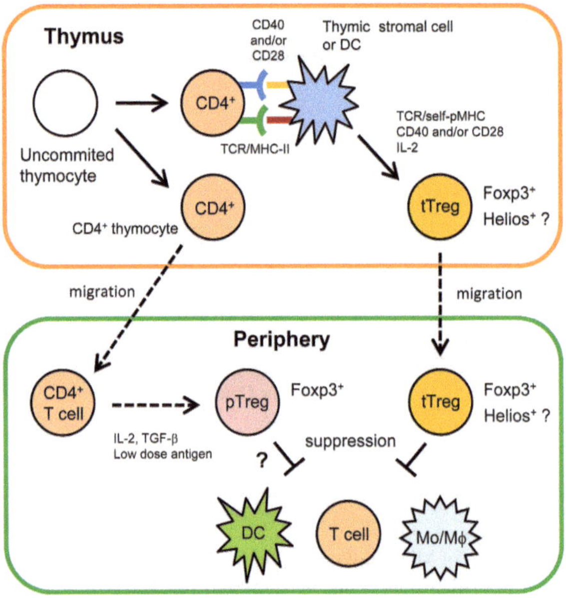

Fig. 22.1 Development of human Treg cells. This figure describes the development of human Treg cells in physiological conditions. Treg cells naturally develop in the thymus from CD4$^+$ thymocytes, termed tTreg cells. tTreg cells then migrate into periphery. Alternatively, Treg cells are induced from CD4$^+$ T cells, termed pTreg cells, in periphery by low-dose antigen, IL-2 and TGF-β1 stimulation. Both tTreg and pTreg cells may contribute to peripheral immune suppression. However, detailed mechanisms of their actions are still being investigated. *DC* dendritic cell, *Mo* monocyte, *MΦ* macrophage

mouse and human. Mouse Tc cells readily convert to FOXP3+ Treg cells following in vitro stimulation with transforming growth factor (TGF)-β and retinoic acid [9]. In contrast to mouse pTreg cells, human pTreg cells which are induced from naïve T cells by in vitro TCR stimulation and TGF-β show little in vitro suppressive activity but they can secrete proinflammatory cytokines [10, 11]. It has been reported that Helios, an Ikaros transcription factor family member, can differentiate tTreg cells from pTreg cells [12, 13]. However, a recent report showed that tTreg cells consist of Helios+ and Helios- subpopulations [14]. This suggests that it is still controversial to use Helios expression to identify tTreg cells. Other markers may need to be investigated to discriminate tTreg from pTreg cells.

22.2 Treg Cells in HIV-1 Infection

Treg cells have been reported to be susceptible to Human Immunodeficiency Virus type 1 (HIV-1) infection and replication due to high expression of CD4 and the HIV-1 coreceptor CCR5 [15]. Accumulating reports have tried to clarify the homeostasis and function of Treg cells in HIV/Simian immunodeficiency virus (SIV) infection [16]. However, some conflicting results about Treg levels upon infection have been reported, especially in the acute phase of infection, due to differences in the pathogenesis among HIV-1 strains, Treg cell definition and methods used to measure Treg levels. We and others have shown that Treg cells were severely depleted in the gut-associated lymphoid tissue [17] and PB [18] during acute phase of infection in SIV/Macaque and humanized mouse models [19] (Table 22.1). These data suggest that Treg cells may be primary target cells in HIV-1/SIV infection and may regulate anti-HIV immune responses in the acute phase. This issue will be further discussed later in this chapter. On the other hand, based on recent findings, Treg cell frequency is likely to be increased in PB [23–27] and lymphoid tissue (LT) [30–33] during chronic HIV-1 infection (Table 22.1). However, the absolute number of circulating Treg cells was reduced [22, 24, 26, 34] because HIV-1 infection progressively depletes the whole CD4+ T cell population (Table 22.1). These data suggest that Treg cells are relatively resistant to cell death or their proliferation is induced by HIV-1 infection in the chronic phase [23, 41].

Several mechanisms have been postulated for Treg-mediated regulation of HIV-1 replication and immune responses. Elahi et al. [42] showed that HIV-specific CD8+ T cells restricted by protective human leukocyte antigen (HLA) alleles are not suppressed by Treg cells, whereas T cells restricted by nonprotective HLA alleles are highly suppressed ex vivo within the same individual. The abrogation of Treg-mediated suppression of HIV-specific CD8+ T cells restricted by protective alleles allows more robust antiviral responses, thus maintaining increased control of HIV-1 infection. Also, the CD39/adenosine pathway has recently shown to be involved in acquired immune deficiency syndrome (AIDS) progression. HIV-1 positive patients have a significant increase of Treg-associated expression of CD39 [23, 43]. Interestingly, the expansion of CD39+ Tregs correlates with the level of immune activation

Table 22.1 Change of Treg cell levels in HIV-1 or SIV infection

Status	Quantitation	Tissues	Levels (References)
Acute HIV/no ART	Frequency	PB	↓[19, 20]
		LT	↓[17, 19]
Chronic HIV/no ART	Frequency	PB	→[21, 22], ↑ [23–27], ↓ [28, 29]
		LT	↑[30–33]
	Number	PB	↓[22, 24, 26, 34]
		LT	↓ [22]
Chronic HIV/ART	Frequency	PB	↑ [25, 32, 35], ↓ [23]
		LT	↑[36], ↓ [32]
	Number	PB	↑[24, 32]
		LT	↓[32]
Acute SIV/Mac	Frequency	PB	↑[37–39], ↓ [18]
		LT	↑ [39, 40], ↓ [17]
	Number	PB	↓ [39]
Chronic SIV/Mac	Frequency	PB	↓ [37]
		LT	↓[17, 18, 37]
Acute SIV/AGM or SM	Frequency	PB	↓[37], ↑ [39]
	Number	PB	↑[39]
Chronic SIV/AGM or SM	Frequency	PB	→[37]

↑ increased, ↓ decreased, → unchanged, *PB* peripheral blood, *LT* lymphoid tissue, *Mac* macaque, *AGM* African green monkey, *SM* sooty mangabey

and lower $CD4^+$ T cell counts in HIV-1-infected persons. In addition, a CD39 gene polymorphism is associated with downregulation of CD39 expression and a slower AIDS progression [43]. Moreno-Fernandez et al [44] described a possible mechanism of cyclic adenosine mono phosphate (cAMP)-mediated control of HIV-1 viral replication by Treg cells. Using a Tc- and Treg-cell coculture system, they showed that Treg cells might suppress HIV-1 replication in Tc cells by transferring cAMP from Treg to Tc cells through gap junctions. Treg cells also have been reported to inhibit HIV-1 release from infected bone marrow-derived macrophages and to induce cell death by apoptosis and pyroptosis. In addition, Treg cells transform virus-infected macrophages from an M1 to an M2 phenotype [45]. These data suggest that Treg-mediated suppression of HIV-1 replication in Tc cells and macrophages may play a role in Treg-mediated control of HIV-1 replication and immunopathogenesis in either acute or chronic phases of HIV-1 infection. Therefore, Treg cells play important roles in regulating viral replication and immune responses at different steps and stages of HIV-1 infection. Further studies will be needed to determine precise mechanisms, and to develop Treg cell-based immunotherapy for treatment of HIV-1 infection.

22.3 Development and Functional Properties of Treg Cells in Humanized Mouse Models

Humanized mice in the most frequently used current models are created by transplanting human hematopoietic stem cells (hu HSC mice) or HSC/Liver/Thymus (BLT mice). Efficient multilineage hemetopoiesis in these humanized mice has been commonly generated in mouse strains such as *rag2*$^{-/-}$ *Il2rg*^{null} (DKO) or NOD/LtSZ-SCID *Il2rg*^{null} (NSG) mice. These humanized mice mount functional humoral and cellular immune responses and are useful for studying human immune cell development.

Jiang et al. [19] first reported that functional Tregs cells developed in the DKO-hu HSC mouse model. Treg cells were detected in the PB and LT at similar frequencies to human or mouse LTs, as either $CD4^+CD25^+$ or $CD4^+FOXP3^+$ cells by multicolor flow cytometry. Similar to human PB mononuclear cells, approximately 60 % of human $CD3^+CD4^+CD25^+$ cells expressed FOXP3 and 40–50 % of $CD3^+CD4^+FOXP3^+$ cells expressed CD25. In a suppression assay, $CD4^+CD25^{+/hi}$ Treg cells were able to suppress proliferation of $CD4^+CD25^-$ responder T cells upon activation with anti-CD3/28 beads or dendritic cells plus staphylococcal enterotoxin B superantigen. Therefore, Treg cells with normal phenotypes develop in all lymphoid organs of the DKO-hu HSC mouse, and these Treg cells are functional with suppressive activity.

Onoe et al. [46] also performed a detailed analysis of Treg cells in the NSG-BLT mouse model. They analyzed Treg cell populations in PB and LTs with CD25, CD127, FOXP3 and Helios expression. $FOXP3^+$ $Helios^+$ tTreg cells develop normally in the human fetal thymic grafts and are present in PB, spleen (Sp), and lymph nodes of NSG-BLT mice. They also suggested that CD45 isoform expression was normally reversed in association with thymic egress, postthymic phenotypic conversion and suppressive function in the NSG-BLT mice.

Billerbeck et al. [47] created NSG mice expressing human stem cell factor, granulocyte-macrophage colony-stimulating factor, and IL-3, called NSG-SGM3 mice. NSG-SGM3-hu HSC mice are expected to improve certain hematopoietic lineages such as myeloid lineages. Although PB $CD19^+$ human B cell frequency is significantly decreased in NSG-SGM3-hu HSC mice, $CD45^+$ human leukocyte, $CD3^+$ human T cell and $CD33^+$ human myeloid cell frequencies in PB are significantly increased in NSG-SGM3-hu HSC mice compared to NSG-hu HSC mice. Consistent with decreased $CD8^+$ T cells, increased $CD4^+$ T cells were observed in multiple LTs of NSG-SGM3-hu HSC mice. In addition, the $CD4^+FOXP3^+$ Treg cell population was significantly increased in PB and LTs of NSG-SGM3-hu HSC mice. These Treg cells exhibit similar phenotype and suppressive activity as human PB- or tissue-derived Treg cells. In summary, Treg cells develop normally and distribute in all lymphoid organs in currently used humanized mouse models. Their phenotype and suppressive activity are similar to human PB- or tissue-derived Treg cells. Therefore, current humanized mouse models are useful for studying Treg development and function in vivo in physiological condition as well as microbial or viral infections.

22.4 Role of Treg Cells in HIV-1 Infection in a Humanized Mouse Model

As we described in this chapter, humanized mice provide a robust model for studying human Treg cell development and function, as well as immunopathogenesis of HIV-1 infection. We have developed an acute HIV-1 infection model [48] using DKO-hu HSC mice infected with the highly pathogenic HIV-1 R3A strain [49]. In the HIV-R3A acute infection model, HIV-1 viral load reaches its peak around 1 week post infection and then significantly decreases with substantial $CD4^+$ T cell depletion. High levels of productive infection occur in the LTs. Using this model, we utilized multicolor flow cytometry to analyze Treg cells from HIV-R3A infected DKO-hu HSC mice. We found that $FOXP3^+$ human Treg cells were preferentially infected and depleted by apoptosis. We measured levels of active caspase-3^+ cells to determine whether HIV infection induced apoptosis in $FOXP3^+$ Treg cells from HIV-infected DKO-hu HSC mice. We found that a high percentage of $CD3^+CD8^-FOXP3^+$ Treg cells was caspase-3^+ compared with $CD3^+CD8^-FOXP3^-$ T cells isolated from LTs. In addition, significant levels of caspase-3 expression were induced in both $p24^+$ and $p24^-$ populations; however, a higher percentage of the $p24^+FOXP3^+$ Treg cells expressed active caspase-3. We also performed Treg cell depletion from humanized mice to determine the role of Treg cells in HIV-1 infection by injecting ONTAK, which is a fusion protein composed of an IL-2 fragment and a toxin. When Treg cells were depleted before HIV-1 infection by injecting ONTAK, HIV-1 infection was significantly decreased as observed by intracellular p24 staining of T cells and plasma HIV-1 viral load. These results indicate that Treg cells are primary target cells for HIV-1 infection and are preferentially depleted by apoptosis induced by HIV-1 infection [19]. In addition, Treg cells may contribute to the suppression of anti-HIV immunity and to the efficiency of HIV-1 acute infection. Interestingly, HIV-1 infection of Treg-depleted humanized mice leads to severe tissue injury and liver fibrosis [50]. Therefore, Tregs modulate anti-HIV immunity to protect against tissue injury from overreactive anti-HIV immune responses. HIV-1, on the other hand, exploits this host self-protective mechanism to its own benefit during acute infection.

Determining the role of Treg cells is also of great importance for understanding the immunopathogenesis of chronic HIV-1 infection. Our preliminary data showed that Treg-cell depletion from HIV-JRCSF-chronically infected humanized mice by ONTAK injection induced enhanced human T cell activation, suggesting that Treg cells actively suppressed immune activation during the chronic phase of HIV-1 infection. Also, Treg-cell depletion resulted in a significant increase in HIV-1 viral infection and replication [51], suggesting that immune activation during chronic HIV-1 infection enhanced HIV-1 replication, or that Treg cells directly suppressed HIV-1 replication.

Therefore, Treg cells play a distinct role in the acute and chronic phases of HIV-1 infection (Fig. 22.2). Treg cells are likely to serve as primary target cells and to suppress anti-HIV immunity, contributing to robust viral replication in the acute

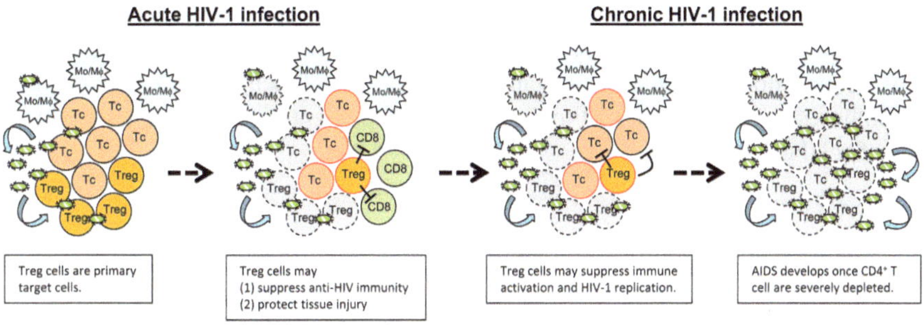

Fig. 22.2 A model of Treg cell function in HIV-1 infection and immunopathogenesis. Treg cells are likely to serve as primary target cells for primary HIV-1 infection. Treg cells may suppress anti-HIV immunity to protect tissue injury from overreactive immune responses. On the other hand, Treg cells may suppress immune activation and HIV-1 replication during chronic HIV-1 infection. In addition, Treg cells also serve as HIV-1 producing cells. These multifaceted roles of Tregs likely make multiple important contributions to establishing and/or sustaining persistent HIV infection, immune activation and CD4+ T cell depletion

phase. However, Treg cells may be important in controlling immune activation and HIV-1 replication during chronic HIV-1 infection. Some recent reports mechanistically support the hypothesis that Treg cells directly suppress HIV-1 viral replication [43–44]. Thus, Treg cells may actively control HIV-1 replication and immune activation, and consequently may slow down disease progression during the chronic phase of infection.

22.5 Perspectives on Future Studies of Treg Cells in HIV-1 Infection

Many studies have been reported that investigate the role of Treg cells in HIV-1 infection. However, some of their results are conflicting. These discrepancies may result from differences in the markers used for Treg cell identification, the compartments and the stage of HIV-1 disease in which Treg cells are examined, and the methods of Treg cell quantification [52]. More reliable Treg-specific markers are needed for identifying and purifying the Treg cell subset. Speaking to this point, the role of pTreg in HIV-1 infection is largely unknown due to the lack of an identifying cellular marker. In addition, human Treg cells have been reported to consist of heterogeneous populations [53], but how these Treg subsets contributes to HIV-1 infection and immunopathogenesis is not determined. Humanized mouse models will help to determine the role of these Treg cell subsets in the different compartments and HIV-1 disease settings. Although recent studies have also shown that Treg cells play multiple roles during HIV-1 infection and pathogenesis, the precise mechanisms should be determined in the future studies. On the other hand, modulating Treg cells may result in medical benefit for persons with HIV-1 infection. For example, manipulation or induction of Treg cells in chronic HIV-1 infection may slow down HIV-1 disease progression with or without antiretroviral therapy. Various animal models including humanized mouse models will allow us to test these hypotheses and will be useful to develop novel preventative or therapeutic strategies for HIV-1 infection.

References

1. Sakaguchi S, et al. Immunologic self-tolerance maintained by activated T cells expressing IL-2 receptor alpha-chains (CD25). Breakdown of a single mechanism of self-tolerance causes various autoimmune diseases. J Immunol. 1995;155(3):1151–64.
2. Hori S, et al. Control of regulatory T cell development by the transcription factor Foxp3. Science. 2003;299(5609):1057–61.
3. Allan SE, et al. Inducible reprogramming of human T cells into Treg cells by a conditionally active form of FOXP3. Eur J Immunol. 2008;38(12):3282–9.
4. Hoffmann P, et al. Loss of FOXP3 expression in natural human CD4+CD25+ regulatory T cells upon repetitive in vitro stimulation. Eur J Immunol. 2009;39(4):1088–97.
5. Liu W, et al. CD127 expression inversely correlates with FoxP3 and suppressive function of human CD4+ T reg cells. J Exp Med. 2006;203(7):1701–11.
6. Seddiki N, et al. Expression of interleukin (IL)-2 and IL-7 receptors discriminates between human regulatory and activated T cells. J Exp Med. 2006;203(7):1693–700.
7. Furtado GC, et al. Interleukin 2 signaling is required for CD4(+) regulatory T cell function. J Exp Med. 2002;196(6):851–7.
8. Thornton AM, et al. Cutting edge: IL-2 is critically required for the in vitro activation of CD4+CD25+ T cell suppressor function. J Immunol. 2004;172(11):6519–23.
9. Coombes JL, et al. A functionally specialized population of mucosal CD103+DCs induces Foxp3+ regulatory T cells via a TGF-beta and retinoic acid-dependent mechanism. J Exp Med. 2007;204(8):1757–64.
10. Walker MR, et al. De novo generation of antigen-specific $CD4^{+}CD25^{+}$ regulatory T cells from human $CD4^{+}CD25^{-}$ cells. Proc Natl Acad Sci U S A. 2005;102(11):4103–8.
11. Tran DQ, et al. Induction of FOXP3 expression in naive human $CD4^{+}FOXP3$ T cells by T-cell receptor stimulation is transforming growth factor-beta dependent but does not confer a regulatory phenotype. Blood. 2007;110(8):2983–90.
12. Gottschalk RA, et al. Expression of Helios in peripherally induced $Foxp3^{+}$ regulatory T Cells. J Immunol. 2011.
13. Thornton AM, et al. Expression of Helios, an Ikaros transcription factor family member, differentiates thymic-derived from peripherally induced Foxp3+ T regulatory cells. J Immunol. 2010;184(7):3433–41.
14. Himmel ME, et al. Helios+ and Helios- cells coexist within the natural FOXP3+ T regulatory cell subset in humans. J Immunol. 2013;190(5):2001–8.
15. Oswald-Richter K, et al. HIV infection of naturally occurring and genetically reprogrammed human regulatory T-cells. PLoS Biol. 2004;2(7):E198.
16. Moreno-Fernandez ME, et al. Homeostasis and function of regulatory T cells in HIV/SIV infection. J Virol. 2012.
17. Chase AJ, et al. Severe depletion of $CD4^{+}CD25^{+}$ regulatory T cells from the intestinal lamina propria but not peripheral blood or lymph nodes during acute simian immunodeficiency virus infection. J Virol. 2007;81(23):12748–57.
18. Qin S, et al. Chemokine and cytokine mediated loss of regulatory T cells in lymph nodes during pathogenic simian immunodeficiency virus infection. J Immunol. 2008;180(8):5530–6.

19. Jiang Q, et al. FoxP3+CD4+ regulatory T cells play an important role in acute HIV-1 infection in humanized Rag2-/-gammaC-/- mice in vivo. Blood. 2008;112(7):2858–68.
20. Kared H, et al. HIV-specific regulatory T cells are associated with higher CD4 cell counts in primary infection. AIDS. 2008;22(18):2451–60.
21. Kinter AL, et al. CD25(+)CD4(+) regulatory T cells from the peripheral blood of asymptomatic HIV-infected individuals regulate CD4(+) and CD8(+) HIV-specific T cell immune responses in vitro and are associated with favorable clinical markers of disease status. J Exp Med. 2004;200(3):331–43.
22. Angin M, et al. Preserved function of regulatory T cells in chronic HIV-1 infection despite decreased numbers in blood and tissue. J Infect Dis. 2012;205(10):1495–500.
23. Presicce P, et al. Frequency of circulating regulatory T cells increases during chronic HIV infection and is largely controlled by highly active antiretroviral therapy. PLoS ONE. 2011;6(12):e28118.
24. Schulze ZWJ, et al. Comprehensive analysis of frequency and phenotype of T regulatory cells in HIV infection: CD39 expression of FoxP3+ T regulatory cells correlates with progressive disease. J Virol. 2011;85(3):1287–97.
25. Weiss L, et al. Human immunodeficiency virus-driven expansion of CD4+CD25+ regulatory T cells, which suppress HIV-specific CD4T-cell responses in HIV-infected patients. Blood. 2004;104(10):3249–56.
26. Suchard MS, et al. FOXP3 expression is upregulated in CD4T cells in progressive HIV-1 infection and is a marker of disease severity. PLoS ONE. 2010;5(7):e11762.
27. Zhou H, et al. Excessive conversion and impaired thymic output contribute to disturbed regulatory T-cell homeostasis in AIDS patients with low CD4 cell counts. AIDS. 2013;27(7):1059–69.
28. Garcia-Saez AJ, et al. Peptides derived from apoptotic Bax and Bid reproduce the poration activity of the parent full-length proteins. Biophys J. 2005;88(6):3976–90.
29. Baker CA, et al. Peripheral CD4 loss of regulatory T cells is associated with persistent viraemia in chronic HIV infection. Clin Exp Immunol. 2007;147(3):533–9.
30. Andersson J, et al. The prevalence of regulatory T cells in lymphoid tissue is correlated with viral load in HIV-infected patients. J Immunol. 2005;174(6):3143–7.
31. Bandera A, et al. CD4+ T cell depletion, immune activation and increased production of regulatory T cells in the thymus of HIV-infected individuals. PLoS ONE. 2010;5(5):e10788.
32. Epple HJ, et al. Mucosal but not peripheral FOXP3+ regulatory T cells are highly increased in untreated HIV infection and normalize after suppressive HAART. Blood. 2006;108(9):3072–8.
33. Nilsson J, et al. HIV-1-driven regulatory T-cell accumulation in lymphoid tissues is associated with disease progression in HIV/AIDS. Blood. 2006;108(12):3808–17.
34. Simonetta F, et al. Early and long lasting alteration of effector CD45RA-Foxp3high regulatory T cells homeostasis during HIV infection. J Infect Dis. 2012;205(10):1510–9.
35. Mendez-Lagares G, et al. Severe immune dysregulation affects CD4+CD25hiFoxP3+ Regulatory T-cells in HIV-infected patients with low-level CD4 T-cell repopulation despite suppressive HAART. J Infect Dis. 2012;205(10):1501–9.
36. Krathwohl MD, et al. Abnormal presence of semimature dendritic cells that induce regulatory T cells in HIV-infected subjects. J Infect Dis. 2006;193(4):494–504.
37. Pereira LE, et al. Simian immunodeficiency virus (SIV) infection influences the level and function of regulatory T cells in SIV-infected rhesus macaques but not SIV-infected sooty mangabeys. J Virol. 2007;81(9):4445–56.
38. Li SY, et al. Dynamics and functions of CD4(+)CD25 (high) regulatory T lymphocytes in Chinese rhesus macaques during the early stage of infection with SIVmac239. Arch Virol. 2012;157(5):961–7.
39. Favre D, et al. Critical loss of the balance between Th17 and T regulatory cell populations in pathogenic SIV infection. PLoS Pathog. 2009;5(2):e1000295.
40. Estes JD, et al. Premature induction of an immunosuppressive regulatory T cell response during acute simian immunodeficiency virus infection. J Infect Dis. 2006;193(5):703–12.

41. Xing S, et al. Increased turnover of FoxP3high regulatory T cells is associated with hyperactivation and disease progression of chronic HIV-1 infection. J Acquir Immune Defic Syndr. 2010;54(5):455–62.
42. Elahi S, et al. Protective HIV-specific CD8(+) T cells evade T(reg) cell suppression. Nat Med. 2011;17(8):989–95.
43. Nikolova M, et al. CD39/adenosine pathway is involved in AIDS progression. PLoS Pathog. 2011;7(7):e1002110.
44. Moreno-Fernandez ME, et al. Regulatory T cells control HIV replication in activated T cells through a cAMP-dependent mechanism. Blood. 2011;117(20):5372–80.
45. Huang X, et al. Functional proteomics analysis for regulatory T cell surveillance of the HIV-1-infected macrophage. J Proteome Res. 2010;9(12):6759–73.
46. Onoe T, et al. Human natural regulatory T cell development, suppressive function, and post-thymic maturation in a humanized mouse model. J Immunol. 2011;187(7):3895–903.
47. Billerbeck E, et al. Development of human $CD4^+$ $FoxP3^+$ regulatory T cells in human stem cell factor-, granulocyte-macrophage colony-stimulating factor-, and interleukin-3-expressing NOD-SCID IL2Rgamma(null) humanized mice. Blood. 2011;117(11):3076–86.
48. Zhang L, et al. HIV-1 infection and pathogenesis in a novel humanized mouse model. Blood. 2007;109(7):2978–81.
49. Meissner EG, et al. Characterization of a thymus-tropic HIV-1 isolate from a rapid progressor: role of the envelope. Virology. 2004;328(1):74–88.
50. Nunoya J, et al. Regulatory T cells prevent liver fibrosis during HIV type 1 infection in a humanized mouse model. J Infect Dis. 2013;209(7):1039–44.
51. Holmes D, et al. Foxp3 and Treg cells in HIV-1 infection and immuno-pathogenesis. Immunol Res. 2008;41(3):248–66.
52. Imamichi H, Lane HC. Regulatory T cells in HIV-1 infection: the good, the bad, and the ugly. J Infect Dis. 2012;205(10):1479–82.
53. Miyara M, et al. Functional delineation and differentiation dynamics of human $CD4^+$ T cells expressing the FoxP3 transcription factor. Immunity. 2009;30(6):899–911.

Chapter 23
Role of Toll-Like Receptor (TLR) Signaling in HIV-1-Induced Adaptive Immune Activation

J. Judy Chang and Marcus Altfeld

Abbreviations

DC	Dendritic cells
EBV	Epstein Barr virus
HCV	Hepatitis C virus
HIV	Human immunodeficiency virus
huCD34$^+$	Human CD34$^+$ hematopoietic cells
mDC	Myeloid dendritic cells
NK	Natural killer
NSG	NOD/SCID/IL-2r$\gamma^{-/-}$
pDC	Plasmacytoid dendritic cells
TLR	Toll-like receptor

23.1 Immune Activation in HIV-1 Infection—Pathogenesis and Priming

Human immunodeficiency virus (HIV)-1 infection induces activation of the innate and adaptive immune responses, which can mediate both beneficial viral control and detrimental immune pathology. At first encounter with the virus, innate immunity

J. J. Chang (✉)
Department of Infectious Diseases, Monash University, Level 7, The Alfred Centre, 99 Commercial Road, Melbourne, VIC 3004, Australia
e-mail: ju-hui.chang@monash.edu

J. J. Chang · M. Altfeld
Ragon Institute of Massachusetts General Hospital, Massachusetts Institute of Technology and Harvard, Harvard Medical School, Boston, MA, USA

M. Altfeld
Heinrich-Pette-Institute, Leibniz Institute for Experimental Virology, Hamburg, Germany

L. Y. Poluektova et al. (eds.), *Humanized Mice for HIV Research*,
DOI 10.1007/978-1-4939-1655-9_23

can contribute to the initial restriction of viral replication and plays an important role in the activation of the adaptive immune response. Despite these initial benefits of activating the immune response, the level of ongoing $CD8^+$ T cell activation during chronic HIV-1 infection has been revealed to be a better predictor of HIV-1 disease progression than HIV-1 viral load alone [1]. Furthermore, T cell activation has also been associated with slower $CD4^+$ T cell recovery and enhanced $CD4^+$ T cell apoptosis even when HIV-1 viral load is suppressed by antiretroviral therapy [2, 3]. This elevated immune activation in chronic HIV-1 infection can be driven directly by HIV-1 replication and indirectly through changes in the microbiome, microbial translocation, coinfections, immune deregulation, and lymphoid tissue fibrosis [4–7]. The mechanisms that drive chronic immune activation in HIV-1, therefore, need to be examined and targeted to improve disease management in HIV-1-infected individuals. Examination of HIV-1 has largely been limited to in vitro work with human samples or to studies in nonhuman primate models. The development of a small animal model will allow for easier dissection of the individual mechanisms contributing to HIV-1-associated immune activation, in part through the possibility to perform interventional studies blocking and enhancing individual pathways.

23.2 Innate Immune System in Humanized Mouse Model

The use of humanized mouse model in examining human immune responses has been carried out in cancer, autoimmunity, viral infections, and vaccinations [8, 9]. The advantage of this model is that it allows for the in vivo manipulation of human immune response through the depletion of specific cell populations [10], administration of inhibiting and activating compounds [10–14] and inoculation with infectious agents [11, 12, 15]. Over 20+ years, the humanized mouse model itself has been modified and adapted, and several different models are now available. The dominant mouse strains recently used for the humanized mouse model can include NOD/SCID[16, 18], NOD/SCID/$\beta2m^{-/-}$ [18], and NOD/SCID/IL-$2r\gamma^{-/-}$ (NSG) mice [19, 20], as well as other mouse strains deficient in both Rag2 and IL-2rγ [21, 22]. Furthermore, the method of "humanizing" the mice with a human immune system can also vary and include adoptive transfer of human PBMC (Hu-PBL mice) [23], human $CD34^+$ hematopoietic cells ($huCD34^+$) isolated from cord blood [21] or fetal liver tissue (Hu-HSC mice) [22], co-engraftment with human fetal thymus and liver (SCID-Hu mice) [24], and a combination of human fetal thymus and liver engraftment with injection of $huCD34^+$ cells (BLT mice) [11, 25]. The reconstituted human immune system can be further enhanced by the administration of cytokines or cytokine-expressing plasmids [26].

These variations of the mouse models result in differing human immune system developments. Those mice with the NOD/SCID-based backgrounds and $Rag2^{-/-}$/IL-$2r\gamma^{-/-}$-based backgrounds have different reconstitution of immune cells at different tissue sights [27, 28]. However, further enhancements can be used to assist the development of the human immune system and overcome some of the defects in each model. In both the NOD/SCID/β2m$^{-/-}$ and NSG models, human T cells, B cells and antibodies can successfully develop [29]. Human fetal thymus and liver engraftment allowed for development of the major human T cell subsets and can be improved by the administration of huCD34^{+} cells (BLT mouse model), which also have better development of B cells and monocytes [30, 31]. The presence of dendritic cells (DC)s in different humanized mouse models has been demonstrated in bone marrow, blood, and spleen [10]. HuCD34^{+}-transplanted NSG and NOD/SCID/β2m$^{-/-}$ mice develop myeloid dendritic cells (mDCs) [10, 13], plasmacytoid dendritic cells (pDCs) [10, 32], and natural killer (NK) cells [13]. DCs and monocyte development in NSG mice reconstituted with huCD34^{+} cells can further be enhanced by the administration of GM-CSF and IL-4 [33]. Presence of GM-CSF and IL-4 can also help the development of human T cells, B cells, as well as IgG antibody production [33]. Using mice with the NOD/SCID/β2m$^{-/-}$ or NSG backgrounds also helps to reduce murine NK cell development as compared to NOD/SCID models but unfortunately also result in poor human NK cell reconstitution [34–36]. Administration of human IL-15 and Flt-3 via DNA vectors or recombinant proteins assists in the development of human NK cells in both NOD/SCID, NSG and Balb/c/Rag2$^{-/-}$/IL-2rγ$^{-/-}$ mouse models reconstituted with huCD34^{+} cells [26, 34, 37]. Furthermore, the manipulation of the genetic background of the mouse can also be used to enhance development of specific human cell subsets. Knock-ins of human IL-3 and GM-CSF in Rag2$^{-/-}$/IL-2rγ$^{-/-}$ mice and CSF1 in Balb/c/Rag2$^{-/-}$/IL-2rγ$^{-/-}$ mice allowed for better macrophage reconstitution following administration with huCD34^{+} cells [38, 39]. Overall, the genetic background of mice, tissue used for human reconstitution, and any subsequent cytokine treatment can all vary between different humanized mouse models described, each with its strengths and weaknesses. The specific humanized mouse model to be used should, therefore, be carefully chosen and depends on the experimental question and the specific immune cell population of interest.

The various improvements to the humanized mouse model allow for the examination of the interplay between the innate and adaptive immune system during viral infections or vaccinations, and provide avenues of intervention. BLT, Hu-HSC, and Hu-PBL mice can be successfully infected with Epstein Barr virus (EBV), which activates and matures human DCs, and in turn, induces both a virus-specific T cell response as well as NK cell response [11, 15, 28, 40]. Utilizing the ability to deplete specific cell populations, depletion of pDCs demonstrated their importance in delaying EBV-induced mortality [15] . Rag2$^{-/-}$IL-2rγ$^{-/-}$ and NSG mice reconstituted with huCD34^{+} cells can be infected with influenza A virus and produce inflammatory cytokines and neutralizing antibodies in response to the virus. This demonstrates the ability of the humanized mouse model to imitate the natural multifaceted immune responses that is observed in vivo as compared to in vitro models [33, 38]. HuCD34^{+}-transplanted Balb/c/Rag2$^{-/-}$/IL-2rγ$^{-/-}$ mouse model can be infected with Hepatitis C virus (HCV), which not only induced HCV-specific T cell responses but infiltration of the mouse liver with human innate and adaptive immune cells [41]. The same mouse model can also be infected with HSV-2, and when the virus is used in a vaginal challenge model, T cells and NK cells were detected at the mucosa [27].

Both of these viral infection models demonstrated the ability to use these mice to examine local immune responses at specific sites of infection. In addition, the use of attenuated HSV-2 in humanized mice demonstrated protection against subsequent challenge infections [27]. These examples of viral infections in different humanized mouse models demonstrate the possibilities of using these models to examine interactions between human immune cell subsets with each other and the virus, tissue and mucosa specific immunity, and the effects of vaccination. Similar to other viruses, HIV-1 can also successfully infect human immune cells in the humanized mice, thus providing a small animal model alternative to nonhuman primates. Like several other viruses, HIV-1 can infect a range of humanized mouse models including huCD34^{+}-transplanted Rag2$^{-/-}$/IL-2rγ$^{-/-}$ [42], NSG BLT mice [30, 31], and NOD/SCID BLT mice [31]. Following HIV-1 infection, a decline of CD4^{+} T cell is observed [30, 31, 42], CD4^{+} and CD8^{+} T cell activation develops [30, 31], and antigen-specific T cell and B cell responses are induced [31, 43]. Furthermore, the viral reservoir can be assessed, and initial studies have tested approaches to eliminate the HIV-1 reservoir using the humanized mouse model [44]. This highlighted the possibility of using the BLT humanized mouse model to examine not only the role of innate immunity in HIV-1 infections but also its consequences on the development of the adaptive immunity and subsequent disease outcome.

23.3 Direct Activation of TLR7/8 by HIV-1 and IFNα-Mediated Immune Activation

HIV-1 ssRNA encodes for multiple TLR7/8 ligands which have been shown to activate pDCs, mDCs, and other immune cells in the in vitro system [45–47]. In pDCs, activation of the TLR7 pathway by HIV-1 leads to the production of IFNα [45–48] and induction of the adaptive T cell response [46, 49]. This has potentially beneficial outcomes as the stimulation of TLRs have demonstrated importance not only in the priming of CD4^{+} and CD8^{+} T cells but also in directly modulating the type of adaptive immune response induced [50, 51]. However, the persistent production of IFNα might be detrimental in the setting of a chronic persistent infection [52, 53]. In SIV infection of nonhuman primates, pDCs in rhesus macaques, the pathogenic host of SIV, produce large quantities of IFNα [54], while in sooty mangabeys, the natural host with reduced pathogenesis [55], lower IFNα levels are observed in chronic infection [54, 56, 57]. This is also observed in the expression of interferon-stimulated genes (ISGs), which are elevated in SIV-infected rhesus macaques [58]. Likewise, in human HIV-1 infections, pDCs also express high levels of IFNα, resulting in the upregulation of ISGs, which may contribute to the overall systemic immune activation [48, 59, 60]. Consistent expression of IFNα can also contribute to the apoptosis of CD4^{+} T cells and hence, exacerbate the immune deficiency observed in HIV-1-infected individuals [61]. In addition, pDC activation by HIV-1 has also been shown to induce an increase in indoleamine 2,3-dioxygenase expression which, through modulating regulatory T cells, can suppress T cell responses and

Table 23.1 TLRs in humanized mouse model—Determined by expression and function

Mouse model	Humanization	TLR	References
NOD/SCID/IL-2r$\gamma^{-/-}$	huCD34^{+}	TLR2 on monocytes	[13]
		on mDCs	[13]
		TLR3 on mDCs	[10]
		TLR4 on monocytes	[13]
		on mDCs	[10, 13]
		TLR7 on pDCs	[10]
		TLR8 on mDCs	[10]
NOD/SCID/β2m$^{-/-}$	huCD34^{+}	TLR7 on pDCs	[32]
NOD/SCID	huCD34^{+}	TLR3 on mDCs	[36]
		TLR4	[89]
		TLR4 on mDCs	[89]
Rag2$^{-/-}$IL2r$\gamma_c^{-/-}$	huCD34^{+}	TLR4[a]	[38]
		TLR7 on pDCs	[42]
		TLR9 on pDCs	[42]

[a] Mice also have human IL-3 and GM-CSF knock-in

proliferation [49, 62–64]. Therapeutic manipulation to suppress the pDC/TLR7/IFNα responses have led to conflicting results. Treatments with chloroquine, which can inhibit endosomal acidification and therefore inhibit signaling through TLR3, 7, 8 and 9, have shown to help reduce HIV-1 replication [65, 66] and immune activation [67–69]. However, other studies have also demonstrated that chloroquine can increase HIV-1 infectivity [70, 71]. Further studies in adequate in vivo models are needed to assess the repercussions of suppressing specific arms of the innate immune response for HIV-1 pathogenesis and immune activation, and also to examine other venues that may more specifically target isolated pathways of innate immunity. These studies will also provide important insights into the contribution of individual pathways to the overall immune activation associated with HIV-1 infection.

The humanized mouse model is one such model that could be utilized to examine responses by specific innate immune pathways and to manipulate these pathways to alter the resulting immune response. NSG BLT mice and Hu-HSC mice all have demonstrated reconstitution of human pDCs, which can respond to TLR7 stimulation (Table 23.1) [10, 42]. Even sex-based differences observed in human studies of TLR7-associated IFNα production [46] have been duplicated in huCD34^{+}-transplanted NOD/SCID/β2m$^{-/-}$ mouse models [32]. Furthermore, TLR7 stimulation of pDCs and its contribution to immune activation and deregulation can be mimicked in mice, where chronic stimulation of TLR7 and TLR9 by synthetic ligands has been shown to result in immune activation and lymphoid follicle destruction similar to that observed in chronic HIV-1 infection in humans [30, 72, 73]. Humanized mice can be infected with HIV-1 and following such infection in huCD34^{+}-transplanted Rag2$^{-/-}$IL2r$\gamma_c^{-/-}$ and NSG BLT mice, pDC activation and production of IFNα can be observed, demonstrating that the humanized mouse model can be used to examine the roles of pDCs, TLR7, and IFNα in HIV-1 infection [30, 42]. Using the NSG BLT mouse model, the administration of recombinant IFNα2b not only induced immune activation levels comparable to that induced by HIV-1 infection, but also induced infiltration of activated T cells in secondary lymphoid organs, adding to the growing evidence that IFNα contributes to immune activation in HIV-1 infection [30]. Although IFNα has potentially detrimental properties in inducing chronic immune activation, it also stimulates the expression of ISGs which possess antiviral activities [74–76]. Cholesterol-25-hydrolase (CH25H), an ISG which can inhibit HIV-1 entry in in vitro cell cultures, can also lead to reduced HIV-1 viral replication in vivo when used to treat HIV-1-infected NOD/Rag2$^{-/-}$IL2r$\gamma_c^{-/-}$ humanized mice [77]. Overall, these early studies using different humanized mouse models of HIV-1 infection suggest that they allow for a reasonable assessment of TLR pathways, the type I IFN axis, and DC function. Future studies will determine the validity of this model to study pathways leading to HIV-1-associated immune activation and to assess interventions aimed at reducing HIV-1-associated immune activation.

23.4 Activation of TLRs by Microbe and Microbial Products

Although HIV-1 directly encodes for TLR7/8 ligands, it can also stimulate directly and indirectly through the other TLR pathways. It has been well established that HIV-1 infection leads to depletion of gut-associated CD4$^+$ T cells and therefore compromises the gut mucosal barrier allowing for microbial translocation [4]. One microbial product, lipopolysaccharide (LPS), which can stimulate through the TLR4 pathway has been associated with HIV-1 associated T cell activation [4, 7, 78], monocyte activation [79], lower CD4$^+$ T cell count [80], and CD4$^+$ T cell recovery [7]. In addition to stimulation of TLR4, stimulation of other TLR pathways including TLR3, 5, and 9 also activates T cells and leads to CD4$^+$ T cell apoptosis [78]. Furthermore, gram-positive bacterial stimulation via the TLR2 pathway has been described to result in increased HIV-1 susceptibility of skin-resident DCs [81] as well as transmission of HIV-1 from DCs to CD4$^+$ T cells [82]. Presence of microbial products resulting from microbial translocation or coinfection can not only cause deregulation of the immune system by themselves, but the activating signal might be further amplified by the presence of HIV-1 [83]. Preexposure of monocytes with HIV-1 and HIV-1-encoded TLR8 ligands enhanced subsequent TLR4 stimulation, exacerbating the microbial-induced immune activation [83], and the stimulation of TLR2, 4, and 9 can increase HIV-1 replication in chronically infected cells [84]. In addition, HIV-1 can also lead to increase in TLR2, 3, 4, 6, 7, and 8 expression, all of which can amplify the activation of these immune cells [85–87]. All these factors can contribute to and amplify HIV-1-associated immune activation, which suggest that suppression of HIV-1 or microbial stimulation of TLRs might therefore be beneficial in reducing HIV-1 pathogenesis. These studies have demonstrated the complexities of the various immune cell populations and TLR pathways that play a role in HIV-1 infection and pathogenesis. The question of how each pathway contributes to progressive HIV-1 disease and methods of therapeutically modulating these pathways still needs to be examined.

The role of microbial translocation and the stimulation of the other TLR pathways have also been examined in humanized mice (Table 23.1). Although limited, the expression and function of TLRs in different humanized mouse models have now been started to be studied. TLR2 and TLR4 expressions have been observed on monocytes in huCD34^{+}-transplanted NSG mice [13], and TLR3 expression on mDCs in huCD34^{+}-transplanted NOD/SCID and both huCD34^{+} and human PBMC-transplanted NSG mice [10, 36, 88]. Stimulation through TLRs can lead to DC activation and maturation [10, 33, 36, 88, 89], production of cytokines [10, 38, 39], and recruitment and priming of T cells [10, 36, 88, 89]. HIV-1 infection of huCD34^{+}-transplanted Rag2$^{-/-}$IL2$\gamma_c^{-/-}$ mice also results in depletion of gut CD4^{+} T cells and subsequent microbial translocation [90]. The use of humanized mice also allowed for the administration of dextran sodium sulfate, which can cause gut barrier dysfunction in the absence of HIV-1 infection. Treatment in huCD34^{+}-transplanted Rag2$^{-/-}$IL2$\gamma_c^{-/-}$ mice demonstrates that dextran sodium sulfate disruption only induces microbial translocation while HIV-1 additionally inhibits microbial clearance potentially reflecting the in vivo consequences of HIV-1 disruption of TLR responses highlighted above [90]. Although the presence of murine immune cells has been reduced in these different humanized mouse models, it is still possible that they may play a role in responding to microbial products. Examination of sepsis in the BLT mouse model, however, suggests that any remnant mouse innate or adaptive immunity that may respond to bacterial products are not enough to induce an inflammatory response on their own and that the reconstituted human immune cells are essential [12]. The humanized mouse models therefore can be utilized to examine not only TLR pathways directly stimulated by HIV-1 but also those induced by microbial translocation or coinfections. However, the expression and function of the various TLR pathways still needs to be more comprehensively studied, particularly the TLR expression patterns and function on the reconstituted human immune cells at various tissue sites.

23.5 Future Direction and Development of the Model

Significant progress has been made in the development of humanized mouse models to more accurately replicate the in vivo human immune system. The establishment of stable human T cell populations has allowed it to be used as a small animal model for HIV-1 infection. Although there have only been a very limited number of studies to date, initial studies assessing immune responses in different humanized mouse models have demonstrated that the TLRs respond to HIV-1 and other microbial pathogens in humanized mice similar to that observed in humans. Furthermore, studies employing in vivo stimulation and blocking of TLRs in humanized mice have already started to dissect how various TLR pathways can contribute to HIV-

1-associated immune activation. The expression and functionality of TLRs in the various humanized mouse models still needs to be fully examined in addition to the role of residual mouse TLR responses, which might represent a confounding factor in these models. Although these early studies are limited, they provide a potential avenue to evaluate activators and inhibitors of individual innate pathways for both dissecting the contribution of these pathways to HIV-1-induced immune activation as well as their therapeutic effect in reducing HIV-1 pathogenesis.

References

1. Giorgi JV, Hultin LE, McKeating JA, Johnson TD, Owens B, Jacobson LP, et al. Shorter survival in advanced human immunodeficiency virus type 1 infection is more closely associated with T lymphocyte activation than with plasma virus burden or virus chemokine coreceptor usage. J Infect Dis. 1999;179(4):859–70.
2. Hunt PW, Martin JN, Sinclair E, Bredt B, Hagos E, Lampiris H, et al. T cell activation is associated with lower CD4+ T cell gains in human immunodeficiency virus-infected patients with sustained viral suppression during antiretroviral therapy. J Infect Dis. 2003;187(10):1534–43.
3. Groux H, Torpier G, Monte D, Mouton Y, Capron A, Ameisen JC. Activation-induced death by apoptosis in CD4+ T cells from human immunodeficiency virus-infected asymptomatic individuals. J Exp Med. 1992;175(2):331–40.
4. Brenchley JM, Price DA, Schacker TW, Asher TE, Silvestri G, Rao S, et al. Microbial translocation is a cause of systemic immune activation in chronic HIV infection. Nat Med. 2006;12(12):1365–71.
5. Biancotto A, Grivel JC, Iglehart SJ, Vanpouille C, Lisco A, Sieg SF, et al. Abnormal activation and cytokine spectra in lymph nodes of people chronically infected with HIV-1. Blood. 2007;109(10):4272–9.
6. Gonzalez VD, Falconer K, Blom KG, Reichard O, Morn B, Laursen AL, et al. High levels of chronic immune activation in the T-cell compartments of patients coinfected with hepatitis C virus and human immunodeficiency virus type 1 and on highly active antiretroviral therapy are reverted by alpha interferon and ribavirin treatment. J Virol. 2009;83(21):11407–11.
7. Jiang W, Lederman MM, Hunt P, Sieg SF, Haley K, Rodriguez B, et al. Plasma levels of bacterial DNA correlate with immune activation and the magnitude of immune restoration in persons with antiretroviral-treated HIV infection. J Infect Dis. 2009;199(8):1177–85 (Clinical trial Research Support, N.I.H., Extramural).
8. Shultz LD, Brehm MA, Garcia-Martinez JV, Greiner DL. Humanized mice for immune system investigation: progress, promise and challenges. Nat Rev Immunol. 2012;12(11):786–98.
9. Ito R, Takahashi T, Katano I, Ito M. Current advances in humanized mouse models. Cell Mol Immunol. 2012;9(3):208–14.
10. Meixlsperger S, Leung CS, Ramer PC, Pack M, Vanoaica LD, Breton G, et al. CD141+ dendritic cells produce prominent amounts of IFN-alpha after dsRNA recognition and can be targeted via DEC-205 in humanized mice. Blood. 2013;121(25):5034–44.
11. Melkus MW, Estes JD, Padgett-Thomas A, Gatlin J, Denton PW, Othieno FA, et al. Humanized mice mount specific adaptive and innate immune responses to EBV and TSST-1. Nat Med. 2006;12(11):1316–22 (Research Support, N.I.H., Extramural).
12. Ye C, Choi JG, Abraham S, Wu H, Diaz D, Terreros D, et al. Human macrophage and dendritic cell-specific silencing of high-mobility group protein B1 ameliorates sepsis in a humanized mouse model. Proc Natl Acad Sci U S A. 2012;109(51):21052–7.

13. Tanaka S, Saito Y, Kunisawa J, Kurashima Y, Wake T, Suzuki N, et al. Development of mature and functional human myeloid subsets in hematopoietic stem cell-engrafted NOD/SCID/IL2rgammaKO mice. J Immunol. 2012;188(12):6145–55.
14. Seung E, Dudek TE, Allen TM, Freeman GJ, Luster AD, Tager AM. PD-1 blockade in chronically HIV-1-infected humanized mice suppresses viral loads. PLoS ONE. 2013;8(10):e77780.
15. Lim WH, Kireta S, Russ GR, Coates PT. Human plasmacytoid dendritic cells regulate immune responses to Epstein–Barr virus (EBV) infection and delay EBV-related mortality in humanized NOD-SCID mice. Blood. 2007;109(3):1043–50.
16. Hesselton RM, Greiner DL, Mordes JP, Rajan TV, Sullivan JL, Shultz LD. High levels of human peripheral blood mononuclear cell engraftment and enhanced susceptibility to human immunodeficiency virus type 1 infection in NOD/LtSz-scid/scid mice. J Infect Dis. 1995;172(4):974–82.
17. Lowry PA, Shultz LD, Greiner DL, Hesselton RM, Kittler EL, Tiarks CY, et al. Improved engraftment of human cord blood stem cells in NOD/LtSz-scid/scid mice after irradiation or multiple-day injections into unirradiated recipients. Biol Blood Marrow Transplant. 1996;2(1):15–23.
18. Christianson SW, Greiner DL, Hesselton RA, Leif JH, Wagar EJ, Schweitzer IB, et al. Enhanced human CD4+ T cell engraftment in beta2-microglobulin-deficient NOD-scid mice. J Immunol. 1997;158(8):3578–86.
19. Shultz LD, Lyons BL, Burzenski LM, Gott B, Chen X, Chaleff S, et al. Human lymphoid and myeloid cell development in NOD/LtSz-scid IL2R gamma null mice engrafted with mobilized human hemopoietic stem cells. J Immunol. 2005;174(10):6477–89.
20. Ito M, Hiramatsu H, Kobayashi K, Suzue K, Kawahata M, Hioki K, et al. NOD/SCID/gamma(c)(null) mouse: an excellent recipient mouse model for engraftment of human cells. Blood. 2002;100(9):3175–82.
21. Traggiai E, Chicha L, Mazzucchelli L, Bronz L, Piffaretti JC, Lanzavecchia A, et al. Development of a human adaptive immune system in cord blood cell-transplanted mice. Science. 2004;304(5667):104–7.
22. Gimeno R, Weijer K, Voordouw A, Uittenbogaart CH, Legrand N, Alves NL, et al. Monitoring the effect of gene silencing by RNA interference in human CD34+ cells injected into newborn RAG2−/− gammac−/− mice: functional inactivation of p53 in developing T cells. Blood. 2004;104(13):3886–93.
23. Mosier DE, Gulizia RJ, Baird SM, Wilson DB. Transfer of a functional human immune system to mice with severe combined immunodeficiency. Nature. 1988;335(6187):256–9.
24. McCune JM, Namikawa R, Kaneshima H, Shultz LD, Lieberman M, Weissman IL. The SCID-hu mouse: murine model for the analysis of human hematolymphoid differentiation and function. Science. 1988;241(4873):1632–9.
25. Lan P, Tonomura N, Shimizu A, Wang S, Yang YG. Reconstitution of a functional human immune system in immunodeficient mice through combined human fetal thymus/liver and CD34+ cell transplantation. Blood. 2006;108(2):487–92.
26. Chen Q, Khoury M, Chen J. Expression of human cytokines dramatically improves reconstitution of specific human-blood lineage cells in humanized mice. Proc Natl Acad Sci U S A. 2009;106(51):21783–8.
27. Kwant-Mitchell A, Ashkar AA, Rosenthal KL. Mucosal innate and adaptive immune responses against herpes simplex virus type 2 in a humanized mouse model. J Virol. 2009;83(20):10664–76.
28. Strowig T, Gurer C, Ploss A, Liu YF, Arrey F, Sashihara J, et al. Priming of protective T cell responses against virus-induced tumors in mice with human immune system components. J Exp Med. 2009;206(6):1423–34.
29. Ishikawa F, Yasukawa M, Lyons B, Yoshida S, Miyamoto T, Yoshimoto G, et al. Development of functional human blood and immune systems in NOD/SCID/IL2 receptor {gamma} chain(null) mice. Blood. 2005;106(5):1565–73.
30. Long BR, Stoddart CA. Alpha interferon and HIV infection cause activation of human T cells in NSG-BLT mice. J Virol. 2012;86(6):3327–36.

31. Brainard DM, Seung E, Frahm N, Cariappa A, Bailey CC, Hart WK, et al. Induction of robust cellular and humoral virus-specific adaptive immune responses in human immunodeficiency virus-infected humanized BLT mice. J Virol. 2009;83(14):7305–21.
32. Seillet C, Laffont S, Tremollieres F, Rouquie N, Ribot C, Arnal JF, et al. The TLR-mediated response of plasmacytoid dendritic cells is positively regulated by estradiol in vivo through cell-intrinsic estrogen receptor alpha signaling. Blood. 2012;119(2):454–64.
33. Chen Q, He F, Kwang J, Chan JK, Chen J. GM-CSF and IL-4 stimulate antibody responses in humanized mice by promoting T, B, and dendritic cell maturation. J Immunol. 2012;189(11):5223–9.
34. Kalberer CP, Siegler U, Wodnar-Filipowicz A. Human NK cell development in NOD/SCID mice receiving grafts of cord blood CD34+ cells. Blood. 2003;102(1):127–35.
35. Rajesh D, Zhou Y, Jankowska-Gan E, Roenneburg DA, Dart ML, Torrealba J, et al. Th1 and Th17 immunocompetence in humanized NOD/SCID/IL2rgammanull mice. Hum Immunol. 2010;71(6):551–9.
36. Vuckovic S, Abdul Wahid FS, Rice A, Kato M, Khalil D, Rodwell R, et al. Compartmentalization of allogeneic T-cell responses in the bone marrow and spleen of humanized NOD/SCID mice containing activated human resident myeloid dendritic cells. Exp Hematol. 2008;36(11):1496–506.
37. Huntington ND, Legrand N, Alves NL, Jaron B, Weijer K, Plet A, et al. IL-15 trans-presentation promotes human NK cell development and differentiation in vivo. J Exp Med. 2009;206(1):25–34.
38. Willinger T, Rongvaux A, Takizawa H, Yancopoulos GD, Valenzuela DM, Murphy AJ, et al. Human IL-3/GM-CSF knock-in mice support human alveolar macrophage development and human immune responses in the lung. Proc Natl Acad Sci U S A. 2011;108(6):2390–5.
39. Rathinam C, Poueymirou WT, Rojas J, Murphy AJ, Valenzuela DM, Yancopoulos GD, et al. Efficient differentiation and function of human macrophages in humanized CSF-1 mice. Blood. 2011;118(11):3119–28.
40. Sato K, Misawa N, Nie C, Satou Y, Iwakiri D, Matsuoka M, et al. A novel animal model of Epstein–Barr virus-associated hemophagocytic lymphohistiocytosis in humanized mice. Blood. 2011;117(21):5663–73.
41. Washburn ML, Bility MT, Zhang L, Kovalev GI, Buntzman A, Frelinger JA, et al. A humanized mouse model to study hepatitis C virus infection, immune response, and liver disease. Gastroenterology. 2011;140(4):1334–44.
42. Zhang L, Jiang Q, Li G, Jeffrey J, Kovalev GI, Su L. Efficient infection, activation, and impairment of pDCs in the BM and peripheral lymphoid organs during early HIV-1 infection in humanized rag2(-)/(-)gamma C(-)/(-) mice in vivo. Blood. 2011;117(23):6184–92.
43. Dudek TE, No DC, Seung E, Vrbanac VD, Fadda L, Bhoumik P, et al. Rapid evolution of HIV-1 to functional CD8+ T-cell responses in humanized BLT mice. Sci Transl Med. 2012;4(143):143ra98.
44. Hauber I, Hofmann-Sieber H, Chemnitz J, Dubrau D, Chusainow J, Stucka R, et al. Highly significant antiviral activity of HIV-1 LTR-specific tre-recombinase in humanized mice. PLoS Pathog. 2013;9(9):e1003587.
45. Meier A, Alter G, Frahm N, Sidhu H, Li B, Bagchi A, et al. MyD88-dependent immune activation mediated by human immunodeficiency virus type 1-encoded Toll-like receptor ligands. J Virol. 2007;81(15):8180–91.
46. Meier A, Chang JJ, Chan ES, Pollard RB, Sidhu HK, Kulkarni S, et al. Sex differences in the Toll-like receptor-mediated response of plasmacytoid dendritic cells to HIV-1. Nat Med. 2009;15(8):955–9.
47. Beignon AS, McKenna K, Skoberne M, Manches O, DaSilva I, Kavanagh DG, et al. Endocytosis of HIV-1 activates plasmacytoid dendritic cells via Toll-like receptor-viral RNA interactions. J Clin Invest. 2005;115(11):3265–75.
48. Lehmann C, Harper JM, Taubert D, Hartmann P, Fatkenheuer G, Jung N, et al. Increased interferon alpha expression in circulating plasmacytoid dendritic cells of HIV-1-infected patients. J Acquir Immune Defic Syndr. 2008;48(5):522–30.

49. Boasso A, Shearer GM. Chronic innate immune activation as a cause of HIV-1 immunopathogenesis. Clin Immunol. 2008;126(3):235–42 (Review).
50. Lombardi V, Van Overtvelt L, Horiot S, Moingeon P. Human dendritic cells stimulated via TLR7 and/or TLR8 induce the sequential production of Il-10, IFN-gamma, and IL-17A by naive CD4+ T cells. J Immunol. 2009;182(6):3372–9.
51. Agrawal S, Agrawal A, Doughty B, Gerwitz A, Blenis J, Van Dyke T, et al. Cutting edge: different Toll-like receptor agonists instruct dendritic cells to induce distinct Th responses via differential modulation of extracellular signal-regulated kinase-mitogen-activated protein kinase and c-Fos. J Immunol. 2003;171(10):4984–9.
52. Odorizzi PM, Wherry EJ. Immunology. An interferon paradox. Science. 2013;340(6129):155–6.
53. Chang JJ, Lacas A, Lindsay RJ, Doyle EH, Axten KL, Pereyra F, et al. Differential regulation of toll-like receptor pathways in acute and chronic HIV-1 infection. AIDS. 2012;26(5):533-41 (Research Support, N.I.H., Extramural Research Support, Non-U.S. Gov't).
54. Mandl JN, Barry AP, Vanderford TH, Kozyr N, Chavan R, Klucking S, et al. Divergent TLR7 and TLR9 signaling and type I interferon production distinguish pathogenic and nonpathogenic AIDS virus infections. Nat Med. 2008;14(10):1077–87.
55. Silvestri G, Sodora DL, Koup RA, Paiardini M, O'Neil SP, McClure HM, et al. Nonpathogenic SIV infection of sooty mangabeys is characterized by limited bystander immunopathology despite chronic high-level viremia. Immunity. 2003;18(3):441–52.
56. Sodora DL, Allan JS, Apetrei C, Brenchley JM, Douek DC, Else JG, et al. Toward an AIDS vaccine: lessons from natural simian immunodeficiency virus infections of African nonhuman primate hosts. Nat Med. 2009;15(8):861–5.
57. Harris LD, Tabb B, Sodora DL, Paiardini M, Klatt NR, Douek DC, et al. Downregulation of robust acute type I interferon responses distinguishes nonpathogenic simian immunodeficiency virus (SIV) infection of natural hosts from pathogenic SIV infection of rhesus macaques. J Virol. 2010;84(15):7886–91.
58. Bosinger SE, Li Q, Gordon SN, Klatt NR, Duan L, Xu L, et al. Global genomic analysis reveals rapid control of a robust innate response in SIV-infected sooty mangabeys. J Clin Invest. 2009;119(12):3556–72.
59. Bosinger SE, Hosiawa KA, Cameron MJ, Persad D, Ran L, Xu L, et al. Gene expression profiling of host response in models of acute HIV infection. J Immunol. 2004;173(11):6858–63.
60. Chang JJ, Woods M, Lindsay RJ, Doyle EH, Griesbeck M, Chan ES, et al. Higher expression of several interferon-stimulated genes in HIV-1-infected females after adjusting for the level of viral replication. J Infect Dis. 2013;208(5):830–8.
61. Herbeuval JP, Nilsson J, Boasso A, Hardy AW, Kruhlak MJ, Anderson SA, et al. Differential expression of IFN-alpha and TRAIL/DR5 in lymphoid tissue of progressor versus nonprogressor HIV-1-infected patients. Proc Natl Acad Sci U S A. 2006;103(18):7000–5.
62. Boasso A, Herbeuval JP, Hardy AW, Anderson SA, Dolan MJ, Fuchs D, et al. HIV inhibits CD4+ T-cell proliferation by inducing indoleamine 2,3-dioxygenase in plasmacytoid dendritic cells. Blood. 2007;109(8):3351–9 (Clinical Trial, Comparative Study Multicenter Study Research Support, N.I.H., Intramural Research Support, Non-U.S. Gov't).
63. Malleret B, Maneglier B, Karlsson I, Lebon P, Nascimbeni M, Perie L, et al. Primary infection with simian immunodeficiency virus: plasmacytoid dendritic cell homing to lymph nodes, type I interferon, and immune suppression. Blood. 2008;112(12):4598–608.
64. Manches O, Munn D, Fallahi A, Lifson J, Chaperot L, Plumas J, et al. HIV-activated human plasmacytoid DCs induce Tregs through an indoleamine 2,3-dioxygenase-dependent mechanism. J Clin Invest. 2008;118(10):3431–9.
65. Chiang G, Sassaroli M, Louie M, Chen H, Stecher VJ, Sperber K. Inhibition of HIV-1 replication by hydroxychloroquine: mechanism of action and comparison with zidovudine. Clin Ther. 1996;18(6):1080-92 (Comparative Study Research Support, Non-U.S. Gov't Research Support, U.S. Gov't, P.H.S.).
66. Naarding MA, Baan E, Pollakis G, Paxton WA. Effect of chloroquine on reducing HIV-1 replication in vitro and the DC-SIGN mediated transfer of virus to CD4+ T-lymphocytes. Retrovirology. 2007;4:6 (Research Support, Non-U.S. Gov't).

67. Murray SM, Down CM, Boulware DR, Stauffer WM, Cavert WP, Schacker TW, et al. Reduction of immune activation with chloroquine therapy during chronic HIV infection. J Virol. 2010;84(22):12082–6 (Randomized Controlled Trial Research Support, N.I.H., Research Support, Non-U.S. Gov't).
68. Martinson JA, Montoya CJ, Usuga X, Ronquillo R, Landay AL, Desai SN. Chloroquine modulates HIV-1-induced plasmacytoid dendritic cell alpha interferon: implication for T-cell activation. Antimicrob Agents Chemother. 2010;54(2):871–81.
69. Piconi S, Parisotto S, Rizzardini G, Passerini S, Terzi R, Argenteri B, et al. Hydroxychloroquine drastically reduces immune activation in HIV-infected, antiretroviral therapy-treated immunologic nonresponders. Blood. 2011;118(12):3263–72 (Research Support, Non-U.S. Gov't).
70. Fredericksen BL, Wei BL, Yao J, Luo T, Garcia JV. Inhibition of endosomal/lysosomal degradation increases the infectivity of human immunodeficiency virus. J Virol. 2002;76(22):11440–6 (Research Support, U.S. Gov't, P.H.S.).
71. Vijaykumar TS, Nath A, Chauhan A. Chloroquine mediated molecular tuning of astrocytes for enhanced permissiveness to HIV infection. Virology. 2008;381(1):1–5 (Research Support, N.I.H., Extramural).
72. Baenziger S, Heikenwalder M, Johansen P, Schlaepfer E, Hofer U, Miller RC, et al. Triggering TLR7 in mice induces immune activation and lymphoid system disruption, resembling HIV-mediated pathology. Blood. 2009;113(2):377–88.
73. Heikenwalder M, Polymenidou M, Junt T, Sigurdson C, Wagner H, Akira S, et al. Lymphoid follicle destruction and immunosuppression after repeated CpG oligodeoxynucleotide administration. Nat Med. 2004;10(2):187–92.
74. Baca-Regen L, Heinzinger N, Stevenson M, Gendelman HE. Alpha interferon-induced antiretroviral activities: restriction of viral nucleic acid synthesis and progeny virion production in human immunodeficiency virus type 1-infected monocytes. J Virol. 1994;68(11):7559–65.
75. Cullen BR. Role and mechanism of action of the APOBEC3 family of antiretroviral resistance factors. J Virol. 2006;80(3):1067–76.
76. Okumura A, Lu G, Pitha-Rowe I, Pitha PM. Innate antiviral response targets HIV-1 release by the induction of ubiquitin-like protein ISG15. Proc Natl Acad Sci U S A. 2006;103(5):1440–5 (Research Support, N.I.H., Extramural).
77. Liu SY, Aliyari R, Chikere K, Li G, Marsden MD, Smith JK, et al. Interferon-inducible cholesterol-25-hydroxylase broadly inhibits viral entry by production of 25-hydroxycholesterol. Immunity. 2013;38(1):92–105.
78. Funderburg N, Luciano AA, Jiang W, Rodriguez B, Sieg SF, Lederman MM. Toll-like receptor ligands induce human T cell activation and death, a model for HIV pathogenesis. PLoS ONE. 2008;3(4):e1915 (Research Support, N.I.H., Extramural).
79. Ancuta P, Kamat A, Kunstman KJ, Kim EY, Autissier P, Wurcel A, et al. Microbial translocation is associated with increased monocyte activation and dementia in AIDS patients. PLoS ONE. 2008;3(6):e2516 [Research Support, N.I.H., Extramural Research Support, Non-U.S. Gov't).
80. Balagopal A, Philp FH, Astemborski J, Block TM, Mehta A, Long R, et al. Human immunodeficiency virus-related microbial translocation and progression of hepatitis C. Gastroenterology. 2008;135(1):226–33.
81. Ogawa Y, Kawamura T, Kimura T, Ito M, Blauvelt A, Shimada S. Gram-positive bacteria enhance HIV-1 susceptibility in Langerhans cells, but not in dendritic cells, via Toll-like receptor activation. Blood. 2009;113(21):5157–66.
82. Thibault S, Fromentin R, Tardif MR, Tremblay MJ. TLR2 and TLR4 triggering exerts contrasting effects with regard to HIV-1 infection of human dendritic cells and subsequent virus transfer to CD4+ T cells. Retrovirology. 2009;6:42 (Research Support, Non-U.S. Gov't).
83. Mureith MW, Chang JJ, Lifson JD, Ndung'u T, Altfeld M. Exposure to HIV-1-encoded Toll-like receptor 8 ligands enhances monocyte response to microbial encoded Toll-like receptor 2/4 ligands. AIDS. 2010;24(12):1841–8 (Research Support, N.I.H., Extramural Research Support, Non-U.S. Gov't).

84. Nordone SK, Ignacio GA, Su L, Sempowski GD, Golenbock DT, Li L, et al. Failure of TLR4-driven NF-kappa B activation to stimulate virus replication in models of HIV type 1 activation. AIDS Res Hum Retroviruses. 2007;23(11):1387–95.
85. Heggelund L, Muller F, Lien E, Yndestad A, Ueland T, Kristiansen KI, et al. Increased expression of toll-like receptor 2 on monocytes in HIV infection: possible roles in inflammation and viral replication. Clin Infect Dis. 2004;39(2):264–9.
86. Song Y, Zhuang Y, Zhai S, Huang D, Zhang Y, Kang W, et al. Increased expression of TLR7 in CD8(+) T cells leads to TLR7-mediated activation and accessory cell-dependent IFN-gamma production in HIV type 1 infection. AIDS Res Hum Retroviruses. 2009;25(12):1287–95.
87. Lester RT, Yao XD, Ball TB, McKinnon LR, Kaul R, Wachihi C, et al. Toll-like receptor expression and responsiveness are increased in viraemic HIV-1 infection. AIDS. 2008;22(6):685–94.
88. Spranger S, Frankenberger B, Schendel DJ. NOD/scid IL-2Rg(null) mice: a preclinical model system to evaluate human dendritic cell-based vaccine strategies in vivo. J Transl Med. 2012;10:30.
89. Cravens PD, Melkus MW, Padgett-Thomas A, Islas-Ohlmayer M, Del PMM, Garcia JV. Development and activation of human dendritic cells in vivo in a xenograft model of human hematopoiesis. Stem Cells. 2005;23(2):264–78.
90. Hofer U, Schlaepfer E, Baenziger S, Nischang M, Regenass S, Schwendener R, et al. Inadequate clearance of translocated bacterial products in HIV-infected humanized mice. PLoS Pathog. 2010;6(4):e1000867.

Chapter 24
Latent HIV-1 Infection of Resting CD4$^+$ T cells: Testing Approaches to Overcome HIV Latency

Shailesh K. Choudhary

Abbreviations

ALT	Antilatency therapy
ART	Antiretroviral therapy
BET	Bromodomain and extraterminal
BLT	Bone marrow–liver–thymus
Brd4	Bromodomain containing 4
CTD	C-terminal domain
CMV	Cytomegalovirus
CTL	Cytotoxic T-lymphocyte
DC	Dendritic cell
EBV	Epstein–Barr virus
HSC	Hematopoietic stem cells
HDAC	Histone deacetylases
HDACis	HDAC inhibitors
HMBA	Hexamethylene bisacetamide
hu	Humanized
IT	Immunotoxin
IUPM	Infected cells per million
LTR	Long terminal repeat
NK	Natural killer
NELF	Negative elongation factor
NSG	NOD/scid-$\gamma c^{-/-}$
PKC	Protein Kinase C
P-TEFb	Positive transcription elongation factor b
RCI	Resting CD4$^+$ T-cell infection
VOR	Vorinostat

S. K. Choudhary (✉)
Department of Medicine, Gene Therapy Center, University of North Carolina at Chapel Hill, CB 7352, Chapel Hill, NC 27599, USA
e-mail: shailesh_choudhary@med.unc.edu

L. Y. Poluektova et al. (eds.), *Humanized Mice for HIV Research*,
DOI 10.1007/978-1-4939-1655-9_24

24.1 Introduction

HIV-1 infection persists in patients even after years of ART. Although ART suppresses viremia to a clinically undetectable level, the use of ultrasensitive assays shows that low-level viremia often persists [1]. Furthermore, intensification of ART does not result in a decline in low-level viral replication suggesting that persistently infected long-lived cells are the major source of residual viremia [2, 3]. Indeed, latently infected memory $CD4^+$ T cells have been characterized as a long-term reservoir of persistent infection in the peripheral blood of humans [4, 5].

In addition to latent $CD4^+$ T cells, suboptimal concentration of antiretrovirals in certain anatomical compartments such as the central nervous system, and the lower level of active drugs within cells have been suggested to contribute to low-level viremia and the maintenance of persistent infection [6]. Because the majority of lymphocytes (~98 %) reside in lymphoid tissues such as the spleen, lymph nodes, and gut-associated lymphoid tissue, it is imperative to measure drug concentration, residual viremia, and latent infection in those specific anatomical compartments to address HIV-1 eradication. Such studies cannot be conducted ethically or safely in patients and must be addressed in a representative and tractable animal model of HIV-1 persistence in the context of suppressive ART. The data generated in such a system has the potential to clarify the mechanism of viral persistence, the effect of combination strategies in different cellular and anatomical compartments for viral clearance and ultimately should help optimize rational advancement of eradication therapies in clinical studies.

24.2 Humanized Mice as the Models of Latent HIV-1 Infection in Resting $CD4^+$ T Cells

$Rag2^{-/-}\gamma c^{-/-}$ mice transplanted with human $CD34^+$ hematopoietic stem cells (HSC) and NOD/scid-$\gamma c^{-/-}$ (NSG) mice transplanted with HSC in combination with a fetal liver/thymus implant, known as bone marrow–liver–thymus (BLT) mice, show stable reconstitution of human T, B, natural killer (NK), and dendritic cells in both primary and secondary lymphoid organs [7, 8]. These humanized (hu) mice support viral replication and $CD4^+$ T-cell depletion after infection with both CCR5 and CXCR4 tropic HIV-1 [9–12]. It has been demonstrated by others and ourselves that plasma viremia in these mice can be suppressed below the limit of detection with ART, and that discontinuation of ART results in viral rebound, consistent with the presence of persistent infection [12–16]. We further identified the presence of resting $CD4^+$ T-cell infection (RCI) in the context of suppressive ART and quantified the replication-competent but quiescent HIV-1 within resting $CD4^+$ T cells [14, 17]. The frequency of RCI in the hu-$Rag2^{-/-}\gamma c^{-/-}$ and BLT mice ranged from one to twelve infectious units per million (IUPM) similar to that observed in humans. These results support the relevance of this model for studying HIV-1 latency and ALT, and lay the groundwork for using this model to pursue detailed studies of residual viral expression within specific tissues, parallel measures of ART concentrations at these sites, and quantitative studies of persistently infected resting $CD4^+$ T cells.

24.3 Approaches to Overcome HIV-1 Latency in Hu-Mouse Models

HIV-1 preferentially integrates into the introns of transcriptionally active genes in resting $CD4^+$ T cells [18]. Therefore, the establishment and maintenance of HIV-1 latency in most cases is due to either active regulation of HIV- transcription by corepressor complexes or a consequence of transcriptional interference. HIV-1 latency, however, is amenable to therapeutic intervention: targeting restrictions at transcriptional initiation with cytokines [19], or at transcriptional elongation with hexamethylene bisacetamide (HMBA) [20] leads to viral expression. Additionally, several signaling pathways have been identified that are critical to the maintenance of HIV-1 latency in chronically infected cell lines and primary $CD4^+$ T cells (Table 24.1) [19–36] and might be exploited to study HIV-1 eradication in the murine models (Figure 24.1). Because HIV-1 latency is a cumulative result of numerous restrictions that are imposed by cellular and viral factors, combination strategies are likely to be required for complete elimination of HIV-1 [37]. As empirical combination testing

Table 24.1 Signaling pathways that allow proviral expression in latently infected cells

Stimulus	Intervention	Physiological activity	Effects on HIV-1 reactivation
HDAC inhibitors	Vorinostat, panobinostat, belinostat, givinostat, entinostat	Histone and nonhistone acetylation/chromatin remodeling	21–23
Protein lysine methyltransferases inhibitor	BIX-01294	Chromatin remodeling	24
DNA cytosine methylation inhibitor	5-aza-2′-deoxycytidine	Removal of DNA CpG methylation	25, 26
P-TEFb release	HMBA	PKC, PI3K/AKT signaling	20, 27
Tat binding to P-TEFb	JQ1	Inhibits bromodomain reader BRD4 to recruit P-TEFb to cellular promoter	28
NFκB activators	Anti-CD3/CD28	T- cell receptor engagement	29
	Prostratin Bryostatin	PKC activation	30–33
	TNFα	TRAF recruitment	34
γcR engagement	IL-7	JAK/STAT signaling	19, 35, 36

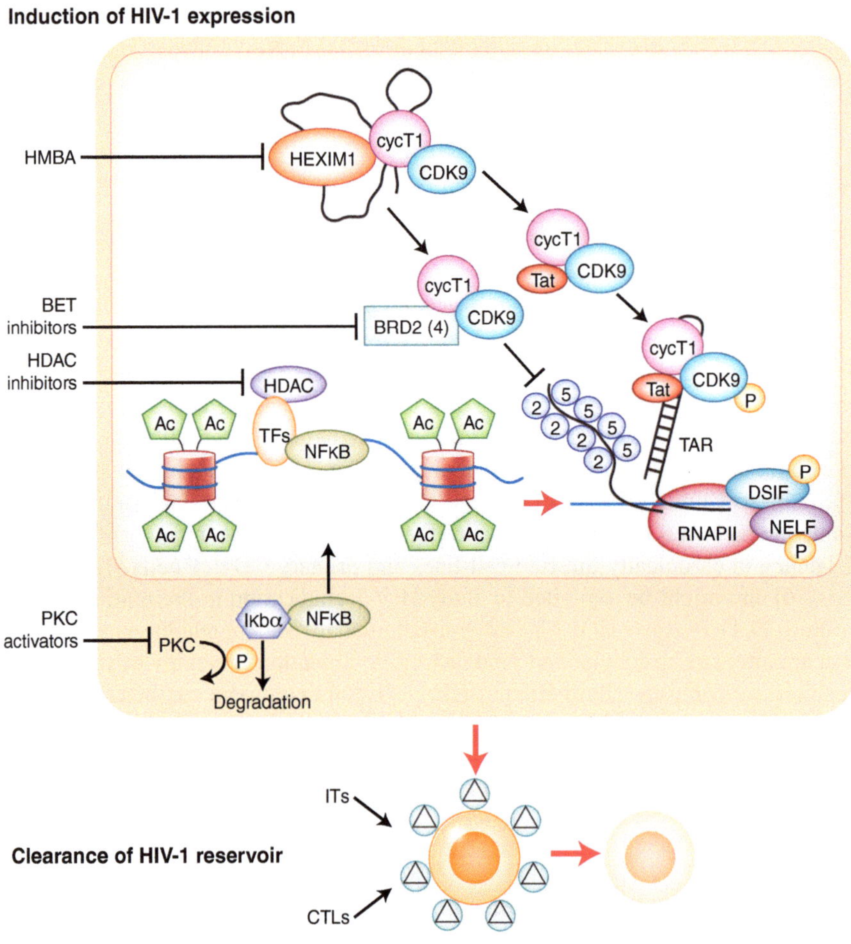

Fig. 24.1 Strategies to eradicate latent HIV-1 infection in resting CD4⁺ T cell. **a** Induction of HIV-1 expression. Class I histone deacetylases (*HDACs*) recruited by various transcription factors (*TFs*) inhibit the transcription from HIV-1 long terminal repeat (*LTR*). HDAC inhibitors facilitate histone acetyltransferases recruitment, histone acetylation, and chromatin remodeling at HIV-1 LTR allowing increased viral expression. The positive transcription elongation factor b (*P-TEFb*), composed of cyclin T1 (*cycT1*) and cyclin-dependent kinase 9 (*CDK9*), is sequestered in an inhibitory complex by hexamethylene bisacetamide-inducible protein (HEXIM1) as part of eukaryotic gene regulation. Hexamethylene bisacetamide (*HMBA*) mediated signaling allows release of P-TEFb from HEXIM1 complex. Both HIV-1 and cellular genes compete to recruit free P-TEFb to their promoter by interaction with Tat (viral transactivator) and BET protein family members BRD2/BRD4, respectively. BET inhibitor such as JQ1 enhances the availability of P-TEFb to Tat, although a Tat-independent effect of BET inhibitor on transcription elongation is also reported. The recruitment of P-TEFb to HIV-1 TAR (trans-activation-response region) allows phosphorylation of negative elongation factor (*NELF*) and DRB (5,6-dichloro-1-β-D-ribofuranosylbenzimidazole) sensitivity-inducing factor (*DSIF*), causing release of NELF and converting DSIF to a positive transcription factor. Furthermore, P-TEFb phosphorylates serine 2 (*Ser2*) and serine 5 (*Ser5*) of the carboxy-terminal domain (*CTD*) of RNA polymerase II (*RNAPII*), increasing its processivity and therefore viral transcription. NF-κB binds to DNA more avidly and induces latent HIV-1

cannot be rapidly or ethically performed in humans, insights gained from preclinical testing of antilatency strategies in a relevant animal model of HIV-1 latency should facilitate the implementation of the most promising therapies in humans.

24.3.1 Targeting Epigenetic Restrictions

HIV-1 latency is established and maintained largely due to the recruitment of class 1 selective histone deacetylases (HDACs) to the HIV-1 long terminal repeat (LTR). The recruitment of HDACs leads to the removal of acetyl groups from lysine residues within the histone tail turning off the transcription. HDAC inhibitors (HDACis) allow histone acetylation and facilitate recruitment of transcriptional factors to initiate transcription (Figure 24.1). Among all the epigenetic reagents tested, class I specific HDACis are the most potent inducers of latent HIV-1 from patients' resting $CD4^+$ T cells ex vivo [21, 23, 38, 39]. Recently, Archin and colleagues have shown that administration of a single dose of the class I specific HDACi vorinostat (VOR) in patients is sufficient to induce HIV-1 transcription in resting $CD4^+$ T in vivo [22]. More clinical trials are currently underway (NCT01365065, NCT01319383) to study the effects of multiple doses of VOR, but these assessments are limited to the peripheral blood. The humanized mouse model in contrast would allow comprehensive evaluation of the ability of VOR to deplete RCI in different cellular and anatomical reservoirs and address whether combination ALT will be needed for HIV-1 eradication.

VOR has been shown to suppress CWR22 prostate tumors and ameliorate motor deficits in a mouse model of Huntington's disease [40, 41]. Doses of VOR up to 100 mg/kg are well tolerated in mice and could be evaluated for histone H3 acetylation and viral expression in human cells. To this end, we propose a framework to test ALT in humanized mouse models (Figure 24.2). Within this framework, a single dose and ultimately multiple VOR doses could be tested to study the kinetics of HIV-1 RNA expression in resting $CD4^+$ T cells. The failure to detect replication-competent virus and HIV-1 RNA in the VOR-treated group in early study cohorts together with the absence of rebound of plasma viremia in late study cohorts will serve as a measure of the success of this therapeutic intervention.

Partial achievement of decline in RCI, however, will suggest either the existence of additional mechanisms of HIV-1 persistence or the possibility of differential penetration of the drug within various anatomical compartments reducing the effectiveness of VOR. The former issue may be addressed through combinatorial interventions of VOR and non-HDACi drugs, whereas the latter may be remedied through testing of alternative, and possibly more penetrant, HDACis, including

expression. PKC activators such as prostratin, tumor necrosis factor-α, and bryostatin phosphorylates IκBα causing release and translocation of NF-κB to the nucleus promoting HIV-1 transcription. **b** Clearance of HIV-1 reservoir. Following induction of proviral expression, targeting infected cells with cytotoxic T-lymphocytes (*CTLs*) and HIV-1 gp120 specific immunotoxins (*ITs*) might facilitate clearance of the latent reservoir

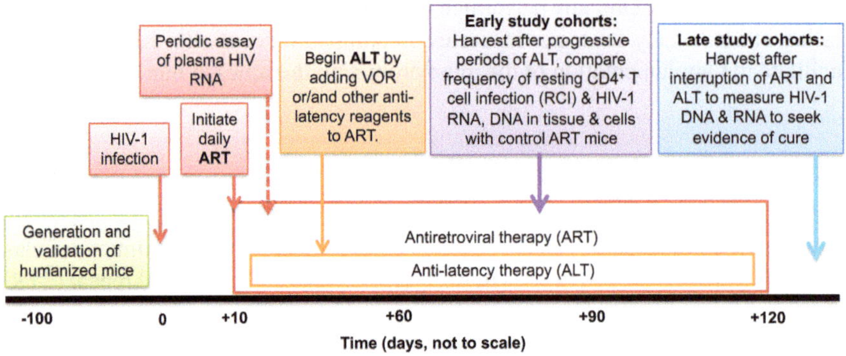

Fig. 24.2 Schematic for modeling eradication approaches in humanized mice

panobinostat, belinostat, givinostat, and entinostat, each of which has been shown to be effective at inducing latent HIV-1 infection in cultured cells [42, 43]. Indeed, many of these drugs are currently undergoing phase II clinical trials to treat cancers or juvenile arthritis and demonstration of their ability to clear quiescent provirus in an animal model should accelerate clinical testing for HIV-1 eradication.

24.3.2 Removing the Block at Transcription Elongation

Despite initiation of transcription, blocks at transcriptional elongation in resting cells in the absence of active positive transcription elongation factor b (P-TEFb), a cyclin-dependent kinase controlling elongation by RNA polymerase II, can contribute to the maintenance of HIV-1 latency (Figure 24.1) [20, 44, 45]. The recruitment of P-TEFb to the promoter region triggers phosphorylation of the negative elongation factor (NELF) and their dissociation, as well as phosphorylation of the C-terminal domain (CTD) of RNA polymerase, which triggers efficient initiation and elongation [46, 47]. Because P-TEFb is critical to the regulation of mammalian gene expression, its availability is tightly regulated by its sequestration in the large inhibitory hexamethylene bisacetamide-inducible protein 1 (HEXIM1) complex [48–50]. The HIV-1 transactivator Tat competes with HEXIM1 for P-TEFb binding and recruits active P-TEFb to its own promoter, allowing efficient transcription of HIV-1 genes [51]. Because the quantity of active P-TEFb in resting $CD4^+$ T cells is low, its availability to the HIV-1 promoter may be limited, effectively restricting robust HIV-1 transcription [45]. HMBA, a potent inducer of cell differentiation and HIV production in chronically infected cells, facilitates Tat-independent release of active P-TEFb from the inhibitory complex [52] and its recruitment to the HIV-1 LTR [20, 27]. It is important to note that we have observed HMBA-induced HIV-1 viral outgrowth from RCI [20], which further supports P-TEFb activity at the HIV-1 promoter as a critical component of latency release.

Recently, it was discovered that the cellular bromodomain and extraterminal (BET) family of human bromodomain containing 4 (Brd4) competes with Tat to recruit active P-TEFb to their respective promoters through their interaction with mediators and chromatin [53–55]. The bromodomain inhibitor JQ1, which inhibits Brd4-mediated recruitment of P-TEFb, predictably causes a significant increase in viral transcription in latently infected $CD4^+$ T cells [56]. Whereas HMBA triggers the transient release of the active P-TEFb complex, JQ1 enhances its availability to the HIV-1 promoter by antagonizing HIV-1 competitive factors and thereby allowing efficient transcription elongation [27, 54]. It has also been reported that Brd4 and related family member Brd2 inhibit HIV-1 transcription in the absence of Tat and the BET inhibitor allows transcription elongation in a Tat-independent manner [57]. Notably, both HMBA and JQ1 induce HIV-1 expression without T-cell activation, the most desired property of antilatency reagents [56].

Despite their promise, the therapeutic potential of HMBA and JQ1 cannot be addressed in in vivo systems. Humanized mice provide an ideal platform to characterize the antilatency effects of HMBA and JQ1 either alone or in combination. The biological activity of HMBA/JQ1 correlates with the availability of active P-TEFb complex as indicated by phosphorylation of CDK9 at threonine 186 and this could be measured by flow cytometry [37]. The measurement of the kinetics of HIV-1 RNA expression in resting $CD4^+$ T cells together with the quantitation of RCI will allow assessing the ability of HMBA and JQ1 to deplete latent HIV-1 infection.

24.3.3 Induction of NF-κB by Protein Kinase C (PKC) Signaling

The enhancer region of the HIV-1 promoter contains binding sites for the transcription factor NF-κB. NF-κB is postulated to be a potent inducer of HIV-1 transcription in part due to its high affinity for its binding site on DNA and therefore promotes proviral expression even when latency is maintained due to transcriptional interference. The PKC agonists prostratin and bryostatin-1 activate NF-κB and induce latent HIV expression alone and enhance the antilatency potential of VOR synergistically [33, 58]. Interestingly, each of these drugs downregulates CD4 and coreceptors CXCR4 and CCR5, thereby limiting de novo HIV-1 infection [59, 60]. Several analogs of both bryostatin and prostratin have recently been synthesized and tested for induction of latent HIV expression [61, 62] and their successful testing in an animal model will allow optimization of better PK parameters, higher activity, and less toxicity for human testing.

Bryostatin-1 induces expression of CD69, an early activation marker of T cells and a surrogate marker for induction of HIV-1 latency, in more than 50 % of cells up to 24 h following a single intravenous injection [63]. To verify NF-κB activation following bryostatin exposure, electrophoretic mobility shift assays are routinely employed to demonstrate upregulation of active p50/p65 heterodimers. The remainder of the experimental approach should be similar to that outlined in Figure 24.2. In the end, if only partial depletion of the viral reservoir is observed, it will bolster the argument for an adjunctive approach targeting different mechanisms of viral persistence to achieve HIV-1 eradication.

24.4 Adjunctive Approaches to Clear Persistent Infection

The disruption of HIV latency via the induction of proviral expression is a critical research goal. However, it is not certain that proviral expression alone will allow the host immune response or viral cytopathic effects to clear of persistent infection. Recently Shan and colleagues have shown that following induction of HIV-1 expression, latently infected cells survive even in the presence of autologous cytotoxic T-lymphocytes (CTLs) but can be eliminated if autologous CTLs are prestimulated in vitro with HIV-1 peptides [64]. The immune response in chronic HIV-1 infection is generally weak due to impaired CTL responses and is unable to eradicate HIV-1 infected cells [65, 66]. Furthermore, memory $CD4^+$ T cells are resistant to apoptotic death largely due to higher expression of Bcl-2 [67]. Combining HIV-1 induction with adjunctive therapies such as anti-HIV immunotoxin (IT) or CTL, which facilitates the clearance of HIV-1 expressing cells, may be more effective in eradication of the latent reservoir in vivo.

24.4.1 Selective Killing of Latently Infected Cells with Immunotoxins (ITs)

ITs are potent cytotoxic agents containing a targeting moiety, typically an antibody or ligand, conjugated to a toxin. HIV-1 expressing cells can be directly targeted by ITs against the HIV-1 gp120 proteins that are expressed on the surface of HIV-1 infected cells [68]. Brooks and colleagues have demonstrated that anti-gp120 IT is effective at killing latently infected T cells from SCID-hu thy/liv mice after stimulation ex vivo [69]. Moreover, treatment of infected mice with the immunotoxins CD4-PE40 and 3B3(Fv)-PE38 caused sustained suppression of viremia even after discontinuation of ART, suggesting depletion of the latent viral reservoir [70]. Despite its success, CD4-PE–based IT have been shown to cause some liver toxicity in rhesus macaques, which is associated with the high isoelectric point of the CD4 moiety. However, if CD4-PE is shown to be effective in depleting latently infected cells in a mouse model, the isoelectric point could be lowered for future studies without affecting its biological activity [71]. 3B3-PE exhibits no hepatotoxicity in rhesus macaques, although viral clearance cannot be accurately assessed in macaques on ART as the SIV Env in RT-SHIVs (chimeric SIVs harboring the HIV-1 reverse transcriptase) are not reactive with the 3B3 antibody. The murine model of HIV-1 latency could provide an alternative platform for rapidly testing the safety and efficacy of a proof-of-principle eradication strategy that involves: (a) induction of HIV-1 expression in latently infected cells in the context of ART, coupled with (b) targeted killing of HIV-1 expressing cells with an anti-HIV immunotoxin.

24.4.2 Selective Killing of Latently Infected Cells with Cytotoxic T-Lymphocyte (CTL)

CTL therapy targeting viral antigens along with ALT are considered as a potential therapeutic strategy to eradicate HIV-1 infection in humans. The feasibility of this approach has been strengthened by the success of autologous T-cell therapies in treating Epstein–Barr virus (EBV) positive lymphoma and of stem cell transplantation of virus-specific CTLs targeting cytomegalovirus (CMV) and EBV positive tumors in patients [72, 73]. The humanized mouse models could provide an ideal platform for tailoring effector T-cell responses to eliminate HIV-1 expressing cells. The implantation of human fetal liver and thymus SCID-hu mice with $CD34^+$ HSCs transduced with lentival vector to express an HLA-A2.01-restricted anti-Gag TCR target has been demonstrated to kill cells expressing HIV antigens ex vivo [74]. NSG-BLT mice when reconstituted with HIV-1 gag 77–85 SLYNTVALT (SL9) specific TCR allows suppression of HIV-1 replication in the plasma and in multiple lymphoid organs in vivo [75]. HIV-1-specific immunoglobulin sequences linked to the signaling domain of the T-cell receptor zeta chain, when transduced into $CD8^+$ cells, also target infected cells expressing surface HIV-1 gp120 but in an HLA-independent manner [76]. In a proof-of-principle study, those TCRs could be transduced into $CD8^+$ T cells isolated from the spleen of humanized mice and reintroduced into infected animals receiving optimized HIV-1 inductive therapy without issues of immune rejection. Labeling of cells before reintroduction in the animals would allow the distribution of these cells to be studies in various anatomic compartments. Cells transduced with a construct lacking the SL9 or/and universal T-cell receptors can serve as a control to TCR-mediated killing. Alternatively, $CD8^+$ T cells could be stimulated with antigen-presenting cells such as monocyte-derived dendritic cells (DCs) pulsed with overlapping peptides for an antigen of interest to derive CTL for the clearance of latent infection.

24.4.3 Killing of Latently Infected Cells with Bcl-2 Inhibitors

The cytokine interleukin 7 (IL-7) plays a critical role in thymopoiesis, peripheral expansion, survival, and function of T cells and is therefore undergoing pilot trials as an immune adjuvant therapy in HIV-1 infection [77, 78]. IL-7 has also been shown to induce latent HIV-1 infection in resting $CD4^+$ T cells from aviremic patients ex vivo [19, 36]. Induction of NFAT via the JAK/STAT signaling plays a key role in HIV-1 expression from latently infected resting $CD4^+$ T cells. However, IL-7 has also been implicated in the maintenance of the latent HIV-1 reservoir by inducing homeostatic proliferation in the setting of low $CD4^+$ T-cell counts [79]. Administration of IL-7 to HIV-1 infected patients on ART resulted in an increase in the number of $CD4^+$ T cells harboring integrated HIV-1 DNA in peripheral circulation [80]. Bosque and colleagues report that IL-7 driven homeostatic proliferation did not result in depletion of the latent reservoir in memory $CD4^+$ T cells despite their partial reactivation [81]. Because homeostatic mechanisms contribute to the maintenance of human memory $CD4^+$ T cells in hu-$Rag2^{-/-}\gamma c^{-/-}$ mice, this model could address the role of IL-7 in the maintenance of HIV latency [82]. In the setting of ART, increases in RCI following IL-7 therapy compared to control would suggest an increase in the frequency of integrated HIV within resting $CD4^+$ T cells, as such cells may undergo mitosis and duplicate the proviral integrant without viral expression. This result would illuminate the potential effects of using IL-7, which would fill a pressing and clinically relevant knowledge gap, as the cytokine is currently being tested for its ability to enhance T-cell reconstitution and function in HIV infection [77, 78, 80].

IL-7 promotes survival by upregulating the expression of antiapoptotic proteins such as Bcl-2 and Mcl-1 [67]. It is tempting to speculate that sensitization of cells to apoptosis with a Bcl-2 inhibitor simultaneously with induction of HIV-1 expression might clear latent infection in memory cells undergoing homeostatic proliferation. The Bcl-2 inhibitors ABT-737 and obatoclax are in the early stages of clinical development to treat cancer and have shown promising results in some patients at tolerable doses [83, 84]. The use of Bcl-2 inhibitors is unlikely to be approved for ALT in humans due to limited safety data but could easily be tested as a proof-of-concept study in a humanized mouse model to augment the clearance of viral reservoirs in $CD4^+$ T cells.

24.5 Summary and Future Perspectives

Exploiting humanized mice to study HIV-1 latency and test new strategies to clear persistent infection to its fullest extent should accelerate the development of effective therapies to eradicate HIV-1 infection. In the setting of highly potent intensified ART, HDAC inhibitors such as VOR either alone or together with other inducers of HIV-1 expression such as bryostatin-1 may disrupt proviral quiescence, and allow clearance of infection. Interventions that increase the clearance of infected cells, such as chimeric toxins targeted to the HIV-1 envelope glycoprotein or HIV-1 specific CTLs together with HIV-1 induction may speed the eradication of chronic HIV-1 infection.

The complete eradication of HIV-1, however, will not be achieved until cellular and anatomic reservoirs can also be targeted. The comprehensive analysis of antiretrovirals and antilatency reagents, including a determination of compartmental and intracellular drug concentration along with measurements of viral replication, will clarify the contributions of specific tissue compartments in the maintenance of persistent infection, and will help define the most promising and effective approaches to eradicating persistent HIV-1 infection.

References

1. Palmer S, Wiegand AP, Maldarelli F, Bazmi H, Mican JM, Polis M, et al. New real-time reverse transcriptase-initiated PCR assay with single-copy sensitivity for human immunodeficiency virus type 1 RNA in plasma. J Clin Microbiol. 2003;41(10):4531–6. (Epub 2003/10/9).
2. Dinoso JB, Kim SY, Wiegand AM, Palmer SE, Gange SJ, Cranmer L, et al. Treatment intensification does not reduce residual HIV-1 viremia in patients on highly active antiretroviral therapy. Proc Natl Acad Sci U S A. 2009;106(23):9403–8. (Epub 2009/5/28).
3. Gandhi RT, Bosch RJ, Aga E, Albrecht M, Demeter LM, Dykes C, et al. No evidence for decay of the latent reservoir in HIV-1-infected patients receiving intensive enfuvirtide-containing antiretroviral therapy. J Infect Dis. 2010;201(2):293–6. (Epub 2009/12/17).
4. Siliciano JD, Kajdas J, Finzi D, Quinn TC, Chadwick K, Margolick JB, et al. Long-term follow-up studies confirm the stability of the latent reservoir for HIV-1 in resting CD4+ T cells. Nat Med. 2003;9(6):727–8. (Epub 2003/5/20).
5. Finzi D, Hermankova M, Pierson T, Carruth LM, Buck C, Chaisson RE, et al. Identification of a reservoir for HIV-1 in patients on highly active antiretroviral therapy. Science. 1997;278(5341):1295–300. (Epub 1997/11/21).
6. Gisolf EH, Enting RH, Jurriaans S, de Wolf F, van der Ende ME, Hoetelmans RM, et al. Cerebrospinal fluid HIV-1 RNA during treatment with ritonavir/saquinavir or ritonavir/saquinavir/stavudine. AIDS. 2000;14(11):1583–9. (Epub 2000/9/13).
7. Traggiai E, Chicha L, Mazzucchelli L, Bronz L, Piffaretti JC, Lanzavecchia A, et al. Development of a human adaptive immune system in cord blood cell-transplanted mice. Science. 2004;304(5667):104–7. (Epub 2004/4/6).
8. Melkus MW, Estes JD, Padgett-Thomas A, Gatlin J, Denton PW, Othieno FA, et al. Humanized mice mount specific adaptive and innate immune responses to EBV and TSST-1. Nat Med. 2006;12(11):1316–22. (Epub 2006/10/24).
9. Zhang L, Kovalev GI, Su L. HIV-1 infection and pathogenesis in a novel humanized mouse model. Blood. 2007;109(7):2978–81. (Epub 2006/11/30).
10. Baenziger S, Tussiwand R, Schlaepfer E, Mazzucchelli L, Heikenwalder M, Kurrer MO, et al. Disseminated and sustained HIV infection in CD34+ cord blood cell-transplanted Rag2-/-gamma c-/- mice. Proc Natl Acad Sci U S A. 2006;103(43):15951–6. (Epub 2006/10/14).
11. Berges BK, Akkina SR, Remling L, Akkina R. Humanized Rag2(-/-)gammac(-/-) (RAG-hu) mice can sustain long-term chronic HIV-1 infection lasting more than a year. Virology. 2010;397(1):100–3. (Epub 2009/11/20).
12. Dash PK, Gendelman HE, Roy U, Balkundi S, Alnouti Y, Mosley RL, et al. Long-acting NanoART ElicitsPotent antiretroviral and neuroprotective responses in HIV-1 Infected Humanized Mice. AIDS. 2012;26:2135–44. (Epub 2012/7/25).
13. Choudhary SK, Rezk NL, Ince WL, Cheema M, Zhang L, Su L, et al. Suppression of human immunodeficiency virus type 1 (HIV-1) viremia with reverse transcriptase and integrase inhibitors, CD4+T-cell recovery, and viral rebound upon interruption of therapy in a new model for HIV treatment in the humanized Rag2-/-{gamma}c-/- mouse. J Virol. 2009;83(16):8254–8. (Epub 2009/6/6).
14. Denton PW, Olesen R, Choudhary SK, Archin NM, Wahl A, Swanson MD, et al. Generation of HIV latency in humanized BLT mice. J Virol. 2012;86(1):630–4. (Epub 2011/10/21).
15. Sango K, Joseph A, Patel M, Osiecki K, Dutta M, Goldstein H. Highly active antiretroviral therapy potently suppresses HIV infection in humanized Rag2-/-gammac-/- mice. AIDS Res Hum Retroviruses. 2010;26(7):735–46. (Epub 2010/7/14).
16. Marsden MD, Kovochich M, Suree N, Shimizu S, Mehta R, Cortado R, et al. HIV latency in the humanized BLT mouse. J Virol. 2012;86(1):339–47. (Epub 2011/11/11).
17. Choudhary SK, Archin NM, Cheema M, Dahl NP, Garcia JV, Margolis DM. Latent HIV-1 infection of resting CD4(+) T cells in the humanized Rag2(-)/(-) gammac(-)/(-) mouse. J Virol. 2012;86(1):114–20. (Epub 2011/10/21).

18. Han Y, Lassen K, Monie D, Sedaghat AR, Shimoji S, Liu X, et al. Resting CD4+ T cells from human immunodeficiency virus type 1 (HIV-1)-infected individuals carry integrated HIV-1 genomes within actively transcribed host genes. J Virol. 2004;78(12):6122–33. (Epub 2004/5/28).
19. Wang FX, Xu Y, Sullivan J, Souder E, Argyris EG, Acheampong EA, et al. IL-7 is a potent and proviral strain-specific inducer of latent HIV-1 cellular reservoirs of infected individuals on virally suppressive HAART. J Clin Invest. 2005;115(1):128–37. (Epub 2005/1/5).
20. Choudhary SK, Archin NM, Margolis DM. Hexamethylbisacetamide and disruption of human immunodeficiency virus type 1 latency in CD4(+) T cells. J Infect Dis. 2008;197(8):1162–70. (Epub 2008/4/19).
21. Archin NM, Espeseth A, Parker D, Cheema M, Hazuda D, Margolis DM. Expression of latent HIV induced by the potent HDAC inhibitor suberoylanilide hydroxamic acid. AIDS Res Hum Retrovir. 2009;25(2):207–12. (Epub 2009/2/26).
22. Archin NM, Liberty AL, Kashuba AD, Choudhary SK, Kuruc JD, Crooks AM, et al. Administration of vorinostat disrupts HIV-1 latency in patients on antiretroviral therapy. Nature. 2012;487(7408):482–5. (Epub 2012/7/28).
23. Contreras X, Schweneker M, Chen CS, McCune JM, Deeks SG, Martin J, et al. Suberoylanilide hydroxamic acid reactivates HIV from latently infected cells. J Biol Chem. 2009;284(11):6782–9. (Epub 2009/1/13).
24. Imai K, Togami H, Okamoto T. Involvement of histone H3 lysine 9 (H3K9) methyltransferase G9a in the maintenance of HIV-1 latency and its reactivation by BIX01294. J Biol Chem. 2010;285(22):16538–45. (Epub 2010/3/26).
25. Blazkova J, Trejbalova K, Gondois-Rey F, Halfon P, Philibert P, Guiguen A, et al. CpG methylation controls reactivation of HIV from latency. PLoS Pathog. 2009;5(8):e1000554. (Epub 2009/8/22).
26. Kauder SE, Bosque A, Lindqvist A, Planelles V, Verdin E. Epigenetic regulation of HIV-1 latency by cytosine methylation. PLoS Pathog. 2009;5(6):e1000495. (Epub 2009/6/27).
27. Contreras X, Barboric M, Lenasi T, Peterlin BM. HMBA releases P-TEFb from HEXIM1 and 7SK snRNA via PI3K/Akt and activates HIV transcription. PLoS Pathog. 2007;3(10):1459–69. (Epub 2007/10/17).
28. Banerjee C, Archin N, Michaels D, Belkina AC, Denis GV, Bradner J, et al. BET bromodomain inhibition as a novel strategy for reactivation of HIV-1. J Leukoc Biol. 2012;92(6):1147–54. (Epub 2012/7/18).
29. Wong JK, Hezareh M, Gunthard HF, Havlir DV, Ignacio CC, Spina CA, et al. Recovery of replication-competent HIV despite prolonged suppression of plasma viremia. Science. 1997;278(5341):1291–5. (Epub 1997/11/21).
30. Korin YD, Brooks DG, Brown S, Korotzer A, Zack JA. Effects of prostratin on T-cell activation and human immunodeficiency virus latency. J Virol. 2002;76(16):8118–23. (Epub 2002/7/23).
31. Kulkosky J, Culnan DM, Roman J, Dornadula G, Schnell M, Boyd MR, et al. Prostratin: activation of latent HIV-1 expression suggests a potential inductive adjuvant therapy for HAART. Blood. 2001;98(10):3006–15. (Epub 2001/11/8).
32. Reuse S, Calao M, Kabeya K, Guiguen A, Gatot JS, Quivy V, et al. Synergistic activation of HIV-1 expression by deacetylase inhibitors and prostratin: implications for treatment of latent infection. PLoS ONE. 2009;4(6):e6093. (Epub 2009/7/1).
33. Perez M, de Vinuesa AG, Sanchez-Duffhues G, Marquez N, Bellido ML, Munoz-Fernandez MA, et al. Bryostatin-1 synergizes with histone deacetylase inhibitors to reactivate HIV-1 from latency. Curr HIV Res. 2010;8(6):418–29. (Epub 2010/7/20).
34. Wolschendorf F, Duverger A, Jones J, Wagner FH, Huff J, Benjamin WH, et al. Hit-and-run stimulation: a novel concept to reactivate latent HIV-1 infection without cytokine gene induction. J Virol. 2010;84(17):8712–20. (Epub 2010/6/12).
35. Chun TW, Engel D, Mizell SB, Ehler LA, Fauci AS. Induction of HIV-1 replication in latently infected CD4+ T cells using a combination of cytokines. J Exp Med. 1998;188(1):83–91. (Epub 1998/7/7).

36. Lehrman G, Ylisastigui L, Bosch RJ, Margolis DM. Interleukin-7 induces HIV type 1 outgrowth from peripheral resting CD4+ T cells. J Acquir Immune Defic Syndr. 2004;36(5):1103–4. (Epub 2004/7/13).
37. Choudhary SK, Margolis DM. Curing HIV: pharmacologic approaches to target HIV-1 latency. Annu Rev Pharmacol Toxicol. 2011;51:397–418. (Epub 2011/1/8).
38. Archin NM, Keedy KS, Espeseth A, Dang H, Hazuda DJ, Margolis DM. Expression of latent human immunodeficiency type 1 is induced by novel and selective histone deacetylase inhibitors. AIDS. 2009;23(14):1799–806. (Epub 2009/7/11).
39. Burnett JC, Lim KI, Calafi A, Rossi JJ, Schaffer DV, Arkin AP. Combinatorial latency reactivation for HIV-1 subtypes and variants. J Virol. 2010;84(12):5958–74. (Epub 2010/4/2).
40. Butler LM, Agus DB, Scher HI, Higgins B, Rose A, Cordon-Cardo C, et al. Suberoylanilide hydroxamic acid, an inhibitor of histone deacetylase, suppresses the growth of prostate cancer cells in vitro and in vivo. Cancer Res. 2000;60(18):5165–70. (Epub 2000/10/4).
41. Hockly E, Richon VM, Woodman B, Smith DL, Zhou X, Rosa E, et al. Suberoylanilide hydroxamic acid, a histone deacetylase inhibitor, ameliorates motor deficits in a mouse model of Huntington's disease. Proc Natl Acad Sci U S A. 2003;100(4):2041–6. (Epub 2003/2/11).
42. Matalon S, Rasmussen TA, Dinarello CA. Histone deacetylase inhibitors for purging HIV-1 from the latent reservoir. Mol Med. 2011;17(5–6):466–72. (Epub 2011/3/23).
43. Wightman F, Ellenberg P, Churchill M, Lewin SR. HDAC inhibitors in HIV. Immunol Cell Biol. 2012;90(1):47–54. (Epub 2011/11/16).
44. Lassen KG, Bailey JR, Siliciano RF. Analysis of human immunodeficiency virus type 1 transcriptional elongation in resting CD4+ T cells in vivo. J Virol. 2004;78(17):9105–14. (Epub 2004/8/17).
45. Wolschendorf F, Bosque A, Shishido T, Duverger A, Jones J, Planelles V, et al. Kinase control prevents HIV-1 reactivation in spite of high levels of induced NF-κB activity. J Virol. 2012;86(8):4548–58. (Epub 2012/2/22).
46. Ping YH, Rana TM. DSIF and NELF interact with RNA polymerase II elongation complex and HIV-1 Tat stimulates P-TEFb-mediated phosphorylation of RNA polymerase II and DSIF during transcription elongation. J Biol Chem. 2001;276(16):12951–8. (Epub 2000/12/22).
47. Zhou M, Halanski MA, Radonovich MF, Kashanchi F, Peng J, Price DH, et al. Tat modifies the activity of CDK9 to phosphorylate serine 5 of the RNA polymerase II carboxyl-terminal domain during human immunodeficiency virus type 1 transcription. Mol Cell Biol. 2000;20(14):5077–86. (Epub 2000/6/24).
48. Barboric M, Kohoutek J, Price JP, Blazek D, Price DH, Peterlin BM. Interplay between 7SK snRNA and oppositely charged regions in HEXIM1 direct the inhibition of P-TEFb. EMBO J. 2005;24(24):4291–303. (Epub 2005/12/20).
49. Yik JH, Chen R, Nishimura R, Jennings JL, Link AJ, Zhou Q. Inhibition of P-TEFb (CDK9/Cyclin T) kinase and RNA polymerase II transcription by the coordinated actions of HEXIM1 and 7SK snRNA. Mol Cell. 2003;12(4):971–82. (Epub 2003/10/29).
50. Markert A, Grimm M, Martinez J, Wiesner J, Meyerhans A, Meyuhas O, et al. The La-related protein LARP7 is a component of the 7SK ribonucleoprotein and affects transcription of cellular and viral polymerase II genes. EMBO Rep. 2008;9(6):569–75. (Epub 2008/5/17).
51. Barboric M, Yik JH, Czudnochowski N, Yang Z, Chen R, Contreras X, et al. Tat competes with HEXIM1 to increase the active pool of P-TEFb for HIV-1 transcription. Nucleic Acids Res. 2007;35(6):2003–12. (Epub 2007/3/8).
52. Klichko V, Archin N, Kaur R, Lehrman G, Margolis D. Hexamethylbisacetamide remodels the human immunodeficiency virus type 1 (HIV-1) promoter and induces Tat-independent HIV-1 expression but blunts cell activation. J Virol. 2006;80(9):4570–9. (Epub 2006/4/14).
53. Zhu J, Gaiha GD, John SP, Pertel T, Chin CR, Gao G, et al. Reactivation of latent HIV-1 by inhibition of BRD4. Cell Rep. 2012;2(4):807–16. (Epub 2012/10/9).
54. Bartholomeeusen K, Xiang Y, Fujinaga K, Peterlin BM. Bromodomain and extra-terminal (BET) bromodomain inhibition activate transcription via transient release of positive transcription elongation factor b (P-TEFb) from 7SK small nuclear ribonucleoprotein. J Biol Chem. 2012;287(43):36609–16. (Epub 2012/9/7).

55. Li Z, Guo J, Wu Y, Zhou Q. The BET bromodomain inhibitor JQ1 activates HIV latency through antagonizing Brd4 inhibition of Tat-transactivation. Nucleic Acids Res. 2013;41(1):277–87. (Epub 2012/10/23).
56. Banerjee C, Archin N, Michaels D, Belkina AC, Denis GV, Bradner J, et al. BET bromodomain inhibition as a novel strategy for reactivation of HIV-1. J Leukoc Biol. 2012;92:1147–54. (Epub 2012/7/18).
57. Boehm D, Calvanese V, Dar RD, Xing S, Schroeder S, Martins L, et al. BET bromodomain-targeting compounds reactivate HIV from latency via a Tat-independent mechanism. Cell Cycle. 2013;12(3):452–62. (Epub 2012/12/21).
58. Williams SA, Chen LF, Kwon H, Fenard D, Bisgrove D, Verdin E, et al. Prostratin antagonizes HIV latency by activating NF-κB. J Biol Chem. 2004;279(40):42008–17. (Epub 2004/7/31).
59. Boto WM, Brown L, Chrest J, Adler WH. Distinct modulatory effects of bryostatin 1 and staurosporine on the biosynthesis and expression of the HIV receptor protein (CD4) by T cells. Cell Regul. 1991;2(2):95–103. (Epub 1991/2/1).
60. Biancotto A, Grivel JC, Gondois-Rey F, Bettendroffer L, Vigne R, Brown S, et al. Dual role of prostratin in inhibition of infection and reactivation of human immunodeficiency virus from latency in primary blood lymphocytes and lymphoid tissue. J Virol. 2004;78(19):10507–15. (Epub 2004/9/16).
61. Beans EJ, Fournogerakis D, Gauntlett C, Heumann LV, Kramer R, Marsden MD, et al. Highly potent, synthetically accessible prostratin analogs induce latent HIV expression in vitro and ex vivo. Proc Natl Acad Sci U S A. 2013;110(29):11698–703. (Epub 2013/7/3).
62. DeChristopher BA, Loy BA, Marsden MD, Schrier AJ, Zack JA, Wender PA. Designed, synthetically accessible bryostatin analogues potently induce activation of latent HIV reservoirs in vitro. Nat Chem. 2012;4(9):705–10. (Epub 2012/8/24).
63. Bear HD, McFadden AW, Kostuchenko PJ, Lipshy KA, Hamad GG, Turner AJ, et al. Bryostatin 1 activates splenic lymphocytes and induces sustained depletion of splenocyte protein kinase C activity in vivo after a single intravenous administration. Anticancer Drugs. 1996;7(3):299–306. (Epub 1996/5/1).
64. Shan L, Deng K, Shroff NS, Durand CM, Rabi SA, Yang HC, et al. Stimulation of HIV-1-specific cytolytic T lymphocytes facilitates elimination of latent viral reservoir after virus reactivation. Immunity. 2012;36(3):491–501. (Epub 2012/3/13).
65. Rosenberg ES, Altfeld M, Poon SH, Phillips MN, Wilkes BM, Eldridge RL, et al. Immune control of HIV-1 after early treatment of acute infection. Nature. 2000;407(6803):523–6. (Epub 2000/10/12).
66. Chouquet C, Autran B, Gomard E, Bouley JM, Calvez V, Katlama C, et al. Correlation between breadth of memory HIV-specific cytotoxic T cells, viral load and disease progression in HIV infection. AIDS. 2002;16(18):2399–407. (Epub 2002/12/4).
67. Lees JR, Farber DL. Generation, persistence and plasticity of CD4 T-cell memories. Immunology. 2010;130(4):463–70. (Epub 2010/5/15).
68. Berger EA. Targeted cytotoxic therapy: adapting a rapidly progressing anticancer paradigm for depletion of persistent HIV-infected cell reservoirs. Curr Opin HIV AIDS. 2011;6(1):80–5. (Epub 2011/1/19).
69. Brooks DG, Hamer DH, Arlen PA, Gao L, Bristol G, Kitchen CM, et al. Molecular characterization, reactivation, and depletion of latent HIV. Immunity. 2003;19(3):413–23. (Epub 2003/9/23).
70. Goldstein H, Pettoello-Mantovani M, Bera TK, Pastan IH, Berger EA. Chimeric toxins targeted to the human immunodeficiency virus type 1 envelope glycoprotein augment the in vivo activity of combination antiretroviral therapy in thy/liv-SCID-Hu mice. J Infect Dis. 2000;181(3):921–6. (Epub 2000/3/18).
71. Onda M, Nagata S, Tsutsumi Y, Vincent JJ, Wang Q, Kreitman RJ, et al. Lowering the isoelectric point of the Fv portion of recombinant immunotoxins leads to decreased nonspecific animal toxicity without affecting antitumor activity. Cancer Res. 2001;61(13):5070–7. (Epub 2001/6/30).

72. Savoldo B, Huls MH, Liu Z, Okamura T, Volk HD, Reinke P, et al. Autologous Epstein-Barr virus (EBV)-specific cytotoxic T cells for the treatment of persistent active EBV infection. Blood. 2002;100(12):4059–66. (Epub 2002/10/24).
73. Feuchtinger T, Opherk K, Bethge WA, Topp MS, Schuster FR, Weissinger EM, et al. Adoptive transfer of pp65-specific T cells for the treatment of chemorefractory cytomegalovirus disease or reactivation after haploidentical and matched unrelated stem cell transplantation. Blood. 2010;116(20):4360–7. (Epub 2010/7/14).
74. Kitchen SG, Bennett M, Galic Z, Kim J, Xu Q, Young A, et al. Engineering antigen-specific T cells from genetically modified human hematopoietic stem cells in immunodeficient mice. PLoS ONE. 2009;4(12):e8208. (Epub 2009/12/10).
75. Kitchen SG, Levin BR, Bristol G, Rezek V, Kim S, Aguilera-Sandoval C, et al. In vivo suppression of HIV by antigen specific T cells derived from engineered hematopoietic stem cells. PLoS Pathog. 2012;8(4):e1002649. (Epub 2012/4/19).
76. Yang OO, Tran AC, Kalams SA, Johnson RP, Roberts MR, Walker BD. Lysis of HIV-1-infected cells and inhibition of viral replication by universal receptor T cells. Proc Natl Acad Sci U S A. 1997;94(21):11478–83. (Epub 1997/10/23).
77. Hodge JN, Srinivasula S, Hu Z, Read SW, Porter BO, Kim I, et al. Decreases in IL-7 levels during antiretroviral treatment of HIV infection suggest a primary mechanism of receptor-mediated clearance. Blood. 2011;118(12):3244–53. (Epub 2011/7/23).
78. Sereti I, Dunham RM, Spritzler J, Aga E, Proschan MA, Medvik K, et al. IL-7 administration drives T cell-cycle entry and expansion in HIV-1 infection. Blood. 2009;113(25):6304–14. (Epub 2009/4/22).
79. Chomont N, El-Far M, Ancuta P, Trautmann L, Procopio FA, Yassine-Diab B, et al. HIV reservoir size and persistence are driven by T cell survival and homeostatic proliferation. Nat Med. 2009;15(8):893–900. (Epub 2009/6/23).
80. Vandergeeten C, Fromentin R, DaFonseca S, Lawani MB, Sereti I, Lederman MM, et al. Interleukin-7 promotes HIV persistence during antiretroviral therapy. Blood. 2013;121(21):4321–9. (Epub 2013/4/17).
81. Bosque A, Famiglietti M, Weyrich AS, Goulston C, Planelles V. Homeostatic proliferation fails to efficiently reactivate HIV-1 latently infected central memory CD4+ T cells. PLoS Pathog. 2011;7(10):e1002288. (Epub 2011/10/15).
82. Legrand N, Huntington ND, Nagasawa M, Bakker AQ, Schotte R, Strick-Marchand H, et al. Functional CD47/signal regulatory protein alpha (SIRP(alpha)) interaction is required for optimal human T- and natural killer- (NK) cell homeostasis in vivo. Proc Natl Acad Sci U S A. 2011;108(32):13224–9. (Epub 2011/7/27).
83. Mitchell C, Yacoub A, Hossein H, Martin AP, Bareford MD, Eulitt P, et al. Inhibition of MCL-1 in breast cancer cells promotes cell death in vitro and in vivo. Cancer Biol Ther. 2010;10(9):903–17. (Epub 2010/9/22).
84. O'Brien SM, Claxton DF, Crump M, Faderl S, Kipps T, Keating MJ, et al. Phase I study of obatoclax mesylate (GX15-070), a small molecule pan-Bcl-2 family antagonist, in patients with advanced chronic lymphocytic leukemia. Blood. 2009;113(2s):299–305. (Epub 2008/10/22).

Chapter 25
Brain HIV-1 Infection Modeling in Humanized Mice

Larisa Y. Poluektova, Adrian A. Epstein and Santhi Gorantla

Abbreviations

CSF-1	Monocyte-colony stimulating factor
DNA-PKcs	Catalytic subunit of the DNA-dependent protein kinase
HAND	HIV-1-associated neurologic disorders
HSC	Hematopoietic stem cells
IL-2Rγ_c	Interleukin-2 receptor common gamma chain
NSG/NOG	NOD/Shi LtJ-*scid/IL2R*γ^{null} mice
scid	Severe combined immune deficiency
SIRP- α	Signal regulatory protein-α

25.1 Introduction

The HIV-1-associated damage of central and peripheral nervous systems became evident very early in the HIV pandemic [1]. The three clinical forms, now considered as HIV-1-associated neurologic disorders (HAND): asymptomatic, which do not affect human daily life, but could be found by rigorous neuropsychological evaluation; mild cognitive and motor deficit; and dementia. In the era of highly active antiretroviral therapy (HAART), the incidence of dementia reduced and the character of mild neurocognitive impairment changed to the memory/learning and executive control rather than the motor and cognitive speed deficits seen pre-HAART [2]. However, up to now several questions remain unanswered. Is the

L. Y. Poluektova (✉) · A. A. Epstein · S. Gorantla
Department of Pharmacology & Experimental Neuroscience, University of Nebraska Medical Center, 985880 Nebraska Medical Center, Omaha, NE 68198-5880, USA
e-mail: lpoluekt@unmc.edu

L. Y. Poluektova et al. (eds.), *Humanized Mice for HIV Research*,
DOI 10.1007/978-1-4939-1655-9_25

damage of nervous system occurring primarily due to persistent productive infection within the brain? Is HAND a combination of several connected pathological processes, such as peripheral lymphoid tissue infection or immune suppression? Is it a combination of both events and therefore related to total levels of viral replication? Are neurological symptoms directly related to the neurotoxicity of viral proteins or indirect-metabolic effects of inflammation? Does chronic immune activation associated with increased permeability of gut contributes to neuronal dysfunction? Does HIV-1 tropism to different cell types (lymphocytes or macrophages) or viral envelope protein coreceptor usage (CCR5 or CXCR4) and ability to bind different types of G-protein-coupled chemokine receptors important for neuronal dysfunction?

Humanized mice on NSG/NOG backgrounds with hemato-lymphoid reconstitution produce human host cells for HIV-1 infection and the infected-lymphocytes and macrophages, are able to enter mouse brain.

Thus it was very attractive to investigate if it is possible to elicit HIV-1-associated neuropathology in humanized animals.

25.2 Human HIV-1 Infected Cells in the Mouse Brain

HIV-1 virus belongs to *Lentiviradae* [3, 4]. Encephalopathy as well as encephalitis driven by persistent infection of brain resident microglia/macrophages are the cause of neurocognitive deficits, neurodegeneration and dementia. For two other members of the family that induce encephalitis without immune suppression, like the visna-maedi virus and the caprine arthritis encephalitis virus, the major target in the brain are macrophages/microglia [5]. The suggested common features of the pathogenesis of *Lentiviruses* are their ability to integrate DNA into monocytes/dendritic cells. When these cells reach their final tissue destinations, including brain, they mature to macrophages in the presence of CSF-1 (monocyte-colony stimulating factor, M-CSF) and disseminate viral particles and spread infection to microglia/macrophages. Short-term modeling of HIV-1 encephalitis in C.B-17/*scid* and NOD/*scid* mice was achieved by transplantation of HIV-1-infected bone marrow-derived human macrophages, with consequential neuropathologies characterized, and this model was applied for antiretroviral and adjunctive therapeutic development [6, 7]. However, this model was lacking important components of the peripheral HIV-1 infection of lymphoid tissues, such as loss of $CD4^+$ cells that contribute to disease progression. Moreover, infection within the brain is secondary to systemic infection and appropriate animal modeling should utilize blood-brain transmission of cell-free virus or infected cells. Perivascular distribution of infected macrophages within the highly vascularized brain are uniquely found in simian immunodeficiency viruses (SIV) infected animals, but for human encephalitic brains up to two third of infected cells are parenchymal microglia [8]. For HIV-1, not all laboratory variants and primary isolates of different viral clades are able to productively infect

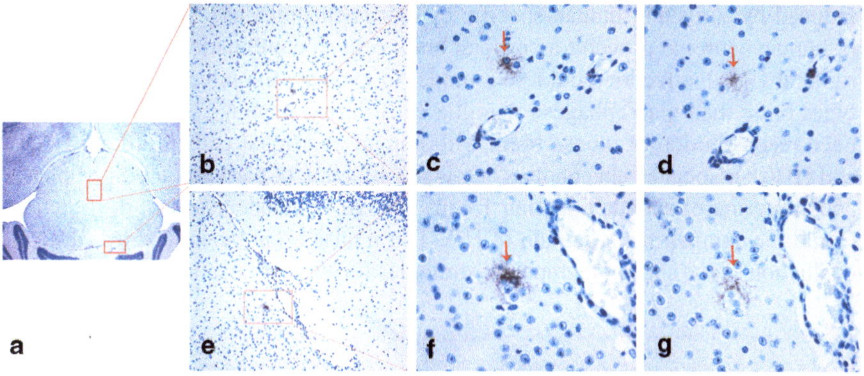

Fig. 25.1 HIV-1p24-positive cells in humanized mouse encephalitic brain. **a** Representative 5 μ thick horizontal sections of paraffin-embedded mouse brain tissue stained with antibodies to human HLA-DR antigen. Human cells (*brown*) are distributed around the vessels, along the meninges and fissures. Magnification 1×. *Red squares* are areas magnified in panels and **b** and **e,** stained for HIV-1p24 antigen. Positive cells are shown in perivascular spaces and white-matter tracks. Magnification 10×. **c-d**, and **f-g** are corresponding further magnified views of two adjacent sections confirming the microglia-like morphology of HIV-1p24-positive cells indicated with *red arrows*. Magnification 40×

monocyte/macrophages, and not all viral strains will be able to induce brain pathology [9]. The selection of viruses for humanized mice infection is important, as well as the time of analysis. We were able to observe scattering HIV-1p24+ human lymphocytes through the white-matter tracks, striatum, olfactory bulbs, and their interaction with macrophages in the meninges mostly in mice sacrificed at peak of viremia at 5–7 weeks [10]. Depletion of human $CD8^+$ cells in HIV-1-infected humanized mice also significantly increased the entry of infected cells into the brain and incidence of meningitis [11]. In a rare case of encephalitis, developed in HIV-1-infected humanized mouse after $CD8^+$ cell depletion, HIV-1p24 positive cells with microglial morphology were found. The origin of these cells remains unknown leaving with two predictions: mouse microglia phagocytized human HIV-1 infected cells, or human microglia-like cells became infected (Fig. 25.1). As of today there is no evidence that HIV-1 is productively infecting neurons, astrocytes, oligodendrocytes, epindemal cells of choroid plexus, and brain microvasculature of murine origin. Animals sacrificed at 16–17 weeks after infection occasionally had infected meningeal or perivascular cells. Productive infection of mouse microglial cells is possible, if virus will be delivered by different mechanisms (not human CD4, CCR5, or CXCR4), such as vesicular stomatitis virus (VSV)-pseudotyped [12, 13]. HIV-1 infection of human brain endothelial cells, oligodendrocytes, astrocytes, immature neuroblasts, and mature neurons was suggested and discussed in review [8].

For the humanized-mouse model for HAND, several questions are unanswered. Are the human cell products associated with chronic HIV-1 infection affecting the mouse endothelial cells and blood-brain-barrier? On the other hand, do we need to transplant human neurons and astrocytes into the mouse brain to confirm the effects induced by virus and human-specific cytokines/chemokines? The transplantation of human neuronal progenitors or fetal brain cells in combination with chronic peripheral HIV-1 infection could be a model to evaluate the infection of neural cell types other than macrophages/microglia cells during human CNS disease. Do we need reconstitution of mouse brain with human microglial cells to mimic HAND and HIV-1-associated encephalitis? Or is the infection of perivascular macrophages and scattering of activated-human T cells enough to induce neuronal dysfunction? Could the compartmentalization of HIV-1 in mouse brain be created and used for evaluation of the special therapeutic strategies for viral eradication [14]?

25.3 Repopulation of Mouse Brain with Human Microglia

Establishment of a stable viral brain-reservoir is an unmet challenge due to the requirement of a substantial population of human microglial cells within the mouse brain. The nature of mouse microglia as a yolk-sac-born cell and a unique lineage of macrophages was confirmed by several recent studies [15–18]. These cells are developmentally dependent on the transcription factor PU.1 and interferon regulatory factor 8 (IRF8), as well independent of Myb (*my*elo*b*lastosis) family of transcription factors, DNA-binding protein inhibitor Id2, basic leucine zipper transcription factor, ATF-like 3 (Batf3), and Kruppel-like factor 4 (Klf4) that delineate bone marrow-derived monocytic lineage. Microglia also differentiate independently of CSF-1 and are present in CSF-1 KO mice as a result of the IL-34 action. The human microglia seed brain during the embryonic stage. Limited studies about microglia about microglia during human embryonic development detect microglial cells as early as in the fifth week of gestation [19]. Microglia are expected to accumulate parallel to neurovasculogenesis, but the precise markers to distinguish human microglial cells from the later hematopoietic monocytes/macrophages remain unknown.

In mice, fate-mapping studies showed that by one year of age ~30% of microglia-like cells in the brain will be bone-marrow-derived cells [15]. This observation is in concordance with data obtained earlier on NOD/*scid*-humanized mice when stable and significant human myelopoiesis was established [20]. Accordingly, humanized mice should have some extent of human origin microglia-like cells in the brain. Animals carrying human hematopoiesis survive over a year with maintained meylopoiesis.

Analysis of human myelopoiesis in a one-year-old animal engrafted neonatally with cord-blood-isolated $CD34^+$ cells is shown in Fig. 25.2. The lifelong development of monocytes by retaining lineage-negative $CD34^+$ HSCs in the BM of humanized animals ensure that human monocytes/macrophages are generated and able to enter mouse brain.

These may not be “true” microglial cells and have different properties. Development of these cells requires, in addition to CSF-1, a species-specific cytokine IL-34 that also reacts via CSF-1R [17]. Taken together and resolving around these

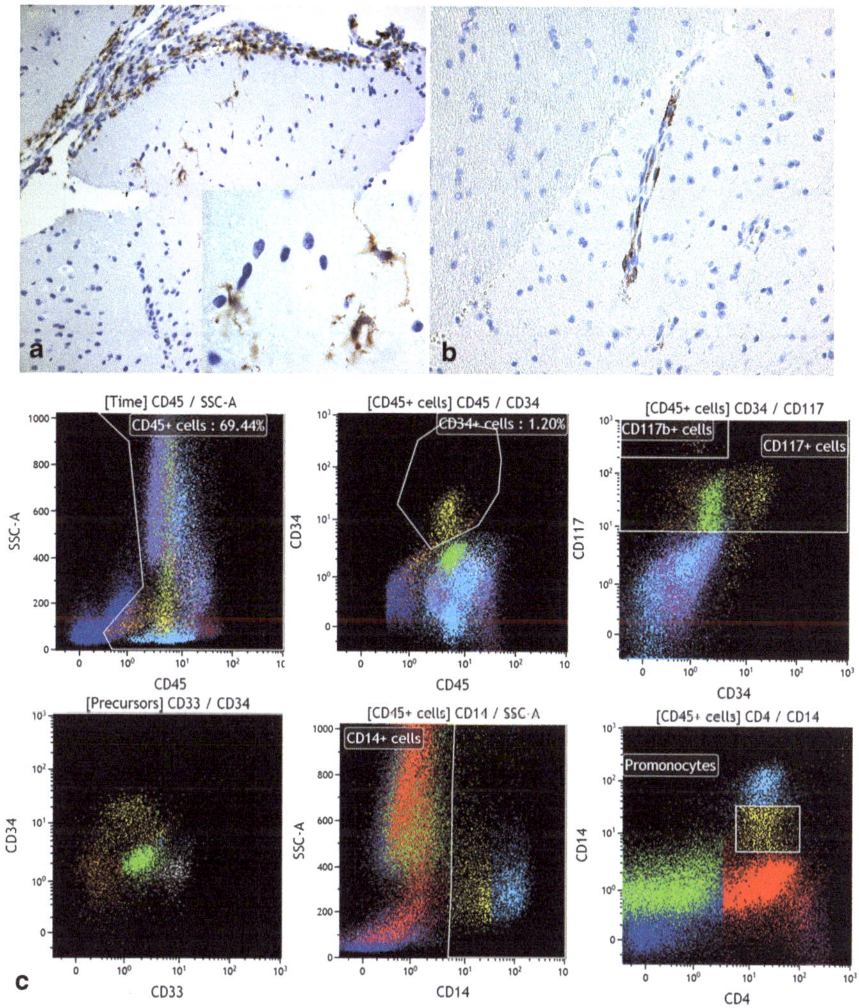

Fig. 25.2 Morphology of human microglia and perivascular macrophages in mouse brain and development of human monocytes in mouse bone marrow. **a** and **b**, Representative 5 μ thick sections of paraffin-embedded mouse brain tissue stained for CD14 showing microglia-like cells with different morphologies found in humanized NSG mice at 6–12 months of age. Cells are close to meninges They are ramified and expressed low levels of CD14 (**a**), or had elongated shape and located around the vessel (**b**). Images were captured under objective 20× and insert under objective 100×. **c** FACS analysis of the bone marrow cells obtained from a 1 year old humanized mouse created by transplantating cord-blood-derived CD34$^+$ cells at birth. Bone marrow contained 69 % of human CD45$^+$ cells. Several sets of antibodies were used to characterize progenitors and monocyte development. *Upper line* panels: *left* panel shows human CD45$^+$, *middle* panel -CD34$^+$, and *right* panel -CD34+/CD117+ progenitors and CD117b+ mast cells. *Low line* panels: *left* represents myeloid progenitors CD33/CD34$^+$, *middle* -CD14$^+$ cells from CD45$^+$ gated cells, and *right* panel highlights the proportion of CD45/CD14/CD4$^+$ promonocytes. (Data were collected using a Beckman Coulter FC500 flow cytometer (Beckman Coulter, Miami, FL) and analyzed using Beckman Coulter Cytomics CXP software (Applied Cytometry Systems, Dinnington, United Kingdom). Results are expressed as percentages of total human CD45$^+$ cells. Courtesy of Professor Samuel Pirrucello, Pathology/Microbiology, UNMC)

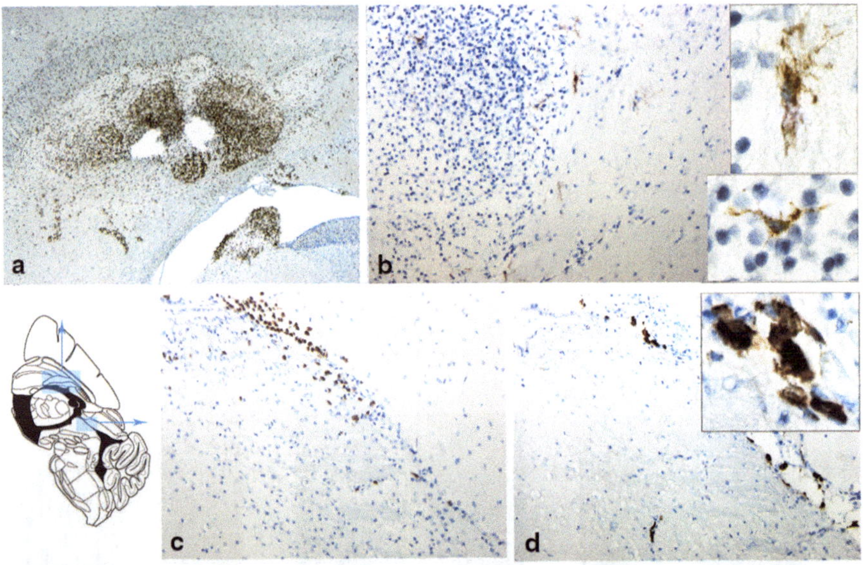

Fig. 25.3 Pattern of the distribution of human microglia and macrophages derived from fetal brain cells. NSG mice were injected at birth with human fetal brain derived mixed cell population cultured for 1 week under serum-free conditions. At 3 months of age, human cells of macrophage/microglia lineage were found without peripheral transplantation of human CD34$^+$ HSC. **a** and **c**, Representative distribution of human DNA-PKcs-positive cells around lateral ventricle and along the aqueduct. **b** Human microglia inside and around the cluster of human immature neurons/astrocytes stained for CD45 with characteristic ramified morphology. **d** Human CD163-positive cells with round macrophages morphology. Inserts represent magnified views of a single or group of cells. *Black* and *white* schematic representation of sagittal brain section at the levels of the captured images depicted as *blue* rectangles. **a** magnification 4×, **b**, **c**, and **d** magnification 20×, inserts—magnification 100×. Sections counterstained with hematoxylin

uncertainties, populating the mouse brain with human microglia requires human microglial precursors and a niche for them in mouse brain.

We confirmed the feasibility of generating human brain-resident microglia and macrophages by transplantation of human fetal brain cells into the brain of newborn irradiated NSG mouse without peripheral transplantation of human CD34+ cells. As shown in Fig. 25.3, human cells with microglial (CD45^{low}) or macrophage (CD163$^+$) morphology are present and distributed outside the clusters of human cells and distinguished by nuclei stained with antibodies to human DNA-PKcs, a molecule that has truncated and reduced expression in *scid/scid* animals. This observation suggests that appropriate cytokine support (like IL-34), appropriate selection of microglial precursors, and development of strategies to oblate mouse microglia at birth followed by human cell transplantation, are needed to create a stable pool of human cells resident in the mouse brain. Such achievement will provide a novel animal model system susceptible to HIV-1 infection to study viral compartmentalization and CNS penetrating antiretroviral drugs. Although the introduction of HAART has been able to reverse some of the clinical manifestations, pathological alterations persist within the CNS of infected patients, current drug regimens hardly penetrate the CNS. Therefore limited in controlling HIV replication within the brain.

25.4 Conclusion and Future Directions

Humanized mice are life-long carriers of human hematopoiesis and human-cell-populated lymphoid tissues susceptible to HIV-1 infection. Limited in the spectrum of brain-resident populations of human cells (macrophages/microglia), they are nonetheless attractive and important tools for HAND research. Introduction of strategies to evaluate in vivo longitudinal changes of HIV-1-infected mouse brain by clinically translatable tools such as magnetic resonance imaging (MRI), magnetic resonance spectroscopy (MRS) [21], diffusion-tensor MR imaging [22, 23], and new experimental imaging methods such as manganese-enhanced MRI (MEMRI) [24] will open new horizons in our understanding of brain pathology and facilitate the development of a new prevention and treatment strategies.

Acknowledgments This contribution was supported by the Chair of the Department of Pharmacology and Experimental Neuroscience Dr. Howard E. Gendelman and grants P01 NS043985 DHHS/NIH/NINDS, P01 DA028555 DHHS/NIH/NIDA, R01AG043540. The authors would like to thank Edward Makarov, Jaclyn Knibbe-Hollinger, Tanuja Gutti, Prashanta Dash, Aditya Bade, and Sidra Akhter for ongoing contributions in humanized mice for NeuroAIDS projects.

References

1. Epstein LG, Sharer LR, Cho ES, Myenhofer M, Navia B, Price RW. HTLV-III/LAV-like retrovirus particles in the brains of patients with AIDS encephalopathy. AIDS Res. 1984;1(6):447–54.
2. Heaton RK, Franklin DR, Ellis RJ, McCutchan JA, Letendre SL, Leblanc S, et al. HIV-associated neurocognitive disorders before and during the era of combination antiretroviral therapy: differences in rates, nature, and predictors. J Neurovirol. 2010;17(1):3–16.
3. Koenig S, Gendelman HE, Orenstein JM, Dal Canto MC, Pezeshkpour GH, Yungbluth M, et al. Detection of AIDS virus in macrophages in brain tissue from AIDS patients with encephalopathy. Science. 1986;233(4768):1089–93.
4. Epstein LG, Sharer LR, Gajdusek DC. Hypothesis: AIDS encephalopathy is due to primary and persistent infection of the brain with a human retrovirus of the lentivirus subfamily. Med Hypotheses. 1986;21(1):87–96.
5. Blacklaws BA. Small ruminant lentiviruses: immunopathogenesis of visna-maedi and caprine arthritis and encephalitis virus. Comp Immunol Microbiol Infect Dis. 2012;35(3):259–69.
6. Tyor WR, Power C, Gendelman HE, Markham RB. A model of human immunodeficiency virus encephalitis in scid mice. Proc Natl Acad Sci U S A. 1993;90(18):8658–62.
7. Persidsky Y, Limoges J, McComb R, Bock P, Baldwin T, Tyor W, et al. Human immunodeficiency virus encephalitis in SCID mice. Am J Pathol. 1996;149(3):1027–53.
8. Bissel SJ, Wiley CA. Human immunodeficiency virus infection of the brain: pitfalls in evaluating infected/affected cell populations. Brain Pathol. 2004;14(1):97–108.
9. Koppensteiner H, Brack-Werner R, Schindler M. Macrophages and their relevance in human immunodeficiency virus type I infection. Retrovirology. 2012;9:82.

10. Gorantla S, Makarov E, Finke-Dwyer J, Castanedo A, Holguin A, Gebhart CL, et al. Links between progressive HIV-1 infection of humanized mice and viral neuropathogenesis. Am J Pathol. 2010;177(6):2938–49.
11. Gorantla S, Makarov E, Finke-Dwyer J, Gebhart CL, Domm W, Dewhurst S, et al. CD8 + cell depletion accelerates HIV-1 immunopathology in humanized mice. J Immunol. 2010;184(12):7082–91.
12. Dou H, Morehead J, Bradley J, Gorantla S, Ellison B, Kingsley J, et al. Neuropathologic and neuroinflammatory activities of HIV-1-infected human astrocytes in murine brain. Glia. 2006;54(2):81–93.
13. Gorantla S, Liu J, Sneller H, Dou H, Holguin A, Smith L, et al. Copolymer-1 induces adaptive immune anti-inflammatory glial and neuroprotective responses in a murine model of HIV-1 encephalitis. J Immunol. 2007;179(7):4345–56.
14. Forsman A, Weiss RA. Why is HIV a pathogen? Trends Microbiol. 2008;16(12):555–60.
15. Ginhoux F, Greter M, Leboeuf M, Nandi S, See P, Gokhan S, et al. Fate mapping analysis reveals that adult microglia derive from primitive macrophages. Science. 2010;330(6005):841–5.
16. Horiuchi M, Wakayama K, Itoh A, Kawai K, Pleasure D, Ozato K, et al. Interferon regulatory factor 8/interferon consensus sequence binding protein is a critical transcription factor for the physiological phenotype of microglia. J Neuroinflammation. 2012;9:227.
17. Wang Y, Szretter KJ, Vermi W, Gilfillan S, Rossini C, Cella M, et al. IL-34 is a tissue-restricted ligand of CSF1R required for the development of Langerhans cells and microglia. Nat Immunol. 2012;13(8):753–60.
18. Kierdorf K, Erny D, Goldmann T, Sander V, Schulz C, Perdiguero EG, et al. Microglia emerge from erythromyeloid precursors via Pu.1- and Irf8-dependent pathways. Nat Neurosci. 2013;16(3):273–80.
19. Verney C, Monier A, Fallet-Bianco C, Gressens P. Early microglial colonization of the human forebrain and possible involvement in periventricular white-matter injury of preterm infants. J Anat. 2010;217(4):436–48.
20. Asheuer M, Pflumio F, Benhamida S, Dubart-Kupperschmitt A, Fouquet F, Imai Y, et al. Human CD34 + cells differentiate into microglia and express recombinant therapeutic protein. Proc Natl Acad Sci U S A. 2004;101(10):3557–62.
21. Dash PK, Gorantla S, Gendelman HE, Knibbe J, Casale GP, Makarov E, et al. Loss of neuronal integrity during progressive HIV-1 infection of humanized mice. J Neurosci. 2011;31(9):3148–57.
22. Towgood KJ, Pitkanen M, Kulasegaram R, Fradera A, Kumar A, Soni S, et al. Mapping the brain in younger and older asymptomatic HIV-1 men: frontal volume changes in the absence of other cortical or diffusion tensor abnormalities. Cortex. 2012;48(2):230–41.
23. Muller-Oehring EM, Schulte T, Rosenbloom MJ, Pfefferbaum A, Sullivan EV. Callosal degradation in HIV-1 infection predicts hierarchical perception: a DTI study. Neuropsychologia. 2010;48(4):1133–43.
24. Liu Y, Sajja BR, Gendelman HE, Boska MD. Mouse brain fixation to preserve In vivo manganese enhancement for ex vivo manganese-enhanced MRI. J Magn Reson Imaging. 24 Jan 2013. doi:10.1002/jmri.24005.

Part IV
Humanized Mice for HIV-1-Specific Adaptive Cellular and Humoral Immune Responses

Chapter 26
Antibody-based Protection Against HIV Infection

Brian Moldt and Dennis R. Burton

26.1 Introduction

Neutralizing antibodies have shown to be a correlate of protection for many viral vaccines and are, therefore, a highly sought after property in a vaccine against Human Immunodeficiency Virus (HIV) [1, 2]. The development of such a vaccine has been elusive for a variety of reasons that are beyond the scope of this review (reviewed elsewhere, [3–5]). Nevertheless, a rapidly increasing number of highly potent broadly neutralizing antibodies (bNAbs) have been isolated from chronically infected HIV patients, and in an attempt to learn how to elicit such antibodies in a vaccine, an extensive characterization has been carried out [6–12]. To investigate the protective potency of these neutralizing antibodies against viral challenge in vivo, passive antibody transfer experiments have been performed in rhesus macaques, chimpanzees, and humanized mice [13–23]. The macaque has been the model of choice for most studies. However, it does have some limitations, notably the use of a chimeric simian-human immunodeficiency virus (SHIV) rather than HIV-1 as a challenge virus, the possibility of antihuman antibodies being raised in the macaque and potential suboptimal interaction of human antibodies with macaque effector cells and critical molecules such as Fc receptors. The chimpanzee is an endangered species and was only used in early animal studies related to AIDS research.

Several different types of humanized mice have been used in vaccine-related studies including passive transfer of antibodies, and it is worth noting that the

B. Moldt (✉) · D. R. Burton
Department of Immunology and Microbial Science, International AIDS Vaccine Initiative Neutralizing Antibody Center, and Center for HIV/AIDS Vaccine Immunology and Immunogen Discovery, The Scripps Research Institute, La Jolla, CA 92037, USA
e-mail: bmoldt@scripps.edu

D. R. Burton
Ragon Institute of Massachusetts General Hospital, Massachusetts Institute of Technology, and Harvard University, Boston, MA 02129, USA

L. Y. Poluektova et al. (eds.), *Humanized Mice for HIV Research*,
DOI 10.1007/978-1-4939-1655-9_26

existing types have differing levels of "humanization" and that each model will have its own set of advantages and limitations. Susceptibility to mucosal infection for example is probably most relevant for vaccine-related studies as the initial local viral propagation and expansion in mucosal tissue is believed to be the stage where the virus is most vulnerable [24, 25]. For treatment studies, the route of infection may be less important but establishment of viral reservoirs in specific tissues that mimics the pathogenesis seen in human infections may be crucial.

The first humanized mouse to be used in antibody-mediated protection studies was the hu-PBL-SCID mouse. However, mucosal transmission of HIV in this model is low and highly variable and these studies therefore relied on intraperitoneal challenge [19, 22, 23, 26–28]. More recent models such as the Rag-hu, the NSG-hu, and the BLT mouse have been shown to be susceptible to mucosal challenge, and the BLT mouse appears especially well-suited for mucosal transmission as the female reproductive tract and the rectum are efficiently populated with cells of human origin [29–31]. Key findings regarding neutralizing antibodies from humanized mouse studies are: (i) protection against primary HIV challenge by passive transfer of neutralizing antibodies [19, 22, 23], (ii) sustained delivery of protective neutralizing antibodies by viral vectors (vectored immunoprophylaxis) [32], and (iii) HIV therapy by neutralizing antibodies [33]. The main focus here will be on passive transfer protection studies but short sections on vectored immunoprophylaxis and HIV therapy will also be included. Figure 26.1 illustrates the three concepts.

26.2 Passive Antibody Protection Studies in hu-PBL-SCID Mice

This model is rarely used today, however, the hu-PBL-SCID mouse was instrumental in early findings related to antibody-mediated protection against HIV in vivo. The first direct evidence demonstrating a role for neutralizing antibodies was obtained from passive transfer studies in chimpanzees but was shortly after followed by similar experiments in the hu-PBL-SCID mouse. In these proof-of-concept studies, it was shown that a strain specific neutralizing antibody recognizing the gp120 V3 region could protect against challenge of a homologous T-cell line-adapted (TCLA) virus [19, 20]. However, a subsequent mouse study also showed that the strain specific antibody was unable to protect against challenge of primary heterologous HIV strains emphasizing that a broad neutralizing antibody response would be needed in an effective vaccine [21].

The first demonstration of protection against a primary HIV stain was performed in the hu-PBL-SCID mice. Here, the authors showed that the CD4-binding site antibody b12, one of the first bNAbs to be isolated, protected against both challenge of TCLA viruses and challenge of a primary strain [22, 23]. One previous attempt had been done in chimpanzees using the gp41 membrane-proximal external region (MPER)-specific bNAb 2F5. However, the antibody only induced a delay in seroconversion and not sterilizing immunity [34, 35]. The lack of protection may be related to the potency of 2F5 against the particular challenge virus as a later study performed in macaques demonstrated efficient protection against a SHIV version of another primary isolate [17].

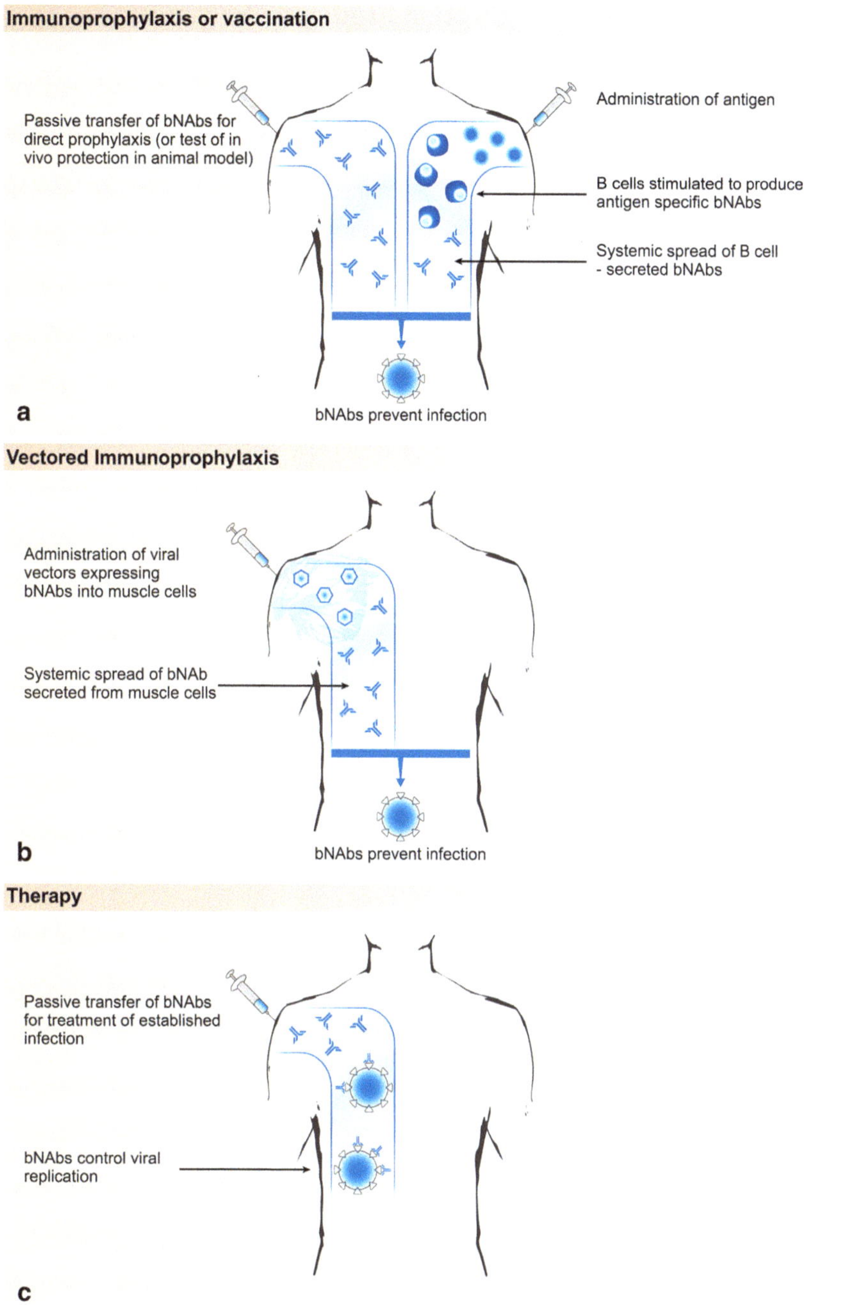

Fig. 26.1 Schematic representation of prophylaxis and treatment strategies against HIV using bNAbs. **a** Passive administration of antibodies has demonstrated the potency of bNAbs in protection against HIV. A conventional vaccination approach is shown on the *right*. **b** Vectored immunoprophylaxis as an alternative to conventional vaccination. bNAbs are delivered by AAV vectors and the approach thereby bypass the difficulties in developing an antigen that elicits bNAbs. **c** bNAbs as therapeutics. Passive administration of bNAbs represses viral replication in an established infection. (Figure adapted from Balazs and West [45])

26.3 Passive Antibody Protection Studies in Newer Humanized Mice Models

From these early studies in the hu-PBL-SCID mouse, the use of humanized mice for passive transfer studies waned in place of the macaque/SHIV model. However, there has been a renewed interest in parallel with the development of new models such as the Rag-hu, the NSG-hu, and the BLT mouse, as well as with the isolation of newer more potent broadly neutralizing antibodies. In contrast to the hu-PBL-SCID mouse, most of the new mouse models enable mucosal infection making them more suitable for evaluation of strategies designed to prevent mucosal HIV transmission [26, 29–31]. Two studies of particular interest related to antibody-mediated protection will be discussed below [36, 37].

The first study investigated the effect of antibody isotypes by comparing IgG1 b12 with monomeric and polymeric IgA2 b12 [37]. Interestingly, despite having comparable neutralization potency in vitro, it was found that polymeric IgA2 showed higher potency than IgG1 and monomeric IgA2 in protection against vaginal challenge of NSG-hu mice. The difference between monomeric and polymeric IgA2 could be related to the unique ability of polymeric IgA to form secretory IgA, however, it should be noted that in order to reach equivalent level of antibody in plasma and genital secretion at the time of challenge, the authors administered tenfold more IgA than IgG. A recent study using the macaque/SHIV model also showed enhanced protection of IgA in vivo compared to IgG [38]. However, in the macaque study, the enhanced protective effect was specific for dimeric IgA1 and not for dimeric IgA2. In addition, it was shown that the protection correlated with the antibody isotype's ability to mediate virion capture and inhibition of transcytosis in vitro [38]. The results from the two studies strongly advocate for additional experiments investigating the potency of IgA in vivo, especially because most previous passive antibody transfer studies have been done with IgG.

The second study investigated bNAbs as topical preexposure prophylaxis (PrEP) against HIV [36]. Here, using vaginal challenge of RAG-hu mice, it was shown that a cocktail of the bNAbs b12, 2G12, 4E10, and 2F5 were fully protective, and that the more recently isolated NAb VRC01 protected up to 80 % of the mice as a single antibody. In light of the encouraging results from the recent CAPRISA004 trial investigating the use of topically applied microbicides to prevent infection in humans, these findings are highly relevant [39]. The current microbicides against HIV are small-molecule drugs such as maraviroc and tenofovir [40]. However, if proven sufficiently effective, the use of biological molecules such as antibodies may represent an attractive alternative without the side effects commonly associated with antiviral drugs, although cost may be a substantial barrier. A previous study performed in macaques using the b12 antibody also provided support for pursuing the development of bNAbs as microbicides, and the mouse model in particular would allow for an quick and relatively inexpensive method of evaluating the potency of various bNAb combinations [36, 41].

In contrast to the conventional humanized mice models, an alternative approach is to generate transgenic mice expressing human genes facilitating susceptibility to HIV infection. Two related models based on bioluminescence were recently described in which human CD4 and CCR5 were either expressed in the liver or ubiquitously expressed in all tissues [42, 43]. Using these models, several of the newer broadly neutralizing antibodies, including 3BNC117, 3BC176, and VRC01, were shown to protect against pseudovirus challenge. Further studies will be needed to determine the level of translatability between the results generated here compared to results from the more established animal models.

26.4 Vectored Immunoprophylaxis

As highlighted in the previous sections, passive transfer studies in humanized mice and nonhuman primates have demonstrated highly efficient protection of bNAbs against HIV in vivo. However, the continuous difficulties in the development of an effective vaccine eliciting bNAbs have prompted an interest in pursuing alternative preventive strategies. One of the most promising is vectored immunoprophylaxis in which adeno-associated virus (AAV) vectors are used to deliver genes encoding bNAbs. The AAV vector has been demonstrated in various preclinical and clinical gene therapy trials to support long-term expression of transgenes in nondividing tissues such as muscle [44]. Two seminal studies have demonstrated the efficacy of AAV-based immunoprophylaxis against SIV and HIV [32, 45]. The first was performed in macaques and showed that a neutralizing immunoadhesin (chimeric antibody-like molecule) specific for SIV gp120, delivered by intramuscular injection of an AAV expression vector, mediated protection against an intravenous SIV challenge [45]. Importantly, serum levels of the immunoadhesin was shown to be stable for more than 1 year and still measurable 5 years postinjection with no adverse effects for the monkeys [46]. The mouse study, using NSG-hu mice, expanded upon these results by generating AAV vectors expressing full-length human antibodies including the bNAbs b12 and VRC01 [32]. Three key findings were achieved. Firstly, injection of AAV vectors expressing b12 or VRC01 into muscle tissue resulted in stable expression of human antibodies for an extended time period. Secondly, the antibodies were highly efficient in protecting against a high-dose intravenous HIV challenge. Thirdly, the observed protection was antibody dose-dependent and correlated with the potency of the bNAb as VRC01 protected at lower serum concentration than b12. Interestingly, in a separate study the authors showed that implementing the same AAV-based strategy resulted in efficient protection against influenza suggesting that the approach may be pursued for various infectious diseases if proven sufficiently safe for human use [47].

26.5 HIV Therapy

From the initial isolation of bNAbs against HIV, there has been an interest in developing these as antiviral drugs. It is important to note that the bNAbs do not control the infection in patients from whom they were isolated because of a continuing emergence of viral escape variants [48]. For therapeutic use of bNAbs, as seen for conventional antiretroviral therapy, a combination of antibodies with multiple epitope specificity may therefore be needed to inhibit the emergence of viable escape variants for long-term treatment. The first study to investigate the potential of bNAbs against an established infection was performed in hu-PBL-SCID mice [49]. The authors found that treating HIV infected mice with a single antibody (b12) was ineffective in controlling viral replication. Treatment with a cocktail of bNAbs (b12, 2G12, and 2F5) did suppress viral replication but only transiently. In light of this, the modest therapeutic effects observed when treating human HIV-infected patients following cessation of conventional antiviral therapy with a cocktail of 2G12, 2F5, and 4E10 may have been disappointing but perhaps anticipated [50]. In the human trial, the antibody cocktail did not control the virus but it did delay the viral rebound in a subpopulation of the treated patients. Interestingly, acutely infected individuals seemed more prone to respond to the treatment than chronically infected patients indicating that the virus diversification occurring in the establishment of chronic infection could restrict the antibody potency. It should also be noted that analysis of the post-therapy virus only showed escape from the 2G12 antibody indicating that more optimal antibody combinations may be more effective [50].

Despite the limited success in previous investigations, the recent isolation of more potent bNAbs inspired a renewed interested in using bNAbs as therapeutics and three studies have shown more promising results [33, 51, 52]. Here, using humanized NRG mice, the authors observed viral escape from treatment by a single antibody and a cocktail of three bNAbs (in line with what was seen in the previous mouse experiment). However, using a cocktail of five bNAbs (3BC176, PG16, 45-46G54W, PGT128, and 10-1074) the authors showed efficient repression of viral replication. The repression lasted as long as the antibodies were present, as viral rebound was not observed before discontinuing the antibody administrations. Interestingly, in follow-up studies a cocktail of three antibodies specifically designed to restrict the possibility of viral escape and a single antibody in the combination of antiretroviral therapy were found to be sufficient to control the infection [51, 52]. Additional animal (and human) studies will be needed to evaluate the impact of these encouraging results. However, the possibility of using bNAbs in therapy may be a new emerging field in line with the treatment of other diseases such as specific cancers and autoimmune disorders [53].

26.6 Summary

In conclusion, the early studies performed in the hu-PBL-SCID mouse provided critical evidence for protection against HIV by neutralizing antibodies supporting an important role of the humoral system in a vaccine. Newer types of humanized mice may be better suited to represent human infections and are gaining much interest from the field as valuable animal models. Alternative strategies to curb the HIV/AIDS pandemic such as vectored immunoprophylaxis and treatment of established infections by neutralizing antibodies have shown to be highly effective in several mouse studies and support further development.

Acknowledgments We thank Christina Corbaci for help with graphic design. The authors acknowledge the financial support of the National Institute of Allergy and Infectious Diseases, the Center for HIV/AIDS Vaccine Immunology and Immunogen Discovery; the International AIDS Vaccine Initiative; the Ragon Institute of MGH, MIT, and Harvard; and the Lundbeck Foundation.

References

1. Amanna IJ, Slifka MK. Contributions of humoral and cellular immunity to vaccine-induced protection in humans. Virology. 2011;411(2):206–15. (PubMed PMID: 21216425. Pubmed Central PMCID: Journal—In Process. Epub 2011/01/11. eng.).
2. Plotkin SA. Correlates of protection induced by vaccination. Clin Vaccine Immunol. 2010;17(7):1055–65. (PubMed PMID: 20463105. Pubmed Central PMCID: 2897268. Epub 2010/05/14. eng.).
3. Burton DR, Ahmed R, Barouch DH, Butera ST, Crotty S, Godzik A, et al. A blueprint for HIV vaccine discovery. Cell host microbe. 2012;12(4):396–407. (PubMed PMID: 23084910. Pubmed Central PMCID: 3513329).
4. Virgin HW, Walker BD. Immunology and the elusive AIDS vaccine. Nature. 2010;464(7286):224–31. (PubMed PMID: 20220841. Pubmed Central PMCID: Journal—In Process. Epub 2010/03/12. eng.).
5. Fauci AS, Johnston MI, Dieffenbach CW, Burton DR, Hammer SM, Hoxie JA, et al. HIV vaccine research: the way forward. Science. 2008;321(5888):530–2. (PubMed PMID: 18653883. Epub 2008/07/26. eng.).
6. Walker LM, Phogat SK, Chan-Hui PY, Wagner D, Phung P, Goss JL, et al. Broad and potent neutralizing antibodies from an African donor reveal a new HIV-1 vaccine target. Science. 2009;326(5950):285–9. (PubMed PMID: 19729618. Pubmed Central PMCID: 2916884. Epub 2009/09/05. Eng).
7. Walker LM, Huber M, Doores KJ, Falkowska E, Pejchal R, Julien JP, et al. Broad neutralization coverage of HIV by multiple highly potent antibodies. Nature. 2011;477(7365):466–70.
8. Wu X, Yang ZY, Li Y, Hogerkorp CM, Schief WR, Seaman MS, et al. Rational design of envelope surface identifies broadly neutralizing human monoclonal antibodies to HIV-1. Science. 2010;329(5993):856–61. (Pubmed Central PMCID: 2965066).
9. Huang J, Ofek G, Laub L, Louder MK, Doria-Rose NA, Longo NS, et al. Broad and potent neutralization of HIV-1 by a gp41-specific human antibody. Nature. 2012; In Press.

10. Scheid JF, Mouquet H, Ueberheide B, Diskin R, Klein F, Olivera TYK, et al. Sequence and structural convergence of broad and potent HIV antibodies that mimic CD4 binding. Science. 2011;333(6049):1633–7.
11. Bonsignori M, Hwang KK, Chen X, Tsao CY, Morris L, Gray E, et al. Analysis of a clonal lineage of HIV-1 envelope V2/V3 conformational epitope-specific broadly neutralizing antibodies and their inferred unmutated common ancestors. J Virol. 2011;85(19):9998–10009. (PubMed PMID: 21795340. Epub 2011/07/29. Eng.).
12. Liao HX, Lynch R, Zhou TQ, Gao F, Alam SM, Boyd SD, et al. Co-evolution of a broadly neutralizing HIV-1 antibody and founder virus. Nature. 2013;496(7446):469–76. (PubMed PMID: WOS:000317984400034. English).
13. Moldt B, Rakasz EG, Schultz N, Chan-Hui PY, Swiderek K, Weisgrau KL, et al. Highly potent HIV-specific antibody neutralization in vitro translates into effective protection against mucosal SHIV challenge in vivo. Proc Natl Acad Sci U S A. 2012;109(46):18921–5. (PubMed PMID: WOS:000311576300062. English).
14. Parren PW, Marx PA, Hessell AJ, Luckay A, Harouse J, Cheng-Mayer C, et al. Antibody protects macaques against vaginal challenge with a pathogenic R5 simian/human immunodeficiency virus at serum levels giving complete neutralization in vitro. J Virol. 2001;75(17):8340–7. (PubMed PMID: 11483779).
15. Mascola JR, Stiegler G, VanCott TC, Katinger H, Carpenter CB, Hanson CE, et al. Protection of macaques against vaginal transmission of a pathogenic HIV-1/SIV chimeric virus by passive infusion of neutralizing antibodies. Nat Med. 2000;6(2):207–10. (PubMed PMID: 10655111).
16. Hessell AJ, Rakasz EG, Poignard P, Hangartner L, Landucci G, Forthal DN, et al. Broadly neutralizing human anti-HIV antibody 2G12 is effective in protection against mucosal SHIV challenge even at low serum neutralizing titers. PLoS Pathog. 2009;5(5):e1000433. (PubMed PMID: 19436712. Pubmed Central PMCID: 2674935. Epub 2009/05/14. eng).
17. Hessell AJ, Rakasz EG, Tehrani DM, Huber M, Landucci G, Forthal DN, et al. Broadly neutralizing monoclonal antibodies 2F5 and 4E10 directed against the human immunodeficiency virus type 1 gp41 membrane-proximal external region protect against mucosal challenge by simian-human immunodeficiency virus SHIVBa-L. J Virol. 2010;84(3):1302–13. (Pubmed Central PMCID: 2812338).
18. Mascola JR, Lewis MG, Stiegler G, Harris D, VanCott TC, Hayes D, et al. Protection of macaques against pathogenic simian/human immunodeficiency virus 89.6PD by passive transfer of neutralizing antibodies. J Virol. 1999;73(5):4009–18. (PubMed PMID: 10196297).
19. Safrit JT, Fung MSC, Andrews CA, Braun DG, Sun WNC, Chang TW, et al. Hu-Pbl-Scid Mice Can Be Protected from Hiv-1 Infection by Passive Transfer of Monoclonal-Antibody to the Principal Neutralizing Determinant of Envelope Gp120. AIDS. 1993;7(1):15–21. (PubMed PMID: WOS:A1993KG97000002. English).
20. Emini EA, Schleif WA, Nunberg JH, Conley AJ, Eda Y, Tokiyoshi S, et al. Prevention of Hiv-1 Infection in chimpanzees by Gp120 V3 domain-specific monoclonal-antibody. Nature. 1992;355(6362):728–30. (PubMed PMID: WOS:A1992HE60400064. English).
21. Gauduin MC, Safrit JT, Weir R, Fung MSC, Koup RA. Preexposure and postexposure protection against human-immunodeficiency-virus type-1 infection mediated by a monoclonal-antibody. J Infect Dis. 1995;171(5):1203–9. (PubMed PMID: WOS:A1995QU64000018. English).
22. Gauduin MC, Parren PW, Weir R, Barbas CF, 3rd, Burton DR, Koup RA. Passive immunization with a human monoclonal antibody protects hu-PBL-SCID mice against challenge by primary isolates of HIV-1. Nat Med. 1997;3(12):1389–93. (PubMed PMID: 9396610).
23. Parren PW, Ditzel HJ, Gulizia RJ, Binley JM, Barbas CF, 3rd, Burton DR, et al. Protection against HIV-1 infection in hu-PBL-SCID mice by passive immunization with a neutralizing human monoclonal antibody against the gp120 CD4-binding site. AIDS. 1995;9(6):F1–6. (PubMed PMID: 7662189).

24. Haase AT. Early events in sexual transmission of HIV and SIV and opportunities for interventions. Annu Rev Med. 2011;62:127–39. (PubMed PMID: 21054171. Pubmed Central PMCID: Journal—In Process. Epub 2010/11/09. eng.).
25. Haase AT. Targeting early infection to prevent HIV-1 mucosal transmission. Nature. 2010;464(7286):217–23. (PubMed PMID: 20220840. Pubmed Central PMCID: Journal—In Process. Epub 2010/03/12. eng.).
26. D'cruz OJ, Uckun FM. Limitations of the human-PBL-SCID mouse model for vaginal transmission of HIV-1. Am J Reprod Immunol. 2007;57(5):353–60. (PubMed PMID: WOS:000245674000002. English).
27. Mosier DE, Gulizia RJ, Baird SM, Wilson DB. Transfer of a functional human immune-system to mice with severe combined immunodeficiency. Nature. 1988;335(6187):256–9. (PubMed PMID: WOS:A1988Q047700061. English).
28. Mosier DE, Gulizia RJ, Baird SM, Wilson DB, Spector DH, Spector SA. Human immunodeficiency virus infection of human-PBL-SCID mice. Science. 1991;251(4995):791–4. (PubMed PMID: 1990441).
29. Sun ZF, Denton PW, Estes JD, Othieno FA, Wei BL, Wege AK, et al. Intrarectal transmission, systemic infection, and CD4(+) T cell depletion in humanized mice infected with HIV-1. J Exp Med. 2007;204(4):705–14. (PubMed PMID: WOS:000245920600003. English).
30. Stoddart CA, Maidji E, Galkina SA, Kosikova G, Rivera JM, Moreno ME, et al. Superior human leukocyte reconstitution and susceptibility to vaginal HIV transmission in humanized NOD-scid IL-2R gamma(-/-) (NSG) BLT mice. Virology. 2011;417(1):154–60. (PubMed PMID: WOS:000293820300018. English).
31. Denton PW, Garcia JV. Mucosal HIV-1 transmission and prevention strategies in BLT humanized mice. Trends Microbiol. 2012;20(6):268–74. (PubMed PMID: WOS:000305502500003. English).
32. Balazs AB, Chen J, Hong CM, Rao DS, Yang L, Baltimore D. Antibody-based protection against HIV infection by vectored immunoprophylaxis. Nature. 2012;481(7379):81–4. (PubMed PMID: 22139420. Pubmed Central PMCID: 3253190. Epub 2011/12/06. eng).
33. Klein F, Halper-Stromberg A, Horwitz JA, Gruell H, Scheid JF, Bournazos S, et al. HIV therapy by a combination of broadly neutralizing antibodies in humanized mice. Nature. 2012;492(7427):118–22. (PubMed PMID: 23103874).
34. Conley AJ, Kessler JA, II, Boots LJ, McKenna PM, Schleif WA, Emini EA, et al. The consequence of passive administration of an anti-human immunodeficiency virus type 1 neutralizing monoclonal antibody before challenge of chimpanzees with a primary virus isolate. J Virol. 1996;70(10):6751–8. (PubMed PMID: 8794312).
35. Muster T, Steindl F, Purtscher M, Trkola A, Klima A, Himmler G, et al. A conserved neutralizing epitope on gp41 of human immunodeficiency virus type 1. J Virol. 1993;67(11):6642–7. (PubMed PMID: 7692082).
36. Veselinovic M, Neff CP, Mulder LR, Akkina R. Topical gel formulation of broadly neutralizing anti-HIV-1 monoclonal antibody VRC01 confers protection against HIV-1 vaginal challenge in a humanized mouse model. Virology. 2012;432(2):505–10. (PubMed PMID: 22832125. Pubmed Central PMCID: 3652800).
37. Hur EM, Patel SN, Shimizu S, Rao DS, Gnanapragasam PN, An DS, et al. Inhibitory effect of HIV-specific neutralizing IgA on mucosal transmission of HIV in humanized mice. Blood. 2012;120(23):4571–82. (PubMed PMID: 23065154. Pubmed Central PMCID: 3512234).
38. Watkins JD, Sholukh AM, Mukhtar MM, Siddappa NB, Lakhashe SK, Kim M, et al. Anti-HIV IgA isotypes: differential virion capture and inhibition of transcytosis are linked to prevention of mucosal R5 SHIV transmission. AIDS. 2013;27(9):F13–20. (PubMed PMID: 23775002).
39. Abdool Karim Q, Abdool Karim SS, Frohlich JA, Grobler AC, Baxter C, Mansoor LE, et al. Effectiveness and safety of tenofovir gel, an antiretroviral microbicide, for the prevention of HIV infection in women. Science. 2010;329(5996):1168–74. (PubMed PMID: 20643915. Pubmed Central PMCID: 3001187).

40. Shattock RJ, Rosenberg Z. Microbicides: topical prevention against HIV. Cold Spring Harb Perspect Med. 2012;2(2):a007385. (PubMed PMID: 22355798. Pubmed Central PMCID: 3281595).
41. Veazey RS, Shattock RJ, Pope M, Kirijan JC, Jones J, Hu Q, et al. Prevention of virus transmission to macaque monkeys by a vaginally applied monoclonal antibody to HIV-1 gp120. Nat Med. 2003;9(3):343–6. (PubMed PMID: 12579198).
42. Pietzsch J, Gruell H, Bournazos S, Donovan BM, Klein F, Diskin R, et al. A mouse model for HIV-1 entry. Proc Natl Acad Sci U S A. 2012;109(39):15859–64. (PubMed PMID: 23019371. Pubmed Central PMCID: 3465400).
43. Gruell H, Bournazos S, Ravetch JV, Ploss A, Nussenzweig MC, Pietzsch J. Antibody and antiretroviral pre-exposure prophylaxis prevent cervicovaginal HIV-1 infection in a transgenic mouse model. J Virol. 2013 Aug;87(15):8535–44. (PubMed PMID: 23720722).
44. Asokan A, Schaffer DV, Samulski RJ. The AAV vector toolkit: poised at the clinical crossroads. Mol Ther. 2012;20(4):699–708. (PubMed PMID: 22273577. Pubmed Central PMCID: 3321598).
45. Johnson PR, Schnepp BC, Zhang J, Connell MJ, Greene SM, Yuste E, et al. Vector-mediated gene transfer engenders long-lived neutralizing activity and protection against SIV infection in monkeys. Nat Med. 2009;15(8):901–6. (PubMed PMID: 19448633. Pubmed Central PMCID: 2723177. Epub 2009/05/19. eng.).
46. Balazs AB, West AP, Jr. Antibody gene transfer for HIV immunoprophylaxis. Nat Immunol. 2013;14(1):1–5. (PubMed PMID: 23238748).
47. Balazs AB, Bloom JD, Hong CM, Rao DS, Baltimore D. Broad protection against influenza infection by vectored immunoprophylaxis in mice. Nat Biotechnol. 2013;31(7):647–52. (PubMed PMID: 23728362).
48. Wei X, Decker JM, Wang S, Hui H, Kappes JC, Wu X, et al. Antibody neutralization and escape by HIV-1. Nature. 2003;422(6929):307–12. (PubMed PMID: 12646921).
49. Poignard P, Sabbe R, Picchio GR, Wang M, Gulizia RJ, Katinger H, et al. Neutralizing antibodies have limited effects on the control of established HIV-1 infection in vivo. Immunity. 1999;10(4):431–8. (PubMed PMID: 10229186).
50. Trkola A, Kuster H, Rusert P, Joos B, Fischer M, Leemann C, et al. Delay of HIV-1 rebound after cessation of antiretroviral therapy through passive transfer of human neutralizing antibodies. Nat Med. 2005;11:615–22. (PubMed PMID: 15880120).
51. Diskin R, Klein F, Horwitz JA, Halper-Stromberg A, Sather DN, Marcovecchio PM, et al. Restricting HIV-1 pathways for escape using rationally designed anti-HIV-1 antibodies. J Exp Med. 2013;210(6):1235–49. (PubMed PMID: 23712429. Pubmed Central PMCID: 3674693).
52. Horwitz JA, Halper-Stromberg A, Mouquet H, Gitlin AD, Tretiakova A, Eisenreich TR, et al. HIV-1 suppression and durable control by combining single broadly neutralizing antibodies and antiretroviral drugs in humanized mice. Proc Natl Acad Sci U S A. 2013 Oct 8;110(41):16538–43. (PubMed PMID: 24043801).
53. Nelson AL, Dhimolea E, Reichert JM. Development trends for human monoclonal antibody therapeutics. Nat Rev Drug Discov. 2010;9(10):767–74. (PubMed PMID: 20811384).

Chapter 27
B-Cell Responses in Humanized Mice: The Glass is Half Full

Edward Seung and Andrew M. Tager

27.1 Introduction

For the past several decades, investigators have created "humanized" mice to bridge the gap between small-animal models and human studies and allow examination of the human immune system in an experimental setting. Humanized mice were first created by transplanting human cells and/or tissues into mice with severe deficiencies in their own immune systems [1–3], and this has remained the most common approach used to generate mice with more and more components of the human immune system. Advances in these humanized mouse models have come mainly from gradual improvements in the immunodeficient mice used, rendering them more receptive to donor human cells and tissues, as has been recently reviewed [4–6]. One of the latest breakthroughs in this area has been the development of immunodeficient mice with targeted deletion of the interleukin-2 (IL-2) receptor common γ-chain locus (IL2rγ_c); the use of these mice has resulted in far higher levels of human cell engraftment than previously possible [7]. This chapter will focus on human B-cell responses in two of the newer humanized-mouse models that have taken advantage of IL2rγ_c null immunodeficient mice: BLT (Bone marrow-Liver-Thymus) and Hu-HSC (Hematopoietic Stem Cell) mice. BLT mice are generated by the cotransplantation of human fetal liver and thymus fragments under the renal capsule

E. Seung (✉) · A. M. Tager
Center for Immunology and Inflammatory Diseases, Massachusetts General Hospital, Harvard Medical School, 149 13th Street #8301, Charlestown, MA 02129, USA
e-mail: seung.edward@mgh.harvard.edu

A. M. Tager
Pulmonary and Critical Care Unit, Massachusetts General Hospital, Harvard Medical School, 149 13th Street #8301, Charlestown, MA 02129, USA
e-mail: amtager@mgh.harvard.edu

L. Y. Poluektova et al. (eds.), *Humanized Mice for HIV Research*,
DOI 10.1007/978-1-4939-1655-9_27

of sublethally-irradiated mice, followed by injection of $CD34^+$ HSCs isolated from the same fetal liver [8, 9]. Hu-HSC mice, also referred to as HIS (Human Immune System) mice, are generated by injecting HSCs from umbilical cord blood [10, 11], from the blood of granulocyte colony-stimulating factor-treated adults [12], or from fetal liver [13], into sublethally-irradiated mice (often newborn pup). We will review the capacity of these models to recapitulate the human humoral immune system, focusing on B cell ontogeny, immunoglobulin production, and generation of functional antibodies.

27.2 B-Cell Ontogeny

27.2.1 Normal B-Cell Development

Subsets of peripheral B cells in mice include: (1) follicular (FO) B cells, which home to follicles in the secondary lymphoid organs (i.e., spleen and lymph nodes), recirculate to a niche in the bone marrow, and make up the bulk of mature B cells in the periphery and (2) marginal-zone (MZ) B cells, which reside in the outer-white pulp of the spleen between the marginal sinus and the red pulp. FO B cells in secondary lymphoid organs are responsible for T cell-dependent humoral immune responses to protein antigens. B-cell follicles in these organs are found adjacent to T-cell zones, which allows FO B cells to interact with activated-T-helper cells at the interface of these regions [14]. In contrast, FO B cells in the bone marrow, present in aggregates around the sinusoids, respond in a T cell-independent manner to blood-borne pathogens. Most mature FO B cells are follicular type I (FO-I) B cells, which are recognized by $IgD^{hi}IgM^{lo}CD21^{mid}$ surface expression; a smaller population of follicular type II (FO-II) B cells are recognized by $IgD^{hi}IgM^{hi}CD$-21^{mid} expression [15]. MZ B cells express high levels of CD1d, which helps in their presentation of lipid antigens, and also respond strongly to blood-borne pathogens [16]. They express high levels of CD21 as well, which helps them transport immune complexes from the blood to follicles in the spleen [17].

Mouse B cell maturation in the bone marrow and spleen has been well described. Early precursors, known as pro- and pre-B cells, commit to the B cell lineage in the bone marrow. While in the marrow, these cells rearrange their immunoglobulin heavy-chain gene and their light-chain gene loci, respectively, as they differentiate into immature B cells. Immature B cells then emigrate to the spleen, where their subsequent maturation is referred to as "peripheral B cell development." In the spleen, B-cell precursors first mature through T1 and T2 transitional stages, and then into follicular B cells or MZ B cells (Fig. 27.1). Although B-cell maturation in humans has yet to be elucidated as well as it has been in mice, human transitional, follicular, and MZ B-cell subsets corresponding to those in mice have all been described [14].

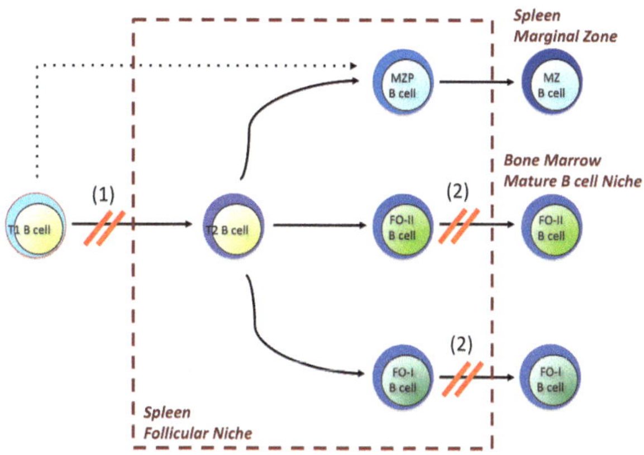

Fig. 27.1 Blocks in peripheral B-cell development in humanized mice. (Modified from reference [14] with permission of Dr. Shiv Pillai). In the normal development of mouse B cells, early precursors commit to the B-cell lineage in the bone marrow where they differentiate into immature B cells. The immature B cells then emigrate to the spleen, where their subsequent maturation is referred to as "peripheral B cell development." During this phase of their development, B-cell precursors first mature through T1 and T2 transitional stages, and then into follicular (FO) B cells or marginal zone precursors (MZP) and marginal zone (MZ) B cells. Multiple investigators have found human B-cell maturation to be incomplete in both BLT and Hu-HSC mice [18–22], and we have found evidence for blocks in B-cell maturation in BLT mice in at least two stages of peripheral development: (1) entry of transitional T1 B cells into splenic follicles is impaired, resulting in an abnormal accumulation of T1 B cells in the peripheral blood and (2) recirculation of mature FO B cells (type I and type II) to the bone marrow is impaired, resulting in a paucity of FO-I and FO-II B cells in the bone marrow mature B cell niche. Courtesy of Dr. Shiv Pillai

27.2.2 Blocks in Human B-Cell Maturation in Humanized Mice

Multiple investigators have documented incomplete B-cell development in both BLT and Hu-HSC mice [18–22]. The B-cell markers CD10 and CD20 are classically used to define the three major $CD19^+$ B-cell maturation stages: (1) very immature Pro-B cells are $CD19^+$ $CD20^-$ $CD10^+$, (2) immature B cells are $CD19^+$ $CD20^+$ $CD10^+$, and (3) mature B cells are $CD19^+$ $CD20^+$ $CD10^-$. In BLT mice, we found that even though mature human follicular B cells, MZ B cells, and plasmablasts were all present, a higher than expected proportion of the $CD19^+$ B cells in the bone marrow and spleen were $CD10^+$, suggesting skewing toward immaturity [18]. Incomplete B-cell maturation in BLT mice is also suggested by CD5 expression by the majority of circulating human B cells [19]. Although CD5 expression does not necessarily denote B-cell immaturity, most circulating mature naïve B lymphocytes in humans are $CD19^+$ $CD5^-$ follicular B cells [19]. B-cell development has similarly been reported to be incomplete in Hu-HSC mice, in which human B cells are predominantly immature [20], and a larger than expected percentage of $CD19^+$ B cells express CD5 [23].

We also found that human transitional B cells ($CD19^+$ $CD10^{hi}CD24^{hi}CD38^{hi}$) were present in much greater frequencies in the spleens of BLT mice than adult humans [18]. T1 versus T2 subclasses of transitional B cell can be distinguished by their surface expressions of IgM and IgD: T1 cells are $IgM^{hi}IgD^{lo}$, whereas cells with a T2 phenotype are $IgM^{hi}IgD^{hi}$ [14]. We found that most of the transitional cells in the spleens of BLT mice had a T1 phenotype; cells with a T2 phenotype were present, but at lower frequencies [18]. In more recent studies, we also found that human transitional B cells were more abundant in the peripheral blood of BLT mice than adult humans, with most having a T1 phenotype as well (unpublished data). As shown in Fig. 27.1, the maturation of T1 cells into T2 cells and then into more mature B cells occurs with T1 cell entry into the B cell follicular niche of the spleen. The accumulation of T1 B cells in the blood and spleen therefore suggests a defect in their entry into splenic follicles. This defect may result from an inability of human B cells to drive the development of proper follicular architecture in the BLT mouse spleen.

Along with the maturation defects seen in splenic and circulating B cells, we found that the majority of the human $CD19^+$ B cells in BLT mouse bone marrow (approximately 90%) were immature $CD10^+$ $CD27^-$ B cells (unpublished data). These $CD10^+$ B cells were largely $CD24^{hi}CD38^{hi}$, and either expressed no surface IgM, as do pro- and pre-B cells, or IgM alone without IgD, as do immature B cells (unpublished data). Compared to the bone marrow compartment of humans, there was consequently a paucity of mature FO B cells, suggesting that the recirculation of these cells to the bone marrow is also defective in BLT mice (Fig. 27.1). This second defect in B-cell maturation may reflect lack of maturity of B cells developing in the spleen, but may also result from incomplete development of B cell niche architecture in the BLT bone marrow.

27.2.3 Improving Human B-Cell Maturation in Humanized Mice

A recent study by Lang and colleagues suggests that human B-cell maturity in humanized mice improves with increasing time following the transplantation of human cells [24]. These investigators demonstrated that while the majority of Hu-HSC mice evaluated prior to 18 weeks following human cell transplant contained mostly immature B cells as others have previously shown, the percentage of mice with mostly mature $CD20^+$ $CD10^-$ B cells increased substantially with age, and exceeded 60% after 24 weeks of reconstitution [24]. In this same study, improving the quality of T cell help in Hu-HSC mice by the adoptive transfer of autologous mature human T cells hastened the development of mature B cells [24]. In vivo depletion of the human T cells of Hu-HSC mice conversely further delayed B cell maturation [24]. Other investigators have noted the presence of mature naïve B cells in Hu-HSC mice after 6 months of reconstitution [25]. Increased numbers of mature B cells have also been reported in Hu-HSC mice generated with the use of the alkylating agent busulfan instead of irradiation as the pre-cell transplant conditioning regimen [26]. In BLT mice, we noted that the percentages of B cells in the spleen that were mature follicular type II (FO-II) cells increased with increasing time following human stem cell and tissue transplantation [18], suggesting that B cell maturation also increases in BLT mice with longer times allowed for human immune reconstitution.

27.3 Lymphoid Organ Microarchitecture

The lymphoid organs of both mice and humans contain B and T cells organized into distinct zones to optimize their effectiveness. One of the most important structures for the development of humoral immunity in lymphoid tissues is the B-cell follicle, an aggregate of B cells containing naïve cells, as well as activated, proliferating and maturing B cells in germinal centers [14]. T cells in zones adjacent to B-cell follicles participate in T cell-dependent humoral immune responses. In contrast, a number of immunohistochemistries showed a lack of distinct B- and T-cell zones in the lymphoid organs of BLT [18] and Hu-HSC mice [11, 24, 27]. Human B cells in the spleens of humanized mice were found interspersed with T cells and scattered throughout the organ. Some investigators have noted organization of human B cells into discrete follicle-like structures in Hu-HSC mice, although adjacent T-cell zones were lacking [24, 27]. Some investigators have also detected B-cell follicles with adjacent T-cell areas in the spleens of BLT mice [28], which can be accentuated by immunization of BLT mice with the classic T-cell-dependent antigen sheep red blood cells [9]. Gut-associated lymphoid tissue (GALT) structures, which are essential parts of the gastrointestinal (GI) tract immune system, have also been identified in BLT mice constructed from NOD/SCID mice, but not from NSG mice that lack IL2rγ_c [29]. Nochi and colleagues hypothesized this presence or absence of GALT in BLT mice generated from these two different recipient mouse strains results from the presence of cryptopatches containing lymphoid tissue inducer (LTi) cells in NOD/SCID mice, versus the absence of these structures in NSG mice. Consistent with their capacity to generate GALT, BLT mice generated from NOD/SCIDs demonstrate greater numbers of intestinal IgA-secreting plasma cells than IgG-secreting plasma cells. In contrast, BLT mice generated from NSGs demonstrate few plasma cells in general, and fewer plasma cells secreting IgA.

Follicular dendritic cells (FDCs), which play an essential role in organization of B cell follicular structures [30–32], are of nonhematopoietic lineage [33, 34], and consequently, if present in the lymphoid tissues of Hu-HSC and BLT models, will be of mouse origin. A loose network of mouse follicular dendritic cells has been detected in the spleen of Hu-HSC mice [27]. The observation of B cell follicle-like structures in humanized mice by some investigators, and the ability of FDCs to provide costimulation across species [35, 36], suggests that these mouse FDCs can interact with human lymphocytes. This raises the possibility that manipulating mouse FDCs may be able to improve B–T cell organization and microarchitecture in the lymphoid organs of humanized mice in the future. Such architectural improvements would be expected to lead to functional improvements in humoral immunity as well.

27.4 Immunoglobulin Production

Given the issues of incomplete B-cell development and poor lymphoid architecture discussed previously, immunoglobulin production in humanized mice has not surprisingly often been reported to be poor. Figure 27.2 shows the variability in total immunoglobin G (IgG) and immunoglobin M (IgM) serum levels in naïve Hu-HSC and BLT mice that have been reported, which overall have been substantially lower than those found in humans, particularly in adults [37]. Immunization of humanized mice has generally not induced strong humoral responses. BLT mice immunized with tetanus toxoid followed by a boost, which typically produces strong antibody responses in humans, elicited low-level antigen-specific IgM antibodies (Abs) and barely detectable IgG Abs in one of three mice tested [38]. Biswas and colleagues also reported weak antigen-binding human IgM and IgG responses in BLT mice immunized with recombinant HIVgp140 or WNV-E protein complexed to IC31® adjuvant [19]. However, Gorantla and colleagues were able to detect antigen-specific IgG after vaccinating Hu-HSC mice with a tetanus toxoid conjugate (ActHIB), although HIV infection produced detectable HIV-specific IgM by enzyme-linked immunosorbent assay (ELISA; personal correspondence) but usually no IgG humoral responses [21].

Studies measuring human humoral immune responses of BLT or Hu-HSC mice in response to infections have been more encouraging, though they also have produced mixed results. Hu-HSC mice infected with Epstein–Barr virus (EBV) have been reported to have no detectable EBV-specific IgM or IgG responses against the

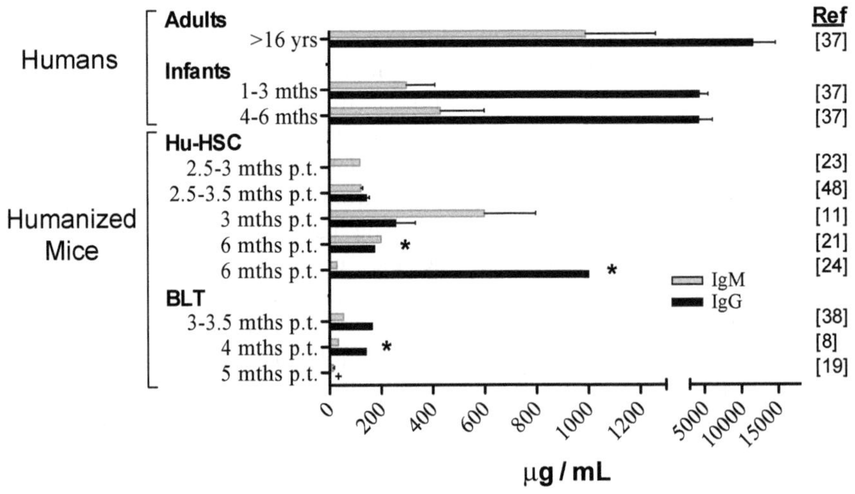

Fig. 27.2 Baseline IgG and IgM levels in humans and humanized mice. Total immunoglobin M (IgM, *grey bars*) and immunoglobin G (IgG, *black bars*) levels in the blood of human adults and infants, and naïve Hu-HSC and BLT mice. Multiple studies indicate that humanized mice generally have substantially lower levels of IgM and IgG than are found in humans, particularly in adults. + no data, *p.t.* posttransplantation, * approximate values estimated from figures

viral capsid antigen [39], and Hu-HSC mice infected with HIV-1 demonstrated little or no HIV-specific IgM or IgG titers 3 to 10 weeks postinfection [21, 40, 41]. Other studies in Hu-HSC mice infected with HIV have detected some HIV-specific class-switched Abs by Western blot analysis [42] and ELISA [43], though these responses were limited. In BLT mice, Sun et al. [44] were able to detect HIV-specific IgG Abs against at least two HIV proteins by 8 weeks postinfection in three of four mice. We found that all BLT mice HIV-infected for more than 12 weeks demonstrated human Abs to at least seven HIV proteins by Western blot analysis [18]. In subsequent experiments, we have found that class-switched HIV-specific human IgG Abs, as detected by ELISA, develop in BLT mice, but can take 8 to 18 weeks postinfection to develop (unpublished data). The occurrence of Ig class-switching in these mice suggests that activation-induced cytidine deaminase (AID) is present and functional in their human B cells, as this enzyme is required for Ig diversification by both class-switch recombination and somatic hypermutation [45, 46]. B cells isolated from Hu-HSC mice have also been shown to undergo somatic hypermutation [47], further suggesting the induction of AID activity. Recently, the presence of AID has been detected in the GALT structures of BLT mice, consistent with the detection of class-switched IgA and IgG Abs in their GI tract [29].

27.5 Development of Functional Antibodies

Most studies of human Abs in both Hu-HSC mice [10, 21, 23, 24, 47, 48] and BLT mice [8, 38], have reported the antigen-binding specificity of these antibodies developing after immunization or infection [10, 11, 18, 19, 21, 24, 28, 42, 43, 48–50], rather than the functionality of these Abs. The first reports that humanized mice can generate *functional* antibodies able to neutralize a pathogen came from studies of dengue virus (DENV) infection in Hu-HSC [49] and BLT mice [50]. Kuruvilla and colleagues found that DENV-infected Hu-HSC mice produced anti-DENV IgM, followed by IgG at 6 weeks postinfection, and that almost a third of these mice produced Abs that could neutralize DENV infection [49]. Jaiswal and colleagues demonstrated that upon ex vivo restimulation, splenocytes from BLT mice that had been infected with a live-attenuated DENV vaccine strain produced DENV-specific IgM (but not IgG) Abs that possessed neutralizing activity against the virus [50]. While reports of neutralizing antibodies being produced by humanized Hu-HSC or BLT mice are currently few, ongoing experiments in our laboratory and those of other investigators cited above indicate that these mice can produce Abs with more robust functional activities against infecting viruses than previously thought. Current efforts to further improve the B-cell compartment of humanized-mouse models will likely further improve the frequency of functional Abs.

27.6 Conclusions

There are substantial grounds for optimism about the state of B-cell responses in humanized mice. Investigators are continuing to improve human immunity in these models [7, 51], and several strategies to improve humoral immune responses in humanized mice to both infections and immunizations have recently been reported. Chen and colleagues demonstrated that hydrodynamic injection of Hu-HSC mice with plasmids encoding GM-CSF and IL-4 enhanced total IgG levels by 12-fold compared to vector controls. In this study, Hu-HSC mice receiving these plasmids were the only group to generate detectable levels of antigen-specific IgG following immunization, as well as Abs with modest neutralizing activity against H5N1 avian influenza virus [23]. Transgenic expression of human signal regulatory protein alpha (SIRPα), a receptor that negatively regulates phagocytosis, enhanced human $CD45^+$ cell engraftment in humanized BALB/c $Rag2^{-/-}\gamma_c^{-/-}$ mice by 3.5-fold, and improved antigen-specific IgM titers (~2.5-fold) and IgG frequency (6-fold) after immunization compared to nontransgenic control mice [52]. The human cell engraftment in these human $SIRP\alpha^+$ mice reached levels similar to those reported for humanized NSG mice, but direct comparisons of the humoral responses of humanized mice generated with these two strains of mice have yet to be made. Transgenic expression of human HLA molecules in Hu-HSC mice has also been shown to produce higher titers of antigen-specific IgG Abs upon immunization. Considerable variability was noted, but increases up to 3000 to 6000-fold were noted in these transgenic mice compared to control mice [53, 54]. Other strategies have been shown to improve B cell maturation in humanized mice, including increasing the length of time between immune cell transplantation and use of the mice [18, 21, 24], and improving T cell help by adoptive transfer of autologous mature T cells [24]. Recent reports indicate that the immunoglobulin gene repertoires of the human B cells in humanized mice closely recapitulate those in humans, creating further optimism about the potential of humanized mice to model human humoral immune responses [48, 55]. Given (1) the progress that investigators have already made in generating humoral immune responses in humanized mice, (2) the ongoing efforts of investigators to improve B-cell maturation and lymphoid architecture in these models, and (3) the breadth and diversity of the human immunoglobulin gene repertoires that these models already contain, we believe the current state of human B-cell responses in humanized mice is a case of "the glass being half full." We are optimistic that ongoing improvements in humoral immunity in humanized-mouse models will ultimately allow investigators to use these models for both the development and preclinical assessment of vaccines and immunomodulatory therapies that focus on B-cell responses.

Acknowledgments The authors gratefully acknowledge the support by a Harvard University CFAR (HU CFAR, NIH/NIAID P30-AI060354) Scholar Award to E.S. and by NIH/NIAID P01-AI104715 (Project 2 and Core B), HU CFAR (NIH/NIAID P30-AI060354) Core support, and a Ragon Institute Initiative to A.M.T.

References

1. Lapidot T, Pflumio F, Doedens M, Murdoch B, Williams DE, Dick JE. Cytokine stimulation of multilineage hematopoiesis from immature human cells engrafted in SCID mice. Science. 1992;255(5048):1137–41.
2. McCune JM, Namikawa R, Kaneshima H, Shultz LD, Lieberman M, Weissman IL. The SCID-hu mouse: murine model for the analysis of human hematolymphoid differentiation and function. Science. 1988;241(4873):1632–9.
3. Mosier DE, Gulizia RJ, Baird SM, Wilson DB. Transfer of a functional human immune system to mice with severe combined immunodeficiency. Nature. 1988;335(6187):256–9.
4. Shultz LD, Ishikawa F, Greiner DL. Humanized mice in translational biomedical research. Nat Rev Immunol. 2007;7(2):118–30.
5. Willinger T, Rongvaux A, Strowig T, Manz MG, Flavell RA. Improving human hemato-lymphoid-system mice by cytokine knock-in gene replacement. Trends Immunol. 2011;32(7):321–7.
6. Ito R, Takahashi T, Katano I, Ito M. Current advances in humanized mouse models. Cell Mol Immunol. 2012;9(3):208–14.
7. Shultz LD, Brehm MA, Garcia-Martinez JV, Greiner DL. Humanized mice for immune system investigation: progress, promise and challenges. Nat Rev Immunol. 2012;12(11):786–98.
8. Lan P, Tonomura N, Shimizu A, Wang S, Yang YG. Reconstitution of a functional human immune system in immunodeficient mice through combined human fetal thymus/liver and CD34 + cell transplantation. Blood. 2006;108(2):487–92.
9. Melkus MW, Estes JD, Padgett-Thomas A, Gatlin J, Denton PW, Othieno FA, et al. Humanized mice mount specific adaptive and innate immune responses to EBV and TSST-1. Nat Med. 2006;12(11):1316–22.
10. Traggiai E, Chicha L, Mazzucchelli L, Bronz L, Piffaretti JC, Lanzavecchia A, et al. Development of a human adaptive immune system in cord blood cell-transplanted mice. Science. 2004;304(5667):104–7.
11. Ishikawa F, Yasukawa M, Lyons B, Yoshida S, Miyamoto T, Yoshimoto G, et al. Development of functional human blood and immune systems in NOD/SCID/IL2 receptor {gamma} chain(null) mice. Blood. 2005;106(5):1565–73.
12. Lepus CM, Gibson TF, Gerber SA, Kawikova I, Szczepanik M, Hossain J, et al. Comparison of human fetal liver, umbilical cord blood, and adult blood hematopoietic stem cell engraftment in NOD-scid/gammac-/-, Balb/c-Rag1-/-gammac-/-, and C.B-17-scid/bg immunodeficient mice. Hum Immunol. 2009;70(10):790–802.
13. Gimeno R, Weijer K, Voordouw A, Uittenbogaart CH, Legrand N, Alves NL, et al. Monitoring the effect of gene silencing by RNA interference in human CD34 + cells injected into newborn RAG2-/- gammac-/- mice: functional inactivation of p53 in developing T cells. Blood. 2004;104(13):3886–93.
14. Pillai S, Cariappa A. The follicular versus marginal zone B lymphocyte cell fate decision. Nat Rev Immunol. 2009;9(11):767–77.
15. Cariappa A, Boboila C, Moran ST, Liu H, Shi HN, Pillai S. The recirculating B cell pool contains two functionally distinct, long-lived, posttransitional, follicular B cell populations. J Immunol. 2007;179(4):2270–81.
16. Pillai S, Cariappa A, Moran ST. Marginal zone B cells. Annu Rev Immunol. 2005;23:161–96.
17. Guinamard R, Okigaki M, Schlessinger J, Ravetch JV. Absence of marginal zone B cells in Pyk-2-deficient mice defines their role in the humoral response. Nat Immunol. 2000;1(1):31–6.
18. Brainard DM, Seung E, Frahm N, Cariappa A, Bailey CC, Hart WK, et al. Induction of robust cellular and humoral virus-specific adaptive immune responses in human immunodeficiency virus-infected humanized BLT mice. J Virol. 2009;83(14):7305–21.

19. Biswas S, Chang H, Sarkis PT, Fikrig E, Zhu Q, Marasco WA. Humoral immune responses in humanized BLT mice immunized with West Nile virus and HIV-1 envelope proteins are largely mediated via human CD5+ B cells. Immunology. 2011;134(4):419–33.
20. Vuyyuru R, Patton J, Manser T. Human immune system mice: current potential and limitations for translational research on human antibody responses. Immunol Res. 2011;51(2–3):257–66.
21. Gorantla S, Sneller H, Walters L, Sharp JG, Pirruccello SJ, West JT, et al. Human immunodeficiency virus type 1 pathobiology studied in humanized BALB/c-Rag2-/-gammac-/- mice. J Virol. 2007;81(6):2700–12.
22. Watanabe Y, Takahashi T, Okajima A, Shiokawa M, Ishii N, Katano I, et al. The analysis of the functions of human B and T cells in humanized NOD/shi-scid/gammac(null) (NOG) mice (hu-HSC NOG mice). Int Immunol. 2009;21(7):843–58.
23. Chen Q, He F, Kwang J, Chan JK, Chen J. GM-CSF and IL-4 stimulate antibody responses in humanized mice by promoting T, B, and dendritic cell maturation. J Immunol. 2012;189(11):5223–9.
24. Lang J, Kelly M, Freed BM, McCarter MD, Kedl RM, Torres RM, et al. Studies of lymphocyte reconstitution in a humanized mouse model reveal a requirement of T cells for human B cell maturation. J Immunol. 2013;190(5):2090–101.
25. Chang H, Biswas S, Tallarico AS, Sarkis PT, Geng S, Panditrao MM, et al. Human B-cell ontogeny in humanized NOD/SCID gammac(null) mice generates a diverse yet auto/poly- and HIV-1-reactive antibody repertoire. Genes Immun. 2012 Jul;13(5):399–410.
26. Choi B, Chun E, Kim M, Kim ST, Yoon K, Lee KY, et al. Human B cell development and antibody production in humanized NOD/SCID/IL-2Rgamma(null) (NSG) mice conditioned by busulfan. J Clin Immunol. 2011;31(2):253–64.
27. Watanabe S, Terashima K, Ohta S, Horibata S, Yajima M, Shiozawa Y, et al. Hematopoietic stem cell-engrafted NOD/SCID/IL2Rgamma null mice develop human lymphoid systems and induce long-lasting HIV-1 infection with specific humoral immune responses. Blood. 2007;109(1):212–8.
28. Tonomura N, Habiro K, Shimizu A, Sykes M, Yang YG. Antigen-specific human T-cell responses and T cell-dependent production of human antibodies in a humanized mouse model. Blood. 2008;111(8):4293–6.
29. Nochi T, Denton PW, Wahl A, Garcia JV. Cryptopatches are essential for the development of human GALT. Cell reports. 2013;3(6):1874–84.
30. Allen CD, Cyster JG. Follicular dendritic cell networks of primary follicles and germinal centers: phenotype and function. Semin immunol. 2008;20(1):14–25.
31. Burton GF, Conrad DH, Szakal AK, Tew JG. Follicular dendritic cells and B cell costimulation. J Immunol. 1993;150(1):31–8.
32. Grouard G, de Bouteiller O, Banchereau J, Liu YJ. Human follicular dendritic cells enhance cytokine-dependent growth and differentiation of CD40-activated B cells. J Immunol. 1995;155(7):3345–52.
33. Yoshida K, Kaji M, Takahashi T, van den Berg TK, Dijkstra CD. Host origin of follicular dendritic cells induced in the spleen of SCID mice after transfer of allogeneic lymphocytes. Immunology. 1995;84(1):117–26.
34. Humphrey JH, Grennan D, Sundaram V. The origin of follicular dendritic cells in the mouse and the mechanism of trapping of immune complexes on them. Eur J Immunol. 1984;14(9):859–64.
35. Heath SL, Tew JG, Tew JG, Szakal AK, Burton GF. Follicular dendritic cells and human immunodeficiency virus infectivity. Nature. 1995;377(6551):740–4.
36. Tew JG, Wu J, Qin D, Helm S, Burton GF, Szakal AK. Follicular dendritic cells and presentation of antigen and costimulatory signals to B cells. Immunol Rev. 1997;156:39–52.
37. Stiehm ER, Fudenberg HH. Serum levels of immune globulins in health and disease: a survey. Pediatrics. 1966;37(5):715–27.

38. Rajesh D, Zhou Y, Jankowska-Gan E, Roenneburg DA, Dart ML, Torrealba J, et al. Th1 and Th17 immunocompetence in humanized NOD/SCID/IL2rgammanull mice. Hum Immunol. 2010;71(6):551–9.
39. Strowig T, Gurer C, Ploss A, Liu YF, Arrey F, Sashihara J, et al. Priming of protective T cell responses against virus-induced tumors in mice with human immune system components. J Exp Med. 2009;206(6):1423–34.
40. Baenziger S, Tussiwand R, Schlaepfer E, Mazzucchelli L, Heikenwalder M, Kurrer MO, et al. Disseminated and sustained HIV infection in CD34+ cord blood cell-transplanted Rag2-/-gamma c-/- mice. Proc Natl Acad Sci U S A. 2006;103(43):15951–6.
41. Berges BK, Rowan MR. The utility of the new generation of humanized mice to study HIV-1 infection: transmission, prevention, pathogenesis, and treatment. Retrovirology. 2011;8:65.
42. Sato K, Nie C, Misawa N, Tanaka Y, Ito M, Koyanagi Y. Dynamics of memory and naive CD8 + T lymphocytes in humanized NOD/SCID/IL-2Rgammanull mice infected with CCR5-tropic HIV-1. Vaccine. 2010;28 (Suppl 2):B32–7.
43. Sango K, Joseph A, Patel M, Osiecki K, Dutta M, Goldstein H. Highly active antiretroviral therapy potently suppresses HIV infection in humanized Rag2-/-gammac-/- mice. AIDS Res Hum Retroviruses. 2010;26(7):735–46.
44. Sun Z, Denton PW, Estes JD, Othieno FA, Wei BL, Wege AK, et al. Intrarectal transmission, systemic infection, and CD4+ T cell depletion in humanized mice infected with HIV-1. J Exp Med. 2007;204(4):705–14.
45. Muramatsu M, Kinoshita K, Fagarasan S, Yamada S, Shinkai Y, Honjo T. Class switch recombination and hypermutation require activation-induced cytidine deaminase (AID), a potential RNA editing enzyme. Cell. 2000;102(5):553–63.
46. Revy P, Muto T, Levy Y, Geissmann F, Plebani A, Sanal O, et al. Activation-induced cytidine deaminase (AID) deficiency causes the autosomal recessive form of the Hyper-IgM syndrome (HIGM2). Cell. 2000;102(5):565–75.
47. Scheeren FA, Nagasawa M, Weijer K, Cupedo T, Kirberg J, Legrand N, et al. T cell-independent development and induction of somatic hypermutation in human IgM+ IgD+ CD27+ B cells. J Exp Med. 2008;205(9):2033–42.
48. Becker PD, Legrand N, van Geelen CM, Noerder M, Huntington ND, Lim A, et al. Generation of human antigen-specific monoclonal IgM antibodies using vaccinated "human immune system" mice. PLoS ONE 5(10): e13137.
49. Kuruvilla JG, Troyer RM, Devi S, Akkina R. Dengue virus infection and immune response in humanized RAG2(-/-)gamma(c)(-/-) (RAG-hu) mice. Virology. 2007;369(1):143–52.
50. Jaiswal S, Pazoles P, Woda M, Shultz LD, Greiner DL, Brehm MA, et al. Enhanced humoral and HLA-A2-restricted dengue virus-specific T-cell responses in humanized BLT NSG mice. Immunology. 2012;136(3):334–43.
51. Akkina R. New generation humanized mice for virus research: comparative aspects and future prospects. Virology. 435(1):14–28.
52. Strowig T, Rongvaux A, Rathinam C, Takizawa H, Borsotti C, Philbrick W, et al. Transgenic expression of human signal regulatory protein alpha in Rag2-/-gamma(c)-/- mice improves engraftment of human hematopoietic cells in humanized mice. Proc Natl Acad Sci U S A. 2011;108(32):13218–23.
53. Danner R, Chaudhari SN, Rosenberger J, Surls J, Richie TL, Brumeanu TD, et al. Expression of HLA class II molecules in humanized NOD. Rag1KO.IL2RgcKO mice is critical for development and function of human T and B cells. PLoS ONE. 2011;6(5):e19826.
54. Suzuki M, Takahashi T, Katano I, Ito R, Ito M, Harigae H, et al. Induction of human humoral immune responses in a novel HLA-DR-expressing transgenic NOD/Shi-scid/gammacnull mouse. Int Immunol. 2012;24(4):243–52.
55. Ippolito GC, Hoi KH, Reddy ST, Carroll SM, Ge X, Rogosch T, et al. Antibody repertoires in humanized NOD-scid-IL2Rgamma(null) mice and human B cells reveals human-like diversification and tolerance checkpoints in the mouse. PLoS ONE. 2012;7(4):e35497.

Part V
Therapeutics Developments for HIV-1

Chapter 28
Species Similarities and Differences in Pharmacokinetics and Distribution of Antiretroviral Drugs

Andrew Owen and Paul Curley

28.1 Background

There are important species differences relating to drug-handling that must be considered when interpreting pharmacokinetic and pharmacodynamic data generated in mouse models. In general, smaller species exhibit higher pharmacokinetic clearance of xenobiotics than larger animals [1]. The genetic basis for these species differences stem from differences in the expression, distribution and function of metabolic enzymes and transporters, as well as important physiological and anatomical differences between rodents and humans. This chapter focuses on key mechanisms that underpin species differences affecting the pharmacokinetics and disposition of antiretroviral drugs. Wherever possible, direct comparisons between mice and humans have been made. However, for some factors, there is a paucity of pharmacokinetic and/or mechanistic knowledge in mice and areas for future research are indicated. In these cases, data from rats have been used as a surrogate to form a preliminary position on the likely differences.

As with all xenobiotics, there are numerous factors that influence the pharmacokinetics, distribution and pharmacodynamics of antiretroviral drugs. These include a plethora of proteins involved in phase I metabolism, phase II metabolism, drug influx transport and drug efflux transport, as well as the ligand-responsive nuclear receptor type transcription factors that regulate them. Some of the key proteins that influence disposition of antiretroviral drugs and their rodent orthologues are presented in Table 28.1. The human and murine proteins that underpin pharmacokinetic exposure to antiretroviral drugs are discussed in the following sections of this chapter with a focus on differences in function, substrate specificity, expression and distribution. A diagrammatical representation of the general mechanisms by which humans and mice can differ is presented in Fig. 28.1.

A. Owen (✉) · P. Curley
Department of Molecular and Clinical Pharmacology, University of Liverpool,
70 Pembroke Place, Liverpool L69 3GF, UK
e-mail: aowen@liv.ac.uk

L. Y. Poluektova et al. (eds.), *Humanized Mice for HIV Research*,
DOI 10.1007/978-1-4939-1655-9_28

Table 28.1 Summary of some of the key proteins effecting antiretroviral pharmacokinetics in humans and murine orthologues

Common name	Human gene	Murine gene	Antiretrovirals affected	References
CYP3A4	*CYP3A4*	*Cyp3a11*	Ritonavir, Saquinavir, Darunavir, Lopinavir, Atazanavir, Nelfinavir, Indinavir, Efavirenz, Nevirapine, Etravirine, Rilpivirine, Delavirdine,Maraviroc	[2–10]
CYP2B6	*CYP2B6*	*Cyp2b10*	Efavirenz, nevirapine	[11–14]
CYP2D6	*CYP2D6*	*Cyp2d22*	Ritonavir, Saquinavir, Darunavir, Lopinavir, Atazanavir, Nelfinavir, Indinavir, Delavirdine	[4, 15, 16]
UGT2B7	*UGT2B7*	ND	Zidovudine, efavirenz	[17–20]
P-GP	*ABCB1*	*Abcb1a* *Abcb1b*	Ritonavir, Saquinavir, Darunavir, Lopinavir, Atazanavir, Nelfinavir, Indinavir, Maraviroc	[21–29]
BCRP	*ABCG2*	*Abcg2*	Abacavir, Zidivudine, Lamivudine, Atazanavir, Lopinavir	[23, 30–32]
MRP1	*ABCC1*	*Abcc1*	Ritonavir, Saquinavir, Lopinavir, Nelfinavir, Indinavir	[24, 33–35]
MRP2	*ABCC2*	*Abcc2*	Saquinavir, Lopinavir, Nelfinavir	[24, 33, 36]
MRP4	*ABCC4*	*Abcc4*	Tenofovir	[37, 38]
MRP7	*ABCC10*	*Abcc10*	Tenofovir, nevirapine	[39–41]
OATP1A2	*SLCO1A2*	ND	Saquinavir, darunavir, lopinavir	[42]
OATP1B1	*SLCO1B1*	ND	Saquinavir, darunavir, lopinavir	[42]
OAT1	*SLCO22A6*	*Slco22a6*	Zidovudine, Tenofovir	[105–107]
OAT2	*SLCO22A7*	*Slco22a7*	Zidovudine	[105–107]
OAT3	*SLCO22A8*	*Slco22a8*	Zidovudine, Tenofovir	[105, 106, 108]
OAT4	*SLCO22A11*	*Slco22a11*	Zidovudine	[106]
OCT1	*SLC22A1*	*Slc22a1*	Lamivudine	[42, 109]
OCT2	*SLC22A2*	*Slc22a2*	Lamivudine	[42, 109]
PXR	*NR1I2*	*Nr1i2*	Ritonavir, Darunavir, Lopinavir, Atazanavir, Efavirenz	[116, 121]
CAR	*NR1I3*	*Nr1i3*	Abacavir, Efavirenz	[116, 132]

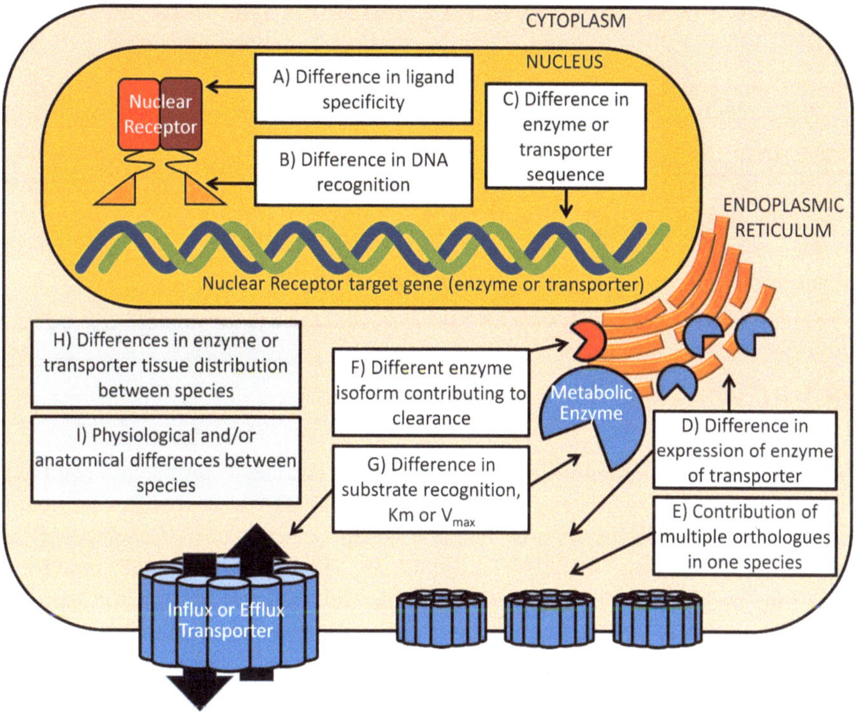

Fig. 28.1 Diagrammatical representation of the general mechanisms discussed in this chapter for differences in drug disposition between humans and mice both at a cellular level (*white boxes*) and systemically (*grey boxes*). Differences in molecular recognition of ligands (**a**) or DNA response elements (**b**) by transcription factors can result in differences in constitutive or inducible target-gene expression. Nuclear receptor target-genes include enzymes and transporters that influence antiretroviral pharmacokinetics and distribution. Where orthologues for human enzymes and transporters have been identified, there are often significant differences in DNA sequence identity (**c**) and differences in relative expression (**d**). In some cases the precise orthologue is difficult to accurately predict and/or multiple murine orthologues exist that co-ordinate to provide the metabolic function (**e**). Similarly, there are cases were distinct isoforms contribute to metabolism in one species but not the other (**f**). When there is confidence that the correct orthologue has been identified, there are sometimes differences in the kinetics of transport or metabolism between species (**g**). There are paradigms for differences in the tissue distribution between human and murine orthologues (**h**) and clear physiological and anatomical differences (**i**) that contribute to differences in pharmacokinetic and pharmacodynamic response

28.2 Species Differences in Key Metabolic Enzymes

Phase I enzymes such as the cytochrome P450 (CYP) family serve to introduce polar groups, increasing water solubility and thus facilitate clearance of their substrates from the systemic circulation. Phase II enzymes such as UDP-glucuronyl transferases (UGTs) conjugate the more reactive phase I metabolic products to glucuronic acid. While there are also other phase I and phase II enzymes, this chapter focuses on CYPs and UGTs because of their importance in the pharmacokinetics of antiretroviral drugs.

28.3 CYP3A

All of the protease inhibitors, including ritonavir, darunavir, lopinavir, saquinavir and atazanavir are substrates for CYP3A [2–5]. CYP3A also provides the major route for metabolic clearance of maraviroc and contributes to clearance of the non-nucleoside reverse transcriptase inhibitors, efavirenz, nevirapine, etravirine, rilpivirine and delavirdine [6–8]. Of particular relevance is the specific isoform, CYP3A4, which is inhibited by ritonavir to 'boost' the pharmacokinetics of protease inhibitors [43]. Due to the importance of CYP3A4 in metabolic clearance of its substrates it is also implicated in a number of adverse drug–drug interactions [44]. It is therefore of critical importance to understand differences between human and murine CYP3A.

In humans, the CYP3A family consists of four enzymes, CYP3A4, CYP3A5, CYP3A7 and CYP3A43, but CYP3A4 represents the most abundant isoform in the human liver [9]. The murine CYP3a subfamily is comprised of six isoforms, Cyp3a11, Cyp3a13, Cyp3a16, Cyp3a25, Cyp3a41 and Cyp3a44. The greatest similarity is found between CYP3A4 and Cyp3a11, which share 76 % sequence identity [9, 10]. There have been a number of reports demonstrating general similarities with protease inhibitors and maraviroc being substrates for both human and murine CYP3A/3a [44, 45]. Despite these similarities, there are important differences in CYP3A function in the mouse model. For example, lopinavir is extensively metabolized by human and mouse liver microsomes at similar rates [46]. However, species differences in the formed metabolites exist, with 12 metabolic products in humans compared to only 11 in mice, which do not generate the M9 metabolite [46]. Conversely, nevirapine phase I metabolism forms four hydroxylated metabolites (2-, 3-, 8- and 12-hydroxynevirapine), all of which are also generated in mice [47, 48]. However, this should be interpreted in the context that nevirapine is metabolized by CYP2B (see further) as well as CYP3A and the contribution of each CYP was not determined in these studies [49]. Significant differences in metabolite formation have also been reported for non-antiretroviral xenobiotics. For example, the bioactive metabolite of vitamin D ($1\alpha,25(OH)_2D_3$) undergoes hydroxylation by CYP3A and in human liver microsomes yielding four monohydroxylated products (M1–M4) with M4 being the major product. However, the major metabolite from mouse liver microsomes is a unique metabolite (M5), with minor formation of M1 and M2 [50].

It is also important to consider inhibition of CYP3A. Ritonavir has been shown to inhibit murine clearance of protease inhibitors [51, 52] as well as non-antiretroviral drugs such as docetaxel [53]. Similarities between murine and human have also been reported for other inhibitors. Indeed, no difference in K_I and K_{inact} for inhibition by midazolam of triazolam metabolism by CYP3A versus Cyp3a was reported [54]. While this is encouraging, it should be noted that a higher concentration ratio of ritonavir to protease inhibitor is used for 'boosting' in rodents [51, 52] compared to what is used clinically in humans [55], and it should also be noted that the in vivo pharmacokinetic response to ritonavir is not due to CYP3A inhibition alone [56]—see also further.

There are other important examples for differences in substrate affinity and inhibition for non-antiretroviral drugs. For example, midazolam is metabolized to 4-OH-midazolam and α-OH-midazolam in both species but while ketoconazole inhibits formation of both metabolites in human liver microsomes, only the 4-OH midazolam is completely inhibited in mouse liver microsomes [57]. Importantly, subsequent analysis revealed that Cyp2c11 contributes to the metabolism of midazxolam in mice but not in humans [57]. Thus, a different compliment of CYP enzymes can contribute to substrate turn-over in mice compared to humans due to differences in substrate recognition for multiple CYP isoforms, and single isoforms cannot be considered in isolation.

28.4 CYP2B

The non-nucleoside reverse transcriptase inhibitors, efavirenz and nevirapine, are both substrates of CYP2B6 but whereas efavirenz metabolic clearance is almost exclusively via CYP2B6, CYP3A4 also plays a major role for nevirapine [13, 14]. In humans, there are two CYP2B isoforms, CYP2B6 and CYP2B7, with CYP2B7 having an almost identical coding sequence to CYP2B6 with high likelihood of being a pseudogene [58]. However, the murine Cyp2b subfamily consists of five members, Cyp2b9, Cyp2b10, Cyp2b13, Cyp2b19 and Cyp2b23 [9], and the murine orthologue is thought to be Cyp2b10, which shares 74 % sequence identity [11, 12]. In humans, CYP2B6 has been reported in liver, peripheral blood mononuclear cells and brain [59, 60], but not in small intestine [9]. Conversely, in mice Cyp2b10 is expressed in the liver but is also expressed in small intestine with expression in the duodenum exceeding that of the liver [9]. The metabolic consequences of this have not been studied in detail but this constitutes a potentially important species difference for CYP2B/Cyp2b.

Currently there is a paucity of information relating to metabolism of efavirenz and nevirapine in mice. However, the rate of efavirenz clearance is faster in rats (plasma half life [$t_{1/2}$] = 7.95 h) than humans ($t_{1/2}$ = 40–76 h) [61, 62]. Pentoxyresorufin has been used as a probe substrate for CYP2B6 and CYP2b10 in isolated mouse hepatocytes, mouse liver microsomes and human liver microsomes [63–65] and thus both species do have substrates in common.

28.5 CYP2D

CYP2D is of minor importance for antiretroviral drugs, contributing only to clearance of ritonavir, nelfinavir and delavirdine. Metabolism of ritonavir is primarily via CYP3A4 to form M1 and M11 metabolites but CYP2D6 contributes to formation of the M2 metabolite [4]. Similarly, nelfinavir is metabolized via CYP3A4, CYP2C19 and CYP2D6 [15], and the major route of delavirdine metabolism is via CYP3A4 with a minor (~20%) contribution from CYP2D6 [16].

The genes coding for CYP2D are markedly different between mouse and humans. In humans there is one functional gene (*CYP2D6*), whereas in mice there are nine (*Cyp2d22, 2d26, 2d40, 2d13, 2d34, 2d11, 2d10, 2d12* and *2d9*) [66]. Although, similar substrate profiles have been reported in some cases, metabolism is not identical. For example, both CYP2D6 and Cyp2d9 metabolize bufuralol but only Cyp2d9 metabolizes testosterone to 16α-hydroxytestosterone [67]. There is currently limited data regarding the affinity of antiretroviral drugs for mouse Cyp2d isoforms but the Km and V_{max} of delavirdine for the rat isoform is similar to that of human [68].

There are also similarities and differences in the inhibition of CYP2D isoforms between species. For example, paroxetine inhibits CYP2D-mediated metabolism of berberine in both species [69] but hydroxylation of bufuralol via CYP2D is strongly inhibited by quinidine in humans (60–80%) but not mice (0–15%) [70]. Indeed, the IC_{50} of 31 compounds, showed marked differences in inhibition of CYP2D6 and Cyp2d22 [71].

28.6 Uridine 5′-Diphosphate-Glucuronosyltransferases (UGTs)

Phase I metabolism of pharmaceutical agents can often produce highly reactive metabolites [72] and conjugation of phase I metabolic products usually produce less active and more water soluble phase II products [72]. The human UGT2B subfamily consists of six members and UGT2B7 plays a particularly important role for efavirenz and zidovudine [17–20].

The mouse Ugt2b subfamily consists of seven members [73] and human UGT2B7 shares relatively low sequence identity with all the possible mouse orthologues; Ugt2b1 (70%), Ugt2b5 (66%), Ugt2b37 (64%), Ugt2b38 (66%), Ugt2b34 (69%), Ugt2b35 (67%) and Ugt2b36 (67%) [74]. Therefore, activity of Ugt2b in mouse compared to human is difficult to study [75, 76]. UGT2B is highly expressed in the liver of both species. UGT2B7 is expressed in human small intestine, colon, kidney, bladder and uterus, and in mouse liver, GI, brain, lung, male and female genital tract [73, 77]. Despite complexities regarding ortholgues, there are strong indications of species differences in UGT2B. For example, zidovudine undergoes extensive phase

II metabolism in humans with 60–70% inactivation through glucuronidation [78] but is excreted largely unchanged (70%) in the urine of mice [79].

28.7 Species Differences in Key Drug Transporters

Influx and efflux drug transporters profoundly influence the pharmacokinetics and distribution of antiretroviral and other drugs. Transporters within the intestinal epithelium can limit or facilitate the entry of xenobiotics into the systemic circulation whereas expression in liver or kidney can dramatically influence clearance of their substrates from the body. Moreover, transporter expression at other sites can influence drug penetration of substrates into cells or anatomical sites and sometimes contribute to 'sanctuary sites' for HIV where drug concentrations are insufficient to stem viral replication [80]. The following sections provide detail on the key drug transporters known to be involved in disposition of antiretroviral drugs.

28.8 P-glycoprotein (ABCB1)

P-glycoprotein (coded by the *ABCB1* gene) has been shown to affect the pharmacokinetics of numerous antiretroviral (and other) drugs, particularly the protease inhibitors [21–27] and maraviroc [28]. P-glycoprotein performs ATP-dependent efflux from cells and it is expressed on the apical membrane in many excretory and barrier tissues [81]. In humans, P-glycoprotein is expressed in liver, kidney, adrenal gland, intestine, blood–brain barrier, placenta, blood–testis and blood–ovarian barriers [81]. The rodent equivalent transport system is coded by two genes, *Abcb1a* and *Abcb1b* [29], the products of which play the role of the single human protein. Expression of *Abcb1a* mRNA occurs in intestine, brain, testis, liver and heart, and *Abcb1b* is found in adrenal gland, ovaries, placenta, kidneys, liver and heart [82]. The two mouse isoforms of *Abcb1* share 90 and 80% sequence identity with the human *ABCB1* [83]. Though structurally similar and expressed in a similar range of tissues, the relative expression may vary between species. For example, mouse brain microvessels were found to express >2.5-fold higher protein than human [84].

The uptake of protease inhibitors is significantly reduced in cell lines overexpressing P-glycoprotein compared to control cells [85] and they are also substrates for murine Abcb1 indicating overlap between species. Furthermore, the coadministration of LY335979 (P-glycoprotein inhibitor) increased nelfinavir concentrations in brain (37-fold) and testes (4-fold) of mice [86] and the accumulation of saquinavir (3-fold) and lopinavir (9-fold) was higher in Abcb1 knockout mice compared to wild type mice [87, 88]. Similarly, indinavir and nelfinavir show a 2–5-fold increase in plasma concentrations in Abcb1 knockout mice [89]. The use of *Abcb1a*/*Abcb1b* knockout mice and human cell systems has also demonstrated maraviroc is a substrate human and mouse P-glycoprotein [28]. Despite these similarities, there are data for non-antiretroviral drugs indicating differences in substrate affinity across species. For example, risperidone has a greater Km and V_{max} in mice (Km=26.3 μM, V_{max}=71.4 nMol/mg/min) compared to humans (Km=12.4 μM, V_{max}=16.8 nMol/mg/min)[90, 91].

With respect to inhibition, concentration-dependent inhibion of berberine efflux by amprenavir, ritonavir and lopinavir in murine macrophages (RAW267.7), human derived macrophages (THP-1) and human *ABCB1*-transfected MDCK cells has been reported [92]. Inhibition of the protease inhibitors was found to be similar in both species, the rank order was lopinavir > ritonavir > amprenavir [92]. Ritonavir has also been shown to inhibit intestinal P-glycoprotein in both human and mouse models [25]. Incubation of Caco-2 monolayers with darunavir (100 μM) demonstrated the apparent permeability was significantly increased in the presence of P-glycoprotein inhibitors, including ritonavir (75 μM). This was further demonstrated in *Abcb1a/Abcb1b* knockout mice. Intestinal absorbtion of darunavir (100 μM) was significantly increased by 2.7 fold in wild type mice in the presence of ritonavir (75 μM) [25]. Furthermore, administration of darunavir to *Abcb1a/Abcb1b* knockout mice showed a similar increase in the apparent permiability of darunavir compared to wild type mice [25].

28.9 Breast Cancer Resistance Protein (BCRP; *ABCG2*)

A number of antiretroviral drugs such as abacavir, zidovudine, lamivudine and atazanavir are substrates for BCRP [23, 30, 31]. Lamivudine and zidovudine have been identified as substrates for both human BCRP [31] and mouse Bcrp [30, 32] but detailed studies of the kinetics were not conducted in order to compare potential differences in affinity across species. BCRP is coded by *ABCG2* in humans and *Abcg2* in mice [82]. BCRP is expressed in a variety of tissues including intestine, liver, kidney, brain, testis and placenta [93], and there is some sequence similarity between the coding sequence (81 %) and protein (86 %) between mouse and human [94]. However, there is an apparent species difference in the quantity of expression with BCRP being ~2-fold higher in brain microvessels and 8-fold lower in hepatocytes from human versus mouse [95]. Generally speaking, there is a paucity of information pertaining to cross-species comparisons for BCRP/Bcrp and further work in this area is warranted.

28.10 Multidrug Resistance Proteins (MRPs; *ABCCs*)

The MRP family of transporters consists of ten members [82] and many antiretrovirals interact either as substrates or inhibitors. The protease inhibitors ritonavir, lopinavir, nelfinavir, saquinavir and indinavir are substrates for multiple MRP transporters. Ritonavir and indinavir are substrates for MRP1 only while lopinavir, nelfinavir and saquinavir are substrates for both MRP1 and MRP2 [24, 33, 34, 36]. Additionally, tenofovir is a substrate for MRP4 (*ABCC4*) and MRP7 (*ABCC10*) [38, 41] and other nucleos(t)ide reverse transcriptase inhibitors are substrates for MRP4 (*ABCC4*) and MRP5 (*ABCC5*) [42]. Nevirapine is also a substrate for MRP7 [40].

The sequence identity of mouse and human MRP's shows a high degree of similarity for MRP1 (88 %), MRP2 (78 %), MRP4 (87 %) and MRP5 (95 %) [35, 37, 96, 97]. MRP1/Mrp1 mRNA is expressed in multiple tissues in both humans and mice including intestine, liver, kidney, brain and genital tract. MRP2/Mrp2 mRNA expression is found to be highly expressed in the liver and kidneys of both species [98]. However, in some cases expression differs greatly, with human MRP1 mRNA being moderately expressed in placenta and ovaries and murine Mrp1 being highly expressed [98]. Similarly, MRP4/Mrp4 mRNA is highly expressed in the kidney of mice and humans but Mrp4 mRNA is also moderately expressed in the stomach and ovary of mice and most highly expressed in the lung of humans [98]. Finally, murine Mrp5 is highly expressed in the brain, gonads, placenta, lung and stomach, whereas human MRP5 is ubiquitously expressed, with high levels in skeletal muscle and brain [98].

There is evidence that human and murine MRP's show similar substrate overlap but there are also some important differences in the efficiency of transport. For example, both orthologues transport vinblastine, docetaxel, saquinavir and etoposide but murine Mrp2 transport of saquinavir (2-fold) and docetaxel (3-fold) is more efficient than human MRP2 [99]. Conversely, vinblastine transport is 1.6-fold more efficient by human MRP2 yet etoposide transport is equivalent between species [98]. Similarly, human and mouse MRP1/Mrp1 confer resistance to etoposide and vincristine whereas only human MRP1 confers resistance to doxorubicin [100]. Species differences in the kinetics of transport have also been reported with estradiol-17β-glucuronide having higher affinity for mouse MRP2 (Km = 71 μM, V_{max} = 2621 pMol/mg/min) than the human orthologue (Km = 98 μM, V_{max} = 9683 pMol/mg/min) [99]. Some differences in interaction potential have emerged also with sulfinpyrazone stimulating human MRP2-mediated transport of estradiol-17β-glucuronide, while strongly inhibiting mouse Mrp2 [82].

28.11 Solute Carrier (SLC) Transporters

Many antiretroviral drugs have been shown to interact with members of the SLC superfamily of transporters. Multiple protease inhibitors (lopinavir, saquinavir and darunavir) are substrates for the organic anion-transporting polypeptides (OATP1A2 and 1B1 coded by *SLCO1A2* and *SLCO1B1*, respectively) while tenofovir (OAT1 and OAT3) and zidovudine are substrates for the organic anion transporters (OAT1, OAT2, OAT3 and OAT4 coded by *SLC22A6*, *SLC22A7*, *SLC22A8* and *SLC22A9*, respectively) [42].

The expression of OATP's can be highly organ-specific in humans. For example, OATP1A2 is found predominantly in brain and retina while OATP1B1 is found almost exclusively in liver [101]. The mouse Oatp family consists of five members, Oatp1a1, Oatp1a4, Oatp1a5, Oatp1a6 and Oatp1b2, and identification of mouse orthologues of OATP1A2 and OATP1B1 has proven difficult due to the tissue distribution and low sequence identity [102]. For example, OATP1B1 is only found in human liver whereas mouse liver expresses Oatp1A1, Oatp1a4, Oatp1b2 and Oatp2b1. These transporters are also expressed in small intestine (Oatp1a4, Oatp1a5 and Oatp2b1), kidney (Oatp1a1, Oatp1a4 and Oatp2b1) and brain (Oatp1a4 and Oatp2b1). Only Oatp1b2 is expressed exclusively in the liver but the sequence identity with OATP1B1 is only 64 % [103, 104].

OAT1-4 are all highly expressed in human kidney, supporting their role in renal excretion, OAT1, OAT2 and OAT4 are expressed in the liver, and OAT2 is highly expressed in both human kidney and liver [101]. Oat1, Oat2 and Oat3 expression in mice is similar to humans with mRNA expressed in the liver and kidneys. Interestingly, Oat2 mRNA expression is sexually dimorphic in mice with males, only expressing Oat2 in kidneys while females express in both liver and kidney [105, 101], and OAT4 is expressed in humans but not in mice [106]. Tenofovir transport displays similarities across species, being a substrate for both human and mouse OAT1 and OAT3 [106, 107]. However, zidovudine is a substrate for human OAT1, OAT2 and OAT3 but only Oat1 and Oat3 in mice [105, 108].

The organic cation transporters (OCT1 and OCT2), which are coded by *SLC22A1* and *SLC22A2*, are expressed in numerous tissues and lamivudine is a substrate [42]. Inhibition of OCT1 has been shown to be species specific for non-antiretroviral drugs. For example, the inhibition constant (Ki) of 1-methyl-4-phenylpyridinium by tetramethylammonium, tetraethylammonium, tetrapropylammonium and tetrabutylammonium is significantly lower in *Xenopus laevis* oocytes transfected with the mouse compared to the human orthologue [109].

28.12 Four Other Transport Proteins of Interest

The transporters detailed so far in this chapter are of particular relevance for antiretroviral drugs. However, the International Transporter Consortium (ITC) have listed a total of 12 transporters recommended for evaluation during drug development ([110, 111] and many of these now feature in EMA and FDA regulatory documents. Therefore, a brief description of species similarities and differences for transporters appearing on this list and not discussed previously is given here.

The bile salt export pump (BSEP, coded by *ABCB11*) facilitates the excretion of lipophilic xenobiotics and there appear to be only small differences in the transport of substrates between mouse and human; with no significant differences in Km [112]. However, differences between human and mouse have been reported for the inhibition of transport of taurocholate, glycocholate, taurochenodeoxycholate and glycochenodeoxycholate by cyclosporine A [112]. Multidrug and toxin extrusion proteins (MATE), coded by *SLC47A1* and *SLC47A2*, share some overlap of substrate specificity with the OCT transporters [82] and are involved in both hepatic and renal clearance [82]. In humans there are two MATE's expressed, MATE1 and MATE2-K [82] whereas mice do not express MATE2-K [113] and its absence has been suggested to result in misinterpretation of murine data for high affinity substrates (e.g. oxaliplatin)[113].

28.13 Species Differences in Ligand-Activated Transcription Factors that Transcriptionally Regulate Metabolic Enzymes and Transporters

The enzymes and transporters that influence pharmacokinetics and disposition of antiretroviral drugs are regulated by a complex interplay between numerous transcription factors. Of particular importance are the nuclear receptor family that sense and respond to endobiotic and xenobiotic stimuli. These pathways have evolved to provide the organism with a mechanism to respond to potential toxins in the environment and diet. Some subtle but important species differences exist in these networks, presumably due to differences in the evolutionary environment between species. Two of the most well studied nuclear receptors, the pregnane X receptor (PXR; coded by *NR1I2*) and the constitutive androstane receptor (CAR; coded by *NR1I3*), are involved in the constitutive and inducible regulation of numerous CYP's and transporters (e.g. CYP3A4 and P-glycoprotein) [114, 115].

28.14 Pregnane X Receptor (PXR)

PXR is activated by multiple antiretroviral drugs including amprenavir, atazanavir, darunavir, ritonavir, lopinavir and efavirenz [116]. PXR also regulates many of the enzymes and transporters that effect disposition of antiretroviral drug's including CYP2B, CYP2C, CYP3A, MDR1 MRP2 and MRP3 [117]. Importantly, PXR can mediate ligand-dependent regulation but even in the absence of enzyme inducers the expression of PXR correlates with enzymes and transporters in various tissues [118, 119]. Therefore, baseline as well as inducible expression of enzymes and transporters is influenced by PXR. PXR is most highly expressed in liver, small intestine and colon but at lower levels in multiple tissues including kidney, stomach and lung [120]. There is a high degree of sequence identity in the DNA binding domain (DBD) of human and mouse PXR (96 %) but not the ligand binding domain (LBD; 76 %) [121].

Endogenous hormones have been demonstrated to be activators of both human and mouse PXR including pregnenolone, oestradiol and progesterone [122]. In addition to hormones certain exogenous drugs have also been shown to activate PXR in both species. Dexamethasone, phenobarbital and clotrimazole have all been shown to activate human and mouse PXR, although clotrimazole activates mouse

PXR only weakly [123]. However, rifampicin and SR12813 activate human but not mouse PXR, and pregnenolone 16α-carbonitrile strongly activates murine but not human PXR [64, 124]. Similarly, ritonavir has been demonstrated to activate human PXR in vivo while showing little effect on mouse PXR [123] indicating important differences in ligands between species. As a consequence of these molecular differences, there are similarities and differences between human and rodent hepatocytes in their response to PXR ligands [125, 126].

28.15 Constitutive Androstane Receptor (CAR)

Activation of CAR has been shown to regulate numerous enzymes and transporters including CYP2B, CYP2C, CYP3A, UGT's, MRP2, MRP4 [117] and UGT2B [127]. CAR responds to abacavir, efavirenz and nevirapine [116], and both efavirenz and nevirapine have been shown to induce their own metabolism through CAR effects on CYP2B6 [128]. Like PXR, CAR correlates with target gene expression in the absence of ligands [129, 130]. CAR is primarily expressed in the liver in both human and mice, but is also expressed in the small intestine [123, 131]. CAR has limited sequence identity between human and mouse in the DBD (88 %) and LBD (72 %) [132].

CAR from both species responds to activation from paracetamol (acetominophen), phenobarbitol, phenytoin and TCPOBOP [123, 133]. However, chlorpromazine and clotrimazole only activate mouse CAR [123], and human but not mouse CAR responds to CITCO. Conversely, meclizine is an agonist of mouse CAR but an inverse agonist of human CAR [134]. Similarities and differences between human and rodent hepatocytes in their response to CAR ligands have been reported [125, 126].

28.16 Differences in Murine and Human Physiology

It is important to recognize that in addition to the metabolic differences outlined above, differences in tissue size and blood flow may also influence species differences in pharmacokinetics and thus pharmacodynamics. In some cases considering physiological differences can be used to normalize for species differences using allometric scaling [135]. For example, methotrexate is cleared much faster in mice than in humans, but by accounting for dose per kilogram, the clearance is 133 % of creatinine clearance in both species [1]. Similar approaches have been used to investigate species differences in tissue distribution. For example, the pharmacokinetics of atomoxetine and duloxetine in the brain extracellular fluid in rats were recently interpreted using scaling factors, such as brain weight and volume of extracellular fluid [136]. The physiology of other compartments may also contribute subtle differences. For example, the pH of the gastrointestinal tract significantly impacts on the absorption of some antiretroviral drugs such as raltegravir [137] and mice have a more acidic gastrointestinal tract compared to humans (human: stomach 6.6, duodenum 6.6, jejunum 6.63, illeum 7.49, proximal colon 6.6 and distal colon 7.5; mice: stomach 4.04, duodenum 4.74, jejunum 5.01, illeum 5.24, proximal colon 5.02 and distal colon 4.72) [138, 139].

28.17 Progress with Murine Transgenic and Humanized Liver Models

Mice transplanted with human hepatocytes have been shown to display a more human-like metabolic response to drugs and represent an interesting avenue for future research. For example, human CYP3A4 was shown to be induced by rifampicin in urokinase-type plasminogen activator-transgenic SCID mice (uPA/SCID mice) transplanted with human hepatocytes [140]. Transgenic mice also represent an interesting avenue of research. Recently transgenic mice expressing human PXR and CYP3A4 were developed, and displayed a more human-like response with rifampicin impacting upon darunavir [141]. A review on the use of transgenic humanized mice was recently published [142].

28.18 Discussion and Future Directions

This chapter has discussed some key similarities and differences in factors influencing the disposition of antiretroviral drugs in mice compared to humans. There are clearly some critical species differences and the complex interplay between distinct pathways influencing drug response complicates thorough evaluation of differences. In some cases, compensatory mechanisms appear to exist and the overall phenotype can sometimes be similar even when molecular differences are evident. Currently there is a paucity of information relating to differences between humans and mice in the context of some key mechanisms known to influence the pharmacokinetics of antiretroviral and other drugs. This lack of substantive data taken collectively with different rates clearance and clear genetic differences that exist between species indicates that pharmacokinetic and pharmacodynamics data generated in mice should be interpreted with caution. Progress is being made with humanized models for the immune system as well as for the liver and there is at least potential for convergence of these models in the future (Fig. 28.2). In the meantime, a better understanding of mechanistic differences will aid in the interpretation of pharmacokinetic and pharmacodynamics data generated in mice for existing as well as development antiretroviral medicines.

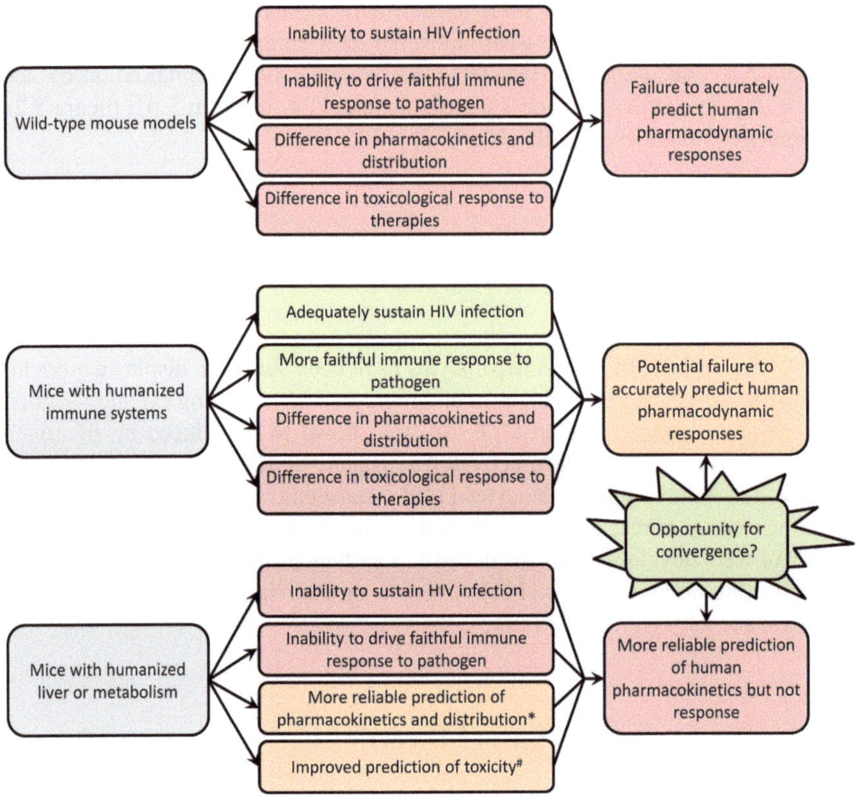

Fig. 28.2 Schematic representation of the limitations of current murine model systems for studying antiretroviral pharmacokinetics and pharmacodynamics. Wild-type mice have limitations for studying virology, immunology and pharmacology of HIV. While mice with humanized immune systems are able to sustain HIV infection and provide tools for studying immunology, data generated for drug-response should be interpreted with caution. Parallel research has resulted in progress for development of humanized liver or metabolism models and there is the potential for future convergence. *notwithstanding extra-hepatic factors that are important in mediating differences in pharmacokinetics and pharmacodynamics. #relates to concentration-dependent toxicities or toxicities that manifest through generation of a reactive metabolite generated in liver and this does not consider other extra-hepatic genes and proteins may be important mediators of the toxicological phenotype

References

1. Boxenbaum H. Interspecies scaling, allometry, physiological time, and the ground plan of pharmacokinetics. J Pharmacokinet Biopharm. 1982;10(2):201–27.
2. Back D, Sekar V, Hoetelmans R. Darunavir: pharmacokinetics and drug interactions. Antivir Ther. 2008;13(1):1–13.
3. Taburet A-M, Piketty C, Chazallon C, Vincent I, Gérard L, Calvez V, et al. Interactions between atazanavir-ritonavir and tenofovir in heavily pretreated human immunodeficiency virus-infected patients. Antimicrob Agents Chemother. 2004;48(6):2091–6.

4. Kumar GN, Rodrigues AD, Buko AM, Denissen JF. Cytochrome P450-mediated metabolism of the HIV-1 protease inhibitor ritonavir (ABT-538) in human liver microsomes. J Pharmacol Exp Ther. 1996;277(1):423–31.
5. Lim ML, Min SS, Eron JJ, Bertz RJ, Robinson M, Gaedigk A, et al. Coadministration of lopinavir/ritonavir and phenytoin results in two-way drug interaction through cytochrome P-450 induction. J Acquir Immune Defic Syndr. 2004;36(5):1034–40.
6. Smith PF, DiCenzo R, Morse GD. Clinical pharmacokinetics of non-nucleoside reverse transcriptase inhibitors. Clin Pharmacokinet. 2001;40(12):893–905.
7. Kakuda T, Leopold L, Nijs S, Vandevoorde A, Crauwels H, Bertelsen K, et al., Editors. Pharmacokinetic interaction between etravirine or rilpivirine and telaprevir in healthy volunteers: a randomised, two-way crossover trial. Program and abstracts of the 13th International Workshop on Clinical Pharmacology of HIV Therapy; 2012.
8. MacArthur RD, Novak RM. Maraviroc: the first of a new class of antiretroviral agents. Clin Infect Dis. 2008;47(2):236–41.
9. Martignoni M, Groothuis GM, de Kanter R. Species differences between mouse, rat, dog, monkey and human CYP-mediated drug metabolism, inhibition and induction. Expert Opin Drug Metab Toxicol. 2006;2(6):875–94.
10. Nelson DR, Zeldin DC, Hoffman SM, Maltais LJ, Wain HM, Nebert DW. Comparison of cytochrome P450 (CYP) genes from the mouse and human genomes, including nomenclature recommendations for genes, pseudogenes and alternative-splice variants. Pharmacogenet Genomics. 2004;14(1):1–18.
11. Tompkins LM, Wallace AD. Mechanisms of cytochrome P450 induction. J Biochem Mol Toxicol. 2007;21(4):176–81.
12. Davies R, Schuurman A, Barker CR, Clothier B, Chernova T, Higginson FM, et al. Hepatic gene expression in protoporphyic Fech mice is associated with cholestatic injury but not a marked depletion of the heme regulatory pool. Am J Pathol. 2005;166(4):1041–53.
13. Erickson DA, Mather G, Trager WF, Levy RH, Keirns JJ. Characterization of the in vitro biotransformation of the HIV-1 reverse transcriptase inhibitor nevirapine by human hepatic cytochromes P-450. Drug Metab Dispos. 1999;27(12):1488–95.
14. Ogburn ET, Jones DR, Masters AR, Xu C, Guo Y, Desta Z. Efavirenz primary and secondary metabolism in vitro and in vivo: identification of novel metabolic pathways and cytochrome P450 2A6 as the principal catalyst of efavirenz 7-hydroxylation. Drug Metab Dispos. 2010;38(7):1218–29
15. Baede-van Dijk PA, Hugen PW, Verweij-van Wissen CP, Koopmans PP, Burger DM, Hekster YA. Analysis of variation in plasma concentrations of nelfinavir and its active metabolite M8 in HIV-positive patients. AIDS. 2001;15(8):991–8.
16. Voorman RL, Maio SM, Hauer MJ, Sanders PE, Payne NA, Ackland MJ. Metabolism of delavirdine, a human immunodeficiency virus type-1 reverse transcriptase inhibitor, by microsomal cytochrome P450 in humans, rats, and other species: probable involvement of CYP2D6 and CYP3A. Drug Metab Dispos. 1998;26(7):631–9.
17. Bae S, Jeong Y-J, Lee C, Liu K-H. Identification of human UGT isoforms responsible for glucuronidation of efavirenz and its three hydroxy metabolites. Xenobiotica. 2011;41(6):437–44.
18. Bélanger A-S, Caron P, Harvey M, Zimmerman PA, Mehlotra RK, Guillemette C. Glucuronidation of the antiretroviral drug efavirenz by UGT2B7 and an in vitro investigation of drug-drug interaction with zidovudine. Drug Metab and Dispos. 2009;37(9):1793–6.
19. Michaud V, Bar-Magen T, Turgeon J, Flockhart D, Desta Z, Wainberg MA. The dual role of pharmacogenetics in HIV treatment: mutations and polymorphisms regulating antiretroviral drug resistance and disposition. Pharmacol Rev. 2012;64(3):803–33.
20. Turgeon D, Carrier J-S, Lévesque É, Hum DW, Bélanger A. Relative enzymatic activity, protein stability, and tissue distribution of human steroid-metabolizing UGT2B subfamily members. Endocrinology. 2001;142(2):778–87.
21. Rittweger M. Clinical pharmacokinetics of darunavir. Clin pharmacokinet. 2007;46(9):739–56.

22. Zhang X, Tierney C, Albrecht M, Demeter LM, Morse G, DiFrancesco R, et al. Discordant associations between SLCO1B1 521T→ C and plasma levels of ritonavir-boosted protease inhibitors in AIDS clinical trials group study A5146. Ther drug monit. 2013;35(2):209–16.
23. Bousquet L, Roucairol C, Hembury A, Nevers M-C, Creminon C, Farinotti R, et al. Comparison of ABC transporter modulation by atazanavir in lymphocytes and human brain endothelial cells: ABC transporters are involved in the atazanavir-limited passage across an in vitro human model of the blood-brain barrier. AIDS Res Hum Retroviruses. 2008;24(9):1147–54.
24. Chandler B, Detsika M, Owen A, Evans S, Hartkoorn RC, Cane PA, et al. Short communication effect of transporter modulation on the emergence of nelfinavir resistance in vitro. Antivir Ther. 2007;12:831–4.
25. Holmstock N, Mols R, Annaert P, Augustijns P. In situ intestinal perfusion in knockout mice demonstrates inhibition of intestinal p-glycoprotein by ritonavir causing increased darunavir absorption. Drug Metab Dispos. 2010;38(9):1407–10.
26. Lee CG, Gottesman MM, Cardarelli CO, Ramachandra M, Jeang K-T, Ambudkar SV, et al. HIV-1 protease inhibitors are substrates for the MDR 1 multidrug transporter. Biochemistry. 1998;37(11):3594–601.
27. Shaik N, Giri N, Pan G, Elmquist WF. P-glycoprotein-mediated active efflux of the anti-HIV1 nucleoside abacavir limits cellular accumulation and brain distribution. Drug Metab Dispos. 2007;35(11):2076–85.
28. Walker DK, Abel S, Comby P, Muirhead GJ, Nedderman AN, Smith DA. Species differences in the disposition of the CCR5 antagonist, UK-427,857, a new potential treatment for HIV. Drug Metab Dispos. 2005;33(4):587–95.
29. Devault A, Gros P. Two members of the mouse mdr gene family confer multidrug resistance with overlapping but distinct drug specificities. Mol Cell Biol. 1990;10(4):1652–63.
30. Pan G, Giri N, Elmquist WF. Abcg2/Bcrp1 mediates the polarized transport of antiretroviral nucleosides abacavir and zidovudine. Drug Metab Dispos. 2007;35(7):1165–73.
31. Wang X, Furukawa T, Nitanda T, Okamoto M, Sugimoto Y, Akiyama S-I, et al. Breast cancer resistance protein (BCRP/ABCG2) induces cellular resistance to HIV-1 nucleoside reverse transcriptase inhibitors. Mol Pharmacol. 2003;63(1):65–72.
32. Giri N, Shaik N, Pan G, Terasaki T, Mukai C, Kitagaki S, et al. Investigation of the role of breast cancer resistance protein (Bcrp/Abcg2) on pharmacokinetics and central nervous system penetration of abacavir and zidovudine in the mouse. Drug Metab Dispos. 2008;36(8):1476–84.
33. Janneh O, Hartkoorn R, Jones E, Owen A, Ward S, Davey R, et al. Cultured CD4T cells and primary human lymphocytes express hOATPs: intracellular accumulation of saquinavir and lopinavir. Br J Pharmacol. 2008;155(6):875–83.
34. Srinivas RV, Middlemas D, Flynn P, Fridland A. Human immunodeficiency virus protease inhibitors serve as substrates for multidrug transporter proteins MDR1 and MRP1 but retain antiviral efficacy in cell lines expressing these transporters. Antimicrob Agents Chemother. 1998;42(12):3157–62.
35. Ensembl. Gene: ABCC1 ENSG00000103222. [cited 2013 23rd July]. http://www.ensembl.org/Homo_sapiens/Gene/Compara_Ortholog?g=ENSG00000103222;r=16:16043434-16236931. Accessed 23 July 2013.
36. Huisman MT, Smit JW, Crommentuyn KM, Zelcer N, Wiltshire HR, Beijnen JH, et al. Multidrug resistance protein 2 (MRP2) transports HIV protease inhibitors, and transport can be enhanced by other drugs. AIDS. 2002;16(17):2295–301.
37. Ensembl. Gene: ABCC4 ENSG00000125257. 2013 [cited 2013 23rd July]. http://www.ensembl.org/Homo_sapiens/Gene/Compara_Ortholog?g=ENSG00000125257;r=13:95672083-95953687. Accessed 23 July 2013.
38. Kearney BP, Flaherty JF, Shah J. Tenofovir disoproxil fumarate. Clin Pharmacokinet. 2004;43(9):595–612.
39. Ensembl. Gene: ABCC10 ENSG00000124574. 2013 [cited 2013 31st July]. http://www.ensembl.org/Homo_sapiens/Gene/Compara_Ortholog?g=ENSG00000124574;r=6:43395104-43418168.

40. Liptrott NJ, Pushpakom S, Wyen C, Fätkenheuer G, Hoffmann C, Mauss S, et al. Association of ABCC10 polymorphisms with nevirapine plasma concentrations in the German Competence Network for HIV/AIDS. Pharmacogenet Genomics. 2012;22(1):10–9.
41. Pushpakom SP, Liptrott NJ, Rodríguez-Nóvoa S, Labarga P, Soriano V, Albalater M, et al. Genetic variants of ABCC10, a novel tenofovir transporter, are associated with kidney tubular dysfunction. J Infect Dis. 2011;204(1):145–53.
42. Kis O, Robillard K, Chan GN, Bendayan R. The complexities of antiretroviral drug–drug interactions: role of ABC and SLC transporters. Trends Pharm Sci. 2010;31(1):22–35.
43. Weiss J, Rose J, Storch CH, Ketabi-Kiyanvash N, Sauer A, Haefeli WE, et al. Modulation of human BCRP (ABCG2) activity by anti-HIV drugs. J Antimicrob Chemother. 2007;59(2):238–45.
44. Back DJ. Drug-drug interactions that matter. Top HIV Med. 2006;14(2):88–92. (Epub 2006/07/13).
45. Abel S, Back DJ, Vourvahis M. Maraviroc: pharmacokinetics and drug interactions. Antivir Ther. 2009;14(5):607–18. (Epub 2009/08/26).
46. Kumar GN, Jayanti V, Lee RD, Whittern DN, Uchic J, Thomas S, et al. In vitro metabolism of the HIV-1 protease inhibitor ABT-378: species comparison and metabolite identification. Drug Metab Dispos. 1999;27(1):86–91.
47. Riska PS, Joseph DP, Dinallo RM, Davidson WC, Keirns JJ, Hattox SE. Biotransformation of nevirapine, a non-nucleoside HIV-1 reverse transcriptase inhibitor, in mice, rats, rabbits, dogs, monkeys, and chimpanzees. Drug Metab Dispos. 1999;27(12):1434–47.
48. Riska P, Lamson M, MacGregor T, Sabo J, Hattox S, Pav J, et al. Disposition and biotransformation of the antiretroviral drug nevirapine in humans. Drug Metab Dispos. 1999;27(8):895–901.
49. Dellamonica P, Di Perri G, Garraffo R. NNRTIs: pharmacological data. Med Mal Infect. 2012;42(7):287–95.
50. Deb S, Pandey M, Adomat H, Guns EST. Cytochrome P450 3A-mediated microsomal biotransformation of 1α, 25-dihydroxyvitamin D3 in mouse and human liver: drug-related induction and inhibition of catabolism. Drug Metab Dispos. 2012;40(5):907–18.
51. van Waterschoot RA, ter Heine R, Wagenaar E, van der Kruijssen CM, Rooswinkel RW, Huitema AD, et al. Effects of cytochrome P450 3A (CYP3A) and the drug transporters P-glycoprotein (MDR1/ABCB1) and MRP2 (ABCC2) on the pharmacokinetics of lopinavir. Br J Pharmacol. 2010;160(5):1224–33. (Epub 2010/07/02).
52. Merry C, Barry MG, Mulcahy F, Ryan M, Heavey J, Tjia JF, et al. Saquinavir pharmacokinetics alone and in combination with ritonavir in HIV-infected patients. AIDS. 1997;11(4):F29–33.
53. van Herwaarden AE, Wagenaar E, van der Kruijssen CM, van Waterschoot RA, Smit JW, Song J-Y, et al. Knockout of cytochrome P450 3A yields new mouse models for understanding xenobiotic metabolism. J Clin Invest. 2007;117(11):3583–92.
54. Kuze J, Mutoh T, Takenaka T, Oda N, Hanioka N, Narimatsu S. Evaluation of animal models for intestinal first-pass metabolism of drug candidates to be metabolized by CYP3A enzymes via in vivo and in vitro oxidation of midazolam and triazolam. Xenobiotica. 2013;43(0):1–9.
55. Schipani A, Egan D, Dickinson L, Davies G, Boffito M, Youle M, et al. Estimation of the effect of SLCO1B1 polymorphisms on lopinavir plasma concentration in HIV-infected adults. Antivir Ther. 2012;17(5):861–8. (Epub 2012/04/06).
56. Tomaru A, Takeda-Morishita M, Banba H, Takayama K. Analysis of the pharmacokinetic boosting effects of ritonavir on oral bioavailability of drugs in mice. Drug metabolism and pharmacokinetics. 2013;28(2):144–52. PubMed PMID: 22971642.
57. Perloff MD, von Moltke LL, Kotegawa T, Shader RI, Greenblatt DJ. Midazolam and triazolam biotransformation in mouse and human liver microsomes: relative contribution of CYP3A and CYP2C isoforms. J Pharmacol Exp Ther. 2000;292(2):618–28.
58. Sheng J, Hua Z, Guo J, Caggana M, Ding X. Identification of a new human CYP2A gene fragment with close linkage to CYP2GP1. Drug Metab Dispos. 2001;29(1):4–7. (Epub 2000/12/22).

59. Liptrott N, Khoo S, Back D, Owen A. Detection of ABCC2, CYP2B6 and CYP3A4 in human peripheral blood mononuclear cells using flow cytometry. J Immunol Methods. 2008;339(2):270–4.
60. Miksys S, Lerman C, Shields PG, Mash DC, Tyndale RF. Smoking, alcoholism and genetic polymorphisms alter CYP2B6 levels in human brain. Neuropharmacology. 2003;45(1):122–32.
61. Adkins JC, Noble S. Efavirenz. Drugs. 1998;56(6):1055–64.
62. Nirogi R, Bhyrapuneni G, Kandikere V, Muddana N, Saralaya R, Komarneni P, et al. Pharmacokinetic profiling of efavirenz–emtricitabine–tenofovir fixed dose combination in pregnant and non-pregnant rats. Biopharm Drug Dispos. 2012;33(5):265–77.
63. Micuda S, Mundlova L, Anzenbacherova E, Anzenbacher P, Chladek J, Fuksa L, et al. Inhibitory effects of memantine on human cytochrome P450 activities: prediction of in vivo drug interactions. Eur J Clin Pharmacol. 2004;60(8):583–9.
64. Posti K, Leinonen S, Tetri S, Kottari S, Viitala P, Pelkonen O, et al. Modulation of murine phenobarbital-inducible CYP2A5, CYP2B10 and CYP1A enzymes by inhibitors of protein kinases and phosphatases. Eur J Biochem. 1999;264(1):19–26.
65. Scheer N, Ross J, Kapelyukh Y, Rode A, Wolf CR. In vivo responses of the human and murine pregnane X receptor to dexamethasone in mice. Drug Metab Dispos. 2010;38(7):1046–53.
66. Ingelman-Sundberg M. Genetic polymorphisms of cytochrome P450 2D6 (CYP2D6): clinical consequences, evolutionary aspects and functional diversity. Pharmacogenomics J. 2004;5(1):6–13.
67. Smith G, Modi S, Pillai I, Lian L-Y, Sutcliffe MJ, Pritchard MP, et al. Determinants of the substrate specificity of human cytochrome P-450 CYP2D6: design and construction of a mutant with testosterone hydroxylase activity. Biochem J. 1998;331(Pt 3):783–92.
68. Voorman RL, Maio SM, Hauer MJ, Sanders PE, Payne NA, Ackland MJ. Metabolism of delavirdine, a human immunodeficiency virus type-1 reverse transcriptase inhibitor, by microsomal cytochrome P450 in humans, rats, and other species: probable involvement of CYP2D6 and CYP3A. Drug Metab Dispos. 1998;26(7):631–9. (Epub 1998/07/14).
69. Guo Y, Li F, Ma X, Cheng X, Zhou H, Klaassen CD. CYP2D plays a major role in berberine metabolism in liver of mice and humans. Xenobiotica. 2011;41(11):996–1005.
70. Bogaards J, Bertrand M, Jackson P, Oudshoorn M, Weaver R, Van Bladeren P, et al. Determining the best animal model for human cytochrome P450 activities: a comparison of mouse, rat, rabbit, dog, micropig, monkey and man. Xenobiotica. 2000;30(12):1131–52.
71. McLaughlin LA, Dickmann LJ, Wolf CR, Henderson CJ. Functional expression and comparative characterization of nine murine cytochromes P450 by fluorescent inhibition screening. Drug Metab Dispos. 2008;36(7):1322–31.
72. Rang H, Dale M, Ritter J, Moore P. Pharmacology. 5th edn. New Delhi: Charchill Livingstone; 2006.
73. Buckley DB, Klaassen CD. Tissue-and gender-specific mRNA expression of UDP-glucuronosyltransferases (UGTs) in mice. Drug Metab Dispos. 2007;35(1):121–7.
74. Ensembl. Gene: UGT2B7 ENSG00000171234. 2013 [cited 2013 23rd July]. https://www.ebi.ac.uk/s4/jump;jsessionid=9ED08AF0587C08212C07263C27DD2519?from=aHR0cDovL3d3dy5lYmkuYWMudWsvczQvc3VtbWFyeS9tb2xlY3VsYXI7anNlc3Npb25pZD05RUQwOEFGMDU4N0MwODIxMkMwNzI2M0MyN0REMjUxOT90ZXJtPVVHVDJC-NyZjbGFzc2lmaWNhdGlvbj05NjA2&hash=07237&url=http://www.ensembl.org/Homo_sapiens/Gene/Compara_Ortholog?g%3DENSG00000171234%3Bdb%3Dcore.
75. Trottier J, Milkiewicz P, Kaeding J, Verreault M, Barbier O. Coordinate regulation of hepatic bile acid oxidation and conjugation by nuclear receptors. Mol Pharm. 2006;3(3):212–22.
76. Pretheeban M, Hammond G, Bandiera S, Riggs W, Rurak D. Ontogenesis of UDP-glucuronosyltransferase enzymes in sheep. Comp Biochem Physiol A Mol Integr Physiol. 2011;159(2):159–66.
77. Nakamura A, Nakajima M, Yamanaka H, Fujiwara R, Yokoi T. Expression of UGT1A and UGT2B mRNA in human normal tissues and various cell lines. Drug Metab Dispos. 2008;36(8):1461–4.
78. Veal GJ, Back DJ. Metabolism of zidovudine. Gen Pharm. 1995;26(7):1469–75.

79. Chow HH. A physiologically based pharmacokinetic model of zidovudine (AZT) in the mouse: model development and scale-up to humans. J Pharm Sci. 1997;86(11):1223–8.
80. Owen A, Khoo SH. Intracellular pharmacokinetics of antiretroviral agents. J HIV Ther. 2004;9(4):97–101. (Epub 2005/02/26).
81. Marchetti S, Mazzanti R, Beijnen JH, Schellens JH. Concise review: clinical relevance of drug–drug and herb–drug interactions mediated by the ABC transporter ABCB1 (MDR1, P-glycoprotein). Oncologist. 2007;12(8):927–41.
82. Chu X, Bleasby K, Evers R. Species differences in drug transporters and implications for translating preclinical findings to humans. Expert Opin Drug Metab Toxicol. 2013;9(0):1–16.
83. Mason BL, Pariante CM, Thomas SA. A revised role for P-glycoprotein in the brain distribution of dexamethasone, cortisol, and corticosterone in wild-type and ABCB1A/B-deficient mice. Endocrinology. 2008;149(10):5244–53.
84. Uchida Y, Ohtsuki S, Katsukura Y, Ikeda C, Suzuki T, Kamiie J, et al. Quantitative targeted absolute proteomics of human blood–brain barrier transporters and receptors. J Neurochem. 2011;117(2):333–45.
85. Jones K, Bray PG, Khoo SH, Davey RA, Meaden ER, Ward SA, et al. P-Glycoprotein and transporter MRP1 reduce HIV protease inhibitor uptake in CD4 cells: potential for accelerated viral drug resistance? AIDS. 2001;15(11):1353.
86. Choo EF, Leake B, Wandel C, Imamura H, Wood AJ, Wilkinson GR, et al. Pharmacological inhibition of P-glycoprotein transport enhances the distribution of HIV-1 protease inhibitors into brain and testes. Drug Metab Dispos. 2000;28(6):655–60.
87. Washington CB, Wiltshire HR, Man M, Moy T, Harris SR, Worth E, et al. The disposition of saquinavir in normal and P-glycoprotein deficient mice, rats, and in cultured cells. Drug Metab Dispos. 2000;28(9):1058–62.
88. van Waterschoot R, ter Heine R, Wagenaar E, van der Kruijssen C, Rooswinkel R, Huitema A, et al. Effects of cytochrome P450 3A (CYP3A) and the drug transporters P-glycoprotein (MDR1/ABCB1) and MRP2 (ABCC2) on the pharmacokinetics of lopinavir. Br J Pharmacol. 2010;160(5):1224–33.
89. Kageyama M, Namiki H, Fukushima H, Terasaka S, Togawa T, Tanaka A, et al. Effect of chronic administration of ritonavir on function of cytochrome P450 3A and P-glycoprotein in rats. Biol Pharm Bull. 2005;28(1):130–7.
90. Ejsing TB, Pedersen AD, Linnet K. P-glycoprotein interaction with risperidone and 9-OH-risperidone studied in vitro, in knock-out mice and in drug–drug interaction experiments. Hum Psychopharmacol. 2005;20(7):493–500.
91. Boulton DW, DeVane CL, Liston HL, Markowitz JS. In vitro P-glycoprotein affinity for atypical and conventional antipsychotics. Life Sci. 2002;71(2):163–9.
92. Zha W, Wang G, Xu W, Liu X, Wang Y, Zha BS, et al. Inhibition of P-glycoprotein by HIV protease inhibitors increases intracellular accumulation of berberine in murine and human macrophages. PloS one. 2013;8(1):e54349.
93. König J, Müller F, Fromm MF. Transporters and drug-drug interactions: important determinants of drug disposition and effects. Pharmacol Rev. 2013;65(3):944–66.
94. Breedveld P, Beijnen JH, Schellens JH. Use of P-glycoprotein and BCRP inhibitors to improve oral bioavailability and CNS penetration of anticancer drugs. Trends Pharmacol Sci. 2006;27(1):17–24.
95. Lai Y. Identification of interspecies difference in hepatobiliary transporters to improve extrapolation of human biliary secretion. Expert opinion on drug metabolism & toxicology. 2009;5(10):1175–87. PubMed PMID: 19611403.
96. Ensembl. Gene: ABCC2 ENSG00000023839. 2013 [cited 2013 23rd July]. http://www.ensembl.org/Homo_sapiens/Gene/Compara_Ortholog?g=ENSG00000023839;r=10:101542489-101611949.
97. Ensembl. Gene: ABCC5 ENSG00000114770. 2013 [cited 2013 23rd July]. http://www.ensembl.org/Homo_sapiens/Gene/Compara_Ortholog?g=ENSG00000114770;r=3:183637722-183735803. Accessed 23 July 2013.

98. Maher JM, Slitt AL, Cherrington NJ, Cheng X, Klaassen CD. Tissue distribution and hepatic and renal ontogeny of the multidrug resistance-associated protein (Mrp) family in mice. Drug Metab Dispos. 2005;33(7):947–55.
99. Zimmermann C, van de Wetering K, van de Steeg E, Wagenaar E, Vens C, Schinkel AH. Species-dependent transport and modulation properties of human and mouse multidrug resistance protein 2 (MRP2/Mrp2, ABCC2/Abcc2). Drug Metab Dispos. 2008;36(4):631–40.
100. Takayanagi Si, Ishikawa T. Molecular identification and characterization of rat Abcc1 cDNA: existence of two splicing variants and species difference in drug-resistance profile. J Exp Ther Oncol. 2003;3(3):136–46.
101. Bleasby K, Castle J, Roberts C, Cheng C, Bailey W, Sina J, et al. Expression profiles of 50 xenobiotic transporter genes in humans and pre-clinical species: a resource for investigations into drug disposition. Xenobiotica. 2006;36(10–11):963–88.
102. Ensembl. Gene: SLCO1B1 ENSG00000134538. 2013. http://www.ensembl.org/Homo_sapiens/Gene/Compara_Ortholog?g=ENSG00000134538;r=12:21284136-21392180;t=ENST00000256958. Accessed 25 July 2013.
103. Iusuf D, van de Steeg E, Schinkel AH. Functions of OATP1A and 1B transporters in vivo: insights from mouse models. Trends Pharmacol Sci. 2012;33(2):100–8.
104. Ensembl. Gene: SLCO1B1 ENSG00000134538. 2013 [cited 2013 25th July]. http://www.ensembl.org/Homo_sapiens/Gene/Compara_Ortholog?g=ENSG00000134538;r=12:21284136-21392180;t=ENST00000256958. Accessed 25 July 2013.
105. Burckhardt G. Drug transport by Organic Anion Transporters (OATs). Pharmacology & therapeutics. 2012;136(1):106–30. PubMed PMID: 22841915.
106. Rizwan AN, Burckhardt G. Organic anion transporters of the SLC22 family: biopharmaceutical, physiological, and pathological roles. Pharm Res. 2007;24(3):450–70.
107. Truong DM, Kaler G, Khandelwal A, Swaan PW, Nigam SK. Multi-level analysis of organic anion transporters 1, 3, and 6 reveals major differences in structural determinants of antiviral discrimination. J Biol Chem. 2008;283(13):8654–63.
108. Kobayashi Y, Ohshiro N, Shibusawa A, Sasaki T, Tokuyama S, Sekine T, et al. Isolation, characterization and differential gene expression of multispecific organic anion transporter 2 in mice. Mol pharmacol. 2002;62(1):7–14.
109. Dresser MJ, Gray AT, Giacomini KM. Kinetic and selectivity differences between rodent, rabbit, and human organic cation transporters (OCT1). J Pharmacol Exp Ther. 2000;292(3):1146–52.
110. Tweedie D, Polli JW, Berglund EG, Huang SM, Zhang L, Poirier A, et al. Transporter studies in drug development: experience to date and follow-up on decision trees from the international transporter consortium. Clin Pharmacol Ther. 2013;94(1):113–25. (Epub 2013/04/17).
111. Hillgren KM, Keppler D, Zur AA, Giacomini KM, Stieger B, Cass CE, et al. Emerging transporters of clinical importance: an update from the international transporter consortium. Clin Pharmacol Ther. 2013;94(1):52–63. (Epub 2013/04/17).
112. Kis E, Ioja E, Nagy T, Szente L, Herédi-Szabó K, Krajcsi P. Effect of membrane cholesterol on BSEP/Bsep activity: species specificity studies for substrates and inhibitors. Drug Metab Dispos. 2009;37(9):1878–86.
113. Chu X, Bleasby K, Evers R. Species differences in drug transporters and implications for translating preclinical findings to humans. Expert Opin Drug Metab Toxicol. 2013;9(3):237–52.
114. Kliewer SA, Goodwin B, Willson TM. The nuclear pregnane X receptor: a key regulator of xenobiotic metabolism. Endocr Rev. 2002;23(5):687–702.
115. Wu B, Li S, Dong D. 3D structures and ligand specificities of nuclear xenobiotic receptors CAR, PXR and VDR. Drug discovery today. 2013;18(11-12):574–81. PubMed PMID: 23299080.
116. Chan GN, Patel R, Cummins CL, Bendayan R. Induction of P-glycoprotein by antiretroviral drugs in human brain microvessel endothelial cells. Antimicrob Agents Chemother. 2013.

117. Scheer N, Roland Wolf C. Xenobiotic receptor humanized mice and their utility. Drug Metab Rev. 2013;45(1):110–21.
118. Albermann N, Schmitz-Winnenthal FH, Z'Graggen K, Volk C, Hoffmann MM, Haefeli WE, et al. Expression of the drug transporters MDR1/ABCB1, MRP1/ABCC1, MRP2/ABCC2, BCRP/ABCG2, and PXR in peripheral blood mononuclear cells and their relationship with the expression in intestine and liver. Biochem Pharmacol. 2005;70(6):949–58. (Epub 2005/08/02).
119. Owen A, Chandler B, Back DJ, Khoo SH. Expression of pregnane-X-receptor transcript in peripheral blood mononuclear cells and correlation with MDR1 mRNA. Antivir Ther. 2004;9(5):819–21. (Epub 2004/11/13).
120. Kodama S, Negishi M. PXR cross-talks with internal and external signals in physiological and pathophysiological responses. Drug Metab Rev. 2013;45(0):1–11.
121. LeCluyse EL. Pregnane X receptor: molecular basis for species differences in CYP3A induction by xenobiotics. Chem Biol Interact. 2001;134(3):283–9.
122. Gong H, Xie W. Orphan nuclear receptors, PXR and LXR: new ligands and therapeutic potential. Expert Opin Ther Targets. 2004;8(1):49–54.
123. Chang TK, Waxman DJ. Synthetic drugs and natural products as modulators of constitutive androstane receptor (CAR) and pregnane X receptor (PXR). Drug Metab Rev. 2006;38(1–2):51–73.
124. Holmstock N, Gonzalez FJ, Baes M, Annaert P, Augustijns P. PXR/CYP3A4-humanized mice for studying drug-drug interactions involving intestinal P-glycoprotein. Mol Pharm. 2013.
125. Martin P, Riley R, Back DJ, Owen A. Comparison of the induction profile for drug disposition proteins by typical nuclear receptor activators in human hepatic and intestinal cells. Br J Pharmacol. 2008;153(4):805–19. (Epub 2007/11/27).
126. Martin P, Riley R, Thompson P, Williams D, Back D, Owen A. Effect of prototypical inducers on ligand activated nuclear receptor regulated drug disposition genes in rodent hepatic and intestinal cells. Acta Pharmacol Sin. 2010;31(1):51–65. (Epub 2010/01/06).
127. Shelby MK, Klaassen CD. Induction of rat UDP-glucuronosyltransferases in liver and duodenum by microsomal enzyme inducers that activate various transcriptional pathways. Drug Metab Dispos. 2006;34(10):1772–8. (Epub 2006/07/21).
128. Faucette SR, Zhang T-C, Moore R, Sueyoshi T, Omiecinski CJ, LeCluyse EL, et al. Relative activation of human pregnane X receptor versus constitutive androstane receptor defines distinct classes of CYP2B6 and CYP3A4 inducers. J Pharmacol Exp Ther. 2007;320(1):72–80.
129. Chang TK, Bandiera SM, Chen J. Constitutive androstane receptor and pregnane X receptor gene expression in human liver: interindividual variability and correlation with CYP2B6 mRNA levels. Drug Metab Dispos. 2003;31(1):7–10.
130. Wortham M, Czerwinski M, He L, Parkinson A, Wan YJ. Expression of constitutive androstane receptor, hepatic nuclear factor 4 alpha, and P450 oxidoreductase genes determines interindividual variability in basal expression and activity of a broad scope of xenobiotic metabolism genes in the human liver. Drug Metab Dispos. 2007;35(9):1700–10. (Epub 2007/06/20).
131. Scheer N, Ross J, Rode A, Zevnik B, Niehaves S, Faust N, et al. A novel panel of mouse models to evaluate the role of human pregnane X receptor and constitutive androstane receptor in drug response. J Clin Invest. 2008;118(9):3228–39.
132. Choi H-S, Chung M, Tzameli I, Simha D, Lee Y-K, Seol W, et al. Differential transactivation by two isoforms of the orphan nuclear hormone receptor CAR. J Biol Chem. 1997;272(38):23565–71.
133. Sueyoshi T, Kawamoto T, Zelko I, Honkakoski P, Negishi M. The repressed nuclear receptor CAR responds to phenobarbital in activating the human CYP2B6 gene. J Biol Chem. 1999;274(10):6043–6.
134. Huang W, Zhang J, Wei P, Schrader WT, Moore DD. Meclizine is an agonist ligand for mouse constitutive androstane receptor (CAR) and an inverse agonist for human CAR. Mol Endocrinol. 2004;18(10):2402–8.

135. Agutter PS, Tuszynski JA. Analytic theories of allometric scaling. J Exp Biol. 2011;214(7):1055–62.
136. Kielbasa W, Stratford RE. Exploratory translational modeling approach in drug development to predict human brain pharmacokinetics and pharmacologically relevant clinical doses. Drug Metab Dispos. 2012;40(5):877–83.
137. Moss DM, Siccardi M, Murphy M, Piperakis MM, Khoo SH, Back DJ, et al. Divalent metals and pH alter raltegravir disposition in vitro. Antimicrob Agents Chemother. 2012;56(6):3020–6.
138. McConnell EL, Basit AW, Murdan S. Measurements of rat and mouse gastrointestinal pH, fluid and lymphoid tissue, and implications for in-vivo experiments. J Pharm Pharmacol. 2008;60(1):63–70.
139. Evans D, Pye G, Bramley R, Clark A, Dyson T, Hardcastle J. Measurement of gastrointestinal pH profiles in normal ambulant human subjects. Gut. 1988;29(8):1035–41.
140. Tateno C, Yoshizane Y, Saito N, Kataoka M, Utoh R, Yamasaki C, et al. Near completely humanized liver in mice shows human-type metabolic responses to drugs. Am J Pathol. 2004;165(3):901–12.
141. Holmstock N, Gonzalez FJ, Baes M, Annaert P, Augustijns P. PXR/CYP3A4-Humanized Mice for Studying Drug–Drug Interactions Involving Intestinal P-Glycoprotein. Mol Pharm. 2013;10(3):1056–62.
142. Scheer N, Wolf CR. Genetically humanized mouse models of drug metabolizing enzymes and transporters and their applications. Xenobiotica; the fate of foreign compounds in biological systems. 2014 Jan;44(2):96–108. PubMed PMID: 23845026.

Chapter 29
Antiretroviral Treatment Testing in HIV-Infected Humanized Mice

Roberto F. Speck

Preclinical testing in animal models is a critical step in the drug discovery and development process. However, cross-species differences may confound preclinical testing results. For example, genomic responses in mouse models poorly mimic human inflammatory diseases [1], and a large number of pathogens are specific only for humans (e.g., human immunodeficiency virus (HIV) type-1 and hepatitis C and B viruses). Importantly, mice are normally not permissive to these pathogens.

Nevertheless, nonhuman primates have served well as models for the study of human pathogens. Nonhuman primates (e.g., sooty mangabey, macaques) are primarily permissive to infection by simian-immunodeficiency virus (SIV), which is closely related to HIV [2]. Other than humans, the chimpanzee is the only primate species that is susceptible to HIV, but even there, HIV replicates poorly in chimps. Costs, ethics, and availability limit the use of nonhuman primate models and, particularly, chimpanzees. In addition, differences in SIV and HIV complicate extrapolation of results from SIV-based monkey models to the clinical setting of HIV-infected patients. For example, non-nucleoside reverse transcriptase inhibitors (NNRTI) are the most frequently used drugs for treating HIV-infected patients. They bind to a hydrophobic pocket close to the active catalytic site of HIV-1 reverse transcriptase (RT), but not of RT from SIVmac or to a lesser extent of RT from SIVagm, and have little or no inhibitory activity against SIV [3]. Thus, drug testing may be hampered by subtle differences in drug targets.

Humanized (hu) mice are a valuable alternative to SIV-based monkey models. They are highly susceptible to HIV infection [4] and allow investigations of HIV pathogenesis and novel drug treatment approaches. "Humanized mice" is a general term for mice transplanted with human tissue or cells. The first descriptions of humanized mice date back to the 1980s with the seminal publications from two laboratories. First, the fetal thymus/liver severe combined immunedeficiency (SCID) mouse model (SCID-hu Thy/Liv) was developed by the McCune laboratory [5]. In

R. F. Speck (✉)
Division of Infectious Diseases and Hospital Epidemiology, University Hospital of Zurich, University of Zurich, Raemistrasse 100, 8091 Zurich, Switzerland
e-mail: Roberto.speck@usz.ch

L. Y. Poluektova et al. (eds.), *Humanized Mice for HIV Research*,
DOI 10.1007/978-1-4939-1655-9_29

this model, fetal thymus/liver tissue is surgically placed under the renal capsule, and a conjoint organoid forms that resembles human thymus. The main disadvantages of this model are that studies and, in particular, drug testing are limited mainly to the conjoint organ, and monitoring requires repeated biopsies of the organ. Second, the hu-PBL-SCID mouse model, developed by Mosier et al. [6], is based on the intraperitoneal injection of human peripheral blood lymphocytes (PBL). Unfortunately, this model shows graft-vs-host disease (GvHD) within weeks after injection, and thus, the human PBL are "supraphysiologically" permissive to HIV and might need dosages of compounds to inhibit HIV replication that are higher than those for infected humans. Notably, GvHD is accompanied by immune activation, which is a main trigger for HIV replication. Furthermore, the progressive GvHD limits experimentation to a couple of weeks. Long-term studies with anti-HIV compounds are not feasible.

Newer generations of humanized mice are based on the transplantation of human CD34$^+$ cells into immunocompromised mice with or without implantation of human thymic/liver tissue [7–10]. After 3–4 months, the mice develop a lymphoid system with multilineage hematopoiesis. Critically for studying HIV, the lymphatic cells display a phenotype with quiescent/activated and naive/memory cells that is similar to humans.

In this chapter, we present the research and drug testing in SCID and humanized mice over the past three decades. We use the term humanized mice exclusively to these newer versions of mice. Ideally, experiments begin on humanized mice at the age of 3–4 months. These mice have a normal life span, and HIV infection can be monitored up to 18 months [11].

We focus mainly on currently available drugs for treating HIV-infected patients and the extent to which antiretroviral therapy (ART) in these models mirrors the human setting. A very important aspect herein is the pharmacokinetic analysis of the compounds before any efficacy testing in HIV-infected mice. A thorough characterization of those features is indispensable for defining a gold standard for comparing investigational compounds.

29.1 Antiretroviral Testing in SCID-hu Thy/Liv and hu-PBL-SCID Mice

The efficacy of any compound depends on the blood levels that are attained without significant toxicity (therapeutic range). The metabolism of compounds intended for use in humans must be diligently characterized. The most fundamental of these characteristics is dosage. Unfortunately, the literature provides a wide range of dosages of the human anti-HIV drugs in HIV mouse models (Table 29.1).

The first anti-HIV drug on the market, AZT, is a good example. This nucleoside analogue blocks the reverse transcription (RT) step of HIV, and thus it is called nucleoside reverse transcriptase inhibitor (NRTI). In humans, but not in mice, it is extensively metabolized [12] to, among other compounds, a highly toxic catabolite,

Table 29.1 Compilation of ART used in SCID-hu Thy/LIV mice, hu-PBL mice, humanized mice, and BLT mice

Mouse model	Treatment start/application mode of ART/mode of infection	AZT	Other ART compounds	Commentary	Reference
SCID-hu Thy/Liv mice	1 day prior to HIV infection and thereafter for 2 weeks AZT in drinking water Mode: intrathymic HIV injection	160–200 mg/kg/d	–	Transient HIV suppression	[18]
SCID-hu Thy/Liv mice	AZT started 0.5–48 h post-HIV challenge 1st dose i.p., then orally Mode: i.v. HIV injection	125–250 mg/kg/d	–	HIV suppression was dependent upon start of AZT after HIV infection	[20]
SCID-hu implanted with human fetal lymph node	1 day prior to HIV infection AZT in drinking water Mode: i.v. HIV injection	~200 mg/kg/d	–	Investigation of novel NNRTI compared to AZT	[43]
SCID-hu Thy/Liv mice	1 day prior to HIV infection and thereafter for 2 weeks AZT in drinking water, ddI i.p. Mode: i.v. HIV injection	6.6–165 mg/kg/d	ddI: up to 40 mg/kg/d	50 % infection rate when treated with AZT 40 mg/kg/d and ddI 13.7 mg/kg/d	[90]
Modified hu-PBL-SCID	1st dose of AZT 6 h post-infection for total 7 d AZT i.p. Mode: i.p. HIV injection	0.5, 5 and 50 mg/kg/d once daily	–	Dose-dependent suppressive effect of AZT	[19]
SCID-hu Thy/Liv	1 day prior to HIV infection AZT and nevirapine by gavage, ddI i.p. Mode: intraimplant infection	up to 180 mg/kg/d	ddI: up to ~150 mg/kg/d Nevirapine: up to 100 mg/kg/d	Study performed for developing standardization procedure	[38]
SCID-hu Thy/Liv	1 day prior to HIV infection AZT orally, ddI i.p. Mode: intrathymic HIV injection	5 mg/kg/d b.i.d given orally	ddI: 1 mg/kg/d	The investigational drug, bicyclam SDZ SID 791 was studied in combination with AZT and/or ddI	[51]

Table 29.1 (continued)

Mouse model	Treatment start/application mode of ART/mode of infection	AZT	Other ART compounds	Commentary	Reference
hu-PBL-SCID	2 days prior to HIV infection AZT for 2 days i.p. and then delivered by osmotic mini-pumps for 7 days Mode: i.p. HIV infection	20 mg/kg/d	–	MDL 74,968 (novel acycloun-cleotide) derivative of guanine compared to PMEA (adefovir) and AZT	[40]
hu-PBL-SCID	Either ART given immediately after HIV infection or for estab-lished HIV infection Saquinavir admixed to the food vs. given by gavage Model: intraimplant HIV infection	–	Saquinavir: 250 mg/kg/d or 2500 mg/kg/d vs. 2000 mg/kg/d by gavage b.i.d	Similar PK data as in human Significant decrease of HIV infection with Saquinavir on established HIV infection No prevention of HIV infection when started immediately after HIV challenge	[27]
SCID-hu Thy/Liv	Either ART for established HIV infection or given just immedi-ately after HIV infection Mode: intrathymic infection	66 mg/kg/d in drinking water	ddI: 50 mg/kg/d Saquinavir: 24–30 mg/kg/d in drinking water[a] Indinavir 180–225 mg/kg/d	Study suggests very nicely that ART might allow *de novo* pro-duction of T-lymphocytes	[29]
hu-PBL SCID	Immediately prior to HIV infection AZT in drinking water Mode: intraperitoneal injection of HIV	200 mg/kg/d	–	Comparison of TXU (anti-CD7)-pokeweed antiviral protein with AZT	[54]

Table 29.1 (continued)

Mouse model	Treatment start/application mode of ART/mode of infection	AZT	Other ART compounds	Commentary	Reference
SCID-hu Thy/Liv	ART for prevention or for treatment of established infection Added to the food Model: intraimplant HIV infection	10 and 100 mg/kg/d	3TC: 10 and 100 mg/kg/d Saquinavir: 2500 mg/kg/d (± Ritonavir 200 mg/kg/d) Ritonavir: 10 and 200 mg/kg/d	Efficient HIV suppression of established infection as well as prevention of HIV infection with triple ART even at low doses such as AZT/3TC/Ritonavir 10/10/20 mg/kg/d Study discussed potential suppressive effect of 3TC/AZT on hematopoietic activity	[28]
SCID-hu Thy/Liv	8 weeks postinfection AZT and indinavir in drinking water Mode: intraimplant HIV infection	60 mg/kg/d	ddI: 50 mg/kg/d i.p. Indinavir 225 mg/kg/d	Viral breakthrough despite ART with three active compounds	[37]
hu-PBL SCID	1 day prior to HIV infection for 14 days AZT given i.p. Mode: i.p. HIV infection	300–480 mg/kg/d i.p.	–	Type I IFN compared to AZT	[91]
SCID-hu Thy/Liv	1 day prior to HIV challenge 3TC given for 13 days Mode: intraimplant HIV infection	–	3TC: 30/mg/kg/d b.i.d. gavage	3TC compared to the structurally related nucleoside analog 2′deoxy-3′-oxa-4′-thiocytidine	[42]
SCID mouse model of HIVE (HIV encephalitis)	ART start immediately after intracerebral injection of HIV-infected macrophages ART was given i.p. Mode: intracerebral injection of HIV	45 mg/kg/d t.i.d., ip.	3TC: 45 mg/kg/d t.i.d., i.p. Indinavir : 45 mg/kg/d t.i.d., i.p.	ART decreased the amount of astrogliosis and viral load in treated mice compared with controls	[74]

Table 29.1 (continued)

Mouse model	Treatment start/application mode of ART/mode of infection	AZT	Other ART compounds	Commentary	Reference
SCID-hu Thy/Liv	Drugs given either on day − 1, day + 1, day + 3 and day + 7 All drugs but Enfuvirtide[b] given by gavage b.i.d,; Enfuvirtide s.c. b.i.d. Mode: intraimplant infection	–	3TC: 10–300 mg/kg/d (gavage) FTC[a]: 10–100 mg/kg/d Nevirapine 100–1000 mg/kg/d Efavirenz 30–100 mg/kg/d Atazanavir 100–1000 mg/kg/d Indinavir 100–1000 mg/kg/d Enfuvirtide 10–100 mg/kg/d	2nd generation antiretrovirals more potent than than their first-generation predecessors (first generation antivirals: 3TC, Nevirapine, Indinavir)	[30]
Humanized mice	ART for established HIV infection Mode: i.v. HIV infection	–	TDF 60 mg/kg/d i.p. FTC 60 mg/kg/d i.p. T20[b] 100 mg/kg s.c.	Study shows the efficacious anti-HIV effects of TDV/FTC/T20 in combination with investigational integrase inhibitor L-870812	[72, 75]
hu-PBL NSG mice	One single dose prior to HIV infection Mode: i.p. HIV infection	–	Atazanavir 250 mg/kg gavage Ritonavir 250 mg/kg gavage	Comparison to nanoformulated Atazanavir, Ritonavir ± Efavirenz	[55]

Table 29.1 (continued)

Mouse model	Treatment start/application mode of ART/mode of infection	AZT	Other ART compounds	Commentary	Reference
BLT mice	ART for established HIV infection	–	FTC 140–200 mg/kg/d i.p. TDF 146–208 mg/kg/d i.p. Raltegravir 56–80 mg/kg/d i.p.	very efficient ART regimen to suppress HIV	[77]
Humanized mice	ART for established HIV infection 3TC, Tenofovir, AZT, Rritonavir admixed to the food Mode: i.p. HIV infection	60–90 mg/kg/d	3TC: 60–90 mg/kg/d AZT: ~60 mg/kg/d Ritonavir: ~120 mg/kg/d	PK done over time for all compounds; AZT proved to be very toxic; ritonavir below the therapeutic range Assessment of a novel long-acting nano-based form of Rilpivirine	[4]
SCID-hu Thy/Liv	One dose Tenofovir/Emtricitabine[a] given 24 h prior to HIV infection Tenofovir/Emtricitabine given by gavage Mode: implant HIV infection	–	Tenofovir/Emtricitabine 200/130 mg/kg/d Tenofovir/Emtricitabine 60/40 mg/kg/d	Studying the effect of PC-1505 (fusion inhibitor) for its preexposure prophylactic anti-HIV effect as compared to one dose of Tenofovir/Emtricitabine	[47])

[a] Emtricitabine = FTC

[b] T20 = Enfuvirtide

3′-amino-3′-deoxythymidine [13]. In experiments with HIV-infected mice, a dosage of AZT of 60–480 mg/kg/d was given either by i.p. injections, orally by gavage or by adding AZT to the food pellets or drinking water, or by osmotic minipumps placed under the skin (Table 29.1). The toxic effects and potentially AZT-associated laboratory anomalies were rarely reported. Pharmacokinetics (PK) and toxicity also depend on the mouse strain used: AZT produced only a mild macrocytic anemia in C57Bl/6 mice, even when administered by gavage at 400 mg/kg/d for 30 days [14], whereas much lower doses were toxic in Balb/c mice [15]. Indeed, AZT provided in the drinking water displayed significant myelotoxicity at doses of 240–350 mg/kg/d given over 30 days in uninfected Balb/c mice [15] and of ~146 mg/kg/d given over 55 days in mice infected with the Rauscher murine leukemia complex [16]. In the latter case, drug toxicity necessitated stopping the study. Most data related to AZT's anti-HIV effect were generated in HIV-infected hu-PBL-SCID and SCID-hu Thy/Liv mice. These strains are similar to BALB/c except they carry the Igh-1b allele from the C57BL/Ka strain and, thus, will most likely be equally susceptible to higher doses of AZT.

We observed a striking toxicity in HIV-infected humanized NOD/SCID γ-chain−/− (NSG) mice on AZT (~66 mg/kg/d) when AZT was admixed to freely accessible food pellets [4]; no toxicity was found in a pilot experiment with uninfected mice. This discrepancy might be explained by assuming that HIV-1 Tat protein potentiates AZT-induced cellular toxicity [17].

Irrespective of dosing and toxicity issues, the SCID-hu Thy/Liv and the hu-PBL SCID mouse models have proved very useful for assessing AZT's anti-HIV inhibitory potential. AZT was clearly efficacious as monotherapy in suppressing HIV replication when given either before or within 1 day after HIV challenge but could not prevent HIV infection [18, 19]. In the SCID-hu mouse, AZT showed a time-dependent suppression of HIV with 100% efficacy when given within 2 h after viral challenge but gradual loss thereafter [20]. These data were reminiscent of data reported in clinical trials showing a substantial decrease of HIV-p24 antigen and a dose-dependent toxicity: reduction of the AZT dose in human was associated with increase of HIV-p24 Ag [21, 22]. Thus, the data in HIV-infected mice substantially supplemented our knowledge of AZT from clinical trials. Notably, extent to which toxic effects may have confounded AZT's HIV inhibitory effects in the various mouse models is not known. Other nucleoside or nucleotide analogues have not been studied as extensively as AZT (Table 29.1): studies examining PK and toxicity of those compounds are limited [14, 23, 24].

The discovery of protease inhibitors (PIs) of HIV and their clinical introduction in 1996 was a milestone in HIV medicine. ART consisting of a combination of two NRTIs and a PI showed a dramatic decrease of HIV-associated mortality and morbidity [25] and remains a first-line therapy today [26]. The first PIs were ritonavir and saquinavir. Goldstein et al. made an exemplary effort to document the PK and toxicity of ritonavir and Saquinavir before investigating its anti-HIV efficacy in SCID-hu Thy/Liv mice [27, 28](Table 29.1). PIs, in general, have a very short half-life. Since more-than-twice-daily gavage was too invasive for the mice, the Goldstein group incorporated saquinavir into the animal's diet. In this way, they attained plasma levels of 1000–2000 ng/ml, which are equal to those in humans (i.e., dose: 2500 mg/kg/d for mice; 7200 mg in humans) [27]. Like AZT monotherapy, saquinavir alone reduced HIV replication in a dose-dependent manner in established HIV infection but did not prevent HIV infection [27]. The Goldstein group also convincingly showed the efficacy of a triple ART for treating established HIV infection and indeed preventing HIV infection when given immediately after viral exposure [28]. Another study used a tenfold lower saquinavir dosage based on the information provided by the manufacturer [29]. Since saquinavir was given in this study in concert with ddI and AZT, the overall anti-HIV efficacy of the regimen did not appear to be affected: it resulted in $CD4^{+}$ thymocytes resurgence in HIV-infected mice, compared to controls.

Recently, Stoddart et al. completed an extensive study to validate the SCID-hu Thy/Liv model by examining the potency of four classes of licensed anti-HIV drugs, including 3TC, emtricitabine, nevirapine, efavirenz, indinavir, atazanavir and enfuvirtide [30]. The dosages were adjusted for difference in the ratios of surface area to body weight between mice and humans. Their major conclusion was that second-generation anti-HIV drugs were more potent than the first-generation anti-HIV drugs. While they carefully assessed the dose range of the drugs, they omitted more detailed PK of the various compounds.

To summarize, SCID-hu Thy/Liv and the hu-PBL SCID mouse models and their congeners closely recapitulate ART in humans. To better appreciate their value and positioning, additional information would be desirable, including standardization of the various mouse models, infection procedures, and in depth characterization of PK/PD (PD: pharmacodynamics) and toxicity of compounds. Most studies relied on information by the manufacturer or on conversion factors for estimating the dosage in mice [31–33]. However, the drug dosage should be verified by PK and toxicity analysis (i.e., general status and detailed laboratory analyses). Importantly, the plasma level measurements of the NRTIs are only an approximation of the concentration of their active intracellular di- or tri-phosphorylated metabolites [34]. Also when drugs are given in combination, unexpected toxicity is possible. An excellent illustration is provided by the hematopoietic toxicity for AZT and ddI when given together to normal C57BL/6 female mice [35]. Guidelines are made available by the FDA and EMEA for testing combination drugs in preclinical studies [36]. Furthermore, the majority of studies administered the anti-HIV compounds either before or immediately after HIV infection (Table 29.1), which mimics pre- or post-exposure prophylaxis somewhat. Such a drug dosing protocol may overestimate the drug's antiviral potency for treating established HIV infection. In fact, Amado et al. reported viral breakthrough linked to a progressive decline in thymocytes numbers when treating established HIV infection in thy/liv SCID mice with triple ART of AZT, ddI, and Indinavir [37].

The mode of infection must also be considered in evaluating the potential of novel compounds, particularly if they are developed for pre- or postexposure prophylaxis. Direct inoculation of HIV into the Thy/Liv conjoint organ in SCID mice or intravenously in hu-PBL mice with a highly activated immune system may result in a number of HIV-infected cells that is over a threshold amenable to anti-HIV drugs to prevent HIV infection.

Preclinical drug testing must take into account the different HIV strains. Antiviral compounds vary in their potency against different strains, pointing to the need to test compounds with various HIV strains to fully characterize them [38]. Unfortunately, this has been overlooked in most published reports. Thus, defined criteria for preclinical testing—a gold standard—would be a very valuable reference when testing novel treatment approaches.

Nevertheless, the SCID-hu Thy/Liv and the hu-PBL SCID mouse models have been valuable systems for assessing investigational compounds including (i) various reverse transcriptase inhibitors, such as a derivative of stavudine, stampidine [39]; a novel acyclonucleotide derivative of guanine MDL 74,968 [40]; a novel NRTI 4′-ethynyl-2-fluoro-2′deoxyadenosine [41], the nucleoside analog 2′-deoxy-3′-oxa-4′thiocytidine [42]; nonnucleoside RT inhibitors, such as BHAPs [43] and 2′-beta-fluoro-2′,3′dideoxyadenosine [44, 45]; (ii) fusion inhibitors such as an albumin-conjugated CD34 peptide HIV-1 fusion inhibitor [46] and PC-1505 [47]; (iii) CCR5 or CXCR4 antagonists, such as RANTES-analogs [48], the CCR5 antagonist SCH-C [49], a CXCR4 antagonist KRH-3955 [50]; and (iv) various other therapeutic modalities, such as the bicyclam SDZ SID 791 (JM 3100) [51], the oligonucleotide ISIS5320 [52], lovastatin [53], TXU-(anti-CD7)-pokeweed antiviral protein [54], long-acting nanoformulated atazanavir and ritonavir [55], rapamycin [56], and the maturation inhibitor Bevirimat [57].

A number of studies compared the activity of the investigational compound to AZT [43, 54]; such a comparison is only admissible when the optimal dosing, including dosing schedules, is known. Again, key for extrapolating data from mouse to man is the thorough characterization of the investigational compounds, including PK analysis that also covers the concentration attained in the various body compartments in the synopsis with their inhibitory activities against HIV.

29.2 ART Testing in Humanized Mice

Humanized mice are generated by direct transplantation of human hematopoietic progenitor cells (HPCs) intrahepatically [7] or intravenously [58] or in which a conjoint organ of liver thymus tissue acts, after total body irradiation, as a scaffold for the subsequent engraftment of HPCs [59]. The latter kind of humanized mice are known as BLT mice (BLT = bone marrow liver thymus). T-cells generated in the BLT mouse model are "educated" in a human environment and can mount HLA-restricted immune responses. Simpler humanized mice have less prominent responses.

The main advantage of humanized mice is the multilineage hematopoiesis with the generation of primary lymphoid organs of human origin (i.e., bone marrow, thymus, and lymph nodes) and secondary lymphoid organs (i.e., spleen and lymphoid tissue of the gastrointestinal (GI) tract) [8]. Notably, the phenotypes and the ratios of lymphocytes in the humanized mice are very similar to those in humans [7, 59]. Very importantly, humanized mice reconstitute the GI [9, 59, 60] and female reproductive tracts [61, 62]. In particular, Sun et al. showed that the small intestine of BLT mice is repopulated with substantial numbers of intraepithelial and lamina propria lymphocytes and that the rectum and colon are repopulated with $CD4^+$ T-cells in abundant clusters within the follicular aggregates and throughout the lamina propria [9]. Similarly, $CD4^+$ T cells, macrophages and dendritic cells are found throughout the vagina, ectocervix, endocervix, and uterus in the BLT mice [61, 62].

Here, particularly, the background strain and the kind of humanization make a difference in the level of mucosal reconstitution [62, 63]. In very nice work, the Garcia laboratory reported that the IL-2 receptor γ-chain molecule is critical for the intestinal T-cell reconstitution, resulting in the homing of $CD8^+CD4^+$ cells with two CD8 alpha/alpha chains into the GI tract, which is typical for human intestinal lymphocytes [60, 64]. They also showed that NOD/SCID γ-chain−/− (NSG) mice have better GI reconstitution than the Rag2−/−γ-chain−/− mice. Obviously, these properties make humanized mice very attractive for studying mucosal transmission either by the vaginal or rectal route. Indeed, the successful HIV transmission via vaginal [65, 66] or rectal routes [9, 62] have been reported.

Humanized mice are highly susceptible to HIV infection resulting in high peripheral viral loads and disseminated infection [4, 9, 67–70]. HIV replication is monitored by measuring HIV RNA copy numbers in the peripheral blood over time. This is a vast improvement over the SCID-hu Thy/Liv mouse model, in which HIV monitoring requires repeated biopsies of the conjoint thy/liv organ. Also notably, humanized mice show sustained HIV replication over 1 year [11], permitting long-term drug evaluation. In contrast, hu-PBL SCID mice have a rather tight window for experimentation due to xenogeneic reaction [71]. Thus, humanized mice are rather straightforward for preclinical testing of novel compounds/interventions, including the urgent assessment of microbicides.

ART in Humanized Mice for Treating Disseminated HIV Infection or as Preexposure Prophylaxis Choudhary et al. [72] and Sango et al. [73] reported the first trials of ART in humanized mice (Table 29.1). Overall, the ARTs appeared to be efficient. However, only 2/6 mice in the study by Choudhary et al. showed sustained viral load suppression; the other mice died prematurely or had viral breakthrough. These results might have been due to intraperitoneal injection of the drugs, which is poorly tolerated by the mice or the dosages might not have been optimal. PK analyses in the mice of the drug administered showed a very wide range of the maximal concentration (Cmax) and the areas under the curve (AUC), and the median values were much higher than in humans. ART by Sango et al. was also very efficient and was done with 45 mg/kg/d t.i.d. (ter in die; three times daily) of AZT, 3TC, and Indinavir each and was based on work that examined drug penetration in a SCID mouse model of HIV encephalopathy [74](Table 29.1). ART was also given in studies examining latency and its reactivation [75–77]. All of the reports convincingly demonstrated the efficacy of the chosen regimens: Choudhary et al. [75] administered the regimen as discussed above, Marsden et al. [76] applied the regimen of

AZT, ddI, and Indinavir as described [29] and Denton et el. used a regimen of i.p. injection of FTC (=emtricitabine), tenofovir, and raltegravir [77] (Table 29.1). No PK data were presented in the latter two papers.

In the context of evaluating a long-acting formulation of the novel NNRTI, Rilpivirin, we also completed a PK analysis of tenofovir, 3TC, AZT, and ritonavir [4]. Similar to the Goldstein's laboratory, we considered daily gavage or i.p. injections over weeks too invasive for the mice strains used for humanization. Thus, we added the anti-HIV drugs to the food. The drug dosages used in humans were converted for mice by the formula of Reagan-Shaw et al. [78]. The predicted dosages for TDF and 3TC were corroborated by their plasma measurements (Table 29.1). Thus, the dosage of tenofovir given i.p. by Denton et al. [77] is very likely above the therapeutic range for humans but comparable to that needed to suppress HIV in nonhuman primates. The calculated conversions were not correct for AZT and ritonavir: AZT was dosed too high with corresponding toxicity, and ritonavir was underdosed. These data underline the need to validate drug dosages chosen by in vivo PK analyses.

To conclude, the humanized mice are a straightforward preclinical small-animal model for testing novel HIV drugs, particularly for high-titer disseminated viral infections. Two papers, one by our laboratory, evaluated long-acting nanoformulated antiretroviral compounds in humanized mice [4, 79]. Both papers demonstrated the suppression of HIV RNA during treatment with those compounds and viral rebound after their interruption. Dash et al. applied atazanavir in concert with ritonavir as nanodrugs [79], and we added nanobased rilpivirine to a regimen of tenofovir/emtirictabine [4]. Both studies provided PK and toxicity data. These papers illustrate nicely the value of humanized mice for the preclinical proof of concept analyses of novel investigational therapeutic compounds. The targeting of monocyte-macrophages by the nanodrugs developed and their subsequent slow release from this cellular reservoir presented by Dash et al. is really worth mentioning [79]. It would have been nice to include a separate cohort of mice treated with an ART gold standard for comparison. Unfortunately, till now, such a gold standard is lacking.

Mucosal Transmission/Microbicide Testing Obviously, the presence of human lymphoid cells in the GI tract and the female reproductive tract makes humanized mice very attractive for testing microbicides.

Denton et al. reported rather complete prevention of HIV transmission in BLT mice when preexposure prophylaxis was given with emtricitabine/tenofovir. This differed from an 88% transmission rate in control mice in the setting of vaginal (100% protection) [61], rectal (100% protection), or intravenous (88% protection) challenge with HIV [80]. Taking advantage of reconstitution of the oral cavity and the upper GI tract with leucocytes in BLT mice, the same group investigated the efficacy of this drug combination for preventing oral HIV transmission. Again, mice given preexposure prophylaxis were protected against HIV transmission [81]. In all these studies, emtricitabine/tenofovir was given over 7 days, starting 2 days before HIV challenge. The doses of emtricitabine/tenofovir chosen were based on the previous publication [77] and were rather high (i.e., 140–175 mg/kg/d and 208–260 mg/

kg/d i.p.), compared to those used by Stoddart et al. [30] and Nischang et al. [4]. In a very similar fashion, Neff et al. evaluated the integrase inhibitor raltegravir and the CCR5 entry inhibitor maraviroc for preexposure prophylaxis [82]. Mice that received preexposure prophylaxis were protected, while all other mice became HIV positive. Raltegravir and maraviroc doses were calculated by an interspecies allometric scaling factor [32, 78] as we did in our study [4].

Topical application of tenofovir [83, 84], maraviroc [85] and some investigational compounds [83] efficiently prevented HIV mucosal transmission. For example, when 1 % tenofovir gel was applied to BLT mice 4 h before and 4 h after vaginal exposure to HIV, HIV infection was prevented in 7/8 mice [83]. The protocol was chosen to ensure full compliance with the Caprisa 004 protocol, which resulted in a 54 % lower HIV incidence in high-adherers women [86]. For other inhibitors, a simpler protocol applied each inhibitor vaginally 30 min before HIV challenge: three peptide-based inhibitors prevented HIV infection completely. In the other study, maraviroc was given vaginally 1 h before HIV challenge [85]. Finally, 1 % tenofovir gel given 30 min before rectal challenge was also efficient with prevention rate of ~80 % [84].

In summary, reconstituting the rectal and female reproductive tracts with human lymphatic tissue permits, for the first time, evaluation of the efficacy of topically applied compounds for preventing HIV transmission. Again, we lack PK data of ART in the male and female genital tract of mice. For a preclinical proof of concept, it would be very desirable to know how tissue concentrations compare in humans and mice. Data in human show a preferential accumulation of NRTIs in the male and female genital tract while there is much lesser accumulation of PIs in this body compartment. The PK data of most currently available anti-HIV drugs in these anatomical sanctuary sites are very nicely summarized by Else et al. [87].

Novel Approaches for Interfering with HIV in Humanized Mice A detailed discussion of novel therapeutic approaches is beyond the scope of this review and I like to refer to recently published reviews [88, 89].

29.3 Dosing of ARTs and Investigational Drugs

> All things are poison, and nothing is without poison; only the dose permits something not to be poisonous. (Paracelsus 1493–1541)

Importantly, the long-term goal of preclinical testing in mice is to prepare drugs for use in clinical practice. Thus, for every approved compound tested in mice, we should approximate the PK in human clinical practice as accurately as possible. Otherwise, we may generate misleading data. For investigational compounds that are in preclinical stage of evaluation, the criterion for evaluating therapeutic doses is a dose titration to define the highest dose without or with acceptable side effects. The primary endpoint of interest in this scenario will define the threshold of severity of side effects acceptable.

Where do we stand with these criteria in anti-HIV testing in humanized mice? As noted here, in the majority of studies, we lack a thorough characterization of the PK and toxicity profile of the corresponding compound. A large number of studies were based on converting the dose administered based on calculations to the dose in mice. PK studies in mice are very labor intensive, specimens are small, and consequently analyses difficult. Furthermore, drug administration is cumbersome, and mice are fragile and HIV-infected mice present the risk of accidental needle stick.

The modes of administration include i.p. or s.c. injection, adding the drug to the drinking water or to the food, osmotic minipumps and, for mucosal transmission, local application of drugs. All these have pros and cons. For example, i.p. injection is invasive for humanized mice and includes some biohazard; s.c. administration is less invasive but, for a number of drugs, not the ideal mode. For the comfort of mice and personnel, adding compounds to freely accessible drinking water or food has obvious advantages. It also permits the mice to reach a fairly stable drug level within the therapeutic range over time. However, this does not recapitulate the PK in humans with a peak concentration usually within 1–2 h of ingestion and subsequent disappearance of the compound according to its half-life. The relatively high steady state of compound when provided by food or water with unlimited access may also be toxic and is certainly not the ideal solution. To better mimic the PK in humans, we might consider limiting access to food containing the drugs overall to one or two time periods of 2–3 h a day.

Of course, the purpose of the experiments defines how diligent the dosing should be. Preclinical testing of a compound needs a greater effort than latency studies in humanized mice where the primary requirement is suppressed HIV RNA irrespective of the dosing. However, even in the latter case, it might create confounding results when doses are so high as to affect the reticulo-lymphatic system. It might be very difficult to recognize an artifact generated by these high dosages.

29.4 Conclusion

Humanized mice represent a very important advancement in HIV research. Nevertheless, standardization and defining a gold standard would be highly desirable. This would include conditions for pre- and postexposure prophylaxis, disseminated infection, infection with various HIV strains, mode of infection, level of detection of HIV RNA, dose of infection, PK analysis of compounds, and toxicity analyses. Such a standardization/gold standard would clearly facilitate the interexperimental comparison of studies and thus moving research in HIV drugs forward. *A workshop for defining the minimum requirements for preclinical testing would be very welcome. That workshop could take into account the previous work, the technical challenges and feasibilities of working with humanized mice, and the rapidly evolving field of humanized mice.*

Acknowledgments I thank Dr. Cheryl A. Stoddart (Division of Experimental Medicine, University of California, San Francisco) for her critical review of the manuscript. This work was funded by the Swiss National Science Foundation (#31003A_135682/1) and the Clinical Research Focus Program "Human Hemato-Lymphatic Diseases" of the University of Zurich.

References

1. Seok J, Warren HS, Cuenca AG, Mindrinos MN, Baker HV, Xu W, et al. Genomic responses in mouse models poorly mimic human inflammatory diseases. Proc Natl Acad Sci U S A. 2013;110(9):3507–12.
2. Pandrea I, Silvestri G, Apetrei C. AIDS in African nonhuman primate hosts of SIVs: a new paradigm of SIV infection. Curr HIV Res. 2009;7(1):57–72.
3. Isaka Y, Miki S, Kawauchi S, Suyama A, Sugimoto H, Adachi A, et al. A single amino acid change at Leu-188 in the reverse transcriptase of HIV-2 and SIV renders them sensitive to non-nucleoside reverse transcriptase inhibitors. Arch Virol. 2001;146(4):743–55.
4. Nischang M, Sutmuller R, Gers-Huber G, Audige A, Li D, Rochat MA, et al. Humanized mice recapitulate key features of HIV-1 infection: a novel concept using long-acting antiretroviral drugs for treating HIV-1. PLoS ONE. 2012;7(6):e38853.
5. Namikawa R, Weilbaecher KN, Kaneshima H, Yee EJ, McCune JM. Long-term human hematopoiesis in the SCID-hu mouse. J Exp Med. 1990;172(4):1055–63.
6. Mosier DE, Gulizia RJ, Baird SM, Wilson DB. Transfer of a functional human immune system to mice with severe combined immunodeficiency. Nature. 1988;335(6187):256–9.
7. Traggiai E, Chicha L, Mazzucchelli L, Bronz L, Piffaretti JC, Lanzavecchia A, et al. Development of a human adaptive immune system in cord blood cell-transplanted mice. Science. 2004;304(5667):104–7.
8. Rongvaux A, Takizawa H, Strowig T, Willinger T, Eynon EE, Flavell RA, et al. Human hemato-lymphoid system mice: current use and future potential for medicine. Annu Rev Immunol. 2013;31:635–74.
9. Sun Z, Denton PW, Estes JD, Othieno FA, Wei BL, Wege AK, et al. Intrarectal transmission, systemic infection, and CD4+ T cell depletion in humanized mice infected with HIV-1. J Exp Med. 2007;204(4):705–14.
10. Denton PW, Garcia JV. Humanized mouse models of HIV infection. AIDS Rev. 2011;13(3):135–48.
11. Berges BK, Akkina SR, Remling L, Akkina R. Humanized Rag2(−/−)gammac(−/−) (RAG-hu) mice can sustain long-term chronic HIV-1 infection lasting more than a year. Virology. 2010;397(1):100–3.
12. Resetar A, Spector T. Glucuronidation of 3′-azido-3′-deoxythymidine: human and rat enzyme specificity. Biochem Pharmacol J. 1989;38(9):1389–93.
13. Cretton EM, Xie MY, Bevan RJ, Goudgaon NM, Schinazi RF, Sommadossi JP. Catabolism of 3′-azido-3′-deoxythymidine in hepatocytes and liver microsomes, with evidence of formation of 3′-amino-3′-deoxythymidine, a highly toxic catabolite for human bone marrow cells. Mol Pharmacol. 1991;39(2):258–66.
14. Luster MI, Germolec DR, White KL Jr, Fuchs BA, Fort MM, Tomaszewski JE, et al. A comparison of three nucleoside analogs with anti-retroviral activity on immune and hematopoietic functions in mice: in vitro toxicity to precursor cells and microstromal environment. Toxicol Appl Pharmacol. 1989;101(2):328–39.
15. Calabresi P, Falcone A, St Clair MH, Wiemann MC, Chu SH, Darnowski JW. Benzylacyclouridine reverses azidothymidine-induced marrow suppression without impairment of anti-human immunodeficiency virus activity. Blood. 1990;76(11):2210–5.
16. Ruprecht RM, O'Brien LG, Rossoni LD, Nusinoff-Lehrman S. Suppression of mouse viraemia and retroviral disease by 3′-azido-3′-deoxythymidine. Nature. 1986;323(6087):467–9.

17. Prakash O, Teng S, Ali M, Zhu X, Coleman R, Dabdoub RA, et al. The human immunodeficiency virus type 1 Tat protein potentiates zidovudine-induced cellular toxicity in transgenic mice. Arch Biochem Biophys. 1997;343(2):173–80.
18. McCune JM, Namikawa R, Shih CC, Rabin L, Kaneshima H. Suppression of HIV infection in AZT-treated SCID-hu mice. Science. 1990;247(4942):564–6.
19. Alder J, Hui YH, Clement J. Efficacy of AZT therapy in reducing p24 antigen burden in a modified SCID mouse model of HIV infection. Antiviral Res. 1995;27(1-2):85–97.
20. Shih CC, Kaneshima H, Rabin L, Namikawa R, Sager P, McGowan J, et al. Postexposure prophylaxis with zidovudine suppresses human immunodeficiency virus type 1 infection in SCID-hu mice in a time-dependent manner. J Infect Dis. 1991;163(3):625–7.
21. Parks WP, Parks ES, Fischl MA, Leuther MD, Allain JP, Nusinoff-Lehrman S, et al. HIV-1 inhibition by azidothymidine in a concurrently randomized placebo-controlled trail. J Acquir Immune Defic Syndr. 1988;1(2):125–30.
22. Jackson GG, Paul DA, Falk LA, Rubenis M, Despotes JC, Mack D, et al. Human immunodeficiency virus (HIV) antigenemia (p24) in the acquired immunodeficiency syndrome (AIDS) and the effect of treatment with zidovudine (AZT). Ann Intern Med. 1988;108(2):175–80.
23. Manouilov KK, Xu ZS, Boudinot FD, Schinazi RF, Chu CK. Lymphatic targeting of anti-HIV nucleosides: distribution of 2′,3′-dideoxyinosine after intravenous and oral administration of dipalmitoylphosphatidyl prodrug in mice. Antiviral Res. 1997;34(3):91–9.
24. Russell JW, Klunk LJ. Comparative pharmacokinetics of new anti-HIV agents: 2′,3′-dideoxyadenosine and 2′, 3′-dideoxyinosine. Biochem Pharmacol. 1989;38(9):1385–8.
25. Palella FJ Jr, Delaney KM, Moorman AC, Loveless MO, Fuhrer J, Satten GA, et al. Declining morbidity and mortality among patients with advanced human immunodeficiency virus infection. HIV outpatient study investigators. N Engl J Med. 1998;338(13):853–60.
26. Thompson MA, Aberg JA, Hoy JF, Telenti A, Benson C, Cahn P, et al. Antiretroviral treatment of adult HIV infection: 2012 recommendations of the International Antiviral Society-USA panel. JAMA. 2012;308(4):387–402.
27. Pettoello-Mantovani M, Kollmann TR, Raker C, Kim A, Yurasov S, Tudor R, et al. Saquinavir-mediated inhibition of human immunodeficiency virus (HIV) infection in SCID mice implanted with human fetal thymus and liver tissue: an in vivo model for evaluating the effect of drug therapy on HIV infection in lymphoid tissues. Antimicrob Agents Chemother. 1997;41(9):1880–7.
28. Pettoello-Mantovani M, Kollmann TR, Katopodis NF, Raker C, Kim A, Yurasov S, et al. thy/liv-SCID-hu mice: a system for investigating the in vivo effects of multidrug therapy on plasma viremia and human immunodeficiency virus replication in lymphoid tissues. J Infect Dis. 1998;177(2):337–46.
29. Withers-Ward ES, Amado RG, Koka PS, Jamieson BD, Kaplan AH, Chen IS, et al. Transient renewal of thymopoiesis in HIV-infected human thymic implants following antiviral therapy. Nat Med. 1997;3(10):1102–9.
30. Stoddart CA, Bales CA, Bare JC, Chkhenkeli G, Galkina SA, Kinkade AN, et al. Validation of the SCID-hu Thy/Liv mouse model with four classes of licensed antiretrovirals. PLoS ONE. 2007;2(7):e655.
31. Mordenti J. Man versus beast: pharmacokinetic scaling in mammals. J Pharm Sci. 1986;75(11):1028–40.
32. Freireich EJ, Gehan EA, Rall DP, Schmidt LH, Skipper HE. Quantitative comparison of toxicity of anticancer agents in mouse, rat, hamster, dog, monkey, and man. Cancer Chemother Rep. 1966;50(4):219–44.
33. Bast RCHJ, Frei E. American Cancer Society (2000) Cancer medicine e5 online 5th ed Hamilton, Ont AtlantaGA: BC Decker; American Cancer.
34. Barry MG, Khoo SH, Veal GJ, Hoggard PG, Gibbons SE, Wilkins EG, et al. The effect of zidovudine dose on the formation of intracellular phosphorylated metabolites. AIDS. 1996;10(12):1361–7.

35. Gallicchio VS, Scott KW, Hughes NK, Tse KF, Gaines H, Kirk PR, et al. Increased hematopoietic toxicity following administration of interferon-a with combination dideoxynucleoside therapy (zidovudine plus ddI) administered in normal mice. Life Sci. 1995;56(3):PL71–81.
36. Lodola A. Developing combination drugs in preclinical studies. Methods Mol Biol. 2011;691:3–16.
37. Amado RG, Jamieson BD, Cortado R, Cole SW, Zack JA. Reconstitution of human thymic implants is limited by human immunodeficiency virus breakthrough during antiretroviral therapy. J Virol. 1999;73(8):6361–9.
38. Rabin L, Hincenbergs M, Moreno MB, Warren S, Linquist V, Datema R, et al. Use of standardized SCID-hu Thy/Liv mouse model for preclinical efficacy testing of anti-human immunodeficiency virus type 1 compounds. Antimicrob Agents Chemother. 1996;40(3):755–62.
39. Uckun FM, Qazi S, Pendergrass S, Lisowski E, Waurzyniak B, Chen CL, et al. In vivo toxicity, pharmacokinetics, and anti-human immunodeficiency virus activity of stavudine-5′-(p-bromophenyl methoxyalaninyl phosphate) (stampidine) in mice. Antimicrob Agents Chemother. 2002;46(11):3428–36.
40. Bridges CG, Taylor DL, Ahmed PS, Brennan TM, Hornsperger JM, Nave JF, et al. MDL 74,968, a new acyclonucleotide analog: activity against human immunodeficiency virus in vitro and in the hu-PBL-SCID.beige mouse model of infection. Antimicrob Agents Chemother. 1996;40(5):1072–7.
41. Hattori S, Ide K, Nakata H, Harada H, Suzu S, Ashida N, et al. Potent activity of a nucleoside reverse transcriptase inhibitor, 4′-ethynyl-2-fluoro-2′-deoxyadenosine, against human immunodeficiency virus type 1 infection in a model using human peripheral blood mononuclear cell-transplanted NOD/SCID Janus kinase 3 knockout mice. Antimicrob Agents Chemother. 2009;53(9):3887–93.
42. Stoddart CA, Moreno ME, Linquist-Stepps VD, Bare C, Bogan MR, Gobbi A, et al. Antiviral activity of 2′-deoxy-3′-oxa-4′-thiocytidine (BCH-10652) against lamivudine-resistant human immunodeficiency virus type 1 in SCID-hu Thy/Liv mice. Antimicrob Agents Chemother. 2000;44(3):783–6.
43. Romero DL, Busso M, Tan CK, Reusser F, Palmer JR, Poppe SM, et al. Nonnucleoside reverse transcriptase inhibitors that potently and specifically block human immunodeficiency virus type 1 replication. Proc Natl Acad Sci U S A. 1991;88(19):8806–10.
44. Ruxrungtham K, Boone E, Ford H Jr, Driscoll JS, Davey RT Jr, Lane HC. Potent activity of 2′-beta-fluoro-2′,3′-dideoxyadenosine against human immunodeficiency virus type 1 infection in hu-PBL-SCID mice. Antimicrob Agents Chemother. 1996;40(10):2369–74.
45. Boyle MJ, Connors M, Flanigan ME, Geiger SP, Ford H Jr, Baseler M, et al. The human HIV/peripheral blood lymphocyte (PBL)-SCID mouse. A modified human PBL-SCID model for the study of HIV pathogenesis and therapy. J Immunol. 1995;154(12):6612–23.
46. Stoddart CA, Nault G, Galkina SA, Thibaudeau K, Bakis P, Bousquet-Gagnon N, et al. Albumin-conjugated C34 peptide HIV-1 fusion inhibitor: equipotent to C34 and T-20 in vitro with sustained activity in SCID-hu Thy/Liv mice. J Biol Chem. 2008;283(49):34045–52.
47. Stoddart CA, Nault G, Galkina SA, Bousquet-Gagnon N, Bridon D, Quraishi O. Preexposure prophylaxis with albumin-conjugated C34 peptide HIV-1 fusion inhibitor in SCID-hu Thy/Liv mice. Antimicrob Agents Chemother. 2012;56(4):2162–5.
48. Hartley O, Gaertner H, Wilken J, Thompson D, Fish R, Ramos A, et al. Medicinal chemistry applied to a synthetic protein: development of highly potent HIV entry inhibitors. Proc Natl Acad Sci U S A. 2004;101(47):16460–5.
49. Strizki JM, Xu S, Wagner NE, Wojcik L, Liu J, Hou Y, et al. SCH-C (SCH 351125), an orally bioavailable, small molecule antagonist of the chemokine receptor CCR5, is a potent inhibitor of HIV-1 infection in vitro and in vivo. Proc Natl Acad Sci U S A. 2001;98(22):12718–23.
50. Murakami T, Kumakura S, Yamazaki T, Tanaka R, Hamatake M, Okuma K, et al. The novel CXCR4 antagonist KRH-3955 is an orally bioavailable and extremely potent inhibitor of human immunodeficiency virus type 1 infection: comparative studies with AMD3100. Antimicrob Agents Chemother. 2009;53(7):2940–8.

51. Datema R, Rabin L, Hincenbergs M, Moreno MB, Warren S, Linquist V, et al. Antiviral efficacy in vivo of the anti-human immunodeficiency virus bicyclam SDZ SID 791 (JM 3100), an inhibitor of infectious cell entry. Antimicrob Agents Chemother. 1996;40(3):750–4.
52. Stoddart CA, Rabin L, Hincenbergs M, Moreno M, Linquist-Stepps V, Leeds JM, et al. Inhibition of human immunodeficiency virus type 1 infection in SCID-hu Thy/Liv mice by the G-quartet-forming oligonucleotide, ISIS 5320. Antimicrob Agents Chemother. 1998;42(8):2113–5.
53. del Real G, Jimenez-Baranda S, Mira E, Lacalle RA, Lucas P, Gomez-Mouton C, et al. Statins inhibit HIV-1 infection by down-regulating Rho activity. J Exp Med. 2004;200(4):541–7.
54. Uckun FM, Chelstrom LM, Tuel-Ahlgren L, Dibirdik I, Irvin JD, Langlie MC, et al. TXU (anti-CD7)-pokeweed antiviral protein as a potent inhibitor of human immunodeficiency virus. Antimicrob Agents Chemother. 1998;42(2):383–8.
55. Roy U, McMillan J, Alnouti Y, Gautum N, Smith N, Balkundi S, et al. Pharmacodynamic and antiretroviral activities of combination nanoformulated antiretrovirals in HIV-1-infected human peripheral blood lymphocyte-reconstituted mice. J Infect Dis. 2012;206(10):1577–88.
56. Nicoletti F, Lapenta C, Donati S, Spada M, Ranazzi A, Cacopardo B, et al. Inhibition of human immunodeficiency virus (HIV-1) infection in human peripheral blood leucocytes-SCID reconstituted mice by rapamycin. Clin Exp Immunol. 2009;155(1):28–34.
57. Stoddart CA, Joshi P, Sloan B, Bare JC, Smith PC, Allaway GP, et al. Potent activity of the HIV-1 maturation inhibitor bevirimat in SCID-hu Thy/Liv mice. PLoS ONE. 2007;2(11):e1251.
58. Singh M, Singh P, Gaudray G, Musumeci L, Thielen C, Vaira D, et al. An improved protocol for efficient engraftment in NOD/LTSZ-SCIDIL-2Rgammanull mice allows HIV replication and development of anti-HIV immune responses. PLoS ONE. 2012;7(6):e38491.
59. Melkus MW, Estes JD, Padgett-Thomas A, Gatlin J, Denton PW, Othieno FA, et al. Humanized mice mount specific adaptive and innate immune responses to EBV and TSST-1. Nat Med. 2006;12(11):1316–22.
60. Denton PW, Nochi T, Lim A, Krisko JF, Martinez-Torres F, Choudhary SK, et al. IL-2 receptor gamma-chain molecule is critical for intestinal T-cell reconstitution in humanized mice. Mucosal Immunol. 2012;5(5):555–66.
61. Denton PW, Estes JD, Sun Z, Othieno FA, Wei BL, Wege AK, et al. Antiretroviral pre-exposure prophylaxis prevents vaginal transmission of HIV-1 in humanized BLT mice. PLoS Med. 2008;5(1):e16.
62. Berges BK, Akkina SR, Folkvord JM, Connick E, Akkina R. Mucosal transmission of R5 and X4 tropic HIV-1 via vaginal and rectal routes in humanized Rag2−/− gammac−/− (RAG-hu) mice. Virology. 2008;373(2):342–51.
63. Hofer U, Baenziger S, Heikenwalder M, Schlaepfer E, Gehre N, Regenass S, et al. RAG2−/− gamma(c)−/− mice transplanted with $CD34^+$ cells from human cord blood show low levels of intestinal engraftment and are resistant to rectal transmission of human immunodeficiency virus. J Virol. 2008;82(24):12145–53.
64. Nochi T, Denton PW, Wahl A, Garcia JV. Cryptopatches are essential for the development of human GALT. Cell Rep. 2013;3(6):1874–84.
65. Stoddart CA, Maidji E, Galkina SA, Kosikova G, Rivera JM, Moreno ME, et al. Superior human leukocyte reconstitution and susceptibility to vaginal HIV transmission in humanized NOD-scid IL-2Rgamma(−/−) (NSG) BLT mice. Virology. 2011;417(1):154–60.
66. Akkina R, Berges BK, Palmer BE, Remling L, Neff CP, Kuruvilla J, et al. Humanized Rag1−/− gammac−/− mice support multilineage hematopoiesis and are susceptible to HIV-1 infection via systemic and vaginal routes. PLoS ONE. 2011;6(6):e20169.
67. Baenziger S, Tussiwand R, Schlaepfer E, Mazzucchelli L, Heikenwalder M, Kurrer MO, et al. Disseminated and sustained HIV infection in $CD34^+$ cord blood cell-transplanted Rag2−/−gamma c−/− mice. Proc Natl Acad Sci U S A. 2006;103(43):15951–6.
68. Berges BK, Wheat WH, Palmer BE, Connick E, Akkina R. HIV-1 infection and CD4 T cell depletion in the humanized Rag2−/−gamma c−/− (RAG-hu) mouse model. Retrovirology. 2006;3:76.

69. Zhang L, Kovalev GI, Su L. HIV-1 infection and pathogenesis in a novel humanized mouse model. Blood. 2007;109(7):2978–81.
70. Gorantla S, Sneller H, Walters L, Sharp JG, Pirruccello SJ, West JT, et al. Human immunodeficiency virus type 1 pathobiology studied in humanized BALB/c-Rag2−/−gammac−/− mice. J Virol. 2007;81(6):2700–12.
71. Hoffmann-Fezer G, Gall C, Zengerle U, Kranz B, Thierfelder S. Immunohistology and immunocytology of human T-cell chimerism and graft-versus-host disease in SCID mice. Blood. 1993;81(12):3440–8.
72. Choudhary SK, Rezk NL, Ince WL, Cheema M, Zhang L, Su L, et al. Suppression of human immunodeficiency virus type 1 (HIV-1) viremia with reverse transcriptase and integrase inhibitors, $CD4^+$ T-cell recovery, and viral rebound upon interruption of therapy in a new model for HIV treatment in the humanized Rag2−/−{gamma}c−/− mouse. J Virol. 2009;83(16):8254–8.
73. Sango K, Joseph A, Patel M, Osiecki K, Dutta M, Goldstein H. Highly active antiretroviral therapy potently suppresses HIV infection in humanized Rag2−/−gammac−/− mice. AIDS Res Hum Retroviruses. 2010;26(7):735–46.
74. Cook JE, Dasgupta S, Middaugh LD, Terry EC, Gorry PR, Wesselingh SL, et al. Highly active antiretroviral therapy and human immunodeficiency virus encephalitis. Ann Neurol. 2005;57(6):795–803.
75. Choudhary SK, Archin NM, Cheema M, Dahl NP, Garcia JV, Margolis DM. Latent HIV-1 infection of resting CD4 T cells in the humanized Rag2−/− gammac−/− mouse. J Virol. 2012;86(1):114–20.
76. Marsden MD, Kovochich M, Suree N, Shimizu S, Mehta R, Cortado R, et al. HIV latency in the humanized BLT mouse. J Virol. 2012;86(1):339–47.
77. Denton PW, Olesen R, Choudhary SK, Archin NM, Wahl A, Swanson MD, et al. Generation of HIV latency in humanized BLT mice. J Virol. 2012;86(1):630–4.
78. Reagan-Shaw S, Nihal M, Ahmad N. Dose translation from animal to human studies revisited. FASEB J. 2008;22(3):659–61.
79. Dash PK, Gendelman HE, Roy U, Balkundi S, Alnouti Y, Mosley RL, et al. Long-acting nanoformulated antiretroviral therapy elicits potent antiretroviral and neuroprotective responses in HIV-1-infected humanized mice. AIDS. 2012;26(17):2135–44.
80. Denton PW, Krisko JF, Powell DA, Mathias M, Kwak YT, Martinez-Torres F, et al. Systemic administration of antiretrovirals prior to exposure prevents rectal and intravenous HIV-1 transmission in humanized BLT mice. PLoS ONE. 2010;5(1):e8829.
81. Wahl A, Swanson MD, Nochi T, Olesen R, Denton PW, Chateau M, et al. Human breast milk and antiretrovirals dramatically reduce oral HIV-1 transmission in BLT humanized mice. PLoS Pathog. 2012;8(6):e1002732.
82. Neff CP, Ndolo T, Tandon A, Habu Y, Akkina R. Oral pre-exposure prophylaxis by anti-retrovirals raltegravir and maraviroc protects against HIV-1 vaginal transmission in a humanized mouse model. PLoS ONE. 2010;5(12):e15257.
83. Denton PW, Othieno F, Martinez-Torres F, Zou W, Krisko JF, Fleming E, et al. One percent tenofovir applied topically to humanized BLT mice and used according to the CAPRISA 004 experimental design demonstrates partial protection from vaginal HIV infection, validating the BLT model for evaluation of new microbicide candidates. J Virol. 2011;85(15):7582–93.
84. Chateau ML, Denton PW, Swanson MD, McGowan I, Garcia JV. Rectal transmission of transmitted/founder HIV-1 is efficiently prevented by topical 1 % tenofovir in BLT humanized mice. PLoS ONE. 2013;8(3):e60024.
85. Neff CP, Kurisu T, Ndolo T, Fox K, Akkina R. A topical microbicide gel formulation of CCR5 antagonist maraviroc prevents HIV-1 vaginal transmission in humanized RAG-hu mice. PLoS ONE. 2011;6(6):e20209.
86. Abdool Karim Q, Abdool Karim SS, Frohlich JA, Grobler AC, Baxter C, Mansoor LE, et al. Effectiveness and safety of tenofovir gel, an antiretroviral microbicide, for the prevention of HIV infection in women. Science. 2010;329(5996):1168–74.

87. Else LJ, Taylor S, Back DJ, Khoo SH. Pharmacokinetics of antiretroviral drugs in anatomical sanctuary sites: the male and female genital tract. Antivir Ther. 2011;16(8):1149–67.
88. Nischang M, Gers-Huber G, Audige A, Akkina R, Speck RF. Modeling HIV infection and therapies in humanized mice. Swiss Med Wkly. 2012;142:13618.
89. Akkina R. New generation humanized mice for virus research: comparative aspects and future prospects. Virology. 2013;435(1):14–28.
90. Kaneshima H, Shih CC, Namikawa R, Rabin L, Outzen H, Machado SG, et al. Human immunodeficiency virus infection of human lymph nodes in the SCID-hu mouse. Proc Natl Acad Sci U S A. 1991;88(10):4523–7.
91. Lapenta C, Santini SM, Proietti E, Rizza P, Logozzi M, Spada M, et al. Type I interferon is a powerful inhibitor of in vivo HIV-1 infection and preserves human CD4(+) T cells from virus-induced depletion in SCID mice transplanted with human cells. Virology. 1999;263(1):78–88.

Chapter 30
Humanized Mice as a Platform for the Development of Long-Acting Nanoformulated Antiretroviral Therapy

JoEllyn M. McMillan and Howard E. Gendelman

30.1 Introduction

The delivery of medicines through nanoparticles is emerging as a means to enhance drug efficacy, reduce drug toxicities, and enhance delivery of medicines to target sites of disease [1–3]. Such technologies are likely to see substantive improvements in therapeutics over the next decade and have already shown the delivery of nucleic acids, proteins, enzymes, and antibodies to specific cells and tissues for improvements in bioavailability and plasma half-life [1, 2, 4]. For targeted delivery, such as cancer chemotherapy [5–7], nanoparticles are commonly designed to evade the innate immune system and as such positively affect the drug's therapeutic index [3]. However, cells of the immune system may also be targets for disease and most notably infectious diseases. As such, nanomedicines may also take advantage of the predilection of natural clearance mechanisms for the particles themselves [8]. Early liposomal formulations of amphotericin B were designed to take advantage of the normal clearance function of macrophages to improve treatment of leishmaniasis and for cryptococcal and other fungal infections [9–11]. Liposomal amphotericin is not only preferentially taken up by the same cells infected by the Leishmania parasite or the fungus but also provides a decrease in systemic toxicity, allowing for administration of higher doses and an increase in efficacy [9, 12]. Indeed, during the past half decade, targeting of nanoparticles to mononuclear phagocytes (MP; monocytes, macrophages, and dendritic cells) acting as Trojan horses for delivery of anti-inflammatory and anti-infective medicines was developed [2, 13]. This includes works in our own and others' laboratories designed to improve adherence and ease of access to antiretroviral therapy (ART) for human immunodeficiency virus (HIV) infections [14–20].

H. E. Gendelman (✉)· J. M. McMillan
Department of Pharmacology and Experimental Neuroscience,
University of Nebraska Medical Center, 985880 Nebraska Medical Center,
Omaha, NE 68198-5880, USA
e-mail: hegendel@unmc.edu

L. Y. Poluektova et al. (eds.), *Humanized Mice for HIV Research*,
DOI 10.1007/978-1-4939-1655-9_30

The need for improved drug delivery schemes for HIV disease is immediate. The reasons bridge potential elimination of virus in reservoirs of infection to ease of access and reduction of viral resistance [21]. Moreover, HIV infection remains among the most significant widespread diseases worldwide [22]. As of 2011, there were over 34 million people worldwide infected with HIV with a yearly toll of more than 2 million newly infected individuals and deaths [22, 23]. The advent of combination ART (cART) has reduced the comorbid complications and death from infection. However, cART has not succeeded in eradicating virus from reservoir sites such as gut, lymphoid tissues, and brain [24]. Other major limitations include significant adverse side effects, need for lifelong dosing regimens, and the emergence of drug-resistant viral strains [25–29]. Recently, single dose multi-class cART, including Atripla™ (emtricitabine/tenofovir/efavirenz), Complera™ (emtricitabine/tenofovir/rilpivirine), and Stribild™ (Quad; elvitegravir/cobicistat/emtricitabine/tenofovir) was developed to counter multiple daily dosing regimens [30–32]. However, toxicities are associated with these medicines, including lactic acidosis, hepatotoxicity, exacerbation of hepatitis after sudden discontinuance, peripheral neuropathy, and renal impairments [33–37]. Thus, there is a critical need for long-acting and slow-release (once-a-month) formulations of standard or new ART regimens that can positively impact current therapeutic limitations [38–40]. With this in mind, the development of ART nanoparticles (nanoART) that are carried within circulating and tissue MP and delivered to virus target tissues and hard to reach viral depots such as the CNS and lymphoid tissue was initiated [15, 16, 41–44]. Cell-based nanoART can travel to sites of inflammation, extend drug half-life by bypassing hepatic metabolism, and release drug slowly from tissue macrophage stores with limited tissue toxicities [2, 13, 45]. In developing this model system for clinical use, there are a number of delivery limitations that need to be overcome [2, 13, 46]. These include the following: (1) the drug-laden cells must migrate to the site of disease and deliver sufficient amount of drug to provide effective therapy without compromising the cell's normal function; (2) predictive cell-based assays need be developed to screen nanoformulations for appropriate cell and antiviral activity; (3) appropriate animal models need to be used to evaluate drug distribution, pharmacokinetic (PK), and pharmacodynamic (PD) patterns together with potential inherent toxicities for the drug [47–50]. Specifically, small animal models are of distinct advantage in generating statistically significant data sets for reduction of residual HIV infection. This is required if the full potential for nanoART can be realized in order to provide sustained plasma and tissue drug levels and long-term antiviral efficacy [51].

30.2 Promise of Cell-Based Drug Delivery

Targeted drug delivery systems to reduce toxicities and improve PK have been most widely studied for antineoplastic drugs [5–7]. More recent studies have sought to develop this paradigm for delivery of drugs targeted to other diseases, including infectious, inflammatory, and neurodegenerative diseases, and for the delivery of vaccines [52–56]. The intent has often been to reduce uptake by cells of the reticuloendothelial system in order to reduce dose and improve delivery to the target site; thus, many nanoparticle delivery systems for drugs contain polyethylene glycol (PEG) coating, which reduces recognition by phagocytic cells [13].

Targeted cell-based delivery is also being developed as a means of carrying drug nanoparticles across biologic barriers, such as the blood–brain barrier. Such systems are being developed to take advantage of the phagocytic and chemotactic capabilities of MPs to deliver drugs to sites of CNS disease and lymphoid tissue [2, 13]. The feasibility of this concept has been demonstrated for "nanozyme" delivery to the CNS in a mouse model of Parkinson's disease [57] and for delivery of nanoART to a site of HIV infection in the CNS in a mouse model of HIV encephalitis [15]. Thus, by targeting drugs to macrophages, delivery can be achieved to sites of disease or infection that are normally inaccessible to free drug in circulation.

30.3 NanoART

30.3.1 HIV Therapeutics

ART has greatly reduced the morbidity and mortality associated with HIV-1 infection. Among the reduced comorbid conditions are opportunistic infections and the cognitive and behavioral impairments associated with advanced viral infection commonly referred to as HIV-associated neurocognitive disorders (HAND) [58, 59]. For the infected patient, ART is started early in disease and the initial and subsequent maintenance therapy includes combinations of reverse transcriptase and protease inhibitors [60]. Newer drugs such as integrase strand transfer inhibitors (raltegravir and elvitegravir), entry inhibitors (enfuvirtide), and CCR5 antagonists (maraviroc) provide improved virologic control with reduced toxicities [60–64]. Of importance, viral resistance may develop more slowly in patients [65–67]. Preexposure prophylaxis regimens have recently been approved by the Food and Drug Administration (tenofovir/emtricitabine) that show great effectiveness in reducing HIV transmission in adherent individuals [68, 69].

Drug maintenance therapies commonly include a combination of protease inhibitors [60]. Indeed, lopinavir, atazanavir (ATV), or fosamprenavir are used in combination with a much lower, boosting dose of ritonavir (RTV) to inhibit metabolism of the therapeutic component by hepatic cytochrome P450 and extend plasma half-life [38]. However, these drugs suffer from a low penetration into viral reservoirs such as the CNS, complicated by the lack of medication adherence. High plasma protein binding, low lipophilicity, and efflux by p-glycoprotein on the luminal surface of brain microvascular endothelial cells serve to reduce CNS penetration [70–74]. In addition, renal, cardiovascular, and peripheral neuropathic complications, especially in light of coinfections such as hepatitis C and tuberculosis, can affect regimen continuation [75–81]. In spite of the available therapeutics, eradication of virus from protected sites, such as lymphoid tissue and brain, is still not possible and secondary drug toxicities, adherence, and viral resistance remain common place [24].

30.3.2 Nanoparticle-Based Systems for ART Delivery

Nanoparticle-based systems have been developed with the goal of improving antiretroviral therapy. For improvements in oral delivery, efavirenz (EFV)–poloxamine/poloxamer polymeric micelles were produced ,and displayed promising physical stability and PK [82]. EFV-loaded poly(ethylene oxide)-poly(propylene oxide) polymeric micelles administered intranasally exhibited fourfold enhanced drug bioavailability to the CNS and fivefold increased relative exposure index (ratio between the area under curve in the CNS and plasma) with respect to the same system administered intravenously [83]. Increased CNS delivery of ART by drug polybutylcyanoacrylate and methylmethacrylate-sulfopropymethacrylate nanoparticles has also been reported [84]. To overcome some of the disadvantages of polymeric nanoparticles that include residual contamination with polymer initiators and toxic monomers and stability issues, Shibata et al. [20] produced biodegradable combination antiretroviral poly(DL-lactide-co-glycolic acid) (PLGA) nanoparticles containing EFV and lopinavir by high-pressure homogenization. These nanoparticles efficiently inhibited in vitro HIV-1 infection and transduction. PLGA-chitosan nanoparticles of tenofovir, a nucleotide analog reverse transcriptase inhibitor, demonstrated sustained in vitro drug release suggesting that these particles could be effective and attractive drug carriers [85].

30.3.3 NanoART for Cell-Based Delivery

For cell-based antiretroviral drug delivery to be a clinically viable therapeutic strategy, there are developmental limitations that need to be overcome. These include effective delivery of drug in sufficient quantities to the disease site, development of predictive in vitro screening assays, and development and use of predictive and specific animal models [86]. Many of the limitations for cell-based drug delivery can be overcome by the careful and systematic development of drug nanoformulations. The physicochemical characteristics of a nanoparticle drug carrier will influence its suitability for cell-based drug delivery. Nanocarriers for drugs usually consist of a hydrophobic polymeric shell surrounding a core for drug carriage. Drug loading capacity, particle stability, and interaction with the surrounding environment are determined by the composition of the polymer shell [13, 87]. Polymer composition, surface charge, size, and shape of the nanoparticle, all contribute to uptake and retention by the target cell carrier. The most efficient nanocarriers for drug delivery have a high drug to carrier ratio. Interaction of the nanoparticle with cell surfaces is greatly influenced by the surface charge of the nanocarrier [13], with highly charged carriers being taken up by MPs to a greater extent than more neutral particles [88–90]. Nanoparticles in the range of 200–1000 nm were readily taken up by MPs [91] and rod-shaped particles were internalized by HeLa cells up to four times faster than more uniformly shaped particles [92–94]. To be effective for drug delivery, once internalized, the nanoparticle must be stored in nondegrading compartments for long-term carriage and subsequent drug release [13].

For development of cell-based drug delivery for treatment of HIV infection, normal MP functions can be harnessed, including phagocytosis, intercellular communication, and migration to sites of injury, and to maximize macrophage engulfment and location in protective subcellular organelles. With these goals in mind, our laboratories developed long-acting polymeric crystalline nanoparticles of antiretroviral drugs for cell-based drug delivery. Polymeric crystalline formulations of ATV, RTV, indinavir (IDV), or EFV were prepared by fractionating crystalline drug in the presence of polymer surfactants, such as poloxamers 188 and 407 (P188 and P407, respectively) by wet-milling and high-pressure homogenization [17, 43, 95]. The antiretroviral drug nanoparticles prepared by these methods were in the range of 200–500 nm in size with a loading capacity of >70 % (w/w). NanoART were screened in cell-based assays using human monocyte-derived macrophages (MDM) for cell uptake, retention, release, and antiretroviral efficacy. Particles that were rod-shaped (ATV and RTV) were internalized by MDM more rapidly and to a greater extent than the ellipsoid particles (IDV and EFV) [17, 43, 95]. Preloading MDM with nanoformulated ATV, RTV, or IDV and then infecting the cells up to 15 days later with HIV-1_{ADA} provided 85, 80, and 40 % inhibition of progeny virion production and HIV-1 p24 staining, while treatment with an equivalent concentration of EFV nanoparticles resulted in nearly complete inhibition of viral infection [43, 95]. NanoART were internalized by clathrin-mediated endocytosis and stored in nondegrading recycling Rab 11^+ and Rab 14^+ endosomes and released generally intact at the cell surface [17, 96]. Of particular importance, active targeting of nanoART to specific cell compartments could not only enhance cell storage but also enable the direction of the drug to the cell compartments where HIV replicates [97].

Cell-based nanoparticle delivery can be improved by actively targeting nanoformulations to the carrier cells. Nanoformulations of antiretroviral drugs are readily taken up by MPs via general phagocytic and endocytic mechanisms [2, 13]. The addition of targeting moieties on the surface of nanoparticles can direct the particles to certain cells and specific subcellular compartments for storage and protection from degradation [98–102]. In recent studies, nanoART targeted to folate receptors on macrophages were synthesized and demonstrated over twofold enhanced MDM uptake, retention, and antiretroviral efficacy in MDM infected with HIV-1_{ADA} [17]. In addition, liposomal formulations of stavudine and zidovudine targeted with mannose and galactose provided increased macrophage uptake and enhanced liver, spleen, lung, and lymph node drug levels in rodents [103, 104].

For effective cell-based nanoformulated drug delivery, uptake of the drug by the carrier must not affect the normal functions of the cell. The internalized nanoparticle must be carried intact inside the cell for effective delivery to the site of disease and for the cell to serve as a reservoir for long-term release of drug [99]. For generally predictive means of determining global functional changes as a result of nanoART carriage, proteomics has been of use [105, 106]. The specific proteomics technique of pulsed stable isotope labeling of amino acids in cell culture (pSILAC) was used to determine dynamic global proteomic changes in MDM loaded with nanoART [106]. Proteome changes were substantiated by functional assays and cytokine/chemokine secretion and indicated that nanoART carriage induces a unique activated macrophage phenotype that is primed for further nanoART uptake and storage and that enhances cell migration [106]. These targeting strategies were aimed at macrophages, however, active targeting of nanoART to T lymphocytes may provide improved antiviral therapy and penetrance into viral reservoir sites, such as lymph nodes, with the ultimate goal of viral eradication.

30.4 Mouse Models for NanoART Pharmacokinetics and Efficacy

30.4.1 Normal Mice for Pharmacokinetics, Biodistribution, and Pharmacodynamics

Determination of PK and biodistribution of nanoformulations for cell-based nanoART delivery is easily achieved using normal mouse strains, such as Balb/cJ [107]. Uptake of parenterally injected nanoformulations by MPs and nanoART sequestration inside cells of the reticuloendothelial system can be monitored by drug quantitation and by using dye-labeled formulations, followed by flow cytometry and confocal microscopy to isolate and visualize MPs [107]. Drug metabolism and enzyme induction and their contribution to ART PK, biodistribution, and toxicity have been well studied in rodents and are generally predictive of human events [47, 108–111]. For a nanoformulated drug to be successful, it is essential that a desired PK and PD profile be achieved. The standard assessments designed for small molecule PK may not be applicable to PK of nanomedicines. Because of the unique behavior of nanoparticles, including nanoparticle absorption, drug release kinetics, and nanoparticle clearance using blood and plasma drug levels as a determination of PK, may not provide a complete picture of nanoparticle distribution [86]. In particular, for targeted and cell-based nanoparticle delivery, determination of levels of drug in plasma may not be an accurate indication of the PD of the nanoformulated drug [112]. Of significance, determination of drug concentrations at the target site or in target cells may be more predictive of therapeutic efficacy [61]. In addition, when using MPs as carriers for drug nanoparticles, the activation state of the cells may influence their uptake of the nanoparticles and the delivery of drug to target sites. Thus, to develop an effective dosing regimen, determine optimal route of administration, and determine tissue and cell distribution, careful in vivo assessments of drug nanoformulations are required.

NanoART developed for cell-based drug delivery were characterized for PK and biodistribution in normal mice [107]. Balb/cJ mice treated subcutaneously with a single administration of nanoART (P188-ATV/RTV) had plasma AUC (area under the curve) and tissue drug levels 14 days after injection that were 10- and 40-fold higher than levels following administration of nonformulated drugs. Dose-escalation studies revealed a nonlinear PK behavior that has been reported previously for ATV and RTV and attributed to induction/inhibition of the enzymes and transporters involved in elimination of the drugs (in particular, cytochrome P450 3A and p-glycoprotein) [113, 114]. Multiple doses provided up to 270-fold higher plasma and tissue levels after 6 weeks of weekly nanoART administration. Localization and storage of the nanoART in nonlysosomal compartments of liver Kupffer cells were demonstrated, providing evidence for MPs as nanoART reservoirs [107]. Extended and enhanced plasma and tissue drug levels were found when nanoART targeted to MP folate receptors were administered intramuscularly to mice [17]. Importantly, drug levels in lymph nodes of up to 70 ng/g tissue were detected in these mice. These results suggest that by targeting nanoART to macrophage and even T-cell receptors, penetration of drug into previously inaccessible HIV viral reservoirs can be achieved.

30.4.2 Human Peripheral Blood Lymphocyte Reconstituted Mouse Models for Acute Antiviral Efficacy and Toxicity Studies

While PK studies in normal mice can demonstrate enhanced and extended plasma and tissue ART levels after a single administration of nanoART, PD and efficacy of the nanoformulations need to be demonstrated. The challenge, however, in determining PD and efficacy of nanoART is the specificity of HIV-1 infection for humans. The development of "humanized" mouse models that support the survival of human immune cells and recapitulate HIV-1 pathogenesis has allowed the study of new ART therapies and mechanisms of HIV-1 pathogenesis [47, 48, 115]. Severe combined immunodeficient (SCID) mice contain an autosomal recessive mutation in the DNA dependent protein kinase (*prkdc*) gene that causes a deficiency in mature T and B lymphocytes [116]. Transfer of the SCID mutation onto the nonobese diabetic (NOD) background resulted in a mouse strain (NOD/SCID) with very few occurrences of spontaneous lymphocyte production (leakiness) [49]. Because of the lack of functional lymphocytes, transplantation of foreign tissues and cells is possible in these mice [50]. Mouse models have been developed using transplanted human thymus, fetal liver, and peripheral blood lymphocytes, wherein immune cells from the immunocompetent human tissues are able to reconstitute the NOD/SCID mouse immune system [117–119]. Immune-deficient mice transplanted with human immune cells are especially useful for assessing the PK and biodistribution of nanoART targeted to human immune cells for cell-based delivery.

A number of rodent systems based upon the SCID mouse have been developed for the study of HIV-1 infection and evaluation of new therapeutics. In the HIV-1 encephalitis model, SCID mice injected with HIV-1 infected human MDM demonstrate key features of HIV-1 neuropathology [120, 121]. Using this mouse model, Dou et al. demonstrated that mouse bone marrow macrophages, loaded with indinavir nanoparticles and administered to CB-17/SCID mice injected intracranially with HIV-1 infected human MDM, would migrate to the site of viral infection and release drug at sufficient concentrations to reduce viral infection [15]. The mouse model has also proven effective in examining the interaction of the innate and adaptive immune systems in overall CNS response to HIV infection [122, 123]. In this model, NOD/SCID mice are reconstituted with human peripheral blood lymphocytes (huPBL) and then injected intracranially with HIV-1 infected MDM. HuPBL-reconstituted NOD/SCID mice infected with HIV-1 were used to evaluate PK, biodistribution, and antiviral efficacy of nanoART in a system of activated immune cells [18]. In this study, subcutaneous administration of 250 mg/kg nanoART (P188-ATV/RTV) provided sustained plasma and tissue drug levels 9 days after dosing compared with undetectable drug levels with free ART administration. Drug levels were highest in spleen and liver and reached levels of up to 10 μg/g tissue 7 days after two, weekly doses of 250 mg/kg nanoART (P188-ATV/RTV). In addition, nanoART suppressed HIV-1 infection to the limit of detection, when given either 24 h before or 12 h after virus administration [18]. This model was also used to demonstrate that folate-targeted nanoART administration provided drug levels in brain, three to four times that observed with nontargeted nanoART, and that this was associated with decreased viral load and suppressed CD11b-associated glial activation in the brain [42]. However, because of the development of graft versus host disease [108, 109, 118, 124, 125], the adult huPBL reconstitution model is useful for only short-term PK, PD, and efficacy studies.

30.4.3 Humanized Mouse Models for Chronic Dosing and Preexposure Prophylaxis in HIV Infection

For chronic treatment and HIV infection studies, there are a couple of models of long-term human immune cell reconstitution of NOD/SCID mice (i.e., "humanized" mice). In the hu-HSC model, newborn NOD/SCID mice are irradiated to deplete their normal bone marrow stem cells and then given human $CD34^+$ hematopoietic stem cells (HSC) from human fetal liver or human cord blood. Reconstitution of the mouse immune system with human immune cells occurs over a period of 12 weeks [124, 125]. In response to infection with HIV-1, these "humanized" mice demonstrate hallmark lymphocytic and humoral immune responses and support chronic HIV-1 infection. Because of the long-term survival of these mice and the stability of the immune reconstitution, chronic ART treatment regimens can be evaluated (Fig. 30.1). In humanized mice chronically treated with HIV-1, weekly subcutaneous doses of nanoART (P188-ATV/RTV, 250 mg/kg) provided sustained plasma, liver, and spleen drug levels, above the minimal effective plasma concentration for

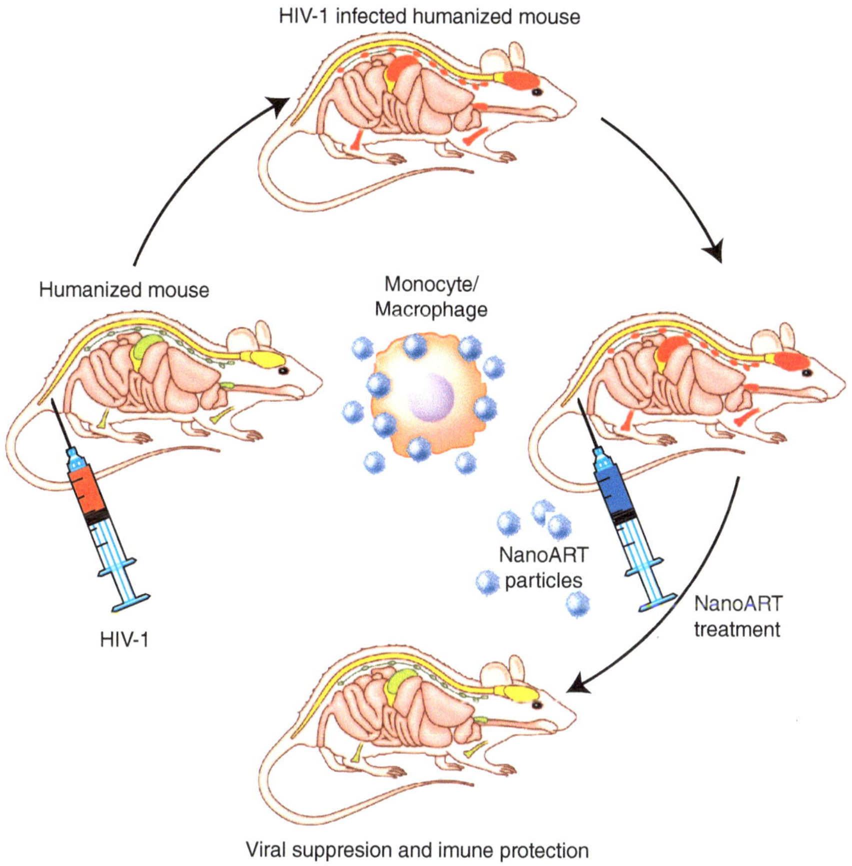

Fig. 30.1 Schematic model illustrating the use of HIV-1-infected "humanized" NOD/SCID mice for testing nanoformulated antiretroviral therapies (*nanoART*). This schematic reflects the potential of long-acting antiretrovirals to reduce viral load and protect both $CD4^+$ T cells and the CNS against HIV-1-associated injuries. The colors reflect normal immune tissue homeostasis (*green* and *yellow*) and tissues damaged by chronic viral replication in mice (*red*). $CD4^+$ T lymphocyte decline is observed following HIV-1 infection. In addition, infected and immune activated lymphocytes and macrophages have been observed in lymphoid organs and peripheral blood. Infected animals are treated with nanoART which can gain entry into monocyte-macrophages (*center* of picture), captured by blood-filtrating liver/splenic tissue macrophages, and serve as a long-term drug depot leading to the suppression of viral replication and protection of $CD4^+$ T cell numbers (*green* and *yellow*). The nanoformulation may also facilitate ART delivery to the nervous system, although, this is yet to be demonstrated experimentally. (Figure and legend reprinted with permission from [115])

ATV (150 mg/ml), and suppressed HIV-1 viral infectivity to below the limit of detection during the course of nanoART treatment [14]. Upon cessation of treatment, however, plasma viral load rebounded and the presence of infected T cells in lymph nodes was not suppressed. In this same model brain metabolite changes in response to chronic HIV infection correlated with microgliosis and were partially reversed with weekly nanoART treatment [126].

In the BLT mouse model, immune-deficient mice, reconstituted at 6–8 weeks of age with human bone marrow $CD34^+$ stem cells and implanted with pieces of liver and thymus tissue, are important and useful for studying several tissue pathologies associated with HIV infection [118, 127–130]. Human cell engraftment occurs in many tissues, including the female reproductive tract. Generation of T and B cells, macrophages, NK cells, and dendritic cells is seen along with a functional human thymus [131, 132]. ART treatment regimens, including nanoART and preexposure prophylaxis to reduce vaginal transmission of HIV-1 and latency, are being explored using BLT mice [118, 119, 129, 133–136]. The problem of HIV persistence during ART therapy can also be studied in these mice. Using BLT mice infected with HIV and on cART therapy, Denton et al. [137] were able to greatly reduce viral RNA levels in reservoir sites by coadministration of an HIV-specific immunotoxin. These studies provided "proof-of-concept" that viable "kick and kill" therapeutic strategies [138] can be developed to aid HIV eradication efforts.

30.5 Conclusion

The development of nanoART as an effective therapeutic modality for HIV-1 infection offers many promises. The ability to maintain extended plasma drug levels from a single intramuscular injection would improve patient adherence and reduce the potential for development of resistant viral strains. The ability to specifically target cell and tissue reservoirs and to direct the nanoformulations to the same subcellular compartments where viral replication occurs, offers for the first time, the potential for viral eradication. As a means of screening new nanoformulations of ART, humanized mice offer many advantages. These mice recapitulate most aspects of the human immune system, and are thus, ideal models for determining the effects of new ART formulations on HIV-1 viral infection, progression, latency, and transmission and developing "kick and kill" strategies for HIV eradication.

References

1. Kim BY, Rutka JT, Chan WC. Nanomedicine. N Engl J Med. 2010;363(25):2434–43.
2. McMillan J, Batrakova E, Gendelman HE. Cell delivery of therapeutic nanoparticles. Prog Mol Biol Transl Sci. 2011;104:563–601.
3. Wagner V, Dullaart A, Bock AK, Zweck A. The emerging nanomedicine landscape. Nat Biotechnol. 2006;24(10):1211–7.
4. Pautler M, Brenner S. Nanomedicine: promises and challenges for the future of public health. Int J Nanomedicine. 2010;5:803–9.
5. Misra R, Acharya S, Sahoo SK. Cancer nanotechnology: application of nanotechnology in cancer therapy. Drug Discov Today. 2010;15(19–20):842–50.
6. Sampathkumar SG, Yarema KJ. Targeting cancer cells with dendrimers. Chem Biol. 2005;12(1):5–6.

7. Seigneuric R, Markey L, Nuyten DS, Dubernet C, Evelo CT, Finot E, et al. From nanotechnology to nanomedicine: applications to cancer research. Curr Mol Med. 2010;10(7):640–52.
8. Armstead AL, Li B. Nanomedicine as an emerging approach against intracellular pathogens. Int J Nanomedicine. 2011;6:3281–93.
9. New RR, Chance ML, Heath S. Antileishmanial activity of amphotericin and other antifungal agents entrapped in liposomes. J Antimicrob Chemother. 1981;8(5):371–81.
10. Graybill JR, Craven PC, Taylor RL, Williams DM, Magee WE. Treatment of murine cryptococcosis with liposome-associated amphotericin B. J Infect Dis. 1982;145(5):748–52.
11. Taylor RL, Williams DM, Craven PC, Graybill JR, Drutz DJ, Magee WE. Amphotericin B in liposomes: a novel therapy for histoplasmosis. Am Rev Respir Dis. 1982;125(5):610–1.
12. Heath S, Chance ML, New RR. Quantitative and ultrastructural studies on the uptake of drug loaded liposomes by mononuclear phagocytes infected with Leishmania donovani. Mol Biochem Parasitol. 1984;12(1):49–60.
13. Batrakova EV, Gendelman HE, Kabanov AV. Cell-mediated drug delivery. Expert Opin Drug Deliv. 2011;8(4):415–33.
14. Dash PK, Gendelman HE, Roy U, Balkundi S, Alnouti Y, Mosley RL, et al. Long-acting nanoformulated antiretroviral therapy elicits potent antiretroviral and neuroprotective responses in HIV-1-infected humanized mice. AIDS. 2012;26(17):2135–44.
15. Dou H, Grotepas CB, McMillan JM, Destache CJ, Chaubal M, Werling J, et al. Macrophage delivery of nanoformulated antiretroviral drug to the brain in a murine model of neuroAIDS. J Immunol. 2009;183(1):661–9.
16. Nowacek AS, McMillan J, Miller R, Anderson A, Rabinow B, Gendelman HE. Nanoformulated antiretroviral drug combinations extend drug release and antiretroviral responses in HIV-1-infected macrophages: implications for neuroAIDS therapeutics. J Neuroimmune Pharmacol. 2010;5(4):592–601.
17. Puligujja P, McMillan J, Kendrick L, Li T, Balkundi S, Smith N, et al. Macrophage folate receptor-targeted antiretroviral therapy facilitates drug entry, retention, antiretroviral activities and biodistribution for reduction of human immunodeficiency virus infections. Nanomedicine. 2013;9(8):1263–73.
18. Roy U, McMillan J, Alnouti Y, Gautum N, Smith N, Balkundi S, et al. Pharmacodynamic and antiretroviral activities of combination nanoformulated antiretrovirals in HIV-1-infected human peripheral blood lymphocyte-reconstituted mice. J Infect Dis. 2012;206(10):1577–88.
19. Destache CJ, Belgum T, Christensen K, Shibata A, Sharma A, Dash A. Combination antiretroviral drugs in PLGA nanoparticle for HIV-1. BMC Infect Dis. 2009;9:198.
20. Shibata A, McMullen E, Pham A, Belshan M, Sanford B, Zhou Y, et al. Polymeric nanoparticles containing combination antiretroviral drugs for HIV Type 1 treatment. AIDS Res Hum Retroviruses. 2013;29(5):746–54.
21. Palmer S, Josefsson L, Coffin JM. HIV reservoirs and the possibility of a cure for HIV infection. J Intern Med. 2011;270(6):550–60.
22. WHO. HIV/AIDS Fact Sheet NË360 2012 (accessed 2013 June). http://www.who.int/mediacentre/factsheets/fs360/en/.
23. CDC. Diagnoses of HIV Infection in the United States and Dependent Areas, 2011 (accessed 2013 June). http://www.cdc.gov/hiv/library/reports/surveillance/2011/surveillance_Report_vol_23.html.
24. Valcour V, Sithinamsuwan P, Letendre S, Ances B. Pathogenesis of HIV in the central nervous system. Curr HIV/AIDS Rep. 2011;8(1):54–61.
25. Chen RY, Westfall AO, Raper JL, Cloud GA, Chatham AK, Acosta EP, et al. Immunologic and virologic consequences of temporary antiretroviral treatment interruption in clinical practice. AIDS Res Hum Retroviruses. 2002;18(13):909–16.
26. Chulamokha L, DeSimone JA, Pomerantz RJ. Antiretroviral therapy in the developing world. J neurovirol. 2005;11(Suppl 1):76–80.
27. Fellay J, Boubaker K, Ledergerber B, Bernasconi E, Furrer H, Battegay M, et al. Prevalence of adverse events associated with potent antiretroviral treatment: Swiss HIV Cohort Study. Lancet. 2001;358(9290):1322–7.

28. Hawkins T. Appearance-related side effects of HIV-1 treatment. AIDS Patient Care STDS. 2006;20(1):6–18.
29. Shehu-Xhilaga M, Tachedjian G, Crowe SM, Kedzierska K. Antiretroviral compounds: mechanisms underlying failure of HAART to eradicate HIV-1. Curr Med Chem. 2005;12(15):1705–19.
30. Llibre JM, Clotet B. Once-daily single-tablet regimens: a long and winding road to excellence in antiretroviral treatment. AIDS Rev. 2012;14(3):168–78.
31. Permpalung N, Putcharoen O, Avihingsanon A, Ruxrungtham K. Treatment of HIV infection with once-daily regimens. Expert Opin Pharmacother. 2012;13(16):2301–17.
32. Wegzyn CM, Wyles DL. Antiviral drug advances in the treatment of human immunodeficiency virus (HIV) and chronic hepatitis C virus (HCV). Curr Opin Pharmacol. 2012;12(5):556–61.
33. Kontorinis N, Dieterich D. Hepatotoxicity of antiretroviral therapy. AIDS Rev. 2003;5(1):36–43.
34. Kranick SM, Nath A. Neurologic complications of HIV-1 infection and its treatment in the era of antiretroviral therapy. Continuum (Minneap Minn). 2012;18(6 Infectious Disease):1319–37.
35. Luetkemeyer AF, Havlir DV, Currier JS. Complications of HIV disease and antiretroviral therapy. Top Antivir Med. 2012;20(2):48–60.
36. McCance-Katz EF. Treatment of opioid dependence and coinfection with HIV and hepatitis C virus in opioid-dependent patients: the importance of drug interactions between opioids and antiretroviral agents. Clin Infect Dis. 2005;41(Suppl 1):S89–95.
37. Rather ZA, Chowta MN, Raju GJ, Mubeen F. Evaluation of the adverse reactions of antiretroviral drug regimens in a tertiary care hospital. Indian J Pharmacol. 2013;45(2):145–8.
38. McKinnon JE, Mellors JW, Swindells S. Simplification strategies to reduce antiretroviral drug exposure: progress and prospects. Antivir Ther. 2009;14(1):1–12.
39. Swindells S, Flexner C, Fletcher CV, Jacobson JM. The critical need for alternative antiretroviral formulations, and obstacles to their development. J Infect Dis. 2011;204(5):669–74.
40. Williams J, Sayles HR, Meza J, Sayre P, Sandkovsky U, Gendelman HE, et al. Long-acting parenteral nanoformulated antiretroviral therapy: interest and attitudes of HIV-infected patients. Nanomedicine. 2013;8(11):1807–13.
41. Dou H, Destache CJ, Morehead JR, Mosley RL, Boska MD, Kingsley J, et al. Development of a macrophage-based nanoparticle platform for antiretroviral drug delivery. Blood. 2006;108(8):2827–35.
42. Kanmogne GD, Singh S, Roy U, Liu X, McMillan J, Gorantla S, et al. Mononuclear phagocyte intercellular crosstalk facilitates transmission of cell-targeted nanoformulated antiretroviral drugs to human brain endothelial cells. Int J Nanomedicine. 2012;7:2373–88.
43. Nowacek AS, Balkundi S, McMillan J, Roy U, Martinez-Skinner A, Mosley RL, et al. Analyses of nanoformulated antiretroviral drug charge, size, shape and content for uptake, drug release and antiviral activities in human monocyte-derived macrophages. J Control Release. 2011;150(2):204–11.
44. Nowacek AS, Miller RL, McMillan J, Kanmogne G, Kanmogne M, Mosley RL, et al. NanoART synthesis, characterization, uptake, release and toxicology for human monocyte-macrophage drug delivery. Nanomedicine (Lond). 2009;4(8):903–17.
45. Nowacek A, Gendelman HE. NanoART, neuroAIDS and CNS drug delivery. Nanomedicine (Lond). 2009;4(5):557–74.
46. Lewin SR, Rouzioux C. HIV cure and eradication: how will we get from the laboratory to effective clinical trials? AIDS. 2011;25(7):885–97.
47. Denton PW, Garcia JV. Humanized mouse models of HIV infection. AIDS Rev. 2011;13(3):135–48.
48. Shultz LD, Brehm MA, Garcia-Martinez JV, Greiner DL. Humanized mice for immune system investigation: progress, promise and challenges. Nat Rev Immunol. 2012;12(11):786–98.
49. Van Duyne R, Pedati C, Guendel I, Carpio L, Kehn-Hall K, Saifuddin M, et al. The utilization of humanized mouse models for the study of human retroviral infections. Retrovirology. 2009;6:76.

50. Watanabe S, Terashima K, Ohta S, Horibata S, Yajima M, Shiozawa Y, et al. Hematopoietic stem cell-engrafted NOD/SCID/IL2Rgamma null mice develop human lymphoid systems and induce long-lasting HIV-1 infection with specific humoral immune responses. Blood. 2007;109(1):212–8.
51. Mahajan SD, Aalinkeel R, Law WC, Reynolds JL, Nair BB, Sykes DE, et al. Anti-HIV-1 nanotherapeutics: promises and challenges for the future. Int J Nanomedicine. 2012;7:5301–14.
52. Chen J, Li Z, Huang H, Yang Y, Ding Q, Mai J, et al. Improved antigen cross-presentation by polyethyleneimine-based nanoparticles. Int J Nanomedicine. 2011;6:77–84.
53. Kasturi SP, Skountzou I, Albrecht RA, Koutsonanos D, Hua T, Nakaya HI, et al. Programming the magnitude and persistence of antibody responses with innate immunity. Nature. 2011;470(7335):543–7.
54. Klippstein R, Pozo D. Nanotechnology-based manipulation of dendritic cells for enhanced immunotherapy strategies. Nanomedicine. 2010;6(4):523–9.
55. Reddy ST, Swartz MA, Hubbell JA. Targeting dendritic cells with biomaterials: developing the next generation of vaccines. Trends Immunol. 2006;27(12):573–9.
56. Xiang SD, Selomulya C, Ho J, Apostolopoulos V, Plebanski M. Delivery of DNA vaccines: an overview on the use of biodegradable polymeric and magnetic nanoparticles. Wiley Interdiscip Rev Nanomed Nanobiotechnol. 2010;2(3):205–18.
57. Batrakova EV, Li S, Reynolds AD, Mosley RL, Bronich TK, Kabanov AV, et al. A macrophage-nanozyme delivery system for Parkinson's disease. Bioconjug Chem. 2007;18(5):1498–506.
58. Hogan C, Wilkins E. Neurological complications in HIV. Clin Med. 2011;11(6):571–5.
59. Spudich SS, Ances BM. Neurologic complications of HIV infection. Top Antivir Med. 2012;20(2):41–7.
60. Arts EJ, Hazuda DJ. HIV-1 antiretroviral drug therapy. Cold Spring Harb Perspect Med. 2012;2(4):a007161.
61. Allen TM. Ligand-targeted therapeutics in anticancer therapy. Nat Rev Cancer. 2002;2(10):750–63.
62. Lee FJ, Carr A. Tolerability of HIV integrase inhibitors. Curr Opin HIV AIDS. 2012;7(5):422–8.
63. Malet I, Calvez V, Marcelin AG. The future of integrase inhibitors of HIV-1. Curr Opin Virol. 2012;2(5):580–7.
64. Sharma AK, George V, Valiathan R, Pilakka-Kanthikeel S, Pallikkuth S. Inhibitors of HIV-1 entry and integration: recent developments and impact on treatment. Recent Pat Inflamm Allergy Drug Discov. 2013;7(2):151–61.
65. De Feo CJ, Weiss CD. Escape from human immunodeficiency virus type 1 (HIV-1) entry inhibitors. Viruses. 2012;4(12):3859–911.
66. Quashie PK, Mesplede T, Wainberg MA. Evolution of HIV integrase resistance mutations. Curr Opin Infect Dis. 2013;26(1):43–9.
67. Wainberg MA, Mesplede T, Quashie PK. The development of novel HIV integrase inhibitors and the problem of drug resistance. Curr Opin Virol. 2012;2(5):656–62.
68. De Man J, Colebunders R, Florence E, Laga M, Kenyon C. What is the place of pre-exposure prophylaxis in HIV prevention? AIDS Rev. 2013;15(2):102–11.
69. Underhill K, Morrow KM, Operario D, Mayer KH. Could FDA approval of pre-exposure Prophylaxis make a difference? A qualitative study of PrEP acceptability and FDA perceptions among men who have sex with men. AIDS Behav. 2013;18(2):241–9.
70. Wire MB, Shelton MJ, Studenberg S. Fosamprenavir: clinical pharmacokinetics and drug interactions of the amprenavir prodrug. Clin Pharmacokinet. 2006;45(2):137–68.
71. van Heeswijk RP, Veldkamp A, Mulder JW, Meenhorst PL, Lange JM, Beijnen JH, et al. Combination of protease inhibitors for the treatment of HIV-1-infected patients: a review of pharmacokinetics and clinical experience. Antivir Ther. 2001;6(4):201–29.
72. Bazzoli C, Jullien V, Le Tiec C, Rey E, Mentre F, Taburet AM. Intracellular pharmacokinetics of antiretroviral drugs in HIV-infected patients, and their correlation with drug action. Clin Pharmacokinet. 2010;49(1):17–45.
73. Perez-Valero I, Bayon C, Cambron I, Gonzalez A, Arribas JR. Protease inhibitor monotherapy and the CNS: peace of mind? J Antimicrob Chemother. 2011;66(9):1954–62.

74. Wynn HE, Brundage RC, Fletcher CV. Clinical implications of CNS penetration of antiretroviral drugs. CNS Drugs. 2002;16(9):595–609.
75. Gray JM, Cohn DL. Tuberculosis and HIV coinfection. Semin Respir Crit Care Med. 2013;34(1):32–43.
76. Idemyor V. HIV and tuberculosis coinfection: inextricably linked liaison. J Natl Med Assoc. 2007;99(12):1414–9.
77. Stephan C. Virological efficacy and safety of antiretroviral therapy-switch to atazanavir-based regimen: a review of the literature. Expert Opin Pharmacother. 2012;13(16):2355–67.
78. Scarpino M, Pinzone MR, Di Rosa M, Madeddu G, Foca E, Martellotta F, et al. Kidney disease in HIV-infected patients. Eur Rev Med Pharmacol Sci. 2013;17(19):2660–7.
79. Garg H, Joshi A, Mukherjee D. Cardiovascular complications of HIV infection and treatment. Cardiovasc Hematol Agents Med Chem. 2013;11(1):58–66.
80. Nachega JB, Trotta MP, Nelson M, Ammassari A. Impact of metabolic complications on antiretroviral treatment adherence: clinical and public health implications. Curr HIV/AIDS Rep. 2009;6(3):121–9.
81. Parfieniuk-Kowerda A, Czaban SL, Grzeszczuk A, Jaroszewicz J, Flisiak R. Assessment of serum IGF-1 and adipokines related to metabolic dysfunction in HIV-infected adults. Cytokine. 2013;64(1):97–102.
82. Seremeta KP, Chiappetta DA, Sosnik A. Poly(epsilon-caprolactone), Eudragit® RS 100 and poly(epsilon-caprolactone)/Eudragit® RS 100 blend submicron particles for the sustained release of the antiretroviral efavirenz. Colloids Surf B Biointerfaces. 2013;102:441–9.
83. Chiappetta DA, Hocht C, Opezzo JA, Sosnik A. Intranasal administration of antiretroviral-loaded micelles for anatomical targeting to the brain in HIV. Nanomedicine (Lond). 2013;8(2):223–37.
84. Mulik RS, Monkkonen J, Juvonen RO, Mahadik KR, Paradkar AR. ApoE3 mediated polymeric nanoparticles containing curcumin: apoptosis induced in vitro anticancer activity against neuroblastoma cells. Int J Pharm. 2012;437(1–2):29–41.
85. Belletti D, Tosi G, Forni F, Gamberini MC, Baraldi C, Vandelli MA, et al. Chemico-physical investigation of tenofovir loaded polymeric nanoparticles. Int J Pharm. 2012;436(1–2):753–63.
86. Desai N. Challenges in development of nanoparticle-based therapeutics. AAPS J. 2012;14(2):282–95.
87. Kanwar JR, Sun X, Punj V, Sriramoju B, Mohan RR, Zhou SF, et al. Nanoparticles in the treatment and diagnosis of neurological disorders: untamed dragon with fire power to heal. Nanomedicine. 2012;8(4):399–414.
88. Fujiwara M, Baldeschwieler JD, Grubbs RH. Receptor-mediated endocytosis of poly(acrylic acid)-conjugated liposomes by macrophages. Biochim Biophys Acta. 1996;1278(1):59–67.
89. Miller CR, Bondurant B, McLean SD, McGovern KA, O'Brien DF. Liposome-cell interactions in vitro: effect of liposome surface charge on the binding and endocytosis of conventional and sterically stabilized liposomes. Biochemistry. 1998;37(37):12875–83.
90. Nishikawa K, Arai H, Inoue K. Scavenger receptor-mediated uptake and metabolism of lipid vesicles containing acidic phospholipids by mouse peritoneal macrophages. J Biol Chem. 1990;265(9):5226–31.
91. Tabata Y, Ikada Y. Effect of the size and surface charge of polymer microspheres on their phagocytosis by macrophage. Biomaterials. 1988;9(4):356–62.
92. Chithrani BD, Ghazani AA, Chan WC. Determining the size and shape dependence of gold nanoparticle uptake into mammalian cells. Nano Lett. 2006;6(4):662–8.
93. Gratton SE, Ropp PA, Pohlhaus PD, Luft JC, Madden VJ, Napier ME, et al. The effect of particle design on cellular internalization pathways. Proc Natl Acad Sci U S A. 2008;105(33):11613–8.
94. Huang X, Teng X, Chen D, Tang F, He J. The effect of the shape of mesoporous silica nanoparticles on cellular uptake and cell function. Biomaterials. 2010;31(3):438–48.
95. Balkundi S, Nowacek AS, Veerubhotla RS, Chen H, Martinez-Skinner A, Roy U, et al. Comparative manufacture and cell-based delivery of antiretroviral nanoformulations. Int J Nanomedicine. 2011;6:3393–404.
96. Kadiu I, Nowacek A, McMillan J, Gendelman HE. Macrophage endocytic trafficking of antiretroviral nanoparticles. Nanomedicine (Lond). 2011;6(6):975–94.

97. Kadiu I, Gendelman HE. Macrophage bridging conduit trafficking of HIV-1 through the endoplasmic reticulum and Golgi network. J Proteome Res. 2011;10(7):3225–38.
98. Bale SS, Kwon SJ, Shah DA, Banerjee A, Dordick JS, Kane RS. Nanoparticle-mediated cytoplasmic delivery of proteins to target cellular machinery. ACS Nano. 2010;4(3):1493–500.
99. Duncan R, Richardson SC. Endocytosis and intracellular trafficking as gateways for nanomedicine delivery: opportunities and challenges. Mol Pharm. 2012;9(9):2380–402.
100. Huang K, Voss B, Kumar D, Hamm HE, Harth E. Dendritic molecular transporters provide control of delivery to intracellular compartments. Bioconjug Chem. 2007;18(2):403–9.
101. Liu Y, Li K, Pan J, Liu B, Feng SS. Folic acid conjugated nanoparticles of mixed lipid monolayer shell and biodegradable polymer core for targeted delivery of Docetaxel. Biomaterials. 2010;31(2):330–8.
102. Micheli MR, Bova R, Magini A, Polidoro M, Emiliani C. Lipid-based nanocarriers for CNS-targeted drug delivery. Recent Pat CNS Drug Discov. 2012;7(1):71–86.
103. Garg M, Asthana A, Agashe HB, Agrawal GP, Jain NK. Stavudine-loaded mannosylated liposomes: in-vitro anti-HIV-I activity, tissue distribution and pharmacokinetics. J Pharm Pharmacol. 2006;58(5):605–16.
104. Garg M, Dutta T, Jain NK. Reduced hepatic toxicity, enhanced cellular uptake and altered pharmacokinetics of stavudine loaded galactosylated liposomes. Eur J Pharm Biopharm. 2007;67(1):76–85.
105. Kraft-Terry SD, Engebretsen IL, Bastola DK, Fox HS, Ciborowski P, Gendelman HE. Pulsed stable isotope labeling of amino acids in cell culture uncovers the dynamic interactions between HIV-1 and the monocyte-derived macrophage. J Proteome Res. 2011;10(6):2852–62.
106. Martinez-Skinner AL, Veerubhotla RS, Liu H, Xiong H, Yu F, McMillan JM, et al. Functional proteome of macrophage carried nanoformulated antiretroviral therapy demonstrates enhanced particle carrying capacity. J Proteome Res. 2013;12(5):2282–94.
107. Gautam N, Roy U, Balkundi S, Puligujja P, Guo D, Smith N, et al. Preclinical pharmacokinetics and tissue distribution of long-acting nanoformulated antiretroviral therapy. Antimicrob Agents Chemother. 2013;57(7):3110–20.
108. Nischang M, Gers-Huber G, Audige A, Akkina R, Speck RF. Modeling HIV infection and therapies in humanized mice. Swiss Med Wkly. 2012;142:13618.
109. Nischang M, Sutmuller R, Gers-Huber G, Audige A, Li D, Rochat MA, et al. Humanized mice recapitulate key features of HIV-1 infection: a novel concept using long-acting anti-retroviral drugs for treating HIV-1. PloS one. 2012;7(6):e38853.
110. Garcia-Lerma JG, Heneine W. Animal models of antiretroviral prophylaxis for HIV prevention. Curr Opin HIV AIDS. 2012;7(6):505–13.
111. Van Rompay KK. Evaluation of antiretrovirals in animal models of HIV infection. Antiviral Res. 2010;85(1):159–75.
112. Emerich DF, Thanos CG. The pinpoint promise of nanoparticle-based drug delivery and molecular diagnosis. Biomol Eng. 2006;23(4):171–84.
113. Foisy MM, Yakiwchuk EM, Hughes CA. Induction effects of ritonavir: implications for drug interactions. Ann Pharmacother. 2008;42(7):1048–59.
114. Havlir DV, O'Marro SD. Atazanavir: new option for treatment of HIV infection. Clin Infect Dis. 2004;38(11):1599–604.
115. Gorantla S, Poluektova L, Gendelman HE. Rodent models for HIV-associated neurocognitive disorders. Trends Neurosci. 2012;35(3):197–208.
116. Bosma MJ, Carroll AM. The SCID mouse mutant: definition, characterization, and potential uses. Annu Rev Immunol. 1991;9:323–50.
117. Wege AK, Melkus MW, Denton PW, Estes JD, Garcia JV. Functional and phenotypic characterization of the humanized BLT mouse model. Curr Top Microbiol Immunol. 2008;324:149–65.
118. Akkina R. New generation humanized mice for virus research: comparative aspects and future prospects. Virology. 2013;435(1):14–28.
119. Akkina R. Human immune responses and potential for vaccine assessment in humanized mice. Curr Opin Immunol. 2013;25(3):403–9.

120. Tyor WR, Power C, Gendelman HE, Markham RB. A model of human immunodeficiency virus encephalitis in scid mice. Proc Natl Acad Sci U S A. 1993;90(18):8658–62.
121. Persidsky Y, Gendelman HE. Murine models for human immunodeficiency virus type 1-associated dementia: the development of new treatment testing paradigms. J Neurovirol. 2002;8(Suppl 2):49–52.
122. Gorantla S, Che M, Gendelman HE. Isolation, propagation, and HIV-1 infection of monocyte-derived macrophages and recovery of virus from brain and cerebrospinal fluid. Methods Mol Biol. 2005;304:35–48.
123. Poluektova L, Gorantla S, Faraci J, Birusingh K, Dou H, Gendelman HE. Neuroregulatory events follow adaptive immune-mediated elimination of HIV-1-infected macrophages: studies in a murine model of viral encephalitis. J Immunol. 2004;172(12):7610–7.
124. Gorantla S, Makarov E, Finke-Dwyer J, Castanedo A, Holguin A, Gebhart CL, et al. Links between progressive HIV-1 infection of humanized mice and viral neuropathogenesis. Am J Pathol. 2010;177(6):2938–49.
125. Dash PK, Gorantla S, Gendelman HE, Knibbe J, Casale GP, Makarov E, et al. Loss of neuronal integrity during progressive HIV-1 infection of humanized mice. J Neurosci. 2011;31(9):3148–57.
126. Epstein AA, Narayanasamy P, Dash PK, High R, Bathena SP, Gorantla S, et al. Combinatorial assessments of brain tissue metabolomics and histopathology in rodent models of human immunodeficiency virus infection. J Neuroimmune Pharmacol. 2013;8(5):1224–38.
127. Denton PW, Garcia JV. Mucosal HIV-1 transmission and prevention strategies in BLT humanized mice. Trends Microbiol. 2012;20(6):268–74.
128. Deruaz M, Luster AD. BLT humanized mice as model to study HIV vaginal transmission. J Infect Dis. 2013;208 Suppl 2:S131–6.
129. Marsden MD, Kovochich M, Suree N, Shimizu S, Mehta R, Cortado R, et al. HIV latency in the humanized BLT mouse. J Virol. 2012;86(1):339–47.
130. Lavender KJ, Pang WW, Messer RJ, Duley AK, Race B, Phillips K, et al. BLT-humanized C57BL/6 Rag2-/-gammac-/-CD47-/- mice are resistant to GVHD and develop B- and T-cell immunity to HIV infection. Blood. 2013;122(25):4013–20.
131. Lan P, Tonomura N, Shimizu A, Wang S, Yang YG. Reconstitution of a functional human immune system in immunodeficient mice through combined human fetal thymus/liver and $CD34^+$ cell transplantation. Blood. 2006;108(2):487–92.
132. Melkus MW, Estes JD, Padgett-Thomas A, Gatlin J, Denton PW, Othieno FA, et al. Humanized mice mount specific adaptive and innate immune responses to EBV and TSST-1. Nat Med. 2006;12(11):1316–22.
133. Chateau ML, Denton PW, Swanson MD, McGowan I, Garcia JV. Rectal transmission of transmitted/founder HIV-1 is efficiently prevented by topical 1 % tenofovir in BLT humanized mice. PloS one. 2013;8(3):e60024.
134. Dudek TE, No DC, Seung E, Vrbanac VD, Fadda L, Bhoumik P, et al. Rapid evolution of HIV-1 to functional $CD8^+$ T cell responses in humanized BLT mice. Sci Transl Med. 2012;4(143):143ra98.
135. Denton PW, Olesen R, Choudhary SK, Archin NM, Wahl A, Swanson MD, et al. Generation of HIV latency in humanized BLT mice. J Virol. 2012;86(1):630–4.
136. Denton PW, Othieno F, Martinez-Torres F, Zou W, Krisko JF, Fleming E, et al. One percent tenofovir applied topically to humanized BLT mice and used according to the CAPRISA 004 experimental design demonstrates partial protection from vaginal HIV infection, validating the BLT model for evaluation of new microbicide candidates. J Virol. 2011;85(15):7582–93.
137. Denton PW, Long JM, Wietgrefe SW, Sykes C, Spagnuolo RA, Snyder OD, et al. Targeted cytotoxic therapy kills persisting HIV infected cells during ART. PLoS pathog. 2014;10(1):e1003872.
138. Deeks SG. HIV: shock and kill. Nature. 2012;487(7408):439–40.

Chapter 31
Targeted Delivery of Aptamers and siRNAs for HIV Prevention and Therapies in Humanized Mice

Charles Preston Neff and Ramesh Akkina

31.1 Introduction

Viral genome integration into host cell chromosome and propensity for establishing latency are formidable obstacles for a complete cure for HIV [1]. However, with the advent of highly active antiretroviral therapy (HAART) using drugs that target various stages of the viral life cycle, HIV-1 infected patients are now routinely treated to sustain substantial reduction in viral load and maintain healthy CD4 T cell levels. This treatment combines up to three antiretroviral drugs that effectively prevent AIDS progression; as a result, HIV-1 infection is no longer a death sentence. However, HAART does not eradicate the virus, thus requiring indefinite treatment that is dependent on patient compliance. This matter is further complicated by lack of wide drug availability, side effects, and intricacy of the treatment schedule that includes a strict regimen with tailored drug prescriptions. In this regard, new forms of treatment are necessary to replace life-long drug use or as an adjunct therapy to increase efficacy.

31.2 RNA-based Treatment Strategies for HIV

Due to high specificity and novel mechanisms of action, RNA-based antiviral constructs are promising alternatives to current drug therapy [2]. These include ribozymes and small interfering RNAs (siRNAs) that cleave viral or host transcripts

R. Akkina (✉)
Department of Microbiology, Immunology and Pathology,
Colorado State University, 1619 Campus delivery, Fort Collins, CO 80523, USA
e-mail: akkina@colostate.edu

C. P. Neff
Department of Medicine, Division of Allergy and Clinical Immunology Aurora, University of Colorado, Denver, Research Complex 2, B164, 12700 E 19th Ave, Aurora, CO 80045, USA

L. Y. Poluektova et al. (eds.), *Humanized Mice for HIV Research*,
DOI 10.1007/978-1-4939-1655-9_31

essential for viral infection and replication, RNA aptamers and decoys that inhibit viral protein activity and/or block viral-cell protein interactions [3]. There are advantages to applying these anti-HIV nucleic acid based approaches. First, unlike therapeutic proteins, they are nonimmunogenic molecules, and thus can be used on a prolonged basis. Second, they are highly specific allowing specific targeting of desired viral and/or cell specific RNA transcripts or critical proteins involved in viral infection and replication. Furthermore, they can be designed to augment present HAART strategies and lower drug doses to reduce side effects.

31.3 SiRNAs Against HIV and Host Factors

Inhibition of gene expression by RNA interference (RNAi) is mediated by two pathways [3]. In the first, posttranscriptional gene silencing (PTGS), translational suppression and degradation of mRNAs is mediated by micro-RNAs (miRNAs) in which the guide strand (antisense) displays imperfect complementarity to the target. PTGS by sequence specific cleavage of target mRNA is seen with exogenously supplied siRNAs or shRNAs with perfect or near perfect complementarity. Similar to the host miRNA pathway, siRNAs and shRNAs engage many of the endogenous host factors to mediate gene silencing. The second RNAi pathway, transcriptional gene silencing (TGS), involves inhibitory action of the siRNAs on the cell's transcriptional apparatus [4]. Whereas the TGS pathway is the subject of new investigations on silencing the HIV genome permanently leading to viral eradication, previous studies focused on PTGS with many important advances of which some have gone on to clinical trials [5] as potential functional cures.

Anti-HIV activity by PTGS is induced either by transfecting siRNAs into cells in a dsRNA form or as endogenously transcribed shRNAs which are subsequently processed by cellular machinery into mature siRNAs [3]. The activity and longevity of synthetic siRNAs have been substantially increased by improved designs and synthetic approaches. Precursor dsRNAs are designed asymmetrically with a 3′ overhang of 2nt at one end, with a blunt end at the other, resulting in a Dicer substrate in which there is a preferential generation of the guide strand. Chemical modifications of the nucleotides increase the half-life of these molecules. Since the discovery of RNAi, many siRNAs with a range of efficacies that target the HIV life cycle have been examined (Table 31.1). A recent report developed a Web-based database detailing numerous constructs thus facilitating the search process and documentation [25]. The viral targets included transcripts encompassing env, tat, rev, and RT whereas the cellular transcripts included those coding for cell surface receptors CD4, CCR5, and CXCR4 as well as cellular factors exemplified by TNPO3. For in vitro studies most used transfection of chemically synthesized siRNAs. For endogenous expression of siRNAs/shRNAs into HIV susceptible cells, many reports utilized lentiviral vectors for gene transduction [26]. The siRNAs comprised of those targeting coreceptor molecules such as CCR5, and against viral transcripts exemplified by tat and rev. Since use of a single siRNA

Table 31.1 siRNAs against HIV and host factors

	Target gene	Delivery method	Efficacy	Reference
Viral	LTR	Transfection, lentiviral vector	10–87 %	[6–8]
	Gag/Pol	Transfection	20–98 %	[9–11]
	Integrase	Transfection, lentiviral vector	43–96 %	[6, 12]
	Vif	Transfection	69–96 %	[13, 14]
	Tat/rev	Aptamer conjugate, nanoparticle, lentiviral vector	17–100 %	[15–17]
Host	CD4	Aptamer conjugate, nanoparticle	85–95 %	[18–20]
	CCR5	Lentiviral vector	20–97 %	[5, 21]
	CXCR4	Lentiviral vector, transposon	14–93 %	[22, 23]
	TNPO3	Aptamer conjugate, nanoparticle	85–95 %	[15, 24]

to a single target would invariably lead to the generation of viral escape mutants, anti-HIV siRNAs were combined with other RNA-based constructs. A combinatorial construct in a lentiviral vector employed tat-rev siRNA together with a tar decoy and a CCR5 ribozyme. Its efficacy was evaluated both in vitro and in vivo in the SCID-hu mouse system and subsequently in a human clinical trial [27]. More recent combinatorial constructs also included host restriction factors such as TRIM5α variants. Testing these in the new generation $Rag1^{-/-}\gamma c^{-/-}$ (RAG-hu) mice showed promising results [28].

31.4 Aptamers Against HIV and Host Factors

Aptamers are synthetic single-stranded nucleic acid molecules, either RNA or DNA, that bind target molecules with high specificity and affinity similar to antibodies [29, 30]. Aptamers to desired targets are selected and identified by a process called systematic evolution of ligands by exponential enrichment (SELEX) [31]. Aptamers can be chemically synthesized in bulk quantities and chemical modifications are possible that improve nuclease resistance and prolong in vivo half-life. Other advantages of these molecules are their apparent lack of immunogenicity and systemic toxicity. A number of aptamers have been developed to inhibit HIV, which are either directed to host cell receptors or viral proteins (Table 31.2), deriving a potential alternative functional cure. Aptamers against HIV RT were shown to act by competing with primer/template access, inhibiting DNA polymerization or RNase

Table 31.2 Anti-HIV aptamers against host and viral targets

	Target	Location	Nucleic acid	Efficacy	Reference
Viral	RT	Intracellular	RNA	30–60 %	[32, 33]
	Integrase	Intracellular	RNA	–	[34]
	Rev	Intracellular	RNA	–	[35]
	GP120	Extracellular	RNA	80–99 %	[15, 36, 37]
Host	CD4	Extracellular	RNA, DNA	60–70 %	[38, 39]

activity [40]. When expressed in cells, viral replication was suppressed. Based on these successes, aptamers against other HIV proteins such as viral integrase are currently being developed. Aptamers against the cellular CD4 receptor were shown to inhibit HIV entry whereas anti-gp120 aptamers bind to HIV envelop resulting in viral neutralization [15, 41]. These later aptamers have found novel applications for cell specific delivery of anti-HIV siRNAs into naive cells for protection and into HIV-infected cells to inhibit viral replication (Fig 31.1)

31.5 Systemic Delivery of siRNAs and Aptamers

RNA is an inherently labile and unstable molecule in relative contrast to DNA. While in vitro delivery of siRNAs and aptamers use standard methods of transfection for routine testing, in vivo delivery of these molecules has posed severe challenges. A major limitation has been their degradation in vivo by host nucleases [3, 30]. This is partially overcome by chemical modifications of the phosphate backbone (e.g., phosphorothioate) and/or sugar moieties (e.g., 2′-fuoro). However, such modifications of siRNAs may lead to inefficient processing by the cellular Dicer and/or their loading into the cellular RISC complex mediating RNAi. This necessitates empirical testing.

For delivery, noncovalent assemblies of siRNAs in general employ cationic molecules such as protamine or encapsulation into a nanoparticle using a cationic lipid. These complexes can be targeted to specific cells via antibodies or peptides. Covalent assemblies of siRNAs employ direct conjugation to molecules such as aptamers, cholesterol, and antibody-protamine fusion proteins (see below). Irrespective of the delivery system, the siRNAs need to find their way into cell cytoplasm to mediate the RNAi effect. In some studies involving HIV suppression, encapsulating the siRNAs into nanoparticles enabled efficient and lasting delivery. In this regard, humanized mice that permit HIV infection and display helper CD4 T-cell depletion have played an important role [42]. In a recent study, Zhou et al. used dendrimer formulated siRNAs against HIV in RAG-hu mice [24, 43].

The fifth generation PAMAM dendrimers belong to a class of highly branched polymers bearing 96 cationic primary amine groups on their spherical surface with a triethanolamine core with a flexible structure. Stable and uniform nanoparticle complexes are formed with nucleic acids via electrostatic interactions. Mixing of dendrimer with siRNAs leads to their encapsulation and thus protection from nucleases. Upon internalization of these particles via cellular macropinocytosis, the precursor siRNA molecules are processed by cell Dicer enzyme into mature molecules that trigger RNAi response to specific RNA transcripts in the cell. To determine their in vivo efficacy, dendrimer formulated siRNAs (against host cell factors CD4 and TNPO3 and viral transcripts tat-rev) were administered to previously HIV infected viremic humanized $Rag2^{-/-}\gamma c^{-/-}$ mice [24, 43].

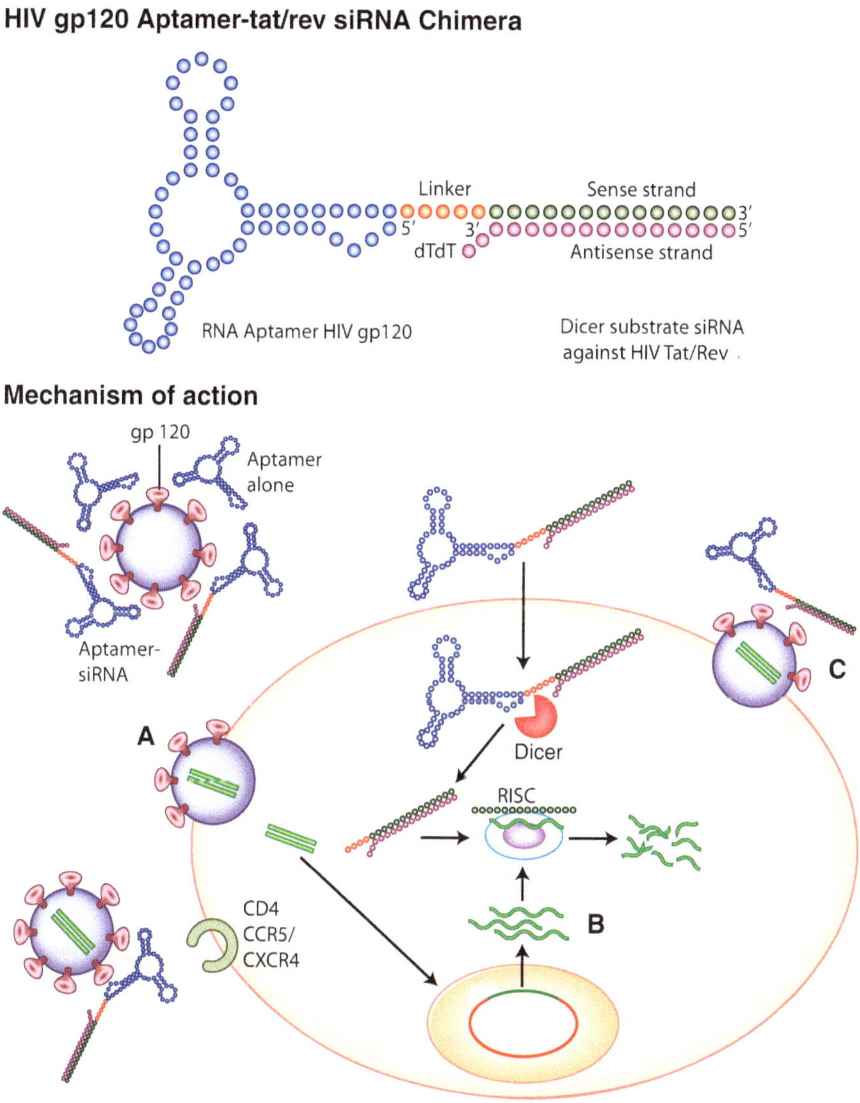

Fig. 31.1 Schematics of aptamer-siRNA chimera and mechanisms of antiviral action: gp120 aptamer is conjugated by a linker segment to a tat/rev siRNA resulting in an apatmer-siRNA chimera (*Top*). Mechanisms of action (*Bottom*). Binding of free floating apatmer alone or aptamer-siRNA chimera via the aptamer moiety to HIV gp120 on cell free plasma virus surface leads to neutralization via interference with viral binding to cell surface receptors (*A*). Binding of the chimera to gp120 expressed on infected cells leads to internalization of the chimera and cytoplasmic processing by Dicer to release the anti-HIV siRNA. siRNA is incorporated into the RISC complex resulting in subsequent cleavage of viral transcripts (*B*). Viral budding is inhibited by apatmer interactions with gp120 on infected cell's surface

In these mice, suppression of HIV viremia by several orders of magnitude was seen together with protection from CD4 T cell loss. Dendrimer-siRNAs were shown to preferentially accumulate in PBMCs and in liver with no discernible toxicity. Treatment cessation led to viral rebounds. However, retreatment suppressed viral loads again further confirming the selectivity of viral suppression. These proof-of-concept studies showed that siRNAs can be effectively delivered in vivo to HIV susceptible cells in lasting manner providing a therapeutic benefit but does not result in viral eradication.

31.6 Cell Specific Targeting of Aptamers, siRNAs, and Aptamer-siRNA Chimeras In Vivo

The nonspecific delivery methods described above require large amounts of siRNAs to over saturate the system to reach the HIV susceptible cells in vivo. This approach has several drawbacks that include exorbitant cost as well as having potentially immense off target effects in vivo. To overcome this major drawback Kumar et al. used an innovative approach [44]. A single chain antibody (scFV) to pan T cell marker CD7 was used to deliver siRNAs into T cells. To permit binding, a modified scFVCD7Cys was conjugated to a nona-d-arginine (9R) peptide. SiRNAs to CCR5, HIV Vif, and tat were tested using hu-HSC NSG mice. It was found that in vivo administration of these conjugates delivered siRNAs to the naive HIV susceptible cells resulted in marked suppression of HIV viremia. While impressive, even non-CD4$^+$ helper T cells were targeted using this approach, and due to possible immunogenicity of the modified antibody-peptide conjugates, long-term repeated dosing of these constructs is not possible. In a different study integrin-targeted and stabilized nanoparticles (I-tsNP) incorporating an LFA-1 integrin-targeted antibody were employed for delivery of anti-HIV siRNAs to human lymphocytes and monocytes in humanized bone marrow liver thymic (BLT) mice [45]. A CCR5 siRNA was administered to humanized mice followed by HIV challenge. The treated mice exhibited increased resistance to HIV infection and reduction in CD4 T cell decline. However this approach targets all T cells including CD8 T cells and therefore is not selective for CD4 T cells which are primary viral targets.

Therefore, methods that specifically target HIV susceptible cells and in particular, virus infected cells are necessary for a more focused therapeutic approach in HIV infected individuals. Gp120 is a subunit of the viral envelop molecule that is expressed on the surface of HIV infected cells. Taking this into consideration, a single chain gp120 antibody tagged siRNAs was used successfully in vitro to target infected cells [46]. However, host immune responses to modified antibodies used for cell targeting remains to be an issue. To circumvent this drawback, Neff et al. used a novel approach that employed an HIV gp120 binding aptamer [15]. To deliver an anti-HIV tat-rev siRNA specifically into HIV infected cells, the anti-gp120 aptamer was conjugated to the siRNA resulting in a chimeric aptamer-siRNA chimera. To reduce degradation by the serum nucleases and to increase their biological half-life in vivo, chimeric RNA constructs were 2′-fuoro-modified during synthesis. In vitro testing confirmed its specificity in entering only HIV infected but not the neighboring noninfected cells. The in vivo efficacy of this chimera was evaluated in RAG-hu mice that are permissive to HIV infection and show CD4 T cell loss.

Viremic mice were treated on a weekly basis for 5 weeks. Viremia and $CD4^+$ T cell levels were measured for several weeks to determine the efficacy. Drastic reduction in viral loads to below detectable levels (<40 RNA copies/ml plasma) was seen within a week and viral suppression persisted throughout the treatment period. Furthermore, viral loads were undetectable even up to 3 weeks after cessation of treatment in majority of mice, indicating sustained efficacy of this approach. Importantly, there was also protection from $CD4^+$ T cell loss in treated infected mice showing that aptamer-siRNA treatment achieved the primary goal of targeted anti-HIV therapy. The antiviral effect of the chimera is due to three different mechanisms operating simultaneously. First, the aptamer portion of the chimera would neutralize the free floating virus in the plasma of viemic animals thus preventing infection of new cells. Second, the aptamer delivered tat-rev siRNA into infected cells would interfere with virus replication and third, the gp120 binding aptamer on cell surface would interfere with viral budding and/or release. RT-PCR and RACE PCR assays confirmed the activity of the intracellularly delivered tat-rev siRNAs on their targets as shown by a drastic reduction in tat-rev transcript levels as well as detection of their specific cleavage products. Undesirable nonspecific innate immune activation may result with the use of RNA-based constructs. However, no such off target effects were noted. Interestingly, while drastic viral suppression was seen in the majority of treated micc, viral rebound was detected in some mice suggesting viral escape to the treatment. Therefore, simultaneous use of multiple aptamer-siRNA constructs with different specificities for treatment is likely to prevent generation of viral escape mutants in a principle similar to the HAART drug treatment approach wherein a combination of drugs are used. Accordingly, in an extension of the above studies a combinatorial siRNA delivery was successfully used exploiting a stick-aptamer approach in which different siRNAs targeted to HIV tat-rev, cellular CD4, and TNPO3 were conjugated to a gp120-aptamer [43].

The methods described above were geared for therapeutic purposes in HIV infected patients but are ineffective in preventing infection in naive individuals. Using the NSG mouse, recent studies of Wheeler et al. employed CD4 aptamers conjugated to siRNAs targeted to host cell and viral transcripts addressing prevention strategies [41, 47]. It was shown that CD4 aptamer-CCR5, HIV vif, and gag siRNA conjugates were specifically taken up by $CD4^+$ T cells and macrophages leading to HIV suppression in vitro. Furthermore, when tested in BLT mice permissible to HIV vaginal transmission, these chimeras inhibited HIV transmission. While a CD4 aptamer by itself was effective to a certain extent in disease prevention, the chimeric constructs were found to be even more effective. As can be seen, aptamer targeted approaches for siRNA delivery in vivo has shown to be a very promising strategy for HIV therapy and prevention and thus paving the way for future human clinical trials.

31.7 Summary and Future Directions

While numerous studies have shown the potential of siRNAs and aptamers for anti-HIV therapies, efficient systemic delivery of these molecules to HIV susceptible or infected cells has proven to be a major hurdle until recently. Chemical modifications of these molecules to increase their in vivo half-life and novel designs permitting their cell specific targeting and cellular uptake are beginning to yield promising results. Future in-depth pharmaco-kinetic (PK) studies that focus on in vivo tissue distribution, biological half-life, clearance and elimination characteristics, and potential long-term toxicities are necessary to derive important preclinical data. Importantly, humanized mice have proven to be highly suitable for in vivo testing of these novel constructs and are expected to play a continued crucial role.

References

1. Richman DD, et al. The challenge of finding a cure for HIV infection. Science. 2009;323(5919):1304–7.
2. Akkina R, et al. siRNAs, ribozymes and RNA decoys in modeling stem cell-based gene therapy for HIV/AIDS. Anticancer Res. 2003;23(3A):1997–2005.
3. Burnett JC, Rossi JJ. RNA-based therapeutics: current progress and future prospects. Chem Biol. 2012;19(1):60–71.
4. Turner AM, et al. Characterization of an HIV-targeted transcriptional gene-silencing RNA in primary cells. Hum gene Ther. 2012;23(5):473–83.
5. DiGiusto DL, et al. RNA-based gene therapy for HIV with lentiviral vector-modified CD34(+) cells in patients undergoing transplantation for AIDS-related lymphoma. Sci Transl Med. 2010;2(36):36ra43.
6. Chang LJ, Liu X, He J. Lentiviral siRNAs targeting multiple highly conserved RNA sequences of human immunodeficiency virus type 1. Gene Ther. 2005;12(14):1133–44.
7. Yamagishi M, et al. Retroviral delivery of promoter-targeted shRNA induces long-term silencing of HIV-1 transcription. Microbes Infect. 2009;11(4):500–8. (Institut Pasteur).
8. Yoshinari K, Miyagishi M, Taira K. Effects on RNAi of the tight structure, sequence and position of the targeted region. Nucleic Acids Res. 2004;32(2):691–9.
9. Capodici J, Kariko K, Weissman D. Inhibition of HIV-1 infection by small interfering RNA-mediated RNA interference. J Immunol. 2002;169(9):5196–201.
10. ter Brake O, et al. Silencing of HIV-1 with RNA interference: a multiple shRNA approach. Mol Ther. 2006;14(6):883–92.
11. von Eije KJ, ter Brake O, Berkhout B. Stringent testing identifies highly potent and escape-proof anti-HIV short hairpin RNAs. J Gene Med. 2009;11(6):459–67.
12. Lau TS, et al. Suppression of HIV replication using RNA interference against HIV-1 integrase. FEBS Lett. 2007;581(17):3253–9.
13. Jacque JM, Triques K, Stevenson M. Modulation of HIV-1 replication by RNA interference. Nature. 2002;418(6896):435–8.
14. McIntyre GJ, et al. 96 shRNAs designed for maximal coverage of HIV-1 variants. Retrovirology. 2009;6:55.
15. Neff CP, et al. An aptamer-siRNA chimera suppresses HIV-1 viral loads and protects from helper CD4(+) T cell decline in humanized mice. Sci Transl Med. 2011;3(66):66ra6.
16. Scherer LJ, et al. Rapid assessment of anti-HIV siRNA efficacy using PCR-derived Pol III shRNA cassettes. Mol Ther. 2004;10(3):597–603.

17. Zhou J, et al. Novel dual inhibitory function aptamer-siRNA delivery system for HIV-1 therapy. Mol Ther. 2008;16(8):1481–9.
18. Berkhout B, ter Brake O. Towards a durable RNAi gene therapy for HIV-AIDS. Expert Opin Biol Ther. 2009;9(2):161–70.
19. Brass AL, et al. Identification of host proteins required for HIV infection through a functional genomic screen. Science. 2008;319(5865):921–6.
20. Zhou J, et al. Functional in vivo delivery of multiplexed anti-HIV-1 siRNAs via a chemically synthesized aptamer with a sticky bridge. Mol Ther. 2013;21(1):192–200.
21. Liang M, et al. Inhibition of HIV-1 infection by a unique short hairpin RNA to chemokine receptor 5 delivered into macrophages through hematopoietic progenitor cell transduction. J Gene Med. 2010;12(3):255–65.
22. Anderson J, Banerjea A, Akkina R. Bispecific short hairpin siRNA constructs targeted to CD4, CXCR4, and CCR5 confer HIV-1 resistance. Oligonucleotides. 2003;13(5):303–12.
23. Tamhane M, Akkina R. Stable gene transfer of CCR5 and CXCR4 siRNAs by sleeping beauty transposon system to confer HIV-1 resistance. AIDS Res Ther. 2008;5:16.
24. Zhou J, et al. Systemic administration of combinatorial dsiRNAs via nanoparticles efficiently suppresses HIV-1 infection in humanized mice. Mol Ther. 2011;19(12):2228–38.
25. Tyagi A, et al. HIVsirDB: a database of HIV inhibiting siRNAs. PloS ONE. 2011;6(10):e25917.
26. Kitchen SG, Shimizu S, An DS. Stem cell-based anti-HIV gene therapy. Virology. 2011;411(2):260–72.
27. Anderson J, et al. Safety and efficacy of a lentiviral vector containing three anti-HIV genes-CCR5 ribozyme, tat-rev siRNA, and TAR decoy-in SCID-hu mouse-derived T cells. Mol Ther. 2007;15(6):1182–8.
28. Walker JE, et al. Generation of an HIV-1-resistant immune system with CD34(+) hematopoietic stem cells transduced with a triple-combination anti-HIV lentiviral vector. J Virol. 2012;86(10):5719–29.
29. Keefe AD, Pai S, Ellington A. Aptamers as therapeutics. Nat Rev Drug Discov. 2010;9(7):537–50.
30. Thiel KW, Giangrande PH. Intracellular delivery of RNA-based therapeutics using aptamers. Ther Deliv. 2010;1(6):849–61.
31. Tuerk C, Gold L. Systematic evolution of ligands by exponential enrichment: RNA ligands to bacteriophage T4 DNA polymerase. Science. 1990;249(4968):505–10.
32. Nickens DG, Patterson JT, Burke DH. Inhibition of HIV-1 reverse transcriptase by RNA aptamers in *Escherichia coli*. RNA. 2003;9(9):1029–33.
33. Tuerk C, MacDougal S, Gold L. RNA pseudoknots that inhibit human immunodeficiency virus type 1 reverse transcriptase. Proc Natl Acad Sci U S A. 1992;89(15):6988–92.
34. Allen P, Worland S, Gold L. Isolation of high-affinity RNA ligands to HIV-1 integrase from a random pool. Virology. 1995;209(2):327–36.
35. Giver L, et al. Selection and design of high-affinity RNA ligands for HIV-1 Rev. Gene. 1993;137(1):19–24.
36. Khati M, et al. Neutralization of infectivity of diverse R5 clinical isolates of human immunodeficiency virus type 1 by gp120-binding 2'F-RNA aptamers. J Virol. 2003;77(23):12692–8.
37. Mufhandu HT, et al. UCLA1, a synthetic derivative of a gp120 RNA aptamer, inhibits entry of human immunodeficiency virus type 1 subtype C. J Virol. 2012;86(9):4989–99.
38. Kraus E, James W, Barclay AN. Cutting edge: novel RNA ligands able to bind CD4 antigen and inhibit CD4+T lymphocyte function. J Immunol. 1998;160(11):5209–12.
39. Zhu Q, et al. Inhibition of HIV-1 protease expression in T cells owing to DNA aptamer-mediated specific delivery of siRNA. Eur J Med Chem. 2012;56:396–9.
40. Whatley AS, et al. Potent inhibition of HIV-1 reverse transcriptase and replication by nonpseudoknot, "UCAA-motif" RNA aptamers. Mol Ther Nucleic Acids. 2013;2:e71.
41. Wheeler LA, et al. Inhibition of HIV transmission in human cervicovaginal explants and humanized mice using CD4 aptamer-siRNA chimeras. J Clin Invest. 2011;121(6):2401–12.
42. Akkina R. New generation humanized mice for virus research: comparative aspects and future prospects. Virology. 2013;435(1):14–28.

43. Zhou J, et al. Functional in vivo delivery of multiplexed anti-HIV-1 siRNAs via a chemically synthesized aptamer with a sticky bridge. Mol Ther. 2013;**21**(1):192–200.
44. Kumar P, et al. T cell-specific siRNA delivery suppresses HIV-1 infection in humanized mice. Cell. 2008;134(4):577–86.
45. Kim SS, et al. RNAi-mediated CCR5 silencing by LFA-1-targeted nanoparticles prevents HIV infection in BLT mice. Mol Ther. 2010;18(2):370–6.
46. Song E, et al. Antibody mediated in vivo delivery of small interfering RNAs via cell-surface receptors. Nat Biotechnol. 2005;23(6):709–17.
47. Wheeler LA, et al. Durable knockdown and protection from HIV transmission in humanized mice treated with gel-formulated CD4 aptamer-siRNA chimeras. Mol Ther. 2013;21(7):1378–89.

Chapter 32
Zinc Finger Nuclease Editing of Hematopoietic Stem Cells as an Anti-HIV Therapy

Nathalia G. Holt, Colin M. Exline, Orla Mulhern, Ursula Hofer, Kathleen A. Burke, Jill E. Oldenburg and Paula M. Cannon

32.1 Introduction

Although the term "humanized mice" can broadly refer to any situation where immune-deficient mice are transplanted with a graft of human primary cells, it is frequently used in the context of mice transplanted with human hematopoietic stem and progenitor cells (HSC). Following introduction of these cells into sublethally irradiated immune-deficient strains such as the NOD/LtSz-scid IL2Rgamma (null) (NSG) mouse [1], the human HSC engraft and differentiate, thereby recapitulating many aspects of human hematopoiesis and giving rise to mature cells of the human immune system [1, 2]. As a result, HSC-transplanted humanized mice are proving useful for the evaluation of therapies based on manipulations of human HSC, including those aimed at providing an anti-HIV effect. The model represents a rigorous test of whether the treatments in any way impact HSC function and can thereby provide meaningful preclinical data to support eventual human clinical trials. In this chapter we review the use of humanized mice in the development of zinc finger nuclease (ZFN) editing of human HSC as an anti-HIV therapy.

Kathleen A. Burke is deceased.

P. M. Cannon (✉)
Department of Molecular Microbiology & Immunology, Keck School of Medicine, University of Southern California, 2011 Zonal Avenue, HMR502, Los Angeles, CA 90033, USA
e-mail: pcannon@usc.edu

N. G. Holt · C. M. Exline · O. Mulhern · U. Hofer · K. A. Burke · J. E. Oldenburg
University of Southern California, Los Angeles, CA, USA

L. Y. Poluektova et al. (eds.), *Humanized Mice for HIV Research*,
DOI 10.1007/978-1-4939-1655-9_32

32.2 CCR5: A Nonessential Human Gene That Makes a Good Anti-HIV Target

ZFNs are being developed for anti-HIV gene therapies based on their ability to knockout specific gene targets, including CCR5, the major coreceptor used by HIV-1 [3]. During entry, the virus interacts with both CD4 and a coreceptor such as CCR5 or CXCR4, and in this way gains access to $CD4^+$ T cells or macrophages. Although this fact alone makes CCR5 an important target for anti-HIV therapies, it has received special attention as a gene knockout target because the lack of functional CCR5 is a relatively common and well tolerated genotype in humans. Specifically, individuals homozygous for the nonfunctional CCR5Δ32 allele represent approximately 1 % of the population, and are profoundly resistant to CCR5-tropic strains of HIV-1 [4]. These observations have encouraged both the development of drugs that target CCR5 [5], as well as genetic strategies aiming to down regulate or prevent its expression [6].

A proof of principle of the capability of CCR5-negative HSC to suppress an established HIV-1 infection was provided, most dramatically, by the case of the "Berlin patient." This HIV+ individual underwent HSC transplantation as part of a treatment for leukemia, but was transplanted with HSC obtained from a CCR5Δ32 homozygous donor. Subsequent control and eventual eradication of HIV-1 from his body has been documented [7–9]. While it is acknowledged that other factors may also have played a role in this outcome, such as the extensive anti-leukemia conditioning that may have depleted the burden of HIV-infected cells, the fundamental observation of an HIV cure is galvanizing efforts to develop gene therapies aimed at recreating this situation, through the modification of a patients' own T cells or HSC to become CCR5-negative [10].

32.3 Engineered Nucleases to Knockout CCR5

CCR5 expression can be prevented in two broad ways. First, as will be reviewed elsewhere in this volume, technologies such as RNA interference (RNAi) can be used to effectively suppress CCR5 gene expression. For example, RNAi has been successfully used to reduce expression of CCR5 in the T-cell progeny derived from modified human HSC [6]. However, RNAi creates only temporary effects, thereby requiring either the continuous delivery of siRNA oligonucleotides to target cells, or the long-term expression of an shRNA precursor from a stable vector such as an integrating lentiviral vector. These requirements add the risk of vector-mediated insertional mutagenesis, immunogenicity, or gene silencing developing, all of which could limit the long-term effectiveness of such approaches.

An alternative strategy to limit CCR5 gene expression is based on the use of engineered nucleases. These reagents can be used to promote the specific and permanent disruption of a gene, but require only transient expression of the nuclease to achieve this. Four main classes of nucleases are currently being developed—ZFNs,

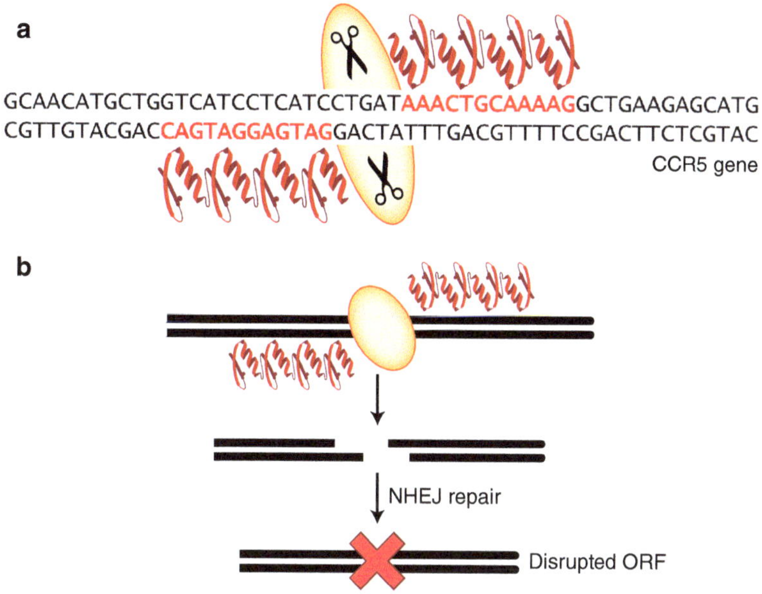

Fig. 32.1 **a** A pair of zinc finger nucleases binding to their target sequence (*red*) in the human CCR5 gene and positioning an endonuclease to cut the DNA. **b** The action of the endonuclease introduces a staggered (*5 bp*) double-stranded break. Following error-prone repair by the NHEJ pathway, insertions or deletions can be introduced that disrupt the open-reading frame

homing endonucleases, TALENs, and CRISPR/Cas9—and their varying characteristics suggest that they will find different applications [11]. What they all have in common is that they can be engineered to bind to a specific DNA sequence, such as the CCR5 gene, where the action of an endogenous or linked endonuclease introduces a site-specific double-stranded break at that DNA site. Subsequent repair of the break by the host cell frequently occurs through the error-prone nonhomologous end joining (NHEJ) pathway, with the result that small insertions or deletions are introduced at the break site that ultimately disrupt the open-reading frame (Fig. 32.1). In this way, the transient expression of an engineered nuclease can lead to the permanent disruption of a gene in the modified cell and all of its progeny. This ability makes the reagents particularly applicable to the modification of long-lived cells such as HSC.

ZFNs engineered to disrupt CCR5 were first shown to function in cells of relevance for HIV-1 infection using primary human $CD4^+$ T cells [12]. For this cell type, a modified adenovirus vector, Ad5/F35 [13], was found to be an effective way to deliver the ZFNs to the cells, and disruption of approximately 50 % of the CCR5 alleles in the cell population was achieved [12]. In contrast, human HSC have proven more difficult to engineer, in part because the cells have only a limited life-span ex vivo before suffering loss of viability and "stemness." Various gene delivery strategies can be employed with HSC, but since a key advantage of ZFN therapy is the fact that only transient expression of the nuclease is required, preferred strategies include electroporation of ZFN-encoding DNA or mRNA, and transduction by nonpermanent viral vector systems such as adenovirus, adeno-associated virus, or modified, nonintegrating, lentiviral vectors [14–19].

We previously described an optimized plasmid DNA electroporation (Nucleofection) protocol to deliver ZFNs to human cord blood-derived CD34$^+$ HSC, where we achieved a mean rate of CCR5 disruption of 17% of the alleles in the HSC population [14]. Our ongoing work with a larger team of investigators is aimed at developing strategies to modify HSC purified from the blood of patients whose stem cells have been mobilized from their bone marrow niche using treatments such as G-CSF, since this represents the source of HSC that is most commonly used for autologous gene therapy applications [20, 21]. For example, the same Ad5/F35 adenoviral vectors used to engineer the T cells described above can also be used with mobilized HSC, where a transient preincubation of the mobilized HSC with protein kinase activators significantly increases the resulting levels of CCR5 disruption [19]. Our current favored protocols use ZFN mRNA electroporation, since this can result in high levels of gene disruption in the range of 50% of the alleles [22], without the toxicity that can occur when using plasmid DNA, or the higher titers of adenoviral vectors needed to achieve such gene disruption rates [19]. Overall, procedures to achieve high levels of CCR5 disruption in human HSC isolated from mobilized blood without overt toxicity or loss of function are needed to pave the way for human clinical trials.

32.4 Using Humanized Mice to Evaluate CCR5 ZFN Treated Cells for Anti-HIV Effects

The ability of HSC or T cell transplanted humanized mice to support HIV-1 replication has revolutionized animal studies of HIV-1, which were previously limited to primate models and SIV or SHIV infections. We and others have reported that NSG, or other immune-deficient strains of mice, such as BALB/c-Rag2(null) IL2Rgamma(null) mice [23] show robust human CD4$^+$ T cell development when transplanted with human HSC, and support relatively long-lasting infections by both CCR5 (R5) and CXCR4 (X4)-tropic strains of HIV [14, 24, 25]. In addition, the human HSC give rise to other lineages of cells that HIV-1 infects, such as macrophages, and different T cell subsets of importance for HIV infection and persistence can be identified, such as central memory T cells [26, 27]. Improvements to the mouse models that promote more authentic human hematopoiesis, including the maturation and organization of cells of importance for HIV-1 infection, have been achieved by either the cotransplantation of fetal thymic tissue [28, 29], or by engineering the mice to also express human immunomodulatory molecules [30–34].

Humanized mice infected with HIV-1 also recapitulate the loss of CD4$^+$ T cells that is the hallmark of the disease in humans, and which underlies HIV's pathogenicity. This means that gene therapies based on ZFN-engineered HSC can be evaluated in the mice for anti-HIV efficacy, both by monitoring HIV-1 viremia in the animals, as well as the rates of human $CD4^+$ T cell depletion. Combined with the ability of the model to evaluate the impact of ZFN modification on HSC function, the humanized mouse is proving to be an excellent small animal model for preclinical studies addressing both the efficacy and safety of HSC-based anti-HIV therapies [6, 35–39].

An in vivo anti-HIV effect of CCR5 ZFN engineering was first demonstrated in the studies of Perez et al. [12], following transplantation of ZFN-modified T cells into NSG mice and subsequent challenge with a R5-tropic strain of HIV-1. Here, the ZFN-treated population of T cells preferentially survived and expanded two to threefold compared to unmodified cells in matched cohorts of mice, and the ZFN-treated mice had a statistically significant reduction in HIV-1 levels in peripheral blood. These demonstrations of efficacy supported the initiation of human clinical trials [40], where large-scale methods have been developed that allow both ZFN editing and ex vivo expansion of $CD4^+$ T cells, generating yields of $>10^{10}$ CCR5-edited $CD4^+$ T cells from individual patients [41].

Despite the relative ease of ZFN editing and ex vivo expansion of $CD4^+$ T cells, the finite life-span of the cells, together with the fact that HIV-1 also infects $CD4^+$ cells of myeloid lineages, supports the development of HSC engineering as a way to provide a more long-term and comprehensive protection. Using HSC purified by $CD34^+$ selection from cord blood, we are able to deliver anti-CCR5 ZFNs by either plasmid DNA nucleofection or adenovirus vector delivery, and subsequently engraft the modified HSC into 1-day old neonatal NSG mice (Fig 32.2a). HSC engraftment is promoted by irradiation of the mice with a sublethal dose of 150 cGy total radiation, and the cells can be delivered by facial vein or intrahepatic injection. Meaningful information about the kinetics, quantity, and subsets of human cells that develop can then be achieved by peripheral blood analysis, starting at 8 weeks of age. Typically, we find that human $CD4^+$ T cells appear by 8–12 weeks, and that the animals support R5-tropic HIV-1 infection when injected at 12 weeks (Fig. 32.2a). By following the levels of virus and human CD4+ and CD8+ T cells in the blood of the animals over time, an anti-HIV activity can be observed for the ZFN-modified cells, by both the marked inhibition of R5-tropic HIV-1 replication, as well as preservation of a CD4:CD8 T cell ratio of about 1. This contrasts with the situation in control mice receiving unmodified human HSC, where the detrimental effects of HIV-1 replication on T cell subsets are very apparent (Fig. 32.2b). At the end of such experiments, the human cells that persist in the ZFN cohorts are found to have undergone strong selection for CCR5-negative cells [14]. This suggests that HIV-1 replication itself could be used to increase the percentage of HIV-resistant, CCR5-negative cells in the recipients of these therapies, and temporary antiretroviral drug treatment interruptions have been included in clinical trials of ZFN-modified T cells [40].

Using control and ZFN Nucleofected HSC mice, we have also demonstrated that this modification is effective against multiple different strains of R5-tropic HIV-1 but, as expected, has no impact against an X4-tropic virus (Fig. 32.2c). Consequently, the selection for a X4-tropic strain of HIV-1 is a possible limitation of this type of therapy, which can be managed by careful selection of patients that do not have

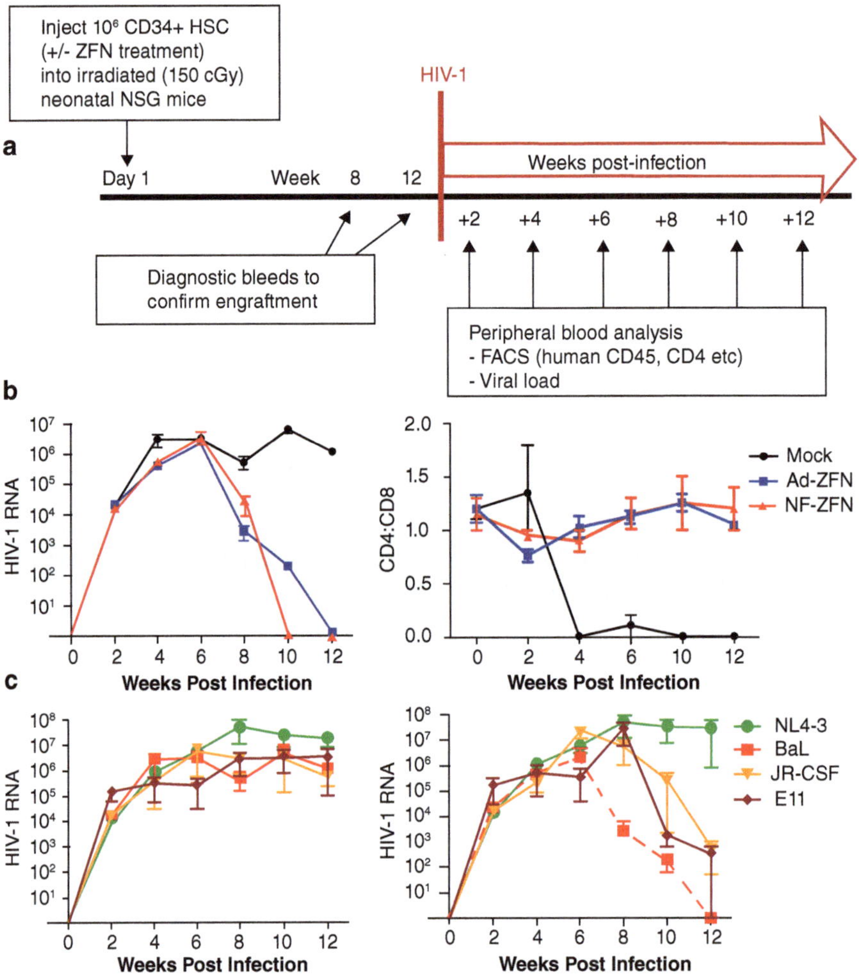

Fig. 32.2 **a** Time-line for generation of humanized NSG mice and infection by HIV-1. The kinetics of HIV-1 infection are monitored in serial blood samples, using FACS analysis of human cells and quantitative RT-PCR to measure virus levels, as described [14]. **b** Humanized NSG mice were engrafted with either unmodified cord blood HSC (Mock, $n=4$), or cells treated with CCR5-ZFNs, delivered by either nucleofection of plasmid DNA (NF, $n=3$) or following transduction with an adenovirus vector at an MOI of 600 (Ad, $n=4$). Following challenge with R5-tropic HIV-1 strain BaL, both ZFN cohorts were able to suppress virus replication (*left*) and maintain normal ratios of CD4:CD8 cells (*right*). In contrast, Mock HSC mice maintained high levels of HIV-1 and became highly depleted for $CD4^+$ T cells. **c** HIV-1 challenges were performed on mock HSC engrafted mice (*left panel*) or mice engrafted with CCR5 ZFN nucleofected cells (*right panel*), using three different R5-tropic strains of HIV-1 (*BaL, JR-CSF, and E11*) and one X4-tropic strain, NL4-3. The ZFN-treated mice were able to suppress all three R5-tropic strains, but not NL-3. $n=4$ mice per group

preexisting X4 virus, and ongoing monitoring. In addition, other groups are evaluating the use of CXCR4 targeted ZFNs in humanized mouse models [42, 43], and although the importance of the CXCR4 molecule for HSC biology means that such engineering could not be used with HSC, it may be possible to combine CXCR4 ZFN editing of $CD4^+$ T cells with CCR5 ZFN editing of either T cells and/or HSC.

The experiments described above can evaluate the anti-HIV efficacy of these genetic modifications, but do so in the rather artificial situation of the treatment occurring before HIV-1 infection. This is a limitation of the model, which does not mimic the human therapeutic scenario, where HSC or T-cell engineering would be performed on an already infected individual, and probably in the setting of antiretroviral drugs that have suppressed virus replication. These limitations stem from the requirement to irradiate and transplant the mice twice in order to (1) engraft a human immune system with which to establish an HIV-1 infection and (2) to allow replacement with the engineered immune system. Multiple rounds of radiation and transplantation will be toxic, and the combined time needed to allow mature $CD4^+$ T cells to develop, for an HIV-1 infection to establish, and possibly also for antiretroviral therapy to suppress HIV-1 replication prior to the therapeutic transplantation, would also make such a model a practical challenge.

32.5 Using Humanized Mice to Evaluate the Safety of ZFN-Treated HSC

Studies to evaluate anti-HIV activities, such as those described above, are typically performed using humanized mice engrafted with HSC purified from fetal liver or cord blood. However, for autologous cell therapy applications, the clinically relevant HSC is more usually purified from mobilized blood. Therefore, it is also important that the effect of ZFNs on the function of mobilized HSC is evaluated as part of the preclinical safety studies.

In our experience using NSG mice, mobilized HSC do not readily engraft neonatal mice, so that a more reliable adult (8 week old) engraftment model needs to be used [19, 44]. These mice are delayed in the development of T cells compared to neonatally engrafted mice, and robust and long-lived HIV infections are not as easy to establish or maintain. In addition, we do not observe the rapid $CD4^+$ T cell depletion following HIV-1 infection that is a feature of mice transplanted as neonates with cord blood or fetal liver derived HSC (Fig. 32.2b). However, the mice provide an adequate model to assess HSC function, by evaluating both the quantity of human cells that engraft, as well as the quality of hematopoiesis that they support, defined as both the kinetics and absolute levels of subsequent lineage differentiation. In this way, mice receiving unmodified or ZFN-treated HSC from the same donor can be compared in cohorts containing multiple animals, to allow statistical analyses [19]. In previous studies using adenovirus-treated mobilized HSC, we were able to use the mice to identify a dose and treatment protocol that provided levels of CCR5 disruption of $>5\,\%$ without an adverse impact on either the absolute levels of engraftment of the human cells, or the quality (kinetics, sub-sets) of human hematopoiesis. In this way humanized mouse studies can complement and extend the more typical in vitro assays of HSC function such as CFU analyses.

In addition, the ability of mobilized blood HSC to engraft, differentiate and undergo considerable expansion in NSG mice means that they can also be used for at-scale toxicology studies. For example, using the calculation that a whole patient dose of CD34$^+$ HSC is 1.5×10^8CD34$^+$ cells ($\geq 2.0 \times 10^6$ cells/kg, 70 kg person) [45], and our experience that 1–2 million HSC per adult NSG mouse reliably engraft and are well tolerated, then a whole patient dose equivalent could be evaluated by engrafting 75–150 animals. Importantly, because the model is incompletely characterized in terms of the spontaneous development of either mouse or human hematopoietic tumors, a control group receiving irradiation and unmodified human HSC will also be necessary. Both groups of mice can be monitored over several weeks for general health and weight and at necropsy, gross macroscopic examination, organ weights, and histopathological examinations can be conducted to assess whether any tumors arise. The model has the advantage that any tumors that may develop can easily be determined to be of murine or human-cell origin and, in the case of ZFN-treated cells, any insertion or deletion signatures of modification at the CCR5 target site could also be used as markers to identify monoclonal outgrowths of engineered cells. Similar biotoxicity studies were undertaken using mice engrafted with ZFN-edited T cells [41].

32.6 Summary

HSC or T cell transplanted humanized mice are an exceptionally well-suited model for the evaluation of anti-HIV gene therapies, and are providing preclinical safety and efficacy data to support the first-in-human uses of engineered nucleases. For both T cell and HSC engineering, a key advantage of the model is the ability to use the authentic human target cell and clinical ZFN reagent formulations, and to assess the impact the edited cells have on an HIV-1 infection in a complex, in vivo environment. However, a limitation of these models is that the ZFN engineering is performed prior to the establishment of an HIV-1 infection, and the mice cannot at present be used to evaluate the effect of these therapies in the context of a preestablished infection, as would be the situation in patients. However, as more therapies based on ZFNs and other classes of engineered nucleases are developed, it is likely that humanized mice will continue to play an important role in preclinical evaluations.

Acknowledgments This work was supported by the National Institutes of Health research grant HL073104 and the California HIV/AIDS Research Program (CHRP) grant ID12-USC-245 the James B. Pendleton Charitable Trust. CE was supported by a fellowship from the CHRP, F10-USC-207, UH was supported by a fellowship from the Swiss National Science Foundation, and OM was supported by a fellowship from the California Institute for Regenerative Medicine. The article is dedicated to the memory of our friend and colleague, Kathy Burke.

References

1. Shultz LD, et al. Human lymphoid and myeloid cell development in NOD/LtSz-scid IL2R gamma null mice engrafted with mobilized human hematopoietic stem cells. J Immunol. 2005;174:6477–89.
2. Ishikawa F, et al. Development of functional human blood and immune systems in NOD/SCID/IL2 receptor γ chain^{null} mice. Blood. 2005;106:1565–73.
3. Wu L, et al. CD4-induced interaction of primary HIV-1 gp120 glycoproteins with the chemokine receptor CCR-5. Nature. 1996;384:179–83.
4. Samson M, et al. Resistance to HIV-1 infection in Caucasian individuals bearing mutant alleles of the CCR-5 chemokine receptor gene. Nature. 1996;382:722–5.
5. Wood A, Armour D. The discovery of the CCR5 receptor antagonist, UK-427,857, a new agent for the treatment of HIV infection and AIDS. In: King FD, Lawton G, editors. Prog Med Chem; 2005;43:239–71.
6. Shimizu S, et al. A highly efficient short hairpin RNA potently down-regulates CCR5 expression in systemic lymphoid organs in the hu-BLT mouse model. Blood. 2010;115:1534–44.
7. Allers K, et al. Evidence for the cure of HIV infection by CCR5Δ32/Δ32 stem cell transplantation. Blood. 2010;117:2791–99.
8. Hütter G, et al. Long-term control of HIV by CCR5 Delta32/Delta32 stem-cell transplantation. N Engl J Med. 2009;360:692–8.
9. Yukl SA, et al. Challenges in detecting HIV persistence during potentially curative interventions: a study of the Berlin patient. PLoS Pathog. 2013;9:e1003347.
10. Cannon P, June C. Chemokine receptor 5 knockout strategies. Curr Opin HIV AIDS. 2011;6:74–9.
11. Gaj T, et al. ZFN, TALEN, and CRISPR/Cas-based methods for genome engineering. Trends Biotechnol. 2013;31:397–405.
12. Perez EE, et al. Establishment of HIV-1 resistance in CD4+ T cells by genome editing using zinc-finger nucleases. Nat Biotechnol. 2008;26:808–16.
13. Shayklmetov DM, et al. Efficient gene transfer into human CD34+ cells by retargeted adenovirus vector. J Virol. 2000;74:2567–83.
14. Holt N, et al. Human hematopoietic stem/progenitor cells modified by zinc-finger nucleases targeted to CCR5 control HIV-1 in vivo. Nat Biotechnol. 2010;28:839–47.
15. Joglekar AV, et al. Integrase-defective lentiviral vectors as a delivery platform for targeted modification of the adenosine deaminase locus. Mol Ther. 2013;21:1705–17.
16. Lei Y, et al. Gene editing of human embryonic stem cells via an engineered baculoviral vector carrying zinc-finger nucleases. Mol Ther. 2011;19:942–50.
17. Li H, et al. In vivo genome editing restores haemostasis in a mouse model of haemophilia. Nature. 2011;475:217–21.
18. Lombardo A, et al. Gene editing in human stem cells using zinc finger nucleases and integrase-defective lentiviral vector delivery. Nat Biotechnol. 2007;25:1298–306.
19. Li L, et al. Genomic editing of the HIV co-receptor CCR5 in adult hematopoietic stem and progenitor cells using zinc finger nucleases. Mol Ther. 2013;21:1259–69.
20. Körbling M, Anderlini P. Peripheral blood stem cell versus bone marrow allotransplantation: does the source of hematopoietic stem cells matter? Blood. 2001;98:2900–08.
21. Sica S, et al. G-CSF and peripheral blood progenitor cells. Lancet. 1992;339:1411.
22. Cannon PM, et al. Electroporation of ZFN mRNA enables efficient CCR5 gene disruption in mobilized blood hematopoietic stem cells at clinical scale. Mol Ther. 2013;21:S71–S2.
23. Traggiai E, et al. Development of a human adaptive immuse system in cord blood cell-transplanted mice. Science. 2004;304:104–7.
24. Baenziger S, et al. Disseminated and sustained HIV infection in CD34+ cord blood cell-transplanted Rag2-/-gamma c-/- mice. Proc Natl Acad Sci U S A. 2006;103:15951–56.
25. Watanabe S, et al. Hematopoietic stem cell-engrafted NOD/SCID/IL2Rγ^{null} mice develop human lymphoid systems and induce long-lasting HIV-1 infection with specific humoral immune responses. Blood. 2007;109:212–8.

26. Berges B, et al. HIV-1 infection and CD4 T cell depletion in the humanized Rag2$^{-/-}\gamma c^{-/-}$ (RAG-hu) mouse model. Retrovirology. 2006;3:76.
27. Nie C, et al. Selective infection of CD4+ effector memory T lymphocytes leads to preferential depletion of memory T lymphocytes in R5 HIV-1-infected humanized NOD/SCID/IL2-Rγ^{null}mice. Virology. 2009;394:64–72.
28. McCune J, et al. The SCID-hu mouse: murine model for the analysis of human hematolymphoid differentiation and function. Science. 1988;241:1632–9.
29. Melkus MW, et al. Humanized mice mount specific adaptive and innate immune responses to EBV and TSST-1. Nat Med. 2006;12:1316–22.
30. Billerbeck E, et al. Development of human CD4+ FoxP3+ regulatory T cells in human stem cell factor–, granulocyte-macrophage colony-stimulating factor–, and interleukin-3–expressing NOD-SCID IL2Rγ^{null} humanized mice. Blood. 2011;117(11):3076–86.
31. Brehm MA. Engraftment of human HSCs in nonirradiated newborn NOD-scid IL2rγ null mice is enhanced by transgenic expression of membrane-bound human SCF. Blood. 2012;119:2778–88.
32. Rongvaux A, et al. Human thrombopoietin knockin mice efficiently support human hematopoiesis in vivo. Proc Natl Acad Sci U S A. 2011;108:2378–83.
33. Shultz LD, et al. Generation of functional human T-cell subsets with HLA-restricted immune responses in HLA class I expressing NOD/SCID/IL2rγ^{null} humanized mice. Proc Natl Acad Sci U S A. 2010;107:13022–7.
34. Strowig T, et al. Transgenic expression of human signal regulatory protein alpha in Rag2$^{-/-}\gamma_c^{-/-}$ mice improves engraftment of human hematopoietic cells in humanized mice. Proc Natl Acad Sci U S A. 2011;108:13218–23.
35. Anderson J, et al. Safety and efficacy of a lentiviral vector containing three anti-HIV genes-CCR5 ribozyme, Tat-rev siRNA, and TAR decoy-in SCID-hu mouse-derived T cells. Mol Ther. 2007;15:1182–8.
36. Hauber I, et al. Highly significant antiviral activity of HIV-1 LTR-specific Tre-recombinase in humanized mice. PLoS Pathog. 2013;9:e1003587.
37. Joseph A, et al. Inhibition of in vivo HIV infection in humanized mice by gene therapy of human hematopoietic stem cells with a lentiviral vector encoding a broadly neutralizing anti-HIV antibody. J Virol. 2010;84:6645–53.
38. Kitchen SG, et al. In vivo suppression of HIV by antigen specific T cells derived from engineered hematopoietic stem cells. PLoS Pathog. 2012;8:e1002649.
39. Walker JE, et al. Generation of an HIV-1 resistant immune system with CD34+ hematopoietic stem cells transduced with a triple-combination anti-HIV lentiviral vector. J Virol. 2012;86:5719–29.
40. Tebas P, et al. Gene editing of CCR5 in autologous CD4 T cells of persons infected with HIV. N Engl J Med. 2014;370:901–10.
41. Maier DA, et al. Efficient clinical scale gene modification via zinc finger nuclease-targeted disruption of the HIV co-receptor CCR5. Hum Gene Ther. 2013;24:245–58.
42. Wilen CB, et al. Engineering HIV-resistant human CD4+ T cells with CXCR4-specific zinc-finger nucleases. PLoS Pathog. 2011;7:e1002020.
43. Yuan J, et al. Zinc-finger nuclease editing of human CXCR4 promotes HIV-1 CD4+ T cell resistance and enrichment. Mol Ther. 2012;20:849–59.
44. Hofer U, et al. Pre-clinical modeling of CCR5 knockout in human hematopoietic stem cells by zinc finger nucleases using humanized mice. J Infect Dis. 2013;208:S160–4.
45. Tricot G, et al. Peripheral blood stem cell transplants for multiple myeloma: identification of favorable variables for rapid engraftment in 225 patients. Blood. 1995;85:588–96.

Chapter 33
Hematopoietic Progenitor Cell Transduction by a Unique Short Hairpin RNA to Chemokine Receptor 5

Saki Shimizu, Erica Eggers and Dong Sung An

33.1 Introduction

Highly active anti retroviral therapy (HAART), the current treatment for HIV-infected individuals, has made significant improvement in the survival and quality of life for HIV-infected individuals [1]. However, daily administration of multiple drugs can only control infection, not provide eradication of virus [2, 3]. If the drug treatment is stopped, the majority of well-controlled individuals can experience a rapid rebound in viral load due to HIV reactivation from chronically infected viral reservoirs [4]. Thus far, the first clinical case of a functional HIV cure in an adult was demonstrated by transplantation of allogeneic bone marrow cells with homozygous CCR5 Δ32/Δ32 gene mutation in an HIV-infected patient with leukemia [5]. HIV RNA and DNA have continued to be undetectable for more than 5 years even in the absence of anti-HIV drug treatment [6]. Because of the first case of cure in the HIV-infected patient, investigation of novel therapeutic strategies to stably control or potentially cure HIV infection without continuous treatment has become an intensive research subject [7–11]. One strategy is to replicate the cure by using CCR5 Δ32/Δ32 homozygous bone marrow transplants in other patients. However, CCR5 Δ32/Δ32 exists in only 1 % of the Caucasian population [12–14]

D. Sung An (✉) · S. Shimizu
Division of Hematology-Oncology, The David Geffen School of Medicine at University of California, Los Angeles, USA
e-mail: an@ucla.edu

D. Sung An · E. Eggers
School of Nursing, University of California, 188(06) BSRB, 615 Charles E. Young Drive South, Los Angeles, CA 90095, USA

D. Sung An · S. Shimizu · E. Eggers
The UCLA AIDS Institute, Los Angeles, USA

L. Y. Poluektova et al. (eds.), *Humanized Mice for HIV Research*,
DOI 10.1007/978-1-4939-1655-9_33

and identification of human leukocyte antigen (HLA) matched CCR5 Δ32/Δ32 bone marrow donors may therefore be extremely challenging. To overcome this limitation, we and others have investigated genetic engineering of human HSPCs to replicate the function of the CCR5 Δ32/Δ32 mutation to potentially produce a self-renewing reservoir of HIV resistant cells that would persist for the entire life of the patient [15–22]. Successful genetic engineering of HSPCs would enable us to use autologous patient-derived HSPCs to maximize engraftment efficiency and reduce transplant complications. Our group investigates RNA interference (RNAi) to downregulate CCR5 expression [23–29]. RNAi is a natural defense mechanism in which a small double-stranded RNA induces sequence-specific degradation of its target mRNA [30]. Anti-HIV small interfering RNA (siRNA)/short hairpin RNA (shRNA) have been developed as gene therapeutic reagents due to their small size, ease of manipulation, and ability to combine with other reagents [23]. We have identified a potent and nontoxic shRNA against CCR5 capable of protecting human primary T lymphocytes from CCR5 tropic HIV infection [24]. To achieve long-term stable CCR5 downregulation by an shRNA, we extensively investigated a lentiviral vector-mediated transduction of this potent and nontoxic shRNA against CCR5 into human primary CD4+ T lymphocytes, CD34+ HSPC-derived monocyte/macrophages in vitro [23, 27, 28]. We were the first group to report multi-lineage differentiation of shRNA-transduced CD34+ HSPC and stable, efficient CCR5 downregulation in HIV target cells including CD4+ T lymphocytes and monocyte/macrophages in primary and secondary lymphoid organs in vivo in the humanized BLT mouse model [26].

RNA Interference to Efficiently Knock Down CCR5 Expression to Protect CD4+ T Lymphocytes from HIV Infection In Vitro We originally reported CCR5 downregulation by RNAi in human peripheral blood mononuclear cells (PBMC) and CCR5 tropic HIV inhibition in vitro [29]. To efficiently introduce shRNAs into PBMC, we utilized a vesicular stomatitis virus G protein (VSV-G) pseudotyped lentiviral vector. The expression of shRNA resulted in tenfold inhibition of CCR5 expression. Blocking CCR5 expression by the shRNA-inhibited CCR5-tropic HIV infection by three- to sevenfold in vitro. These initial in vitro studies established that specific shRNA homologous to CCR5 mRNA could downregulate CCR5 in PBMC and protect PBMC from CCR5-tropic HIV infection.

Optimization of Stable shRNA Expression in Human PBMC Without Causing Cytotoxic Effects In our initial study, we selected an shRNA against CCR5 from a total of eight candidate shRNAs [29]. High levels of shRNA expression were required to efficiently downregulate CCR5 expression in PBMC. To this end, we utilized a transcriptionally strong RNA polymerase III U6 promoter to drive shRNA expression from a lentiviral vector. Despite our initial success in downregulating CCR5 by shRNAs, we and others have found that the majority of shRNAs exhibit cytotoxicity in human PBMC [25, 31, 32]. We reported a twofold decrease in shRNA-transduced PBMC over a 2-week culture [25]. The cytotoxicity was not directly related to CCR5 downregulation, as we observed similar results with other shRNAs directed to irrelevant targets such as luciferase and LacZ [25]. The mechanical understandings of shRNA-induced cytotoxic effect have been published by other investigators and may be caused by competition with endogenous microRNA biogenesis, induction of interferon responses, and/or off-targeting effects [33–39]. To avoid this cytotoxicity, we expressed shRNAs using the transcriptionally weaker H1 promoter to reduce shRNA expression and subsequent cytotoxicity in PBMC in vitro. Importantly, unlike shRNAs expressed from the strong U6 promoter, expression of shRNAs from the H1 promoter did not alter the growth of shRNA-transduced PBMC in vitro. These results demonstrated that the level of shRNA expression is important for determining whether shRNAs will be cytotoxic or not and that optimization of those levels can yield shRNAs that are both effective and minimally cytotoxic. We reported that the choice of promoters and vector designs were important for efficient, stable, and safe expression of shRNA [25].

Identification of a Potent shRNA Against CCR5 (sh1005) from a Random shRNA Library Our optimization study for shRNA expression using the transcriptionally weaker H1 promoter alleviated the cytotoxic effects, but lowering the shRNA expression also reduced the level of CCR5 downregulation. Because the potency of shRNA-mediated gene downregulation is dependent on the given shRNA sequence [40], we decided to screen a large number of shRNAs directed to CCR5 mRNA. We developed a random shRNA library specific to CCR5 mRNA sequence by adopting an enzymatic production of RNAi library (EPRIL), in which cDNAs are converted by a sequence of enzymatic treatments into a RNAi library consisting of a vast array of different shRNA expression lentiviral vector constructs [24, 41]. sh1005, the most potent shRNA to date of more than 380 shRNAs characterized in our laboratory, downregulated CCR5 expression more than 20 fold in human primary PBMC even though it was expressed from a transcriptionally weaker H1 RNA polymerase III promoter [24].

Establishment of the Humanized BLT Mouse Model for In Vivo Evaluation of HSPC-Based Gene Therapy Strategies In vivo investigation of therapeutic strategies for human pathogens including HIV requires HIV-susceptible animal models. A number of humanized mouse models were created and each humanized model contributed significantly to the in vivo investigation of HIV infection [42–44]. The BLT mouse model is a new generation of the improved humanized mouse model [45]. The BLT mouse is transplanted with human CD34+ HSPC and the human fetal thymus-like organoid (thy/liv), which allows a range of human immune cells to be reconstituted at high efficiency, including HIV target cells such as CD4+ T lymphocytes with a diverse T cell receptor repertoire and monocyte/macrophages in primary and secondary lymphoid organs. Among the many humanized mouse models, BLT mice reconstituted with human thymus/liver and CD34+ HSPC have recently been extensively used for in vivo investigations of HIV infection in lymphoid organs [46, 47], mucosal transmission, and functional immune responses against HIV [48, 49]. Therefore, we thought that this mouse model would be an ideal small animal model to investigate the potential for genetically engineered HSPCs to resist HIV infection. Successful development of the model enables us to determine how effectively human HSPCs transduced with lentiviral vectors bearing anti-HIV genes inhabit bone marrow and undergo human thymopoiesis in the human thy/liv implant to differentiate into human lymphocytes and monocytes/macrophages, how they migrate into and survive within the systemic lymphoid organs, the safety of this approach and the mechanisms to control HIV, prior to human application. As described in the following sections, we were the first group to utilize the BLT mouse model to test HSPC-based gene therapy strategies using an shRNA against CCR5 and published in *Blood* [26].

Examination of Engraftment of sh1005 Transduced HSPCs in Lymphoid Tissues of the BLT Mouse To efficiently introduce sh1005 into fetal liver-derived human CD34+ cells, we used VSV-G pseudotyped lentiviral vectors. The efficiency of vector transduction in CD34+ cells was measured by EGFP expression in ex vivo cultured cells and the mean EGFP expression was 50.8 %. As a control, half of CD34+ cells were transduced with non-shRNA control vector marked with mCherry. The mean mCherry expression was 42.9 %. To examine sh1005 in the humanized BLT mouse model, we first transplanted vector-transduced fetal liver-derived CD34+ cells and CD34− cells solidified with matrigel and a fetal thymus segment from the same donor under the kidney capsule to generate a vector-transduced human thymus tissue in NOD.CB17-*Prkdcscid*/J (NOD SCID) mouse (later we used NOD.Cg-*Prkdcscid Il2rgtm1Wjl*/SzJ (NSG) mice) (Fig. 33.1). Three weeks later, frozen fetal liver CD34+ cells from the same donor were thawed and transduced with lentiviral vectors. The efficiency of vector transduction in CD34+ cells measured by mean EGFP expression was 45.9 %. The mean mCherry expression in the control vector-transduced CD34+ cells was 28.3 %. Vector-transduced CD34+ HPSC were injected intravenously into the sub-lethally irradiated mouse. To examine stable CCR5 downregulation by sh1005 in the BLT mice, we co-transplanted an equal mix of sh1005 vector (EGFP+) and non-shRNA control vector (mCherry+) transduced CD34+ cells. Because both vectors are present within the same animal, this experimental design allows us to examine responses specific to the anti-HIV gene as well as to control for mouse-to-mouse variation.

Multi-lineage Hematopoietic Cell Differentiation HSPC-based gene therapy requires efficient multi-lineage hematopoietic system reconstitution with cells stably expressing anti-HIV genes. We therefore evaluate in vivo multi-lineage differentiation by assessing the differentiated phenotypes of vector-transduced cells. Differentiated human hematopoietic cell engraftment was examined from 11 weeks post-CD34+ injection in peripheral blood. Human CD45+ lymphoid cells were detected in a gated lymphocyte population of peripheral blood from transplanted mice by flow cytometric analysis. EGFP and mCherry expression were found in this human CD45+ population in transplanted mice (average %EGFP ± Standard deviation was 24.2 ± 12.8 %, average %mCherry ± Standard deviation was 19.0 ± 11.7 %). In our follow-up studies, human CD3+, CD4+, CD8+ T lymphocytes, CD19+ B lymphocytes as well as CD14+/CD33+ monocyte/macrophage populations were also detected in transplanted mice. These results demonstrated that sh1005 is nontoxic for human HSPCs and their multi-lineage differentiation.

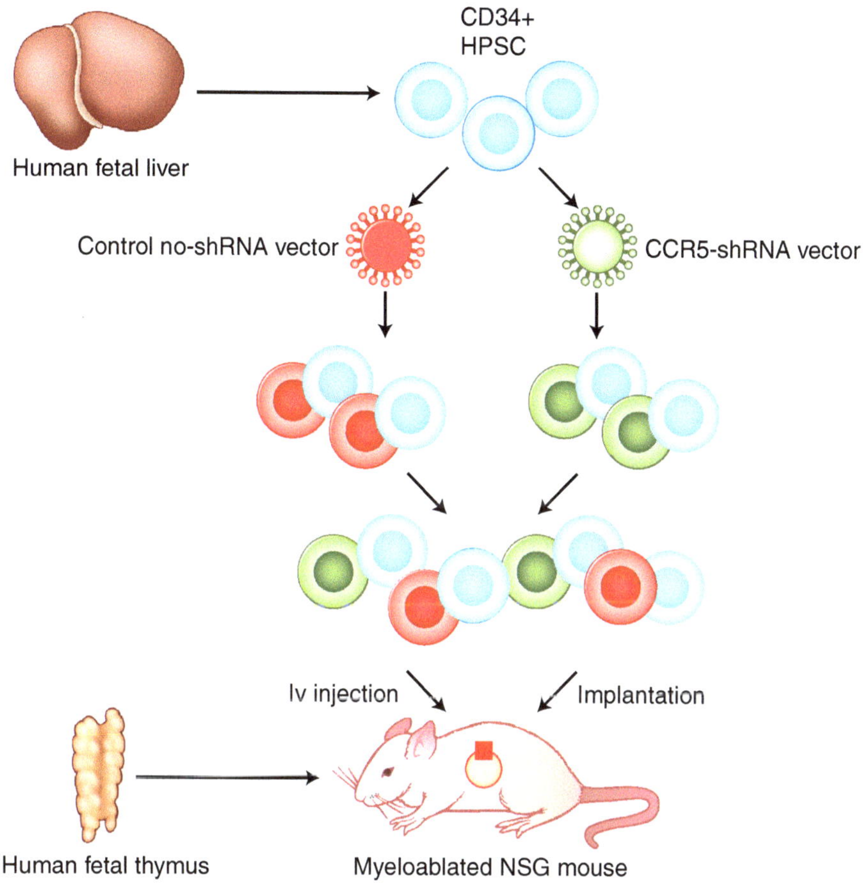

Fig. 33.1 Modeling anti-HIV-1 HSPC gene therapy strategies in hu BLT mouse model. CD34+ hematopoietic stem/progenitor cells (*HSPCs*) isolated from a fetal liver are transduced with a lentiviral vector expressing a short hairpin RNA (*shRNA*) directed to human CCR5. For an internal control, half of CD34+ HSPC are transduced with a control vector without shRNA. CCR5 shRNA vector and the control vector are marked with EGFP or mCherry, respectively, to monitor each vector-transduced cell by multi-color flow cytometry. Vector-transduced CD34+ HSPC are mixed and implanted under the kidney capsule with a piece of thymus. At the same time or 3 weeks later, vector-transduced CD34+ HSPC are intravenously injected in NSG mice after myeloablative conditioning by irradiation or busulfan. HSPC engraftment and appearance of differentiated cell in peripheral blood typically starts at 6–8 weeks post CD34+ cell transplant. Engraftment of anti-HIV-1 gene-transduced HSPCs, multi-linage differentiation, T cell receptor (*TCR*) rearrangement in thymus and spleen, memory T cell development in tissues including the gut associated lymphoid tissue, functional effects of anti-HIV-1 transgene such as CCR5 downregulation in CD4+ T lymphocytes and macrophages, long term CCR5 downregulation in second transplanted mice, CD4+ cell protection, and HIV-1 inhibition can be investigated in the reconstituted hu BLT mice. Adapted from Shimizu et al., a highly efficient short hairpin RNA potently downregulates CCR5 expression in systemic lymphoid organs in the hu-BLT mouse model. BLOOD, 25 FEBRUARY 2010 _ VOLUME 115, NUMBER 8

T Cell Development in Human Thy/liv in the BLT Mice One of the innovative aspects of the BLT mouse model is the ability assess human T cell development in engrafted human thymic tissue. We examined diversity within the T cell receptor repertoire as a marker for T cell development [26]. We examined T cell receptor (TCR) vα and vβ rearrangement using a quantitative PCR-based TCR spectratyping assay in sorted EGFP+ and mCherry+ thymocytes from human thymus organoid. We compared the profile and peak distribution of each TCRv α and β and examined Gaussian distribution in both EGFP+ and mCherry+ sorted populations. These experiments determined polyclonal human TCR development in the BLT mouse model.

Memory T Cell Development in the BLT Mice To examine the effect on naïve and memory T cell differentiation, we compared CD27 and CD45RA expression in EGFP+ and mCherry+ CD3+T cells in multiple lymphoid organs including peripheral blood, spleen, bone marrow, lung, liver, and small intestine. There was no significant difference in naïve and memory T cell differentiation based on CD27 and CD45RA expression between the two populations, suggesting sh1005 did not have apparent adverse effects on T cell differentiation. These experiments demonstrated that sh1005-transduced human HSPCs could differentiate into mature immune cells in T cells in lymphoid organs in the BLT mice.

In Vivo Examination of CCR5 Downregulation in CD4+ T Lymphocytes in Lymphoid Organs CCR5-knockdown was examined in human CD4+ T lymphocytes in various lymphoid tissues in reconstituted animals at 14–20 weeks post CD34+ HPSC transplant. CCR5 expression was efficiently reduced in EGFP+CD4+T-lymphocytes in lymphoid tissues including peripheral blood, human thymus organoid, spleen, bone marrow, lung, lymph nodes, liver, and small intestines. CCR5 reduction was efficient even in the naturally highly CCR5-expressing lamina propria lymphocytes isolated from the gut, a major target of HIV in new infection. In contrast, CCR5 was not reduced in the control mCherry+ human CD4+T-lymphocytes in the same animal. These results along with the above indicated that sh1005 expression did not affect human T-lymphocyte differentiation and migration and effectively induced CCR5 downregulation in systemic lymphoid organs in vivo.

In Vivo Examination of CCR5 Downregulation in Monocyte/Macrophages in Lymphoid Organs Monocyte/macrophage populations are another major target for R5 tropic HIV infection. HSPC-based gene therapy could genetically modify not only CD4 progeny cells but also monocyte/macrophage progeny. CCR5-knockdown was examined in human CD14+/CD33+ monocytes/macrophages in lymphoid tissues in the reconstituted huBLT mice at 14–20 weeks post CD34+HPSC transplant [26]. CCR5 expression was efficiently reduced in EGFP+ human CD14+/CD33+ Monocytes/macrophages in peripheral blood, spleen, and lung. In contrast, CCR5 was not reduced in the control mCherry+ human CD14+/CD33+ Monocytes/macrophages in the same animal. These results along with the above results demonstrated that *shRNA 1005* was capable of downregulating CCR5 expression in the two major HIV target cell populations including monocyte/macrophages and CD4+ T lymphocytes in systemic lymphoid organs in vivo.

Long Term Hematopoietic Repopulation of shRNA 1005-Transduced HSPCs and CCR5 Downregulation in Serially Transplanted Mice In order to determine if this therapeutic strategy can lead to stable control or cure from a single treatment, it is important to investigate the long-term repopulating potential of gene-modified HSPCs. We assessed this using a serial transplantation of bone marrow cells from a primary BLT mouse to a secondary BLT mouse [26]. We isolated bone marrow cells from the femurs of sh1005 vector-transduced HSPCs transplanted EGFP expressing BLT mice. The bone marrow donor cells were then injected both into a thy/liv organoid for transplantation and intravenously in preconditioned secondary recipient mice. We examined the presence of CCR5 downregulated EGFP+ human CD4+ T lymphocytes in multiple tissues of the secondary transplanted mice from 14 weeks post bone marrow transplant. These results demonstrated that lentiviral vector transduction of HSPCs and sh1005 expression is capable of supporting long-term repopulation of the hematopoietic system and stable expression of anti-HIV genes in vivo in serially transplanted BLT mice.

CCR5 Tropic HIV Is Inhibited in Isolated CCR5 Downregulated Splenocytes in Ex Vivo Culture To examine HIV inhibition in CCR5 downregulated cells generated in the BLT mice, isolated EGFP+ and mCherry+ splenocytes purified by cell sorting were infected with either R5 tropic $HIV_{NFNSX\ SL9}$ or X4 tropic HIV_{NL4-3} ex-vivo in culture. There was no increase in p24 HIV antigen capsid protein production in the culture supernatant of EGFP+ splenocytes over the 12-day culture period [26]. In contrast, p24 produced approximately fourfold higher levels in the culture supernatant on days 7 and 12 in mCherry+ splenocytes, suggesting they are susceptible to R5 tropic $HIV_{NFNSX\ SL9}$ (p value = 0.003). These results indicated that CCR5 downregulation effectively inhibited R5 tropic HIV infection. In contrast to the R5 tropic HIV infection, X4 tropic HIV_{NL4-3} infection produced comparable amounts of p24 in both EGFP+ and mCherry+ splenocyte culture supernatants. These results demonstrated that isolated CD4+ T lymphocytes from a BLT mouse spleen with CCR5 downregulation by sh1005 was sufficient to protect from R5 but not X4 tropic HIV challenge ex vivo culture. Current progress examining the in vivo efficacy of protected cells to evade HIV infection demonstrates the protection of cells in the BLT mice and will be submitted for a publication elsewhere.

In summary, we have identified a potent and nontoxic shRNA against CCR5 (sh1005) by developing and screening a CCR5-specific random shRNA library. The transduction of sh1005 into human HSPCs via a lentiviral vector and transplant in the humanized BLT mice resulted in efficient development of sh1005-modified human hematopoietic cells without apparent adverse effects. We have demonstrated stable downregulation of CCR5 in CD4+ T lymphocytes and monocyte/macrophages in lymphoid tissues in the gene modified CD34+ HSPC transplanted humanized BLT mouse model. Therefore, we conclude that sh1005 is efficient for CCR5 knockdown and nontoxic, and we have established the BLT mouse model system to test our therapeutic candidate vectors.

Since reports of the BLT mouse model in 2006 [45], this new generation of improved mouse/human chimera technology has become one of the major small animal models for the investigation of HIV infection, pathogenesis, prevention, and therapeutic interventions. Furthermore, recently Drs. Victor Garcia and Jerome Zack's laboratories reported HIV latently infected cells in the spleen of HIV-infected BLT mice [50, 51]. We are currently investigating if CCR5 knock down CD4+T lymphocytes are protected from establishment of chronic HIV infection in vivo in the HIV-challenged BLT mice. In theory, HIV inhibitors such as sh1005 that work at an early step of the HIV life cycle could effectively protect cells from viral entry, as evidenced by the functional cure by CCR5 Δ32/Δ32 bone marrow transplants. Therefore, our strategy may be able to protect cells from chronic infection and may reduce the number of latently infected cells due to the blockade of viral DNA integration. Since our original report of establishing the humanized BLT mouse model to test the CCR5 shRNA anti-HIV HSPC gene therapy strategy [26], several researchers have utilized the BLT mouse model to test their novel anti-HIV gene therapy strategies [15, 23, 46–49, 52–54]. These include a combination of shRNAs against CCR5 and HIV sequence [23], an engineering of anti-HIV IgA producing B cell immunity in mucosal tissues [52] and engineering anti-HIV T cell receptor expressing T cells through gene modified human CD34+ HSPC [54]. Furthermore, the BLT mice were also used to test for cancer therapeutic reagents using an anti-tumor genetically engineered TCR HSPC gene therapy approach [55]. We foresee that in vivo investigation in the BLT mouse model system will provide critical information for efficacy and safety of newly developed therapeutic strategies for HIV infection as well as other human diseases before their clinical applications.

Acknowledgements We would like to thank the National Institute of Allergy and Infectious Diseases (1R01AI100652-01A1), UCLA Center for AIDS Research (CFAR) (5P30 AI028697) and National Heart, Lung, and Blood Institute (1R01HL086409) for their support.

References

1. Ho DD. Time to hit HIV, early and hard. N Engl J Med. 1995;333(7):450–1. (Epub 1995/08/17).
2. Finzi D, Hermankova M, Pierson T, Carruth LM, Buck C, Chaisson RE, et al. Identification of a reservoir for HIV-1 in patients on highly active antiretroviral therapy. Science. 1997;278(5341):1295–300. (Epub 1997/11/21).
3. Chun TW, Stuyver L, Mizell SB, Ehler LA, Mican JA, Baseler M, et al. Presence of an inducible HIV-1 latent reservoir during highly active antiretroviral therapy. Proc Natl Acad Sci U S A. 1997;94(24):13193–7. (Epub 1997/12/16).
4. Chun TW, Engel D, Mizell SB, Ehler LA, Fauci AS. Induction of HIV-1 replication in latently infected CD4+ T cells using a combination of cytokines. J Exp Med. 1998;188(1):83–91. (Epub 1998/07/07).
5. Hutter G, Nowak D, Mossner M, Ganepola S, Mussig A, Allers K, et al. Long-term control of HIV by CCR5 Delta32/Delta32 stem-cell transplantation. N Engl J Med. 2009;360(7):692–8. (Epub 2009/02/14).
6. Allers K, Hutter G, Hofmann J, Loddenkemper C, Rieger K, Thiel E, et al. Evidence for the cure of HIV infection by CCR5Delta32/Delta32 stem cell transplantation. Blood. 2011;117(10):2791–9. (Epub 2010/12/15).

7. Stone D, Kiem HP, Jerome KR. Targeted gene disruption to cure HIV. Curr Opin HIV AIDS. 2013;8(3):217–23. (Epub 2013/03/13).
8. Burnett JC, Zaia JA, Rossi JJ. Creating genetic resistance to HIV. Curr Opin Immunol. 2012;24(5):625–32. (Epub 2012/09/19).
9. Lai Y. CCR5-targeted hematopoietic stem cell gene approaches for HIV disease: current progress and future prospects. Curr Stem Cell Res Ther. 2012;7(4):310–7. (Epub 2012/04/11).
10. Kiem HP, Jerome KR, Deeks SG, McCune JM. Hematopoietic-stem-cell-based gene therapy for HIV disease. Cell Stem Cell. 2012;10(2):137–47. (Epub 2012/02/07).
11. Kitchen SG, Shimizu S, An DS. Stem cell-based anti-HIV gene therapy. Virology. 2011;411(2):260–72. (Epub 2011/01/21).
12. Huang Y, Paxton WA, Wolinsky SM, Neumann AU, Zhang L, He T, et al. The role of a mutant CCR5 allele in HIV-1 transmission and disease progression. Nat Med. 1996;2(11):1240–3. (Epub 1996/11/01).
13. Lucotte G. Frequencies of the CC chemokine receptor 5 delta 32 allele in various populations of defined racial background. Biomed Pharmacother. 1997;51(10):469–73. (Epub 1997/01/01).
14. Rappaport J, Cho YY, Hendel H, Schwartz EJ, Schachter F, Zagury JF. 32 bp CCR-5 gene deletion and resistance to fast progression in HIV-1 infected heterozygotes. Lancet. 1997;349(9056):922–3. (Epub 1997/03/29).
15. Kim SS, Peer D, Kumar P, Subramanya S, Wu H, Asthana D, et al. RNAi-mediated CCR5 silencing by LFA-1-targeted nanoparticles prevents HIV infection in BLT mice. Mol Ther. 2010;18(2):370–6. (Epub 2009/12/10).
16. Holt N, Wang J, Kim K, Friedman G, Wang X, Taupin V, et al. Human hematopoietic stem/progenitor cells modified by zinc-finger nucleases targeted to CCR5 control HIV-1 in vivo. Nat Biotechnol. 2010;28(8):839–47. (Epub 2010/07/06).
17. Tamhane M, Akkina R. Stable gene transfer of CCR5 and CXCR4 siRNAs by sleeping beauty transposon system to confer HIV-1 resistance. AIDS Res Ther. 2008;5:16. (Epub 2008/08/01).
18. Qureshi A, Zheng R, Parlett T, Shi X, Balaraman P, Cheloufi S, et al. Gene silencing of HIV chemokine receptors using ribozymes and single-stranded antisense RNA. Biochem J. 2006;394(Pt 2):511–8. (Epub 2005/11/19).
19. Feng Y, Leavitt M, Tritz R, Duarte E, Kang D, Mamounas M, et al. Inhibition of CCR5-dependent HIV-1 infection by hairpin ribozyme gene therapy against CC-chemokine receptor 5. Virology. 2000;276(2):271–8. (Epub 2000/10/21).
20. Cagnon L, Rossi JJ. Downregulation of the CCR5 beta-chemokine receptor and inhibition of HIV-1 infection by stable VA1-ribozyme chimeric transcripts. Antisense Nucleic Acid Drug Dev. 2000;10(4):251–61. (Epub 2000/09/13).
21. Arteaga HJ, Hinkula J, van Dijk-Hard I, Dilber MS, Wahren B, Christensson B, et al. Choosing CCR5 or Rev siRNA in HIV-1. Nat Biotechnol. 2003;21(3):230–1. (Epub 2003/03/01).
22. Anderson JS, Walker J, Nolta JA, Bauer G. Specific transduction of HIV-susceptible cells for CCR5 knockdown and resistance to HIV infection: a novel method for targeted gene therapy and intracellular immunization. J Acquir Immune Defic Syndr. 2009;52(2):152–61. (Epub 2009/07/14).
23. Ringpis GE, Shimizu S, Arokium H, Camba-Colon J, Carroll MV, Cortado R, et al. Engineering HIV-1-resistant T-cells from short-hairpin RNA-expressing hematopoietic stem/progenitor cells in humanized BLT mice. PLoS One. 2012;7(12):e53492. (Epub 2013/01/10).
24. An DS, Donahue RE, Kamata M, Poon B, Metzger M, Mao SH, et al. Stable reduction of CCR5 by RNAi through hematopoietic stem cell transplant in non-human primates. Proc Natl Acad Sci U S A. 2007;104(32):13110–5. (Epub 2007/08/03).
25. An DS, Qin FX, Auyeung VC, Mao SH, Kung SK, Baltimore D, et al. Optimization and functional effects of stable short hairpin RNA expression in primary human lymphocytes via lentiviral vectors. Mol Ther. 2006;14(4):494–504. (Epub 2006/07/18).

26. Shimizu S, Hong P, Arumugam B, Pokomo L, Boyer J, Koizumi N, et al. A highly efficient short hairpin RNA potently down-regulates CCR5 expression in systemic lymphoid organs in the hu-BLT mouse model. Blood. 2010;115(8):1534–44. (Epub 2009/12/19).
27. Shimizu S, Kamata M, Kittipongdaja P, Chen KN, Kim S, Pang S, et al. Characterization of a potent non-cytotoxic shRNA directed to the HIV-1 co-receptor CCR5. Genet Vaccines Ther. 2009;7:8. (Epub 2009/06/12).
28. Liang M, Kamata M, Chen KN, Pariente N, An DS, Chen IS. Inhibition of HIV-1 infection by a unique short hairpin RNA to chemokine receptor 5 delivered into macrophages through hematopoietic progenitor cell transduction. J Gene Med. 2010;12(3):255–65. (Epub 2010/02/27).
29. Qin XF, An DS, Chen IS, Baltimore D. Inhibiting HIV-1 infection in human T cells by lentiviral-mediated delivery of small interfering RNA against CCR5. Proc Natl Acad Sci U S A. 2003;100(1):183–8. (Epub 2003/01/09).
30. Fire A, Xu S, Montgomery MK, Kostas SA, Driver SE, Mello CC. Potent and specific genetic interference by double-stranded RNA in Caenorhabditis elegans. Nature. 1998;391(6669):806–11. (Epub 1998/03/05).
31. Castanotto D, Sakurai K, Lingeman R, Li H, Shively L, Aagaard L, et al. Combinatorial delivery of small interfering RNAs reduces RNAi efficacy by selective incorporation into RISC. Nucleic Acids Res. 2007;35(15):5154–64. (Epub 2007/07/31).
32. Boudreau RL, Martins I, Davidson BL. Artificial microRNAs as siRNA shuttles: improved safety as compared to shRNAs in vitro and in vivo. Mol Ther. 2009;17(1):169–75. (Epub 2008/11/13).
33. Grimm D, Wang L, Lee JS, Schurmann N, Gu S, Borner K, et al. Argonaute proteins are key determinants of RNAi efficacy, toxicity, and persistence in the adult mouse liver. J Clin Invest. 2010;120(9):3106–19. (Epub 2010/08/11).
34. Fish RJ, Kruithof EK. Short-term cytotoxic effects and long-term instability of RNAi delivered using lentiviral vectors. BMC Mol Biol. 2004;5:9. (Epub 2004/08/05).
35. Jackson AL, Bartz SR, Schelter J, Kobayashi SV, Burchard J, Mao M, et al. Expression profiling reveals off-target gene regulation by RNAi. Nat Biotechnol. 2003;21(6):635–7. (Epub 2003/05/20).
36. Bridge AJ, Pebernard S, Ducraux A, Nicoulaz AL, Iggo R. Induction of an interferon response by RNAi vectors in mammalian cells. Nat Genet. 2003;34(3):263–4. (Epub 2003/06/11).
37. Kariko K, Bhuyan P, Capodici J, Weissman D. Small interfering RNAs mediate sequence-independent gene suppression and induce immune activation by signaling through toll-like receptor 3. J Immunol. 2004;172(11):6545–9. (Epub 2004/05/22).
38. Scacheri PC, Rozenblatt-Rosen O, Caplen NJ, Wolfsberg TG, Umayam L, Lee JC, et al. Short interfering RNAs can induce unexpected and divergent changes in the levels of untargeted proteins in mammalian cells. Proc Natl Acad Sci U S A. 2004;101(7):1892–7. (Epub 2004/02/11).
39. Sledz CA, Holko M, de Veer MJ, Silverman RH, Williams BR. Activation of the interferon system by short-interfering RNAs. Nat Cell Biol. 2003;5(9):834–9. (Epub 2003/08/28).
40. Mittal V. Improving the efficiency of RNA interference in mammals. Nat Rev Genet. 2004;5(5):355–65. (Epub 2004/05/15).
41. Shirane D, Sugao K, Namiki S, Tanabe M, Iino M, Hirose K. Enzymatic production of RNAi libraries from cDNAs. Nat Genet. 2004;36(2):190–6. (Epub 2004/01/06).
42. Hatziioannou T, Evans DT. Animal models for HIV/AIDS research. Nat Rev Microbiol. 2012;10(12):852–67. (Epub 2012/11/17).
43. Duyne RV, Narayanan A, K KH, Saifuddin M, Shultz L, Kashanchi F. Humanized mouse models of HIV-1 latency. Curr HIV Res. 2011;9(8):595–605. (Epub 2012/01/04).
44. Denton PW, Garcia JV. Humanized mouse models of HIV infection. AIDS Rev. 2011;13(3):135–48. (Epub 2011/07/30).
45. Melkus MW, Estes JD, Padgett-Thomas A, Gatlin J, Denton PW, Othieno FA, et al. Humanized mice mount specific adaptive and innate immune responses to EBV and TSST-1. Nat Med. 2006;12(11):1316–22. (Epub 2006/10/24).

46. Brainard DM, Seung E, Frahm N, Cariappa A, Bailey CC, Hart WK, et al. Induction of robust cellular and humoral virus-specific adaptive immune responses in human immunodeficiency virus-infected humanized BLT mice. J Virol. 2009;83(14):7305–21. (Epub 2009/05/08).
47. Biswas S, Chang H, Sarkis PT, Fikrig E, Zhu Q, Marasco WA. Humoral immune responses in humanized BLT mice immunized with West Nile virus and HIV-1 envelope proteins are largely mediated via human CD5+ B cells. Immunology. 2011;134(4):419–33. (Epub 2011/11/03).
48. Denton PW, Krisko JF, Powell DA, Mathias M, Kwak YT, Martinez-Torres F, et al. Systemic administration of antiretrovirals prior to exposure prevents rectal and intravenous HIV-1 transmission in humanized BLT mice. PLoS One. 2010;5(1):e8829. (Epub 2010/01/26).
49. Denton PW, Estes JD, Sun Z, Othieno FA, Wei BL, Wege AK, et al. Antiretroviral pre-exposure prophylaxis prevents vaginal transmission of HIV-1 in humanized BLT mice. PLoS Med. 2008;5(1):e16. (Epub 2008/01/18).
50. Denton PW, Olesen R, Choudhary SK, Archin NM, Wahl A, Swanson MD, et al. Generation of HIV latency in humanized BLT mice. J Virol. 2012;86(1):630–4. (Epub 2011/10/21).
51. Marsden MD, Kovochich M, Suree N, Shimizu S, Mehta R, Cortado R, et al. HIV Latency in the Humanized BLT Mouse. J Virol. 2012;86(1):339–47. (Epub 2011/11/11).
52. Hur EM, Patel SN, Shimizu S, Rao DS, Gnanapragasam PN, An DS, et al. Inhibitory effect of HIV-specific neutralizing IgA on mucosal transmission of HIV in humanized mice. Blood. 2012;120(23):4571–82. (Epub 2012/10/16).
53. Stoddart CA, Maidji E, Galkina SA, Kosikova G, Rivera JM, Moreno ME, et al. Superior human leukocyte reconstitution and susceptibility to vaginal HIV transmission in humanized NOD-scid IL-2Rgamma(-/-) (NSG) BLT mice. Virology. 2011;417(1):154–60. (Epub 2011/06/21).
54. Kitchen SG, Levin BR, Bristol G, Rezek V, Kim S, Aguilera-Sandoval C, et al. In vivo suppression of HIV by antigen specific T cells derived from engineered hematopoietic stem cells. PLoS Pathog. 2012;8(4):e1002649. (Epub 2012/04/19).
55. Vatakis DN, Koya RC, Nixon CC, Wei L, Kim SG, Avancena P, et al. Antitumor activity from antigen-specific CD8 T cells generated in vivo from genetically engineered human hematopoietic stem cells. Proc Natl Acad Sci U S A. 2011;108(51):E1408–16. (Epub 2011/11/30).

Chapter 34
Cell-Based Approaches for Treating HIV Infection

Scott G. Kitchen and Jerome A. Zack

34.1 Introduction

34.1.1 The Need for Cell-Based Therapeutics

More than three decades have passed since the identification of HIV disease, yet a therapeutic strategy that results in a cure for the majority of individuals infected with HIV remains elusive. The development and success of antiretroviral therapy is a testament to the Herculean effort made to understand and attempt to cure HIV infection over the years. Despite the successful development of drugs that can suppress viral replication for prolonged periods of time, there is currently no viable treatment that results in the complete eradication of the virus from the body. Treatment with antiretroviral drugs is expensive and has many associated problems, including harmful side effects, cellular and organ toxicities, adherence issues, efficacy problems, and drug resistance development. Due to the presence of long-lived viral reservoirs, if therapy is terminated virus replication and disease progression resume, requiring patients to remain on these medications permanently. In addition, the failure to develop an effective vaccine candidate and the continued unabated spread of the HIV in pandemic proportions highlights the importance of developing a therapeutic strategy that can result in viral clearance and cure.

34.1.2 Cell-Based HIV Curative Strategies

To date there has only been a single reported case of cured chronic HIV infection in an adult, via bone marrow transplant from a donor lacking the gene for normal

S. G. Kitchen (✉) · J. A. Zack
Division of Hematology-Oncology, The David Geffen School of Medicine at UCLA, The UCLA AIDS Institute, 173 BSRB, 615 Charles E. Young Dr. S., Los Angeles, CA 90095, USA
e-mail: skitchen@ucla.edu

L. Y. Poluektova et al. (eds.), *Humanized Mice for HIV Research*,
DOI 10.1007/978-1-4939-1655-9_34

CCR5, which is a cell surface receptor required for most strains of HIV to infect cells [1]. However, the mortality rate for this procedure is about 40 %, and matched bone marrow with this genetic profile is rare, rendering this approach impractical for broader clinical applicability. This case of Mr. Timothy Brown, the "Berlin Patient," has provided a much-needed proof of principle that an HIV-infected individual can be "functionally cured" following the establishment of infection. The "cure" in this case utilized a cell-based therapeutic strategy and this result has stimulated a renewed and focused interest in the development of cell-based therapies to treat others with HIV infection. Although there are currently few therapeutic approaches in clinical practice or trial, many new lines of investigation have been recently opened in cell-based therapies. In most of these developments, there are two main approaches in cell-based therapeutic strategies to treat HIV infection: (1) protecting cells from infection through cellular modification; and (2) modifying cells specifically to target HIV infection. It is notable that these approaches utilize gene therapy or gene knockout methodologies, as this is the current state of the art in cellular modification. With these new areas also comes the need for model systems to allow for their investigation and development. Here we discuss the use of humanized mouse models in the development of cell-based therapies to treat and more closely examine HIV infection and disease.

34.2 Humanized Mouse-Based Systems to Study Cell-Based Therapies

34.2.1 Advantages of the Humanized Mouse Model for Cell-Based Therapy Development

Humanized mouse models have undergone a resurgence in interest since their initial conception and utilization [2, 3] with the development of more sophisticated systems that allow greater multilineage hematopoietic reconstitution and enable the study of HIV infection on a more complete human immune system in a surrogate model [4–7]. Historically, in HIV/AIDS-related work, the use of these models has had a significant impact on our understanding of human hematopoiesis [3, 8], HIV pathogenesis [9–11], HIV latency [12], gene transfer and therapy to target HIV infection [13–15], human immune responses to viral infection [7, 16, 17], and has significantly contributed to the continuing development of new antiretroviral therapies and prevention methods [18–22]. Capitalizing on the development of these models, many investigators have focused on efforts to examine cell-based therapeutic strategies to target HIV infection. The humanized mouse model has many advantages over other in vivo systems in the development of therapeutic strategies, namely that the model allows the use of human reagents in the context of spreading HIV infection. Simian or rodent-based models have important roles in the identification of fundamental biologic aspects of viral infection and immunity, but lack the direct application to human therapy that the humanized mouse has as a key strength. In addition, humanized mouse models allow a high degree of experimental manipulation that cannot be performed in other systems or in humans, which can provide greater insight into mechanisms of efficacy and potential safety. The humanized mouse model may not fully recapitulate human immune responses [23], however, there is strong evidence that human immune responses can be detected and engineered against HIV in vivo [24–26]. In addition, humanized mouse models are well documented to allow the sustained examination of the pathogenic effects of HIV infection [4], mimicking the natural history of HIV infection in humans. A major significance in the use of humanized mouse models, in addition to therapeutic feasibility testing and basic development, lies in the use of these models for preclinical safety and toxicity testing. Thus, the humanized mouse models are fast becoming the "gold standard" for preclinical studies to provide early and germane evidence in this regard; and it is anticipated that their continued use and development will facilitate important aspects of the regulatory approval process of new and novel therapeutic approaches to ultimately lead to human clinical trials.

34.2.2 Cell-Based Therapeutic Testing in Humanized Mice: Basic Models

The details of many different humanized mouse models are discussed elsewhere, and several systems have been utilized in the development of cell-based approaches to target HIV infection. The first report of the use of the humanized mouse system that lay the groundwork for subsequent attempts to utilize gene therapy-based approaches to combat HIV infection involved the use of the severe combined immunodeficient (SCID) mouse implanted with human fetal thymus and liver (SCID-hu thy/liv) [13]. This study demonstrated that human CD34+ hematopoietic stem cells (HSCs) carrying a neomycin-resistant transgene delivered by a retroviral vector can differentiate into mature human thymocytes in the Thy/Liv implant [13].

Many studies utilizing the SCID-hu model examining potential cell-based therapies followed (see below) and the development of the humanized Rag 2, common gamma chain knockout (Rag2-/-cγ-/-) and the Non Obese Diabetic (NOD)-SCID mouse or the NOD-SCID, cγ-/- (NSG) mouse, humanized in the bone marrow, with human fetal liver and thymus tissue (BLT) mouse models [6, 27–29], allowed further study of the development of anti-HIV transgene-containing cells in the periphery as well as in different organs of an in vivo model. A general scheme of this model is described in Figure 34.1. In the humanized Rag 2-/-cγ-/- model, a neonatal immunodeficient mouse is irradiated and injected with (genetically modified) human HSCs. These cells develop over time into mature human immune cells capable of carrying the transgene of interest. A key advantage of this system is the long duration of cellular engraftment and the lack of the significant graft-versus-host disease that is seen in the BLT model in the long term. However, this system often produces poor engraftment and involves the manipulation of newborn mice.

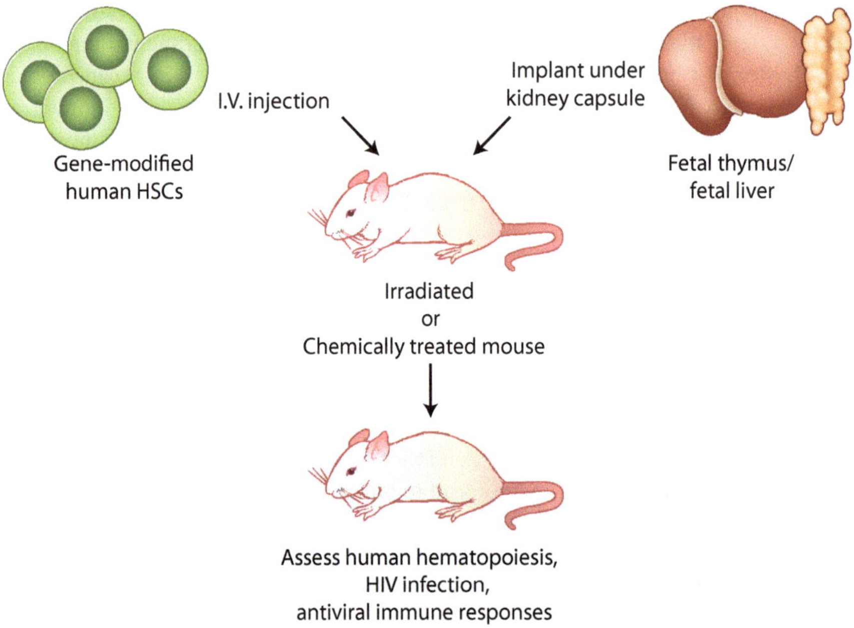

Fig. 34.1 Construction of humanized Rag2-/-cγ-/- or NSG mice with genetically modified HSCs. CD34+ cells are isolated from fetal liver by cell sorting and are then transduced with a lentiviral vector containing the gene of interest. These cells are then intravenously (I.V.) injected into either irradiated or chemically (busulfan, 6TC, or other myeloablative agent) treated Rag2-/-cγ-/- or NSG mice. For BLT mouse construction and better human T-cell reconstitution, human fetal liver and thymus are implanted under the kidney capsule. Six to twelve weeks following injection of cells, human cell reconstitution is analyzed and mice can then be infected with HIV or treated according to the experimental design. Subsequent analysis on mouse blood and organs can then be performed

The BLT model is of high utility in examining cell-based anti-HIV therapeutics in that it usually produces robust engraftment of multilineage human immune cells following the implantation of human fetal liver and thymus and human HSCs. In addition to these models, a key feature of most humanized mouse systems is their adaptability to different types of experimentation. Thus, there is a great deal of ongoing development of other humanized mouse model systems in HIV therapeutic testing and future mention of these models and their utility and importance is highly anticipated.

34.3 Protecting Immune Cells from Infection

The case of Timothy Brown illustrates the use of an adoptive strategy to treat HIV infection involving cells that are protected from natural infection by HIV [1]. In this case, the patient received two rounds of myeloablative therapy (for the treatment of acute myelogenous leukemia) followed by a human leukocyte antigen (HLA) matched allogeneic bone marrow transplant with HSC from a donor that was CCR5-/- (CCR5Δ32/Δ32) [1]. The reconstituting lymphocytes all developed from the donor HSCs and lacked CCR5 expression and were thus refractory to HIV infection by the CCR5-tropic virus. Initial and subsequent follow-up to this patient revealed the absence of detectable replicating virus and this is thought to be the first demonstration that a therapeutic approach could clear replicating HIV from the body [1, 30]. Although this result is of high significance, this strategy is confounded by the fact that the donor cells have to be HLA-matched, the CCR5 Δ32 mutation is only found in less than 1 % of the Caucasian population (which minimizes the potential pool of donors), and the toxic nature of the myeloablative strategy required. However, this observation has stimulated strong interest in the development of strategies that involve the modification of cells to inhibit or protect from HIV infection in a less toxic manner. It is of note that the majority of these strategies involve the use of gene therapy methodologies to alter the ability of the cell to be infected or to target the cell towards HIV. Fortunately, humanized mouse models have a strong track record in preclinical testing regarding new strategies to treat HIV infection (reviewed in [31]).

34.3.1 Utilization of Humanized Mice to Assess the Cellular Targeting of HIV

Strategies have been developed that target multiple stages of the viral life cycle, from entry to expression. Early studies in humanized mice utilizing a transdominant mutant of HIV Rev-1, RevM10, which interferes with Rev function, were among the first to demonstrate that T cells in the Thy/Liv implant and macrophages in the bone marrow can differentiate in vivo in the SCID-hu model and carry a protective anti-HIV gene, which renders the mature cells refractory to HIV infection ex vivo [32, 33]. The SCID-hu model was also used to demonstrate the development of T cells from HSCs transduced with a protective anti-HIV ribozyme in the Thy/Liv implant [34]. The data generated by this and subsequent studies lay the foundation for a phase II and subsequent phase III gene therapy trial [35, 36], which demonstrates the utility of the use of humanized mouse models in preclinical development of cell-based therapies. The SCID-hu model was extremely effective at developing effective progenitor cell transduction protocols [34] and in demonstrating the feasibility of such potential therapies in allowing the characterization of the differentiation of HSCs into mature T cells and other hematopoietic lineages in a surrogate model in vivo.

Other efforts to target HIV RNA have been attempted in humanized mice in different ways. Utilizing the Rag 2-/-cγ-/- model, HSCs transduced with a lentiviral vector encoding a short hairpin (sh)RNA specific to HIV nef were shown to undergo successful multilineage hematopoiesis that resulted in mature cells that were refractory to HIV infection ex vivo [37]. Utilizing a different approach examining peripheral T-cell modification, NSG mice were reconstituted with peripheral CD4+ T cells modified with a vector containing an antisense RNA sequence specific to HIV-1 env [38]. They observed no decrease in viral loads, likely due to the fact that only a minority of cells were modified with the vector; however, they found increased viral evolution when mice were challenged with an HIV variant that imperfectly matched the target sequence in env. This suggests that this approach should be used with caution as it may drive faster viral evolution away from the therapeutic.

Humanized mice have been used to assess the efficacy of targeting "alternative" therapeutic targets in the HIV life cycle, such as the ability of tripartite motif 5alpha (TRIM5α) proteins, which have been shown to inhibit HIV infection of cells through capsid binding, to inhibit HIV pathogenesis in vivo [39, 40]. These studies found that modification of human peripheral CD4+ T cells engrafted in Rag 2-/-cγ-/- mice or CD34+ cells developing in SCID-hu Thy/Liv implants inhibited HIV replication following virus challenge. In a separate study therapeutically targeting HIV DNA integration events, lentiviral vectors expressing the C-terminal part of lens epithelium-derived growth factor/p75 (LEDGGF/$p75_{325-530}$), which competitively inhibits HIV integration by blocking HIV integrase interaction with endogenous LEDGF/p75, were assessed in their ability to block HIV replication and CD4 T-cell depletion in NSG mice [41]. Human T cells were modified with vectors expressing LEDGF/$p75_{325-530}$ or a control, integrase interaction deficient mutant LEDGF/$p75_{325-530}$D366N), infected with HIV, and placed in the periphery of NSG mice. Mice containing cells overexpressing LEDGGF/$p75_{325-530}$ had lower levels of HIV p24 in the plasma and had greater percentages of CD4+ T cells than did mice containing the mutant LEDGF/$p75_{325-530}$D366N, indicating protection of the former cells from HIV infection in vivo. Although the idea of attacking cellular targets such as LEDGF/p75 is relatively new, this study is a clear demonstration of the utility of the humanized mouse system to assess the efficacy of alternative therapeutic targets to treat HIV infection. There is great interest in ongoing and planned efficacy and safety studies assessing additional molecular therapeutic targets in humanized-mouse–based systems.

34.3.2 Humanized Mice in Targeting and Preventing HIV Entry

Following the success of the "Berlin Patient," there is intense interest in the development of therapeutic strategies that prevent HIV entry into cells, particularly strategies that involve the knockdown or interference with HIV binding the co-receptor CCR5. Development of mature thymocytes carrying a CCR5-specific ribozyme was initially demonstrated in SCID-hu mice following development from transduced HSCs [42]. A lentiviral-based, shRNA approach to knockdown CCR5 expression following HSC modification in the BLT humanized mouse models successfully demonstrated the multilineage development of CCR5 knockdown cells and inhibition of CCR5-tropic HIV infection in vivo [14]. Knockout of CCR5 expression in HSC through use of zinc-finger nucleases, has been shown to allow hematopoietic development of HIV-resistant cells in the NSG mouse [43]. In both of these studies, human hematopoiesis is relatively unaffected by CCR5 knockout or knockdown. An attempt at knocking out CXCR4 expression was also performed by using zinc-finger nucleases in human T cells engrafted in the peripheral blood of NSG mice [44]. Protection of cells from CXCR4-tropic variants was observed for a period of time, but was lost following the mutational development of CCR5-tropic variants. In another study, investigators examined the survival advantage conferred by the genetic modification of cells with a different set of protective genes in vivo. It was found that the modification of peripheral human T cells with an HIV-1 entry/viral fusion inhibitor, the membrane-anchored C46 peptide (maC46), conferred selective survival advantage of modified cells versus cells expressing an HIV-1 tat/rev shRNA or an antisense RNA specific for env sequences [45]. In this case, human T cells were modified with lentiviral vectors containing the protective gene or a control vector, infected with HIV, and then placed into the periphery of NOD/Shi-*scid*/cg--/- (NOG mouse; similar to NSG mice).

More recently, a focus has been on a combination approach in targeting both HIV itself and cellular entry receptors [46, 47]. This includes the incorporation of more than one anti-HIV gene that targets different stages of the HIV life cycle or different parts of HIV into a single lentiviral vector. A combination approach utilizing a lentiviral vector encoding an anti-HIV shRNA specific to the R region of the LTR as well as a CCR5-specific shRNA resulted in multilineage hematopoiesis of vector-containing cells following human HSC transduction, engraftment, and development in NSG mice [48]. These resultant vector-containing cells were resistant to HIV infection by both CXCR4- and CCR5-tropic viruses following challenge in vitro and increased in relative levels, as a result of selection, in vivo following virus challenge. In another study, development of mature T cells following HSC transduction with a vector expressing a triple combination of a CCR5-specific ribozyme, a tat-rev specific shRNA, and an HIV TAR decoy, was shown in the SCID-hu system to produce cells that were resistant to HIV infection ex vivo [49]. Subsequent studies demonstrated the multilineage hematopoietic development in multiple organs of cells derived from human HSC in humanized NOD-RAG1-/-cg-/- mice transduced with a vector expressing a human/macaque TRIM5α isoform, a CCR5-specific shRNA, and a TAR decoy [50]. Following in vivo challenge with either a CCR5-tropic or CXCR4-tropic variant of HIV-1, they further observed the selective survival advantage of these gene-modified cells for prolonged periods of time. In all, these studies illustrate the flexibility of the use of humanized mouse models to test the potential in vivo efficacy of a gene therapy anti-HIV strategy.

34.4 Engineering Antiviral Immune Responses

The natural immune response to HIV infection fails to eradicate the virus and there is a strong need to understand the defects that result in this as well as to discover new ways to enhance the antiviral response. As previously discussed, a major focus in investigation in cell-based therapeutics using the humanized mouse model has been in developing ways to protect cells from direct infection by HIV. Protecting developing and mature cells in vivo from infection is likely to be beneficial in reconstituting part of the ablated immune response, however, it may not reconstitute antigen-specific responses to HIV enough to clear the virus. Thus, there is a significant interest in the development of ways to augment or repair defects in the immune response to HIV during infection. In humanized mice, there has been significant development in ways to engineer cells genetically to target HIV infection, in hopes of developing methodologies that can ultimately be carried to the clinic. Two primary approaches have emerged thus far: that of engineering the anti-HIV humoral immunity through the use of molecularly cloned monoclonal antibodies and engineering anti-HIV cell-based immunity through the use of molecularly cloned receptors expressed on effector cells.

34.4.1 Engineering Humoral Immunity to HIV in Humanized Mice

Based on the characterization and putative importance of neutralizing antibodies in HIV transmission and infection, efforts have been made to engineer the development of human B cells to produce molecularly cloned neutralizing antibodies to target HIV and HIV-infected cells. Humanized mice have been utilized to demonstrate that HSCs engineered with a molecularly cloned neutralizing antibody can develop into mature cells that express the antibody [51]. In this case, HSCs were engineered with a lentiviral vector containing the molecularly cloned heavy and light chains of the 2G12 broadly neutralizing anti-HIV antibody and engrafted in NSG mice. Following hematopoietic development in the mouse, they detected lentiviral vector expression in multiple anatomic compartments and found the cloned antibody secreted in the plasma. It is important to note that they observed a marked decrease, or resistance, to HIV infection in terms of the number of cells infected and lower plasma viral RNA levels in mice receiving the cloned antibody, indicating protection of these mice from HIV infection by the antibody. An additional study, in which adeno-associated virus vector was injected intramuscularly into NSG mice, also showed high levels of sustained molecularly cloned neutralizing antibody (b12) production that protected adoptively transferred, previously expanded human peripheral blood mononuclear cells (PBMC) from HIV challenge in vivo [52]. An approach with a single neutralizing antibody would likely result in the development of viral immune escape, however, these studies demonstrate the potential importance of neutralizing antibodies in preventing infection and set the stage for the potential use of this type of approach in the clinical setting. Several other studies are ongoing utilizing a combinatorial approach and/or different types of neutralizing antibodies. Thus, the humanized mouse model has been an important tool in the development of humoral approaches to engineer HIV immunity.

34.4.2 Engineering Cellular Immunity to HIV in Humanized Mice

The cellular immune response has a highly important role in controlling HIV replication during all stages of HIV infection; however, this response fails to eradicate the virus from the body. There is considerable interest in enhancing or reconstituting the antiviral cellular immune response during HIV infection. Two approaches have been examined utilizing humanized mice: an approach that involves "redirecting" peripheral T cells to target HIV and that of "programming" HSCs to produce T cells to target HIV [15, 26, 53]. Both of these approaches utilize a molecularly cloned T-cell receptor directed towards the SL9 (amino acids 77–85; SLYNTVATL) epitope of HIV-1 Gag, which recognize this peptide in the context of human leukocyte antigen (HLA)-A*0201, and which were isolated from clonal cell lines derived from HIV-infected individuals. In the redirection approach, CD8+ T cells transduced with a lentiviral vector containing the molecularly cloned TCR were intrasplenically injected with HIV-infected HLA-A*0201+ PBMC from the same donor [53] and splenic viral loads were assessed. They demonstrated that the redirected CD8+ T cells dramatically decreased the levels of HIV-infected cells in the spleen over those of controls and also demonstrated that the presence of these HIV-specific CTL strongly inhibited infection when uninfected human PBMCs were challenged in vivo. This demonstrates that these mature cells can be redirected to target and kill HIV-infected cells in an in vivo system. However, approaches such as this involving redirection of mature T cells, which are currently utilized in many cancer-related therapeutic trials, could suffer from altered cellular function due to the large amount of ex vivo manipulation involved in transduction with a molecularly cloned TCR.

We have examined a stem-cell–based approach, which would forgo problems with the ex vivo manipulation of mature T cells, to program hematopoietic progenitors (HSCs) to produce mature, functionally naïve T cells specifically targeting HIV in humanized mice [15, 26]. Initially, to demonstrate if cellular modification of HSCs with a molecularly cloned TCR can allow the development of mature T cells expressing the transgenic TCR, we utilized the SCID-hu mouse model. HSCs transduced with the HIV-specific TCR were injected into the Thy/Liv implant in irradiated mice and were found to generate mature T cells that expressed the TCR, responded to SL9 peptide stimulation, and killed HIV antigen coated cells [15]. To further investigate if peripheral cells can be modified and function with this HIV-specific TCR, we utilized a variation of the NSG-BLT mouse model where human fetal thymus and liver were implanted, the mice were irradiated, and then the mice subsequently intravenously injected with HSCs genetically modified with the HIV-specific TCR or a control TCR specific to an unknown antigen [26]. We determined that the presence of the HIV-specific TCR resulted in significantly reduced HIV loads, lower amounts of productively infected cells, and antiviral immune responses by these cells following HIV challenge. Thus, these studies using various humanized mouse models have formed a foundation for further investigation into the use of molecularly cloned, antigen-specific receptors as a cell-based therapeutic strategy to target HIV infection.

34.5 Conclusions

In sum, the various versions of the humanized mouse model have been very useful tools in the investigation into cell-based therapeutic strategies to target HIV infection. The ability of these models to serve as a surrogate system for in vivo infection and pathogenesis of human cells, the high degree to which these models can be manipulated, and the lower cost of utilizing these systems compared to other in vivo model systems (such as nonhuman primates), and the strong track record of these systems in examining human biology in vivo are all factors that have contributed to the expanding interest in the humanized mouse in many aspects of HIV infection and beyond. Due to intense interest in developing a cure for HIV infection, it is highly likely that these systems will continue to be used to further explore new and novel ways to assess preclinical therapeutic strategies.

References

1. Hutter G, Nowak D, Mossner M, Ganepola S, Mussig A, Allers K, et al. Long-term control of HIV by CCR5 Delta32/Delta32 stem-cell transplantation. N Engl J Med. 2009;360(7):692–8.
2. Mosier DE, Gulizia RJ, Baird SM, Wilson DB. Transfer of a functional human immune system to mice with severe combined immunodeficiency. Nature. 1988;335(6187):256–9.
3. McCune JM, Namikawa R, Kaneshima H, Shultz LD, Lieberman M, Weissman IL. The SCID-hu mouse: murine model for the analysis of human hematolymphoid differentiation and function. Science. 1988;241(4873):1632–9.
4. Denton PW, Garcia JV. Humanized mouse models of HIV infection. AIDS Rev. 2011;13(3):135–48.
5. Shultz LD, Lyons BL, Burzenski LM, Gott B, Chen X, Chaleff S, et al. Human lymphoid and myeloid cell development in NOD/LtSz-scid IL2R gamma null mice engrafted with mobilized human hemopoietic stem cells. J Immunol. 2005;174(10):6477–89.
6. Shultz LD, Ishikawa F, Greiner DL. Humanized mice in translational biomedical research. Nat Rev Immunol. 2007;7(2):118–30.
7. Melkus MW, Estes JD, Padgett-Thomas A, Gatlin J, Denton PW, Othieno FA, et al. Humanized mice mount specific adaptive and innate immune responses to EBV and TSST-1. Nat Med. 2006;12(11):1316–22.
8. Galic Z, Kitchen SG, Kacena A, Subramanian A, Burke B, Cortado R, et al. T lineage differentiation from human embryonic stem cells. Proc Natl Acad Sci U S A. 2006;103(31):11742–7.
9. Mosier DE, Gulizia RJ, MacIsaac PD, Torbett BE, Levy JA. Rapid loss of CD4+ T cells in human-PBL-SCID mice by noncytopathic HIV isolates. Science. 1993;260(5108):689–92.
10. Bonyhadi ML, Rabin L, Salimi S, Brown DA, Kosek J, McCune JM, et al. HIV induces thymus depletion in vivo. Nature. 1993;363(6431):728–32.
11. Aldrovandi GM, Feuer G, Gao L, Jamieson B, Kristeva M, Chen IS, et al. The SCID-hu mouse as a model for HIV-1 infection. Nature 1993;363(6431):732–6.
12. Brooks DG, Kitchen SG, Kitchen CM, Scripture-Adams DD, Zack JA. Generation of HIV latency during thymopoiesis. Nat Med. 2001;7(4):459–64.
13. Akkina RK, Rosenblatt JD, Campbell AG, Chen IS, Zack JA. Modeling human lymphoid precursor cell gene therapy in the SCID-hu mouse. Blood. 1994;84(5):1393–8.
14. Shimizu S, Hong P, Arumugam B, Pokomo L, Boyer J, Koizumi N, et al. A highly efficient short hairpin RNA potently down-regulates CCR5 expression in systemic lymphoid organs in the hu-BLT mouse model. Blood. 2010;115(8):1534–44.

15. Kitchen SG, Bennett M, Galic Z, Kim J, Xu Q, Young A, et al. Engineering antigen-specific T cells from genetically modified human hematopoietic stem cells in immunodeficient mice. PLoS ONE. 2009;4(12):e8208.
16. Shultz LD, Saito Y, Najima Y, Tanaka S, Ochi T, Tomizawa M, et al. Generation of functional human T-cell subsets with HLA-restricted immune responses in HLA class I expressing NOD/SCID/IL2r gamma(null) humanized mice. Proc Natl Acad Sci U S A. 2010;107(29):13022–7.
17. Watanabe S, Terashima K, Ohta S, Horibata S, Yajima M, Shiozawa Y, et al. Hematopoietic stem cell-engrafted NOD/SCID/IL2Rgamma null mice develop human lymphoid systems and induce long-lasting HIV-1 infection with specific humoral immune responses. Blood. 2007;109(1):212–8.
18. Denton PW, Krisko JF, Powell DA, Mathias M, Kwak YT, Martinez-Torres F, et al. Systemic administration of antiretrovirals prior to exposure prevents rectal and intravenous HIV-1 transmission in humanized BLT mice. PLoS ONE. 2010;5(1):e8829.
19. Denton PW, Othieno F, Martinez-Torres F, Zou W, Krisko JF, Fleming E, et al. One percent tenofovir applied topically to humanized BLT mice and used according to the CAPRISA 004 experimental design demonstrates partial protection from vaginal HIV infection, validating the BLT model for evaluation of new microbicide candidates. J Virol. 2011;85(15):7582–93.
20. Neff CP, Kurisu T, Ndolo T, Fox K, Akkina R. A topical microbicide gel formulation of CCR5 antagonist maraviroc prevents HIV-1 vaginal transmission in humanized RAG-hu mice. PLoS One. 2011;6(6):e20209.
21. Neff CP, Zhou J, Remling L, Kuruvilla J, Zhang J, Li H, et al. An aptamer-siRNA chimera suppresses HIV-1 viral loads and protects from helper CD4(+) T cell decline in humanized mice. Sci Transl Med. 2011;3(66):66ra66.
22. Neff CP, Ndolo T, Tandon A, Habu Y, Akkina R. Oral pre-exposure prophylaxis by anti-retrovirals raltegravir and maraviroc protects against HIV-1 vaginal transmission in a humanized mouse model. PLoS ONE. 2010;5(12):e15257.
23. Stoddart CA, Maidji E, Galkina SA, Kosikova G, Rivera JM, Moreno ME, et al. Superior human leukocyte reconstitution and susceptibility to vaginal HIV transmission in humanized NOD-scid IL-2R[γ]-/- (NSG) BLT mice. Virology. 2011;417(1):154–60.
24. Brainard DM, Seung E, Frahm N, Cariappa A, Bailey CC, Hart WK, et al. Induction of robust cellular and humoral virus-specific adaptive immune responses in human immunodeficiency virus-infected humanized BLT mice. J Virol. 2009;83(14):7305–21.
25. Singh M, Singh P, Gaudray G, Musumeci L, Thielen C, Vaira D, et al. An improved protocol for efficient engraftment in NOD/LTSZ-SCIDIL-2Rgammanull mice allows HIV replication and development of anti-HIV immune responses. PLoS ONE. 2012;7(6):e38491.
26. Kitchen SG, Levin BR, Bristol G, Rezek V, Kim S, Aguilera-Sandoval C, et al. In vivo suppression of HIV by antigen specific T cells derived from engineered hematopoietic stem cells. PLoS Pathog. 2012;8(4):e1002649.
27. Wege AK, Melkus MW, Denton PW, Estes JD, Garcia JV. Functional and phenotypic characterization of the humanized BLT mouse model. Curr Top Microbiol Immunol. 2008;324:149–65.
28. Melkus MW, Estes JD, Padgett-Thomas A, Gatlin J, Denton PW, Othieno FA, et al. Humanized mice mount specific adaptive and innate immune responses to EBV and TSST-1. Nat Med. 2006;12(11):1316–22.
29. An DS, Poon B, Ho Tsong Fang R, Weijer K, Blom B, Spits H, et al. Use of a novel chimeric mouse model with a functionally active human immune system to study human immunodeficiency virus type 1 infection. Clin Vaccine Immunol. 2007;14(4):391–6.
30. Allers K, Hutter G, Hofmann J, Loddenkemper C, Rieger K, Thiel E, et al. Evidence for the cure of HIV infection by CCR5Delta32/Delta32 stem cell transplantation. Blood. 2011;117(10):2791–9.
31. Berges BK, Rowan MR. The utility of the new generation of humanized mice to study HIV-1 infection: transmission, prevention, pathogenesis, and treatment. Retrovirology. 2011;8:65.
32. Bonyhadi ML, Moss K, Voytovich A, Auten J, Kalfoglou C, Plavec I, et al. RevM10-expressing T cells derived in vivo from transduced human hematopoietic stem-progenitor cells inhibit human immunodeficiency virus replication. J Virol. 1997;71(6):4707–16.

33. Su L, Lee R, Bonyhadi M, Matsuzaki H, Forestell S, Escaich S, et al. Hematopoietic stem cell-based gene therapy for acquired immunodeficiency syndrome: efficient transduction and expression of RevM10 in myeloid cells in vivo and in vitro. Blood. 1997;89(7):2283–90.
34. Amado RG, Symonds G, Jamieson BD, Zhao G, Rosenblatt JD, Zack JA. Effects of megakaryocyte growth and development factor on survival and retroviral transduction of T lymphoid progenitor cells. Hum Gene Ther. 1998;9(2):173–83.
35. Amado RG, Mitsuyasu RT, Symonds G, Rosenblatt JD, Zack J, Sun LQ, et al. A phase I trial of autologous CD34+ hematopoietic progenitor cells transduced with an anti-HIV ribozyme. Hum Gene Ther. 1999;10(13):2255–70.
36. Mitsuyasu R, Merigan T, Carr A, Zack J, Winters M, Workman C, et al. Phase 2 gene therapy trial of an anti-HIV ribozyme in autologous CD34+ cells. Nat Med. 2009;15:285–92.
37. ter Brake O, Legrand N, von Eije KJ, Centlivre M, Spits H, Weijer K, et al. Evaluation of safety and efficacy of RNAi against HIV-1 in the human immune system (Rag-2(-/-)gammac(-/-)) mouse model. Gene Ther. 2009;16(1):148–53.
38. Mukherjee R, Plesa G, Sherrill-Mix S, Richardson MW, Riley JL, Bushman FD. HIV sequence variation associated with env antisense adoptive T-cell therapy in the hNSG mouse model. Mol Ther J Am Soc Gene Ther. 2010;18(4):803–11.
39. Neagu MR, Ziegler P, Pertel T, Strambio-De-Castillia C, Grutter C, Martinetti G, et al. Potent inhibition of HIV-1 by TRIM5-cyclophilin fusion proteins engineered from human components. J Clin Investig. 2009;119(10):3035–47.
40. Anderson J, Akkina R. Human immunodeficiency virus type 1 restriction by human-rhesus chimeric tripartite motif 5alpha (TRIM 5alpha) in CD34(+) cell-derived macrophages in vitro and in T cells in vivo in severe combined immunodeficient (SCID-hu) mice transplanted with human fetal tissue. Hum Gene Ther. 2008;19(3):217–28.
41. Vets S, Kimpel J, Volk A, De Rijck J, Schrijvers R, Verbinnen B, et al. Lens epithelium-derived growth factor/p75 qualifies as a target for HIV gene therapy in the NSG mouse model. Mol Ther J Am Soc Gene Ther. 2012;20(5):908–17.
42. Bai J, Gorantla S, Banda N, Cagnon L, Rossi J, Akkina R. Characterization of anti-CCR5 ribozyme-transduced CD34+ hematopoietic progenitor cells in vitro and in a SCID-hu mouse model in vivo. Mol Ther J Am Soc Gene Ther. 2000;1(3):244–54.
43. Holt N, Wang J, Kim K, Friedman G, Wang X, Taupin V, et al. Human hematopoietic stem/progenitor cells modified by zinc-finger nucleases targeted to CCR5 control HIV-1 in vivo. Nat Biotechnol. 2010;28(8):839–47.
44. Wilen CB, Wang J, Tilton JC, Miller JC, Kim KA, Rebar EJ, et al. Engineering HIV-resistant human CD4+ T cells with CXCR4-specific zinc-finger nucleases. PLoS Pathog. 2011;7(4):e1002020.
45. Kimpel J, Braun SE, Qiu G, Wong FE, Conolle M, Schmitz JE, et al. Survival of the fittest: positive selection of CD4+ T cells expressing a membrane-bound fusion inhibitor following HIV-1 infection. PLoS ONE. 2010;5(8):e12357.
46. Berges BK, Rowan MR. The utility of the new generation of humanized mice to study HIV-1 infection: transmission, prevention, pathogenesis, and treatment. Retrovirology. 2011;8(1):65.
47. Kitchen SG, Shimizu S, An DS. Stem cell-based anti-HIV gene therapy. Virology. 2011;411(2):260–72.
48. Ringpis GE, Shimizu S, Arokium H, Camba-Colon J, Carroll MV, Cortado R, et al. Engineering HIV-1-resistant T-cells from short-hairpin RNA-expressing hematopoietic stem/progenitor cells in humanized BLT mice. PLoS ONE. 2012;7(12):e53492.
49. Anderson J, Li MJ, Palmer B, Remling L, Li S, Yam P, et al. Safety and efficacy of a lentiviral vector containing three anti-HIV genes—CCR5 ribozyme, tat-rev siRNA, and TAR decoy—in SCID-hu mouse-derived T cells. Mol Ther J Am Soc Gene Ther. 2007;15(6):1182–8.
50. Walker JE, Chen RX, McGee J, Nacey C, Pollard RB, Abedi M, et al. Generation of an HIV-1-resistant immune system with CD34(+) hematopoietic stem cells transduced with a triple-combination anti-HIV lentiviral vector. J Virol. 2012;86(10):5719–29.

51. Joseph A, Zheng JH, Chen K, Dutta M, Chen C, Stiegler G, et al. Inhibition of in vivo HIV infection in humanized mice by gene therapy of human hematopoietic stem cells with a lentiviral vector encoding a broadly neutralizing anti-HIV antibody. J Virol. 2010;84(13):6645–53.
52. Balazs AB, Chen J, Hong CM, Rao DS, Yang L, Baltimore D. Antibody-based protection against HIV infection by vectored immunoprophylaxis. Nature. 2012;481(7379):81–4.
53. Joseph A, Zheng JH, Follenzi A, Dilorenzo T, Sango K, Hyman J, et al. Lentiviral vectors encoding human immunodeficiency virus type 1 (HIV-1)-specific T-cell receptor genes efficiently convert peripheral blood CD8 T lymphocytes into cytotoxic T lymphocytes with potent in vitro and in vivo HIV-1-specific inhibitory activity. J Virol. 2008;82(6):3078–89.

Chapter 35
A Conditionally Replicating Human Immunodeficiency Virus in BRG-HIS Mice

Mireille Centlivre, Nicolas Legrand and Ben Berkhout

Vaccination has been a major advance for health care, allowing the eradication or the reduction of incidence and mortality of various infectious diseases [1]. However, there remain major pathogens, e.g., human immunodeficiency virus (HIV), hepatitis C virus, or the causative agent of malaria (plasmodium), for which classical vaccination approaches have failed. Two main classes of vaccines are already in long-term use in humans [2]. The first vaccine type consists of live attenuated pathogens, which mimic the natural infection but in a weakened nonpathogenic fashion. Live-attenuated vaccines have been enormously successful against a variety of viral pathogens such as smallpox, measles, and yellow fever [2]. The second vaccination method comprises subunit vaccines (hepatitis B), inactivated vaccines (diphtheria, tetanus), carbohydrate vaccines (pneumococcus) and conjugate vaccines (meningococcus, haemophilus influenza type B). This vaccine type usually requires adjuvants to enhance the induced immune response. Furthermore, they

B. Berkhout (✉) · M. Centlivre
Laboratory of Experimental Virology, Department of Medical Microbiology, Center for Infection and Immunity Amsterdam (CINIMA), Academic Medical Center, University of Amsterdam (AMC-UvA), Meibergdreef 15, 1105 AZ Amsterdam, The Netherlands
e-mail: b.berkhout@amc.uva.nl

M. Centlivre
Centre d'Immunologie et des Maladies Infectieuses-Paris (CIMI-Paris), INSERM U1135, Sorbonnes Universités, UPMC University Paris 06, UMR_S CR7, 91 Bld de l'Hôpital, 75013 Paris, France

N. Legrand
Department of Cell Biology and Histology, Center for Immunology Amsterdam (CIA), Academic Medical Center, University of Amsterdam (AMC-UvA), Meibergdreef 15, 1105 AZ Amsterdam, The Netherlands

N. Legrand
AXENIS, 28 rue du Docteur Roux, 75015 Paris, France

L. Y. Poluektova et al. (eds.), *Humanized Mice for HIV Research*,
DOI 10.1007/978-1-4939-1655-9_35

depend on a boosting strategy in order to maintain protective immunity, in contrast with live attenuated vaccines that can confer lifelong memory.

Despite this major step ahead for global public health and obvious benefits over the past century, vaccination faces new challenges when dealing with specific patient categories (infants, elderly, or immuno-compromised individuals) or pathogens such as HIV with an extremely high antigenic variability. The use of a live-attenuated HIV strain as an effective HIV vaccine has been proposed, but the idea has been challenged by safety concerns arising from preclinical studies performed with live-attenuated simian immunodeficiency virus (SIV, the simian counterpart of HIV) strains [3]. It was indeed reported that replication of attenuated SIV strains increases after several years in a fraction of infected monkeys, concomitant with the onset of AIDS, and pathogenicity was scored upon inoculation of newborn macaques [4, 5]. Furthermore, although naturally attenuated HIV-1 variants lacking the *nef* gene result in a benign course of infection in humans [6], a decline in $CD4^+$ T cell numbers has been reported for some of these individuals, which is an early sign of AIDS development [7]. These combined results have reinforced the development of subunit or inactivated virus vaccines. Unfortunately, these vaccines have not elicited the potent broad-based immune responses or long-term memory necessary to confer life-long protection in immunized individuals [8–10]. For this reason, live-attenuated HIV vaccine approaches are still being considered, especially since live attenuated SIV vaccines confer robust protection against a broad range of SIV challenge variants, including uncloned pathogenic SIV, SIV-infected cells and simian-human immunodeficiency virus (SHIV) recombinants expressing the HIV-1 envelope protein [11–14]. Attempts to determine the mechanism(s) by which a live-attenuated virus protects against infection by the wild-type virus have failed to clearly identify a decisive role for antibodies or $CD8^+$ T cells [3]. Overall, efforts to define correlates of protection are further confounded by conflicting data as to whether significant protection requires an extended vaccination period [15] or can be achieved in as little as 3 weeks [16]. This knowledge is critical for the design of new vaccine candidates. Thus, novel approaches are required to establish the protective factors induced by vaccination.

35.1 HIV-rtTA: A Drug-Dependent HIV-1 Strain

To address this issue, we decided to explore an innovative strategy of HIV-1 design by imposing control over virus replication by a nontoxic effector molecule. The unique feature of such a virus is that the inserted genetic elements should be essential for viral gene expression and replication and therefore stably maintained in the viral genome. Tight control of virus replication is an attractive asset of an attenuated HIV vaccine to study the correlates of protection. First, by limiting the extent of virus replication and consequently insuring a reduced spectrum of mutants, one reduces the chances of reversion of the attenuated to a pathogenic phenotype. In contrast, rampant virus replication generates a broad mutant spectrum and is thus

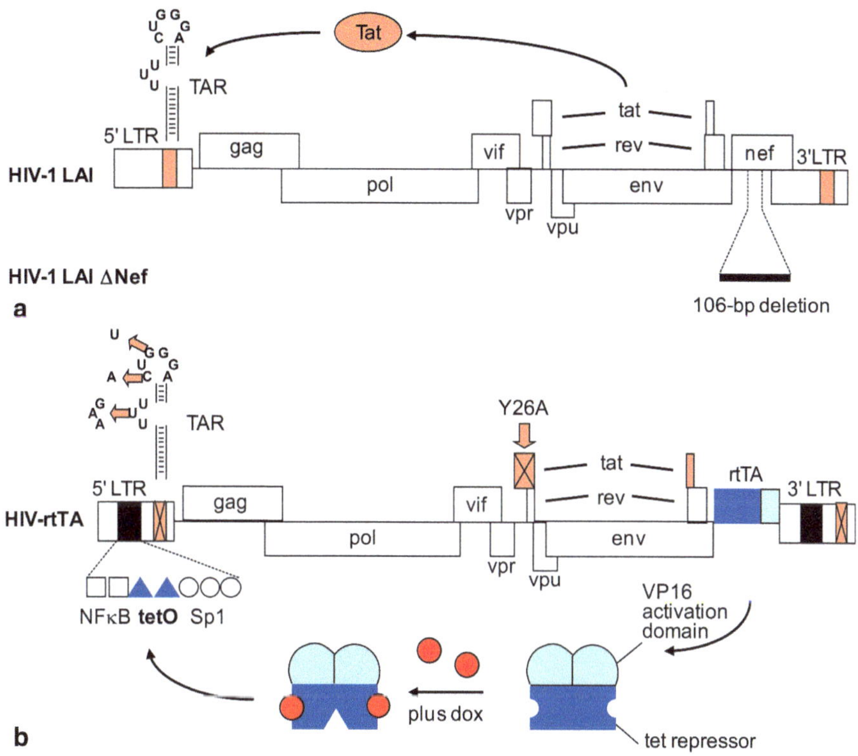

Fig. 35.1 Schematic of the dox-inducible lentiviral vectors [41]. Vectors are based on the third generation self-inactivating lentiviral vectors, containing the central polypurine tract (*cPPT*) and the hepatitis B virus posttranscriptional regulatory element (*PRE*). **a** In the TRECMV-V14 lentiviral vector, reverse tetracycline transactivator (*rtTA*) expression is driven by the constitutive cytomegalovirus (*CMV*) promoter and the green fluorescent protein (*GFP*) from the tetracycline-responsive element *TRE* promoter. **b** In TREAuto-V14, the *TRE* promoter expresses a bicistronic mRNA that encodes both *GFP* and *rtTA,* the latter via an internal ribosomal entry site (*IRES*). *LTR* long terminal repeat

not desirable. Second, HIV-1 preferentially replicates in antigen specific memory $CD4^+$ T-lymphocytes, in particular those that can specifically recognize HIV-infected cells, which are key effector cells of the immune system raised against the viral infection [17]. Control of viral replication should hence limit the loss of cells in this critical HIV-specific $CD4^+$ T cell pool. Last, tight replication control permits strict timing of antigen expression ("boosting on demand"), which provides an attractive tool to decipher whether ongoing vaccine virus replication is needed for durable protection. Overall, it is likely that such a conditional live HIV would significantly contribute to HIV vaccine development, by helping defining the correlates of protection against HIV via the induction of an effective immune response, as generated with standard replication-defective or live-attenuated viral vaccines, but in addition in a time-controllable manner.

To obtain such a drug-dependent HIV-1 strain, we have inserted elements of the inducible Tet-On system into the viral genome (Fig. 35.1). In wild-type HIV-1, transcription of the viral genome is enhanced by the viral Tat protein through interaction with the viral transactivating response (TAR) ribonucleic acid (RNA) element. In HIV reverse tetracycline transactivator (rtTA), Tat and TAR were inactivated by mutations and functionally replaced by the components of the *Escherichia coli*-derived tetracycline-inducible (Tet-On) gene expression system [18]. This system is based on two elements: the tetracycline-inducible repressor protein (TetR) and its tet operator (tetO) deoxyribonucleic acid (DNA) binding site [19]. In HIV-rtTA, the new transcriptional activator is the rtTA protein, a fusion protein between TetR and the activation domain of the herpes simplex virus VP16 protein [20]. The gene coding for the rtTA protein replaces the *nef* gene in HIV-rtTA, and the tetO sequences were inserted in the viral promoter region of the long terminal repeat (LTR). Virus transcription is controlled by the antibiotic dox, a nontoxic and selective effector molecule. Dox binds to rtTA and induces a conformational change that triggers specific binding to the tetO DNA sequence that consequently leads to viral transcription and replication.

Strategies based on the tetracycline operon elements have also been implemented to gain external control over replication of a variety of viruses. Modifying the herpes simplex virus type 1 (HSV-1) genome by incorporation of a Tet-repressor modified promoter to drive the transexpression of a dominant negative mutant of an essential factor for viral replication, it was reported that tetracycline addition strongly reduced HSV-1 production in vitro [21]. More recently, an alternative HSV-1 recombinant demonstrated dose-dependent tetracycline-mediated regulation of viral replication [22]. In a model of conditionally replicative adenovirus, it was shown that addition/withdrawal of dox in vitro or in vivo—after oral administration in immunodeficient nude mice—imposed a tight control on viral replication [23]. Vaccinia virus (VACV) recombinants engineered for the tetracycline-dependent, inducible expression of viral genes essential for VACV replication demonstrated similar replication capacity as wild-type VACV, but exclusively in the presence of dox [24]. These examples indicate that Tet-On-based control over viral replication may provide a general strategy to generate effective virus constructs or vectors for vaccination purposes. Using HIV-rtTA as evolution machine, we have selected greatly improved versions of the Tet-On system that should facilitate such strategies (see below).

We extensively studied the in vitro replication characteristics of the drug-dependent HIV-rtTA variant and demonstrated that it can be turned on and off at will by simple dox addition/withdrawal. The original HIV-rtTA, in which eight tetO-binding sites were inserted in the LTR, replicated poorly as compared to the parental wild-type LAI strain [25, 26]. However, after prolonged culturing of HIV-rtTA, we were able to select optimized virus variants with improved replication capacity, notably due to deletion of six of the eight tetO-binding sites and deletion of 15 base pairs in the spacer between the two remaining tetO-binding sites [25, 26]. Efficient replication of the optimized HIV-rtTA virus has been demonstrated in T cell lines and peripheral blood mononuclear cells (PBMCs), but also ex vivo on human lymphoid tonsil tissue [27].

Additional genetic information introduced into the HIV genome is usually rapidly deleted if it does not encode any essential viral function. However, the inserted Tet-On replication axis was stably maintained in HIV-rtTA over many years of cumulative in vitro replication and therefore proven critical for viral replication. In addition, the inactivated Tat–TAR axis was not functionally repaired over time. In the absence of dox, HIV-rtTA does not replicate. In traditional Tet-On system applications, rtTA is produced from a constitutive promoter and this protein will thus accumulate over time, resulting in some degree of expression leakiness. In contrast, the HIV-rtTA genetic organization ensures that viral gene expression is controlled by an inducible autoregulatory loop. As a consequence, rtTA production is virtually undetectable in the absence of dox, therefore imposing a very strict dox-control over HIV-rtTA replication.

Of note, HIV-rtTA also represents a valuable tool to improve the Tet-On system technology. Indeed, through spontaneous virus evolution, we were able to select rtTA proteins with improved function. For instance, we identified rtTA mutants that can be activated at minimal dox-levels (rtTA-V14; 10 ng/mL dox) [28, 29] as compared to the commercially available rtTA2^{S}-S2 and rtTA2^{S}-M2 variants (1000 ng/mL dox) [30]. The selected mutations enhance both the transcriptional activity and dox-sensitivity of rtTA, which is available as the third generation (3G) Tet-On system from Clontech [28, 29]. In vivo, such rtTA variants may be useful to achieve high-level replication in tissues that may receive suboptimal dox-levels due to bioavailability restrictions, as will be discussed in the following section. We also described rtTA variants that reinforce the long-term dox-dependency [31, 32].

35.2 Evaluating Dox Bioavailability In Vivo

Before analyzing the in vivo behavior of the HIV-rtTA virus, we addressed two important technical issues, namely the bioavailability of dox in various anatomical locations and the efficacy of an rtTA-based gene expression system in this context. For this, we generated BALB/c Rag2$^{-/-}$ IL-2R$\gamma_c^{-/-}$ (BRG) human immune system (HIS) mice by injecting fetal liver CD34^{+} CD38^{-} human hematopoietic progenitor cells (hHPCs) into immunodeficient newborn BRG mice [33–35]. All major subsets of the human innate and adaptive immune system are present in the reconstituted BRG-HIS mice, including T and B cells, conventional dendritic cells (cDCs), plasmacytoid dendritic cells (pDCs), macrophages, and natural killer cells. Similarly to what has been reported in other humanized mouse models of the immune system, productive infection of BRG-HIS mice with HIV-1 is possible (both X4- and R5-tropic strains), with infection parameters being reminiscent of those observed in humans [36–39]. The BRG-HIS mouse model therefore represents an attractive tool to analyze HIV-1 pathogenesis and to test the potential of new antiviral compounds in an in vivo setting.

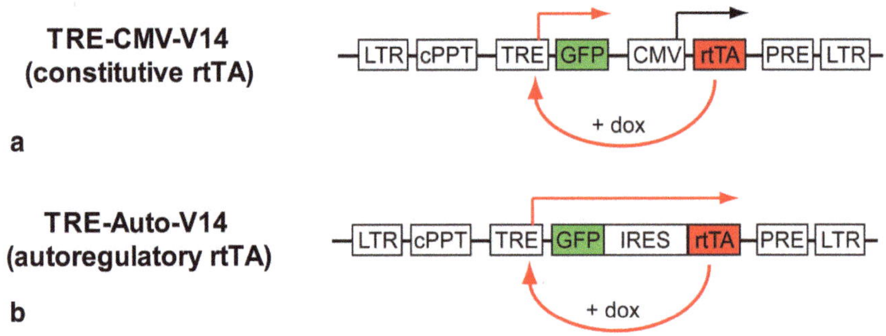

Fig. 35.2 Schematic of the HIV-1 LAI, HIV-1 LAI-ΔNef and HIV reverse tetracycline transactivator (*rtTA*) genomes [49]. **a** In HIV-1 LAI, transcription and replication of the virus is dependent on the *Tat*–transactivating response (*TAR*) axis with Tat binding to the TAR hairpin at the 5'-end of the nascent HIV-1 transcripts. The out-of-frame, 106 bp deletion in the *nef* gene of the HIV-1 LAI-ΔNef variant is indicated. **b** In HIV-rtTA, the *Tat–TAR* axis of transcription regulation has been inactivated by mutations in both Tat (Y26A) and TAR (five mutations in the bulge and loop elements). Transcription and replication of the virus became doxycycline (*dox*) dependent by the introduction of tet operator (*tetO*) elements in the U3 promoter region and replacement of the *nef* gene for the *rtTA* gene. *LTR* long terminal repeat

One important parameter to consider when evaluating HIV-rtTA in vivo is the delivery of dox, its diffusion, and distribution in the animals that will affect the kinetics of on:off dox responses. We addressed this question in the BRG-HIS mice by setting up a biological assay to measure dox concentrations in small samples [40]. We compared two noninvasive methods of dox delivery to the BRG-HIS mice, in the drinking water or the diet of the animals. The possibility to reach high plasmatic dox levels (>1000 ng/mL) and the kinetics of dox build-up/decay in the plasma were similar for the two administration routes, with a dox plateau reached some 3 days after dox-administration [41]. Dox levels became undetectable 2–3 days after dox-withdrawal. Higher dox concentrations were reached with dox-diet than dox-water, which is of importance for an efficient delivery to organs with restricted dox diffusion. For example, hardly any dox could be detected in the brain—that is in the cerebrospinal fluid samples—after delivery by drinking water, whereas a dox concentration of ~400 ng/mL was achieved with a high dose dox-diet (6 g/kg). This observation also highlights the relevance of new, more sensitive rtTA variants, obtained through in vitro viral evolution of HIV-rtTA, for certain in vivo experiments. Indeed, the rtTA-V14 variant, that is activated at reduced dox-levels (10 ng/mL dox) [28, 29], insures better onset of expression/replication even in restricted anatomical locations. Dox-containing food presents several advantages over drinking water, e.g., protection of light-sensitive dox from light, improved dox stability over time and less frequent supply replacements (once a week). When delivered in drinking water, dox has to be refreshed every other day. This delivery route also requires sucrose addition to mask the bitter taste of dox, thus influencing the daily water intake by the animals and limiting the maximal concentration of dox [42–44].

To evaluate dox bioavailability in the lymphoid organs of BRG-HIS mice, we made use of two inducible lentiviral vectors in which green fluorescent protein (GFP) expression is driven from a dox-dependent promoter. In the TRECMV-V14 vector, rtTA was expressed from a constitutive promoter, whereas the TREAuto-V14 vector expressed rtTA from a dox-dependent promoter placed in an autoregulatory loop setting [45], similarly to the design of HIV-rtTA [41]. Both vectors encode the ultra-sensitive rtTA-V14 variant. The hHPCs were transduced with a lentiviral vector and inoculated as a bulk cell population into the newborn BRG mice, thereby generating BRG-HIS mice in which a fraction of the human hematopoietic cells should express GFP upon dox exposure (Fig. 35.2).

The autoregulatory TREAuto-V14 lentiviral vector exhibited enhanced dox-induced gene expression and improved genetic control in the absence of dox, as compared to the constitutive TRECMV-V14 lentiviral vector [41]. Both gene expression systems allowed efficient reporter gene expression—and hence effective dox diffusion—in all analyzed lymphoid organs, i.e., spleen, bone marrow, liver, and thymus, underscoring that dox delivery and diffusion is not a limiting factor in the study of HIV-rtTA replication. Moreover, multiple rounds of dox-dependent gene expression could be induced in human hematopoietic cells in vivo, which is an important feature for gene-therapy applications when toxicity induced by overexpression or long-term exposure to the transgene product has to be avoided [46–48].

35.3 HIV-rtTA Pathogenesis in BRG-HIS Mice

Having established that dox administration and diffusion was not a limiting factor in vivo, an initial evaluation of the HIV-rtTA replication properties and pathogenicity was performed in BRG-HIS mice. Dox-fed BRG-HIS mice (>10 weeks old) were infected with 5.10^4 $TCID_{50}$ of either HIV-rtTA or the parental HIV-1 LAI strain. Furthermore, a ΔNef variant of HIV-1 LAI was included as a control for HIV-rtTA, which carries a deletion in the ΔNef gene to accommodate the rtTA gene insertion. A productive infection by HIV-rtTA was established in vivo in the presence of dox in the HIV-rtTA-infected BRG-HIS mice, as illustrated by the gradual accumulation of viral RNA in the plasma, reaching a plateau of $\sim 10^5$ viral RNA copies/mL at 4–5 weeks postinfection [49]. However, HIV-rtTA viremia was delayed as compared to the parental HIV-1 LAI-infected BRG-HIS mice (peak at 10^6 viral RNA copy/mL at 1 week postinfection followed by a plateau at $\sim 10^4$–10^5 RNA copies/mL). HIV-rtTA established a disseminated infection, as a high frequency of capside (CA)-p24$^+$ human cells were detected in the spleen and bone marrow of the HIV-rtTA-infected animals. However, despite active replication in the presence of dox, HIV-rtTA did not induce human CD4$^+$ T cell depletion in the blood or the lymphoid organs. In contrast, HIV-1 LAI replication in BRG-HIS mice was accompanied by a dramatic depletion of human CD4$^+$ T cells as early as 2 weeks postinfection. Importantly, we reported that HIV-rtTA was genetically stable in vivo and did not show any sign of escape from dox-control for up to 10 weeks of infection.

The fact that we could detect a high frequency of CA-p24$^+$ human cells in the organs of the HIV-rtTA-infected animals, and that more than 95 % of these target cells were human CD4$^+$ T cells, confirmed that HIV-rtTA is able to productively infect these human cells in vivo. Introduction of the Tet-On system in the HIV-1 genome appeared not to change the viral cell tropism, although more sensitive in situ staining is required to address this issue in further detail. We were able to recover replicating virus as evidenced by the capacity of splenocytes of HIV-rtTA-infected animals to spread the infection to co-cultured activated PBMCs ex vivo. Moreover, the absence of human CD4$^+$ T cell depletion in HIV-rtTA-infected animals cannot be explained by the replacement of the *nef* gene by the *rtTA* gene in the viral genome, as HIV-LAI-ΔNef also induced depletion of the human CD4$^+$ T cell in the blood and the lymphoid organs in vivo, but with a 1–2-week delay as compared to HIV-1 LAI [49]. Ten weeks after onset of infection, HIS mice infected with either HIV-1 LAI or HIV-LAI-ΔNef exhibited similar numbers of human cells in central and peripheral lymphoid organs, including the thymus, therefore demonstrating a similar long-term pathogenicity ([49], Centlivre M. personal unpublished data). The attenuated phenotype concomitant with *nef* deletion was also observed in the "bone marrow, liver, thymus" (BLT) humanized mouse model, but unlike in our study, a limited loss of the human CD4$^+$ T cells and thymocytes was reported [50]. This difference in pathogenicity associated with the *nef* deletion may be due to differential deletion strategies in the *nef* gene—an out-of-frame 106 bp deletion upstream of the polypurine tract [49] versus two large deletions flanking the polypurine tract [50]—and/or intrinsic properties of the respective humanized mouse models used in these studies. In particular, human thymopoiesis in the BRG-HIS mice is supported by the murine thymus, whereas human T cell development predominantly takes place in a human thymic organoid in the BLT mice. Overall, pathogenicity observed for HIV-rtTA was less severe than that of the parental HIV-1 LAI virus and the ΔNef control, with evidence of delayed viremia, absence of human CD4$^+$ T cell depletion and accumulation of CA-p24$^+$ human CD4$^+$ T cells.

We hypothesized that the lack of human CD4$^+$ T cell depletion could be attributed to the emergence of an anti-HIV B cell response in the HIV-rtTA infected humanized mice. However, we were not able to detect anti-HIV antibodies in these animals, using ELISA- and Western blot-based human clinical assays [49]. The maintenance of human helper CD4$^+$ T cells did not support the generation of an antibody response, which therefore seems to depend on mechanisms that appear suboptimal in BRG-HIS mice. Considering the rare occurrence of human T cell responses in BRG-HIS mice reported so far, it also seems unlikely that the human T cell response against HIV-rtTA is responsible for the absence of human CD4$^+$ T cell depletion. However, we observed a trend towards enhanced numbers of human CD4$^+$ and CD8$^+$ T cells in the spleen, liver, and blood of HIV-rtTA-infected animals at 10 weeks after infection, as compared to the mock-infected BRG-HIS mice. This does not necessarily reflect an anti-HIV-rtTA T cell response but could be due to a more general immune activation mechanism, mediated by HIV infection and leading to an accumulation of T cells in the context of a nondepleting $CD4^+$ T cell infection.

35.4 Conclusions

We demonstrated the proof-of-concept of dox-dependent gene expression in human hematopoietic cell populations of BRG-HIS mice using dox-inducible HIV-based lentiviral vectors. Furthermore, a bioassay to measure the dox concentration in small biological samples permitted us to monitor drug levels in BRG-HIS mice. HIV-rtTA infection of dox-fed BRG-HIS mice led to a productive and disseminated infection without inducing human $CD4^+$ T cell depletion. Genetic stability of HIV-rtTA was demonstrated, without the emergence of dox-independent virus variants or a reversion towards a functional Tat–TAR axis. Despite vigorous virus replication, we were unable to detect any sign of anti-HIV B cell responses in HIV-rtTA-infected BRG-HIS mice. This research line is currently extended to dox-controlled SIV, which allows testing and safety monitoring of this innovative vaccination strategy in the rhesus macaque model. Comparison of the data obtained in these two models will be of particular interest to judge the intrinsic value of humanized mouse models as preclinical platform for the evaluation of drugs, new antiviral therapeutics strategies, and vaccines. The first data obtained indicate that the conditional-live, attenuated SIV-rtTAΔnef strain exhibits a decent replicative capacity in vivo [51]. The peak viremia obtained with the SIV-rtTAΔnef virus is around two $\log_{10}$ lower than that of the parental SIVΔnef strain. This corroborates the results obtained in the BRG-HIS mice where HIV-rtTA viremia was delayed as compared to the parental HIV-1ΔNef strain. Despite a shift towards increase of effector memory T cell frequency, the capacity of the animals to mount a protective response after a challenge with virulent virus remains to be determined. Pilot challenge experiments indicate that a dox-window of 20 weeks suffices for full protection (Berkhout B., personal unpublished results).

Overall, the proof-of-concept for the in vivo replicative capacity of the dox-dependent HIV-rtTA was obtained. This conditional-live HIV-rtTA virus is a powerful reagent that may help to define the exact correlates of vaccine protection, in particular the importance of ongoing virus replication. Moreover, the ability to control virus replication in a graded manner from the outside by a nontoxic effector molecule will provide a unique tool to study the relationship between virus replication and pathogenicity. However, there are several issues that need to be addressed before HIV-rtTA can be further developed as a safe vaccination strategy. Extensive in vitro analysis of HIV-rtTA has already generated distinct safety measures, including the use of safety-lock codons to prevent the evolution of virus variants with reduced dox-dependency [31]. Potential safety concerns have to be neutralized fully, including the risks associated with inter-HIV recombination events, integration of HIV-rtTA into the human genome, or undesired viral reactivation due to future dox treatment not related to the vaccination protocol. As already mentioned, efficacy after a virulent HIV challenge will be evaluated in nonhuman primates using the

SIV-rtTAΔnef virus. A similar evaluation will have to be performed with HIV-rtTA in humanized mouse models of immunity optimized for immunological functionality. Further in vivo evaluation of conditional live HIV-rtTA/SIV-rtTAΔnef in humanized mice and rhesus macaques will bring us essential knowledge for the generation of a safe and effective HIV vaccine.

References

1. Rappuoli R, Miller HI, Falkow S. Medicine. The intangible value of vaccination. Science. 2002;297(5583):937–9. (Epub 2002/8/10).
2. Pulendran B, Ahmed R. Immunological mechanisms of vaccination. Nat Immunol. 2011;12(6):509–17. (Epub 2011/7/9).
3. Whitney JB, Ruprecht RM. Live attenuated HIV vaccines: pitfalls and prospects. Curr Opin Infect Dis. 2004;17(1):17–26. (Epub 2004/4/20).
4. Baba TW, Liska V, Khimani AH, Ray NB, Dailey PJ, Penninck D, et al. Live attenuated, multiply deleted simian immunodeficiency virus causes AIDS in infant and adult macaques. Nat Med. 1999;5(2):194–203. (Epub 1999/2/4).
5. Wyand MS, Manson KH, Lackner AA, Desrosiers RC. Resistance of neonatal monkeys to live attenuated vaccine strains of simian immunodeficiency virus. Nat Med. 1997;3(1):32–6. (Epub 1997/1/1).
6. Deacon NJ, Tsykin A, Solomon A, Smith K, Ludford-Menting M, Hooker DJ, et al. Genomic structure of an attenuated quasi species of HIV-1 from a blood transfusion donor and recipients. Science. 1995;270(5238):988–91. (Epub 1995/11/10).
7. Greenough TC, Sullivan JL, Desrosiers RC. Declining CD4 T-cell counts in a person infected with nef-deleted HIV-1. N Engl J Med. 1999;340(3):236–7. (Epub 1999/1/23).
8. Sheppard HW. Inactivated- or killed-virus HIV/AIDS vaccines. Curr Drug Targets Infect Disord. 2005;5(2):131–41. (Epub 2005/6/25).
9. Flynn NM, Forthal DN, Harro CD, Judson FN, Mayer KH, Para MF. Placebo-controlled phase 3 trial of a recombinant glycoprotein 120 vaccine to prevent HIV-1 infection. J Infect Dis. 2005;191(5):654–65. (Epub 2005/2/3).
10. Gilbert PB, Peterson ML, Follmann D, Hudgens MG, Francis DP, Gurwith M, et al. Correlation between immunologic responses to a recombinant glycoprotein 120 vaccine and incidence of HIV-1 infection in a phase 3 HIV-1 preventive vaccine trial. J Infect Dis. 2005;191(5):666–77. (Epub 2005/2/3).
11. Ruprecht RM. Live attenuated AIDS viruses as vaccines: promise or peril? Immunol Rev. 1999;170:135–49. (Epub 1999/11/24).
12. Johnson RP, Desrosiers RC. Protective immunity induced by live attenuated simian immunodeficiency virus. Curr Opin Immunol. 1998;10(4):436–43. (Epub 1998/9/2).
13. Tenner-Racz K, Stahl Hennig C, Uberla K, Stoiber H, Ignatius R, Heeney J, et al. Early protection against pathogenic virus infection at a mucosal challenge site after vaccination with attenuated simian immunodeficiency virus. Proc Natl Acad Sci U S A. 2004;101(9):3017–22. (Epub 2004/2/19).
14. Daniel MD, Kirchhoff F, Czajak SC, Sehgal PK, Desrosiers RC. Protective effects of a live attenuated SIV vaccine with a deletion in the nef gene. Science. 1992;258(5090):1938–41. (Epub 1992/12/18).
15. Wyand MS, Manson KH, Garcia-Moll M, Montefiori D, Desrosiers RC. Vaccine protection by a triple deletion mutant of simian immunodeficiency virus. J Virol. 1996;70(6):3724–33. (Epub 1996/6/1).

16. Stebbings R, Berry N, Stott J, Hull R, Walker B, Lines J, et al. Vaccination with live attenuated simian immunodeficiency virus for 21 days protects against superinfection. Virology. 2004;330(1):249–60. (Epub 2004/11/6).
17. Douek DC, Brenchley JM, Betts MR, Ambrozak DR, Hill BJ, Okamoto Y, et al. HIV preferentially infects HIV-specific CD4+T cells. Nature. 2002;417(6884):95–8. (Epub 2002/5/3).
18. Verhoef K, Marzio G, Hillen W, Bujard H, Berkhout B. Strict control of human immunodeficiency virus type 1 replication by a genetic switch: Tet for Tat. J Virol. 2001;75(2):979–87. (Epub 2001/1/3).
19. Hillen W, Berens C. Mechanisms underlying expression of Tn10 encoded tetracycline resistance. Annu Rev Microbiol. 1994;48:345–69. (Epub 1994/1/1).
20. Gossen M, Freundlieb S, Bender G, Muller G, Hillen W, Bujard H. Transcriptional activation by tetracyclines in mammalian cells. Science. 1995;268(5218):1766–9. (Epub 1995/6/23).
21. Yao F, Eriksson E. A novel tetracycline-inducible viral replication switch. Hum Gene Ther. 1999;10(3):419–27. (Epub 1999/2/27).
22. Yao F, Murakami N, Bleiziffer O, Zhang P, Akhrameyeva NV, Xu X, et al. Development of a regulatable oncolytic herpes simplex virus type 1 recombinant virus for tumor therapy. J Virol. 2010;84(16):8163–71. (Epub 2010/6/4).
23. Zhang H, Takayama K, Zhang L, Uchino J, Harada A, Harada T, et al. Tetracycline-inducible promoter-based conditionally replicative adenoviruses for the control of viral replication. Cancer gene therapy. 2009;16(5):415–22. (Epub 2009/1/17).
24. Hagen CJ, Titong A, Sarnoski EA, Verardi PH. Antibiotic-dependent expression of early transcription factor subunits leads to stringent control of vaccinia virus replication. Virus Res. 2014;181:43–52. (Epub 2014/1/8).
25. Marzio G, Verhoef K, Vink M, Berkhout B. In vitro evolution of a highly replicating, doxycycline-dependent HIV for applications in vaccine studies. Proc Natl Acad Sci U S A. 2001;98(11):6342–7. (Epub 2001/5/17).
26. Marzio G, Vink M, Verhoef K, de Ronde A, Berkhout B. Efficient human immunodeficiency virus replication requires a fine-tuned level of transcription. J Virol. 2002;76(6):3084–8. (Epub 2002/2/28).
27. Kiselyeva Y, Ito Y, Lima RG, Grivel JC, Das AT, Berkhout B, et al. Depletion of CD4 T lymphocytes in human lymphoid tissue infected ex vivo with doxycycline-dependent HIV-1. Virology. 2004;328(1):1–6. (Epub 2004/9/24).
28. Das AT, Zhou X, Vink M, Klaver B, Verhoef K, Marzio G, et al. Viral evolution as a tool to improve the tetracycline-regulated gene expression system. J Biol Chem. 2004;279(18):18776–82. (Epub 2004/2/6).
29. Zhou X, Vink M, Klaver B, Berkhout B, Das AT. Optimization of the Tet-On system for regulated gene expression through viral evolution. Gene Ther. 2006;13(19):1382–90. (Epub 2006/5/26).
30. Urlinger S, Baron U, Thellmann M, Hasan MT, Bujard H, Hillen W. Exploring the sequence space for tetracycline-dependent transcriptional activators: novel mutations yield expanded range and sensitivity. Proc Natl Acad Sci U S A. 2000;97(14):7963–8. (Epub 2000/6/22).
31. Zhou X, Vink M, Berkhout B, Das AT. Modification of the Tet-On regulatory system prevents the conditional-live HIV-1 variant from losing doxycycline-control. Retrovirology. 2006;3:82. (Epub 2006/11/11).
32. Zhou X, Vink M, Klaver B, Verhoef K, Marzio G, Das AT, et al. The genetic stability of a conditional live HIV-1 variant can be improved by mutations in the Tet-On regulatory system that restrain evolution. J Biol Chem. 2006;281(25):17084–91. (Epub 2006/4/22).
33. Gimeno R, Weijer K, Voordouw A, Uittenbogaart CH, Legrand N, Alves NL, et al. Monitoring the effect of gene silencing by RNA interference in human CD34+ cells injected into newborn RAG2−/− gammac−/− mice: functional inactivation of p53 in developing T cells. Blood. 2004;104(13):3886–93. (Epub 2004/8/21).
34. Legrand N, Huntington ND, Nagasawa M, Bakker AQ, Schotte R, Strick-Marchand H, et al. Functional CD47/signal regulatory protein alpha (SIRP(alpha)) interaction is required for

optimal human T- and natural killer- (NK) cell homeostasis in vivo. Proc Natl Acad Sci U S A. 2011;108(32):13224–9. (Epub 2011/7/27).
35. Traggiai E, Chicha L, Mazzucchelli L, Bronz L, Piffaretti JC, Lanzavecchia A, et al. Development of a human adaptive immune system in cord blood cell-transplanted mice. Science. 2004;304(5667):104–7. (Epub 2004/4/6).
36. Baenziger S, Tussiwand R, Schlaepfer E, Mazzucchelli L, Heikenwalder M, Kurrer MO, et al. Disseminated and sustained HIV infection in CD34+ cord blood cell-transplanted Rag2−/− gamma c−/− mice. Proc Natl Acad Sci U S A. 2006;103(43):15951–6. (Epub 2006/10/14).
37. Berges BK, Wheat WH, Palmer BE, Connick E, Akkina R. HIV-1 infection and CD4 T cell depletion in the humanized Rag2−/−gamma c−/− (RAG-hu) mouse model. Retrovirology. 2006;3:76. (Epub 2006/11/3).
38. Gorantla S, Sneller H, Walters L, Sharp JG, Pirruccello SJ, West JT, et al. Human immunodeficiency virus type 1 pathobiology studied in humanized BALB/c-Rag2−/−gammac−/− mice. J Virol. 2007;81(6):2700–12. (Epub 2006/12/22).
39. Zhang L, Kovalev GI, Su L. HIV-1 infection and pathogenesis in a novel humanized mouse model. Blood. 2007;109(7):2978–81. (Epub 2006/11/30).
40. Kleibeuker W, Zhou X, Centlivre M, Legrand N, Page M, Almond N, et al. A sensitive cell-based assay to measure the doxycycline concentration in biological samples. Hum Gene Ther. 2009;20(5):524–30. (Epub 2009/2/11).
41. Centlivre M, Zhou X, Pouw SM, Weijer K, Kleibeuker W, Das AT, et al. Autoregulatory lentiviral vectors allow multiple cycles of doxycycline-inducible gene expression in human hematopoietic cells in vivo. Gene Ther. 2010;17(1):14–25. (Epub 2009/9/4).
42. Hojman P, Eriksen J, Gehl J. Tet-On induction with doxycycline after gene transfer in mice: sweetening of drinking water is not a good idea. Anim Biotechnol. 2007;18(3):183–8. (Epub 2007/7/7).
43. Cawthorne C, Swindell R, Stratford IJ, Dive C, Welman A. Comparison of doxycycline delivery methods for Tet-inducible gene expression in a subcutaneous xenograft model. J Biomol Tech. 2007;18(2):120–3. (Epub 2007/5/15).
44. Honnorat-Benabbou VC, Lebugle AA, Sallek B, Duffaut-Lagarrigue D. Stability study of tetracyclines with respect to their use in slow release systems. J Mater Sci Mater Med. 2001;12(2):107–10. (Epub 2004/9/7).
45. Markusic D, Oude-Elferink R, Das AT, Berkhout B, Seppen J. Comparison of single regulated lentiviral vectors with rtTA expression driven by an autoregulatory loop or a constitutive promoter. Nucleic Acids Res. 2005;33(6):e63. (Epub 2005/4/6).
46. Stieger K, Mendes-Madeira A, Meur GL, Weber M, Deschamps JY, Nivard D, et al. Oral administration of doxycycline allows tight control of transgene expression: a key step towards gene therapy of retinal diseases. Gene Ther. 2007;14(23):1668–73. (Epub 2007/10/5).
47. Vigna E, Amendola M, Benedicenti F, Simmons AD, Follenzi A, Naldini L. Efficient Tet-dependent expression of human factor IX in vivo by a new self-regulating lentiviral vector. Mol Ther. 2005;11(5):763–75. (Epub 2005/4/27).
48. Herold MJ, van den Brandt J, Seibler J, Reichardt HM. Inducible and reversible gene silencing by stable integration of an shRNA-encoding lentivirus in transgenic rats. Proc Natl Acad Sci U S A. 2008;105(47):18507–12. (Epub 2008/11/20).
49. Legrand N, van der Velden GJ, Fang RH, Douaisi M, Weijer K, Das AT, et al. A doxycycline-dependent human immunodeficiency virus type 1 replicates in vivo without inducing CD4+ T-cell depletion. J Gen Virol. 2012;93(Pt 9):2017–27. (Epub 2012/6/1).
50. Zou W, Denton PW, Watkins RL, Krisko JF, Nochi T, Foster JL, et al. Nef functions in BLT mice to enhance HIV-1 replication and deplete CD4+CD8+ thymocytes. Retrovirology. 2012;9:44. (Epub 2012/5/30).
51. Manoussaka MS, Berry N, Ferguson D, Stebbings R, Robinson M, Ham C, et al. Conditionally-live attenuated SIV upregulates global T effector memory cell frequency under replication permissive conditions. Retrovirology. 2013;10(1):59. (Epub 2013/6/7).

Part VI
New Models for Other Human-Specific or Human Selective Pathogens

Chapter 36
Dual Reconstituted Mice for Hepatotropic Pathogens

Helene Strick-Marchand and Alexander Ploss

36.1 The Need for Animal Models for Human Hepatotropic Pathogens

Hepatotropic pathogens, in particular hepatitis viruses and plasmodial parasites causing malaria, account for more than 800 million infections worldwide and contribute substantially to human morbidity and mortality. *Plasmodium falciparum* and *Plasmodium vivax* infect annually 250 and 80 million people, respectively, and are responsible for about 1 million malaria-related deaths primarily in young children [1]. Five hepatitis viruses are recognized to cause disease in humans. While hepatitis A (HAV) and E viruses (HEV) primarily cause acute, often self-limiting infections, hepatitis C (HCV), B (HBV) and delta viruses (HDV)—the latter only in conjunction with HBV—have a high propensity for chronicity [2–4]. Chronic viral hepatitis frequently results in severe liver disease, which can culminate in hepatocarcinogenesis [5–7]. HBV infection and consequently HDV, a satellite virus that requires an underlying HBV to propagate its genome [8], can be prevented efficiently by prophylactic vaccination. The currently available anti-HBV drugs can efficiently suppress the virus; however, they can only rarely eradicate the virus and completely cure the patient [9]. Therefore, chronically infected patients with an active disease must usually be treated with antiviral agents for lifelong, resulting in significant morbidity and costs, not to mention the risk of emerging mutants and viral reactivation.

A study of human hepatotropic viruses has been hampered by the lack of infectious systems that accurately mimic the unique host environment of the liver [10]. By and large, only a few human cell lines efficiently support replication of hepatitis

H. Strick-Marchand (✉)
Unité Immunité Innée, Institut Pasteur, Inserm U668, Paris, France
e-mail: helene.strick-marchand@pasteur.fr

A. Ploss (✉)
Department of Molecular Biology, Princeton University, Princeton, NJ 08544, USA
e-mail: aploss@princeton.edu

L. Y. Poluektova et al. (eds.), *Humanized Mice for HIV Research*,
DOI 10.1007/978-1-4939-1655-9_36

viruses—in particular HCV and HBV, but their transformed nature does not reflect normal hepatocyte physiology. Primary human hepatocytes replicate hepatitis viruses only inefficiently, in part due to the difficulty in maintaining their highly differentiated phenotype in cell culture [10]). A study of hepatotropic pathogens in the physiological 3D context of an intact liver would be highly desirable, but animal models are scarce [11] because of the narrow host range of plasmodial parasites causing malaria in humans and most of the known human hepatitis viruses. Chimpanzees are the only immunocompetent species besides humans that are naturally susceptible to many of the aforementioned pathogens [12]. However, large apes are not amenable to genetic manipulation, as an outbred species they are genetically highly variable, and liver tissue is not readily accessible. Furthermore, the National Institutes of Health moratorium on "nonessential" chimpanzee research [13] will severely limit future chimpanzee research, and thus alternative animal models are urgently needed.

Humanized mice have emerged as an important technology that has the potential to fill this void. In this chapter, we discuss human liver chimeric mice, which are susceptible to human hepatotropic pathogens. In addition, we focus on approaches for co-engraftment with components of a human immune system, which may facilitate the analysis of human immune responses, pathogenesis, and the preclinical evaluation of vaccine candidates (Fig. 36.1).

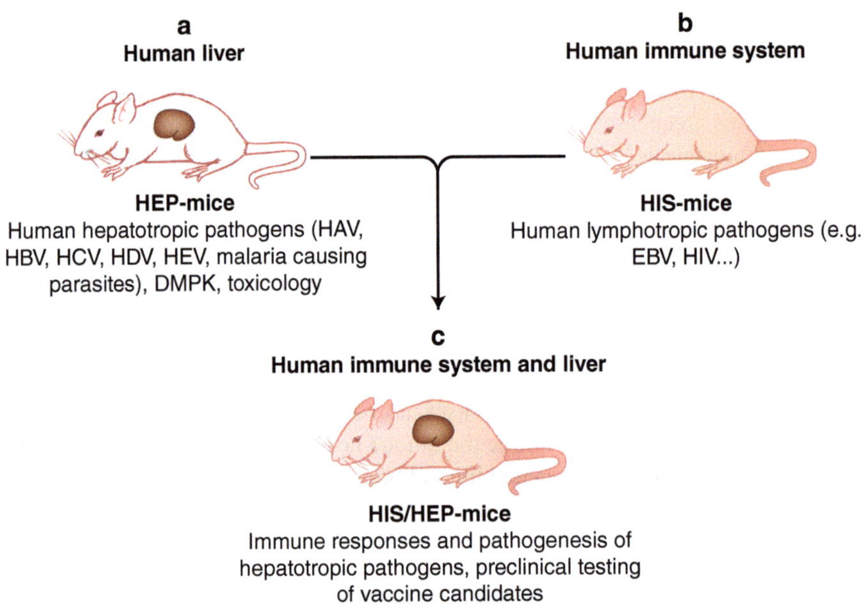

Fig. 36.1 Applications of dually reconstituted mice. *HAV* hepatitis A virus, *HBV* hepatitis B virus, *HCV* hepatitis C virus, *HDV* hepatitis delta virus, *HEV* hepatitis E virus, *DMPK* drug metabolism and pharmacokinetics, *EBV* Epstein Barr virus, *HIS* Human Immune System, *HEP* Human Hepatocyte

36.2 Human Liver Chimeric Mice

To develop a small animal model susceptible to hepatotropic pathogens, human hepatocytes must be robustly engrafted. Although liver fragments can be grafted at ectopic sites (i.e., under the kidney capsule), these models do not take into account the liver's unique architecture. Successful repopulation of the liver with injected cells depends on three parameters: sufficient ablation of the host's immune system to inhibit rejection mechanisms, creation of a niche in which the cells will engraft, and maintenance of a selective advantage for the donor over the host cells (Fig. 36.2).

With the aim of obtaining hosts permissive for transplantation of human cells, immunodeficient mice lacking B, T, (Scid, $Rag1^{-/-}$ or $Rag2^{-/-}$ mice) and natural killer (NK) lymphocytes ($Rag2^{-/-}IL2Rg_c^{-/-}$ mice) were bred with animals harboring severe liver damage to create an environment of competitive liver regeneration. The first model was developed from mice carrying a urokinase plasminogen activator (uPA) transgene under the control of the albumin promoter/enhancer [14]. Originally, uPA transgenic mice were established to study blood clotting mechanisms. However, the transgene caused hepatocyte cytolysis, hypofibrinogenemia, and severe liver damage, thus, creating a niche in which mouse, rat, or human hepatocytes could engraft, proliferate, and regenerate the diseased liver in the context of an immunosuppressed host [15–17].

The second model was initially developed to study a human metabolic disease affecting the liver and kidneys, hereditary tyrosinaemia type I, caused by a deficiency of fumarylacetoacetate hydrolase (FAH). This enzyme is the last step in tyrosine catabolism and its deficiency causes toxic metabolites to accumulate in patients and $FAH^{-/-}$ mice, leading to neonatal death in mice [18]. Treating with the drug 2-(2-nitro-4-trifluoro-methylbenzyol)-1, 3-cyclohexanedione (NTBC) can bypass the lethality, and cycling the drug on and off in mice renders them permissive for exogenous hepatocyte repopulation [19, 20]. Subsequently, efficient human hepatocyte repopulation was demonstrated in immunodeficient $FAH^{-/-}$ mice, although additional liver damage, through the use of an adenoviral vector expressing uPA, was required for human cell engraftment [21, 22].

Recently, a third model based on the expression of the herpes simplex virus thymidine kinase (HSVtk) transgene under control of the albumin promoter/enhancer in immunodeficient NOD/Shi-scid/IL-2Rgamma [null] (NOG) mice was described [23]. In these animals, administration of ganciclovir induces liver damage, creating a niche into which human hepatocytes engraft, resulting in high levels of repopulation.

By using either uPA transgenic, $FAH^{-/-}$, or HSVtk immunodeficient mice, numerous studies showed that these models support high levels of liver chimerism (20–90 %) for several months, producing human serum proteins such as clotting factors and complement proteins, and demonstrating human drug metabolizing activities that are essential for toxicology studies [23–26]. To compare the different models and follow the engraftment longitudinally, the gold standard is an enzyme-linked immunosorbent assay (ELISA) specific for human albumin (hAlb), which quantifies human chimerism and functional hepatocyte engraftment. It is

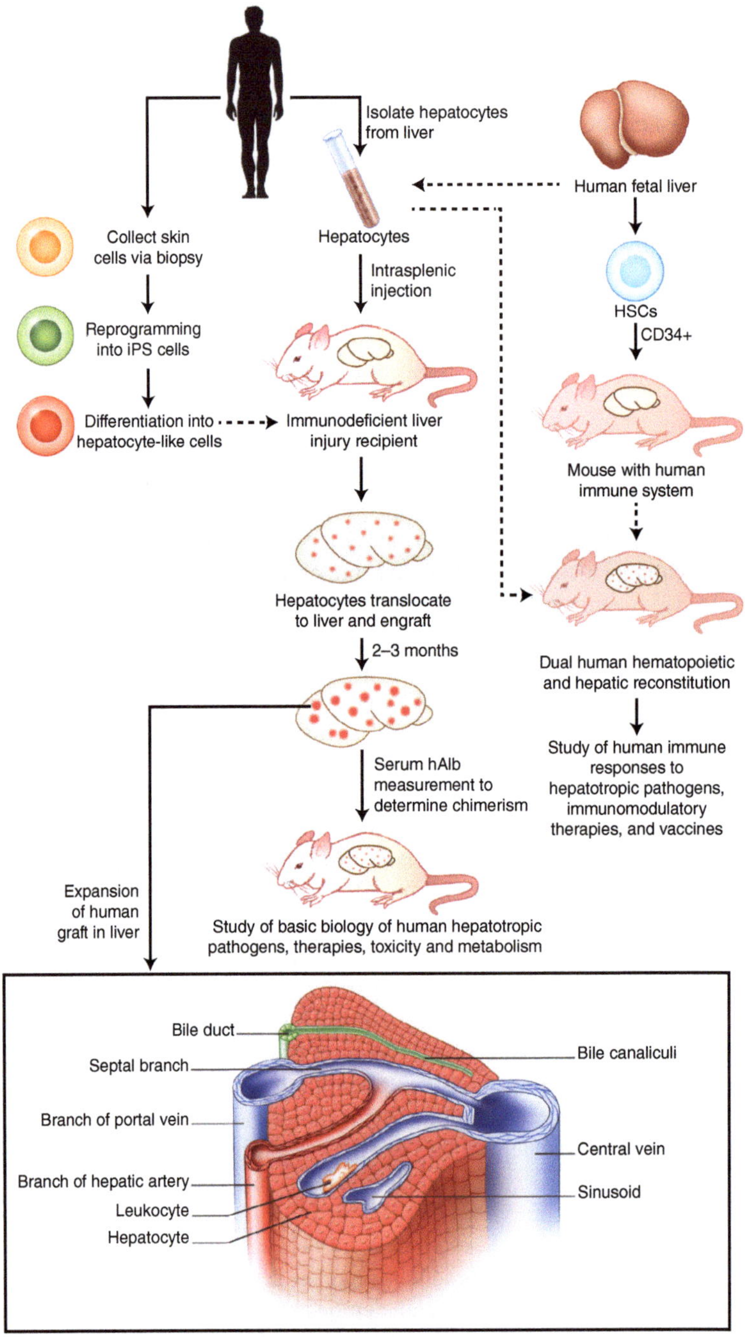

Fig. 36.2 Creation and future improvements of dually reconstituted mice. *NTBC* 2-(2-nitro-4-trifluoro-methylbenzyol)-1, 3-cyclohexanedione, *HSC* hematopoietic stem cell, *HBV* hepatitis B virus, *HCV* hepatitis C virus. ([62], republished with the permission of the American Society for Clinical Investigation, permission conveyed through Copyright Clearance Center, Inc)

Table 36.1 Animal models for liver humanization

Model	Colony management	Hepatocyte engraftment	hES- and hiPS-derived hepatocyte engraftment	Double humanization of the immune system and liver	HBV and HCV infections
Alb-uPA	Homozygous animals are sterile, fertility is restored by transplanting allogeneic hepatocytes	Non-inducible system, mice must be transplanted a few weeks after birth	Yes, low levels of repopulation	Not shown	Yes, high viral titers in plasma
FAH	Normal fertility, continuous NTBC treatment and/or low tyrosine diet	Inducible system with NTBC treatment, mice must be monitored daily during NTBC cycling	Not shown	Not shown	Yes, high viral titers in plasma
TK-NOG	Males are sterile, breeding with TK-NOG females and NOG males	Inducible system with GCV treatment	Not shown	Not shown	Not shown
Caspase 8	Normal fertility	Inducible system, toxicity unknown	Not shown	Yes	Yes for HCV: no viral titers in plasma, RNA only detected in the liver

FAH fumarylacetoacetate hydrolase, *TK-NOG* thymidine kinase- NOD/Shi-scid/IL-2Rgamma (null), *GCV* Ganciclovir, *hES* human embryonic stem cell, *hiPS* human induced pluripotent stem cell, *HBV* hepatitis B virus, *HCV* hepatitis C virus, *NTBC* 2-(2-nitro-4-trifluoro-methylbenzyol)-1, 3-cyclohexanedione, *RNA* ribonucleic acid

noteworthy that all of these models present advantages and caveats, and each model should be chosen to answer a specific question, a side-by-side comparison is presented in Table 36.1 [27–29].

The uPA transgenic and $FAH^{-/-}$ immunodeficient mice harboring human hepatocytes (HUHEP mice) were shown to be productively infected with HBV and HCV, replicating the entire viral life cycle in vivo and resulting in high levels of viremia [17, 30, 31] reviewed in [32]. In both humanized models, HBV is more infectious than HCV, the latter requiring higher levels of human chimerism (> 50 %, equivalent to mg/ml hAlb) to result in a successful infection [31–33]. Differences in viral particle structure and entry mechanisms could account for this observation.

To develop novel therapeutic treatments, HBV-infected HUHEP mice were tested for the efficacy of antiviral compounds such as Lamivudine or BAY 41-4109 and more classical interferon alpha (IFNa) therapy [34–36]. Although these drugs reduced viremia, they did not result in viral eradication, probably due to the inability to covalently closed circular DNA (eliminate cccDNA).

Structural analysis of HCV particles produced by humanized mice showed similarities with those analyzed from chimpanzees and humans, whereas cell culture derived particles were most divergent, further demonstrating the pertinence of these in vivo models [37]. The evaluation of prophylactic or therapeutic compounds for HCV has been extensively conducted in uPA HUHEP mice. For example, antibodies directed against HCV receptors (CD81 and SR-BI) blocked or delayed the onset of infection in HUHEP mice when administered at the time of inoculation [38, 39]. Direct acting antivirals such as a NS3-4A protease inhibitor (telaprevir) or a NS5B polymerase inhibitor (HCV371) demonstrated similar results in HUHEP mice as in chimpanzees [40–42].

Taken together, this body of work demonstrates the potential use of HUHEP mice to analyze the complete viral life cycle, assay virus entry inhibitors or antiviral molecules, and evaluate potential drug toxicities. However, these models do not take into account immune mediated responses, which are intrinsic to the pathophysiology generated by hepatotropic viral infections. Deciphering the cross-talk between the infected hepatocyte and immune cells, as well as the balance between proinflammatory and immunosuppressive responses, remain major challenges in the field.

36.3 Dually Engrafted Mice

With the goal of establishing an animal model humanized for both the immune system and liver, L. Su et al. generated mice with an active Caspase 8 fused to the FK506 binding domain (FKBP) under the control of the albumin promoter in BALB/c $Rag2^{-/-}IL2Rg_c^{-/-}$ mice (AFC8 mice) [43]. Repeated administration of the drug AP20187 led to dimerization of active Caspase 8 and mouse hepatocyte apoptosis. By coinjecting $CD34^+$ human hematopoietic stem cells with human hepatocyte progenitors from a common source of fetal liver, a syngeneic humanized immune system and chimeric liver were generated. Following inoculation with HCV, about half of the animals had low levels of viral RNA in the liver, but no viremia was detectable in the serum. The HCV-infected animals showed hepatic inflammation with a recruitment of Kupffer cells ($CD68^+$), as well as CD4 and CD8 T cell responses following HCV peptide stimulation ex vivo. Portal fibrosis was also observed in the infected animals, with elevated levels of human Tissue inhibitor of metalloproteinase 1 (TIMP1) and collagen 1A1 in the liver of a few animals. Further analysis of the molecular and cellular mediators of the immunopathologies in HCV-infected humanized AFC8 mice would be useful to understand the evolution of the disease.

The AFC8 mice provided a proof-of-principle that a dually humanized small animal model could be generated. The incomplete differentiation of human hepatoblasts could play a role in the low level of human hepatocyte repopulation, and thus an absence of high-level viremia following HCV inoculation. Alternatively, more sustained liver damage could be required to obtain higher levels of human chimerism, but with the risk of creating an inflammatory environment in the liver, which could interfere with interpretations of virus-induced pathologies. Although the humanized immune system and hepatocytes are human leukocyte antigen (HLA) matched in this model, the T cells are educated on mouse Major histocompatibility complex (MHC). The addition of HLA transgenes to this model could enhance the T cell mediated response [44, 45]. The early onset of liver fibrosis, within 3 months following infection, is remarkable and is an unusually rapid progression compared to the relevant clinical setting. In patients, HCV-associated liver fibrosis usually develops only after years, sometimes decades, of persistent infection.

36.4 Utility and Necessary Improvements of Dually Engrafted Animals

Human liver chimeric mice are susceptible to a variety of human hepatotropic pathogens, including HBV [25, 31], HCV [25, 30, 31]; HDV [46] and *P. falciparum* [47–49], which has opened unprecedented opportunities to study their interactions with the human host (Fig. 36.2). Furthermore, liver humanized mice are increasingly used to assess preclinically the efficacy of drug candidates (reviewed in [32]). In this respect, it is particularly attractive that highly engrafted animals exhibit human-like metabolic profiles presumably increasing the predictive value DMPK and toxicological studies (reviewed in [50]; Fig. 36.2). To systematically investigate the impact of human host genetics on human liver disease it would be desirable to engraft animals with patient-specific hepatocytes. Induced pluripotent stem cells (iPSCs) can be routinely generated from easily accessible tissues, such as blood and skin in a process of cellular reprogramming and are easily grown, scalable, and can in a developmentally appropriate and efficient manner be differentiated towards the hepatocyte lineage [28, 51–53] (Fig. 36.2). Unfortunately, the functional engraftment and repopulation of stem-cell-derived hepatocytes in existing chimeric mouse models have been limited to date [29, 53–55]. Furthermore, it will be critical to more closely reproduce the cellular complexity of the liver. Current protocols facilitate only engraftment with human adult hepatocytes. However, other human nonparenchymal cells which shape liver function and modulate immune responses to infection, including stellate cells, liver sinusoidal endothelial cells, and cholangiocytes cannot be detected (Fig. 36.2).

With the advent of better, more immunocompromized xenorecipient strains (reviewed in [44, 56]), mice can now be routinely transplanted with human hematopoietic stem cells, yielding multilineage human cell engraftment. In general, human lymphocyte populations including B, T, and NK cells can be readily detected [57, 58]. However, cells derived from the erythro-myeloid lineage, such as various dendritic cell populations, macrophages, megakaryocytes, and erythrocytes, are underrepresented presumably due to the limited biological cross-reactivity of certain critical lineage-promoting cytokines (reviewed in [59]). These imbalances along with other factors such as the suboptimal architectural organization of lymphoid organs and the lack of human major-histocompatibility complex genes result in a rather limited functional output from the engrafted human immune system (reviewed in [60]). Weak antigen-specific immune responses can be detected in humanized mice following challenge with lymphotropic viruses (e.g., Dengue virus, human immunodeficiency virus (HIV), or Epstein Barr virus), but they are generally not protective (reviewed in [61]). Consequently, existing models need to be further refined to attain stronger human cellular and humoral immune responses.

Undoubtedly, mice harboring both human hepatocytes and immune cells would be very useful for dissecting correlates of protection to human hepatotropic pathogens and for analysis of liver pathogenesis, especially in the context of clinically relevant coinfections (e.g., with HIV; Fig. 36.1). Proof-of-concept has been established that mice can be created that contain both tissue compartments [43]. Following HCV challenge, these dually engrafted AFC8 mice mounted HCV-specific human T cell responses and developed signs of liver disease. The very low human hepatic chimerism may be an inherent problem of the AFC8 model, which would likely be resolved using other, better characterized liver injury models including immunodeficient Alb-uPA transgenic, $FAH^{-/-}$, and/or HSVtk transgenic mice. Future studies need to focus on substantially improving the robustness of the engraftment for both compartments. These optimized doubly engrafted animals will be valuable models to study the pathophysiology of hepatotropic infections, and to evaluate prophylactic and therapeutic therapies in the context of challenge experiments.

References

1. Greenwood BM, Fidock DA, Kyle DE, Kappe SH, Alonso PL, Collins FH, et al. Malaria: progress, perils, and prospects for eradication. J Clin Invest. 2008;118(4):1266–76.
2. Dienstag JL. Hepatitis B virus infection. N Engl J Med. 2008;359(14):1486–500.
3. Hoofnagle JH, Nelson KE, Purcell RH. Hepatitis E. N Engl J Med. 2012;367(13):1237–44.
4. Farci P, Niro GA. Clinical features of hepatitis D. Semin Liver Dis. 2012;32(3):228–36.
5. Yang PL, Althage A, Chung J, Maier H, Wieland S, Isogawa M, et al. Immune effectors required for hepatitis B virus clearance. Proc Natl Acad Sci U S A. 2010;107(2):798–802.
6. Shepard CW, Finelli L, Alter MJ. Global epidemiology of hepatitis C virus infection. Lancet Infect Dis. 2005;5(9):558–67.
7. Feinstone SM. Hepatitis A: epidemiology and prevention. Eur J Gastroenterol Hepatol. 1996;8(4):300–5.
8. Taylor JM. Virology of hepatitis D virus. Semin Liver Dis. 2012;32(3):195–200.
9. Liaw YF. Natural history of chronic hepatitis B virus infection and long-term outcome under treatment. Liver Int. 2009;29(Suppl 1):100–7.
10. Sheahan T, Jones CT, Ploss A. Advances and challenges in studying hepatitis C virus in its native environment. Expert Rev Gastroenterol Hepatol. 2010;4(5):541–50.
11. Billerbeck E, de Jong Y, Dorner M, de la Fuente C, Ploss A. Animal models for hepatitis C. Curr Top Microbiol Immunol. 2013;369:49–86.
12. Bukh J. A critical role for the chimpanzee model in the study of hepatitis C. Hepatology (Baltimore, Md.). 2004;39(6):1469–75.
13. Kaiser J. Congress removes roadblock to chimp Retirement. Science (New York, NY). 2013. http://news.sciencemag.org/funding/2013/11/congress-removes-roadblock-chimp-retirement. Accessed: 15. Nov. 2013.

14. Heckel JL, Sandgren EP, Degen JL, Palmiter RD, Brinster RL. Neonatal bleeding in transgenic mice expressing urokinasetype plasminogen activator. Cell. 1990;62(3):447–56.
15. Rhim JA, Sandgren EP, Degen JL, Palmiter RD, Brinster RL. Replacement of diseased mouse liver by hepatic cell transplantation. Science. 1994;263(5150):1149–52.
16. Rhim JA, Sandgren EP, Palmiter RD, Brinster RL. Complete reconstitution of mouse liver with xenogeneic hepatocytes. Proc Natl Acad Sci U S A. 1995;92:4942–6.
17. Dandri M, Burda MR, Torok E, Pollok JM, Iwanska A, Sommer G, et al. Repopulation of mouse liver with human hepatocytes and *in vivo* infection with hepatitis B virus. Hepatology. 2001;33(4):981–8.
18. Grompe M, al-Dhalimy M, Finegold M, Ou CN, Burlingame T, Kennaway NG, et al. Loss of fumarylacetoacetate hydrolase is responsible for the neonatal hepatic dysfunction phenotype of lethal albino mice. Genes Dev. 1993;7(12A):2298–307.
19. Grompe M, Lindstedt S, al-Dhalimy M, Kennaway NG, Papaconstantinou J, Torres-Ramos CA, et al. Pharmacological correction of neonatal lethal hepatic dysfunction in a murine model of hereditary tyrosinaemia type I. Nat Genet. 1995;10(4):453–60.
20. Overturf K, Al-Dhalimy M, Tanguay R, Brantly M, Ou CN, Finegold M, et al. Hepatocytes corrected by gene therapy are selected in vivo in a murine model of hereditary tyrosinaemia type I. Nat Genet. 1996;12(3):266–73.
21. Azuma H, Paulk N, Ranade A, Dorrell C, Al-Dhalimy M, Ellis E, et al. Robust expansion of human hepatocytes in Fah-/-/Rag2-/-/Il2rg-/- mice. Nat Biotechnol. 2007;25(8):903–10.
22. Bissig KD, Le TT, Woods NB, Verma IM. Repopulation of adult and neonatal mice with human hepatocytes: a chimeric animal model. Proc Natl Acad Sci U S A. 2007;104(51):20507–11.
23. Hasegawa M, Kawai K, Mitsui T, Taniguchi K, Monnai M, Wakui M, et al. The reconstituted 'humanized liver' in TK-NOG mice is mature and functional. Biochem Biophys Res Commun. 2011;405(3):405–10.
24. Tateno C, Yoshizane Y, Saito N, Kataoka M, Utoh R, Yamasaki C, et al. Near completely humanized liver in mice shows human-type metabolic responses to drugs. Am J Pathol. 2004;165(3):901–12.
25. Meuleman P, Libbrecht L, De Vos R, de Hemptinne B, Gevaert K, Vandekerckhove J, et al. Morphological and biochemical characterization of a human liver in a uPA-SCID mouse chimera. Hepatology. 2005;41(4):847–56.
26. Lootens L, Van Eenoo P, Meuleman P, Pozo OJ, Van Renterghem P, Leroux-Roels G, et al. Steroid metabolism in chimeric mice with humanized liver. Drug Test Anal. 2009;1(11–12):531–7.
27. Brezillon NM, DaSilva L, L'Hote D, Bernex F, Piquet J, Binart N, et al. Rescue of fertility in homozygous mice for the urokinase plasminogen activator transgene by the transplantation of mouse hepatocytes. Cell Transplant. 2008;17(7):803–12.
28. Touboul T, Hannan NR, Corbineau S, Martinez A, Martinet C, Branchereau S, et al. Generation of functional hepatocytes from human embryonic stem cells under chemically defined conditions that recapitulate liver development. Hepatology. 2010;51(5):1754–65.
29. Yusa K, Rashid ST, Strick-Marchand H, Varela I, Liu PQ, Paschon DE, et al. Targeted gene correction of alpha1-antitrypsin deficiency in induced pluripotent stem cells. Nature. 2011;478(7369):391–4.
30. Mercer DF, Schiller DE, Elliott JF, Douglas DN, Hao C, Rinfret A, et al. Hepatitis C virus replication in mice with chimeric human livers. Nat Med. 2001;7(8):927–33.
31. Bissig KD, Wieland SF, Tran P, Isogawa M, Le TT, Chisari FV, et al. Human liver chimeric mice provide a model for hepatitis B and C virus infection and treatment. J Clin Invest. 2010;120(3):924–30.
32. Meuleman P, Leroux-Roels G. The human liver-uPA-SCID mouse: a model for the evaluation of antiviral compounds against HBV and HCV. Antivir Res. 2008;80(3):231–8.
33. Vanwolleghem T, Libbrecht L, Hansen BE, Desombere I, Roskams T, Meuleman P, et al. Factors determining successful engraftment of hepatocytes and susceptibility to hepatitis B and C virus infection in uPA-SCID mice. J Hepatol. 2010;53(3):468–76.

34. Tsuge M, Hiraga N, Takaishi H, Noguchi C, Oga H, Imamura M, et al. Infection of human hepatocyte chimeric mouse with genetically engineered hepatitis B virus. Hepatology. 2005;42(5):1046–54.
35. Brezillon N, Brunelle MN, Massinet H, Giang E, Lamant C, DaSilva L, et al. Antiviral activity of Bay 41–4109 on hepatitis B virus in humanized Alb-uPA/SCID mice. PLoS ONE. 2011;6(12):e25096.
36. Dandri M, Lutgehetmann M, Petersen J. Experimental models and therapeutic approaches for HBV. Semin Immunopathol. 2013;35(1):7–21.
37. Lindenbach BD, Meuleman P, Ploss A, Vanwolleghem T, Syder AJ, McKeating JA, et al. Cell culture-grown hepatitis C virus is infectious *in vivo* and can be recultured in vitro. Proc Natl Acad Sci U S A. 2006;103(10):3805–9.
38. Meuleman P, Hesselgesser J, Paulson M, Vanwolleghem T, Desombere I, Reiser H, et al. Anti-CD81 antibodies can prevent a hepatitis C virus infection in vivo. Hepatology. 2008;48(6):1761–8.
39. Meuleman P, Catanese MT, Verhoye L, Desombere I, Farhoudi A, Jones CT, et al. A human monoclonal antibody targeting scavenger receptor class B type I precludes hepatitis C virus infection and viral spread in vitro and *in vivo*. Hepatology. 2012;55(2):364–72.
40. Kneteman NM, Weiner AJ, O'Connell J, Collett M, Gao T, Aukerman L, et al. Anti-HCV therapies in chimeric scid-Alb/uPA mice parallel outcomes in human clinical application. Hepatology. 2006;43(6):1346–53.
41. Vanwolleghem T, Meuleman P, Libbrecht L, Roskams T, De Vos R, Leroux-Roels G. Ultra-rapid cardiotoxicity of the hepatitis C virus protease inhibitor BILN 2061 in the urokinase-type plasminogen activator mouse. Gastroenterology. 2007;133(4):1144–55.
42. Kamiya N, Iwao E, Hiraga N, Tsuge M, Imamura M, Takahashi S, et al. Practical evaluation of a mouse with chimeric human liver model for hepatitis C virus infection using an NS3–4A protease inhibitor. J Gen Virol. 2010;91(Pt 7):1668–77.
43. Washburn ML, Bility MT, Zhang L, Kovalev GI, Buntzman A, Frelinger JA, et al. A humanized mouse model to study hepatitis C virus infection, immune response, and liver disease. Gastroenterology. 2011;140(4):1334–44.
44. Shultz LD, Brehm MA, Garcia-Martinez JV, Greiner DL. Humanized mice for immune system investigation: progress, promise and challenges. Nat Rev Immunol. 2012;12(11):786–98.
45. Billerbeck E, Horwitz JA, Labitt RN, Donovan BM, Vega K, Budell WC, et al. Characterization of human antiviral adaptive immune responses during hepatotropic virus infection in HLA-transgenic human immune system mice. J Immunol. 2013;191(4):1753–64.
46. Lutgehetmann M, Mancke LV, Volz T, Helbig M, Allweiss L, Bornscheuer T, et al. Humanized chimeric uPA mouse model for the study of hepatitis B and D virus interactions and preclinical drug evaluation. Hepatology. 2012;55(3):685–94.
47. Sacci JB Jr., Alam U, Douglas D, Lewis J, Tyrrell DL, Azad AF, et al. Plasmodium falciparum infection and exoerythrocytic development in mice with chimeric human livers. Int J Parasitol. 2006;36(3):353–60.
48. Morosan S, Hez-Deroubaix S, Lunel F, Renia L, Giannini C, Van Rooijen N, et al. Liver-stage development of Plasmodium falciparum, in a humanized mouse model. J Infect Dis. 2006;193(7):996–1004.
49. Vaughan AM, Mikolajczak SA, Wilson EM, Grompe M, Kaushansky A, Camargo N, et al. Complete Plasmodium falciparum liver-stage development in liver-chimeric mice. J Clin Invest. 2012;122(10):3618–28.
50. Peltz G. Can 'humanized' mice improve drug development in the 21st century? Trends Pharmacol Sci. 2013;34(5):255–60.
51. Takahashi K, Tanabe K, Ohnuki M, Narita M, Ichisaka T, Tomoda K, et al. Induction of pluripotent stem cells from adult human fibroblasts by defined factors. Cell. 2007;131(5):861–72.
52. Takahashi K, Yamanaka S. Induction of pluripotent stem cells from mouse embryonic and adult fibroblast cultures by defined factors. Cell. 2006;126(4):663–76.

53. Si-Tayeb K, Noto FK, Nagaoka M, Li J, Battle MA, Duris C, et al. Highly efficient generation of human hepatocyte-like cells from induced pluripotent stem cells. Hepatology. 2010;51(1):297–305.
54. Liu H, Kim Y, Sharkis S, Marchionni L, Jang YY. *In vivo* liver regeneration potential of human induced pluripotent stem cells from diverse origins. Sci Transl Med. 2011;3(82):82ra39.
55. Takebe T, Sekine K, Enomura M, Koike H, Kimura M, Ogaeri T, et al. Vascularized and functional human liver from an iPSC-derived organ bud transplant. Nature. 2013;499(7459):481–4.
56. Rongvaux A, Takizawa H, Strowig T, Willinger T, Eynon EE, Flavell RA, et al. Human hemato-lymphoid system mice: current use and future potential for medicine. Annu Rev Immunol. 2013;31:635–74.
57. Huntington ND, Legrand N, Alves NL, Jaron B, Weijer K, Plet A, et al. IL-15 trans-presentation promotes human NK cell development and differentiation in vivo. J Exp Med. 2009;206(1):25–34.
58. Legrand N, Huntington ND, Nagasawa M, Bakker AQ, Schotte R, Strick-Marchand H, et al. Functional CD47/signal regulatory protein alpha (SIRP(alpha)) interaction is required for optimal human T- and natural killer- (NK) cell homeostasis in vivo. Proc Natl Acad Sci U S A. 2011;108(32):13224–9.
59. Willinger T, Rongvaux A, Strowig T, Manz MG, Flavell RA. Improving human hemato-lymphoid-system mice by cytokine knock-in gene replacement. Trends Immunol. 2011;32(7):321–7.
60. Legrand N, Ploss A, Balling R, Becker PD, Borsotti C, Brezillon N, et al. Humanized mice for modeling human infectious disease: challenges, progress, and outlook. Cell Host Microbe. 2009;6(1):5–9.
61. Leung C, Chijioke O, Gujer C, Chatterjee B, Antsiferova O, Landtwing V, et al. Infectious diseases in humanized mice. Eur J Immunol. 2013;43(9):2246–54.
62. Ype de Jong P, Rice CM, Ploss A. New horizons for studying human hepatropic infections. J Clin Invest. 2010;120(3):650–3.

Chapter 37
Dengue Viral Pathogenesis and Immune Responses in Humanized Mice

Anuja Mathew and Ramesh Akkina

37.1 Introduction

Dengue virus, the causative agent of dengue fever, is transmitted to humans by the bite of an infected *Aedes* mosquito. Over 3 billion people live in endemic areas and therefore are at increased risk to contract the disease with an estimated 50–100 million new infections a year [1, 2]. Dengue virus (DENV) belongs to the family Flaviviridae and is comprised of four closely related serotypes: DENV-1, DENV-2, DENV-3 and DENV-4. In the continental USA, while cases of dengue were nonexistent until 1980, there have been outbreaks of laboratory-confirmed, locally acquired dengue cases along the Texas-Mexico border. More recently, between 2009 and 2011, autochthonous dengue fever was discovered in several Florida counties with a number of cases emerging in Key West, Florida [3]. The viral ribonucleic acid (RNA) genome encodes a single polyprotein which is processed by viral and host proteases to produce three structural proteins (core [C], membrane [M], and envelope [E]) and seven nonstructural (NS1, NS2a, NS2b, NS3, NS4a, NS4b, and NS5) proteins. The study of the molecular characteristics of the virus has provided new insights into its biology [4, 5]. A complex interplay between the virus and the host's immune system is widely hypothesized to precipitate the serious and fatal form of the disease [6].

A. Mathew (✉)
Division of Infectious Diseases and Immunology, Department of Medicine, University of Massachusetts Medical School, Worcester, MA 01655, USA
e-mail: anuja.mathew@umassmed.edu

R. Akkina (✉)
Department of Microbiology, Immunology and Pathology, Colorado State University, 1619 Campus delivery, Fort Collins, CO 80523, USA
e-mail: akkina@colostate.edu

L. Y. Poluektova et al. (eds.), *Humanized Mice for HIV Research*,
DOI 10.1007/978-1-4939-1655-9_37

37.2 Dengue Disease, Immune Correlates of Protection and Pathogenesis

Most DENV infections are asymptomatic; the majority of subjects with symptomatic infections experience an uncomplicated acute illness, dengue fever (DF) that lasts for 3–7 days. This may be accompanied by headache, myalgia, and fatigue after resolution of the illness [7, 8]. Laboratory findings include leucopenia, thrombocytopenia, and mild elevations in serum hepatic transaminases. Less than 3 % of infected individuals present with a more severe form of the disease, dengue hemorrhagic fever (DHF), which is distinguished from DF primarily by the occurrence of vascular leakage. When severe, the leakage can lead to hypotension and circulatory shock. Thrombocytopenia is another hallmark of DHF with a platelet count of δ100,000/mm^3 required to fulfill the case definition of DHF [9, 10]. Volume repletion is a highly successful strategy to treat DHF and the case fatality rate is less than 1 % in endemic areas when experienced clinicians and nursing staff are available to provide care to hospitalized patients [10–12].

Individuals who have been infected for the first time with one dengue virus serotype (primary infection) have long-term protective immunity against reinfection with the same serotype [13]. There is transient resistance to infection with other dengue virus serotypes after which individuals are susceptible to infection with other serotypes (secondary infection). Epidemiological observations indicate that 90 % of the cases of DHF occur during secondary heterologous DENV infections and that the risk of DHF is increased 15–80 times in secondary DENV infections. Memory cellular immune responses and/or antibody-dependent enhancement (ADE) of infection—wherein subneutralizing levels of anti-DENV antibodies that are present from a previous heterologous infection or passively acquired by an infant from its mother—are widely hypothesized to trigger the massive immunological cascade responsible for DHF [6].

Significant progress in our understanding of the immunity to dengue viral infection and the pathogenesis of DHF has come from a wide range of clinical observations. High titers of dengue virus-specific neutralizing antibodies have been associated with a lower likelihood of severe disease during secondary infection [14]. In patients with severe disease, T cell associated cytokines and markers of activation are found to be elevated [6]. Host factors including human leukocyte antigen (HLA) alleles, age, nutrition status, and prior T and B cell immunity are key determinants of susceptibility to DHF. Individuals carrying the HLA-A*0203 and HLA-A*33 alleles have been associated with a more resistant phenotype, whereas in contrast, patients carrying HLA-A*0207 and HLA-A*24 were found to have increased susceptibility as determined in ethnic Thai and Vietnamese populations, respectively [15–17].

Since antibodies and T cells are critical contributors to DHF pathogenesis, characterizing the nature and fine specificity of adaptive immune responses during a second infection with any of the four viral serotypes is critical to understand how these components can exacerbate severe illness. However, there are significant challenges with clinical samples which include identifying the serotype of the previous DENV infection, varying levels of preexisting immunity and transporting patient samples in a timely manner to the laboratory from the endemic areas of the world.

37.3 Animal Models for Dengue

Given the complex pathogenesis of severe dengue disease, a suitable animal model that can mimic clinical disease would be invaluable. The following criteria should be considered for all animal models. The ideal animal model should permit natural insect transmission, elicit classical disease symptoms, generate protective/enhancing antibodies and T cell responses, permit vaccine testing, and finally, aid in drug discovery. Over the years, mouse and primate models have shed light on protective and pathological responses to dengue albeit with some limitations (Table 37.1).

Immunocompetent mice such as C57BL/6 and BALB/c mice require very high doses of input virus ($> 10^8$ Plaque Forming Unit (PFU)) to induce disease, far greater than the amount of virus ($\sim 10^4$ PFU) believed to be injected subcutaneously into the human host by an infected mosquito bite [18, 19]. The intracranial route of inoculation has been used in some studies, which is not the natural route of human infection. DENV replication is not typically detected in extraneural sites or in the cell types (monocytes, dendritic cells, and lymphocytes) thought to be most relevant to DENV infection in humans. To assess the contribution of memory DENV-specific CD8 T cells to the immune response in secondary DENV infection, sequential DENV infections were performed in immunocompetent BALB/c mice [20]. However, major limitations with immunocompetent mice include the lack of disease symptoms after primary or secondary infection and the cells that respond are murine in origin.

Productive replication of laboratory strains of DENV was reported after intravenous (i.v.) and subcutaneous (s.c.) infection of mice deficient for both interferon (IFN)-α, -β, and -γ receptors in a 129 background [21]. Mice developed paralysis and other neurological symptoms which are not cardinal features of dengue disease. Generation of a mouse-adapted strain of DENV, designated D2S10, and a triple-plaque-purified clone of DENV-2 D2S10, designated S221, resulted in a vascular leak syndrome with minimal neurological symptoms [22, 23]. AG129 mice develop a broadly cross-reactive and long-lasting antibody responses to DENV [24]. ADE was demonstrated in AG129 mice by passive transfer of dengue monoclonal antibodies, subneutralizing homotypic serum or cross reactive immune serum. This mouse model has also been used for antiviral testing [19, 25, 26]. An obvious limitation of immunodeficient models is the lack of a critical component of the host antiviral system. Furthermore, the infected cells, antibody, and T cell responses are murine in origin and may not truly reflect human responses to dengue infection. While recent work indicates that selective nonadapted strains of virus can induce disease [27], most studies also require the use of mouse adapted viral strains to cause disease symptoms similar to human dengue.

Table 37.1 Animal models for dengue

	Humanized mice			Nonhuman primates	Standard immunocompetent mice	Immunodeficient mice	
	Hu-HSC NSG	BLT-NSG	Hu-HSC RAG$^{-/-}$ $\gamma c^{-/-}$	NHP	BALB/c C57BL/6	A129	AG129
Functional immune system	+++	+++	+++	+++	+++	±	±
Human cells	+++	+++	+++	n/a	n/a	n/a	n/a
Human antibody response	+++	+++	+++	n/a	n/a	n/a	n/a
Human T cell response	+++	+++	n/t	n/a	n/a	n/a	n/a
Cost	++	+++	++	++++	+	+	+
Disease symptoms	Rash fever weight loss	n/t	Fever weight loss	Only using high dose	Only using high dose or intracerebral inoculation	CNS syndrome with lab DENV strain	Vascular syndrome with mouse DENV strain
Dose required for disease	Moderate	Moderate	Moderate	High	High	Moderate	Low
References	[37–40]	[44]	[38]	[34]	[49, 50]	[18, 19, 51]	[18, 19, 22, 51]

HSC hematopoietic stem cell, *RAG* recombination-activating gene, *DENV* dengue virus, *BLT* bone marrow-liver-thymus, *NHP non human primate, CNS central nervous system, NSG NOD-scid IL2rγ^{null}*, *n/a* not applicable, *n/t* not tested

Human primates are the only vertebrates known to be infected by dengue virus in nature. Infection of chimpanzees and several species of monkeys with physiologic doses of DENV (10^4–10^6 PFU) via the s.c. route resulted in viral replication. NHPs are also used to study ADE and to test the efficacy and safety of candidate vaccines [28–31]. In vaccine studies, antibody titers and T cell responses were measured and protection was indicated by reduced/absent viremia [32, 33]. Inoculation with a higher dose of DENV via an i.v. route recently has been shown to induce hemorrhage and coagulopathy [34]. However, this is not a natural route of dengue infection. Overall, while NHPs develop viremia and neutralizing antibody responses, there is only limited evidence of disease or hematologic abnormalities. In addition, for large-scale vaccine testing NHP models involve significant cost and accessibility.

37.4 Humanized Mouse Models for Dengue

Humanized mouse models with multilineage human hematopoiesis and a capacity for eliciting human immune responses are likely to overcome many of the limitations observed in mice and NHP models [35, 36]. A major advantage of humanized models is the presence of human cells in a physiological setting. Furthermore, dengue-specific humoral and/or cellular immune responses are directed at viral antigenic epitopes recognized by the human immune system in contrast to the murine or primate system. With the advent of improved humanized mouse models, new in vivo experimental strategies are being pursued by several groups (Fig. 37.1). Two leading humanized mouse models currently employed to study dengue are the hu-HSC model in which human CD34+ HSC are engrafted, and the BLT mouse model where human fetal thymus, liver, and HSC are transplanted (Table 37.2).

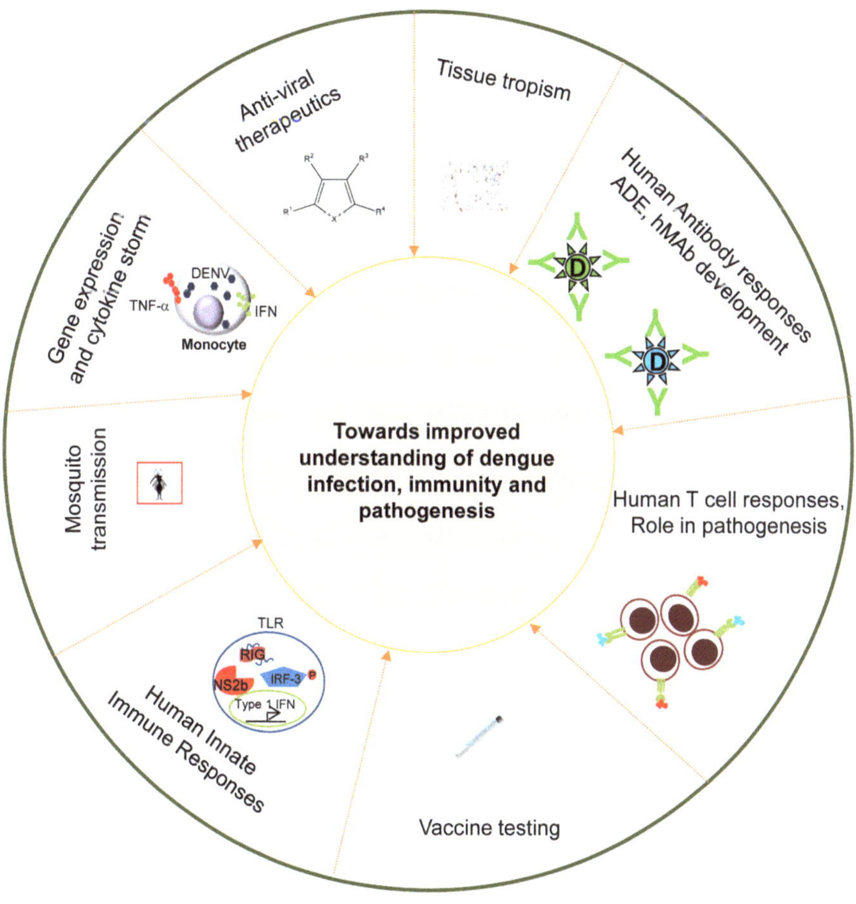

Fig. 37.1 Humanized mice: applications for dengue. *DENV* dengue virus, *IFN* interferon, *ADE* antibody-dependent enhancement, *hMAb* human monoclonal antibodies, *TLR* Toll Like Receptor

Table 37.2 Dengue viral infection in human immune system models

Mouse strain/ model	Human HSC source	Human lymphoid tissues	Dengue viremia/ symptoms	Human cytokine response	Human T cell responses	Human Ab responses	MAbs isolated
Hu-HSC NSG	Cord blood	n/a	+	+	±	+ (weak IgM)	No
Hu-HSC NSG-HLA Tg	A2+ cord blood	n/a	n/t	n/t	DENV-A2 peptides	++ (moderate IgM)	No
Hu-HSC RAG$^{-/-}$ $\gamma c^{-/-}$	Human fetal liver	n/a	+	n/t	n/t	++ (moderate IgM), weak IgG	No
BLT-NSG	Autologous fetal liver	Autologous fetal thymus and liver	±	n/t	Viral non-structural overlapping peptides	++ (moderate IgM)	Yes
BLT-NSG-HLA Tg	Autologous A2+ fetal liver	Autologous fetal thymus and liver	n/t	n/t	DENV-A2 peptides	++ (moderate IgM)	No
References			[37–27, 39]	[40]	[39, 44]	[41, 39, 44, 38]	Unpublished data

HSC hematopoietic stem cell, *RAG* recombination-activating gene, *DENV* dengue virus, *HLA* human leukocyte antigen, *IgM* immunoglobulin M, *IgG* immunoglobulin M, *BLT* bone marrow-liver-thymus, *NSG*, *MAbs* monoclonal antibodies, *n/t* not tested, *n/a* not applicable

37.5 hu-HSC NOD-SCID Mice

Bente et al. used nonobese diabetic–severe combined immune deficient (NOD-SCID) mice transplanted with human CD34+ HSC in their early studies [37]. Mice were inoculated with DENV-2 by the s.c. route. In addition to viremia, clinical signs of DF characterized by fever, rash, and thrombocytopenia were seen. However, due to the lack of sustained multilineage hematopoiesis and paucity of a full complement of human immune cells, these mice were incapable of human immune responses. Therefore, the utility of this model for immunopathogenesis studies has been limited.

37.6 hu-HSC Rag2$^{-/-}$c$\gamma^{-/-}$and NSG Mice

Studies of Kuruvilla et al. employed Balb/c recombination-activating gene (Rag)2$^{-/-}$c$\gamma^{-/-}$ mice which due to more severe immunodeficiency permitted higher human cell reconstitution levels and sustained multilineage human hematopoiesis

[38]. Hu-mice (sometimes referred to as RAG-hu) were prepared by intrahepatic injection of human fetal liver derived CD34 HSC into newborn mice. These mice generated human T cells, B cells, macrophages, natural killer (NK) cells, and dendritic cells which are chief components of innate and adaptive immune responses. Mice were injected with DENV-2 viral strains by the s.c. route. Sustained viremia was detected reaching 10^6 PFU/ml lasting up to three weeks accompanied by fever. Dengue specific antibody responses were detected with IgM appearing in 2 weeks followed by IgG responses in 6 weeks in a minority of infected mice. Most importantly, viral neutralizing antibody responses were seen with reactivity to the principal protective immunity inducing viral surface antigen E protein. However, cell mediated T cell responses were not evaluated in these studies due to lack of human HLA class restriction in this model. To overcome this deficiency, the studies of Jaiswal et al., used NSG (NOD-*scid* $IL2r\gamma^{null}$) transgenic for HLA-A2 which were humanized by transplanting with cord blood human HSC CD34 cells of the corresponding human HLA type [39]. Mice were infected via s.c. or i.p. routes. Productive viral infection was demonstrated by the presence of viral antigens and RNA in plasma and different tissue compartments. Virus-specific IgM Abs was detected 1 week post infection. Of major importance, virus-specific T cell responses were elicited with the secretion of cytokines IFN-γ, interleukin (IL)-2 and Tumor Necrosis Factor (TNF)-α in response to stimulation with A2 restricted dengue viral peptides. Thus, both antibody and cellular responses to DENV are detected in hu-HSC mice. In an extension of studies in hu-HSC NSG mice, Mota et al., used a highly virulent low passage DENV-2 viral strain [40]. Viremia, clinical signs of fever and thrombocytopenia were detected. In addition to monocytes and macrophages, B and T cells were found to be infected. Cytokine detection assays revealed increased levels of IL-6 and TNF-α in infected mice. However, dengue specific antibody and T cell responses were not assessed in this study. Since dengue is an insect transmitted disease, studies incorporating this natural transmission route are likely to increase our understanding of the dynamics of vector–host interactions and to develop ways to interfere with this process. Cox et al., used hu-HSC NSG mice to allow insect-mediated viral transmission by dengue infected *Aedes aegypti* mosquitoes during feeding [41]. More severe signs of disease characterized by higher and more sustained viremia, erythema, and thrombocytopenia were seen in mice bitten by dengue infected mosquitoes versus those infected by the s.c. route. Interestingly, only mice with insect-mediated viral infection produced IgM antibodies compared to mice infected by injection. This is in contrast to a number of other studies wherein antibody production was demonstrated in hu-mice productively infected by either s.c. or i.p. routes [39, 38].

37.7 BLT-NSG Mice

With the exception of the HLA-A2 transgenic NSG mice, standard hu-HSC mice do not permit evaluation of human HLA restricted dengue T cell responses [36]. The use of BLT-NSG mice, where developing human T cells are educated in an autologous human thymic graft, is an important advance to generate authentic human T cell responses during viral infections [42, 43]. Thus, the hypothesis that T cells restricted by multiple HLA alleles expressed by the donor should be able to respond to DENV infection can be tested using this model. Accordingly, in a recent study, overlapping peptide pools that encompass the entire DENV genome were used to assess the breadth, magnitude, and quality of DENV-specific T cell responses [44]. The results demonstrated that nonstructural proteins are the predominant targets of CD8 T cells, which is similar to the findings seen in humans [44, 45]. CD8+ T cells in splenocytes from BLT-NSG A2+ mice engrafted with HLA A2 tissue secreted IFN-γ when stimulated with previously identified HLA-A2-restricted DENV epitopes [46]. These findings set the stage for the exploitation of BLT mice to measure human T cell responses to DENV during controlled primary and secondary homologous and heterologous DENV infections.

37.8 Summary and Future Prospects of Dengue Humanized Mouse Models

Studies from several labs have demonstrated productive DENV infection in various humanized mouse models [37, 39, 44, 38]. The induction of human DENV-specific immune responses, both humoral and cellular, represents a promising first step towards developing an ideal small animal model with a functional human adaptive immune system to study complexities of human DENV infection. Further improvement of these models will likely enable the testing of multiple aspects of the interplay between the virus, host immunity, and pathogenesis of disease (Fig. 37.1). However, there remain several important limitations and challenges in advancing these humanized mouse models to study human dengue disease. Studies performed to date have differed in the immunodeficient mouse strains used, the types of human cells transplanted, and the routes used for DENV challenge. Each of these parameters could influence the differing outcomes of infection. The variable and low IgG responses observed in NSG and BLT-NSG mice have been primarily attributed to a lack of species-cross-reactive cytokines in the xenogenic environment. Recent studies of Lang et al., attributed the inefficient Ig class switch in hu-mouse models to insufficient time allowed for the generation of adequate levels of helper T cells resulting in suboptimal T–B cell interactions [47]. Nevertheless, current ongoing efforts to improve the levels of human-cell engraftment, HLA restricted T cell help, and germinal center formation are likely to lead to more robust humanized mice for dengue studies [36, 48].

References

1. Halstead SB. Pathogenesis of dengue: challenges to molecular biology. Science. 1988;239:476–81.
2. Simmons CP, Farrar JJ, Nguyen v V, Wills B. Dengue. N Engl J Med. 2012;366:1423–32.
3. Radke EG, Gregory CJ, Kintziger KW, Sauber-Schatz EK, Hunsperger EA, Gallagher GR, Barber JM, Biggerstaff BJ, Stanek DR, Tomashek KM, Blackmore CG. Dengue outbreak in Key West, Florida, USA, 2009. Emerg Infect Dis. 2012;18:135–7.
4. Pagni S, Fernandez-Sesma A. Evasion of the human innate immune system by dengue virus. Immunol Res. 2012;54:152–9.
5. Paranjape SM, Harris E. Control of dengue virus translation and replication. Curr Top Microbiol Immunol. 2010;338:15–34.
6. Rothman AL. Immunity to dengue virus: a tale of original antigenic sin and tropical cytokine storms. Nat Rev Immunol. 2011;11:532–43.
7. Halstead SB. Immunological parameters of togavirus disease syndromes. In: Schlesinger RW, editors. The togaviruses. Biology, structure, replication. New York: Academic; 1980. p. 107–173.
8. Kalayanarooj S, Vaughn DW, Nimmannitya S, Green S, Suntayakorn S, Kunentrasai N, Viramitrachai W, Ratanachu-eke S, Kiatpolpoj S, Innis BL, Rothman AL, Nisalak A, Ennis FA. Early clinical and laboratory indicators of acute dengue illness. J Infect Dis. 1997;176:313–21.
9. Anonymous. Dengue haemorrhagic fever: diagnosis, treatment and control. Geneva:World Health Organization; 1986.
10. WHO. Dengue and dengue haemorrhagic fever in the Americas, 1996. Wkly Epidemiol Rec. 1997;72:122–3.
11. Harris E, Perez L, Phares CR, Perez Mde L, Idiaquez W, Rocha J, Cuadra R, Hernandez E, Campos LA, Gonzales A, Amador JJ, Balmaseda A. Fluid intake and decreased risk for hospitalization for dengue fever, Nicaragua. Emerg Infect Dis. 2003;9:1003–6.
12. Ngo NT, Cao XT, Kneen R, Wills B, Nguyen VM, Nguyen TQ, Chu VT, Nguyen TT, Simpson JA, Solomon T, White NJ, Farrar J. Acute management of dengue shock syndrome: a randomized double-blind comparison of 4 intravenous fluid regimens in the first hour. Clin Infect Dis. 2001;32:204–13.
13. Sabin AB. Research on dengue during World War II. Am J Trop Med Hyg. 1952;1:30–50.
14. Endy TP, Nisalak A, Chunsuttitwat S, Vaughn DW, Green S, Ennis FA, Rothman AL, Libraty DH. Relationship of pre-existing dengue virus (DV) neutralizing antibody levels to viremia and disease severity in a prospective cohort study of DV infection in Thailand. J Infect Dis. 2004;189:990–1000.
15. Chiewsilp P, Scott RM, Bhamarapravati N. Histocompatibility antigens and dengue hemorrhagic fever. Am J Trop Med Hyg. 1981;30:1100–5.
16. Loke H, Bethell DB, Phuong CX, Dung M, Schneider J, White NJ, Day NP, Farrar J, Hill AV. Strong HLA class I-restricted T cell responses in dengue hemorrhagic fever: a double-edged sword? J Infect Dis. 2001;184:1369–73.
17. Stephens HA, Klaythong R, Sirikong M, Vaughn DW, Green S, Kalayanarooj S, Endy TP, Libraty DH, Nisalak A, Innis BL, Rothman AL, Ennis FA, Chandanayingyong D. HLA-A and -B allele associations with secondary dengue virus infections correlate with disease severity and the infecting viral serotype in ethnic Thais. Tissue Antigens. 2002;60:309–18.
18. Yauch LE, Shresta S. Mouse models of dengue virus infection and disease. Antiviral Res. 2008;80:87–93.
19. Zompi S, Harris E. Animal models of dengue virus infection. Viruses. 2012;4:62–82.
20. Beaumier CM, Mathew A, Bashyam HS, Rothman AL. Cross-reactive memory CD8(+) T cells alter the immune response to heterologous secondary dengue virus infections in mice in a sequence-specific manner. J Infect Dis. 2008;197:608–17.

21. Shresta S, Kyle JL, Snider HM, Basavapatna M, Beatty PR, Harris E. Interferon-dependent immunity is essential for resistance to primary dengue virus infection in mice, whereas T- and B-cell-dependent immunity are less critical. J Virol. 2004;78:2701–10.
22. Shresta S, Sharar KL, Prigozhin DM, Beatty PR, Harris E. Murine model for dengue virus-induced lethal disease with increased vascular permeability. J Virol. 2006;80:10208–17.
23. Prestwood TR, Prigozhin DM, Sharar KL, Zellweger RM, Shresta S. A mouse-passaged dengue virus strain with reduced affinity for heparan sulfate causes severe disease in mice by establishing increased systemic viral loads. J Virol. 2008;82:8411–21.
24. Yauch LE, Prestwood TR, May MM, Morar MM, Zellweger RM, Peters B, Sette A, Shresta S. CD4+ T cells are not required for the induction of dengue virus-specific CD8+ T cell or antibody responses but contribute to protection after vaccination. J Immunol. 2010;185:5405–16.
25. Chen YL, Yin Z, Duraiswamy J, Schul W, Lim CC, Liu B, Xu HY, Qing M, Yip A, Wang G, Chan WL, Tan HP, Lo M, Liung S, Kondreddi RR, Rao R, Gu H, He H, Keller TH, Shi PY. Inhibition of dengue virus RNA synthesis by an adenosine nucleoside. Antimicrob Agents Chemother. 2011;54:2932–9.
26. Wang QY, Bushell S, Qing M, Xu HY, Bonavia A, Nunes S, Zhou J, Poh MK, Florez de Sessions, P., Niyomrattanakit P, Dong H, Hoffmaster K, Goh A, Nilar S, Schul W, Jones S, Kramer L, Compton T, Shi PY. Inhibition of dengue virus through suppression of host pyrimidine biosynthesis. J Virol. 2011;85:6548–56.
27. Tan GK, Ng JK, Trasti SL, Schul W, Yip G, Alonso S. A non mouse-adapted dengue virus strain as a new model of severe dengue infection in AG129 mice. PLoS Negl Trop Dis. 2010;4:e672.
28. Goncalvez AP, Engle RE, St Claire M, Purcell RH, Lai, CJ. Monoclonal antibody-mediated enhancement of dengue virus infection in vitro and in vivo and strategies for prevention. Proc Natl Acad Sci U S A. 2007;104(22):9422–7
29. Kochel TJ, Watts DM, Gozalo AS, Ewing DF, Porter KR, Russell KL. Cross-serotype neutralization of dengue virus in Aotus nancymae monkeys. J Infect Dis. 2005;191:1000–4.
30. Maves RC, Ore RM, Porter KR, Kochel TJ. Immunogenicity and protective efficacy of a psoralen-inactivated dengue-1 virus vaccine candidate in Aotus nancymaae monkeys. Vaccine. 2011;29:2691–6.
31. Raviprakash K, Porter KR, Kochel TJ, Ewing D, Simmons M, Phillips I, Murphy GS, Weiss WR, Hayes CG. Dengue virus type 1 DNA vaccine induces protective immune responses in rhesus macaques. J Gen Virol. 2000;81:1659–67.
32. Koraka P, Benton S, van Amerongen G, Stittelaar KJ, Osterhaus AD. Efficacy of a live attenuated tetravalent candidate dengue vaccine in naive and previously infected cynomolgus macaques. Vaccine. 2007;25:5409–16.
33. Koraka P, Benton S, van Amerongen G, Stittelaar KJ, Osterhaus AD. Characterization of humoral and cellular immune responses in cynomolgus macaques upon primary and subsequent heterologous infections with dengue viruses. Microbes Infect. 2007;9:940–6.
34. Onlamoon N, Noisakran S, Hsiao HM, Duncan A, Villinger F, Ansari AA, Perng GC. Dengue virus-induced hemorrhage in a nonhuman primate model. Blood. 2010;115:1823–34.
35. Shultz LD, Brehm MA, Garcia-Martinez JV, Greiner DL. Humanized mice for immune system investigation: progress, promise and challenges. Nat Rev Immunol. 2012;12:786–98.
36. Akkina R. Human immune responses and potential for vaccine assessment in humanized mice. Curr Opin Immunol. 2013;25(3):403–9.
37. Bente DA, Melkus MW, Garcia JV, Rico-Hesse R. Dengue fever in humanized NOD/SCID mice. J Virol. 2005;79:13797–9.
38. Kuruvilla JG, Troyer RM, Devi S, Akkina R. Dengue virus infection and immune response in humanized RAG2(−/−)gamma(c)(−/−) (RAG-hu) mice. Virology. 2007;369:143–52.
39. Jaiswal S, Pearson T, Friberg H, Shultz LD, Greiner DL, Rothman AL, Mathew A. Dengue virus infection and virus-specific HLA-A2 restricted immune responses in humanized NOD-scid IL2rgammanull mice. PLoS ONE. 2009;4:e7251.

40. Mota J, Rico-Hesse R. Humanized mice show clinical signs of dengue fever according to infecting virus genotype. J Virol. 2009;83:8638–45.
41. Cox J, Mota J, Sukupolvi-Petty S, Diamond MS, Rico-Hesse R. Mosquito bite delivery of dengue virus enhances immunogenicity and pathogenesis in humanized mice. J Virol. 2012;86:7637–49.
42. Lan P, Tonomura N, Shimizu A, Wang S, Yang YG. Reconstitution of a functional human immune system in immunodeficient mice through combined human fetal thymus/liver and CD34+ cell transplantation. Blood. 2006;108:487–92.
43. Melkus MW, Estes JD, Padgett-Thomas A, Gatlin J, Denton PW, Othieno FA, Wege AK, Haase AT, Garcia JV. Humanized mice mount specific adaptive and innate immune responses to EBV and TSST-1. Nat Med. 2006;12:1316–22.
44. Jaiswal S, Pazoles P, Woda M, Shultz LD, Greiner DL, Brehm MA, Mathew A. Enhanced humoral and HLA-A2-restricted dengue virus-specific T-cell responses in humanized BLT NSG mice. Immunology. 2012;136:334–43.
45. Mathew A, Kurane I, Green S, Stephens HA, Vaughn DW, Kalayanarooj S, Suntayakorn S, Chandanayingyong D, Ennis FA, Rothman AL. Predominance of HLA-restricted cytotoxic T-lymphocyte responses to serotype-cross-reactive epitopes on nonstructural proteins following natural secondary dengue virus infection. J Virol. 1998;72:3999–4004.
46. Bashyam HS, Green S, Rothman AL. Dengue virus-reactive CD8+ T cells display quantitative and qualitative differences in their response to variant epitopes of heterologous viral serotypes. J Immunol. 2006;176:2817–24.
47. Lang J, Kelly M, Freed BM, McCarter MD, Kedl RM, Torres RM, Pelanda R. Studies of lymphocyte reconstitution in a humanized mouse model reveal a requirement of T cells for human B cell maturation. J Immunol. 2013;190(5):2090–101.
48. Rongvaux A, Takizawa H, Strowig T, Willinger T, Eynon EE, Flavell RA, Manz MG. Human hemato-lymphoid system mice: current use and future potential for medicine. Annu Rev Immunol. 2013;31:635–74.
49. Huang KJ, Li SY, Chen SC, Liu HS, Lin YS, Yeh TM, Liu CC, Lei HY. Manifestation of thrombocytopenia in dengue-2-virus-infected mice. J Gen Virol. 2000;81:2177–82.
50. Yen YT, Chen HC, Lin YD, Shieh CC, Wu-Hsieh BA. Enhancement by tumor necrosis factor alpha of dengue virus-induced endothelial cell production of reactive nitrogen and oxygen species is key to hemorrhage development. J Virol. 2008;82:12312–24.
51. Prestwood TR, Morar MM, Zellweger RM, Miller R, May MM, Yauch LE, Lada SM, Shresta S. Gamma interferon (IFN-gamma) receptor restricts systemic dengue virus replication and prevents paralysis in IFN-alpha/beta receptor-deficient mice. J Virol. 2012;86:12561–70.

Chapter 38
HIV-1 and TB: How Humanized Mice Can Help

Antoinette Labuschagné and Muazzam Jacobs

38.1 Introduction

38.1.1 TB and HIV: The Dual Disease

Tuberculosis (TB) and human immunodeficiency virus/acquired immune deficiency syndrome (HIV/AIDS) compose the main burden of infectious disease in resource-limited countries. Estimates by the World Health Organization (WHO) indicate that there were approximately nine million new active cases of TB in 2011 and up to 1.4 million deaths [1]. In addition, 2.5 million new cases of HIV infection and 1.7 million AIDS-related deaths occur per year [2]. Globally, 34 million people were living with HIV at the end of 2011.

Mycobacterium tuberculosis (*M.tb*)-HIV coinfections pose particular diagnostic and therapeutic challenges and exert immense pressure on health care systems in African and Asian countries with large populations of coinfected individuals. In the individual host the two pathogens, *M.tb* and HIV, potentiate one another, accelerating the deterioration of immunological functions, resulting in premature death if untreated. Some 14 million people worldwide are estimated to be dually infected with almost 80% of TB cases among people living with HIV residing in Africa. [1, 3]. TB is the largest single cause of death in the setting of AIDS [4, 5]. HIV coinfection is the most powerful known risk factor for progression of *M.tb* infection to active

A. Labuschagné (✉) · M. Jacobs
Division of Immunology, Department of Clinical and Laboratory Science and Institute of Infectious Diseases and Molecular Medicine, UCT Medical School, University of Cape Town, Anzio Road, Observatory, Werner Beit Building, Cape Town 7925, Western Cape, South Africa
e-mail: toinette.labuschagne@uct.ac.za

M. Jacobs
National Health Laboratory Service, Johannesburg, South Africa

L. Y. Poluektova et al. (eds.), *Humanized Mice for HIV Research*,
DOI 10.1007/978-1-4939-1655-9_38

disease, increasing the risk of latent TB reactivation 20-fold [3, 6]. Patients coinfected with TB and HIV have higher mortality rates in comparison with those with either of the two infections alone [7–13]. During this mutual interaction between HIV and *M.tb* infection: (1) HIV infection predisposes to the development of active TB, and (2) HIV treatment in itself can augment the symptoms of TB or even elicit reactivation as with the immune reconstitution inflammatory syndrome (IRIS)-associated TB, while (3) the course of HIV-related immunodeficiency is worsened by active *M.tb* infection [14]. Very little is known about the underlying mechanisms involved during these associations and there exists a dire need for tractable animal models where these processes can be studied.

38.1.2 Conventional Animal Models of HIV/TB Coinfection

To date, retroviral infection as a TB risk factor can only be effectively modeled in nonhuman primates (NHP). NHP are resistant to infection by HIV; however, they can be infected by simian immunodeficiency virus (SIV), a retrovirus causing immunodeficiency in NHP similar to AIDS. In addition, macaques develop TB similar to humans, characterized by cavitary lung disease and necrotic lesions. Macaques maintain TB latency for years, and only a small proportion of latently infected animals develop reactivation [15]. Coinfection with SIV and bacille Calmette–Guérin (BCG) accelerates progression to AIDS [16] and coincided with reactivation of the clinically latent BCG infection into a TB-like disease [17]. Reactivation of *M.tb* infection with coinfected SIV macaques has also been reported [18]. There are significant differences between SIV and HIV, including genetic heterogeneity and receptors for host entry [19]. Conducting macaque experiments requires relatively high costs with few institutions able to support the facilities and infrastructure to conduct combination SIV and TB research.

While there is no natural homologous murine immunodeficiency virus, as for NHP, HIV-1 virus constructs have been developed that enables infection of immune cells of conventional, immunocompetent mice [20]. The host-specific species range of HIV-1 is converted from human to rodent by replacing the coding region of its surface envelope glycoprotein, gp120, with the envelope-coding region gp80 from that of a mouse virus; ecotropic murine leukemia virus (MLV), restricting replication of the virus to rodents. One single inoculation of this ecotropic HIV-1 (EcoHIV-1) virus establishes spreading infection in conventional immunocompetent mice such as the CB57BL/6 strain, reproduces key characteristics of early HIV-1 infection of human beings, including host cell range and infectious immune responses (Table 38.1). Progression of EcoHIV-1 infection does not result in depletion of $CD4^+$ T cells or any immunodeficiency similar to AIDS, most likely due to mice containing the infection or specific Major Histocompatibility Complex (MHC) requirements. The EcoHIV-1 model has proven useful in preclinical evaluation of antiretroviral drugs, including preexposure antiretroviral therapy during sexual transmission and vaccine challenge studies [21–24], but has not been utilized in conjunction with *M.tb* infection.

Table 38.1 Comparing features of animal models for TB, HIV, and TB/HIV coinfection studies. (Adapted from Dharmadhikari and Nardell [53])

	Granuloma histopathology			Tuberculosis disease state		Retroviral infection			TB/HIV coinfection	
Model	Necrosis	Caseation	Cavitation	Latency	Reactivation	HIV	Susceptibility and replication	Retroviral immunedeficiency	Retroviral reactivation of latent TB	IRIS TB
EcoHIV-1	Rare	No	No	No; Cornell model may do so		No	EcoHIV-1 vaginal susceptibility, replicates systemically	No	No; Cornell model may do so	No
HIV transgenic mice	Minimal; can depend on immune status	Rare	No	No; Cornell model may do so		No	No infection	No; some symptoms of wasting and $CD4^+$ depletion	No; Cornell model may do so	No
Nonhuman primates	Yes	Yes	Yes	Yes	Yes	No: SIV	SIV replicates systemically	Yes	Yes	No
Humanized mice	Yes	Yes	Yes	No; Cornell model may do so		Yes	HIV vaginal and rectal susceptibility replicates systemically	Yes	Not confirmed to date but possible	Not confirmed to date but possible

TB tuberculosis, *HIV* human immunodeficiency virus, *IRIS* immune reconstitution inflammatory syndrome

Mice that express HIV transgenes have been generated using both the full length provirus and individual components of the HIV-1 genome including Nef, Env, Long Terminal Repeat (LTR) and Tat [25–29]. Some mouse strains expressing single HIV proteins developed symptoms of AIDS such as wasting and $CD4^+$ T cell depletion (Table 38.1). However, full length HIV-1 ribonucleic acids (RNAs) are transcribed inefficiently in mouse cells. Human Cyclin T1 (hCyclin T1) interacts with HIV Tat protein in a species-restricted manner to enhance RNA transcription and processing [30]. In JRCSF (R5 tropic HIV-1) and hCyclin T1 double transgenic mice, increased HIV-1 expression correlated with $CD4^+$ T cell depletion [31]. In addition to hCyclin T1, hCD4 and hCCR5 or hCXCR4 are necessary for HIV-1 infection of mouse cells. Transgenic mice that express these human genes have been generated and reported [32, 33]; however, these genes alone are not sufficient to make mice susceptible to HIV-1. No virus spread was observed in vivo for either hCD4/CCR5 or hCD4/CXCR4 transgenic mice. Although none of these transgenic mice have been used to address how specific viral proteins manipulate immunological responses to TB to date, HIV transgenic mice incorporating the entire viral genome have been used to study the effect of *M.tb* infection on the induction of HIV gene expression [34]. In this model, viral gene expression was activated by *M.tb* and suppressed after antimycobacterial chemotherapy. Future TB studies may be able to use these transgenic mice to determine how different HIV-1 component expression changes immunologic responses to TB. Additional perspectives on appropriate animal models to study *M.tb*/HIV coinfection documenting the use of NHP, conventional mice infected with EcoHIV-1, HIV transgenic mice, and humanized mice have also been reviewed elsewhere [35].

38.2 Humanized Mice

The principal step towards generating mice with human immune responses against HIV and TB is to select the best-suited immunodeficient mouse strain to serve as host for human leukocyte engraftment. The use of mouse stocks that are severely immunodeficient, e.g., nonobese diabetic (NOD)-*scid/scid* $Il2rg^{null}$ (NSG), BALB/c-$Rag2^{null}Il2rg^{null}$ (BRG), are becoming the standard. Methods used to reconstitute the most versatile human immune system (HIS) in mice include many factors, such as choice of human tissue and/or cells, route of transplant, age and gender of recipient mice, and preconditioning regime, are all interrelated and interdependent and can have a significant effect on the extent of HIS reconstitution and function.

When considering the pathogenesis of HIV and TB, both are long-term diseases that require a robust and extensive HIS reconstitution, thus eliminating the option of engrafting peripheral blood mononuclear cells (PBMC) to create a human peripheral blood lymphocyte (hu-PBL) mouse. The most commonly used model for infection studies (particularly HIV-1) includes transplantation of human $CD34^+$ HSC cells to create the humanized models (Hu-mice).

38.2.1 Hu-Mice

A recent study from Heuts et al. [36] demonstrated Hu-NSG mice ($CD34^+$ HSC cells from umbilical cord blood) infected with attenuated *M. bovis* BCG or virulent *M.tb* which resulted in organized granuloma formation that resemble lesions observed in human TB. Granulomas in livers and lungs represented with core containing giant cells, human $CD68^+$ macrophages, and high bacilli numbers surrounded by a layer of $CD3^+$ T cells and a fibrotic response encapsulating the lesions from infected humanized mice but not in non-humanized infected controls. Dysfunctional T cell responses and lack of bacterial control was however evident in this model, most likely due to augmented expression of programmed death ligand, PD-1 which inhibits T cell effector functions during human TB [37] and CD57 (a marker for T cell clonal exhaustion). The malfunctioning T cell responses in turn underlie the defective mycobacterial control [36]. A further consideration for the evident dysfunctional response in the hu-NSG model is that human T cells are positively selected by mouse MHC and might not function well in a HLA-restricted manner [38]. The selection choice of using umbilical cord blood derived instead of fetal liver $CD34^+$ cells can also impact on reconstitution and functionality, maturation [39].

38.2.2 BLT-Mice

The model system in which immunocompromised mouse stocks first receive a transplant of human fetal liver (Liv) and thymus (Thy) and subsequently injected with human $CD34^+$ HSC cells derived from the same fetal liver (human leukocyte antigen (HLA)-matched) is designated bone marrow-liver-thymic (BLT) mice. A functional HIS with both systemic human T and B cell repopulation develops [40–42]. Since T cells develop on a syngeneic thymic-organic graft (referred to as a thymic "organoid"), the human T cell maturation and selection is based on human thymic agents, and T cells are capable of mounting HLA-restricted immune responses. The human thymic organoid expands with human lymphoid cells over time, immature $CD4^+CD8^+$ T cells mature to single positive effector cells and improved cell mediated immunity and immunoglobulin G (IgG) and immunoglobulin M (IgM) production [41–43]. The resulting HIS is more robust and functional with respect to carrying an infection such as *M.tb* as is evident in the study produced by Calderon et al. (2013) [44]. Human T cells were functionally competent as determined by proliferative capacity and effector molecule expression (IFN-γ, granulysin, perforin) in response to positive stimuli. BLT-NSG mice were intranasally infected with *M.tb* and presented with progressive bacterial infection in the lung and dissemination to spleen and liver from 2 to 8 weeks post infection. Sites of infection in the lung were characterized by the formation of organized granulomatous lesions, caseous necrosis, bronchial obstruction, and crystallization of cholesterol deposits. Human T cells were distributed throughout the lung, liver, and spleen at the sites of inflammation and bacterial growth and were organized to the periphery of granulomas [44].

38.2.3 Addressing TB/HIV-1

Improvements in the humanized mouse model that are accelerating its application in HIV/AIDS research is clearly documented elsewhere in this book. These advances are also opening the door to investigate disease processes of HIV/*M.tb* coinfection. In particular, areas of study that lack in vivo correlates can now be explored and will aid in the search for biomarkers of disease status that can eventually be used in vaccine and drug studies as well as diagnostic tests. The most significant features of *M.tb* and HIV infections can be reproduced in humanized mice (Table 38.1). The immunological and virological consequences of *M.tb* and HIV-1 copathogenesis within the humanized mouse host can provide a predictable model for the study of the immunopathogenesis in humans. It is increasingly important to understand what problems associated with TB/HIV-1 infection within current humanized models can be addressed:

1. Enhanced interpretation of blood-based data (by drawing parallels between blood samples and tissue responses from timed necropsy)
   Current clinical models that utilize human samples in general fail to effectively address immune responses at local sites of infection. Most studies use peripheral blood cells, serum, plasma, or fluid obtained through lung bronchial alveolar lavage to measure immune parameters in defined cohorts with the inherent limitation of data interpretation associated with extrapolating data obtained in the periphery to active sites of disease. The humanized mouse model allows for a structured approach to investigate immune responses by human cells and overcomes most of the limitation on the sampling associated with clinical studies.
2. Changes in T cell and macrophage function within granulomas
   To elucidate possible mechanisms by which HIV disrupts TB immune pathology, T cells and the macrophage function within granulomas can be examined. Depletion of $CD4^+$ T cells is not the sole cause of impaired *M.tb* control, as HIV has effects on other cells such as macrophages, and the virus influences cytokine production that may also prevent a host from containing an initial or latent *M.tb* infection. The HIV might even disrupt antigen-presenting cells and T cell interactions.
3. Modeling latent TB and reactivation with HIV-1
   Implementing systems such as the Cornell method of TB latency [45–47] within humanized mice can shed light on reactivation and subsequent reactivation of TB through HIV-1 infection. Mechanisms of HIV-1 latency and reactivation during *M.tb* infection in HIV-1 reservoirs and dormancy can *a*lso be explored during *M.tb* coinfection.

4. Mechanisms of TB IRIS
   Although the mechanisms that lead to IRIS-associated TB are not fully understood and have been reviewed more fully elsewhere [48–50], it is clear that this consequence of antiretroviral therapy further complicates HIV/TB coinfection. Information on TB IRIS patients is limited, and there exists a dire need for an appropriate animal model where this phenomenon can be studied. Humanized mice can serve as the platform of simulating IRIS to shed light on these mechanisms.

38.2.4 Model Limitations

Important requirements for a productive HIV/TB coinfection humanized mouse model is that it should: offer a well-functioning human immune system that closely reflects the human condition; support differentiation of human monocytes, dendritic and natural killer (NK) cells, form granulomas that reproduce those seen in humans, characterized by organized TB lesions with central necrosis, cavitation, and caseation; and be created in sufficient numbers to carry out experiments that are standardized, reproducible, and statistically verifiable. According to the data obtained thus far, the humanized mouse models for TB meet these requirements.

Limitations of the current humanized mouse models that have been highlighted thus far include the MHC differences, background graft versus host disease (GVHD), cytokine cross reactivity, Ig class switching, and then the most important for TB/HIV coinfections is the absence of normal lymphoid structure and function. In a normal immune response to TB, antigen-presenting cells (APCs) move from the lung into the lung draining lymph nodes, interacting thereafter specifically with T cells to activate *M.tb*-specific proliferation and differentiation of cells with associated receptors. The *M.tb*-specific T cells traffic out of the draining lymph nodes into distal effector sites. Most of the infrastructure underlying such physiologic responses is not present in any of the humanized mouse models making normal human immune cell trafficking and differentiation unlikely. The effect HIV might have on disrupting APC and T cell interactions will not be able to be studied in current humanized mice, and a better-suited model in this specific case would be the EcoHIV-1 system, for instance. An ideal coinfection model should develop specific humoral and cell-mediated immune activation and responses against TB and HIV-1. Engrafted mice should survive periods sufficient for chronic infection studies, reach T cell levels that can sustain intense and prolonged TB/HIV-1 replication, support development of lymphoid structures such as lymph nodes and thymus. The BLT model may be restricted to investigate aspects of humoral immunity involving B cells [51] as class switching of immunoglobulins is limited and mainly consists of IgM and few IgGs [52].

38.3 Concluding Remarks

The enormity of the TB/HIV-1 syndemic for world health and the bidirectional interaction between the two pathogens clearly highlights the importance of HIV-1 and *M.tb* copathogenesis as an area of intense investigation. With the development of humanized mice that can support and easily allow for monitoring of HIV-1 and TB infection in vivo, the basic understanding of how these pathogens interact will add to the body of knowledge and help uncover possible treatments for coinfected individuals. Producing a mouse model susceptible to diseases caused by pathogens infecting humans, humanized mice may dramatically change the paradigm of the development of new therapeutic methods and drug discovery.

To investigate the pathogenesis of this syndemic in vivo, the SIV and BCG-infected macaque model has been used and accepted. Both the dually infected macaque and humanized mouse models have respective advantages and disadvantages. To reveal an accurate understanding of HIV-1 and TB pathogenesis in vivo, it is important to pool the accumulated knowledge provided from transgenic and conventional mouse models, macaque studies, clinical trials, and novel findings from the studies on humanized mice, together with strategies from multiple other disciplines (such as molecular and mathematical biology) to provide evidence supporting or refuting the various hypotheses of how HIV-1 and TB fuel one another. After the standardization, optimal adjustment, repeated and validated applications, the adoption of the TB/HIV-1 humanized mouse model will rapidly accelerate efforts to understand the complementation between these two pathogens, establish a system that can be used to test interactions between antiretrovirals for HIV and chemotherapies for TB. This will in turn provide a framework for the development of new models of opportunistic infections to address unrevealed but essential aspects of other human-specific pathogens.

References

1. World Health Organization: Global tuberculosis Report 2012. Available: http://www.who.int/tb/publications/global_report/2012/en/index.html. 2012. Accessed: Oct. 2012.
2. UNAIDS: "State of the Epidemic". UNAIDS report on the global AIDS epidemic 2012. http://www.unaids.org/en/media/unaids/contentassets/documents/epidemiology/2012/gr2012/20121120_UNAIDS_Global_Report_2012_en.pdf. 2012. Accessed: 13 March 2013.
3. Getahun H, Gunneberg C, Granich R, Nunn P. HIV infection-associated tuberculosis: the epidemiology and the response. Clin Infect Dis. 2010;50(Suppl 3). p. 201–7. doi:10.1086/651492. PubMed PMID: 20397949.
4. Corbett EL, Watt CJ, Walker N, Maher D, Williams BG, Raviglione MC, Dye C. The growing burden of tuberculosis: global trends and interactions with the HIV epidemic. Arch Intern Med. 2003;163(9):1009–21. PubMed PMID: 12742798.
5. Tabarsi P, Chitsaz E, Moradi A, Baghaei P, Farnia P, Marjani M, Shamai M, Amiri M, Nikaein S, Mansouri D, Masjedi M, Altice F. Treatment outcome, mortality and their predictors among HIV-associated tuberculosis patients. Int J STD AIDS. 2012;23(9):e1–4. doi:10.1258/ijsa.2009.009093. PubMed PMID: 23033530.

6. Selwyn PA, Hartel D, Lewis VA, Schoenbaum EE, Vermund SH, Klein RS, Walker AT, Friedland GH. A prospective study of the risk of tuberculosis among intravenous drug users with human immunodeficiency virus infection. N Engl J Med. 1989;320(9):545–50. PubMed PMID: 2915665.
7. Stoneburner R, Laroche E, Prevots R, Singh T, Blum S, Terry P, Reatrice S, Adler J. Survival in a cohort of human immunodeficiency virus-infected tuberculosis patients in New York City. Implications for the expansion of the AIDS case definition. Arch Intern Med. 1992;152(10):2033–7. PubMed PMID: 1358042.
8. Perriëns JH, Colebunders RL, Karahunga C, Willame JC, Jeugmans J, Kaboto M, Mukadi Y, Pauwels P, Ryder RW, Prignot J, et al. Increased mortality and tuberculosis treatment failure rate among human immunodeficiency virus (HIV) seropositive compared with HIV seronegative patients with pulmonary tuberculosis treated with "standard" chemotherapy in Kinshasa, Zaire. Am Rev Respir Dis. 1991;144(4):750–5. PubMed PMID: 1928943.
9. Nunn P, Brindle R, Carpenter L, Odhiambo J, Wasunna K, Newnham R, Githui W, Gathua S, Omwega M, McAdam K. Cohort study of human immunodeficiency virus infection in patients with tuberculosis in Nairobi, Kenya. Analysis of early (6-month) mortality. Am Rev Respir Dis. 1992;146(4):849–54. PubMed PMID: 1416409.
10. Small PM, Schecter GF, Goodman PC, Sande MA, Chaisson RE, Hopewell PC. Treatment of tuberculosis in patients with advanced human immunodeficiency virus infection. N Engl J Med. 1991;324(5):289–94. PubMed PMID: 1898769.
11. Chaisson RE, Schecter GF, Theuer CP, Rutherford GW, Echenberg DF, Hopewell PC. Tuberculosis in patients with the acquired immunodeficiency syndrome. Clinical features, response to therapy, and survival. Am Rev Respir Dis. 1987;136(3):570–4. PubMed PMID: 3631730.
12. Tabarsi P, Baghaei P, Mirsaeidi M, Amiri M, Alipanah N, Emami H, Novin A, Mansouri D, Masjedi MR, Velayati AA. Treatment outcome of tuberculosis patients diagnosed with human immunodeficiency virus infection in Iran. Saudi Med J. 2008;29(1):148–50. PubMed PMID: 18176695.
13. van der Sande MA, Schim van der Loeff MF, Bennett RC, Dowling M, Aveika AA, Togun TO, Sabally S, Jeffries D, Adegbola RA, Sarge-Njie R, Jaye A, Corrah T, McConkey S, Whittle HC. Incidence of tuberculosis and survival after its diagnosis in patients infected with HIV-1 and HIV-2. AIDS. 2004;18(14):1933–41. PubMed PMID: 15353979.
14. Collins KR, Quiñones-Mateu ME, Toossi Z, Arts EJ. Impact of tuberculosis on HIV-1 replication, diversity, and disease progression. AIDS Rev. 2002;4(3):165–76. PubMed PMID: 12416451.
15. Diedrich CR, Flynn JL. HIV-1/mycobacterium tuberculosis coinfection immunology: how does HIV-1 exacerbate tuberculosis? Infect Immun. 2011;79(4):1407–17. doi:10.1128/IAI.01126-10. PubMed PMID: 21245275. Epub 2011/01/18. eng.
16. Zhou D, Shen Y, Chalifoux L, Lee-Parritz D, Simon M, Sehgal PK, Zheng L, Halloran M, Chen ZW. Mycobacterium bovis bacille Calmette-Guérin enhances pathogenicity of simian immunodeficiency virus infection and accelerates progression to AIDS in macaques: a role of persistent T cell activation in AIDS pathogenesis. J Immunol. 1999;162(4):2204–16. PubMed PMID: 9973496.
17. Shen Y, Shen L, Sehgal P, Huang D, Qiu L, Du G, Letvin NL, Chen ZW. Clinical latency and reactivation of AIDS-related mycobacterial infections. J Virol. 2004;78(24):14023–32. PubMed PMID: 15564509.
18. Diedrich CR, Mattila JT, Klein E, Janssen C, Phuah J, Sturgeon TJ, Montelaro RC, Lin PL, Flynn JL. Reactivation of latent tuberculosis in cynomolgus macaques infected with SIV is associated with early peripheral T cell depletion and not virus load. PLoS ONE. 2010;5(3):e9611. doi:10.1371/journal.pone.0009611. PubMed PMID: 20224771.
19. Brenchley JM, Paiardini M. Immunodeficiency lentiviral infections in natural and non-natural hosts. Blood. 2011;118(4):847–54. doi:10.1182/blood-2010-12-325936. PubMed PMID: 21505193. Epub 2011/04/19.eng.

20. Potash MJ, Chao W, Bentsman G, Paris N, Saini M, Nitkiewicz J, Belem P, Sharer L, Brooks AI, Volsky DJ. A mouse model for study of systemic HIV-1 infection, antiviral immune responses, and neuroinvasiveness. Proc Natl Acad Sci U S A. 2005;102(10):3760–5. PubMed PMID: 15728729. Epub 2005/02/23. eng.
21. Hadas E, Borjabad A, Chao W, Saini M, Ichiyama K, Potash MJ, Volsky DJ. Testing antiretroviral drug efficacy in conventional mice infected with chimeric HIV-1. AIDS. 2007;21(8):905–9. PubMed PMID: 17457083.
22. Hadas E, Chao W, He H, Saini M, Daley E, Saifuddin M, Bentsman G, Ganz E, Volsky DJ, Potash MJ. Transmission of chimeric HIV by mating in conventional mice: prevention by pre-exposure antiretroviral therapy and reduced susceptibility during estrus. Dis Model Mech. 2013;6(5):1292–8. doi:10.1242/dmm.012617. PubMed PMID: 23886803. Epub 2013/07/25.eng.
23. Roshorm Y, Hong JP, Kobayashi N, McMichael AJ, Volsky DJ, Potash MJ, Takiguchi M, Hanke T. Novel HIV-1 clade B candidate vaccines designed for HLA-B*5101(+) patients protected mice against chimaeric ecotropic HIV-1 challenge. Eur J Immunol. 2009;39(7):1831–40. doi:10.1002/eji.200939309. PubMed PMID: 19585509.
24. Saini M, Hadas E, Volsky DJ, Potash MJ. Vaccine-induced protection from infection of mice by chimeric human immunodeficiency virus type 1, EcoHIV/NL4–3. Vaccine. 2007;25(52):8660–3. PubMed PMID: 18023943. Epub 2007/10/31.eng.
25. Hanna Z, Priceputu E, Chrobak P, Hu C, Dugas V, Goupil M, Marquis M, de Repentigny L, Jolicoeur P. Selective expression of human immunodeficiency virus Nef in specific immune cell populations of transgenic mice is associated with distinct AIDS-like phenotypes. J Virol. 2009;83(19):9743–58. doi:10.1128/JVI.00125-09. PubMed PMID: 19605470. Epub 2009/07/15. eng.
26. De SK, Wohlenberg CR, Marinos NJ, Doodnauth D, Bryant JL, Notkins AL. Human chorionic gonadotropin hormone prevents wasting syndrome and death in HIV-1 transgenic mice. J Clin Invest. 1997;99(7):1484–91. PubMed PMID: 9119991.
27. Toggas SM, Masliah E, Rockenstein EM, Rall GF, Abraham CR, Mucke L. Central nervous system damage produced by expression of the HIV-1 coat protein gp120 in transgenic mice. Nature. 1994;367(6459):188–93. PubMed PMID: 8114918.
28. Corboy JR, Buzy JM, Zink MC, Clements JE. Expression directed from HIV long terminal repeats in the central nervous system of transgenic mice. Science. 1992;258(5089):1804–8. PubMed PMID: 1465618.
29. Garza HH Jr, Prakash O, Carr DJ. Aberrant regulation of cytokines in HIV-1TAT72-transgenic mice. J Immunol. 1996;156(10):3631–7. PubMed PMID: 8621896.
30. Wei P, Garber ME, Fang SM, Fischer WH, Jones KA. A novel CDK9-associated C-type cyclin interacts directly with HIV-1 Tat and mediates its high-affinity, loop specific binding to TAR RNA. Cell. 1998;92(4):451–62. PubMed PMID: 9491887.
31. Sun J, Soos T, Kewalramani VN, Osiecki K, Zheng JH, Falkin L, Santambrogio L, Littman DR, Goldstein H. CD4-specific transgenic expression of human cyclin T1 markedly increases human immunodeficiency virus type 1 (HIV-1) production by CD4+ T lymphocytes and myeloid cells in mice transgenic for a provirus encoding a monocyte-tropic HIV-1 isolate. J Virol. 2006;80(4):1850–62. PubMed PMID: 16439541.
32. Browning J, Horner JW, Pettoello-Mantovani M, Raker C, Yurasov S, DePinho RA, Goldstein H. Mice transgenic for human CD4 and CCR5 are susceptible to HIV infection. Proc Natl Acad Sci U S A. 1997;94(26):14637–41. PubMed PMID: 9405665.
33. Sawada S, Gowrishankar K, Kitamura R, Suzuki M, Suzuki G, Tahara S, Koito A. Disturbed CD4+ T cell homeostasis and in vitro HIV-1 susceptibility in transgenic mice expressing T cell line-tropic HIV-1 receptors. J Exp Med. 1998;187(9):1439–49. PubMed PMID: 9565636.
34. Scanga CA, Bafica A, Sher A. Viral gene expression in HIV transgenic mice is activated by Mycobacterium tuberculosis and suppressed after antimycobacterial chemotherapy. J Infect Dis. 2007;195(2):246–54. PubMed PMID: 17191170. Epub 2006/12/11. eng.

35. Pawlowski A, Jansson M, Sköld M, Rottenberg ME, Källenius G. Tuberculosis and HIV co-infection. PLoS Pathog. 2012;8(2):e1002464. doi:10.1371/journal.ppat.1002464. PubMed PMID: 22363214. Epub 2012/02/16. eng.
36. Heuts F, Gavier-Widén D, Carow B, Juarez J, Wigzell H, Rottenberg ME. CD4+ cell-dependent granuloma formation in humanized mice infected with mycobacteria. Proc Natl Acad Sci U S A. 2013;110(16):6482–7. doi:10.1073/pnas.1219985110. Epub 2013 Apr 4. PubMed PMID: 23559373.
37. Jurado JO, Alvarez IB, Pasquinelli V, Martínez GJ, Quiroga MF, Abbate E, Musella RM, Chuluyan HE, García VE. Programmed death (PD)-1:PD-ligand 1/PD-ligand 2 pathway inhibits T cell effector functions during human tuberculosis. J Immunol. 2008;181(1):116–25. PubMed PMID: 18566376.
38. Shultz LD, Saito Y, Najima Y, Tanaka S, Ochi T, Tomizawa M, Doi T, Sone A, Suzuki N, Fujiwara H, Yasukawa M, Ishikawa F. Generation of functional human T-cell subsets with HLA-restricted immune responses in HLA class I expressing NOD/SCID/IL2r gamma(null) humanized mice. Proc Natl Acad Sci U S A. 2010;107(29):13022–7. doi:10.1073/pnas.1000475107. PubMed PMID: 20615947. Epub 2010/07/06.eng.
39. Tanner A, Taylor SE, Decottignies W, Berges BK. Humanized Mice as a Model to Study Human Hematopoietic Stem Cell Transplantation. Stem Cells Dev. 2013;23(1):76–82. [Epub ahead of print] PubMed PMID: 23962058.
40. Lan P, Tonomura N, Shimizu A, Wang S, Yang YG. Reconstitution of a functional human immune system in immunodeficient mice through combined human fetal thymus/liver and CD34+ cell transplantation. Blood. 2006;108(2):487–92. PubMed PMID: 16410443. Epub 2006/01/12.eng.
41. Melkus MW, Estes JD, Padgett-Thomas A, Gatlin J, Denton PW, Othieno FA, Wege AK, Haase AT, Garcia JV. Humanized mice mount specific adaptive and innate immune responses to EBV and TSST-1. Nat Med. 2006;12(11):1316–22. PubMed PMID: 17057712. Epub 2006/10/22.eng.
42. Jaiswal S, Pazoles P, Woda M, Shultz LD, Greiner DL, Brehm MA, Mathew A. Enhanced humoral and HLA-A2-restricted dengue virus-specific T-cell responses in humanized BLT NSG mice. Immunology. 2012;136(3):334–43. doi:10.1111/j.1365-2567.2012.03585.x. PubMed PMID: 22384859.
43. Rajesh D, Zhou Y, Jankowska-Gan E, Roenneburg DA, Dart ML, Torrealba J, Burlingham WJ. Th1 and Th17 immunocompetence in humanized NOD/SCID/IL2rgammanull mice. Hum Immunol. 2010;71(6):551–9. doi:10.1016/j.humimm.2010.02.019. PubMed PMID: 20298731. Epub 2010/03/26.eng.
44. Calderon VE, Valbuena G, Goez Y, Judy BM, Huante MB, Sutjita P, Johnston RK, Estes DM, Hunter RL, Actor JK, Cirillo JD, Endsley JJ. A humanized mouse model of tuberculosis. PLoS ONE. 2013;8(5):e63331. doi:10.1371/journal.pone.0063331. PubMed PMID: 23691024.
45. de Wit D, Wootton M, Dhillon J, Mitchison DA. The bacterial DNA content of mouse organs in the Cornell model of dormant tuberculosis. Tuber Lung Dis. 1995;76(6):555–62. PubMed PMID: 8593379.
46. McCune RM, Feldmann FM, Lambert HP, McDermott W. Microbial persistence. I. The capacity of tubercle bacilli to survive sterilization in mouse tissues. J Exp Med. 1966;123(3):445–68. PubMed PMID: 4957010.
47. Scanga CA, Mohan VP, Joseph H, Yu K, Chan J, Flynn JL. Reactivation of latent tuberculosis: variations on the Cornell murine model. Infect Immun. 1999;67(9):4531–8. PubMed PMID: 10456896.
48. Lawn SD, Wilkinson RJ, Lipman MC, Wood R. Immune reconstitution and "unmasking" of tuberculosis during antiretroviral therapy. Am J Respir Crit Care Med. 2008;177(7):680–5. doi:10.1164/rccm.200709-1311PP. PubMed PMID: 18202347. Epub 2008/01/17. eng.

49. Meintjes G, Rabie H, Wilkinson RJ, Cotton MF. Tuberculosis-associated immune reconstitution inflammatory syndrome and unmasking of tuberculosis by antiretroviral therapy. Clin Chest Med. 2009;30(4):797–810, x. doi:10.1016/j.ccm.2009.08.013. PubMed PMID: 19925968.
50. Sereti I, Rodger AJ, French MA. Biomarkers in immune reconstitution inflammatory syndrome: signals from pathogenesis. Curr Opin HIV AIDS. 2010;5(6):504–10. doi:10.1097/COH.0b013e32833ed774. PubMed PMID: 20966640.
51. Ito R, Takahashi T, Katano I, Ito M. Current advances in humanized mouse models. Cell Mol Immunol. 2012;9(3):208–14. doi:10.1038/cmi.2012.2. PubMed PMID: 22327211. Epub 2012/02/13.eng.
52. Ippolito GC, Hoi KH, Reddy ST, Carroll SM, Ge X, Rogosch T, Zemlin M, Shultz LD, Ellington AD, Vandenberg CL, Georgiou G. Antibody repertoires in humanized NOD-scid-IL2Rγ(null) mice and human B cells reveals human-like diversification and tolerance checkpoints in the mouse. PLoS ONE. 2012;7(4):e35497. doi:10.1371/journal.pone.0035497. PubMed PMID: 22558161. Epub 2012/04/27.eng.
53. Dharmadhikari A, Nardell E. Translational reviews: What animal models teach humans about tuberculosis. Am J Respir Cell Mol Biol. 2008;39:503–8

Chapter 39
Epstein–Barr Virus Infection in Humanized Mice

Shigeyoshi Fujiwara, Go Matsuda and Ken-Ichi Imadome

Abbreviations

EBV	Epstein–Barr virus
IM	Infectious mononucleosis
LPD	Lymphoproliferative disease
RA	Rheumatoid arthritis
LCL	Lymphoblastoid cell line
CTL	Cytotoxic T lymphocyte
EBNA	EBV nuclear antigen
LMP	Latent membrane protein
EBER	EBV-encoded small RNA
BART	BamHI-A rightward transcript
CAEBV	Chronic active EBV infection
PBMC	Peripheral blood mononuclear cells
HSC	Hematopoietic stem cell
HLH	Hemophagocytic lymphohistiocytosis
XLP	X-linked lymphoproliferative disease
HA	Hemagglutinin
NOG	NOD/Shi-*scid* $Il2rg^{null}$
BRG	Balb/c $Rag2^{-/-}Il2rg^{-/-}$
NSG	NOD/LtSz-*scid* $Il2rg^{-/-}$

S. Fujiwara (✉) · G. Matsuda · K.-I. Imadome
Department of Infectious Diseases, National Research Institute for Child Health and Development, Tokyo 157-8535, Japan
e-mail: fujiwara-s@ncchd.go.jp

G. Matsuda
e-mail: matsuda-g@ncchd.go.jp

K.-I. Imadome
e-mail: imadome-k@ncchd.go.jp

L. Y. Poluektova et al. (eds.), *Humanized Mice for HIV Research*,
DOI 10.1007/978-1-4939-1655-9_39

39.1 Introduction

Epstein–Barr virus (EBV), a lymphotropic γ herpesvirus, was first identified in a primary culture of Burkitt lymphoma [1]. Although subsequent studies indicated that EBV is etiologically involved in a number of human malignancies, the virus was found to be a ubiquitous virus latently infecting more than 90% of the adult population worldwide [1]. Primary EBV infection occurs most often in childhood and is usually asymptomatic or accompanied by only nonspecific flu-like symptoms. Between 25 and 75% of primarily EBV infection in adolescence and young adulthood results in the development of infectious mononucleosis (IM) [1]. In a restricted fraction of hosts, EBV is involved in the pathogenesis of a wide variety of diseases, including lymphomas, carcinomas, B-cell lymphoproliferative diseases (LPDs) in immunocompromised hosts, and T- and NK-cell LPDs endemic mainly in east Asia [1]. In addition, EBV has been implicated in various autoimmune diseases including rheumatoid arthritis (RA) [2]. EBV has a unique biological activity to transform B lymphocytes into continuously proliferating lymphoblastoid cell lines (LCLs). EBV-transformed lymphoblastoid cells are normally removed by the virus-specific cytotoxic T lymphocytes (CTL) [3], but in immunocompromised hosts such as transplant recipients and AIDS patients, they may proliferate unlimitedly to cause B-cell LPDs such as posttransplant lymphoproliferative disease (PTLD) and AIDS-associated lymphomas.

EBV has a genome of approximately 170 kbp encoding more than 80 genes. EBV-transformed LCLs express six nuclear proteins (EBV nuclear antigens (EBNAs) 1, 2, 3A, 3B, 3C, and LP), three membrane proteins (latent membrane proteins (LMPs) 1, 2A, and 2B), and two sets of untranslated RNAs (EBV-encoded small RNAs (EBERs) 1 and 2, and BamHI-A rightward transcripts (BARTs)) [4]. EBV-infected cells observed in the virus-associated malignancies and LPDs exhibit various patterns of viral gene expression that are characteristic of the respective diseases. These patterns can be categorized into three groups: latency I, characterized by the expression of EBNA1, EBERs, and BARTs, is seen in Burkitt lymphoma and gastric carcinoma; latency II, characterized by the expression of EBNA1, LMP1, LMP2A, EBERs, and BARTs, is found in CAEBV, Hodgkin lymphoma, and nasopharyngeal carcinoma; latency III, characterized by the expression of all EBV-encoded proteins and RNAs identified in EBV-transformed LCLs, is observed in PTLD and AIDS-associated lymphomas [1].

Humans are the only natural host of EBV, but several new-world monkeys and rabbits can be infected with the virus experimentally [5–8]. Cotton-top tamarins that develop B-cell lymphomas following EBV infection were used for the evaluation of a candidate vaccine against the virus [9]. They are, however, endangered species and cannot be used in large numbers. A rhesus monkey lymphocryptovirus homologous to EBV was shown to reproduce key features of human EBV infection, including oral transmission, atypical lymphocytosis, and lymphadenopathy [10, 11]. Rhesus monkeys are thus a promising animal model, although they have disadvantages common to primate models, including limited accessibility and high costs. Small animal models for EBV infection were not available until the development of humanized mice.

39.2 EBV Pathogenesis in Humanized Mice

39.2.1 B-Cell Lymphoproliferative Disease

EBV studies with humanized mice started with *scid*-hu PBL mice. In 1988, Mosier and others demonstrated that intraperitoneal transplantation of peripheral blood mononuclear cells (PBMC) isolated from EBV-infected healthy carriers induced EBV-positive B-cell lymphomas in C.B-17 *scid* mice [12]. Similar transplantation of PBMC derived from EBV-uninfected donors did not cause lymphomas but subsequent infection with EBV resulted in B-cell lymphomas [13]. Pathological and virological studies demonstrated that these B-cell lymphomas are similar to EBV-associated B-cell LPD in immunocompromised hosts [14, 15]. This early humanized mouse model of EBV-associated B-cell LPD provided various insights into EBV-induced lymphomagenesis. For example, a critical role for $CD4^+$ T cells in in vivo proliferation of EBV-infected B cells was revealed in this model [16, 17]. The *scid*-hu PBL mouse model also revealed the involvement of human IL-10 and CXCL12/CXCR4 signaling in lymphomagenesis by EBV [18, 19]. Analysis on the IFN-γ gene polymorphism indicated that the A/A genotype for the base +874 was more prevalent in donors of PBMC that generated aggressive lymphoproliferation in *scid*-hu PBL mice [20]. Biological studies showed that the IFN-γ allele with adenosine at +874 was associated with inefficient CTL restimulation, probably explaining the above finding [20]. A study with NOD/*scid* mice transplanted with human PBMC revealed an important role of plasmacytoid dendritic cells in cellular immune responses against EBV [21].

Although *scid*-hu PBL mice were the remarkable first small animal model of EBV-associated B-cell LPD, they had disadvantages including inability to mount primary immune responses and a tendency to develop graft-versus-host disease. These disadvantages were removed in new-generation humanized mouse models that were prepared by transplanting human hematopoietic stem cells (HSC) to immunodeficient mice of various strains (e.g. NOD/Shi-*scid* $Il2rg^{null}$ (NOG), Balb/c $Rag2^{-/-}Il2rg^{-/-}$ (BRG), and NOD/LtSz-*scid* $Il2rg^{-/-}$ (NSG)) [22–25]. Models for EBV infection based on new-generation humanized mice are summarized in Table 39.1. Traggiai and others reconstituted human T cells, B cells, and dendritic cells in BRG mice and demonstrated that these mice can be infected with EBV and develop B-cell lymphoproliferation [23]. Characterization of EBV-induced LPD in humanized mice and its comparison to the original human disease was performed by Yajima and others [26] (Figure 39.1). They showed that the development of LPD in EBV-infected humanized NOG (hu-NOG) mice were dependent on viral dose; mice in-

Table 39.1 Humanized mouse models of EBV infection and associated diseases. (Reproduced from pathogens 2, 153–176, 2013. Reproduced under Creative Commons Attribution (CC BY) license)

Mouse strain, age on transplantation, irradiation	Human cells and/or tissues transplanted, route of transplantation	Human immune system components reconstituted	EBV strain used, route of inoculation	Features of EBV infection reproduced	Reference
NOD/*scid*, 8–10 weeks, 325 cGy γ irradiation	$CD34^+$ cells isolated from cord blood, intravenous	B cells, myeloid cells	Akata and EGFP-tagged B95-8, intrasplenic	B-cell LPD in latency II	[31]
Balb/c *Rag2*$^{-/-}$ *IL-2rg*$^{-/-}$ (BRG), newborn, 4 Gy γ irradiation	$CD34^+$ cells isolated from cord blood, intrahepatic	B, cells, T cells, dendritic cells	B95-8, intraperitoneal	B-cell proliferation, presumably EBV-specific T-cell response	[23, 63]
NOD/*scid*, 6–8 weeks, 325 cGy γ irradiation	Fetal thymus, fetal liver, liver-derived HSC (BLT mouse)	B cells, T cells, monocytes/ macrophages, dendritic cells	Akata, intrasplenic	Human MHC-restricted T-cell response to EBV detected by ELISPOT assay	[22]
NOD/Shi-*scid IL-2rg*^{null} (NOG), 6–8 weeks, no irradiation	$CD34^+$ cells isolated from cord blood, intravenous	B cells, T cells, NK cells, monocytes/ macrophages, dendritic cells	Akata, itravenous, intraperitoneal	B-cell LPD, latent infection, erosive arthritis resembling RA, EBV-specific T-cell responses, IgM Ab to p18^{BFRF3}	[26, 50, 51]
NOD/Shi-*scid IL-2rg*^{null} (NOG), newborn, 10 cGy X irradiation	$CD34^+$ cells isolated from cord blood, intrahepatic	B cells, T cells, NK cells, monocytes/ macrophages, dendritic cells	Akata, intravenous	IFN-γ cytokinemia, hemophagocytosis, systemic infiltration of $CD8^+$T cells, signs of HLH	[43]
NOD/LtSz-*scid IL-2rg*$^{-/-}$ (NSG), 2–5 days, 100 cGy irradiation	$CD34^+$ cells isolated from fetal liver, intrahepatic	B cells, T cells, NK cells, monocytes/ macrophages, dendritic cells	Unspecified, intraperitoneal	B-cell LPD, EBV-specific T-cell responses, establishment of EBV-specific T-cell clones	[29]
NOD/LtSz-*scid IL-2rg*$^{-/-}$ (NSG) with HLA-A2 transgene, 2–5 days, 100–150 cGy irradiation	$CD34^+$ cells isolated from fetal liver, intrahepatic	B cells, T cells, NK cells, monocytes/ macrophages, dendritic cells	Unspecified intraperitoneal	EBV-specific T-cell responses restricted by HLA-A2	[29, 58]

Table 39.1 (continued)

Mouse strain, age on transplantation, irradiation	Human cells and/or tissues transplanted, route of transplantation	Human immune system components reconstituted	EBV strain used, route of inoculation	Features of EBV infection reproduced	Reference
NOD/LtSz-*scid IL-2rg*$^{-/-}$ (NSG), 6–10 week old, 2–3 Gy irradiation	Fetal thymus, fetal liver, liver-derived HSC (NSG-BLT mouse)	B cells, T cells, no description of other components	B95-8 recombinants (BZLF1 knocked-out or enhanced BZLF1 expression), intraperitoneal	B cell lymphoma with type I, type IIb, or type III latency, latent infection, EBV-specific T-cell responses	[27, 28]

oculated with more than 10^2 50 % transforming dose (TD_{50}) tended to develop LPD whereas those inoculated with less than 10^1 TD_{50} mostly remained asymptomatic. They demonstrated the latency III type EBV gene expression, expression of B-cell activation markers and germinal center markers, as well as the histology of diffuse large B-cell lymphoma (DLBCL), indicating that EBV-induced LPD closely resembling PTLD and AIDS-associated lymphomas was reproduced (Figure 39.1) [26]. It is interesting that EBV-infected B cells morphologically similar to Hodgkin cells and Reed Sternberg cells were occasionally seen in EBV-induced LPD in hu-NOG mice (Figure 39.1c). EBV-induced B-cell LPD with similar characteristics was also reproduced in humanized NSG (hu-NSG) mice and BLT NSG mice [27–30]. One of the earliest new-generation humanized mice was prepared by transplanting NOD/*scid* mice with human HSC, resulting in reconstitution of B cells and myeloid cells but not T cells [31]. Note that B-cell LPD generated in these mice following EBV infection exhibited the latency II type EBV gene expression [31].

Preclinical studies of experimental therapies for EBV-associated LPD have been carried out so far mainly in *scid*-hu PBL mice. The list of tested regimens includes anti-CTLA-4 antibody [32], rituximab and IL-2 in combination [33], GM-CSF and IL-2 in combination [34], low-dose IL-2 [35], and the combination of CD13/CD19-bispecific antibody, CD28 specific antibody, and autologous T cells [36]. Ganciclovir induced complete regression of B-cell tumors in C.B-17 *scid* mice generated by transplantation of EBV-transformed LCL cells harboring a thymidine kinase gene driven by EBNA2 [37]. Gurer and others prepared a fusion protein of EBNA1 and the heavy chain of antibody against DEC-205, an endocytic receptor on dendritic cells, and showed that vaccination of hu-NSG mice with this fusion protein primed EBNA1-specific T cells and induced anti-EBNA1 antibodies [38].

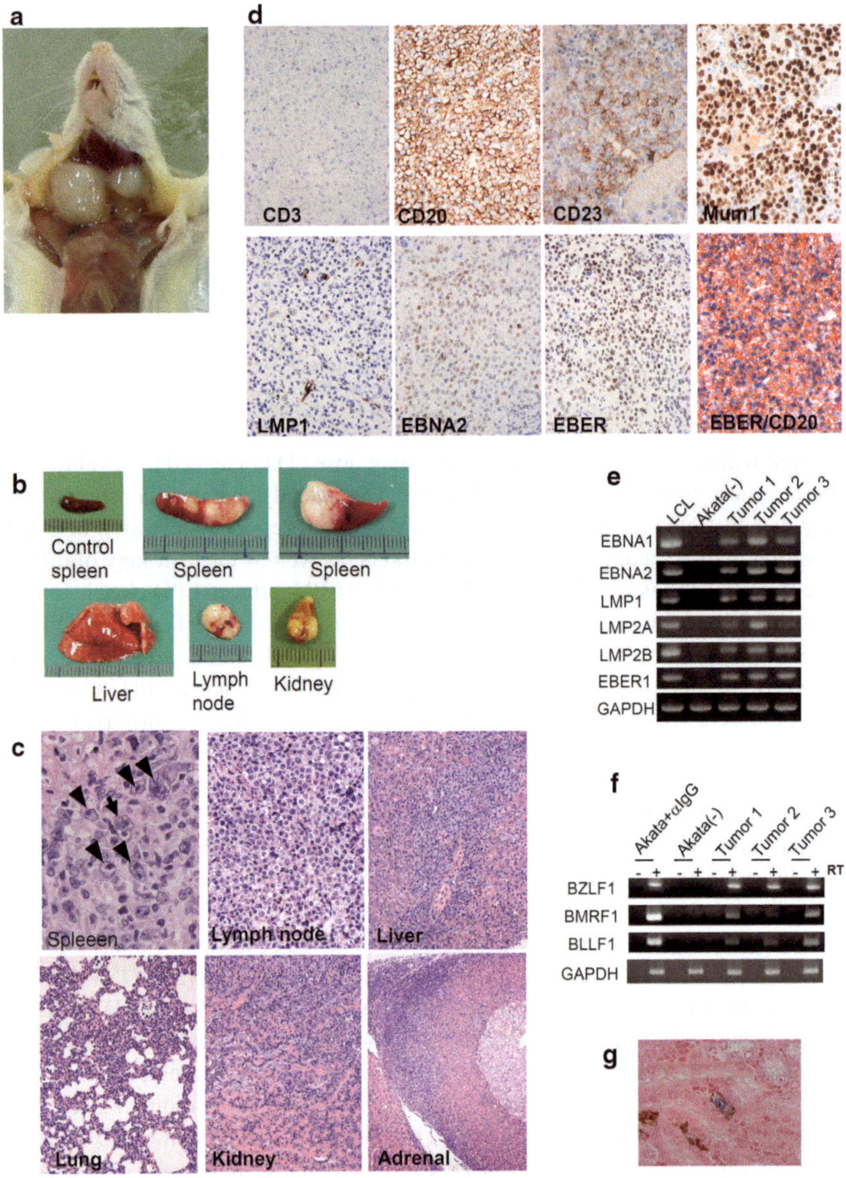

Fig. 39.1 EBV-induced lymphoproliferative disease in humanized NOG mice. **a** Photograph of an EBV-infected mouse showing tumors in the cervical area. **b** Photographs of spleens, liver, lymph node, and kidney from EBV-infected mice with lymphoproliferative disorder. The *upper-left* panel shows the spleen from an uninfected mouse. **c** Photomicrographs of HE-stained tissues of lymphoproliferative disorder. An *arrow* indicates a Reed–Sternberg-like cell and *arrowheads* Hodgkin-like cells. **d** Immunohistochemical staining for lymphocyte surface markers (*CD3*, *CD20*, *CD23*, and *Mum1*) and EBV-encoded proteins (*LMP1* and *EBNA2*), as well as in situ hybridization for *EBER* in a lymph node of a mouse bearing lymphoproliferative disorder. The *bottom-right* panel represents double staining for *EBER* and *CD20*. **e** and **f** RT-PCR detection of latent-cycle (**e**) and

39.2.2 EBV-HLH

Primary EBV infection can be complicated, although in rare occasions, by hemophagocytic lymphohistiocytosis (HLH), a serious hyperinflammatory condition associated with highly activated but ineffective immune responses [39]. Monoclonal proliferation of EBV-infected T or NK (most often $CD8^+$ T) cells have been consistently observed in EBV-HLH [40, 41]. Overproduction of cytokines by EBV-infected T or NK cells as well as by activated macrophages and T cells reacting to the virus is thought to play a central role in the pathogenesis [42]. Sato and others described systemic organ infiltration of activated $CD8^+$ T cells, IFN-γ cytokinemia, normocytic anemia, thrombocytopenia, and hemophagocytosis in EBV-infected hu-NOG mice, where EBV was found restrictedly in B cells [43]. Based on these findings, they proposed that EBV-infected hu-NOG mice may be a useful model of EBV-HLH. EBV-HLH associated with proliferation of the virus-infected B cells has been recently described in patients with X-linked lymphoproliferative disease 1 (XLP-1) and XLP-2 [44]. Using the same EBV-infected hu-NOG mice, Yajima and others observed predominantly B-cell LPD [26]. Because there were a number of differences in the protocols of Sato et al. [43] and Yajima et al. [26], including the sex of NOG mice, age at transplantation of HSC, and the route of transplantation, these differences might explain the different results. As described later, xenogenic transplantation of PBMC obtained from patients with EBV-HLH also reproduced cardinal features of this disease including proliferation of EBV-infected T cells and hypercytokinemia [45].

39.2.3 Rheumatoid Arthritis

Rheumatoid arthritis (RA) is a common autoimmune disease associated with progressive disability and systemic complications [46]. Evidence for the involvement of EBV in the RA pathogenesis includes increased EBV-infected B cells in patients' peripheral blood, infiltration of EBV-specific activated $CD8^+$ T cells in affected joints, and impaired functions of patients' T cells to suppress the outgrowth of EBV-transformed B cells [47]. Furthermore, cynovial EBV infection with the expression of viral proteins including LMP1 has been repeatedly observed in RA lesions [48, 49]. These lines of evidence are, however, rather circumstantial and direct evidence

lytic-cycle (**f**) EBV gene expression in tumors from EBV-infected hNOG mice. Spleen tumors from three different mice were examined for the expression of *EBNA*s *1* and *2*, *LMP*s *1*, *2A* and *2B*, *EBER1*, *BZLF1*, *BMRF1*, and *BLLF1*. RNA samples from LCL **e** and anti-IgG-treated Akata cells **f** were used as positive controls, and those of EBV-negative Akata cells (**e** and **f**) were used as negative controls. Assays were done with (+) or without (−) reverse transcriptase (RT) in **f**. Expression of *GAPDH* was examined as reference. **g** Double staining of *EBER* and *CD20* in the spleen of a hu-NOG mouse persistently infected with EBV without developing lymphoproliferative disorder. *EBER* is stained navy in the nucleus and *CD20* is stained *brown* in the membrane. (Reproduced from [26], by permission of Oxford University Press)

for the causal relationship has been missing, primarily because of the lack of an appropriate animal model. Recently, Kuwana and others reported that EBV-infected hu-NOG mice developed erosive arthritis with massive synovial cell proliferation and infiltration of human CD4⁺ T cells, CD8⁺ T cells, B cells, and macrophages [50]. A histological structure termed pannus, particularly characteristic to RA and involved in the destruction of bone tissues, was found in the arthritis lesions of hu-NOG mice. These results, although restricted to histopathological findings, indicate that EBV triggers the development of erosive arthritis morphologically resembling RA in humanized mice and suggest that the virus is responsible for the development of RA in humans as a trigger. Further studies are required to clarify whether this arthritis in mice is generated by a similar mechanism as that for the original human disease. Effects of antibodies specific to cytokines such as TNF-α, IL-1, and IL-6, that are known to be involved in the pathogenesis of RA, can be readily tested in this model. Inasmuch as there is a strong association between certain HLA types and RA, it will be an interesting experiment to test whether humanized mice prepared with HSC with high-risk HLA types have an increased frequency of arthritis following infection with EBV. The search for signs of other autoimmune diseases such as multiple sclerosis, Sjögren syndrome, and systemic lupus erythematosus in EBV-infected hu-NOG mice is underway.

39.3 Immune Responses to EBV in Humanized Mice

New-generation humanized mice mount a primary immune response to EBV [22, 23]. The spleen of humanized BRG (hu-BRG) mice contained an increased number of T cells after infection with EBV and $CD8^+$ T cells isolated from them proliferated vigorously upon stimulation with autologous EBV-transformed LCL [23]. In BLT mice developed by Melkus et al. [22], transplanted human thymus tissues enabled restriction of T cells by human MHC and hence efficient EBV-specific T-cell responses. These pioneering works were followed by more detailed descriptions of EBV-specific T-cell responses in humanized mice of various strains [26, 29, 43, 51]. EBV-specific $CD4^+$ and $CD8^+$ T-cell clones were established from EBV-infected hu-NSG mice and found to lyse HLA-matched LCLs [29]. Antibodies specific to the human MHC class-I inhibited the production of IFN-γ by $CD8^+$ T cells that were isolated from EBV-infected BLT or hu-NOG mice and stimulated with autologous LCL, indicating that these T cells recognized EBV epitopes presented by the human MHC class I [22, 26]. In accordance with these results, human MHC class I tetramers presenting EBV epitopes identified EBV-specific $CD8^+$ T cells in hu-NOG mice [43]. Depletion of either $CD4^+$ or $CD8^+$ T cells by administration of anti-CD4 or anti-CD8 antibody, respectively, caused more aggressive proliferation of EBV-infected cells and reduced the life span of infected mice, indicating that T-cell responses induced in humanized mice play a protective role [29, 51]. More directly, $CD8^+$ T cells isolated from EBV-infected hu-NOG mice suppressed transformation of autologous B cells by EBV [51]. EBV infection of hu-NOG and hu-NSG mice induced marked proliferation of $CD8^+$ T cells, but only a minor fraction of them appeared to be EBV-specific [26, 29, 43, 51]; the nature of the remaining part of proliferating $CD8^+$ T cells is not clear. One study with NOD/*scid* mice transplanted with human fetal thymic cells demonstrated an important role for EBV-induced $CD8^+$ NKT cells in the suppression of tumorigenesis by EBV-associated Hodgkin lymphoma and nasopharyngeal carcinoma cells [52].

It is a puzzling question how EBV-specific T-cell responses restricted by human MHC were induced in humanized mice that did not have human thymus tissues transplanted [26, 29, 43, 51]. T-cell responses specific to other viruses including HIV-1 [53], an adenovirus vector expressing HCG glycoproteins [54], HSV-2 [55], and influenza virus [56] have also been induced in humanized mice without human thymic tissues. Regarding this issue, experiments with humanized NOG I-$A^{-/-}$ mice performed by Watanabe and others suggested that although murine MHC plays a major role in the positive selection of T cells in humanized mice, human HSC-derived cells, conceivably B cells or dendritic cells, are also involved in the positive selection, possibly explaining T-cell restriction by human MHC observed in humanized mice [57]. This problem of T-cell education was averted in humanized mice expressing human MHC; NSG-derived mouse strains with a human HLA-A2 transgene that were reconstituted with HSC having the same HLA allele mounted efficient EBV-specific T-cell responses restricted by the particular HLA type [29, 58].

Antibody responses to EBV were analyzed in hu-NOG mice and found much less efficient than T-cell responses [26]. Only 3 out of 40 mice produced IgM antibody to $P18^{BFRF3}$, a major component of the viral capsid antigen (VCA). No IgG antibody to EBV-encoded proteins was detected. This inefficiency in antibody responses, especially that in the IgG response, appears common to humanized mouse models of other viruses including HIV-1 [53, 59, 60]. The finding that transfer of functional human T cells expressing TCR specific to hemagglutinin (HA) improved HA-specific IgG responses in hu-NOG mice suggests that suboptimal interactions between T and B cells underlie the inefficient production of antigen-specific IgG antibodies in these mice [57]. Better antibody responses in BLT mice support this notion [61, 62].

39.4 Persistent EBV Infection in Humanized Mice

EBV establishes asymptomatic persistent infection in human hosts that is dependent on T-cell immunosurveillance. In immunocompetent hosts, EBV-transformed cells with expression of highly immunogenic viral proteins are efficiently removed by CTLs and the virus is found persisting in memory B cells where all viral protein expressions are shut down. Yajima and others reported that the majority of hu-NOG mice inoculated with low doses of EBV ($<10^1$ TD_{50}) did not develop LPD and survived up to 6 months without any apparent signs of diseases [26]. EBV DNA was detected in the peripheral blood only for the first several weeks following inoculation and thereafter remained undetectable throughout the course of observation.

Upon autopsy at 6 months postinfection, no macroscopic pathological changes were observed, but occasional EBER-positive cells were found in their spleen and liver, indicating that EBV persisted in the mice. These EBER-positive cells were CD20-positive B cells, but their morphology did not resemble that of resting memory B cells (Figure 39.1g) [26]. mRNAs coding for EBNA1, EBNA2, LMP1, and LMP2A were detected by RT-PCR analyses of RNA obtained from the spleen or liver of these persistently infected mice, indicating the presence of latency III cells (Yajima et al. unpublished results). Persistent EBV infection in hu-NOG mice therefore does not completely reproduce EBV latency in humans. Because proper differentiation of memory B cells requires intricate interactions of B and T cells that have not been reproduced in current humanized mice [57], reproduction of bona fide EBV latency in memory B cells may require more sophisticated humanized mice. Nevertheless, it is an interesting question how immune responses are involved in the induction and maintenance of persistent EBV infection in hu-NOG mice. It is interesting that the blood EBV DNA level fluctuated in a few persistently infected mice and there the rise in the EBV DNA level was immediately followed by the increase in $CD8^+$ T cells and subsequent decline of the EBV DNA level, suggesting an effective T-cell control of EBV-infected cells [51]. If this persistent infection can be disrupted and EBV-positive LPD is induced by certain immunosuppressive regimens, it will be an excellent model of EBV-associated LPD in immunocompromised hosts.

The exact mechanism by which EBV establishes latent infection in memory B cells is not clear. Cocco and others examined EBV gene expression, surface marker expression, and hypermutation of the IG gene variable region in a single cell level in lymphoid tissues of EBV-infected hu-BRG mice [63]. Although they found mainly EBV-infected naïve B cells in the mantle zone, they identified infected cells in latency II that carried mutations in Ig genes in germinal centers. They proposed that these results support the previously presented hypothesis that EBV infects naïve B cells and induces their differentiation into memory B cells via germinal center reactions.

39.5 EBV Reverse Genetics in Humanized Mice

A number of EBV mutants with their particular genes knocked-out by homologous recombination have been prepared and revealed specific functions of these genes [4]. Mutations of certain EBV genes, however, did not generate any new phenotype in in vitro experiments, although strong conservation of these genes among natural EBV isolates suggests that they have important functions. EBNA3B is a representative of such genes; the EBNA3B knock-out virus retained its capacity to transform B cells and its role in the EBV life cycle is not known. Recent work by White and others demonstrated that an EBNA3B-KO EBV mutant induced more aggressive LPD in hu-NSG mice as compared with wild-type EBV, suggesting a tumor suppressor-like function of this gene [30]. B cells transformed by this mutant virus secreted less T-cell chemoattractant CXCL10 and thereby escaped T-cell–mediated killing. EBNA3B might thus function as a safety guard so that the virus should not be a life-threatening harm to the host.

BZLF1 is an immediate-early gene of EBV and acts as a switch from the latent to lytic cycle of EBV infection. Knocking out the BZLF gene did not affect the in vitro transforming activity of EBV and its involvement in lymphomagenesis was first elucidated in experiments with humanized mice. Ma and others generated EBV recombinants with the BZLF1 gene knocked out or with enhanced BZLF1 expression and demonstrated that BZLF1 enhanced EBV-induced lymphomagenesis in NSG-BLT mice, although the exact mechanism was not clear [27, 28].

There are a number of EBV genes, such as *BHRF1* (encoding an Bcl-2-like anti-apoptotic protein) [64], *BXLF1* (encoding EBV thymidine kinase) [65], and *BCRF1* (encoding a cytokine highly homologous to human IL-10) [66], loss-of-function mutants of which exhibit no or only minor phenotype alteration in in vitro studies. Examination of these EBV mutants in humanized mice might reveal their critical roles in the EBV life cycle and pathogenesis.

39.6 Mouse Xenograft Models of EBV-Associated T/NK-Cell LPD

Chronic active EBV infection (CAEBV) is a disease with high morbidity and mortality characterized by prolonged IM-like symptoms and elevated EBV DNA load in the peripheral blood [67–69]. Similar to EBV-HLH, monoclonal or oligoclonal proliferation of EBV-infected T or NK cells is consistently observed in this disease. In EBV-infected hu-NOG mice, however, EBV infection of neither T nor NK cells was recognized. To recapitulate CAEBV and EBV-HLH, Imadome and others transplanted PBMC isolated from patients to NOG mice and recapitulated major features of the two diseases, including systemic monoclonal proliferation of EBV-infected T or NK cells and hypercytokinemia [45] (Figure 39.2). Experiments with these models showed that EBV-infected T and NK cells did not engraft if $CD4^+$ T cells (whether or not infected with EBV) were removed from PBMC. When $CD4^+$ T cells were depleted in vivo just following transplantation of PBMC by administrating the OKT-4 antibody specific to CD4, engraftment of EBV-infected cells was consistently prevented [45]. Furthermore, administration of the antibody after engraftment of EBV-infected cells was also effective and reduced peripheral blood EBV DNA load to an undetectable level, suggesting that therapeutic approaches targeting $CD4^+$ T cells might be possible (Imadome and others, unpublished results).

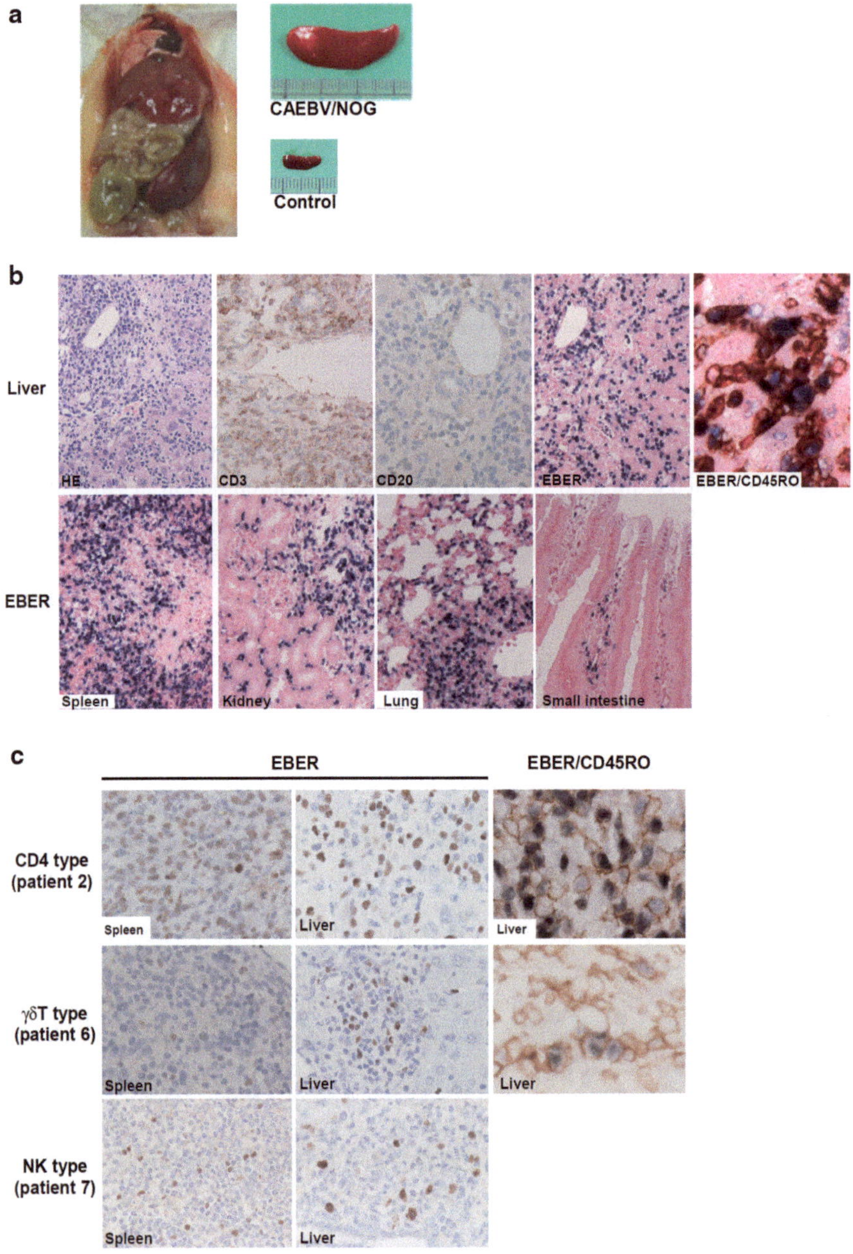

Fig. 39.2 A xenograft mouse model of chronic active EBV infection. **a** Photographs of a model mouse showing splenomegaly and of the excised spleen. This mouse was transplanted with PBMC from a CAEBV patient of the CD8 type. Spleen from a control NOG mouse is also shown. **b** Photomicrographs of various tissues of the mouse shown in **a**. *Upper panels*: liver tissue was stained with hematoxylin-eosin (*HE*), antibodies specific to human *CD3* or *CD20*, or by ISH with an *EBER* probe; the rightmost panel is a double staining with *EBER* and human *CD45RO*. *Bot-*

Acknowledgments This study was supported by grants from the Ministry of Health, Labour and Welfare of Japan (H24-Nanchi-046 and H22-AIDS-I-002), the Grant of National Center for Child Health and Development (25-9), a grant for the Research on Publicly Essential Drugs and Medical Devices from The Japan Health Sciences Foundation (KHD1221), and the Grant-in-Aid for Scientific Research (C) (H22-22590430).

References

1. Rickinson AB, Kieff ED. Epstein–Barr virus. In: Knipe DM, Howley PM, editors. Fields virology. 5. edn. Philadelphia: Lippincott Williams and Wilkins; 2007. pp. 2655–700.
2. Niller HH, Wolf H, Ay E, Minarovits J. Epigenetic dysregulation of Epstein–Barr virus latency and development of autoimmune disease. Adv Exp Med Biol. 2011;711:82–102.
3. Hislop AD, Taylor GS, Sauce D, Rickinson AB. Cellular responses to viral infection in humans: lessons from Epstein–Barr virus. Annu Rev Immunol. 2007;25:587–617.
4. Kieff ED, Rickinson AB. Epstein–Barr virus and its replication. In: Knipe DM, Howley PM, editors. Fields virology. Philadelphia: Lippincott Williams and Wilkins; 2007. pp. 2603–54.
5. Shope T, Dechairo D, Miller G. Malignant lymphoma in cotton top marmosets after inoculation with Epstein–Barr virus. Proc Natl Acad Sci U S A. 1973;70(9):2487–91.
6. Johannessen I, Crawford DH. In vivo models for Epstein–Barr virus (EBV)-associated B cell lymphoproliferative disease (BLPD). Rev Med Virol. 1999;9(4):263–77.
7. Epstein MA. zur Hausen H, Ball G, Rabin H. Pilot experiments with EB virus in owl monkeys (Aotus trivirgatus). III. Serological and biochemical findings in an animal with reticuloproliferative disease. Int J Cancer. 1975;15(1):17–22.
8. Takashima K, Ohashi M, Kitamura Y, Ando K, Nagashima K, Sugihara H, et al. A new animal model for primary and persistent Epstein–Barr virus infection: human EBV-infected rabbit characteristics determined using sequential imaging and pathological analysis. J Med Virol. 2008;80(3):455–66.
9. Epstein MA, Morgan AJ, Finerty S, Randle BJ, Kirkwood JK. Protection of cottontop tamarins against Epstein–Barr virus-induced malignant lymphoma by a prototype subunit vaccine. Nature. 1985;318(6043):287–9.
10. Moghaddam A, Rosenzweig M, Lee-Parritz D, Annis B, Johnson RP, Wang F. An animal model for acute and persistent Epstein–Barr virus infection. Science. 1997;276(5321):2030–3.
11. Wang F. A new animal model for Epstein–Barr virus pathogenesis. Curr Top Microbiol Immunol. 2001;258:201–19.
12. Mosier DE, Gulizia RJ, Baird SM, Wilson DB. Transfer of a functional human immune system to mice with severe combined immunodeficiency. Nature. 1988;335(6187):256–9.
13. Cannon MJ, Pisa P, Fox RI, Cooper NR. Epstein-Barr virus induces aggressive lymphoproliferative disorders of human B cell origin in SCID/hu chimeric mice. J Clin Invest. 1990;85(4):1333–7.
14. Rowe M, Young LS, Crocker J, Stokes H, Henderson S, Rickinson AB. Epstein-Barr virus (EBV)-associated lymphoproliferative disease in the SCID mouse model: implications for the pathogenesis of EBV-positive lymphomas in man. J Exp Med. 1991;173(1):147–58.

---

tom panels: EBER ISH in the spleen, kidney, lung, and small intestine. **c** Photomicrographs of the spleen and liver tissues obtained from NOG mice transplanted with PBMC from CAEBV patients of the *CD4*, *γδT*, and *NK* types, respectively. Tissues were stained by EBER-ISH or by double staining with EBER-ISH and human *CD45RO*. (From [45]. Reproduced under Creative Commons Attribution (CC BY) license.)

15. Mosier DE, Gulizia RJ, Baird SM, Spector S, Spector D, Kipps TJ, et al. Studies of HIV infection and the development of Epstein-Barr virus-related B cell lymphomas following transfer of human lymphocytes to mice with severe combined immunodeficiency. Curr Top Microbiol Immunol. 1989;152:195–9.
16. Veronese ML, Veronesi A, D'Andrea E, Del Mistro A, Indraccolo S, Mazza MR, et al. Lymphoproliferative disease in human peripheral blood mononuclear cell-injected SCID mice. I. T lymphocyte requirement for B cell tumor generation. J Exp Med. 1992;176(6):1763–7.
17. Johannessen I, Asghar M, Crawford DH. Essential role for T cells in human B-cell lymphoproliferative disease development in severe combined immunodeficient mice. Br J Haematol. 2000;109(3):600–10.
18. Baiocchi RA, Ross ME, Tan JC, Chou CC, Sullivan L, Haldar S, et al. Lymphomagenesis in the SCID-hu mouse involves abundant production of human interleukin-10. Blood. 1995;85(4):1063–74.
19. Piovan E, Tosello V, Indraccolo S, Cabrelle A, Baesso I, Trentin L, et al. Chemokine receptor expression in EBV-associated lymphoproliferation in hu/SCID mice: implications for CXCL12/CXCR4 axis in lymphoma generation. Blood. 2005;105(3):931–9.
20. Dierksheide JE, Baiocchi RA, Ferketich AK, Roychowdhury S, Pelletier RP, Eisenbeis CF, et al. IFN-gamma gene polymorphisms associate with development of EBV + lymphoproliferative disease in hu PBL-SCID mice. Blood. 2005;105(4):1558–65.
21. Lim WH, Kireta S, Russ GR, Coates PT. Human plasmacytoid dendritic cells regulate immune responses to Epstein-Barr virus (EBV) infection and delay EBV-related mortality in humanized NOD-SCID mice. Blood. 2007;109(3):1043–50.
22. Melkus MW, Estes JD, Padgett-Thomas A, Gatlin J, Denton PW, Othieno FA, et al. Humanized mice mount specific adaptive and innate immune responses to EBV and TSST-1. Nat Med. 2006;12(11):1316–22.
23. Traggiai E, Chicha L, Mazzucchelli L, Bronz L, Piffaretti JC, Lanzavecchia A, et al. Development of a human adaptive immune system in cord blood cell-transplanted mice. Science. 2004;304(5667):104–7.
24. Ito M, Hiramatsu H, Kobayashi K, Suzue K, Kawahata M, Hioki K, et al. NOD/SCID/gamma(c)(null) mouse: an excellent recipient mouse model for engraftment of human cells. Blood. 2002;100(9):3175–82.
25. Shultz LD, Lyons BL, Burzenski LM, Gott B, Chen X, Chaleff S, et al. Human lymphoid and myeloid cell development in NOD/LtSz-scid IL2R gamma null mice engrafted with mobilized human hemopoietic stem cells. J Immunol. 2005;174(10):6477–89.
26. Yajima M, Imadome K, Nakagawa A, Watanabe S, Terashima K, Nakamura H, et al. A new humanized mouse model of Epstein-Barr virus infection that reproduces persistent infection, lymphoproliferative disorder, and cell-mediated and humoral immune responses. J Infect Dis. 2008;198(5):673–82.
27. Ma SD, Hegde S, Young KH, Sullivan R, Rajesh D, Zhou Y, et al. A new model of Epstein-Barr virus infection reveals an important role for early lytic viral protein expression in the development of lymphomas. J Virol. 2011;85(1):165–77.
28. Ma SD, Yu X, Mertz JE, Gumperz JE, Reinheim E, Zhou Y, et al. An Epstein-Barr Virus (EBV) mutant with enhanced BZLF1 expression causes lymphomas with abortive lytic EBV infection in a humanized mouse model. J Virol. 2012;86(15):7976–87.
29. Strowig T, Gurer C, Ploss A, Liu YF, Arrey F, Sashihara J, et al. Priming of protective T cell responses against virus-induced tumors in mice with human immune system components. J Exp Med. 2009;206(6):1423–34.
30. White RE, Ramer PC, Naresh KN, Meixlsperger S, Pinaud L, Rooney C, et al. EBNA3B-deficient EBV promotes B cell lymphomagenesis in humanized mice and is found in human tumors. J Clin Invest. 2012;122(4):1487–502.
31. Islas-Ohlmayer M, Padgett-Thomas A, Domiati-Saad R, Melkus MW, Cravens PD, Martin Mdel P, et al. Experimental infection of NOD/SCID mice reconstituted with human CD34+ cells with Epstein-Barr virus. J Virol. 2004;78(24):13891–900.

32. May KF Jr, Roychowdhury S, Bhatt D, Kocak E, Bai XF, Liu JQ, et al. Anti-human CTLA-4 monoclonal antibody promotes T-cell expansion and immunity in a hu-PBL-SCID model: a new method for preclinical screening of costimulatory monoclonal antibodies. Blood. 2005;105(3):1114–20.
33. Eisenbeis CF, Grainger A, Fischer B, Baiocchi RA, Carrodeguas L, Roychowdhury S, et al. Combination immunotherapy of B-cell non-Hodgkin's lymphoma with rituximab and interleukin-2: a preclinical and phase I study. Clin Cancer Res. 2004;10(18 Pt 1):6101–10.
34. Baiocchi RA, Ward JS, Carrodeguas L, Eisenbeis CF, Peng R, Roychowdhury S, et al. GM-CSF and IL-2 induce specific cellular immunity and provide protection against Epstein-Barr virus lymphoproliferative disorder. J Clin Invest. 2001;108(6):887–94.
35. Baiocchi RA, Caligiuri MA. Low-dose interleukin 2 prevents the development of Epstein-Barr virus (EBV)-associated lymphoproliferative disease in scid/scid mice reconstituted i.p. with EBV-seropositive human peripheral blood lymphocytes. Proc Natl Acad Sci U S A. 1994;91(12):5577–81.
36. Bohlen H, Manzke O, Titzer S, Lorenzen J, Kube D, Engert A, et al. Prevention of Epstein-Barr virus-induced human B-cell lymphoma in severe combined immunodeficient mice treated with CD3xCD19 bispecific antibodies, CD28 monospecific antibodies, and autologous T cells. Cancer Res. 1997;57(9):1704–9.
37. Franken M, Estabrooks A, Cavacini L, Sherburne B, Wang F, Scadden DT. Epstein-Barr virus-driven gene therapy for EBV-related lymphomas. Nat Med. 1996;2(12):1379–82.
38. Gurer C, Strowig T, Brilot F, Pack M, Trumpfheller C, Arrey F, et al. Targeting the nuclear antigen 1 of Epstein-Barr virus to the human endocytic receptor DEC-205 stimulates protective T-cell responses. Blood. 2008;112(4):1231–9.
39. Henter JI, Horne A, Arico M, Egeler RM, Filipovich AH, Imashuku S, et al. HLH-2004: diagnostic and therapeutic guidelines for hemophagocytic lymphohistiocytosis. Pediatr Blood Cancer. 2007;48(2):124–31.
40. Kikuta H, Sakiyama Y, Matsumoto S, Oh-Ishi T, Nakano T, Nagashima T, et al. Fatal Epstein-Barr virus-associated hemophagocytic syndrome. Blood. 1993;82(11):3259–64.
41. Kawaguchi H, Miyashita T, Herbst H, Niedobitek G, Asada M, Tsuchida M, et al. Epstein-Barr virus-infected T lymphocytes in Epstein-Barr virus-associated hemophagocytic syndrome. J Clin Invest. 1993;92(3):1444–50.
42. Lay JD, Tsao CJ, Chen JY, Kadin ME, Su IJ. Upregulation of tumor necrosis factor-alpha gene by Epstein-Barr virus and activation of macrophages in Epstein-Barr virus-infected T cells in the pathogenesis of hemophagocytic syndrome. J Clin Invest. 1997;100(8):1969–79.
43. Sato K, Misawa N, Nie C, Satou Y, Iwakiri D, Matsuoka M, et al. A novel animal model of Epstein-Barr virus-associated hemophagocytic lymphohistiocytosis in humanized mice. Blood. 2011;117(21):5663–73.
44. Yang X, Wada T, Imadome K, Nishida N, Mukai T, Fujiwara M, et al. Characterization of Epstein-Barr virus (EBV)-infected cells in EBV-associated hemophagocytic lymphohistiocytosis in two patients with X-linked lymphoproliferative syndrome type 1 and type 2. Herpesviridae. 2012;3(1):1.
45. Imadome K, Yajima M, Arai A, Nakazawa A, Kawano F, Ichikawa S, et al. Novel mouse xenograft models reveal a critical role of CD4+T cells in the proliferation of EBV-infected T and NK cells. PLoS Pathog. 2011;7(10):e1002326.
46. McInnes IB, Schett G. The pathogenesis of rheumatoid arthritis. N Engl J Med. 2011;365(23):2205–19.
47. Toussirot E, Roudier J. Pathophysiological links between rheumatoid arthritis and the Epstein-Barr virus: an update. Joint Bone Spine. 2007;74(5):418–26.
48. Takei M, Mitamura K, Fujiwara S, Horie T, Ryu J, Osaka S, et al. Detection of Epstein-Barr virus-encoded small RNA 1 and latent membrane protein 1 in synovial lining cells from rheumatoid arthritis patients. Int Immunol. 1997;9(5):739–43.
49. Takeda T, Mizugaki Y, Matsubara L, Imai S, Koike T, Takada K. Lytic Epstein-Barr virus infection in the synovial tissue of patients with rheumatoid arthritis. Arthritis Rheum. 2000;43(6):1218–25.

50. Kuwana Y, Takei M, Yajima M, Imadome K, Inomata H, Shiozaki M, et al. Epstein-Barr virus induces erosive arthritis in humanized mice. PLoS ONE. 2011;6(10):e26630.
51. Yajima M, Imadome K, Nakagawa A, Watanabe S, Terashima K, Nakamura H, et al. T cell-mediated control of Epstein-Barr virus infection in humanized mice. J Infect Dis. 2009;200(10):1611–5.
52. Yuling H, Ruijing X, Li L, Xiang J, Rui Z, Yujuan W, et al. EBV-induced human CD8+NKT cells suppress tumorigenesis by EBV-associated malignancies. Cancer Res. 2009;69(20):7935–44.
53. Gorantla S, Makarov E, Finke-Dwyer J, Gebhart CL, Domm W, Dewhurst S, et al. CD8+cell depletion accelerates HIV-1 immunopathology in humanized mice. J Immunol. 2010;184(12):7082–91.
54. Marodon G, Desjardins D, Mercey L, Baillou C, Parent P, Manuel M, et al. High diversity of the immune repertoire in humanized NOD.SCID.gamma c-/- mice. Eur J Immunol. 2009;39(8):2136–45.
55. Kwant-Mitchell A, Ashkar AA, Rosenthal KL. Mucosal innate and adaptive immune responses against herpes simplex virus type 2 in a humanized mouse model. J Virol. 2009;83(20):10664–76.
56. Yu CI, Gallegos M, Marches F, Zurawski G, Ramilo O, Garcia-Sastre A, et al. Broad influenza-specific CD8+ T-cell responses in humanized mice vaccinated with influenza virus vaccines. Blood. 2008;112(9):3671–8.
57. Watanabe Y, Takahashi T, Okajima A, Shiokawa M, Ishii N, Katano I, et al. The analysis of the functions of human B and T cells in humanized NOD/shi-scid/gammac(null) (NOG) mice (hu-HSC NOG mice). Int Immunol. 2009;21(7):843–58.
58. Shultz LD, Saito Y, Najima Y, Tanaka S, Ochi T, Tomizawa M, et al. Generation of functional human T-cell subsets with HLA-restricted immune responses in HLA class I expressing NOD/SCID/IL2r gamma(null) humanized mice. Proc Natl Acad Sci U S A. 2010;107(29):13022–7.
59. Baenziger S, Tussiwand R, Schlaepfer E, Mazzucchelli L, Heikenwalder M, Kurrer MO, et al. Disseminated and sustained HIV infection in CD34+ cord blood cell-transplanted Rag2-/-gamma c-/- mice. Proc Natl Acad Sci U S A. 2006;103(43):15951–6.
60. Watanabe S, Terashima K, Ohta S, Horibata S, Yajima M, Shiozawa Y, et al. Hematopoietic stem cell-engrafted NOD/SCID/IL2Rgamma null mice develop human lymphoid systems and induce long-lasting HIV-1 infection with specific humoral immune responses. Blood. 2007;109(1):212–8.
61. Brainard DM, Seung E, Frahm N, Cariappa A, Bailey CC, Hart WK, et al. Induction of robust cellular and humoral virus-specific adaptive immune responses in human immunodeficiency virus-infected humanized BLT mice. J Virol. 2009;83(14):7305–21.
62. Sun Z, Denton PW, Estes JD, Othieno FA, Wei BL, Wege AK, et al. Intrarectal transmission, systemic infection, and CD4+ T cell depletion in humanized mice infected with HIV-1. J Exp Med. 2007;204(4):705–14.
63. Cocco M, Bellan C, Tussiwand R, Corti D, Traggiai E, Lazzi S, et al. CD34+ cord blood cell-transplanted Rag2-/- gamma(c)-/- mice as a model for Epstein-Barr virus infection. Am J Pathol. 2008;173(5):1369–78.
64. Lee MA, Yates JL. BHRF1 of Epstein-Barr virus, which is homologous to human proto-oncogene bcl2, is not essential for transformation of B cells or for virus replication in vitro. J Virol. 1992;66(4):1899–906.
65. Shimizu N, Yoshiyama H, Takada K. Clonal propagation of Epstein-Barr virus (EBV) recombinants in EBV-negative Akata cells. J Virol. 1996;70(10):7260–3.
66. Swaminathan S, Hesselton R, Sullivan J, Kieff E. Epstein-Barr virus recombinants with specifically mutated BCRF1 genes. J Virol. 1993;67(12):7406–13.
67. Okano M. Overview and problematic standpoints of severe chronic active Epstein-Barr virus infection syndrome. Crit Rev Oncol Hematol. 2002;44(3):273–82.
68. Straus SE. Acute progressive Epstein-Barr virus infections. Annu Rev Med. 1992;43:437–49.
69. Kimura H. Pathogenesis of chronic active Epstein-Barr virus infection: is this an infectious disease, lymphoproliferative disorder, or immunodeficiency? Rev Med Virol. 2006;16(4):251–61.

Chapter 40
HTLV-1 Infection of Humanized NOD/SCID IL2 $\gamma c^{-/-}$ and BALB/c-Rag2$^{-/-}\gamma c^{-/-}$ Mouse Models

Madeleine Duc Dodon, Julien Villaudy, Louis Gazzolo and Gerold Feuer

Abbreviations

ATLL	Adult T-cell leukemia/lymphoma
HIS	Human immune system
HP/HSCs	Hematopoietic progenitors/hematopoietic stem cells
HTLV-1	Human T-cell leukemia virus type 1
TSP/HAM	Tropical spastic paraparesis/HTLV-1 associated myelopathy

40.1 Introduction

The first chapter of human retrovirology was written in 1980 with the isolation of HTLV-1 (Human T-lymphotropic Virus type 1) that concluded the long search of a retrovirus etiologically linked to a tumoral process. This retrovirus was identified as the causative agent of adult T-cell Leukaemia/lymphoma (ATLL), an aggressive and clonal lymphoproliferative disorder of mature $CD4^+$ T cells (for a review, see [1]). During the past three decades, numerous epidemiological and clinical studies

M. Duc Dodon (✉) · J. Villaudy · L. Gazzolo
Laboratoire de Biologie Moléculaire de la Cellule,
Unité Mixte de Recherche 5239, Centre National de la Recherche Scientifique,
Ecole Normale Supérieure de Lyon, 69364 Lyon Cedex 7, France
e-mail: madeleine.duc.dodon@ens-lyon.fr

M. Duc Dodon · J. Villaudy · L. Gazzolo
SFR UMS3444 BioSciences Lyon-Gerland-Lyon Sud (UMS3444),
69366 Lyon Cedex 7, France

G. Feuer
Department of Microbiology and Immunology, SUNY Upstate Medical University,
Syracuse, NY 13210, USA

Humurine Technologies, 640 Arrow Highway, LaVerne, CA 91750, USA

L. Y. Poluektova et al. (eds.), *Humanized Mice for HIV Research*,
DOI 10.1007/978-1-4939-1655-9_40

as well as extensive molecular and cellular observations have been performed to understand the pathological mechanisms triggered by HTLV-1 infection. However, experimental approaches carried out in various animal models (rabbits, monkeys, rats, and mice) have failed to recapitulate leukemogenesis [2, 3]. Recently, the growing need of an animal model for the in vivo investigation of human hematopoiesis and immunity culminated with the construction of mice that develop a functional human immune system (HIS) through the engraftment of human tissues and/or hematopoietic progenitor cells in severely immunocompromised mouse strains. The recently developed HIS mice bearing a targeted mutation in the interleukin 2 (IL2) receptor common gamma chain (γc) represent a small and effective animal model, most appropriate to study human infectious diseases and have been used in attempts to in vivo recapitulate HTLV-1-induced leukemogenesis in vivo [4, 5].

40.2 Virological Aspects of HTLV-1

HTLV-1 is a complex retrovirus, which belongs to the genus *Deltaretrovirus* of the subfamily *Orthoretrovirinae*. The HTLV-1 proviral genome has structural genes, *gag*, *pol*, and *env*, bracketed by long terminal repeat (LTR) sequences (Fig. 40.1; [1]). The 5′ LTR including U3/R/U5 serves as the viral promoter for transcription. The pX region, located between *env* and the 3′LTR, contains sequences for regulatory viral proteins, Tax, Rex, p12, p13, p30, and p21. The minus strand of pX encodes an antisense transcript, HTLV-1 basic leucine zipper factor (*Hbz*) [6].

HTLV-1 infects 10–20 million people worldwide, and causes ATLL in a small percentage of infected individuals after a prolonged latency period of up to 20–50 years.

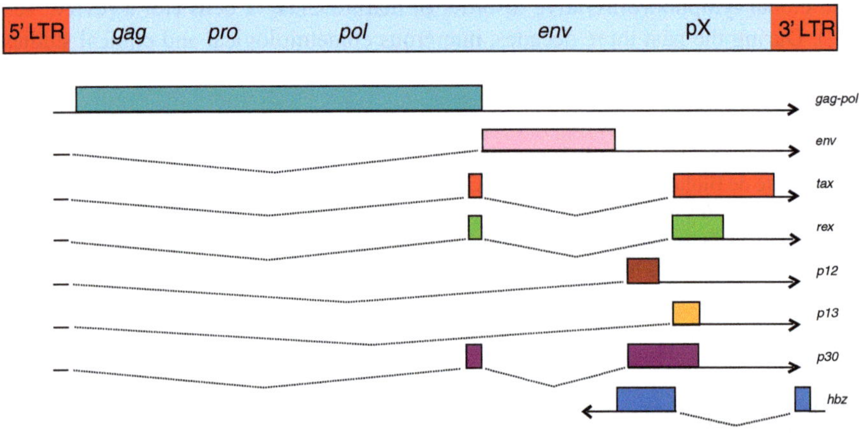

Fig. 40.1 Genome structure of the HTLV-1 provirus. The *gag*, *pol*, and *env* structural genes are flanked by 5′ and 3′LTRs. The *pX* region, Tax, Rex, p12, p13, p30, and the antisense HTLV-1 basic leucine zipper factor (*HBZ*) open reading frame are shown (*colored boxes*)

HTLV-1 can be transmitted through breastfeeding, sexual contact, or parenteral transmission. This malignant disease has unique clinical, pathological, and cytological features, and is classified into four clinical subtypes: smoldering, chronic, acute, and leukaemia/lymphoma. Mild-clinical symptoms characterize smoldering patients, but those with chronic or acute and leukaemia/lymphoma show a rapid progression and common association with lymphadenopathy, hepatosplenomegaly, and hypercalcemia, frequent skin lesions and resistance to treatment with antileukaemia agents. The leukemic T cells, the number of which increases according to the severity of the disease, are polymorphic with multilobulated nuclei (flower cells). They mainly express CD4 and other mature T-cell markers such as CD2, the CD3/T-cell receptor (TCR) molecules on cell surfaces being frequently downregulated. They are activated as ascertained by the expression of CD25, the α chain of the interkeukin-2 (IL-2) receptor. The HTLV-1 provirus is monoclonally integrated in ATLL cells, indicating that these cells arise from a single malignant clone, which displays the same proviral integration pattern. HTLV-1 infection also induces inflammatory manifestations, such as tropical spastic paraparesis or HTLV-1-associated myelopathy (TSP/HAM). The risk to develop either ATLL or TSP/HAM appears to be related to age, route of infection and the immune competency of the host. Thus, the oral route, i.e., breastfeeding, might be critical in the initiation of the long leukemogenic process. The investigation of HTLV-1-mediated leukemogenesis for over three decades has largely been focused on Tax, a 40 kDa HTLV-1 encoded regulatory protein, that functions as a trans-activator of proviral gene expression. It is also a key viral component of HTLV-1-mediated leukemogenesis, primarily due to its ability to modulate the expression of a large variety of cellular genes responsible for cell proliferation, genetic instability, dysregulation of the cell cycle, and apoptosis. However, Tax expression is not detected in a majority of ATLL patients, underlining that Tax expression is not necessary for the maintenance of ATLL. Such a role is associated with expression of HBZ, the other viral regulatory protein. The spliced form of HBZ is expressed in all ATLL and TSP/HAM cases. HBZ protein promotes proliferation of ATLL cells and induces T-cell lymphomas in $CD4^+$ T cells when expressed in transgenic mice, indicating that HBZ is necessary for the proliferation and survival of ATLL cells, for reviews, see [1, 2, 6–8].

40.3 In Search of Target Cells at the Initiation of HTLV-1 Induced-Leukemogenesis

One of the fundamental problems in cancer research concerns the identification of the normal cell in which the malignant events are initiated. This target cell specificity is especially important and noteworthy in hematopoietic cell tumours arising at specific stages of hematopoietic cell development. Thus, leukaemia/lymphoma is defined as the proliferation of a malignant clone in the bone marrow or in peripheral lymphopoietic organs. They are very often characterized by an aberrant expression of transcription factors, explaining the altered expression of numerous cellular genes controlling the maturation of haematopoietic cells. Deciphering these mechanisms has greatly clarified the ways in which leukemia retroviruses affect proliferation and differentiation of the target cells, thus triggering the leukemogenic process. In this regard, molecular and cellular approaches performed with oncogenic chicken and mouse retroviruses, such as avian erythroblastosis virus and murine leukemia virus, have provided evidence that the phenotype of the initial virus-infected target cell is quite different of that of the proliferating leukemic cells [9, 10]. These observations challenge the current belief that the leukemic T cells in ATLL patients originate from mature differentiated $CD4^+$ T cells infected with HTLV-1. Despite considerable progress made in understanding ATLL biology, the exact sequence of events occurring during the initial stages of malignancy, including the types of cells infected with HTLV-1, remain unclear. The primary target cells for HTLV-1 infection may not only influence HTLV-1 pathogenesis, but sequestration of these cells in anatomical sites, such as the bone marrow and spleen may allow the virus to elude the immune system and may contribute to the long-latent period following infection. The observation that neonatal HTLV-1 infection preferentially correlates with the development of ATLL leads to the hypothesis that neonatal transmission of HTLV-1 is favoring viral infection of hematopoietic progenitors/hematopoietic stem cells (HP/HSC) and immature human thymocytes. Such a hypothesis implies that the expression of Tax in these cells may alter the αβT-cell development in the thymus. Interestingly, the role of Tax in T-cell transformation and in the development of HTLV-1-associated lymphoma has been approached using Tax-transgenic mice under the control of either the HTLV-1 LTR or alternative promoters targeting Tax expression in lymphoid compartments [11]. Thus, when transgene expression is placed under the control of the Lck promoter, which restricts Tax expression to developing thymocytes, an ATLL-like phenotype appeared in these transgenic mice [11]. Accordingly, a series of in vitro settings focused on the effect of HTLV-1 infection on progenitor cells as well as on immature thymocytes have supported that hypothesis.

HTLV-1 Infection of $CD34^+$ HP/HSCs Human $CD34^+$ HP/HSCs are maintained in the bone marrow and can differentiate into all requisite mature cell types represented by the immune system. $CD34^+$ cells are able to mature into multiple hematopoietic lineages, yet are concurrently long-lived and have the capacity for self-renewal and maintenance of an undifferentiated state in bone marrow. The ability of HTLV-1 to sustain a productive infection in $CD34^+$ cells identifies hematopoietic progenitor and stem cells as targets that sustain and harbor latent HTLV-1 infection over substantial periods of time [12]. Although HTLV-1 infection of mature T lymphocytes gives rise to immortalization and transformation, infection of $CD34^+$ HP/HSCs in culture induces G0/G1 cell cycle arrest by modulation of the cellular cdk inhibitors $p21^{cip1/waf1}$, $p27^{kip1}$, and surviving by the activity of the Tax1 oncoprotein. This suggests that HTLV-1 exploits the sensitive cell arrest mechanism in $CD34^+$ HP/HSC cells to establish a latent infection [13, 14]. The HTLV-1 provirus was detected in $CD34^+$ cells isolated from peripheral blood lymphocytes of HTLV-1-infected patients has been demonstrated [15], and infection of HP/HSCs may result in skewing of hematopoiesis towards distinct lineages, ultimately giving rise to the outgrowth of malignant clones.

HTLV-1 Infection of Human Immature Thymocytes During early thymocyte differentiation, β-selection is a major checkpoint that controls the proliferation and selection of the double-negative (DN) thymocytes [16, 17]. This process is subsequent to the expression of recombinase Rag1/2 proteins involved in the rearrangement of the T cell receptor (TCR) locus and is mediated by the successful expression of the pre-TCR comprising the TCR βchain, CD3 proteins and an invariant pre-TCR αchain (pTα) at the membrane of human immature thymocytes. Pre-TCR signalling leads to the activation of transcription factors belonging to the rel/NF-κB and AP-1 families involved in the survival, proliferation and differentiation of selected thymocytes into $CD4^+$ $CD8^+$ double-positive (DP) cells. After β-selection, rearrangements of the TCRαlocus are initiated, and after being submitted to either positive or negative selection, the cells become mature functional single-positive (SP) $CD4^+$ and $CD8^+$ cells αβT-cells.

Thus, proliferation and differentiation of αβT lymphocytes are mostly dependent on the events presiding the assembly of the pre-TCR and of those triggered by the β-selection process. Indeed, mutant mice that are lacking the Rag1 or Rag2 proteins, TCRβ or pTα cannot form a pre-TCR complex and as a consequence are displaying a developmental arrest at the DN stage. Genetic ablation or overexpression studies in mice have identified a number of molecules that regulate the transition of DN thymocytes to the DP stage. Among them, transcription factors are critically involved in the molecular programs that control cell proliferation, differentiation, and survival during these early stages of thymopoiesis [18]. They are also intimately linked with the induction of leukemogenic events that result in the transformation of T-cell precursors [19]. HTLV-1 infection within the thymus might therefore represent an essential and critical event of viral leukemogenesis. Moreover, Tax, by silencing E proteins of the basic helix-loop-helix (bHLH) family, downregulates the transcription of pTα during early thymocyte development, indicating that HTLV infection in the thymus is a prerequisite to the induction of the leukemogenic process associated with this retroviral infection [20, 21]. Finally, in in vitro Tax-transduced human immature thymocytes, an increase in the expression of NF-κB and of anti-apoptotic genes was observed, indicating that human T-cell development could be altered by Tax [22].

40.4 HTLV-1 Infection of HIS Mouse Models

During the past few years, remarkable advances have been reported concerning the development of a human immune system in immunodeficient mice, which have neither functional T and B cells nor NK cells. Further efforts at minimizing the response of the innate murine immune system have led to the production of NOD/SCID IL2$\gamma c^{-/-}$ and BALB/c-Rag2$^{-/-}\gamma c^{-/-}$ mouse models [23, 24]. Both models have been used to establish humanized mice after engraftment of human CD34$^+$ hematopoietic stem cells. These strains mice have proven highly suitable for the generation of HIS mice and have been used to study the infection of several human pathogens including HTLV-1.

HTLV-1 Infection of HIS NOD/SCID IL2γc$^{-/-}$ Faithful recapitulation of ATLL in "humanized" mice has been challenging, primarily since HTLV-1 does not readily infect target cells as a cell-free virus and requires cell-to-cell contact for efficient virus transmission. Initial attempts to infect the conjoint Thy/Liv human organ with HTLV-1 in SCID-hu mice failed to induce oncogenesis [12]. HIS NOD/SCID IL2γc$^{-/-}$ mice are generated by inoculation of human CD34$^+$ cells into neonatal NOD/SCID IL2γc$^{-/-}$ mice. These HIS mice support broad and robust maturation of all human hematopoietic lineages and this small animal model has facilitated evaluation of HTLV-1 infection and Tax1 expression on maturation of the human immune system in vivo. Ex vivo HTLV-1 infection of CD34$^+$ HP/HSCs and injection of infected CD34$^+$ cells into NOD/SCID IL2γc$^{-/-}$ or NOD/SCID mice results in maturation of infected HP/HSCs and in significant levels of lymphomagenesis. Malignant cells from infected mice invariably harbor the HTLV-1 provirus and display CD4, and occasionally CD25, both cellular markers of ATLL cells [15]. Transduction of CD34$^+$ cells with a lentivirus vector expressing the Tax oncoprotein and inoculation into NOD/SCID IL2γc$^{-/-}$ mice also resulted in CD4$^+$ lymphomagenesis, although tumors were predominantly monoclonal in origin in contrast to the oligoclonal tumors induced by infection with HTLV-1. It is conceivable that initiation of leukemogenesis/lymphomagenesis by HTLV-1 infection involves the generation of an "infectious leukemic stem cell" that eventually gives rise and allows maturation of malignant mature T cells [25]. The exact molecular processes that contribute and give rise to a full-blown malignant cell types have yet to be precisely identified, and the HIS mouse has become an invaluable tool to delineate molecular and cellular events in HTLV-1 leukemogenesis.

HTLV-1 Infection of HIS BALB/c-Rag2$^{-/-}$γc$^{-/-}$ Mice As indicated above, Tax interferes with ß-selection during early thymopoiesis, indicating that HTLV-1 infection has the potential to perturb thymic human αβT-cell development. To verify that inference, we investigated the in vivo effects of HTLV-1 infection in HIS BALB/c-Rag2$^{-/-}$γc$^{-/-}$ mice [22]. These mice were infected with HTLV-1 through the intraperitoneal inoculation of irradiated HTLV-1-producing cells, at a time when the three main subpopulations of human thymocytes have been detected, i.e., within a period of 1–2 months after transplanting BALB/c-Rag2$^{-/-}$γc$^{-/-}$ immunodeficient animals with human CD34$^+$ HP/HSCs. These mice were analyzed at regular intervals within a 7-month period after inoculation. Significant alterations of human T-cell development have been observed, the extent of which correlated with the proviral load. Specifically, significant differences were observed in the respective percentages of the three main thymocyte subsets (immature, DP and SP) in HIS BALB/c-Rag2$^{-/-}$γc$^{-/-}$ mice with a low-proviral load and in mock-infected animals. Conversely, in the thymus of HTLV-1-infected HIS BALB/c Rag2$^{-/-}$γc$^{-/-}$ mice with

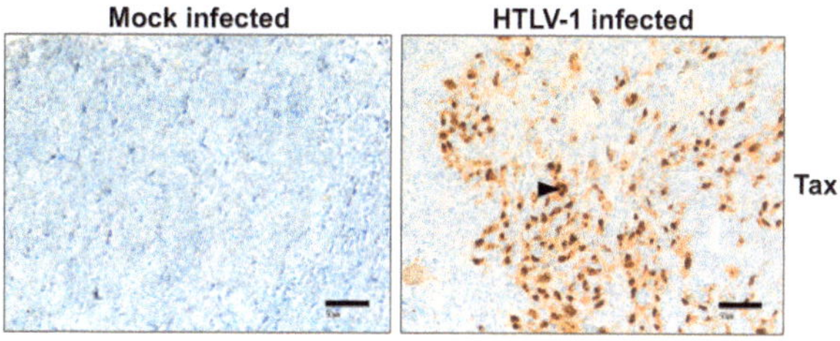

Fig. 40.2 Immuno-histological detection of Tax in the nucleus of large lymphoma cells in the thymus from HTLV-1-infected HIS mice. (From ref. [22]. Reproduced under Creative Commons Attribution (CC BY) license)

a high-proviral load similar to that in ATLL patients, mature SP $CD4^+$ and $CD8^+$ cells were most numerous, at the expense of immature DP thymocytes. These SP cells also accumulated in peripheral organs. Preliminary results obtained with one HTLV-1-infected mouse displaying a thymoma indicate an extraordinary degree of polyclonality of integration sites in the thymus. Interestingly, the three clones that dominated the population in both spleen and lymph nodes were not dominant in the thymus, and the three most abundant clones observed in the thymus were not detected in the periphery.

Human T-cells from thymus and spleen were activated, as shown by the expression of the CD25 marker, that correlated with the presence of *tax* mRNA and with increased expression of NF-κB dependent genes such as *bfl-1*, an antiapoptotic gene, in thymocytes. Finally, hepato-splenomegaly, lymphadenopathy, and T-cell lymphoma/thymoma, in which Tax was detected, were observed in HTLV-1-infected mice, several months after HTLV-1 infection (Fig. 40.2). These data provide strong evidence that in vivo HTLV-1 infection of HIS BALB/c-$Rag2^{-/-}\gamma c^{-/-}$ mice perturbs human T-cell development in the thymus at the level of immature cells, by propelling T-cell development toward the mature stages. They further suggest that the combination of immature target cells in the thymus and the immunodeficient environment of these humanized mice favors the rapid development of a T-cell malignancy. Recently, the induction of HTLV-1-specific adaptive immune responses in some infected mice has been reported [26]. These results demonstrate the potential of this mouse model to elucidate the initial steps of the leukemogenic process induced by HTLV-1.

40.5 Conclusions and Perspectives

The recapitulation of a $CD4^+$ T-cell lymphoma in HTLV-1-infected HIS-mice suggests that HP/HSCs as well as immature thymocytes function as cellular targets for infection and for initiation of leukemogenesis in humans (Fig. 40.3). This animal model of ATLL will provide an important tool for the identification of molecular and cellular events that control the initiation and progression of the lymphoma and potential therapeutic targets to block tumor progression and development. This model will also be helpful for studying the involvement of other HTLV-1 genes

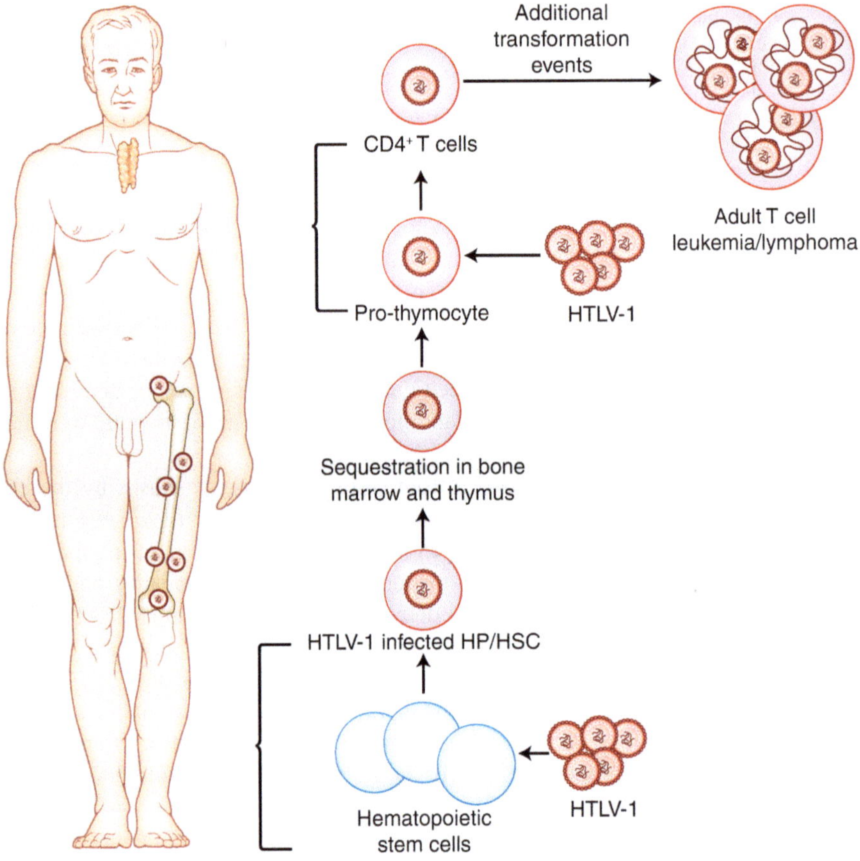

Fig. 40.3 HTLV-1 infection of CD34$^+$ hematopoietic progenitor and stem cells (*HP/HSCs*) and prothymocytes leads to development of adult T-cell leukemia/lymphoma (*ATLL*)

like *hbz* in viral–induced leukemogenesis. The generation of cellular and humoral immune responses in recently developed HIS mice, such as NSG-HLA-A2 [27], may contribute to decipher the multistep progression of leukemia as well as the immunopathogenic mechanisms triggering inflammatory diseases as well as to develop new therapeutic strategies to treat HTLV-1 patients both at an early and a late leukemic stage.

Acknowledgments This contribution was supported in part by INSERM & CNRS, by the European Union Project "Infection and Cancer" (INCA) under the Sixth Research Framework Programme (grant number: LSHC-CT-2005-018704) and by the Fondation de France, comité "Leucémie" (nuRAF09001CCA) to MDD and by grants from the US National Institutes of Health (CA124595) and by the Empire State Stem Cell Fund through New York State Department of Health Contract (NYSTEM #C023059 and #N08G-127) to G.F.

References

1. Matsuoka M, Jeang KT. Human T-cell leukaemia virus type 1 (HTLV-1) infectivity and cellular transformation. Nat Rev Cancer. 2007;7(4):270–80.
2. Duc Dodon M, Villaudy J, Gazzolo L, Haines R, Lairmore M. What we are learning on HTLV-1 pathogenesis from animal models. Front Microbiol. 2012;3:320.
3. Lairmore MD, Silverman L, Ratner L. Animal models for human T-lymphotropic virus type 1 (HTLV-1) infection and transformation. Oncogene. 2005;24(39):6005–15.
4. Manz M, Di Santo J. Renaissance for mouse models of human hematopoiesis and immunobiology. Nat immunol. 2009;10(10):1039–42.
5. Rongvaux A, Takizawa H, Strowig T, Willinger T, Eynon EE, Flavell RA, et al. Human hemato-lymphoid system mice: current use and future potential for medicine. Annu Rev Immunol. 2013;31:635–74.
6. Mesnard JM, Barbeau B, Devaux C. HBZ, a new important player in the mystery of adult T-cell leukemia. Blood. 2006;108(13):3979–82.
7. Tattermusch S, Bangham CR. HTLV-1 infection: what determines the risk of inflammatory disease? Trends Microbiol. 2012;20(10):494–500.
8. Duc Dodon M, Mesnard JM, Barbeau B. Adult T-cell leukemia induced by HTLV-1: before and after HBZ. Med Sci (Paris). 2010;26(4):391–6.
9. Samarut J, Gazzolo L. Target cells infected by avian erythroblastosis virus differentiate and become transformed. Cell. 1982;28(4):921–9.
10. Rulli K, Lenz J, Levy LS. Disruption of hematopoiesis and thymopoiesis in the early premalignant stages of infection with SL3-3 murine leukemia virus. J Virol. 2002;76(5):2363–74.
11. Hasegawa H, Sawa H, Lewis MJ, Orba Y, Sheehy N, Yamamoto Y, et al. Thymus-derived leukemia-lymphoma in mice transgenic for the tax gene of human T-lymphotropic virus type I. Nat Med. 2006;12(4):466–72.
12. Feuer G, Fraser JK, Zack JA, Lee F, Feuer R, Chen IS. Human T-cell leukemia virus infection of human hematopoietic progenitor cells: maintenance of virus infection during differentiation in vitro and in vivo. J Virol. 1996;70(6):4038–44.
13. Banerjee P, Sieburg M, Samuelson E, Feuer G. Human T-cell lymphotropic virus type 1 infection of $CD34^{+}$ hematopoietic progenitor cells induces cell cycle arrest by modulation of p21(cip1/waf1) and survivin. Stem cells. 2008;26(12):3047–58.
14. Tripp A, Banerjee P, Sieburg M, Planelles V, Li F, Feuer G. Induction of cell cycle arrest by human T-cell lymphotropic virus type 1 Tax in hematopoietic progenitor ($CD34^{+}$) cells: modulation of p21cip1/waf1 and p27kip1 expression. J virol. 2005;79(22):14069–78.
15. Banerjee P, Tripp A, Lairmore M, Crawford L, Sieburg M, Ramos J, et al. Adult T-cell leukemia/lymphoma development in HTLV-1-infected humanized SCID mice. Blood. 2010;115(13):2640–8.
16. Blom B, Spits H. Development of human lymphoid cells. Ann Rev Immunol. 2006;24:287–320.
17. Spits H. Development of alphabeta T cells in the human thymus. Nat Rev Immunol. 2002;2(10):760–72.
18. Anderson MK. At the crossroads: diverse roles of early thymocyte transcriptional regulators. Immunol Rev. 2006;209:191–211.
19. Aifantis I, Mandal M, Sawai K, Ferrando A, Vilimas T. Regulation of T-cell progenitor survival and cell-cycle entry by the pre-T-cell receptor. Immunol Rev. 2006;209:159–69.
20. Wencker M, Gazzolo L, Duc Dodon M. The leukemogenic activity of TaxHTLV-1 during human alphabeta T cell development. Front Biosci (Schol Ed). 2009;1:194–204.
21. Wencker M, Sausse C, Derse D, Gazzolo L, Duc Dodon M. Human T-cell leukemia virus type 1 Tax protein down-regulates pre-T-cell receptor alpha gene transcription in human immature thymocytes. J Virol. 2007;81(1):301–8.
22. Villaudy J, Wencker M, Gadot N, Gillet N, Scoazec J-Y, Gazzolo L, et al. HTLV-1 propels thymic human T cell development in "human immune system" Rag2-/- gamma c-/- mice. PLoS pathog. 2011;7(9):e1002231.

23. Akkina R. New generation humanized mice for virus research: comparative aspects and future prospects. Virology. 2012;435(1):14–28.
24. Legrand N, Ploss A, Balling R, Becker P, Borsotti C, Brezillon N, et al. Humanized mice for modeling human infectious disease: challenges, progress, and outlook. Cell Host Microbe. 2009;6(1):5–9.
25. Banerjee P, Crawford L, Samuelson E, Feuer G. Hematopoietic stem cells and retroviral infection. Retrovirology. 2010;7:8.
26. Tezuka K, Xun R, Tei M, Ueno T, Tanaka M, Takenouchi N, et al. An animal model of adult T-cell leukemia-humanized mice with HTLV-1 specific immunity. Blood. 2013;123(3):346–55. (Epub ahead of print).
27. Shultz L, Saito Y, Najima Y, Tanaka S, Ochi T, Tomizawa M, et al. Generation of functional human T-cell subsets with HLA-restricted immune responses in HLA class I expressing NOD/SCID/IL2r gamma(null) humanized mice. Proc Natl Acad of Sci U S A. 2010;107(29):13022–7.

Chapter 41
Plasmodium Falciparum Parasite Development in Humanized Mice: Liver And Blood Stages

Till Strowig and Alexander Ploss

41.1 Introduction

Malaria is a life-threatening disease caused by parasites, which are transmitted through the bites of infected mosquitoes. Nearly half a billion malaria infections occur each year, and close to one million deaths are attributed to the disease annually, mostly of children under the age of 5 years who have frequent attacks and little immunological protection and pregnant women in Sub-Saharan Africa [1]. Although progress has been made in preventing and treating malaria infection, more effective, tolerable, and affordable therapies and vaccines are needed [2–4]. A small animal model for the human parasite could potentially expedite the development of antimalarial drugs and vaccines. Here, we describe progress in the development of humanized mice as a platform to support the life cycle of plasmodial parasites causing diseases in humans.

The Life Cycle of Malaria-causing Plasmodium Parasites Among the four *Plasmodium* species that cause malaria in humans, *P. falciparum* is the most virulent [5]. This species causes the vast majority of deaths from malaria and is also distinguished by its ability to bind to endothelium during the blood stage of the infection and to hide in organs, including the brain. *P. vivax* is less deadly but highly disabling; it is common in tropical areas outside Africa. The ability of *P. vivax* and *P. ovale* to remain dormant for months as hypnozoites in the liver makes infection with these parasites difficult to eradicate. *Plasmodium malariae* does not form hypnozoites, but it can persist for decades as an asymptomatic blood stage infection. A fifth species, *Plasmodium knowlesi*, which was originally described as a malaria parasite of long-tailed macaques, also naturally infects humans in some areas, such as Malaysia.

A. Ploss (✉)
Department of Molecular Biology, Princeton University, Princeton, NJ 08544 USA
e-mail: aploss@princeton.edu

T. Strowig
Helmholtz Center for Infection Research, Braunschweig, Germany

L. Y. Poluektova et al. (eds.), *Humanized Mice for HIV Research*,
DOI 10.1007/978-1-4939-1655-9_41

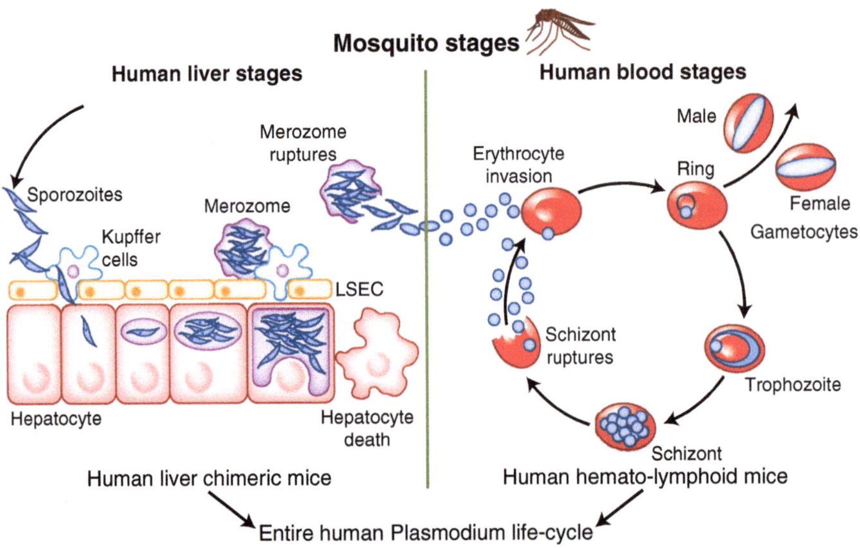

Fig. 41.1 Compartmentalized tissue humanization to model the *P. falciparum* life cycle in humanized mice

Infection of the human host with a *Plasmodium* parasite begins with the bite of an infected *Anopheles* mosquito injecting the individual with sporozoites (Fig. 41.1). These motile forms rapidly access the blood stream and then the liver, where they invade hepatocytes. The asymptomatic liver stage of infection lasts about 6 days, with each sporozoite yielding tens of thousands of merozoites that then invade and multiply within erythrocytes. The blood stages of infection include asexual forms of the parasite that undergo repeated cycles of multiplication as well as male and female sexual forms, called gametocytes that await ingestion by mosquitoes before developing further. Asexual blood stage parasites produce 8–20 new merozoites every 48 h (or 72 h for *P. malariae*), causing parasitic numbers to rise rapidly to levels as high as 10^{13} per host. The asexual stages are pathogenic, and infected individuals can present with diverse sequelae affecting different organ systems. Sexual stage parasites are nonpathogenic but are transmissible to the *Anopheles* vector, where they recombine during a brief period of diploidy and generate genetically distinct sporozoites. The mosquito becomes infectious for its next blood meal donor approximately two weeks after ingesting gametocytes, a time frame that is influenced by external temperature. The development of *P. vivax* within the mosquito can occur at a lower environmental temperature than that required for development of *P. falciparum*, explaining the preponderance of *P. vivax* infections outside tropical and subtropical regions.

Animal Models for Malaria The clinically most relevant malaria parasites, *P. falciparum* and *P. vivax*, display a unique human tropism, and the hallmarks of their pathogenesis cannot be accurately reproduced in animals. While rodents and rodent malaria parasites are commonly used for research and drug/vaccine development, they often do not yield reliable preclinical results that translate into effective human treatments [6]. Both the pathogen and host contribute to this failure. On the microbial side, rodent malaria strains, *P. berghei*, *P. yoelii*, *P. chabaudi*, *P.vinckei*, differ from their highly restricted human counterparts. Most laboratory strains of the parasite have been isolated from African thicket rats, *Thamnomys rutilans*, and none is a natural pathogen of the laboratory mouse. Unlike the human pathogens, there has been no evolutionary adaptation of parasite and the mouse. For this reason, and because the rodent parasites differ in virulence and in the fine details of immune elimination from the host, it has been argued that mouse models are not very relevant to the human disease.

The molecular determinants of *Plasmodium* species tropism are poorly understood. Thus, a genetic adaptation of the murine host to support the parasite life cycle, as has been proposed for other pathogens, is currently not possible. On the host side, mouse and man evolutionarily diverged ca. 65 million years ago, therefore these two species have inhabited different ecological niches and have been challenged with minimally overlapping groups of pathogens. The human and mouse immune systems have evolved to meet these challenges and accumulated many differences, making genes related to immunity, together with genes involved in reproduction and olfaction, the most divergent between the two species [7].

Humanized mice, i.e., engrafted with human tissues and/or expressing human genes, have been proven useful to study a variety of pathogens with a similar narrow host range (reviewed in [8]), although mostly affecting immune cells. Engraftment of the relevant tissue compartments, i.e., liver and red blood cells (RBCs), is also a promising strategy to model human malaria in vivo.

41.2 Humanized Mouse Models for Pre-erythrocytic Stages of *P. Falciparum*

The characterization of liver stages of human malaria is based mostly on the infection of human hepatoma cell lines, which differ substantially in their physiology from primary hepatocytes [9–13]. More recently, sophisticated engineering approaches have culminated in a microscale human liver platform that supports the hepatic stages of *P. falciparum* and *P. vivax* [14]. Such primary human hepatocyte cultures provide a more accurate depiction of the native environment that plasmodial parasites encounter in the liver but suffer from several drawbacks. Infection frequencies are by-and-large very low and, for reasons that have yet to be defined, only specific hepatocyte lots are permissive to infection. Primary hepatocytes—even under advanced culture conditions—are only stable for a few weeks and tend to dedifferentiate, which complicates the study of plasmodial dormancy and reactivation. Furthermore, monolayers of primary hepatocytes lack the cellular complexity and cell–cell contacts dictated by the three-dimensional architecture of the liver, which may influence the susceptibility to infection.

Engraftment of human hepatocytes into suitable xenorecipients has proven to be one solution to study hepatocyte (patho-)biology. To facilitate engraftment, human hepatocytes are usually injected into immunodeficient recipients suffering from liver injury. Suppression of the murine immune system is necessary to prevent graft rejection. The liver injury provides the expansion stimulus to the usually quiescent hepatocytes and gives the transplanted human liver cells a competitive growth advantage over mouse liver cells. Liver injury can be inflicted surgically, e.g., by partial hepatectomy, by treatment with hepatotoxins, such as retrorsine or carbontetrachloride, or with genetic approaches. The latter are most widely used for stable engraftment as they provide a more selective control over the severity of the liver injury, and hepatotoxicity is usually limited to mouse hepatocytes.

Several different liver injury models have been developed over the past few decades (reviewed in [15]). One of the best characterized models is transgenic mice overexpressing the urokinase-type plasminogen activator (uPA) gene [16] under the control of the mouse albumin promoter (alb-uPA), which was originally generated to study coagulopathies. However, transgene expression will lead inadvertently to a severe hepatotoxic effect [17]. In immunodeficient mice carrying homozygously the uPA transgene, the human liver graft is maintained at high levels over long periods of time. There are considerable drawbacks of the alb-uPA model, including hypofertility and high mortality that have made the production of highly engrafted mice laborious and expensive. Some variants of the uPA transgenic model have been developed that may overcome some of these challenges (reviewed in [15]).

To create a more tractable system, mice with targeted disruption in the fumaryl acetoacetate hydrolase (FAH) gene were bred onto immunocompromised backgrounds. FAH deficiency results in acute liver injury, which is neonatally lethal but can be rescued by administration of 2-(2-nitro-4-fluoromethylbenzoyl)-1,3-cyclohexan edione (NTBC) allowing for normal propagation of the line. In turn, withdrawal of this liver protective drug provides easy means to induce liver injury at will. Injection of human adult hepatocytes into the immunodeficient $FAH^{-/-}$ xenorecipients yields high human hepatocyte chimerism [18, 19].

Similarly, transgenic expression of herpes simplex virus thymidine kinase (HSV-TK) in the livers of immunodeficient mice allows for selective ablation of mouse hepatocytes with ganciclovir and facilitates robust engraftment with human hepatocytes [20].

Human liver chimeric mice are a versatile in vivo platform to study human hepatotropic pathogens and have been shown to be susceptible to a variety of human hepatitis viruses, including hepatitis B [11, 22], C [21–23] and delta [24]. Likewise, immunodeficent Alb-uPA and $FAH^{-/-}$ mice, which were highly engrafted with human hepatocytes, supported quantifiable *P. falciparum* liver stage development culminating in the complete maturation of liver stages [25–28]. In immunodeficient liver chimeric $FAH^{-/-}$ mice *P. falciparum,* exoerythrocytic merozoite aggregates budded off after one week suggesting that merosome formation can be modeled in liver chimeric mice [25]. Although it remains to be shown whether human liver chimeric mice can also support hepatic stages of other plasmodial parasites causing diseases in humans, the data is encouraging and will pave the way to studying plasmodial parasite liver stages.

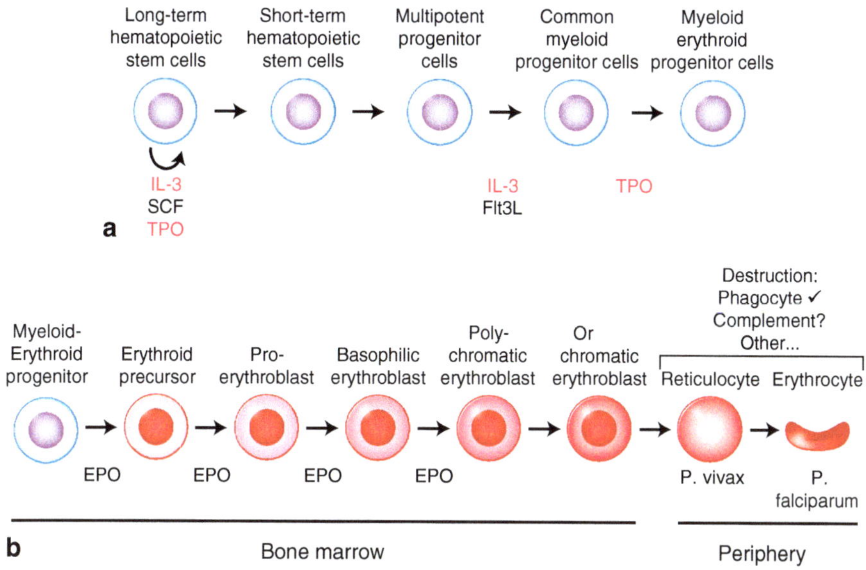

Fig. 41.2 Requirements for erythroid cell differentiation in vivo. **a** Erythropoietin (*EPO*) is crucial in driving differentiation of erythroid progenitor cells to reticulocytes and erythrocytes that can be infected by malaria parasites. Infected and uninfected human RBCs need to avoid destruction by the murine immune system. **b** Cytokine requirements for the efficient development of erythroid precursor (*cytokines in red are non-cross reactive from mouse to man*)

41.3 Blood Stage Malaria Infection

In contrast to plasmodial liver stages, *P. falciparum* blood stages can be routinely propagated in vitro based on seminal work by Jensen and Trager in the late 1970s [29]. This in vitro platform has been instrumental in studying blood stages and also became a valuable tool for functional drug efficacy screens. Despite the utility of such blood stage cell cultures, mouse models for *P. falciparum* blood stage infection have been pursued to analyze pharmacokinetics, pharmacodynamics, and efficacy of novel therapeutics simultaneously [30].

Merozoites from different plasmodial species vary in their preferred target cell. While *P. falciparum* is able to infect mature erythrocytes, *P. vivax* is thought to infect almost exclusively reticulocytes that are relatively scarce (less than 1% in human blood) (Fig. 41.2). Two strategies have been proposed to generate in vivo models varying largely in their utility of malaria parasites and will be discussed in detail below. First, RBCs can be passively infused into immunodeficient mice, providing the relevant tissue compartment for *P. falciparum* infection but not *P. vivax* infection since reticulocytes are lacking. The second model is based on the *de novo* differentiation of human hematopoietic stem cells into human erythroid cells after transfer injection into immunocompromized mice, which would support in principal all malaria parasites but practically has not been established.

Human Red Blood Cell Infusion Models With the advent of a severe combined immunodeficient (SCID) strain with drastically reduced phagocytic activity against it, it became possible to sustain a human erythrocyte graft following repeated injections with packed RBCs [31]. Consequently, these humanized mice were shown to be susceptible to *P. falciparum* blood stage infection. As more severely immunocompromized xenorecipient strains became available, the model was further refined to improve the reproducibility and throughput. Daily injections with relatively large amounts of packed RBCs into non-obese diabetic (NOD) SCID $\beta 2m^{-/-}$ mice yielded a high RBC chimerism exceeding 90 % human erythrocytes, which enabled productive and reproducible infection with in vivo adapted *P. falciparum* strains [30]. Further improvements were made to this model by using NOD SCID IL2Rγ^{NULL} mice, allowing tenfold higher parasitemia [32], and by combining intravenous injections (in lieu of intraperitoneally) with ablation of phagocytic cells using clodronate [33]. Following proof of concept that *P. falciparum* is sensitive to treatment with antimalarial compounds in humanized mice, RBC chimeric mice are now routinely employed to evaluate preclinically the efficacy of lead compounds in vivo. However, the model is labor and resource intensive and has not yet been shown to be susceptible to other *Plasmodium* species, which limits its utility.

De Novo Erythropoiesis in Humanized Mice Attempts to establish *de novo* human erythropoiesis in vivo are based on initial studies by a number of groups (reviewed in [34]) demonstrating that injection of human hematopoietic stem cells into severely immunocompromised mice results in multilineage differentiation, including the detection of erythroid cells in the bone marrow of engrafted mice. However, in these models the frequencies of human erythroid cells in blood are low, suggesting a block in the differentiation or their destruction in the periphery. Several studies have highlighted that differentiation can indeed be enhanced by providing cytokines that support erythropoiesis, in particular, interleukin 3 (IL-3) and erythropoietin (EPO) [35, 36] (Fig. 41.2) and exhibit minimal biological cross-reactivity across species. Moreover, ablation of phagocytes using clodronate liposomes drastically improves the frequencies of human erythrocytes in circulation [36], but none of these studies demonstrated that increases in the frequency of de novo produced RBCs correlate with susceptibility to plasmodial infection. Further studies are needed to test whether human erythropoiesis can be boosted using human cytokine knock-in mice for thrombopoietin and M-CSF, which support myelo-erythropoiesis at earlier steps, as well as providing endogenous human sources of EPO and IL-3 (reviewed in [37]). In addition, other pathways of erythrocyte destruction, such as complement, may have to be considered as contributing to human RBC destruction in vivo.

For all these models, it has to be considered that malaria-induced pathology may not accurate reflect all features of clinical disease in humans. One example is the observation that infected erythrocytes are typically sequestered in blood vessels to avoid clearance in the spleen. This requires specific interactions of parasite-encoded proteins with host proteins that may not be evolutionarily conserved in the mouse. However, this was not observed in initial studies using mice infused with human RBC [30].

41.4 Recapitulation of the Entire *P. Falciparum* Life Cycle in Vivo

Complete maturation of *P. falciparum* liver stages culminating in the release of merozoites in human liver chimeric mice as well as the fact that human RBC containing mice support blood stage infection suggested that the entire *P. falciparum* life cycle can be completed in mice harboring, simultaneously, both tissues. FAH$^{-/-}$ mice on a NOD Rag2$^{-/-}$ IL2Rg^{NULL} background were engrafted with human adult hepatocytes and infected with *P. falciparum* sporozoites. After 7 days when the parasite liver stages had fully matured, mice were infused with human packed RBCs to provide the target cells for schizonts released from the merozome. Subsequent in vitro culture of blood from the infected mice lead to an outgrowth of asexual and sexual *P. falciparum* stages, demonstrating the transition from liver stage to blood stage infection [25]. Recapitulating the entire *P. falciparum* life cycle in humanized mice is a major advance. However, the model will have to be further refined to facilitate robust expansion asexual schizont in vivo following transition from liver to blood stages. Development of gametocytes in vivo would allow us to directly study the transmission from the mammalian host to mosquitoes (Fig. 41.1). Ultimately, it would be desirable to establish robust human hepatocyte engraftment and de novo erythropoiesis in vivo to attempt infections with other malaria parasite species, arguably most importantly with *P. vivax*.

41.5 Summary and Outlook

Humanized mice have greatly helped to study human-tropic pathogens in vivo. Rendering humanized mice susceptible to *P. falciparum* poses unique challenges as the mammalian stages of the parasite depend on the presence of two distinct tissue compartments (Fig. 41.3). Human liver chimeric mice support the complete development of *P. falciparum* liver stages following sporozoite infection. Xenorecipients infused with human RBCs can be used to propagate asexual blood stages in vivo. Engrafting both human liver and erythrocytes in a single recipient offers new avenues to study the entire *P. falciparum* life cycle in vivo. Humanized mice for malaria may enable the dissection of the human stages of malaria pathogenesis as well as allow experimental transmission to mosquitoes. Engraftment of patient-specific hepatocytes and hematopoietic stem cells derived from induced pluripotent stem cells may in the future enable the analysis of human genetic variability on parasite infection and transmission. Humanized mice engrafted with human liver and RBCs lend themselves to perform genetic crosses to identify loci that confer resistance to antimalarials such as choroquine. Similar studies were previously only possible in splenectomized chimpanzees. Animals could be immediately deployed for safety and efficacy assessments of genetically attenuated parasites, antimalarial drug testing, passive immune protection by transfer of antibodies or sera from individuals immunized with vaccine candidates, and direct testing of vaccines.

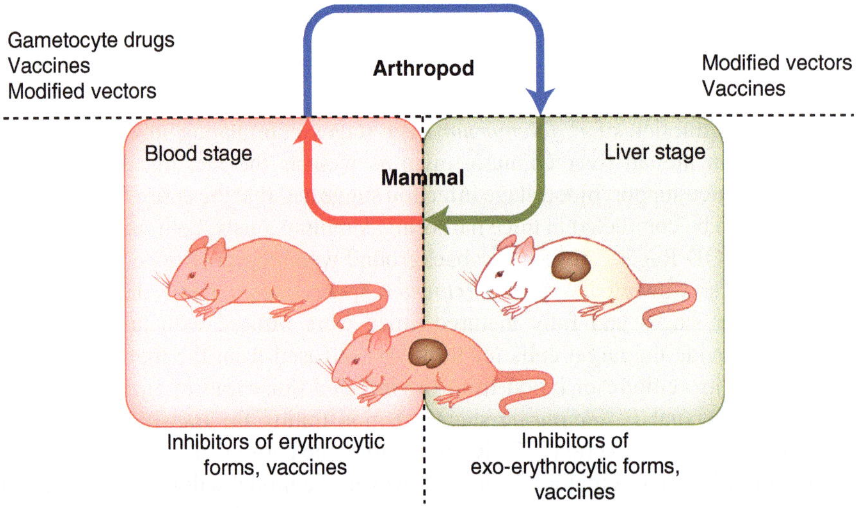

Fig. 41.3 Applications of humanized mice to study and interfere with malaria life-cycle stages

References

1. WHO. Factsheet Malaria. World Health Organization, 2009 Contract No.: 94.
2. Wells TN, Alonso PL, Gutteridge WE. New medicines to improve control and contribute to the eradication of malaria. Nat Rev Drug Discov. 2009;8(11):879–91. (PubMed PMID: 19834482. Epub 2009/10/17. eng).
3. mal ERACGoD. A research agenda for malaria eradication: drugs. PLoS Med. 2011;8(1):e1000402. (PubMed PMID: 21311580. Pubmed Central PMCID: 3026688).
4. mal ERACGoV. A research agenda for malaria eradication: vaccines. PLoS Med. 2011;8(1):e1000398. (PubMed PMID: 21311586. Pubmed Central PMCID: 3026701).
5. Greenwood BM, Fidock DA, Kyle DE, Kappe SH, Alonso PL, Collins FH, et al. Malaria: progress, perils, and prospects for eradication. J Clin Invest. 2008;118(4):1266–76. (PubMed PMID: 18382739. Epub 2008/04/03. eng).
6. Sanni LA, Fonseca LF, Langhorne J. Mouse models for erythrocytic-stage malaria. Methods Mol Med. 2002;72:57–76. (PubMed PMID: 12125155. Epub 2002/07/20. eng).
7. Mestas J, Hughes CC. Of mice and not men: differences between mouse and human immunology. J Immunol. 2004;172(5):2731–8. (PubMed PMID: 14978070).
8. Leung C, Chijioke O, Gujer C, Chatterjee B, Antsiferova O, Landtwing V, et al. Infectious diseases in humanized mice. Eur J Immunol. 2013;43(9):2246–54. (PubMed PMID: 23913412).
9. Chattopadhyay R, Velmurugan S, Chakiath C, Andrews Donkor L, Milhous W, Barnwell JW, et al. Establishment of an in vitro assay for assessing the effects of drugs on the liver stages of *Plasmodium vivax* malaria. PLoS ONE. 2010;5(12):e14275. (PubMed PMID: 21151554. Pubmed Central PMCID: 3000326).
10. Epstein JE, Tewari K, Lyke KE, Sim BK, Billingsley PF, Laurens MB, et al. Live attenuated malaria vaccine designed to protect through hepatic CD8(+) T cell immunity. Science. 2011;334(6055):475–80. (PubMed PMID: 21903775).

11. Hollingdale MR, Leland P, Schwartz AL. In vitro cultivation of the exoerythrocytic stage of *Plasmodium berghei* in a hepatoma cell line. Am J Trop Med Hyg. 1983;32(4):682–4. (PubMed PMID: 6349397).
12. Karnasuta C, Pavanand K, Chantakulkij S, Luttiwongsakorn N, Rassamesoraj M, Laohathai K, et al. Complete development of the liver stage of *Plasmodium falciparum* in a human hepatoma cell line. Am J Trop Med Hyg. 1995;53(6):607–11. (PubMed PMID: 8561262).
13. Sattabongkot J, Yimamnuaychoke N, Leelaudomlipi S, Rasameesoraj M, Jenwithisuk R, Coleman RE, et al. Establishment of a human hepatocyte line that supports in vitro development of the exo-erythrocytic stages of the malaria parasites *Plasmodium falciparum* and *P. vivax*. Am J Trop Med Hyg. 2006;74(5):708–15. (PubMed PMID: 16687667).
14. March S, Ng S, Velmurugan S, Galstian A, Shan J, Logan DJ, et al. A microscale human liver platform that supports the hepatic stages of *Plasmodium falciparum* and *vivax*. Cell Host Microbe. 2013;14(1):104–15. (PubMed PMID: 23870318).
15. Billerbeck E, de Jong Y, Dorner M, de la Fuente C, Ploss A. Animal models for hepatitis C. Curr Top Microbiol Immunol. 2013;369:49–86. (PubMed PMID: 23463197).
16. Heckel JL, Sandgren EP, Degen JL, Palmiter RD, Brinster RL. Neonatal bleeding in transgenic mice expressing urokinase-type plasminogen activator. Cell. 1990;62(3):447–56. (PubMed PMID: 1696178. Epub 1990/08/10. eng).
17. Sandgren EP, Palmiter RD, Heckel JL, Daugherty CC, Brinster RL, Degen JL. Complete hepatic regeneration after somatic deletion of an albumin-plasminogen activator transgene. Cell. 1991;66(2):245–56. (PubMed PMID: 1713128. Epub 1991/07/26. eng).
18. Azuma H, Paulk N, Ranade A, Dorrell C, Al-Dhalimy M, Ellis E, et al. Robust expansion of human hepatocytes in $Fah^{-/-}/Rag2^{-/-}/Il2rg^{-/-}$ mice. Nat Biotechnol. 2007;25(8):903–10. (PubMed PMID: 17664939. Epub 2007/08/01. eng).
19. Bissig KD, Le TT, Woods NB, Verma IM. Repopulation of adult and neonatal mice with human hepatocytes: a chimeric animal model. Proc Natl Acad Sci U S A. 2007;104(51):20507–11. (PubMed PMID: 18077355. Epub 2007/12/14. eng).
20. Hasegawa M, Kawai K, Mitsui T, Taniguchi K, Monnai M, Wakui M, et al. The reconstituted 'humanized liver' in TK-NOG mice is mature and functional. Biochem Biophys Res Commun. 2011;405(3):405–10. (PubMed PMID: 21238430. Epub 2011/01/18. eng).
21. Bissig KD, Wieland SF, Tran P, Isogawa M, Le TT, Chisari FV, et al. Human liver chimeric mice provide a model for hepatitis B and C virus infection and treatment. J Clin Invest. 2010;120(3):924–30. (PubMed PMID: 20179355. Pubmed Central PMCID: 2827952. Epub 2010/02/25. eng).
22. Meuleman P, Libbrecht L, De Vos R, de Hemptinne B, Gevaert K, Vandekerckhove J, et al. Morphological and biochemical characterization of a human liver in a uPA-SCID mouse chimera. Hepatology. 2005;41(4):847–56. (PubMed PMID: 15791625. Epub 2005/03/26. eng).
23. Mercer DF, Schiller DE, Elliott JF, Douglas DN, Hao C, Rinfret A, et al. Hepatitis C virus replication in mice with chimeric human livers. Nat Med. 2001;7(8):927–33.
24. Lutgehetmann M, Mancke LV, Volz T, Helbig M, Allweiss L, Bornscheuer T, et al. Humanized chimeric uPA mouse model for the study of hepatitis B and D virus interactions and preclinical drug evaluation. Hepatology. 2012;55(3):685–94. (PubMed PMID: 22031488).
25. Vaughan A, Mikolajczak SA, Wilson EM, Grompe M, Kaushansky A, Camargo N, et al. Mice with human hepatocyte chimerism support liver stage development and merozoite release of *Plasmodium falciparum*. J Clin Invest. In press 2012.
26. VanBuskirk KM, O'Neill MT, De La Vega P, Maier AG, Krzych U, Williams J, et al. Preerythrocytic, live-attenuated *Plasmodium falciparum* vaccine candidates by design. Proc Natl Acad Sci U S A. 2009;106(31):13004–9. (PubMed PMID: 19625622. Epub 2009/07/25. eng).
27. Sacci JB Jr., Alam U, Douglas D, Lewis J, Tyrrell DL, Azad AF, et al. *Plasmodium falciparum* infection and exoerythrocytic development in mice with chimeric human livers. Int J Parasitol. 2006;36(3):353–60. (PubMed PMID: 16442544. Epub 2006/01/31. eng).

28. Morosan S, Hez-Deroubaix S, Lunel F, Renia L, Giannini C, Van Rooijen N, et al. Liver-stage development of *Plasmodium falciparum*, in a humanized mouse model. J Infect Dis. 2006;193(7):996–1004. (PubMed PMID: 16518762. Epub 2006/03/07. eng).
29. Trager W, Jensen JB. Human malaria parasites in continuous culture. Science. 1976;193(4254):673–5. (PubMed PMID: 781840).
30. Angulo-Barturen I, Jimenez-Diaz MB, Mulet T, Rullas J, Herreros E, Ferrer S, et al. A murine model of falciparum-malaria by in vivo selection of competent strains in non-myelodepleted mice engrafted with human erythrocytes. PLoS ONE. 2008;3(5):e2252. (PubMed PMID: 18493601. Pubmed Central PMCID: 2375113. Epub 2008/05/22. eng).
31. Moore JM, Kumar N, Shultz LD, Rajan TV. Maintenance of the human malarial parasite, *Plasmodium falciparum*, in scid mice and transmission of gametocytes to mosquitoes. J Exp Med. 1995;181(6):2265–70. (PubMed PMID: 7760012. Epub 1995/06/01. eng).
32. Jimenez-Diaz MB, Mulet T, Viera S, Gomez V, Garuti H, Ibanez J, et al. Improved murine model of malaria using *Plasmodium falciparum* competent strains and non-myelodepleted NOD-scid IL2Rgammanull mice engrafted with human erythrocytes. Antimicrob Agents Chemother. 2009;53(10):4533–6. (PubMed PMID: 19596869. Epub 2009/07/15. eng).
33. Arnold L, Tyagi RK, Meija P, Swetman C, Gleeson J, Perignon JL, et al. Further improvements of the *P. falciparum* humanized mouse model. PLoS ONE. 2011;6(3):e18045. (PubMed PMID: 21483851. Pubmed Central PMCID: 3069031).
34. Rongvaux A, Takizawa H, Strowig T, Willinger T, Eynon EE, Flavell RA, et al. Human hemato-lymphoid system mice: current use and future potential for medicine. Annu Rev Immunol. 2013;31:635–74. (PubMed PMID: 23330956).
35. Chen Q, Khoury M, Chen J. Expression of human cytokines dramatically improves reconstitution of specific human-blood lineage cells in humanized mice. Proc Natl Acad Sci U S A. 2009;106(51):21783–8. (PubMed PMID: 19966223. Epub 2009/12/08. eng).
36. Hu Z, Van Rooijen N, Yang YG. Macrophages prevent human red blood cell reconstitution in immunodeficient mice. Blood. 2011;118(22):5938–46. (PubMed PMID: 21926352. Pubmed Central PMCID: 3228505. Epub 2011/09/20. eng).
37. Willinger T, Rongvaux A, Strowig T, Manz MG, Flavell RA. Improving human hemato-lymphoid-system mice by cytokine knock-in gene replacement. Trends Immunol. 2011;32(7):321–7. (PubMed PMID: 21697012. Epub 2011/06/24. eng).

Part VII
Conclusion

Chapter 42
The Future is NOW for Humanized Mouse Models

J. Victor Garcia

Humanized mouse models have matured over the past several years to such a degree that they have become essential components of state-of-the art investigation in a multitude of disciplines. Humanized mouse models are no longer a rarity in the literature or a novelty at major scientific conferences. They have made a mark and they are here to stay. The types of in vivo studies that can be performed on humanized mice have opened new lines of investigation that were clearly out of the realm of possibilities just a few years ago. The availability of these models has completely transformed the landscape in the fields of human immunology and infectious diseases with perhaps the biggest impact in HIV, HCV, HBV, and EBV research. More recent uses in other areas are beginning to demonstrate the utility and the promise of great future potential applications.

The chapters in this book are a clear testament to the significant progress that has been made so far in a multitude of disciplines in a relatively short period of time. It is clear from all these contributions that the only limitation to the use and improvement of humanized mouse models has been our imagination, our determination, and our resources. As the field continues to take bold steps into places where no one has gone before, many more applications and uses for these types of models are likely to be developed. These in turn are likely to continue to produce the kind of cutting edge and highly relevant information that we have become accustomed to seeing in so many outstanding high-impact papers in which humanized mice have contributed in a significant way or in which they themselves have been the focus of the study.

It would be appropriate to indicate some areas of specific importance to the field where focus and resources are needed in order to develop humanized mouse models capable of fulfilling critical gaps in our knowledge. Top among these is our

J. V. Garcia (✉)
Division of Infectious Diseases, Department of Medicine, UNC Center for AIDS Research, University of North Carolina School of Medicine, Genetic Medicine Building 2046, Chapel Hill, NC 27599-7042, USA
e-mail: victor_garcia@med.unc.edu

L. Y. Poluektova et al. (eds.), *Humanized Mice for HIV Research*,
DOI 10.1007/978-1-4939-1655-9_42

understanding of the detailed interplay between the human immune system and mouse tissues. Better understanding of how human cells react to mouse antigens will certainly contribute to models in which cells from both species can coexist in better harmony. There is a substantial amount of information regarding the need for certain cytokines for the development, survival, and maintenance of human hematopoietic lineages. Efforts to date have indicated that unregulated expression of these molecules can improve levels of specific cell types. However, their unregulated expression can have unintended and highly undesirable consequences. Therefore, there is a need for a concerted effort to obtain tightly developmentally regulated tissue specific expression of any human transgenes deemed necessary to improve reconstitution with human cells. This includes the molecules needed to recapitulate key interactions that occur in human tissues between nonhematopoietic and hematopoietic cells, including the formation of germinal centers. Human cell lineages requiring special consideration include follicular dendritic cells, natural killer cells, myeloid cells, phagocytic cells, red blood cells, hepatocytes, and B cells. A humanized mouse model with robust and sustained reconstitution with human red blood cells would be of significant benefit since it would provide a much needed platform for the preclinical evaluation of numerous interventions specifically targeting these cells in a wide variety of contexts ranging from blood disorders to gene therapy to infectious diseases. Phagocytes represent the most abundant cells in blood, yet they are poorly reconstituted in humanized mice. There is a significant need of humanized mouse models that recapitulate the genesis and functional phenotype of these most important cells. Over the years several models have been reported that support engraftment with human hepatocytes. These models have been invaluable in numerous respects. However, the majority are relatively cumbersome and their use has not been widespread. Novel and easy-to-implement models with robust and sustained hepatocyte reconstitution would find wide applicability in numerous vital areas of investigation in drug discovery, cancer, immunology, and infectious diseases.

B cells represent a very interesting paradigm in the field. On the one hand they can repopulate humanized mice at extremely high levels even in the complete absence of human T cells. On the other, they virtually do not produce IgG and for the most part do not mount T-cell dependent immune responses. Even though these limitations are often ignored or minimized, they represent perhaps one of the most significant areas where improvement is needed in all humanized models to date. Specifically, in order for these models to be seriously considered as adequate to study human humoral immune responses and testing of novel vaccines, this limitation has to be at least partially overcome. Unfortunately, investigation of some of the most obvious possibilities has not provided great results and currently there are no good leads to follow. This represents a unique opportunity to learn key details about the intricacies that regulate plasma cell development that will inform how to improve these models while providing novel insight into fundamental aspects of the human immune system.

However, progress in the development and implementation of new and improved models has slowed down and the "new" models that have been described recently have simply been derivatives of other well-established models, making only incremental practical improvements. The time is right for the development of truly novel humanized mouse models that can have broad applicability and the capacity to open lines of investigation in areas thus far underexplored. This is an opportunity to reiterate the need for intellectual, physical, and financial resources to further improve the current models, develop and implement new ones, and broaden the availability of these models to researchers in as many different disciplines as possible.

A model to consider are the National Primate Centers sponsored by the National Institutes of Health of the USA. The National Primate Research Centers (or NPRCs) provide facilities, animals, and expertise for investigators using nonhuman primates (NHP) for biomedical research. In 2011, the NPRCs located throughout the USA facilitated more than 1000 individual research projects involving approximately 2000 researchers. The breadth of the research supported by the NPRCs expands to virtually all areas of biomedicine, including infectious disease, neurobiology, metabolic disease, reproductive biology, aging, etc. The NPRCs provide both physical and intellectual infrastructure to researchers funded by governmental sources, such as the National Health Institute, the Department of Defense, the National Science Foundation, the National Aeronautics and Space Administration as well as researchers funded by foundations and the private sector. Given the similarities between the work currently being performed using NHPs and humanized mice a similar model could be considered in which a National Humanized Mouse Center(s) (NHMRC) could provide physical and intellectual infrastructure to serve virtually all areas of biomedicine. It should be noted that very much like with research involving NHPs, research using humanized mice should continue to be performed and supported at all other types of institutions where individual laboratories, research centers, or other types of research institutes carry out critical research using these models. The broad use of these models could be greatly facilitated by the ability to produce significantly larger numbers of animals for use in studies, by the significantly lower costs associated with their large-scale production and husbandry, and by the fact that they can be readily transported to virtually any facility with adequate resources for their use. The availability of National Humanized Mouse Research Center(s) would therefore continue to improve the current models, develop new ones in response to new technologies and research needs, produce animals for distribution and use, provide on-site facilities and trained personnel to conduct experiments and for the development and implementation of much needed standards for the field. Another consideration would be the development of an International Humanized Mouse Working Group (IHMWG) to facilitate exchange and dissemination of information, the development and implementation of new and improved models, the development and validation of experimental platforms, to facilitate the dialog between investigators in the field, and to identify areas of improvement that would be simplified by the concerted effort of multiple disciplines, unlikely to be met by individual research groups. An IHMWG could also ensure and strengthen the continued availability of appropriate venues for the timely exchange of information in the field currently occurring every 2–3 years at the International Workshop on Humanized Mice.

In summary, the future for humanized mice is bright. With a steadily increasing number of laboratories implementing current and novel humanized mouse models for research in different disciplines, the use of these models will continue to increase and their contributions to biomedical science and translational research will become more evident with time. Humanized mouse models are here to stay and their future is NOW!

Index

L. Y. Poluektova et al. (eds.), *Humanized Mice for HIV Research*,
DOI 10.1007/978-1-4939-1655-9

MIX
Papier aus verantwortungsvollen Quellen
Paper from responsible sources
FSC® C105338

If you have any concerns about our products,
you can contact us on
ProductSafety@springernature.com

In case Publisher is established outside the EU,
the EU authorized representative is:
Springer Nature Customer Service Center GmbH
Europaplatz 3, 69115 Heidelberg, Germany

Printed by Libri Plureos GmbH
in Hamburg, Germany